Earth, Moon, and Sun

Earth
Mass	
Equatorial radius	
Polar radius	
Mean density	5520 kg/m³
Surface gravity	$g = 9.81$ m/s² $= 32.2$ ft/s²
Period of rotation	1 sidereal day $= 23$ h 56 min 4 s $= 8.616 \times 10^4$ s
Moment of inertia:	
about polar axis	$I = 0.331\, M_E R_E^2$
about equatorial axis	$I' = 0.329 M_E R_E^2$
Mean distance from Sun	1.50×10^{11} m
Period of revolution (period of orbit)	1 year $= 365$ days 6 h $= 3.16 \times 10^7$ s
Orbital speed	29.8 km/s

Moon
Mass	7.35×10^{22} kg
Radius	1.74×10^6 m
Mean density	3340 kg/m³
Surface gravity	1.62 m/s²
Period of rotation	27.3 days
Mean distance from Earth	3.84×10^8 m
Period of revolution	1 sidereal month $= 27.3$ days

Sun
Mass	$M_s = 1.99 \times 10^{30}$ kg
Radius	6.96×10^8 m
Mean density	1410 kg/m³
Surface gravity	274 m/s²
Period of rotation	~ 26 days
Luminosity	3.9×10^{26} W

Greek Alphabet

A	α	alpha		N	ν	nu
B	β	beta		Ξ	ξ	xi
Γ	γ	gamma		O	o	omicron
Δ	δ	delta		Π	π	pi
E	ε	epsilon		P	ρ	rho
Z	ζ	zeta		Σ	σ	sigma
H	η	eta		T	τ	tau
Θ	θ	theta		Y	υ	upsilon
I	ι	iota		Φ	ϕ	phi
K	κ	kappa		X	χ	chi
Λ	λ	lambda		Ψ	ψ	psi
M	μ	mu		Ω	ω	omega

Physics

Physics

SECOND EDITION

Hans C. Ohanian

RENSSELAER POLYTECHNIC INSTITUTE

VOLUME TWO, EXPANDED

W·W·Norton & Company
NEW YORK·LONDON

*To Susan Farnsworth Ohanian, writer,
who gently tried to teach me some of her craft.*

Volume 2, Expanded

Copyright © 1989, 1985 by W. W. Norton & Company, Inc.

All rights reserved.

Published simultaneously in Canada by Penguin Books Canada Ltd., 2801 John Street, Markham, Ontario L3R 1B4.
Printed in the United States of America.
Second Edition

Makeup by Roberta Flechner and Ben Gamit

Cover photograph by Robert Houser, courtesy of Comstock © 1988

Photograph credits appear at back of book.

ISBN 0-393-95786-1

W. W. Norton & Company, Inc., 500 Fifth Avenue, New York, N.Y. 10110
W. W. Norton & Company Ltd., 37 Great Russell Street, London WC1B 3NU

1 2 3 4 5 6 7 8 9 0

Contents

PREFACE   xv

22 | Electric Force and Electric Charge   571

22.1	The Electrostatic Force	572
22.2	Coulomb's Law	573
22.3	Charge Quantization and Charge Conservation	577
22.4	Conductors and Insulators; Frictional Electricity	579
Summary		582
Questions		582
Problems		583

23 | The Electric Field   587

23.1	The Superposition of Electric Forces	588
23.2	The Electric Field	590
23.3	Lines of Electric Field	595
23.4	Electric Dipole in an Electric Field	599
Summary		602
Questions		602
Problems		603

24 | Gauss' Law   609

24.1	Electric Flux and the Number of Field Lines	609
24.2	Gauss' Law	611
24.3	Some Examples	613

24.4 Conductors in Electric Fields	616
Summary	620
Questions	620
Problems	621

25 | The Electrostatic Potential — 626

25.1 The Electrostatic Potential	626
25.2 Calculation of the Electrostatic Potential	629
25.3 The Electrostatic Field as a Conservative Field	633
25.4 The Gradient of the Potential	634
25.5 The Potential and Field of a Dipole	637
25.6* The Mean-Value Theorem	639
Summary	642
Questions	643
Problems	644

26 | Electric Energy — 650

26.1 Energy of a System of Point Charges	650
26.2 Energy of a System of Conductors	652
26.3 The Energy Density	657
Summary	657
Questions	657
Problems	658

27 | Capacitors and Dielectrics — 662

27.1 Capacitance	662
27.2 Capacitors in Combination	665
27.3 Dielectrics	667
27.4 Gauss' Law in Dielectrics	672
27.5 Energy in Capacitors	674
Summary	676
Questions	676
Problems	678

28 | Currents and Ohm's Law — 683

28.1 Electric Current	683
28.2 Resistance and Ohm's Law	685
28.3* The Flow of Free Electrons	687
28.4 The Resistivity of Materials	690
28.5 Resistances in Combination	693
Summary	696
Questions	696
Problems	697

* This section is optional.

29 | DC Circuits — 703

29.1	Electromotive Force	703
29.2	Sources of Electromotive Force	705
	Batteries	705
	Electric Generators	706
	Fuel Cells	706
	Solar Cells	707
29.3	Single-Loop Circuits	707
29.4	Multiloop Circuits	710
29.5	Energy in Circuits; Joule Heat	712
29.6*	Electrical Measurements	715
	Ammeter and Voltmeter	715
	Potentiometer	716
	Wheatstone Bridge	717
29.7*	The RC Circuit	717
29.8	The Hazards of Electric Currents	721
Summary		723
Questions		723
Problems		724

INTERLUDE VI. ATMOSPHERIC ELECTRICITY

VI.1	The Fair-Weather Electric Field	VI-1
VI.2	Thunderstorms	VI-3
VI.3	Generation of Electric Charge	VI-5
VI.4	The Electric Field of a Thundercloud	VI-6
VI.5	Lightning	VI-8
Further Reading		VI-11
Questions		VI-11

30 | The Magnetic Force and Field — 730

30.1	The Magnetic Force	730
30.2	The Magnetic Field	735
30.3	The Biot–Savart Law	738
30.4	The Magnetic Dipole	743
Summary		745
Questions		745
Problems		746

31 | Ampère's Law — 753

31.1	Ampère's Law	753
31.2	Solenoids	756
31.3	Motion of Charges in Electric and Magnetic Fields	759
31.4	Crossed Electric and Magnetic Fields; the Hall Effect	765

* This section is optional.

31.5	Force on a Wire	768
31.6	Torque on a Current Loop	769
Summary		772
Questions		772
Problems		773

32 | Electromagnetic Induction — 779

32.1	Motional Emf	779
32.2	Faraday's Law	783
32.3	Some Examples; Lenz' Law	785
32.4	The Induced Electric Field	789
32.5	Inductance	792
32.6	Magnetic Energy	794
32.7*	The RL Circuit	796
Summary		798
Questions		799
Problems		800

INTERLUDE VII. PLASMA

VII.1	The Fourth State of Matter	VII-1
VII.2	Plasmas and the Magnetic Field	VII-3
VII.3	Waves in a Plasma	VII-6
VII.4	Thermonuclear Fusion	VII-8
VII.5	Plasma Confinement	VII-9
Further Reading		VII-11
Questions		VII-12

33 | Magnetic Materials — 806

33.1	Atomic and Nuclear Magnetic Moments	807
33.2	Paramagnetism	809
33.3	Ferromagnetism	811
33.4	Diamagnetism	814
Summary		817
Questions		817
Problems		818

34 | AC Circuits — 823

34.1	Simple AC Circuits with an External Electromotive Force	823
34.2	The Freely Oscillating LC Circuit	828
34.3	The LCR Circuit with an External Emf	832

* This section is optional.

34.4 The Transformer	840
Summary	842
Questions	843
Problems	844

INTERLUDE VIII. SUPERCONDUCTIVITY

VIII.1	Zero Resistance	VIII-1
VIII.2	The Critical Magnetic Field	VIII-3
VIII.3	The Meissner Effect	VIII-3
VIII.4	Superconductors of the Second Kind	VIII-5
VIII.5	The BCS Theory	VIII-5
VIII.6	Technological Applications	VIII-7
	Further Reading	VIII-11
	Questions	VIII-11

35 | The Displacement Current and Maxwell's Equations   849

35.1	The Displacement Current	849
35.2	Maxwell's Equations	853
35.3*	Cavity Oscillations	854
35.4*	The Electric Field of an Accelerated Charge	859
35.5*	The Magnetic Field of an Accelerated Charge	863
	Summary	866
	Questions	867
	Problems	868

36 | Light and Radio Waves   873

36.1	The Plane Wave Pulse	874
36.2	Plane Harmonic Waves	878
36.3	The Generation of Electromagnetic Waves	881
36.4	Energy of a Wave	883
36.5	Momentum of a Wave	885
36.6	The Doppler Shift of Light	888
	Summary	890
	Questions	890
	Problems	892

37 | Reflection, Refraction, and Polarization   899

37.1	Huygens' Construction	899
37.2	Reflection	900
37.3	Refraction	903

---

* This section is optional.

37.4	Polarization	910
Summary		915
Questions		916
Problems		917

38 | Mirrors, Lenses, and Optical Instruments 922

38.1	Spherical Mirrors	922
38.2	Thin Lenses	926
38.3	The Photographic Camera and the Eye	932
38.4	The Magnifier and the Microscope	934
38.5	The Telescope	936
Summary		939
Questions		940
Problems		942

39 | Interference 947

39.1*	The Standing Electromagnetic Wave	948
39.2	Thin Films	950
39.3	The Michelson Interferometer	952
39.4	Interference from Two Slits	954
39.5	Interference from Multiple Slits	958
Summary		963
Questions		963
Problems		965

40 | Diffraction 971

40.1	Diffraction by a Single Slit	971
40.2	Diffraction by a Circular Aperture; Rayleigh's Criterion	976
40.3*	Babinet's Principle	981
Summary		983
Questions		983
Problems		985

41 | The Theory of Special Relativity 989

41.1	The Speed of Light; the Ether	990
41.2	Einstein's Principle of Relativity	992
41.3	The Lorentz Transformations	997
41.4	The Time Dilation	1004
41.5	The Length Contraction	1007

* This section is optional.

41.6 The Combination of Velocities	1008
41.7 Relativistic Momentum and Energy	1010
41.8* Relativity and the Magnetic Field	1014
Summary	1018
Questions	1019
Problems	1020

INTERLUDE IX. GRAVITY AND GEOMETRY

IX.1 The Principle of Equivalence	IX-1
IX.2 The Deflection of Light and the Curvature of Space	IX-3
IX.3 The Theory of General Relativity	IX-6
IX.4 The Gravitational Time Dilation	IX-8
IX.5 Black Holes	IX-10
Further Reading	IX-13
Questions	IX-14

42 | Quanta of Light — 1027

42.1 Blackbody Radiation	1027
42.2 Energy Quanta	1029
42.3 Photons and the Photoelectric Effect	1034
42.4 The Compton Effect	1037
42.5 Wave vs. Particle	1039
Summary	1043
Questions	1043
Problems	1044

43 | Atomic Structure and Spectral Lines — 1049

43.1 Spectral Lines	1050
43.2 The Balmer Series and Other Spectral Series	1052
43.3 The Nuclear Atom	1054
43.4 Bohr's Theory	1057
43.5 The Correspondence Principle	1063
43.6 Quantum Mechanics	1064
Summary	1067
Questions	1067
Problems	1068

INTERLUDE X. LASER LIGHT

X.1 Stimulated Emission	X-1
X.2 Lasers	X-2
X.3 Some Applications	X-5
X.4 Holography	X-9
Further Reading	X-12
Questions	X-13

* This section is optional.

44 | Quantum Structure of Atoms, Molecules, and Solids — 1074

44.1	Principal, Orbital, and Magnetic Quantum Numbers; Spin	1075
44.2	The Exclusion Principle and the Structure of Atoms	1079
44.3	Energy Levels in Molecules	1081
44.4	Energy Bands in Solids	1083
44.5	Semiconductor Devices	1087
	Rectifier (Diode)	1087
	Transistor (Triode)	1088
	Light-Emitting Diode	1089
	Solar Cell	1089
Summary		1090
Questions		1090
Problems		1091

45 | Nuclei — 1095

45.1	Isotopes	1096
45.2	The Strong Force and the Nuclear Binding Energy	1099
45.3	Radioactivity	1103
	Alpha Decay	1104
	Beta Decay	1105
	Gamma Emission	1107
45.4	The Law of Radioactive Decay	1109
45.5	Nuclear Reactions	1111
Summary		1115
Questions		1115
Problems		1116

INTERLUDE XI. NUCLEAR FISSION

XI.1	Fission	XI-1
XI.2	Chain Reactions	XI-3
XI.3	The Bomb	XI-4
XI.4	The Effects of Nuclear Weapons	XI-6
XI.5	Nuclear Reactors	XI-10
Further Reading		XI-12
Questions		XI-13

46 | Elementary Particles — 1121

46.1	The Tools of High-Energy Physics	1121
46.2	The Multitude of Particles	1127
46.3	Interactions and Conservation Laws	1130
46.4	Fields and Quanta	1133

* This section is optional.

46.5	The Unified Electroweak Forces	1136
46.6	Quarks	1138
46.7	Color and Charm	1140
Supplement: Further Reading		1143
Summary		1145
Questions		1146
Problems		1146

Appendix 1:	Index to Tables	A-1
Appendix 2:	Mathematical Symbols and Formulas	A-2
Appendix 3:	Perimeters, Areas, and Volumes	A-4
Appendix 4:	Brief Review of Trigonometry	A-5
Appendix 5:	Brief Review of Calculus	A-10
Appendix 6:	The International System of Units (SI)	A-20
Appendix 7:	Conversion Factors	A-22
Appendix 8:	Best Values of Fundamental Constants	A-27
Appendix 9:	The Chemical Elements and the Periodic Table	A-29
Appendix 10:	Formula Sheets	A-32
Appendix 11:	Answers to Even-Numbered Problems	A-34
Photograph Credits		A-46
Index		Index-1

Preface

In this Second Edition of *Physics,* I have incorporated divers alterations and additions recommended by the many thoughtful users of the First Edition. My objectives in the book remain the same: to present a contemporary, modern view of classical mechanics and electromagnetism, and to offer the student a glimpse of what is going on in physics today. Thus, throughout the book, I encourage students to keep in mind the atomic structure of matter and to think of the material world as a multitude of restless electrons, protons, and neutrons. For instance, in the mechanics chapters, I emphasize that all macroscopic bodies are systems of particles; and in the electricity chapters, I introduce the concepts of positive and negative charge by referring to protons and electrons, not by referring to the antiquated procedure of rubbing glass rods with silk rags (which, according to experts on triboelectricity, can give the wrong sign if the silk has been thoroughly cleaned). I try to make sure that students are always aware of the limitations of the nineteenth-century fiction that matter and electric charge are continua. Blind reliance on this fiction has often been justified by the claim that engineering students need physics as a tool, and that the atomic structure of matter is of little concern to them. But if physics is a tool, it is also a work of art, and its style cannot be dissociated from its function. In this book I give a physicist's view of physics, because I believe that it is fitting that all students should gain some appreciation for the artistic style of the toolmaker.

Core Chapters and Interludes

The book contains two kinds of chapters: *core* chapters and *interlude* chapters. The 41 core chapters cover the essential topics of introductory physics: mechanics of particles, rigid bodies, and fluids; oscillations; wave motion; heat and thermodynamics; electricity and magnetism; optics; and special relativity. An expanded version of the book includes another five core chapters covering quantum physics, nuclei, and elementary particles.

The organization of the core chapters is fairly traditional, with some innovations. For instance, I start the study of magnetism (Chapter 30) with the law for the magnetic force between two moving point charges; this magnetic force is no more complicated than the force between two current elements, and the crucial advantage is that magnetism can be developed from the magnetic-force law in much the same way as electricity is developed from Coulomb's Law. This approach is consistent with the underlying philosophy of the book: particles are primary entities and should always be treated first, whereas macroscopic bodies and currents are composite entities which should be treated later. As another innovation, I include a simple derivation of the electric radiation field of an accelerated charge (Chapter 35); this calculation relies on Richtmeyer and Kennard's clever analysis of the kinks in the electric field lines of an accelerated charge (a set of computer-generated film loops available from the Educational Development Center shows how such kinks propagate along the field lines; these film loops tie in very well with the calculations of Chapter 35).

The interlude chapters (the expanded version contains 11) present some of the fascinating discoveries and applications of physics today: crystal structure and symmetry, the expansion of the universe, automobile safety, ionizing radiation, energy resources, atmospheric electricity, plasmas, superconductivity, general relativity, lasers, and fission. All of these interludes are optional—they rely on the core chapters, but the core chapters do not rely on them. Users of the First Edition will perceive that Chapter 46, on elementary particles, was an interlude in the First Edition. It has now been recast as a core chapter to round out the set of core chapters covering modern physics. However, it remains an interlude in spirit, and can be treated as such.

The inspiration for the interludes grew out of my unhappiness over a paradox afflicting the typical undergraduate physics curriculum: liberal-arts students in a nonmathematical physics course often get to see more of the beauty and excitement of today's physics than do science and engineering students in a calculus-based physics course. While liberal-arts students get a glimpse of gluons, black holes, or the Big Bang, science and engineering students are expected to calculate the motion of blocks on top of other blocks sliding down an inclined plane, or the motion of a baseball thrown in some direction or another by a man (or woman) riding in an elevator. To some extent this is unavoidable—science and engineering students need to learn and practice classical mechanics and electromagnetism, and they have little time left for dabbling in the arcane mysteries of contemporary physics. Nevertheless, most teachers will occasionally find an hour or two to tell their students a little of what is going on in physics today. I wrote the interludes to lend encouragement and support to such excursions to the frontiers of physics.

My choice of topics for the interludes reflect the interests expressed by my students. Over the years, I have often been asked: When will we get to quarks? or Are you going to tell us about gravitational collapse? and I came to feel that such curiosity must not be allowed to wither away. Obviously, in the typical introductory course it will be impossible to cover all of the interludes (I have usually covered two per term), but the broad range of topics will permit teachers to select according to their own tastes. The interludes are mainly descriptive rather than analytic. In them, I try to avoid formulas and instead give students a qualitative feeling for the underlying physics, keeping the discussion simple so that students can read them on their own. Thus, the interludes could be used for supplementary reading, not necessarily accompanied by lectures. For the inquisitive student, each interlude includes a collection of qualitative questions and an extensive annotated list of further readings.

Optional Sections and Chapters

Optional sections and chapters have been indicated by a large asterisk (*). All such optional sections or chapters and all the interludes can be omitted without loss of continuity. Some other sections and chapters could possibly be omitted. For the guidance of teachers, the outline of quintessentials attached to this Preface lists all the sections that are indispensable for the logical coherence of the text.

Mathematical Prerequisites

In order to accommodate students who are taking an introductory calculus course concurrently, derivatives are used slowly and hesitantly at first (Chapter 2), and routinely later on. Likewise, the use of integrals is postponed as far as possible (Chapter 7), and they come into heavy use only in the second volume (after Chapter 21). For students who need a review of calculus, Appendix 5 contains a concise primer on derivatives and integrals.

Examples, Problems, Questions, and Summaries

The core chapters include generous collections of solved examples (about 300 altogether), of qualitative questions for review (about 850 altogether), and of problems (about 2150 altogether). Answers to the even-numbered problems are given in Appendix 11. The problems are grouped by sections, with the most difficult problems at the end of each section. The levels of difficulty of the problems are roughly indicated by no star, one star (*), or two stars (**). No-star problems are easy and straightforward; they are mostly of the plug-in type. One-star problems are of medium difficulty; they contain a few complications requiring the combination of several concepts or the manipulation of several formulas. Two-star problems are difficult and challenging; they demand considerable thought and perhaps some insight, and they occasionally demand substantial mathematical skills.

I have tried to make the problems interesting to the student by drawing on realistic examples from technology, sports, and everyday life. Many of the problems are based on data extracted from engineering handbooks, car-repair manuals, *Jane's Book of Aircraft*, *The Guinness Book of World Records*, newspaper reports, etc. Many other problems deal with atoms and subatomic particles; these are intended to reinforce the atomistic view of the material world. In some cases, cognoscenti will perhaps consider the use of classical physics somewhat objectionable in a problem that really ought to be handled by quantum mechanics. But I believe that the advantages of familiarization with atomic quantities and magnitudes outweigh the disadvantages of a naïve use of classical mechanics.

Each chapter also includes a collection of qualitative questions intended to stimulate thought and to test the grasp of basic concepts (some of these questions are discussion questions that do not have a unique answer). Moreover, each chapter contains a brief summary of the main physical quantities and laws introduced in it. The virtue of these summaries lies in their brevity. They include essential definitions and equations, because the statements in the body of each chapter are adequate.

Units

The SI system of units is used exclusively. In the abbreviations for the units, I follow the dictates of the Conférence Générale des Poids et Mesures of 1971, although I deplore the majestic stupidity of the decision to replace the old, self-explanatory abbreviations amp, coul, nt, sec, °K by an alphabet soup of cryptic symbols A, C, N, s, K, etc. For the sake of clarity, I spell out the names of units in full whenever the abbreviations are likely to lead to ambiguity and confusion.

For reference purposes, the definitions of the British units have been retained. But these units are not used in examples or in problems, with the exception of a handful of problems in the first chapter. In the definitions of the British units, the pound (lb) is taken to be the unit of mass, and the pound-force (lbf) is taken to be the unit of force. This is in accord with the practice approved by the American National Standards Institute (ANSI), the Institute of Electrical and Electronic Engineers (IEEE), and the United States Department of Defense.

Changes from the First Edition

In response to comments from users of the First Edition, I have reorganized some chapters and added some new chapters and sections to provide more thorough and better-balanced coverage. The discussion of angular momentum, which used to be dispersed over two chapters, now appears in one place, in Chapter 12. The discussion of the magnetic force between moving charges in Chapter 30 has been made clearer and simpler. Concise, but self-contained, introductions to dot and cross products of vectors have been included where they are first needed (Chapters 7 and 12, respectively); this makes the book more flexible, since it is now possible to skip the sections on these products in Chapter 3. The chapter on gravitation has been moved forward, so it now precedes the discussion of systems of particles; this move seems sensible since all the simple problems in gravitation assume a fixed center of force, and they can therefore be regarded as single-particle problems. The chapter on special relativity and the chapter on elementary particles have been moved toward the end of the book, in accord with their historical position. However, the prerequisites for these two chapters have not been changed appreciably, and students could read them much sooner, during the first half of the course.

Several new chapters have been added: (14) "Statics and Elasticity"; (38) "Mirrors, Lenses, and Optical Instruments"; (44) "Quantum Structure of Atoms, Molecules, and Solids"; (45) "Nuclei"; and a new Interlude III, "Automobile Collisions and Automobile Structure." The sections that have been added are: 1.7 Significant Figures, Conversion of Units, and Consistency of Units; 19.3 Kinetic Pressure and the Maxwell Distribution; 19.5 The Mean Free Path; 20.3 Thermometers and Thermal Equilibrium; 29.6 Electrical Measurements; 29.7 The RC Circuit; 32.7 The RL Circuit; and 37.4 Polarization.

Besides these additions, the new edition incorporates a multitude of stylistic and pedagogical improvements. For instance, *Comments and Suggestions* have been appended to many of the solved examples in the text; these comments point out generalizations and/or special features of the results, and they provide the student with helpful hints on how to apply (and how not to apply) the methods illustrated in the examples to the solution of problems.

The number of solved examples and of problems has been increased by about 25%, and more attention has been paid to achieving a balance among examples and among problems of diverse levels of difficulty.

Study Guide

An excellent study guide for this book has been written by Professors Van E. Neie (Purdue University) and Peter J. Riley (University of Texas, Austin). This guide includes for every chapter a brief introduction laying out the objectives; a list of key terms for review; detailed commentaries on each of the main ideas; and a large collection of interesting sample problems, which alternate between worked problems (with full solutions) and guided problems (which provide step-by-step schemes that lead students to the solutions). Those with Macintosh computers should also be aware of an exciting new project by Eric Mazur (Harvard University).His software gives access to any of its three component parts: interactively solved problem, demonstrations, and summary notes.

Acknowledgments

I have greatly benefited from comments by reviewers and users of the book. Many of the alterations in this Second Edition originated from users who, to my pleasant surprise, took the trouble to send me their recommendations for improvements, and sometimes even sent me data for the construction of additional realistic and interesting problems. For very detailed, comprehensive reviews, I am indebted to John R. Boccio (Swarthmore College), Roger W. Clapp, Jr. (University of South Florida), A. Douglas Davis (Eastern Illinois University), Anthony P. French (Massachusetts Institute of Technology), J. David Gavenda (University of Texas, Austin), Roger D. Kirby (University of Nebraska), Roland M. Lichtenstein (Rensselaer Polytechnic Institute), Richard T. Mara (Gettysburg College), John T. Marshall (Louisiana

State University), Delo E. Mook (Dartmouth College), Harvey S. Picker (Trinity College), Peter J. Riley (University of Texas, Austin), and Peter L. Scott (University of California, Santa Cruz). For briefer reviews, I am indebted to John R. Albright (Florida State University), I. H. Bailey (Western Australia Institute of Technology), Frank Crawford (University of California, Berkeley), Sumner Davis (University of California, Berkeley), John F. Devlin (University of Michigan, Dearborn), Phil Eastman (University of Waterloo, Canada), Hugh Evans (University of Waterloo, Canada), Frank A. Ferrone (Drexel University), James R. Gaines (Ohio State University), Stephen Gasiorowicz (University of Minnesota), Michael A. Guillen (Harvard University), Lyle Hoffman (Lafayette College), Walter Knight (University of California, Berkeley), Jean P. Krisch (University of Michigan, Ann Arbor), L. B. Meyer (Duke University), Frank Moscatelli (Swarthmore College), Hermann Nann (Indiana University), Mette Owner-Peterson (The Technical University of Denmark), Norman Pearlman (Purdue University), P. Bruce Pipes (Dartmouth College), Jack Prince (Bronx Community College), Malvin Ruderman (Columbia University), Kenneth Schick (Union College), Mark P. Silverman (Trinity College), Gerald A. Smith (Michigan State University), Julia A. Thompson and David Kraus (University of Pittsburgh), Som Tiagi (Drexel University), and Gary A. Williams (University of California, Los Angeles). I am also indebted to the experts who reviewed the interludes: Edmond Brown (Rensselaer Polytechnic Institute; "The Architecture of Crystals"), Priscilla W. Laws (Dickinson College; "Radiation and Life"), Alan H. Guth (Massachusetts Institute of Technology; "The Big Bang and the Expansion of the Universe"), Frank Richardson (National Highway Traffic Safety Administration; "Automobile Collisions and Automobile Structure"), Donald F. Kirwan (University of Rhode Island; "Energy, Entropy, and the Environment"), Irving Kaplan (Massachusetts Institute of Technology; "Nuclear Fission"), Bernard Vonnegut (State University of New York at Albany; "Atmospheric Electricity"), Sam Cohen (Princeton University Plasma Physics Laboratory; "Plasma"), and Margaret L. A. MacVicar (Massachusetts Institute of Technology; "Superconductivity").

I thank Richard D. Deslattes and Barry N. Taylor of the National Bureau of Standards for information on units, standards, and precision measurements. I thank some of my colleagues at Union College: C. C. Jones prepared the beautiful photographs of interference and diffraction by light (Chapters 39 and 40); Barbara C. Boyer gave valuable advice and assistance on photographs dealing with biological material; and the staff of the library helped me find many a tidbit of information needed for an example or a problem.

I thank the editorial staff of W. W. Norton & Co. for their zeal. Drake McFeely, editor, gave this Second Edition the same meticulous attention he gave the First, and he was indefatigable in collecting comments from reviewers and users. Avery Hudson supervised all aspects of the production process; he also improved my sentences, and helped me to attain a high level of stylistic consistency in the text and in the diagrams. And Ruth Mandel provided the detective work needed to find the many splendid new photographs that have been added to the book.

H. C. O.
February 1989

CHAPTER 22

Electric Force and Electric Charge

Ordinary matter — solids, liquids, and gases — consists of atoms, each with a nucleus surrounded by a swarm of electrons. For example, Figure 22.1 shows the structure of an atom of neon. At the center of this atom there is a nucleus made of ten protons and ten neutrons packed very tightly together; the diameter of the nucleus is only about 6×10^{-15} m. Moving around this nucleus there are ten electrons; these electrons are confined to a roughly spherical region about 2×10^{-10} m across. Figure 22.1 shows the electrons at one instant of time. Since the electrons move around the nucleus very quickly, for most purposes the average distribution of electrons is more relevant than the instantaneous distribution. Figure 22.2 is a picture of the average distribution of electrons in the neon atom. This picture gives a good impression of the spherical shape of this atom.

The atom somewhat resembles the Solar System, with the nucleus as Sun and the electrons as planets. In the Solar System, the force that holds a planet near the Sun is the gravitational force. What is the force that holds an electron near the nucleus? This force is the **electric force** of attraction between the electron and the protons in the nucleus. The electric force not only holds electrons inside atoms, but also holds atoms together in molecules or crystals, and holds these building

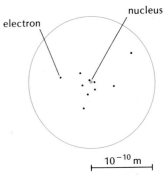

Fig. 22.1 Neon atom as seen at one instant of time with a (hypothetical) gamma-ray microscope.

Electric force

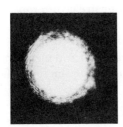

Fig 22.2 Neon atom as seen with an electron-holography microscope. The atom looks like a spherical cloud. The density, or brightness, at any point of this cloud is roughly proportional to the probability for finding an electron at this point. The magnification is $2 \times 10^8 \times$. (Courtesy L. S. Bartell, University of Michigan.)

blocks together in large-scale macroscopic structures — a rock, a tree, a human body, a skyscraper, or a supertanker. All the mechanical "contact" forces of everyday experience — the push of a hand against a door, the pull of an elevator cable, the pressure of water against the hull of a ship — are nothing but the combined electric forces of many atoms. Thus our immediate environment is dominated by electric forces.

In the following chapters we will study electric forces and their effects. For a start (Chapters 22–29), we will assume that the particles exerting these forces are at rest or moving only very slowly (quasistatic). The electric forces exerted under these conditions are called **electrostatic forces.** Later on (Chapters 30–34), we will consider the forces when the particles are moving with uniform velocity or nearly uniform velocity. Then the electric forces are modified — besides the electrostatic force there arises a **magnetic force.** The magnetic force may be regarded as a supplement to the electrostatic force, a supplement depending on the velocities of the particles. The combined electrostatic and magnetic forces are called **electromagnetic forces.** Finally, we will consider the forces when the particles are moving with accelerated motion (Chapters 35 and 36). The electromagnetic forces are then further modified with a drastic consequence, that is, the emission of electromagnetic waves, such as light or radio waves.

Electricity was first discovered through friction. The ancient Greeks noticed that, when rubbed, rods of amber gave off sparks and attracted small bits of straw or feathers. In the nineteenth century, technical applications of electricity were gradually developed; but it was only in the twentieth century that the pervasive presence of electric forces holding together all the matter of our environment was recognized.

We will begin our study of electricity with the fundamental electric force between charged particles rather than following the historical route, because the origin of frictional electricity remains rather mysterious. Even now, physicists have no precise understanding of the detailed mechanism that generates electric charge on rubbed bodies or why some rubbed bodies acquire positive charge and some negative charge.

Electrostatic force

Benjamin Franklin, *1706–1790, American scientist, statesman, and inventor. Although he is most often remembered for his hazardous experiments with a kite in a thunderstorm, which demonstrated that lightning is an electric phenomenon, and for his invention of lightning rods, Franklin also made other significant contributions to the experimental and theoretical studies of electricity, and he was admired and honored by the leading scientific associations in Europe. Among these contributions was his formulation of the law of conservation of electric charge and his introduction of the modern notation for plus and minus charges, which he regarded as an excess or deficiency of "electric fluid."*

22.1 The Electrostatic Force

The electric force between, say, an electron and a proton resembles gravitation in that it decreases in proportion to the inverse square of the distance. But the electric force is much stronger than the gravitational force. The electric attraction between an electron and a proton (at any given distance) is about 2×10^{39} times as strong as the gravitational attraction. Thus the electric force is by far the strongest force felt by an electron in an atom. The other great difference between the gravitational and the electric forces is that gravitation is always attractive, whereas electric forces can be attractive or repulsive. The electric forces between electron and proton, electron and electron, and proton and proton all have the same magnitudes (for the same distances); the electron–proton force is attractive, but the electron–electron and proton–proton forces are repulsive. Table 22.1 gives a qualitative summary of the electric forces between the elementary particles in an atom. The "contact" force between two atoms close together actually arises

Table 22.1 ELECTRIC FORCES (QUALITATIVE)

Particles	Force
Electron and proton	Attractive
Electron and electron	Repulsive
Proton and proton	Repulsive
Neutron and anything	Zero

from a counterplay of the attractive and repulsive electric forces among the electrons of the atoms and the protons in their nuclei. The force between two neighboring atoms depends on the relative location of the electrons and the nuclei. If the distribution of the electrons around the nucleus is somewhat distorted, so the electrons in one atom are closer to the nucleus of the neighboring atom than to its electrons, then the force between these atoms will be attractive. Figure 22.3a shows such a distortion that leads to an attractive force; the distortion may either be intrinsic to the structure of the atom or induced by the presence of the neighboring atom. Figure 22.3b shows a distortion that leads to a repulsive force.

If two macroscopic bodies are separated by some distance, the electric attractions and repulsions between them tend to cancel, provided the bodies are neutral, that is, provided their number of protons matches their number of electrons. For example, if the macroscopic bodies are a man and a woman separated by a distance of 10 m, then each electron in the man will be attracted by the protons in the woman, but simultaneously it will be repelled by the electrons in the woman; these forces cancel each other. Only when the surfaces of the two macroscopic bodies are very near one another ("touching") will the atoms in one surface exert a net force on those in the other surface.

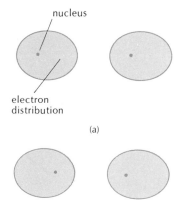

Fig. 22.3 (a) Two neighboring distorted atoms. The colored regions represent the average distributions of the electrons. The electrons of the left atom are closer to the nucleus of the right atom than to its electrons. (b) The nucleus of the left atom is closer to the nucleus of the right atom than to its electrons.

22.2 Coulomb's Law

As we stated in the preceding section, the electric forces between electron and proton, electron and electron, and proton and proton all have the same magnitudes when these particles are placed at the same distances. Obviously, this implies that the force does not depend on the mass of these particles. Instead, it depends on a new quantity: the **electric charge.** For the mathematical formulation of the law of electric force between electrons, protons, and other particles, we assign to each particle an electric charge which is either positive, negative, or zero. The charge on the proton is taken as positive and that on the electron negative. Since experiments show that the absolute values of these charges are exactly the same, we will designate the charges of the proton and electron by $+e$ and $-e$, respectively. The charge on the neutron is zero. Table 22.2 summarizes these values of the charges.

Table 22.2 ELECTRIC CHARGES OF PROTONS, ELECTRONS, AND NEUTRONS

Particle	Charge
Proton, p	$+e$
Electron, e	$-e$
Neutron, n	0

Charles Augustin de Coulomb (koolom), 1736–1806, *French physicist. A military engineer by profession, he retired at age 53 from his post as superintendent of waters and fountains so that he could fully pursue scientific research. With the torsion balance, which he invented, he established that the electric force between small charged balls obeys an inverse-square law.*

In terms of these electric charges, we can then state that the electric force between charges of like sign is repulsive, and that the electric force between charges of unlike sign is attractive.

The precise magnitude of the electric force that one charged particle exerts on another is given by **Coulomb's Law:**

The magnitude of the electric force that a particle exerts on another particle is directly proportional to the product of their charges and inversely proportional to the square of the distance between them. The direction of the force is along the line joining the particles.

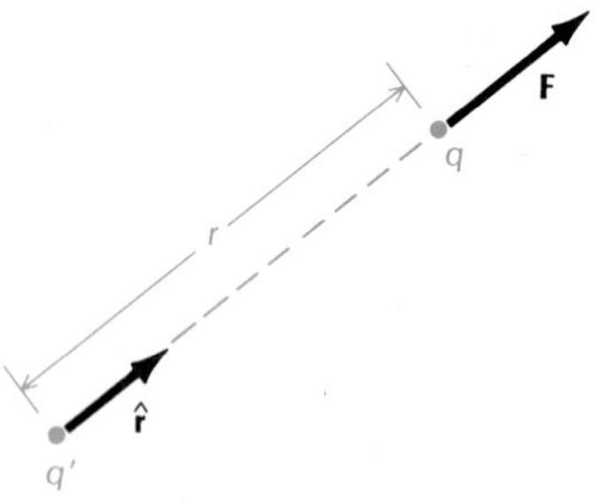

Thus, the magnitude of the force that a particle of charge q' exerts on a particle of charge q at a distance r is

$$F = [\text{constant}] \times \frac{|q'q|}{r^2} \tag{1}$$

To express the direction of the force vectorially, we use the unit vector $\hat{r}$ directed from the charge q' to the charge q (Figure 22.4). With this unit vector, we can write the following equation for the force that q' exerts on q:

$$\mathbf{F} = [\text{constant}] \times \frac{q'q}{r^2} \hat{r} \tag{2}$$

Fig. 22.4 Two charged particles q' and q. The unit vector $\hat{r}$ is directed from q' to q.

This vector equation gives us both the magnitude and the direction of **F**. If the charges are of like sign, then the product $q'q$ is positive and **F** is parallel to $\hat{r}$ (repulsive force). If the charges are of unlike sign, then the product $q'q$ is negative and **F** is antiparallel to $\hat{r}$ (attractive force).

The numerical value of the charge of the proton and the numerical value of the constant in Eqs. (1) and (2) depend on the system of units. In the SI system of units, the electric charge is measured in **coulombs** (C) and the corresponding numerical values are

Coulomb, C

Charge of proton, e

$$\boxed{e = 1.60 \times 10^{-19} \text{ C}} \tag{3}$$

and

$$[\text{constant}] = 8.99 \times 10^9 \text{ N} \cdot \text{m}^2/\text{C}^2 \tag{4}$$

For obscure historical reasons this constant is traditionally written in the complicated form

$$[\text{constant}] = \frac{1}{4\pi\varepsilon_0} \tag{5}$$

with

Permittivity constant

$$\boxed{\varepsilon_0 = 8.85 \times 10^{-12} \text{ C}^2/(\text{N} \cdot \text{m}^2)} \tag{6}$$

The quantity ε_0 is called the **permittivity constant.**[1]

---

[1] The values of e and ε_0 in Eqs. (3) and (6) have been rounded off to three significant figures. The best available values of these constants are listed in Appendix 8.

In terms of the permittivity constant, Coulomb's Law for the force that a particle of charge q' exerts on a particle of charge q becomes

$$\mathbf{F} = \frac{1}{4\pi\varepsilon_0} \frac{q'q}{r^2} \hat{\mathbf{r}} \qquad (7)$$

Coulomb's Law in vector notation

This equation applies to particles — electrons and protons — and also to any small charged bodies, provided that the sizes of these bodies are much less than the distance between them; such bodies are called **point charges.** Equation (7) obviously resembles Newton's Law for the gravitational force (see Section 9.1); the constant $1/4\pi\varepsilon_0$ is analogous to the gravitational constant G, and the electric charges are analogous to the gravitating masses.

Point charge

In the SI system, the coulomb is defined in terms of a standard electric current: one coulomb is the amount of electric charge that a current of one ampere delivers in one second. Unfortunately, the definition of the standard current involves the use of magnetic fields, and we will therefore have to postpone the question of the precise definition of ampere and coulomb to a later chapter. For the time being, we may define the coulomb by the electric repulsive force between two equal charges at a standard distance from one another: the charge on each of two equally charged small bodies is one coulomb if they repel with a force of 8.99×10^9 N when their distance is one meter. (Incidentally: This shows that charge, like everything else, can be defined in terms of the fundamental units of mass, length, and time.)

EXAMPLE 1. Compare the magnitudes of the gravitational force of attraction and of the electric force of attraction between the electron and the proton in a hydrogen atom. According to Newtonian mechanics, what is the acceleration of the electron? Assume that the distance between the two particles is 0.53×10^{-10} m.

SOLUTION: The gravitational force is

$$\frac{GmM}{r^2} = \frac{6.67 \times 10^{-11} \text{ N} \cdot \text{m}^2 \cdot \text{kg}^{-2} \times 9.11 \times 10^{-31} \text{ kg} \times 1.67 \times 10^{-27} \text{ kg}}{(0.53 \times 10^{-10} \text{ m})^2}$$

$$= 3.6 \times 10^{-47} \text{ N}$$

The electric force is

$$\frac{1}{4\pi\varepsilon_0} \frac{e^2}{r^2} = 8.99 \times 10^9 \text{ N} \cdot \text{m}^2/\text{C}^2 \times \frac{(1.60 \times 10^{-19} \text{ C})^2}{(0.53 \times 10^{-10} \text{ m})^2} = 8.2 \times 10^{-8} \text{ N}$$

The ratio of these forces is 8.2×10^{-8} N$/3.6 \times 10^{-47}$ N $= 2.3 \times 10^{39}$.

Since the gravitational force is very small compared with the electric force, it can be neglected. The acceleration of the electron is then

$$a = \frac{F}{m} = \frac{8.2 \times 10^{-8} \text{ N}}{9.1 \times 10^{-31} \text{ kg}} = 9.0 \times 10^{22} \text{ m/s}^2$$

EXAMPLE 2. How much negative and how much positive charge is there on the electrons and protons of a cup of water (0.25 kg)?

SOLUTION: The molecular mass of water is 18 g; hence 250 g of water amounts to 250/18 moles. Each mole has 6.0×10^{23} molecules, giving

$6.0 \times 10^{23} \times (250/18)$ molecules in a cup. Each molecule consists of two hydrogen atoms (one electron apiece) and one oxygen atom (8 electrons). Thus, there are 10 electrons in each molecule, and the total negative charge on all the electrons is $-6.0 \times 10^{23} \times (250/18) \times 10 \times 1.6 \times 10^{-19}$ C $= -1.3 \times 10^7$ C. The total positive charge on the protons is the opposite of this, so it exactly balances the total negative charge on the electrons.

EXAMPLE 3. What is the magnitude of the attractive force exerted by the electrons in a cup of water on the protons in a second cup of water at a distance of 10 m?

SOLUTION: According to the preceding example, the total negative charge on the electrons is -1.3×10^7 C, and the total positive charge on the protons is $+1.3 \times 10^7$ C. If we treat both of these charges (approximately) as point charges, the force is

$$F = \frac{1}{4\pi\varepsilon_0} \frac{qq'}{r^2}$$

$$= 9.0 \times 10^9 \text{ N} \cdot \text{m}^2 \cdot \text{C}^{-2} \times \frac{(-1.3 \times 10^7 \text{ C}) \times (+1.3 \times 10^7 \text{ C})}{(10 \text{ m})^2}$$

$$= -1.5 \times 10^{22} \text{ N}$$

COMMENTS AND SUGGESTIONS: The magnitude of this force is roughly the weight of 10^{18} tons! This enormous attractive force is precisely canceled by an equally large repulsive force exerted by the protons in one cup on the protons in the other cup. Thus, the cups exert no net force on each other.

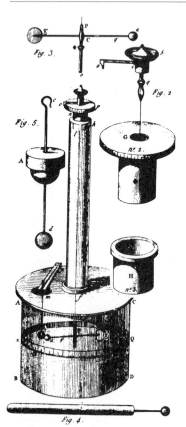

Fig. 22.5 Coulomb's torsion balance.

The inverse-square law for the electric force between pointlike charges was first proposed in the eighteenth century, and it was named after Coulomb because he put it on a firm observational basis by experiments with a torsion balance, which he invented. Figure 22.5 shows Coulomb's torsion balance. The principle of operation of this balance is similar to that of the Cavendish torsion balance used for the measurement of gravitational forces (see Section 9.2); in fact, Cavendish built his balance some years after Coulomb, and he adopted the main features of Coulomb's balance. As shown in Figure 22.5, the beam of the balance is suspended from a fine vertical fiber; a small charged ball is attached at one end of the beam and an uncharged counterweight at the other. When a second small charged ball is brought near the first ball, the electric attraction or repulsion rotates the beam and twists the fiber. The magnitude of the electric force can then be calculated from the measured value of the angular displacement and the (known) value of the torsional constant of the fiber. By repeated measurements with different distances between the balls, Coulomb was able to verify that the force varies as the inverse square of the distance.

Nowadays, the strongest observational evidence for Coulomb's Law comes from atomic mechanics, just as the strongest observational evidence for Newton's Law comes from celestial mechanics. Of course, the motion of an electron inside an atom must be described by quantum mechanics rather than by classical mechanics. The electron has no well-defined orbit, and it is not possible to check the force law by direct measurement of orbital positions; however, orbital energies can be measured and compared with theoretical values calculated from Coulomb's Law. In the best of these calculations, which take into account not only quantum theory but also relativity theory and many subtle effects arising from the interplay of these theories, the orbital energies of the hydrogen atom have been calculated to *nine significant figures.* To within this precision, the theoretical and experimental values agree. This represents one of the greatest triumphs of modern theoretical physics and of Coulomb's Law — even the legendary precision of celestial mechanics and of Newton's Law of gravitation pales by comparison. There is no question that our understanding of the electric force surpasses our understanding of any other force in nature.

22.3 Charge Quantization and Charge Conservation

Not only electrons and protons exert electric forces on each other, but so do many other particles. The magnitudes of these electric forces are given by Eq. (7) with the appropriate values of the electric charges. Table 22.3 lists the electric charges of some particles; a complete list will be found in Section 46.3. The charges of antiparticles are always opposite to those of the corresponding particles; for example, the antielectron (or positron) has charge $+e$, the antiproton has charge $-e$, and the antineutron has charge 0.

Table 22.3 ELECTRIC CHARGES OF SOME PARTICLES

Particle	Charge
Photon, γ	0
Neutrino, ν	0
Electron, e	$-e$
Muon, μ	$-e$
Pion, π^0	0
Pion, π^+	$+e$
Pion, π^-	$-e$
Proton, p	$+e$
Neutron, n	0
Delta, Δ^{++}	$+2e$
Delta, Δ^+	$+e$
Delta, Δ^0	0
Delta, Δ^-	$-e$

All the known particles have charges that are some integer multiple of the fundamental charge e, that is, the charges are always 0, $\pm e$, $\pm 2e$, $\pm 3e$, etc. Why no other charges exist is a mystery for which classical physics offers no explanation. (Recent investigations suggest that the unified theory of fields may supply an explanation; see Chapter 46.) Much effort has been expended on experimental searches for charges of $\frac{1}{3}e$ and $\frac{2}{3}e$, which are the charges of the quarks, the elementary constituents supposedly contained inside protons and neutrons, but neither these fractional charges nor any other fractional charges have ever been found.

Since charges exist only in discrete packets, we say that **charge is quantized** — the fundamental charge e is called the quantum of charge. However, in a description of the charge distribution on macroscopic bodies, the discrete nature of charge can often be ignored, and it is usually sufficient to treat the charge as a continuous "fluid" with a charge density (C/m^3) that varies more or less smoothly as a function of position. This is analogous to describing the mass distribution of a solid, liquid, or gas by a smooth mass density (kg/m^3) which ignores

Quantization of charge

the fact that, on a microscopic scale, the mass is concentrated in atoms. A solid, liquid, or gas seems smooth because there are very many atoms in each cubic millimeter (about 10^{16} atoms/mm³ in a gas under standard conditions), and the distances between atoms are very small. Likewise, charge distributions placed on wires or other conductors will seem smooth because they consist of very many electrons (or protons) in each cubic millimeter.

Conservation of charge

The electric charge is a **conserved quantity**: in any reaction involving charged particles, the total charges before and after the reaction are always the same. Here are some examples of reactions in which particles are destroyed, yet the net electric charge remains constant:

matter–antimatter annihilation:

$$[\text{electron}] + [\text{antielectron}] \rightarrow 2[\text{photons}]$$
$$\textit{charges:} \quad -e \quad + \quad e \quad \rightarrow \quad 0 \qquad (8)$$

radioactive disintegration of a neutron:

$$[\text{neutron}] \rightarrow [\text{proton}] + [\text{electron}] + [\text{antineutrino}]$$
$$\textit{charges:} \quad 0 \quad \rightarrow \quad e \quad + \quad (-e) \quad + \quad 0 \qquad (9)$$

pion creation in a high-energy proton–proton collision:

$$[\text{proton}] + [\text{proton}] \rightarrow [\text{neutron}] + [\text{proton}] + [\text{pion}, \pi^+]$$
$$\textit{charges:} \quad e \quad + \quad e \quad \rightarrow \quad 0 \quad + \quad e \quad + \quad e \qquad (10)$$

No reaction that creates or destroys electric charge has ever been found in nature. In an effort to test the law of charge conservation, physicists have looked for hypothetical reactions, such as

$$[\text{electron}] \rightarrow [\text{neutrino}] + [\text{photon}] \qquad (11)$$

that would destroy electric charge. There is strong experimental evidence that such reactions never happen.

Electric charge is, of course, also conserved in chemical reactions. For instance, in a lead–acid battery (automobile battery), plates of lead and of lead dioxide are immersed in an electrolytic solution of sulfuric acid (Figure 22.6). The reactions that take place on these plates involve sulfate ions (SO_4^{--}) and hydrogen ions (H^+); the reactions release electrons at the lead plate and absorb electrons at the lead-dioxide plate:

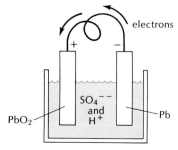

Fig. 22.6 Lead–acid battery.

at lead plate:

$$\text{Pb} + SO_4^{--} \rightarrow \text{PbSO}_4 + 2[\text{electrons}] \qquad (12)$$
$$\textit{charges:} \quad 0 + (-2e) \quad \rightarrow \quad 0 \quad + \quad (-2e)$$

at lead-dioxide plate:

$$\text{PbO}_2 + 4H^+ + SO_4^{--} + 2[\text{electrons}] \rightarrow \text{PbSO}_4 + 2H_2O \qquad (13)$$
$$\textit{charges:} \quad 0 \; + \; 4e \; + \; (-2e) + \quad (-2e) \quad \rightarrow \quad 0 \; + \; 0$$

The plates of such a battery are connected by an external circuit (for example, a wire), and the electrons released by the reaction (12) travel from one plate to the other via this external circuit, forming an electric current.

EXAMPLE 4. A fully "charged" battery contains a large amount of sulfuric acid in the electrolytic solution (H_2SO_4 in the form of SO_4^{--} ions and H^+ ions). As the battery delivers electric charge to the external circuit connecting its terminals, the amount of sulfuric acid in solution gradually decreases. Suppose that while discharging completely, the positive terminal of an automobile battery delivers a charge of 1.8×10^5 C through the external circuit. How many grams of sulfuric acid will be used up in this process?

SOLUTION: Since each electron has a charge of -1.6×10^{-19} C, the number of electrons in -1.8×10^5 C is $1.8 \times 10^5 / 1.6 \times 10^{-19}$, or 1.1×10^{24}. According to the reactions (12) and (13), whenever two electrons are transferred from the lead to the lead-dioxide plate, two sulfate ions are absorbed (one at each plate). Thus, 1.1×10^{24} sulfate ions will be absorbed, that is, 1.1×10^{24} molecules of sulfuric acid will be used up. The required number of moles of sulfuric acid is therefore $1.1 \times 10^{24} / 6.02 \times 10^{23} = 1.9$ and, since the molecular mass of sulfuric acid is 98 g per mole, the required mass of sulfuric acid is 1.9×98 g $= 183$ g.

The conservation of electric charge in chemical reactions, such as the reactions (12) and (13), is a trivial consequence of the conservation of electrons and protons. All such reactions involve nothing but a rearrangement of the electrons and of the nuclei containing the protons; during this rearrangement, the numbers of electrons and protons remain constant. Obviously, the net electric charge must then also remain constant.

The same argument applies to all macroscopic electric processes, such as the operation of electrostatic machines and generators, the flow of currents on wires, the storage of charge in capacitors, and the electric discharge of thunderclouds. All these processes involve nothing but a rearrangement of electrons and protons. Consequently, the net electric charge must remain constant.

22.4 Conductors and Insulators; Frictional Electricity

A **conductor** — such as copper, aluminum, or iron — is a material that permits the motion of electric charge through its volume. An **insulator** — such as glass, porcelain, rubber, or nylon — is a material that does not permit the motion of electric charge. Thus, when we place some electric charge on one end of a conductor, it immediately spreads out over the entire conductor until it finds an equilibrium distribution;[2] when we place some charge on one end of an insulator, it stays in place.

Conductors and insulators

All metals are good conductors. The motion of charge in metals is due to the motion of electrons. In a metal, some of the electrons of each atom are **free;** that is, they are not bound to any particular atom, although they are bound to the metal as a whole. The free electrons come from the outer parts of the atoms. The outer electrons of the atom are not very strongly attached and readily come loose; the inner electrons are firmly bound to the nucleus of the atom and are likely to

Free electrons

---

[2] We will study the conditions for the equilibrium of electric charge on a conductor in Chapter 24. It turns out that when the charges finally reach equilibrium, they will all sit on the surface of the conductor (provided the conductor is homogeneous).

stay put. The free electrons wander through the entire volume of the metal, suffering occasional collisions, but they experience a restraining force only when they hit the surface of the metal. The electrons are held inside the metal in much the same way as the particles of a gas are held inside a container — the particles of gas can wander through the volume of the container, but they are restrained by the walls. In view of this analogy, electrons in a metallic conductor are often said to form a **free-electron gas.** If one end of a metallic conductor has an excess or deficit of electrons, the motion of the free-electron gas will quickly distribute this excess or deficit to other parts of the metallic conductor.

Free-electron gas

The charging of a body of metal is usually accomplished by the removal or addition of electrons. A body will acquire a net positive charge if electrons are removed and a net negative charge if electrons are added. Thus, positive charge on a body of metal is simply a deficit of electrons, and negative charge an excess of electrons.

Liquids containing ions (atoms or molecules with missing electrons or with excess electrons) also are good conductors. For instance, a solution of common salt in water contains ions of Na^+ and Cl^-. The motion of charge through the liquid is due to the motion of these ions. Liquid conductors with an abundance of ions are called **electrolytes.**

Incidentally: Very pure, distilled water is a poor conductor, but ordinary water is a good conductor because it contains some ions contributed by dissolved impurities. The ubiquitous water in our environment makes many substances into conductors. For example, earth (soil) is a reasonably good conductor, mainly because of the presence of water. Furthermore, on a humid day many insulators acquire a microscopic surface film of water, and this permits electric charge to leak away along the insulator; thus, on humid days, it is difficult to store electric charge on bodies supported by insulators.

Fig. 22.7 Lightning.

Ordinary gases are insulators, but ionized gases are good conductors. For example, ordinary air is an insulator, but the ionized air found in the path of a lightning bolt is a good conductor (Figure 22.7). This ionized air contains a mixture of positive ions and free electrons; the motion of charge in such a mixture is due mainly to the motion of the electrons. Such an ionized gas is called a **plasma.**

Plasma

We will end this chapter with a few brief comments on frictional electricity. It is easy to accumulate electric charge on a glass rod merely by rubbing it with a piece of silk. The silk becomes negatively charged and the glass positively charged. The rubbing motion between the surfaces of the silk and the glass rips charges off one of these surfaces and makes them stick to the other, but the detailed mechanism is not well understood. It is believed that what is usually involved is a transfer of ions from one surface to the other. Contaminants residing on the rubbed surfaces play a crucial role in frictional electricity. If glass is rubbed with an absolutely clean piece of silk or other textile material, the glass becomes negatively charged rather than positively charged — ordinary pieces of silk apparently have such a large amount of dirt on their surfaces that the charging process is dominated by the dirt rather than by the silk. Even air can act as contaminant for some surfaces; for instance, careful experiments on the rubbing of platinum with silk show that in vacuum the platinum becomes negatively charged, but that in air it becomes positively charged.

The electric charge that can be accumulated on the surface of a body of ordinary size (a centimeter or more) by rubbing may be as much as 10^{-9} to 10^{-8} coulomb per square centimeter. If the charge concentration on a body is higher than that, it will cause an electric discharge into the surrounding air (a corona discharge; see Section VI.4); a higher charge concentration can subsist only on a small body or on a small spot on a large body.

Once we have accumulated some charge on, say, a rod of glass, we can produce charges on other bodies by a process of **induction,** as follows: First we bring the glass rod near a metallic body supported on an insulating stand. The positive charge on the rod will then attract free electrons to the near side of the body and leave a deficit of free electrons on the far side (Figure 22.8a); thus, the near side will acquire negative charge and the far side positive charge. If we next momentarily connect the far side to the ground, the positive charge will be neutralized by an influx of electrons from the ground (virtually, the positive charge leaks away; Figure 22.8b). This leaves the metallic body with a net negative charge. When we finally withdraw the glass rod, this charge will remain on the metallic body, distributing itself over its entire surface (Figure 22.8c).

Induction

Some old electrostatic machines accumulate large amounts of charge by a repetitive operation of induction. Figure 22.9 illustrates the principle. Two metallic bodies on insulating stands (A and A') initially carry some small positive and negative charges. In order to increase these charges, we place two small metallic bodies B and B' nearby. Charges will then be induced on B and B' as shown in Figure 22.9a. If we now momentarily connect the facing sides of B and B' with a wire, the charges on these sides will cancel, leaving B with a negative charge and B' with a positive charge (Figure 22.9b). Next, we put B' in contact with A and B in contact with A' (Figure 22.9c). The charges then immediately spread out over the contacting bodies; and, consequently, most of the charges of the small bodies are transferred to the larger bodies. The small bodies are then left (almost) uncharged, and we can

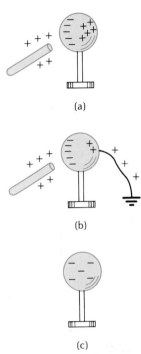

Fig. 22.8 (a) The positively charged glass rod induces a charge distribution on the metallic sphere. (b) When the far side of the sphere is connected to the ground by means of a wire, the positive charge leaks away. (c) Finally, only negative charge remains on the sphere.

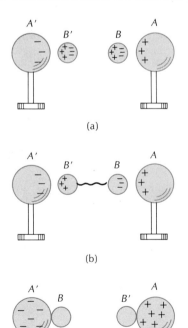

Fig. 22.9 (a) The large spheres induce charge distributions on the small spheres. (b) When the facing sides of the small spheres are connected by means of a wire, the charges of these sides cancel. (c) The large spheres absorb the charge of the small spheres.

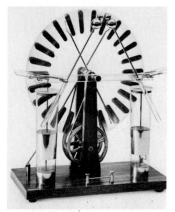

Fig. 22.10 A Wimshurst machine. The positive and negative charges generated by the machine are accumulated on the conductors in the two bottles.

again repeat the sequence of operations. The net result is a stepwise increase of the electric charges on A and A'. In electrostatic machines — such as Wimshurst's machine (Figure 22.10) — the entire sequence of manipulations of the bodies B and B' is carried out automatically when the experimenter turns a crank.

SUMMARY

Electric charge may be positive, negative, or zero; like charges repel, unlike charges attract.

Coulomb's Law: $F = \dfrac{1}{4\pi\varepsilon_0} \dfrac{qq'}{r^2}$

Permittivity constant: $\varepsilon_0 = 8.85 \times 10^{-12}$ C^2/N·m^2

$\dfrac{1}{4\pi\varepsilon_0} = 8.99 \times 10^9$ N·m^2/C^2

Charge of proton: $e = 1.60 \times 10^{-19}$ C

Charge of electron: $-e = -1.60 \times 10^{-19}$ C

Charge conservation: In any reaction the net electric charge remains constant.

Conductor: Permits motion of charge.

Insulator: Does not permit motion of charge.

QUESTIONS

1. Suppose that the Sun has a positive electric charge and each of the planets a negative electric charge, and suppose that there is no gravitational force. In what way will the motions of the planets predicted by this "electric" model of the Solar System differ from the observed motions?

2. The protons in the nucleus of an atom repel each other electrically. What holds the protons (and the neutrons) together, and prevents the nucleus from bursting apart?

3. Describe how you would set up an experiment to determine whether the electric charges on an electron and a proton are exactly the same.

4. Assume that neutrons have a small amount of electric charge, say, positive charge. Discuss some of the consequences of this assumption for the behavior of matter.

5. If we were to assign a positive charge to the electron and a negative charge to the proton, would this affect the mathematical statement [Eq. (7)] of Coulomb's Law?

6. In the cgs, or Gaussian, system of units, Coulomb's Law is written as $F = qq'/r^2$. In terms of grams, centimeters, and seconds, what are the units of electric charge in this system?

7. Could we use the electric charge of an electron as an atomic standard of electric charge to define the coulomb? What would be the advantages and disadvantages of such a standard?

8. Besides electric charge, what other physical quantities are conserved in reactions among particles? Which of these quantities are quantized?

9. Since the free electrons in a piece of metal are free to move any which way, why don't they all fall to the bottom of the piece of metal under the influence of the pull of gravity?

10. If the surface of a piece of metal acts like a container in confining the free electrons, why can't we cause these electrons to spill out by drilling holes in the surface?

11. Water is a conductor, but (dry) snow is an insulator. How can this be?

12. If you rub a plastic comb, it will attract hairs or bits of paper, even though they have no net electric charge. Explain.

13. Some old-fashioned physics textbooks define positive electric charge as the kind of charge that accumulates on a glass rod when rubbed with silk. What is wrong with this definition?

14. When you rub your shoes on a carpet, you sometimes pick up enough electric charge to feel an electric shock if you subsequently touch a radiator or some other metallic body connected to the ground. Why is this more likely to happen in winter than in summer?

15. Some automobile operators hang a conducting strap on the underside of their automobile, so that this strap drags on the street. What is the purpose of this arrangement?

16. An amount of electric charge has been deposited on a Ping-Pong ball. How could you find out whether the charge is positive or negative?

17. Two aluminum spheres of equal radii hang from the ceiling on insulating threads. You have a glass rod and a piece of silk. How can you give these two spheres exactly equal amounts of electric charge?

PROBLEMS

Section 22.2

1. Within a typical thundercloud there are electric charges of -40 C and $+40$ C separated by a vertical distance of 5 km. Treating these charges as pointlike, find the magnitude of the electric force of attraction between them.

2. A crystal of NaCl (common salt) consists of a regular arrangement of ions of Na$^+$ and Cl$^-$. The distance from one ion to its neighbor is 2.82×10^{-10} m. What is the magnitude of the electric force of attraction between the two ions? Treat the ions as point charges.

3. Suppose that the two protons in the nucleus of a helium atom are at a distance of 2×10^{-15} m from each other. What is the magnitude of the electric force of repulsion that they exert on each other? What would be the acceleration of each if this were the only force acting on them? Treat the protons as point charges.

4. An alpha particle (charge $+2e$) is launched at high speed toward a nucleus of uranium (charge $+92e$). What is the magnitude of the electric force on the alpha particle when it is at a distance of 5×10^{-14} m from the nucleus? What is the corresponding instantaneous acceleration of the alpha particle? Treat the alpha particle and the nucleus as point charges.

5. According to recent theoretical and experimental investigations, the subnuclear particles are made of quarks and of antiquarks (see Chapter 46). For example, a positive pion is made of a u quark and a d antiquark. The electric charge on the u quark is $\frac{2}{3}e$ and that on the d antiquark is $\frac{1}{3}e$. Treating the quarks as classical particles, calculate the electric force of attraction between the quarks in the pion if the distance between them is 1.0×10^{-15} m.

6. How many electrons do we need to remove from an initially neutral bowling ball to give it a charge of 1 C?

7. According to recent theoretical speculations, there might exist elementary particles of a mass 10^{-11} kg. If such a particle carries a charge e, what is the gravitational force and what is the electric force it exerts on an identical particle placed at a distance of 10^{-10} m?

8. In the lead atom, the nucleus has an electric charge $82e$. The innermost electron in this atom is typically at a distance of 6.5×10^{-13} m from the nucleus. What is the electric force that the nucleus exerts on such an electron? What is the acceleration that this force produces on the electron? Treat the electron as a classical particle.

9. The electric charge in one mole of protons is called **Faraday's constant.** What is its numerical value?

10. The electric charge flowing through an ordinary 110-volt, 150-watt light bulb is 1.5 C/s. How many electrons per second does this amount to?

11. A maximum electric charge of 7.5×10^{-6} C can be placed on a metallic sphere of radius 15 cm before the surrounding air suffers electric breakdown (sparks). How many excess electrons (or missing electrons) does the sphere have when breakdown is about to occur?

12. How many electrons are in a paper clip of iron of mass 0.3 g?

13. Suppose that you remove all the electrons in a copper penny of mass 2.7 g and place them at a distance of 2.0 m from the remaining copper ions. What is the electric force of attraction on the electrons?

*14. What is the number of electrons and of protons in a human body of mass 73 kg? The chemical composition of the body is given in Problem 1.26.

*15. Deimos is a small moon of Mars, with a mass of 2×10^{15} kg. Suppose that an electron is at a distance of 100 km from Deimos. What is the gravitational attraction acting on the electron? What negative electric charge would have to be placed on Deimos to balance this gravitational attraction? How many electron charges does this amount to? Treat the mass and the charge distributions as pointlike in your calculations.

*16. A small charge of 2×10^{-6} C is at the point $x = 2$ m, $y = 3$ m in the x–y plane. A second small charge of -3×10^{-6} is at the point $x = 4$ m, $y = -2$ m. What is the electric force that the first charge exerts on the second? What is the force that the second charge exerts on the first? Express your answers as vectors, with x and y components.

*17. Two tiny chips of plastic of masses 5×10^{-5} g are separated by a distance of 1 mm. Suppose that they carry equal and opposite electrostatic charges. What must the magnitude of the charge be if the electric attraction between them is to equal their weight?

*18. How many extra electrons would we have to place on the Earth and on the Moon so that the electric repulsion between these bodies cancels their gravitational attraction? Assume that the numbers of extra electrons on the Earth and on the Moon are in the same proportion as the radial dimensions of these bodies (6.38:1.74).

*19. At a place directly below a thundercloud, the induced electric charge on the surface of the Earth is $+10^{-7}$ coulomb per square meter of surface. How many singly charged positive ions per square meter does this represent? The number of atoms on the surface of a solid is typically 3×10^{20} per square meter. What fraction of these atoms must be ions to account for the above electric charge?

*20. Although the best available experimental data are consistent with Coulomb's Law, they are also consistent with a modified Coulomb's Law of the form

$$F = \frac{1}{4\pi\varepsilon_0} \frac{q_1 q_2}{r^2} e^{-r/r_0}$$

where r_0 is a constant with the dimensions of a length and a numerical value which is known to be no less than 10^9 m and is probably much larger. Assuming that $r_0 = 10^9$ m, what is the fractional deviation between Coulomb's Law and the modified Coulomb's Law for $r = 1$ m? For $r = 10^4$ m? (Hint: Use the approximation $e^x \cong 1 + x$ for small x.)

*21. A proton is at the origin of coordinates. An electron is at the point $x = 0.40$ Å, $y = 0.20$ Å, $z = 0.15$ Å. What are the x, y, and z components of the electric force that the proton exerts on the electron? That the electron exerts on the proton?

*22. Precise experiments have established that the magnitudes of the electric charges of an electron and a proton are equal to within an experimental error of $\pm 10^{-21} e$ and that the electric charge of a neutron is zero to within $\pm 10^{-21} e$. Making the worst possible assumption about the combination of errors, what is the largest conceivable electric charge of an oxygen atom consisting of 8 electrons, 8 protons, and 8 neutrons? Treating the atoms as point particles, compare the electric force between two such oxygen atoms with the gravitational force between these atoms. Is the net force attractive or repulsive?

*23. Suppose that under the influence of the electric force of attraction, the electron in a hydrogen atom orbits around the proton on a circle of radius 0.53×10^{-10} m. What is the orbital speed? What is the orbital period?

**24. According to Bohr's theory of the atom, the electron in a hydrogen atom orbits around the nucleus in a circular orbit. The force that holds the electron in this orbit is the Coulomb force. The size of the orbit depends on the angular momentum — the smallest possible orbit has an angular momentum $\hbar = 1.05 \times 10^{-34}$ J·s; the next possible orbit has angular momentum $2\hbar$; the next, $3\hbar$; etc.

(a) Calculate the radius of each of these three possible circular orbits.
(b) In general, show that if a circular orbit has an angular momentum $n\hbar$ (where $n = 1, 2, 3, \ldots$), then its radius is

$$r = \frac{4\pi\varepsilon_0}{m_e e^2} n^2 \hbar^2$$

(c) Evaluate this radius for $n = 1$.

Section 22.3

25. Consider the following hypothetical reactions involving the collision between a high-energy proton (from an accelerator) and a stationary proton (in the nucleus of a hydrogen atom serving as target):

$$p + p \to n + n + \pi^+$$

$$p + p \to n + p + \pi^0$$

$$p + p \to n + p + \pi^+$$

$$p + p \to p + p + \pi^0 + \pi^0$$

$$p + p \to n + p + \pi^0 + \pi^-$$

where the symbols p, n, π^+, π^-, and π^0 stand for proton, neutron, positively charged pion, negatively charged pion, and neutral pion. Which of these reactions are impossible?

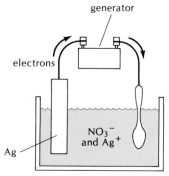

Fig. 22.11

26. Consider the reaction

$$Ni^{++} + 4H_2O \rightarrow NiO_4^{--} + 8H^+ + [electrons]$$

How many electrons does this reaction release?

*27. We can silver-plate a metallic object, such as a spoon, by immersing the spoon and a bar of silver in a solution of silver nitrate ($AgNO_3$). If we then connect the spoon and the silver bar to an electric generator and make a current flow from one to the other, the following reactions will occur at the immersed surfaces (Figure 22.11):

$$Ag^+ + [electron] \rightarrow Ag_{(metal)}$$

$$Ag_{(metal)} \rightarrow Ag^+ + [electron]$$

The first reaction deposits silver on the spoon, and the second removes silver from the silver bar. How many electrons must we make flow from the silver bar to the spoon in order to deposit 1 g of silver on the spoon?

CHAPTER 23

The Electric Field

Up to here, we have taken the view that the gravitational forces and the electric forces between particles are **action-at-a-distance;** that is, a particle exerts a direct gravitational or electric force on another particle even though these particles are not touching. Although such an interpretation of gravitational and electric forces as a ghostly tug-of-war between distant particles is suggested by Newton's Law of gravitation and by Coulomb's Law of electric force, Newton himself expressed considerable misgivings about this interpretation. In his own words:

Action-at-a-distance

> It is inconceivable, that inanimate brute matter, should, without the mediation of something else, which is not material, operate upon and affect other matter without mutual contact. That Gravity should be innate, inherent and essential to Matter so that one Body may act upon another at a Distance thro' a *Vacuum* without the Mediation of anything else, by and through which their Action and Force may be conveyed from one to another, is to me so great an Absurdity that I believe no Man who has in philosophical Matters a competent Faculty of thinking can ever fall into it. Gravity must be caused by an Agent acting constantly according to certain Laws; but whether this Agent be material or immaterial, I have left to the consideration of my readers.

According to the modern view, there is indeed a "material agent" that acts as mediator of force, conveying the force over the distance from one body to another. This agent is the **field.** A gravitating or electrically charged body generates a gravitational or electric field, which permeates the (apparently) empty space around the body and exerts pushes or pulls whenever it comes in contact with another body. Thus, fields convey forces from one body to another through **action-by-contact.**

Field

Action-by-contact

In the present chapter, we will become acquainted with the electric field that conveys the electric force from one charged body to another charged body. But before we discuss the electric field, we will briefly examine what happens to electric forces when many electric point charges interact simultaneously.

23.1 The Superposition of Electric Forces

According to Coulomb's Law, the electric force exerted by a point charge q' on a point charge q is

$$\mathbf{F} = \frac{1}{4\pi\varepsilon_0} \frac{qq'}{r^2} \hat{\mathbf{r}} \tag{1}$$

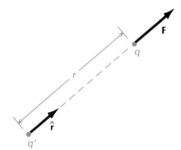

Fig. 23.1 A point charge q' exerts an electric force **F** on the point charge q.

where $\hat{\mathbf{r}}$ is a unit vector directed from the charge q' toward the charge q (Figure 23.1).

If several point charges q_1, q_2, q_3, ..., simultaneously exert electric forces on the charge q, then the collective force on q is obtained by taking the vector sum of the individual forces. Thus, the force on q is

$$\mathbf{F} = \mathbf{F}_1 + \mathbf{F}_2 + \mathbf{F}_3 + \cdots$$

$$= \frac{1}{4\pi\varepsilon_0} \frac{qq_1}{r_1^2} \hat{\mathbf{r}}_1 + \frac{1}{4\pi\varepsilon_0} \frac{qq_2}{r_2^2} \hat{\mathbf{r}}_2 + \frac{1}{4\pi\varepsilon_0} \frac{qq_3}{r_3^2} \hat{\mathbf{r}}_3 + \cdots \tag{2}$$

where r_1, r_2, r_3, ..., and $\hat{\mathbf{r}}_1$, $\hat{\mathbf{r}}_2$, $\hat{\mathbf{r}}_3$, ..., are the distances and unit vectors, respectively (Figure 23.2).

Principle of superposition

Equation (2) expresses the **principle of superposition** of electric forces. According to Eq. (2), the force contributed by each charge is independent of the presence of the other charges. For instance, charge q_2 does not affect the interaction of q_1 with q; it merely adds its own interaction with q. This simple combination law is an important *empirical* fact about electric forces. Since the contact forces (pushes and pulls) of everyday experience are electric forces, they will likewise obey the superposition principle, and they can be combined by simple vector addition. Incidentally: The gravitational forces on the Earth and within the Solar System also obey the superposition principle. Thus, all the forces in our immediate environment obey this principle (see Section 5.3).

The next two examples involve the summation or the integration of the electric forces contributed by discrete or continuous systems of charges.

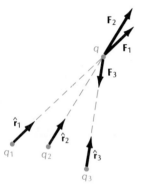

Fig. 23.2 Point charges q_1, q_2, q_3 exert electric forces $\mathbf{F}_1$, $\mathbf{F}_2$, $\mathbf{F}_3$ on the charge q.

EXAMPLE 1. Point charges Q and $-Q$ are separated by a distance d. A point charge q is equidistant from these charges, at a distance x from their midpoint (Figure 23.3). What is the electric force on q?

SOLUTION: With the choice of axes shown in Figure 23.3, the distance between the charges $\pm Q$ and q is $\sqrt{x^2 + d^2/4}$. Hence the magnitudes of the Coulomb forces exerted by $+Q$ and $-Q$ are

$$F_1 = F_2 = \frac{1}{4\pi\varepsilon_0} \frac{qQ}{x^2 + d^2/4} \tag{3}$$

The vector $\mathbf{F}_1$ is directed away from $+Q$ and $\mathbf{F}_2$ is directed toward $-Q$. In the vector sum $\mathbf{F}_1 + \mathbf{F}_2$, the x components obviously cancel and only the z component survives. The latter component has a magnitude

$$F_z = -\frac{1}{4\pi\varepsilon_0}\frac{qQ}{x^2 + d^2/4}\cos\theta - \frac{1}{4\pi\varepsilon_0}\frac{qQ}{x^2 + d^2/4}\cos\theta \tag{4}$$

With $\cos\theta = \frac{1}{2}d/(x^2 + d^2/4)^{1/2}$, this gives

$$F_z = -\frac{1}{4\pi\varepsilon_0}\frac{qQd}{(x^2 + d^2/4)^{3/2}} \tag{5}$$

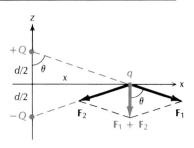

Fig. 23.3 The charges $+Q$ and $-Q$ exert forces $\mathbf{F}_1$ and $\mathbf{F}_2$ on the charge q.

COMMENTS AND SUGGESTIONS: Note that if the charge q is at large distance from the two charges $\pm Q$, then d^2 can be neglected compared with x^2, so $(x^2 + d^2/4)^{3/2} \cong x^3$. The force is then proportional to $F \propto 1/x^3$, that is, the force decreases in inverse proportion to the cube of the distance. Thus, although the force contributed by each point charge $\pm Q$ is an inverse-square force, the net force has a quite different behavior, because at large distance the force contributed by one charge tends to cancel the force generated by the other charge.

EXAMPLE 2. Charge is distributed uniformly along a very long, straight, thin line (say, along a string of silk). If the amount of the charge is λ coulomb per meter of line, what is the electric force on a charge q placed near the line?

SOLUTION: We will pretend that the line of charge is of infinite length; this simplifies the calculation and introduces no appreciable error since, in any case, most of the force on q is contributed by those portions of the line which are nearest to q.

Figure 23.4 shows the line of charge lying along the z axis and the charge q on the x axis, at a distance x from the origin. To find the force, we must perform an integration, regarding the line of charge as made up of infinitesimal line elements, each of which can be treated as a point charge. Before proceeding with such an integration, it is always best to determine the *direction* of the force by a preliminary, qualitative argument. The force generated by the line element dz shown in Figure 23.4 has both x and z components. Upon integration, the z component will cancel against the z component contributed by a line element lying at the same distance above the origin. Therefore the net force will have only an x component.

The line element dz carries a charge $dQ = \lambda\, dz$. Since this line element can be treated as a point charge, it generates a force

$$dF = \frac{1}{4\pi\varepsilon_0}\frac{q\, dQ}{r^2} = \frac{1}{4\pi\varepsilon_0}\frac{q\lambda\, dz}{r^2} \tag{6}$$

This force has an x component:

$$dF_x = \frac{1}{4\pi\varepsilon_0}\frac{q\lambda \cos\theta\, dz}{r^2} \tag{7}$$

The net force is then

$$F_x = \int_{-\infty}^{+\infty}\frac{1}{4\pi\varepsilon_0}\frac{q\lambda \cos\theta}{r^2}\, dz \tag{8}$$

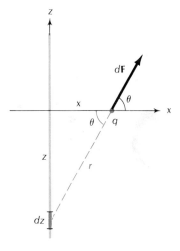

Fig. 23.4 A small segment dz of a line of charge exerts a force on the charge q. The angle θ is reckoned as positive if z is positive, and negative if z is negative.

To evaluate this, it is convenient to express all variables in terms of θ. From Figure 23.4,

$$z = x\tan\theta \tag{9}$$

that is,

$$dz = x \sec^2 \theta \, d\theta \tag{10}$$

Furthermore,

$$r = x \sec \theta \tag{11}$$

With these substitutions we obtain an integral over the angle θ; in terms of this angle, the limits of integration are $\theta = -90°$ and $\theta = 90°$, or, in radians, $\theta = -\pi/2$ and $\theta = \pi/2$:

$$F_x = \frac{1}{4\pi\varepsilon_0} \frac{q\lambda}{x} \int_{-\pi/2}^{\pi/2} \cos\theta \, d\theta \tag{12}$$

Since

$$\int_{-\pi/2}^{\pi/2} \cos\theta \, d\theta = \left[\sin\theta\right]_{-\pi/2}^{\pi/2} = 2$$

we find

$$F_x = \frac{1}{2\pi\varepsilon_0} \frac{q\lambda}{x} \tag{13}$$

COMMENTS AND SUGGESTIONS: This force decreases in inverse proportion to the distance (not the square of the distance). The direction of the force is radially away from the line of charge. The result (13) is exact for an infinite straight line of charge; but it is also a reasonable approximation for a finite line, provided we stay near the line and away from its ends.

The solution of this example illustrates the typical procedure for the evaluation of the electric force exerted by a continuous charge distribution. Such a charge distribution can always be regarded as made up of infinitesimal charge elements, each of which can be treated as a point charge. To perform the evaluation of the net force, begin by making some convenient choice of coordinate axes. Then consider the x, y, and z components of the electric force one by one, and try to decide if any component is zero, because of cancellations brought about by the symmetry of the charge distribution. Components that are not zero must be evaluated by integration over all the infinitesimal charge elements. The choice of variable of integration is often a delicate matter, since the integrand will look very simple for some choices of variable of integration, and very complicated and intractable for others. In the above calculation, the integral turns out to be elementary if θ is used as variable of integration, but rather more difficult if z is used as variable of integration. When faced with an apparently intractable integral, try substitutions of variables (or try finding the integral in a table).

23.2 The Electric Field

As we remarked in the introduction to this chapter, the naive interpretation of Coulomb's Law is that the electric force between charges is action-at-a-distance, that is, a charge q' exerts a direct force on a charge q even though these charges are separated by a large distance and are not touching. However, such an interpretation of electric force leads to serious difficulties in the case of moving charges. Suppose that we suddenly move the charge q' somewhat nearer to the charge q; then the electric force has to increase. But the required increase cannot occur instantaneously — the increase can be regarded as a signal from q' to q, and it is a fundamental principal of physics, based on the theory of Special Relativity, that no signal can propagate faster than the speed of light (this will be discussed in Section 41.3). The time delay in the increase of the electric force suggests that, when we suddenly move the charge q', some kind of disturbance travels through empty space from q' to q at the speed of light and that, when this disturbance reaches q, it adjusts the electric force to the new increased value (Figure 23.5). Thus charges exert forces on one another by means of disturbances that they generate in the space surrounding them. These disturbances are called **electric fields.**

Fields are a form of matter — they are endowed with energy and momentum, and they therefore exist in a material sense. If we think of solids, liquids, gases, and plasmas as the first four states of matter, then fields are the fifth state of matter. In the context of the above example, it is easy to see why the disturbance, or field, generated by the sudden displacement of q' must carry momentum: when we suddenly move q' toward q, the force on q' immediately increases according to Coulomb's Law, but the increase in the force on q will be delayed until a signal has had time to propagate from q' carrying the information regarding the changed position of q'. Thus, action and reaction will be temporarily out of balance. We know from Chapter 10 that the conservation of momentum in a system of particles hinges on the balance of action and reaction — any imbalance implies that the momentum of our system of two charged particles will not be conserved. In order to maintain an overall momentum conservation, the momentum missing from the particles must be transferred to the electric field; that is, the field must acquire momentum. This special case of the two charged particles is representative of the general case. It can be demonstrated rigorously that, when relativity is taken into account, the momentum and energy of a general system of interacting particles cannot be conserved by itself. An extra entity, such as the field, is needed to take up the momentum and energy missing from the particles.

Although the above arguments for the existence of fields arose from the problem of charges in motion, we will now adopt the very natural view that the forces on charges at rest involve the same mechanism. We suppose that each charge at rest generates a permanent, static disturbance in the space surrounding it, and that this disturbance exerts forces on other charges. Thus we take the view that the electric interaction between charges is action-by-contact: a charge q' generates a field which fills the surrounding space and exerts forces on any other charges that it touches. The electric field serves as the mediator of the force according to the scheme

charge q'→electric field of charge q'→force on charge q (14)

The formal mathematical definition of the electric field is as follows: To discover the field at a given position, take a point charge q (a "test charge") and place it at that position. The charge q will then feel an electric force **F**; the electric field **E** is defined as the force **F** divided by the magnitude of charge q,

$$\boxed{\mathbf{E} = \mathbf{F}/q} \qquad (15)$$

Electric field

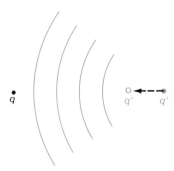

Fig. 23.5 A disturbance emanates from q' and reaches q.

Relation between electric force and field

Thus, the electric field is simply the force per unit positive charge.[1]

The unit of electric field is the newton/coulomb.[2] Table 23.1 gives the magnitudes of some typical electric fields.

Table 23.1 SOME ELECTRIC FIELDS

At surface of uranium nucleus	2×10^{21} N/C
At surface of pulsar	$\sim 10^{14}$ N/C
At orbit of electron in hydrogen atom	6×10^{11} N/C
In X-ray tube	5×10^6 N/C
Electrical breakdown of air	3×10^6 N/C
In Van de Graaff accelerator	2×10^6 N/C
Within lightning bolt	10^4 N/C
Under thundercloud	1×10^4 N/C
Near radar transmitter (FPS-6)	7×10^3 N/C
In sunlight (rms)	1×10^3 N/C
In atmosphere (fair weather)	1×10^2 N/C
In beam of small laser (rms)	1×10^2 N/C
In fluorescent lighting tube	10 N/C
In radio wave	$\sim 10^{-1}$ N/C
Within household wiring	$\sim 3 \times 10^{-2}$ N/C
In thermal radiation in intergalactic space (rms)	3×10^{-6} N/C

The electric field surrounding a charge or a distribution of charges is a function of position. For example, according to Coulomb's Law, a point charge q' exerts a force $(1/4\pi\varepsilon_0)q'q\hat{\mathbf{r}}/r^2$ on a charge q; hence the electric field generated by q' at a distance r is

Electric field of point charge

$$\mathbf{E} = \frac{1}{4\pi\varepsilon_0} \frac{q'}{r^2} \hat{\mathbf{r}} \qquad (16)$$

Note that the electric field is a vector. Its direction must be specified either by unit vectors [as in Eq. (16)] or else by components.

The net electric field generated by any distribution of point charges can be calculated by superposition of the individual fields of the point charges, in much the same way as is done for electric forces.

EXAMPLE 3. A total amount of charge Q is distributed uniformly along the circumference of a thin glass ring of radius R. What is the electric field on the axis of the ring?

SOLUTION: We regard the ring as made up of infinitesimal line elements ds (Figure 23.6), each of which can be treated as a point charge. Before proceeding with the integration of the contributions of all such line elements, let us determine the direction of the net electric field. The field generated by the line element ds shown in Figure 23.6 has both a horizontal and a vertical component. Obviously, for any given line element there is an equal line element on

---

[1] The procedure involved in this definition of the electric field implicitly assumes that all the charges that generate the field **E** remain fixed in their positions while the test charge is brought up. To avoid disturbances to these charges, it is usually convenient to take a very small charge q.

[2] Newton/coulomb is the same thing as volt/meter (see Section 25.1).

the opposite side of the ring's center which contributes an electric field of opposite horizontal component, and all the horizontal components cancel pairwise. The net electric field is therefore vertical.

The charge per unit length along the circumference is $Q/2\pi R$; hence, the charge in the line element ds is $dQ = ds(Q/2\pi R)$. At a height z above the plane of the ring (Figure 23.6), the electric field contributed by the charge element dQ has a magnitude

$$dE = \frac{1}{4\pi\varepsilon_0} \frac{dQ}{r^2} = \frac{1}{4\pi\varepsilon_0} \frac{Q\,ds}{2\pi R} \frac{1}{z^2 + R^2} \tag{17}$$

This field has a vertical component

$$dE_z = \frac{1}{4\pi\varepsilon_0} \frac{Q\,ds}{2\pi R} \frac{\cos\theta}{z^2 + R^2} \tag{18}$$

Consequently, the net electric field is

$$E_z = \int \frac{1}{4\pi\varepsilon_0} \frac{Q}{2\pi R} \frac{\cos\theta}{z^2 + R^2}\,ds \tag{19}$$

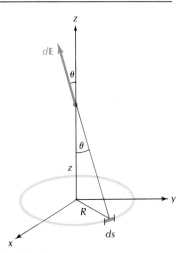

Fig. 23.6 A small segment ds of a ring of charge contributes an electric field $d\mathbf{E}$.

Since the integrand has the same value at all points of the circumference, the integral is of the form [constant] $\times \int ds$; this equals [constant] $\times 2\pi R$, since the length of the circumference is $2\pi R$. Hence, with $\cos\theta = z/\sqrt{z^2 + R^2}$,

$$E_z = \frac{1}{4\pi\varepsilon_0} \frac{Q\cos\theta}{z^2 + R^2} = \frac{1}{4\pi\varepsilon_0} \frac{Qz}{(z^2 + R^2)^{3/2}} \tag{20}$$

In vector notation,

$$\mathbf{E} = \frac{1}{4\pi\varepsilon_0} \frac{Qz}{(z^2 + R^2)^{3/2}}\,\hat{\mathbf{z}} \tag{21}$$

COMMENTS AND SUGGESTIONS: Note that for $z = 0$, $\mathbf{E} = 0$; and for $z \gg R$, $\mathbf{E} \cong Q\hat{\mathbf{z}}/(4\pi\varepsilon_0 z^2)$, which is the electric field of a point charge. These results for these two extreme cases agree with our expectations.

EXAMPLE 4. What is the electric field generated by a large flat sheet, such as a sheet of paper, carrying a uniform charge density of σ coulomb per square meter?

SOLUTION: We will pretend that the sheet is infinitely large. The sheet can be regarded as made up of a collection of many concentric rings. Figure 23.7 shows one of these rings with radius R, width dR; this ring has an area $2\pi R\,dR$ and a charge $dQ = (2\pi R\,dR) \times \sigma$. It produces a vertical electric field [see Eq. (21)],

$$dE = \frac{1}{4\pi\varepsilon_0} \frac{2\pi R\sigma z\,dR}{(z^2 + R^2)^{3/2}} \tag{22}$$

The net electric field is therefore

$$E = \frac{2\pi\sigma z}{4\pi\varepsilon_0} \int_0^\infty \frac{R\,dR}{(z^2 + R^2)^{3/2}} \tag{23}$$

where $R = 0$ is the radius of the smallest ring and $R = \infty$ the radius of the largest.

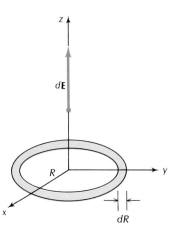

Fig. 23.7 A very large sheet of charge lies in the x–y plane. A thin ring of charge within this sheet produces an electric field $d\mathbf{E}$.

With the substitution of $u = R^2$, the integral becomes

$$\int_0^\infty \frac{R\,dR}{(z^2+R^2)^{3/2}} = \int_0^\infty \frac{(1/2)\,du}{(z^2+u)^{3/2}} = \left[\frac{-1}{(z^2+u)^{1/2}}\right]_0^\infty = \frac{1}{z}$$

and therefore

$$E = \frac{2\pi\sigma z}{4\pi\varepsilon_0}\frac{1}{z} = \frac{\sigma}{2\varepsilon_0} \tag{24}$$

In vector notation,

Electric field of flat sheet

$$\boxed{\mathbf{E} = \frac{\sigma}{2\varepsilon_0}\hat{\mathbf{z}}} \tag{25}$$

COMMENTS AND SUGGESTIONS: This electric field is proportional to the charge density, and it is constant, i.e., it is independent of the distance from the sheet. Although the result is strictly valid only for the case of an infinitely large sheet, it is also a good approximation for a sheet of finite size, provided we stay within a distance much smaller than the size of the sheet and we stay away from the vicinity of the edges. (If we move away from the sheet to distances much larger than its size, then the electric field will resemble that of a point charge.)

In the above calculation we started with the known formula for the electric field of a charged ring, and we built up the charged sheet by integration over many concentric rings. Thus, the rings play the role of charge "elements" in the integration. If, instead of starting with rings, we had started with point charges, the calculation would have been much more difficult. In calculations of the fields of continuous charge distributions, it usually pays to start with the largest possible charge "element" that fits the geometry and has a known formula for its field.

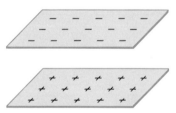

Fig. 23.8 Two very large sheets of charge.

Large charged flat sheets or plates are often used in physics laboratories to produce uniform electric fields. In practice, two parallel charged sheets with charges of opposite signs (see Figure 23.8) are preferred to a single sheet. The electric field generated by two large parallel sheets with uniform charge densities of opposite signs can be obtained by superposing the fields generated by the individual sheets. Each sheet generates an electric field of magnitude $\sigma/2\varepsilon_0$. In the space between the sheets, the two individual fields have the same direction, and they add together, giving a net field

$$E = \sigma/\varepsilon_0 \tag{26}$$

Above and below the pair of sheets, the two individual fields have opposite directions, and they cancel.

This calculation of the electric field generated jointly by two parallel sheets illustrates how the superposition principle can be exploited to combine the electric fields of two (or more) individual charge distributions. The net electric field generated jointly by the entire charge distribution is always the vector sum of the individual electric fields of the individual parts of the charge distribution. Problems involving calculations of electric fields of complicated charge distributions can often be broken up into several individual parts, with simpler individual charge distributions.

EXAMPLE 5. An electron is placed in the uniform electric field between the two charged sheets of Figure 23.8. (a) If the magnitude of the field is 3.0×10^4 N/C, what is the acceleration of the electron? (b) Suppose the electron is initially at rest on the negative sheet and then moves toward the positive sheet under the influence of the electric force. With what speed will it reach the positive sheet? The distance between the sheets is 1.0 cm.

SOLUTION: The electric field E is the force per unit charge. Hence, the magnitude of the force on the electron is $F = eE$ and the acceleration toward the positive sheet is

$$a = \frac{eE}{m_e} = \frac{(1.6 \times 10^{-19} \text{ C}) \times (3 \times 10^4 \text{ N/C})}{9.1 \times 10^{-31} \text{ kg}} = 5.3 \times 10^{15} \text{ m/s}^2$$

For constant acceleration, $v^2 - v_0^2 = 2a(x - x_0)$ [see Eq. (2.25)]. With $v_0 = 0$, $a = 5.3 \times 10^{15}$ m/s², and $x - x_0 = 0.01$ m this gives

$$v = \sqrt{2a(x - x_0)} = 1.0 \times 10^7 \text{ m/s}$$

When we use electric fields for the calculation of electric forces, we must be careful to keep track of what charges generate which electric fields. For instance, if we are dealing with a point charge placed near a uniformly charged flat sheet, then the electric field acting on the point charge is the uniform electric field (26) generated by the sheet, but the electric field acting on the sheet is the Coulomb field (16) generated by the point charge. The former field determines the force acting on the point charge, whereas the latter field determines the force acting on the sheet (these two forces are an action–reaction pair; it can be demonstrated explicitly that these two forces are of equal magnitudes and opposite directions). Of course, we could also inquire, What is the net electric field produced jointly by the uniformly charged sheet and the point charge? This net field would be relevant if we wanted to calculate the force acting on, say, a second point charge placed somewhere near the first charge and the sheet.

To maintain a sharp distinction between the field acting on a charge and the field emanating from this charge, it is useful to introduce the concepts of **external field** and **self-field**. The external field is the field generated by all the other charges that surround the given charge and exert external forces on it. The self-field is the field generated by the charge itself. The latter field exerts only internal forces on the given charge, and does not contribute to the net force acting on the charge from the outside. Thus, the self-field of a charge or of a charge distribution has no effect on the translational or rotational motion of the charge distribution.

External field and self-field

23.3 Lines of Electric Field

The electric field can be represented graphically by drawing, at any given point of space, a vector whose magnitude and direction are those of the electric field at that point. For instance, Figure 23.9 shows the electric field vectors representing the inverse-square electric field of a positive point charge.

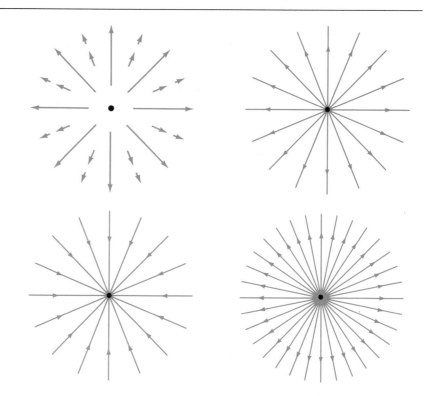

Fig. 23.9 (top left) Electric field vectors surrounding a positive point charge. The magnitudes of these field vectors decrease in inverse proportion to the squares of their distances from the point charge (the distances are reckoned from the tails of the vectors to the point charge).

Fig. 23.10 (top right) Electric field lines of a positive point charge. Note that in three dimensions the lines spread out in all three directions of space, whereas the diagram shows the lines spreading out only in the two dimensions within the page. This gives a misleading impression of the density of field lines as a function of distance. This limitation of two-dimensional diagrams should always be kept in mind when looking at pictures of field lines.

Fig. 23.11 (bottom left) Electric field lines of a negative point charge.

Fig. 23.12 (bottom right) Electric field lines of a positive point charge, twice as large as in Figure 23.10.

Field lines

Alternatively, the electric field can be represented graphically by **field lines.** These lines are drawn in such a way that, at any given point, the tangent to the line has the direction of the electric field. Furthermore, the density of lines is directly proportional to the magnitude of the electric field; that is, where the lines are closely spaced the electric field is strong, and where the lines are far apart the field is weak. Figure 23.10 shows the electric field lines of a positive point charge and Figure 23.11 those of a negative point charge. The arrows on these lines indicate the direction of the electric field along each line.

When we draw a pattern of field lines, we must begin each line on a positive point charge and end on a negative point charge (or continue indefinitely toward infinity). Since the magnitude of the electric field is directly proportional to the amount of electric charge, the number of field lines that we draw emerging from a positive charge must be proportional to the amount of charge. Figure 23.12 shows the electric field lines of a positive charge twice as large as that of Figure 23.10. We will usually adopt the convention that the number of field lines emerging from a charge q is q/ε_0; hence the number of lines emerging from one coulomb of charge is $1/\varepsilon_0 = 1.13 \times 10^{11}$. This normalization is very convenient for making computations with field lines, but it is not always quite practical for making drawings — depending on the magnitude of the charge, it sometimes yields an enormous number of field lines, so that drawing them becomes an unbearable chore, sometimes a fractional number, so that drawing them becomes altogether meaningless. For instance, according to our normalization, a proton with $e = 1.6 \times 10^{-19}$ C has $e/\varepsilon_0 = 1.8 \times 10^{-8}$ field line. Such a number makes good sense in a computation, but no sense at all in a drawing. If we want a draftsman to prepare a drawing of the field lines of a proton, we will first have to alter our normalization. From an artistic point of view, it is desirable that the spacing between the field lines be small

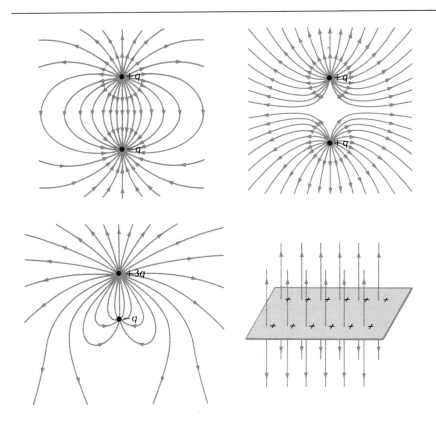

Fig. 23.13 (top left) Field lines generated by positive and negative charges of equal magnitudes.

Fig. 23.14 (top right) Field lines generated by two positive charges of equal magnitudes.

Fig. 23.15 (bottom left) Field lines generated by positive and negative charges of unequal magnitudes. The positive charge has three times the magnitude of the negative charge.

Fig. 23.16 (bottom right) Field lines of a very large sheet of charge.

compared with the distance from the charge or any other relevant distance; this produces a clear picture of the spatial dependence of the electric field. In case of need, we can alter the normalization, but we must be careful to maintain a fixed normalization throughout any given computation or series of drawings.

Figure 23.13 shows the field lines generated jointly by a positive and a negative charge of equal magnitudes. Figure 23.14 shows the field lines of a pair of equal positive charges and Figure 23.15 those of a pair of unequal positive and negative charges. Figure 23.16 shows the field lines of a large, uniformly charged sheet with a positive density of charge.

Note that in all cases, the field lines start on positive charges and end on negative charges — the positive charges are **sources** of field lines and the negative charges are **sinks**. Also, note that field lines never intersect (except where they start or end at point charges). If the lines ever were to intersect, the electric field would have *two* directions at the point of intersection; this is impossible.

The above pictures of field lines help us to develop some intuitive feeling for the spatial dependence of the electric fields surrounding diverse arrangements of electric charges. But we must not fall into the trap of thinking of the field lines as physical objects. The electric field is a form of matter, that is, a physical object, but the field lines are merely mathematical crutches to aid our imagination.

Besides providing us with a pictorial representation of the electric field, the field lines also are useful in some computations of the electric fields of given charge distributions. The next example illustrates how the concept of field lines can be exploited in the computation of the electric field of a large flat charged sheet. But before we can attempt any quantitative computation with field lines, we must give a precise definition of the density of field lines.

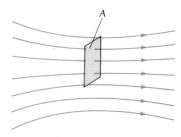

Fig. 23.17 A small area A intercepts some field lines.

Density of lines is the number of lines per unit area, that is, the number of lines intercepted by a small area A erected perpendicularly to the lines (Figure 23.17) divided by the magnitude of this area,

$$[\text{density of lines}] = \frac{[\text{number of intercepted lines}]}{A}$$

The area A used in this equation must be small, but not too small: it must be small compared with the distance over which the electric field varies appreciably, but large compared with the spacing between the field lines, so that it intercepts a fairly large number of field lines. If we were to choose A too small, the density of field lines would not be a well-defined, continuous function of position — the number of intercepted field lines and their calculated density would jump to zero whenever the area A fits between adjacent field lines. By keeping A sufficiently large, we smooth out irrelevant, erratic fluctuations in the density of field lines. Obviously, the restrictions on the choice of the area A are analogous to the restrictions on the volume element used for the computation of the density of a gas of varying density: the volume element must be small compared with the distance over which the density varies appreciably, but large compared with the distance between the molecules.

EXAMPLE 6. Use the concept of field lines and symmetry arguments to obtain the electric field of a large flat sheet carrying a uniform charge density of σ coulombs per square meter.

SOLUTION: In the preceding section, we obtained the electric field of such a sheet by a lengthy integration over the entire sheet. Now we will see how field lines permit us to obtain the answer without integration. Suppose that, as shown in Figure 23.16, the sheet is horizontal and its charge is positive. It is clear that the field lines must start on the charges on the sheet, and some of the field lines must go in the upward direction, some in the downward direction. We can take advantage of symmetry arguments to deduce the arrangement of the field lines. Such arguments are based on the rule that, if the charge distribution makes no distinction between up and down or left and right, then the pattern of field lines must, likewise, not make any such distinction; that is, the pattern of field lines must respect the symmetry of the charge distribution. Since the upper and the lower surfaces of the sheet are physically equivalent, symmetry requires that the pattern of field lines in the space above the sheet be the mirror image of the pattern below the sheet, and therefore one-half of the field lines must leave the sheet in the upward direction, one-half in the downward direction. Next, consider one of the field lines starting on the sheet and going in the, say, upward direction. Since, for an infinite sheet, the portions of the sheet to the right of the field line and the left of the field line are physically equivalent, symmetry requires that the field line bend neither to the right nor to the left. Thus, we conclude that every field line must be a straight vertical line. Finally, since the charge is uniformly distributed, the field lines must also be uniformly distributed. We therefore recognize that the pattern of field lines must consist of uniformly spaced vertical lines, as shown in Figure 23.16. Correspondingly, the electric field is of uniform magnitude throughout all of space, and its direction is vertically upward in the space above the sheet and vertically downward in the space below.

To discover the magnitude of the electric field, consider an area A of the sheet. The amount of charge within this area is σA, and therefore the number of field lines starting on this amount of charge is $\sigma A/\varepsilon_0$. The number of field lines going in the upward direction is half of this, or $\sigma A/2\varepsilon_0$, and the density of field lines is this number divided by the area, or $\sigma/2\varepsilon_0$. This density equals the magnitude of the electric field; consequently, $E = \sigma/2\varepsilon_0$, in agreement with Eq. (25).

The inverse-square law for the electric field of a point charge can be "derived" from the picture of field lines — it is easy to show that the density of lines necessarily obeys an inverse-square law. Consider a point charge q; there will be q/ε_0 lines emerging from this point charge. Since the point charge is spherically symmetric and makes no distinction between one radial direction and another, symmetry arguments tell us that the arrangement of field lines must also be spherically symmetric, and that they must be uniformly distributed over all radial directions. At a distance r from the point charge, the lines are uniformly distributed over the area $4\pi r^2$ of a concentric sphere; consequently, there are $(q/\varepsilon_0)/4\pi r^2$ lines per unit area. This establishes that the density of lines decreases in proportion to the inverse square of the distance. In fact, with our normalization regarding the number of lines per coulomb, the density of lines is not only proportional to, but exactly *equal* to the magnitude of the electric field [compare Eq. (16)]. This "derivation" of Coulomb's Law is really only a consistency check — it is because Coulomb's Law is an inverse-square law that the field can be represented by field lines; any other dependence on distance would make it impossible to draw continuous field lines that start and end only on charges.

23.4 Electric Dipole in an Electric Field

We might expect that, if an electrically neutral body is placed in a given electric field, the body will experience no force. However, this expectation is not always fulfilled. A neutral body may contain within it separate positive and negative charges (of equal magnitudes), and it is possible that the electric force on one of these charges is larger than on the other; the body then experiences a net force. Such an imbalance of the forces on the positive and negative charges will happen if the electric field is stronger at the location of one kind of charge than at the location of the other. For example, the body shown in Figure 23.18 with positive charges on one end and negative charges on the other end will be pushed to the left because the electric field that acts on the body is stronger at the location of the negative charges. Note that this electric field — indicated by field lines in Figure 23.18 — is not the field generated by the body; rather it is an external electric field generated by some other charges (not shown in Figure 23.18).

If the external electric field is uniform, then the forces on the positive and negative charges in a neutral body cancel, and there is no net force. However, there may still be a torque. Figure 23.19 shows a neutral body in a uniform electric field. The body carries equal positive and negative charges $\pm Q$, with the average positions of these charges separated by a distance l. Such a body is called an **electric dipole**. Obviously, there is a torque on the body. The torque of each force about the center of the body is $-\tfrac{1}{2}lQE \sin\theta$, and the net torque of both forces together is

$$\tau = -lQE \sin\theta \qquad (27)$$

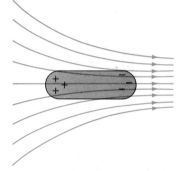

Fig. 23.18 An elongated body with positive charge at one end, negative charge on the other end. The body is placed in a nonuniform electric field.

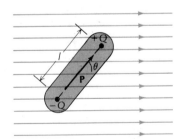

Fig. 23.19 Electric dipole in a uniform electric field. The dots indicate the average positions of the positive and the negative charges, respectively. The vector **p** is the dipole moment.

Electric dipole

where θ is the angle between the direction of the electric field and the line from the negative charge to the positive charge. The minus sign in Eq. (27) indicates that the torque is clockwise, in the sense of negative θ. The torque tends to align the body with the electric field.

We can write Eq. (27) as

$$\tau = -pE \sin\theta \tag{28}$$

where

$$\boxed{p = lQ} \tag{29}$$

Dipole moment

The quantity p is called the **dipole moment** of the body; it is simply the charge multiplied by the separation between the charges. The units of dipole moment are coulomb · meter (C · m).

In vector notation, we can write the torque as

Torque on dipole

$$\boxed{\boldsymbol{\tau} = \mathbf{p} \times \mathbf{E}} \tag{30}$$

where the vector $\mathbf{p}$ is directed from the negative charge toward the positive charge (see Figure 23.19). With the right-hand rule for the cross product, we can verify that $\mathbf{p} \times \mathbf{E}$ yields a torque in the expected direction. For instance, if in Figure 23.19 we place the fingers of the right hand along $\mathbf{p}$ and curl them toward $\mathbf{E}$, our thumb points into the page; such a torque pointing into the plane of the page tends to produce rotation in the clockwise direction.

Corresponding to the torque (28) there exists a potential energy that equals the amount of work that *you* must do against the electric forces to twist the dipole through some angle. If you want to twist the dipole through some angle, you must supply a torque $-\tau = +pE \sin\theta$ [opposite to the torque (28)] and do work[3]

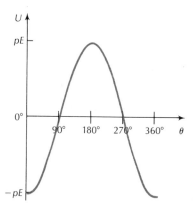

Fig. 23.20 Potential energy of an electric dipole as a function of angle.

$$U = \int_{\theta_0}^{\theta} -\tau\, d\theta' = \int_{\theta_0}^{\theta} pE \sin\theta'\, d\theta'$$
$$= pE\big[-\cos\theta'\big]_{\theta_0}^{\theta} = -pE\cos\theta + pE\cos\theta_0$$

It is customary to take the starting angle as $\theta_0 = 90°$. Then

Potential energy of dipole

$$U = -pE\cos\theta \tag{31}$$

or

$$\boxed{U = -\mathbf{p} \cdot \mathbf{E}} \tag{32}$$

This potential energy has a minimum when $\mathbf{p}$ and $\mathbf{E}$ are parallel, a maximum when they are antiparallel. Figure 23.20 is a plot of the potential energy U vs. the angle θ.

---

[3] The variable of integration in this integral has been written θ' in order to distinguish it from the limit of integration θ.

Many asymmetric molecules have **permanent dipole moments** due to an excess of electrons on one end of the molecule and a corresponding deficit on the other. This means that the molecule has a negative charge on one end and a positive charge on the other. For example, Figure 23.21 shows the structure of a water molecule. In this molecule, the electrons tend to concentrate on the oxygen atom; in Figure 23.21, the left side of the molecule is negatively charged and the right side positively charged. Since the average positions of the positive and the negative charges do not coincide, the water molecule has a dipole moment. For a water molecule in water vapor, $p = 6.1 \times 10^{-30}$ C·m.

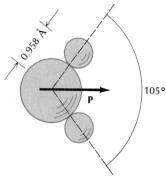

Fig. 23.21 A water molecule.

EXAMPLE 7. A molecule of water with a dipole moment of 6.1×10^{-30} C·m is placed in an electric field of 2.0×10^5 N/C. What is the difference between the potential energies for parallel and antiparallel orientations?

SOLUTION: When the dipole moment is parallel to the electric field, the potential energy is

$$U = -pE \cos 0° = -6.1 \times 10^{-30} \text{ C·m} \times 2.0 \times 10^5 \text{ N/C}$$

$$= -1.2 \times 10^{-24} \text{ J}$$

When the dipole moment is antiparallel,

$$U = -pE \cos 180° = +6.1 \times 10^{-30} \text{ C·m} \times 2.0 \times 10^5 \text{ N/C}$$

$$= 1.2 \times 10^{-24} \text{ J}$$

Hence the energy difference between the parallel and antiparallel orientations is 2.4×10^{-24} J.

Molecules of atoms that do not have a permanent dipole moment may acquire a temporary dipole moment when placed in an electric field. The opposite electric forces on the negative and the positive charges can distort the molecule and produce a charge separation. Such a dipole moment, which lasts only as long as the molecule is immersed in the electric field, is called an **induced dipole moment.** The magnitude of the induced dipole moment is directly proportional to the magnitude of the electric field. (This proportionality holds for electric fields of the strengths we will be concerned with in this text; however, it fails for electric fields of extreme strength.)

The tendency for alignment of a dipole with an electric field can be exploited to make the field lines visible. For this purpose, small bits of thread are mixed into oil in a container placed in some electric field; alternatively, small grass seeds are sprinkled on a sheet of paper placed in the electric field. The electric field induces a dipole moment along the long axis of the bit of thread or the grass seed, and the torque on the dipole then aligns the bit of thread or the grass seed with the electric field. Figures 23.22 and 23.23 show two photographs of field lines made visible by bits of thread (Figures 28.1 and 28.2 show photographs of field lines made visible with grass seeds).

Fig. 23.22 (left) Field lines between a pair of charged parallel plates, made visible by small bits of thread aligned with the field lines. Note the fringing, or spreading, of the field lines near the edges of the plates. (Courtesy H. Waage, Princeton University.)

Fig. 23.23 (right) Field lines between a charged pointed body and a charged flat plate, made visible by small bits of thread aligned with the field lines. Note the strong concentration of field lines at the sharp point, indicating a strong electric field. (Courtesy H. Waage, Princeton University.)

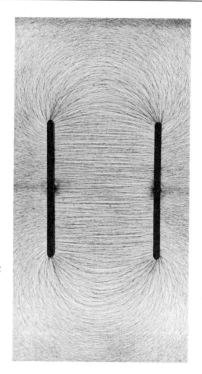
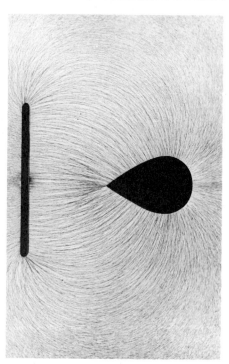

SUMMARY

Superposition principle: Electric forces and fields combine by vector addition.

Definition of electric field: $\mathbf{E} = \mathbf{F}/q$

Electric field of point charge: $\mathbf{E} = \dfrac{1}{4\pi\varepsilon_0} \dfrac{q'}{r^2} \hat{\mathbf{r}}$

Electric field of large uniformly charged plane: $E = \sigma/2\varepsilon_0$

Electric dipole moment: $p = lQ$

Torque on dipole: $\boldsymbol{\tau} = \mathbf{p} \times \mathbf{E}$

Potential energy of dipole: $U = -\mathbf{p} \cdot \mathbf{E}$

QUESTIONS

1. Does it make any difference whether the value of the charge q in the equation defining the electric field [Eq. (15)] is positive or negative?

2. In the cgs, or Gaussian, system of units, Coulomb's Law is written as $F = qq'/r^2$. In terms of grams, centimeters, and seconds, what are the units of the electric field in this system?

3. How would you formally define a gravitational field vector? Is the unit of gravitational field the same as the unit of electric field? According to your definition, what are the magnitude and the direction of the gravitational field at the surface of the Earth?

4. During days of fair weather, the Earth has an atmospheric electric field that points vertically down. This electric field is due to charges on the surface of the Earth. What must be the sign of these charges?

5. A large flat sheet measures $L \times L$; the sheet carries a uniform distribution of charge. Roughly how far from the center of the sheet would you expect the electric field to be markedly different from the uniform electric field of an infinitely large sheet?

6. Figure 23.24 shows diagrams of hypothetical field lines corresponding to some static charge distributions, which are beyond the edge of the diagram. What is wrong with these field lines?

7. If a positive point charge is released from rest in an electric field, will its orbit coincide with a field line? What if the point charge has zero mass?

8. A **tube of force** is the volume enclosed between a bundle of adjacent field lines (Figure 23.25; such a tube of force is analogous to a flow tube in hydrodynamics, and the field lines are analogous to stream lines). Along such a tube of force, the magnitude of the electric field varies in inverse proportion to the cross-sectional area of the tube. Explain.

9. A negative point charge $-q$ sits in front of a very large flat sheet with a uniform distribution of positive charge. Make a rough sketch of the pattern of field lines. Is there any point where the electric field is zero?

10. A very long straight line of positive charge lies along the z axis. A very large flat sheet of positive charge lies in the x–y plane. Sketch a few of the field lines of the net electric field produced by both of these charge distributions acting together.

11. A large, flat, thick slab of insulator has positive charge uniformly distributed over its volume. Sketch the field lines on both sides and inside the slab; pay careful attention to the starting points of the field lines.

12. If our universe is topologically closed, so that a straight line drawn in any direction returns on itself from the opposite direction, can the net charge in the universe be different from zero?

13. How could you build a "compass" that indicates the direction of the electric field?

14. When a neutral metallic body, insulated from the ground, is placed in an electric field, it develops a charge separation, acquiring positive charge on one end and negative charge on the other. This means the body acquires an induced dipole moment. How is the direction of this dipole moment related to the direction of the electric field?

15. By inspection of Figure 23.22, make a rough, qualitative plot of the electric field strength as a function of position along a line parallel to the plates, midway between the plates.

16. One electric dipole is at the origin, oriented parallel to the z axis. Another electric dipole is at some distance on the x axis. The electric field of the first dipole then exerts a torque on the second dipole. For what orientation of the second dipole is the potential energy minimum?

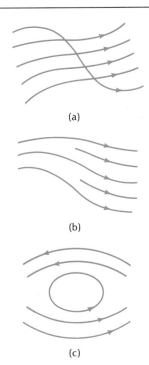

(a)

(b)

(c)

Fig. 23.24

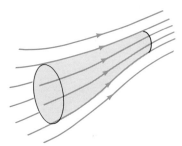

Fig. 23.25

PROBLEMS

Sections 23.1 and 23.2

1. Electric fields as large as 3.4×10^5 N/C have been measured by airplanes flying through thunderclouds. What is the force on an electron exposed to such a field? What is its acceleration?

2. The Earth has not only a magnetic field, but also an atmospheric electric field. During days of fair weather (no thunderclouds), this atmospheric electric field has a strength of about 100 N/C and points down. Taking into account this electric field and also gravity, what will be the acceleration (magnitude and

direction) of a grain of dust of mass 1.0×10^{-18} kg carrying a single *electron* charge?

3. **Millikan's experiment** measures the elementary charge e by the observation of the motion of small oil droplets in an electric field. The oil droplets are charged with one or several elementary charges, and, if the (vertical) electric field has the right magnitude, the electric force on the droplet will balance its weight, holding the drop suspended in midair. Suppose that an oil droplet of radius 1.0×10^{-4} cm carries a single elementary charge. What electric field is required to balance the weight? The density of oil is 0.80 g/cm^3.

4. In an X-ray tube (see Figure IV.1), electrons are exposed to an electric field of 8×10^5 N/C. What is the force on an electron? What is its acceleration?

5. According to a theoretical estimate, at the surface of a neutron star of mass 1.4×10^{30} kg and radius 1.0×10^4 m there is an electric field of magnitude 6×10^3 N/C pointing vertically up. Compare the electric force on a proton with the gravitational force on the proton.

6. A long hair, taken from a girl's braid, has a mass of 1.2×10^{-3} g. The hair carries a charge of 1.3×10^{-9} C distributed along its length. If we want to suspend this hair in midair, what (uniform) electric field do we need?

7. The electric field in the electron gun of a TV tube is supposed to accelerate electrons uniformly from 0 to 3.3×10^7 m/s within a distance of 1.0 cm. What electric field is required?

8. Electric breakdown (sparks) occurs in air if the electric field reaches 3×10^6 N/C. At this field strength, free electrons present in the atmosphere are quickly accelerated to such large speeds that upon impact on atoms they knock electrons off the atom and thereby generate an avalanche of electrons. How far must a free electron move under the influence of the above electric field if it is to attain a kinetic energy of 3×10^{-19} J (which is sufficient to produce ionization)?

9. The nuclei of the atoms in a chunk of metal lying on the surface of the Earth would fall to the bottom of the metal if their weight were the only force acting on them. Actually, within the interior of any metal exposed to gravity there exists a very small electric field that points vertically up. The corresponding electric force on a nucleus just balances the weight of the nucleus. Show that for a nucleus of atomic number Z, mass m, the required field has a magnitude $mg/(Ze)$. What is the numerical value of this electric field in a chunk of iron?

10. The hydrogen atom has a radius of 0.53×10^{-10} m. What is the magnitude of the electric field that the nucleus of the atom (a proton) produces at this radius?

11. What is the strength of the electric field at the surface of a uranium nucleus? The radius of the nucleus is 7.4×10^{-15} m and the electric charge is $92e$. For the purposes of this problem the electric charge may be regarded as concentrated at the center.

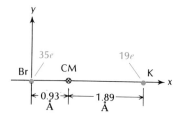

Fig. 23.26 The positive (nuclear) charges in a KBr molecule.

12. Figure 23.26 shows the arrangement of nuclear charges (positive charges) of a KBr molecule. Find the electric field that these charges produce at the center of mass at a distance of 0.93×10^{-10} m from the Br atom.

13. Suppose that in a hydrogen atom the electron is (instantaneously) at a distance of 2.1×10^{-10} m from the proton. What is the net electric field that the electron and the proton produce jointly at a point midway between them?

*14. (a) Equation (23.20) gives the electric field on the axis of a charged ring. Where is the strength of this electric field maximum?
   (b) Roughly sketch the electric field lines in the space surrounding the ring.

*15. The distance between the oxygen nucleus and each of the hydrogen nuclei in an H₂O molecule is 0.958 Å; the angle between the two hydrogen atoms is 105° (Figure 23.27). Find the electric field produced by the nuclear charges (positive charges) at the point P at a distance of 1.2 Å to the right of the oxygen nucleus.

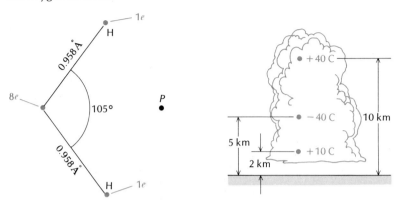

Fig. 23.27 (left) The positive (nuclear) charges in a water molecule.

Fig. 23.28 (right) Charges in a thundercloud.

*16. Figure 23.28 shows the charge distribution within a thundercloud. There is a charge of 40 C at a height of 10 km, −40 C at 5 km, and 10 C at 2 km. Treating these charges as pointlike, find the electric field (magnitude and direction) that they produce at a height of 8 km and a horizontal distance of 3 km.

*17. Suppose that an airplane flies through the thundercloud described in Problem 16 at the 8-km level. Plot the magnitude of the electric field as a function of position along the path of the airplane; start with the airplane 10 km away from the thundercloud.

*18. Suppose that the charge distribution of a thundercloud can be approximated by two point charges, a negative charge of −40 C at a height of 5 km (above ground) and a positive charge of 40 C at a height of 11 km. To find the electric field strength at the ground, we must take into account that the ground is a conductor and that the charge of the thundercloud induces charges on the ground. It can be shown that the effect of the induced charges can be simulated by a point charge of 40 C at 5 km *below ground* and a point charge of −40 C at 11 km below ground (Figure 23.29); these fictitious charges are called *image charges* (described in Section I.4). By adding the electric field of the image charges to that of the two real charges in the thundercloud, calculate the magnitude of the electric field at a point on the ground directly below the thundercloud charges. Similarly, calculate the magnitude of the electric field at horizontal distances of 2, 4, 6, 8, and 10 km. Plot the field vs. the distance.

*19. Consider eight of the ions of Cl⁻ and Na⁺ in a crystal lattice of common salt. The ions are located at the vertices of a cube measuring 2.82×10^{-10} m on an edge. Calculate the magnitude of the electric force that seven of these ions exert on the eight.

*20. Each of two very long, straight, parallel lines carries a positive charge of λ coulombs per meter. The distance between the lines is d. Find the electric field at a point equidistant from the lines, with a distance $2d$ from each line. Draw a diagram showing the direction of the electric field.

*21. Two infinite lines of silk with uniform charge distributions of λ coulombs per meter lie along the x and the y axes, respectively. Find the electric field at a point with coordinates x, y, z; assume that $x > 0$, $y > 0$, $z > 0$.

*22. A semi-infinite line carrying a uniform charge distribution of λ coulombs per meter lies along the positive x axis from $x = 0$ to $x = \infty$. Find the compo-

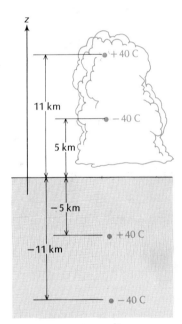

Fig. 23.29 Charges in a thundercloud and their images in the ground.

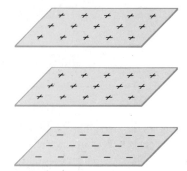

Fig. 23.30 Three parallel sheets.

nents of the electric field at the point with coordinates x, y, with $z = 0$; assume $x > 0, y > 0$.

*23. A semi-infinite line with a uniform charge distribution of $+\lambda$ coulombs per meter lies along the positive x axis from $x = 0$ to $x = \infty$. Another semi-infinite line with a charge distribution of $-\lambda$ coulombs per meter lies along the negative x axis from $x = 0$ to $x = -\infty$. Find the electric field at a point on the y axis.

*24. Electric charge is uniformly distributed over each of three large parallel sheets of paper (Figure 23.30). The charges per unit area on the sheets are 2×10^{-6} C/m², 2×10^{-6} C/m², and -2×10^{-6} C/m², respectively. The distance between one sheet and the next is 1.0 cm. Find the strength of the electric field **E** above the sheets, below the sheets, and in the space between the sheets. Find the direction of **E** at each place.

*25. Each of two very large flat sheets of paper carries a uniform positive charge distribution of 3.0×10^{-4} C/m². The two sheets of paper intersect at an angle of 45° (Figure 23.31). What are the magnitude and the direction of the electric field at a point between the two sheets?

Fig. 23.31 (left) Two sheets intersecting at 45°.

Fig. 23.32 (right) Two sheets intersecting at right angles.

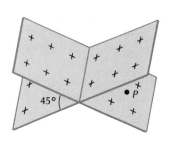
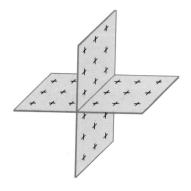

*26. Two large sheets of paper intersect at right angles. Each sheet carries a uniform distribution of positive charge (Figure 23.32). The charge per unit area on the sheets is 3×10^{-6} C/m². Find the magnitude of the electric field in each of the four quadrants. Sketch the field lines in each quadrant.

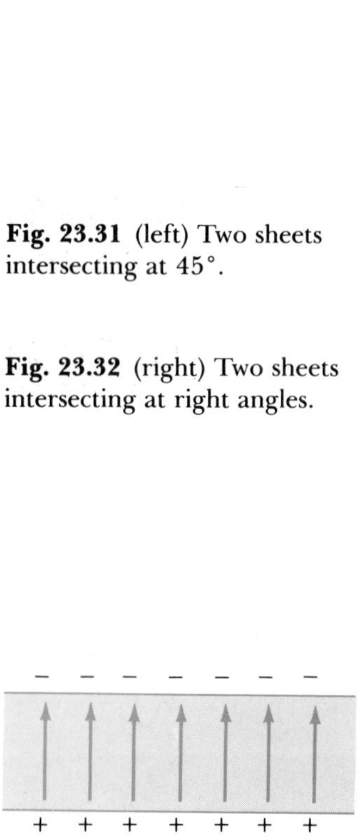

*27. The electric field within a chunk of metal exposed to the Earth's gravity (see Problem 9) is due to a distribution of surface charge. Suppose that we have a slab of iron oriented horizontally (Figure 23.33). What must be the surface charge densities on the upper and lower surfaces?

Fig. 23.33 A horizontal slab of iron.

*28. A very large flat sheet of paper carries charge uniformly distributed over its surface; the amount of charge per unit area is σ. A hole of radius R has been cut out of this paper (see Figure 23.34). Find the electric field on the axis of the hole.

Fig. 23.34 A sheet of paper with a hole.

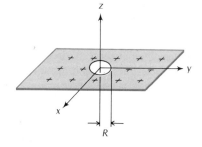

*29. A paper annulus has an inner radius R_1 and an outer radius R_2. An amount of charge Q is uniformly distributed over the surface of the annulus. What is the electric field on the axis of the annulus at a distance z from the center?

*30. A total amount of charge Q is uniformly distributed along a thin, straight plastic rod of length l.
   (a) Find the electric field at the point P, at a distance x from one end of the rod (Figure 23.35).
   (b) Find the electric field at point P', at a distance y from the midpoint of the rod (see Figure 23.35).

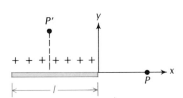

Fig. 23.35 A thin rod, with a uniform distribution of charge.

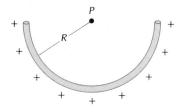

Fig. 23.36 A semicircular rod, with a uniform distribution of charge.

*31. A thin plastic rod is bent so that it has the shape of a semicircle of radius R (Figure 23.36). An amount of charge Q is uniformly distributed along the rod. What is the electric field at the center of the circle?

*32. A Plexiglas square of dimension $l \times l$ has a uniform charge density of magnitude λ coulombs per meter along its edges. Two of the edges are positive and two are negative (Figure 23.37). Find the electric field at the center of the square.

*33. Three thin glass rods carry charges Q, Q, and $-Q$, respectively. The length of each rod is l and the charge is uniformly distributed along each rod. The rods form an equilateral triangle. Calculate the electric field at the center of the triangle.

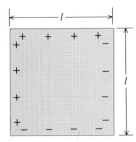

Fig. 23.37 A square, with charge along its edges.

*34. A paper disk of radius R carries an amount of charge Q uniformly distributed over its surface. Find the electric field at a point on the axis of the disk at a distance z from the center. Show that in the limiting case $z \gg R$, the result reduces to that for a point charge.

**35. Two thin, semi-infinite rods lie in the same plane. They make an angle of 45° with each other, and they are joined by another thin rod bent along an arc of circle of radius R, with center at P (see Figure 23.38). All the rods carry a uniform charge distribution of λ coulombs per meter. Find the electric field at the point P.

*36. A thin, semi-infinite rod with a uniform charge distribution of λ coulombs per meter lies along the positive x axis from $x = 0$ to $x = \infty$; a similar rod lies along the positive y axis from $y = 0$ to $y = \infty$. Calculate the electric field at a point in the x–y plane in the first quadrant.

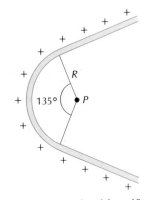

Fig. 23.38 Rods with uniform distributions of charge.

*37. A cylindrical Plexiglas tube of length l, radius R carries a charge Q uniformly distributed over its surface. Find the electric field on the axis of the tube at one of its ends.

*38. Two thin rods of length L carry equal charges Q uniformly distributed over their lengths. The rods are aligned, and their nearest ends are separated by a distance x (Figure 23.39). What is the electric force of repulsion between these rods?

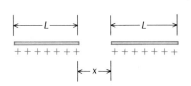

Fig. 23.39 Two aligned rods.

39. A square of paper measuring $l \times l$ carries an amount of charge q uniformly distributed over its surface. The square lies in the x–y plane with its center at the origin and its sides parallel to the x and y axes. Find the electric field at a point on the y axis; assume that the point is outside the square.

Section 23.4

40. A small, straight bit of thread, such as used to make the pictures of field lines in Figure 23.22, has a charge of $+1 \times 10^{-14}$ C at one end and a charge of -1×10^{-14} C at the other. The length of the thread is 2 mm.
    (a) What is the dipole moment?
    (b) What is the torque on this thread if it is placed in an electric field of 6×10^5 N/C at right angles to the field?

41. The two charges of ± 40 C in the thundercloud of Figure 23.28 form a dipole. What is the dipole moment?

42. In a hydrogen atom, the electron is at a distance of 0.53×10^{-10} m from a proton.
    (a) What is the instantaneous dipole moment of this system?
    (b) Taking into account that the electron moves around the proton on a circular orbit, what is the time-average dipole moment of this system?

43. (a) Pretend that the HCl molecule consists of (pointlike) ions of H^+ and Cl^- separated by a distance of 1.0×10^{-10} m. If so, what would be the dipole moment of this system?
    (b) The observed dipole moment is 3.4×10^{-30} C·m. Can you suggest a reason for this discrepancy?

44. The dipole moment of a HCl molecule is 3.4×10^{-30} C·m. Calculate the magnitude of the torque that an electric field of 2.0×10^6 N/C exerts on this molecule when the angle between the electric field and the longitudinal axis of the molecule is $45°$.

45. The dipole moment of the water molecule is 6.1×10^{-30} C·m. In an electric field, a molecule with a dipole moment will tend to settle into an equilibrium orientation such that the dipole moment is parallel to the electric field. If disturbed from this equilibrium orientation, the molecule will oscillate like a torsional pendulum. Calculate the frequency of small oscillations of this kind for a water molecule about an axis through the center of mass (and perpendicular to the plane of the three atoms) when the molecule is in an electric field of 2.0×10^6 N/C. The moment of inertia of the molecule about this axis is 1.93×10^{-47} kg·m².

CHAPTER 24

Gauss' Law

Although the electric field of any given charge distribution can be calculated by means of Coulomb's Law, as in Examples 1–4 of the preceding chapter, this method often involves the evaluation of tedious integrals. Fortunately, there exists another method for calculating the electric field. This method relies on a theorem called Gauss' Law. That law is a consequence of Coulomb's Law, and it therefore contains no new physics. It does, however, contain some new mathematics which supplies an elegant shortcut for calculating the electric field of a given charge distribution, provided that this charge distribution has a certain amount of symmetry. This means that Gauss' Law does not help in every calculation of electric fields, but when it does help it works wonders. Before we present Gauss' Law and some examples of its use, we need to introduce the concept of electric flux.

24.1 Electric Flux and the Number of Field Lines

Consider a mathematical (that is, imagined) surface in the shape of a rectangle of area A. Suppose that this surface is immersed in a constant electric field **E** (Figure 24.1). This electric field makes an angle with the surface; the electric-field vector has a component tangential to the surface and a component normal (that is, perpendicular) to the surface. The **electric flux** Φ through the surface is defined as the product of the area A and the magnitude of the normal component of the electric field,

$$\Phi = E_n A \qquad (1)$$

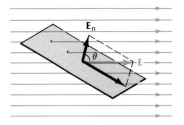

Fig. 24.1 Flat rectangular surface immersed in a uniform electric field. The perpendicular to the surface makes an angle θ with the field lines.

The normal component E_n can also be written as $E \cos \theta$, where θ is the angle between **E** and the perpendicular erected on the surface (Figure 24.1). Hence

$$\Phi = EA \cos \theta \qquad (2)$$

The quantity $A \cos \theta$ can be interpreted as the projection of the area A onto a plane perpendicular to the electric field; that is, $A \cos \theta$ can be regarded as that part of the area A that effectively faces the electric field. According to Section 23.3, the magnitude E of the electric field is numerically equal to the number of field lines intercepted by a unit area facing the electric field. Hence $EA \cos \theta$ must be numerically equal to the number of field lines intercepted by the area A: *the electric flux Φ through an area is equal to the number of field lines intercepted by the area.* Note that this equality hinges on the convention adopted in Section 23.3 — flux and number of intercepted lines are equal if and only if electric field and number of lines per unit area are equal, and the latter is true if and only if we adopt the convention that q/ε_0 lines emerge from each charge q.

More generally, consider a mathematical surface of arbitrary shape immersed in an arbitrary nonuniform electric field (Figure 24.2). Then we can define the electric flux by subdividing the surface into infinitesimal plane areas dS within each of which the electric field is nearly constant; the flux through one such infinitesimal area is $d\Phi = E \cos \theta \, dS$, and the total flux through the surface is the integral obtained by summing all these infinitesimal contributions,

$$\Phi = \int E \cos \theta \, dS \qquad (3)$$

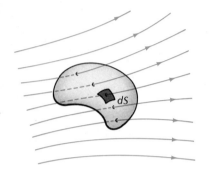

Fig. 24.2 An arbitrary surface immersed in an arbitrary electric field. The flux is $\Phi = 5$.

Electric flux

According to the arguments given above, this electric flux is again numerically equal to the number of field lines intercepted by the surface. Note that lines going through the surface from one side make a positive contribution to the flux; lines going through from the opposite side make a negative contribution (Figure 24.3). This means that the perpendiculars to the surface, with respect to which the angle θ is reckoned, must all be erected on the same side of the surface (for example, on the right side of the surface of Figure 24.3).

Finally, consider a *closed* mathematical surface immersed in an electric field (Figure 24.4). The electric flux through this surface is given by Eq. (3), with the integration extending over all the area of the closed surface. This flux is again equal to the number of field lines intercepted by the surface. The number can be positive or negative. For any closed surface we adopt the convention that the angle θ is reckoned with respect to perpendiculars erected on the *outside* of the

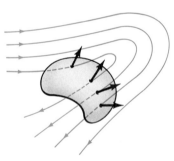

Fig. 24.3 Arbitrary surface immersed in arbitrary electric field. The small black arrows are perpendicular to the surface. The net number of lines going through the surface is $+1-3$, or -2.

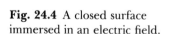

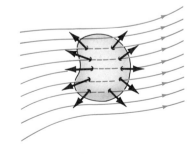

Fig. 24.4 A closed surface immersed in an electric field.

closed surface; then field lines leaving the volume enclosed by the surface make a positive contribution to the flux, and lines entering the volume make a negative contribution.

EXAMPLE 1. A point charge q is inside a closed cubical surface (Figure 24.5). What is the flux of its electric field through this surface?

SOLUTION: A positive charge q is the starting point of q/ε_0 outward field lines, and all of these will emerge from the surface. Hence the flux must be q/ε_0. Note that by appealing to the concept of field lines, we are able to get the answer without explicit calculation of any integral; the explicit integration of the flux integral $\int E \cos\theta\, dS$ for the configuration shown in Figure 24.5 is rather messy.

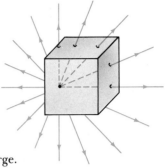

Fig. 24.5 A cubical surface surrounds a positive point charge.

24.2 Gauss' Law

We saw in Example 1 that a point charge inside a closed mathematical surface produces an electric flux q/ε_0 through this surface. This result can be generalized to any arbitrary charge distribution enclosed within an arbitrary surface. The general result is called Gauss' Law:

If the volume within an arbitrary closed mathematical surface holds a net electric charge Q, then the electric flux through the surface is Q/ε_0, that is,

$$\oint E \cos\theta\, dS = Q/\varepsilon_0 \qquad (4)$$

Here, the small circle on the integral sign indicates that the integration is to be performed over a *closed* surface. This closed surface is usually called the **Gaussian surface**.

The proof of Gauss' Law is easy, if we exploit the concept of field lines. The electric field appearing in Eq. (4) is a sum of the individual electric fields of a number of point charges. Some of the point charges are outside the closed surface and some are inside the closed surface. Let us consider the contribution to the flux from each individual electric field of each individual point charge. The individual electric field of a charge outside the closed surface generates no net flux through the closed surface — any field line of this field either does not touch the surface or else enters it at one point and leaves at another; neither case makes any net contribution to the flux (Figure 24.6). However, the individual electric field of a positive or negative charge enclosed within the surface does contribute to the flux. For example, a positive charge q has q/ε_0 outward field lines, all of which will pierce the closed surface — such a charge therefore contributes a flux q/ε_0 (Figure 24.7). Taking into account that positive charges generate positive flux and negative charges negative flux, we see that the net number of field lines piercing the surface, or the net flux through the surface, is equal to the net charge divided by ε_0. This completes the proof of Eq. (4).

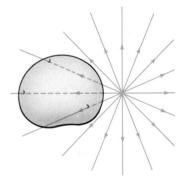

Fig. 24.6 A positive point charge outside the Gaussian surface.

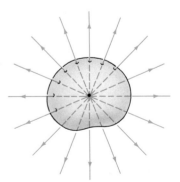

Fig. 24.7 A positive point charge enclosed within the Gaussian surface.

With the notation E_n for the component of **E** normal to the surface, we can put Gauss' Law in the somewhat more compact form

$$\oint E_n \, dS = Q/\varepsilon_0 \qquad (5)$$

An alternative notation frequently used associates a vector $d\mathbf{S}$ with the area dS; this vector has the magnitude of dS and the direction of the outward perpendicular to the surface (Figure 24.8). Then $E \cos\theta \, dS = \mathbf{E} \cdot d\mathbf{S}$, and Eq. (4) takes the form

Gauss' Law in vector notation

$$\boxed{\oint \mathbf{E} \cdot d\mathbf{S} = Q/\varepsilon_0} \qquad (6)$$

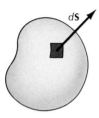

Fig. 24.8 A small area dS with its perpendicular vector $d\mathbf{S}$.

Gauss' Law can be used to calculate the electric field provided the distribution of charge has a high degree of symmetry. Essentially, Gauss' Law can be regarded as a mathematical restriction imposed on the electric field. Symmetry conditions impose further restrictions on the field. By clever combination of all these restrictions, we can often evaluate the electric field without the laborious process of integrating Coulomb's Law.

EXAMPLE 2. Use Gauss' Law to deduce the electric field of a point charge.

SOLUTION: As in the derivation of the electric field of a point charge q by means of field lines, we begin with a symmetry argument: Since the point charge is spherically symmetric, the electric field must also be spherically symmetric. Hence, at all points of a spherical mathematical surface of radius r centered on q, the field is in the radial direction and has the same magnitude (Figure 24.9).

If the Gaussian surface is taken to coincide with this sphere of radius r, then $\cos\theta = 1$, since $\mathbf{E}$ is perpendicular to this surface. Furthermore, $\mathbf{E}$ has a constant magnitude over this surface. Therefore,

$$\oint E \cos\theta \, dS = \oint E \, dS = E \oint dS$$

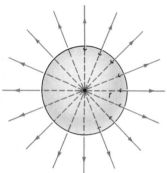

Fig. 24.9 A spherical Gaussian surface surrounds a positive point charge concentrically.

But $\oint dS$ is simply the area of the Gaussian surface, or $4\pi r^2$. Hence Eq. (4) becomes

$$E(4\pi r^2) = q/\varepsilon_0$$

and

$$E = \frac{1}{4\pi\varepsilon_0} \frac{q}{r^2} \qquad (7)$$

The proof of Eq. (4) established that Gauss' Law is a consequence of the field-line concept; that is, Gauss' Law is a consequence of Coulomb's Law. Conversely, Example 2 establishes that Coulomb's Law is a consequence of Gauss' Law. These two laws are therefore equivalent, at least in regard to the electric fields of charges at rest. As we will see in a later chapter, besides such static electric fields, there also exist time-dependent electric fields whose field lines are not attached to charges (the field lines form closed loops). Coulomb's Law does not apply to such time-dependent electric fields, but Gauss' Law does! The latter law is more general than the former. However, this subtle distinction need not concern us as long as all charges are static, as they are in the examples of the next section.

24.3 Some Examples

The following examples illustrate how Gauss' Law can be combined with symmetry requirements and used to evaluate the electric fields of simple charge distributions. The solutions of these examples always involve three steps: First determine the *direction* of the electric field by appealing to the symmetry requirements imposed by the shape of the charge distribution, then choose a suitable Gaussian surface, and then calculate the *magnitude* of the electric field from Gauss' Law. The crucial trick always lies in the choice of Gaussian surface. Keep in mind that the Gaussian surface is a purely mathematical construct. It need not coincide with any of the physical surfaces in the problem, and we are free to choose it in any way we please. A good choice makes the evaluation of the integral $\int E \cos \theta \, dS$ easy, and it also permits us to "solve" Gauss' Law for the unknown value of E.

EXAMPLE 3. Use Gauss' Law to obtain the electric field generated by a very long, thin, charged line, with λ coulombs per meter of line.

SOLUTION: This example has already been worked out via Coulomb's Law, by integration (see Section 23.1). Gauss' Law permits a quicker and more elegant solution. First we need to determine the direction of the electric field. The only direction consistent with the symmetry of the charge distribution is the radial direction (horizontal in Figure 24.10) — if the electric field had any other direction (up or down in Figure 24.10), the field would make an unacceptable distinction between the upper and lower portions of the line of charge. Furthermore, the rotational symmetry of the line of charge tells us that the electric field has constant magnitude over the lateral curved surface of any mathematical cylinder concentric with the line of charge.

Now take a Gaussian surface that coincides with such a cylinder, of radius x and height h (Figure 24.10). On the lateral, curved surface of the cylinder, the electric field is perpendicular to the surface; hence $\cos \theta = 1$ and

$$\int_{\text{curved surface}} E \cos \theta \, dS = \int E \, dS = E \int dS = E \times (2\pi x h) \tag{8}$$

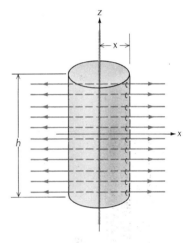

Fig. 24.10 A cylindrical Gaussian surface surrounds a line of charge.

On each of the two circular bases of the cylinder, the electric field is tangent to the surface; hence $\cos \theta = 0$ and

$$\int_{\text{base}} E \cos \theta \, dS = 0 \tag{9}$$

The integral over the entire surface of the cylinder (curved surface plus bases) is then $2\pi x h E$. By Gauss' Law, this must equal the charge contained in the cylinder divided by ε_0. The charge is λh, and therefore

$$2\pi x h E = Q/\varepsilon_0 = \lambda h / \varepsilon_0$$

and

$$E = \frac{1}{2\pi \varepsilon_0} \frac{\lambda}{x} \tag{10}$$

COMMENTS AND SUGGESTIONS: Note that the Gaussian surface in this calculation was carefully chosen so that, over one part of the surface, the electric field is perpendicular to the surface ($\cos \theta = 1$) and constant in magnitude; and so that, over the other parts, the electric field is tangential to the surface ($\cos \theta = 0$). This permits a trivial evaluation of the integral $\int E \cos \theta \, dS$. In the other examples involving Gauss' Law, we will find it convenient to make simi-

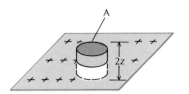

Fig. 24.11 A cylindrical Gaussian surface intersects a very large sheet of charge.

EXAMPLE 4. Use Gauss' Law to deduce the electric field of a large flat sheet with a uniform charge density of σ coulombs per square meter.

SOLUTION: This example was also worked out earlier by a laborious integration procedure; this labor can be bypassed by means of Gauss' Law. Symmetry tells us that the electric field is everywhere perpendicular to the sheet of charge, and that it has a constant magnitude over any surface parallel to this sheet. We take a Gaussian surface in the shape of a cylinder of height $2z$ and base area A (Figure 24.11). One base is above the surface, and one below. Note that since the bases are at the same distance from the sheet, the magnitude of E will be the same on both of them. On each base $\cos\theta = 1$ and, integrating over both, we find

$$\int_{\text{bases}} E\cos\theta\, dS = \int E\, dS = E \times (2A) \tag{11}$$

On the curved lateral surface of the cylinder, $\cos\theta = 0$ and therefore

$$\int_{\substack{\text{curved}\\\text{surface}}} E\cos\theta\, dS = 0 \tag{12}$$

By Gauss' Law, the integral of $E\cos\theta$ over the complete surface must equal the charge within the cylinder divided by ε_0. Since the charge is σA, we obtain

$$2AE = Q/\varepsilon_0 = \sigma A/\varepsilon_0$$

and

$$E = \sigma/2\varepsilon_0 \tag{13}$$

EXAMPLE 5. A sphere of radius R has a total charge q which is uniformly distributed over its volume. (a) What is the electric field at points inside the sphere? (b) What is the electric field at points outside the sphere?

SOLUTION: (a) The charge density within the sphere is

$$\rho = [\text{charge}]/[\text{volume}] = q/(4\pi R^3/3) = 3q/4\pi R^3 \tag{14}$$

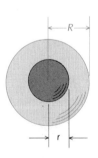

Fig. 24.12 A sphere (colored) with a uniform distribution of charge. The Gaussian surface (gray) has a radius $r < R$.

Since the charge distribution is spherically symmetric, the electric field must be radial and of constant magnitude over any concentric spherical mathematical surface of given radius. To find the magnitude of the electric field inside the charge distribution, take a spherical Gaussian surface of radius r, where $r \leq R$ (Figure 24.12). On this surface $\cos\theta = 1$, so that

$$\oint E\cos\theta\, dS = \oint E\, dS = E\oint dS = 4\pi r^2 E \tag{15}$$

The charge inside this Gaussian surface is

$$Q = [\text{charge density}] \times [\text{volume}] = \rho \times \frac{4\pi}{3} r^3$$

$$= q\frac{r^3}{R^3} \tag{16}$$

Then Gauss' Law gives

$$4\pi r^2 E = \frac{q}{\varepsilon_0}\frac{r^3}{R^3}$$

that is,

$$E = \frac{1}{4\pi\varepsilon_0}\frac{qr}{R^3} \quad \text{for } r \leq R \qquad (17)$$

(b) To find the electric field outside the sphere, take a spherical Gaussian surface of radius r where $r \geq R$ (Figure 24.13). With the usual symmetry arguments, the flux integral again has the form $4\pi r^2 E$ and Gauss' Law gives

$$4\pi r^2 E = q/\varepsilon_0$$

that is,

$$E = \frac{1}{4\pi\varepsilon_0}\frac{q}{r^2} \quad \text{for } r \geq R \qquad (18)$$

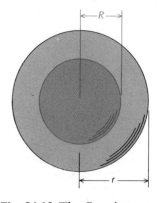

Fig. 24.13 The Gaussian surface (gray) has a radius $r > R$.

COMMENTS AND SUGGESTIONS: According to these results, the electric field outside the region containing the charge is an inverse-square field, exactly the same as it would be if all the charge were located at the center.

The electric field inside the region containing the charge increases linearly with the distance from the center and reaches a maximum value of $E = (1/4\pi\varepsilon_0)q/R^2$ when $r = R$. Figure 24.14 is a plot of the electric field vs. distance.

The above calculation relies on the implicit assumption that the material of the sphere to which the electric charge q is attached exerts no effect on the electric field; that is, the material of the sphere has the same electric properties as a vacuum. For the sake of simplicity, we will ignore the electric properties of materials throughout this chapter and the next two chapters. Later, in Chapter 27, we will see how to take into account the effects of materials on electric fields.

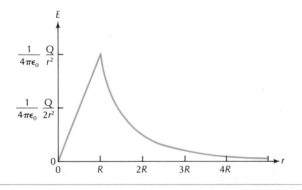

Fig. 24.14 Electric field as a function of radius for a uniformly charged sphere.

It is obvious that the argument of part (b) of the preceding example does not depend on the uniformity of the charge distribution — it depends only on its spherical symmetry. Hence, outside any charge distribution that consists of a sequence of concentric shells of charge — so that the charge density is a function of the radius, but not of the angular direction — the electric field will mimic that of a point charge. This result is similar to Newton's famous theorem concerning the gravitational forces exerted by a planet: the gravitational force exerted by a spherically symmetric planet can be calculated as if all the mass were concentrated in a point at the planetary center. This similarity between electricity and gravitation reflects the similarity of the laws of force. Because of this similarity, we can use the same action–reaction argument as in Section 9.6 to show that a spherical charge distribution mimics a point charge not only in regard to the electric field that it produces but also in regard to the force that it experiences when

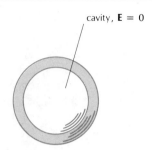

Fig. 24.15 A charged spherical shell (a sphere with a spherical cavity). The electric field produced by the shell is zero within the cavity.

placed in some arbitrary external electric field. Thus, the electric force that an arbitrary electric field exerts on a spherical charge distribution can be calculated as if all of the charge were concentrated in a point at the center.

Note that Eq. (17) can be put in the form

$$E = \frac{1}{4\pi\varepsilon_0} \frac{Q(r)}{r^2} \qquad (19)$$

where $Q(r)$ is the amount of charge inside the sphere of radius r. If the charge distribution is uniform, then $Q(r)$ is as given by Eq. (16); however, the expression (19) does not depend on the radial uniformity of the charge distribution — it depends only on spherical symmetry. According to Eq. (19), only the charge *inside* a radius less than r contributes to the electric field *at* the radius r. As a consequence, the electric field that a spherically symmetric shell of charge produces within the empty cavity inside it is exactly zero (Figure 24.15).

EXAMPLE 6. A proton is (approximately) a spherical ball of charge of radius 1.0×10^{-15} m. What is the electric field at the surface of the proton? If a second proton is brought within touching distance of the first, what is the repulsive electric force?

SOLUTION: For $q = e = 1.6 \times 10^{-19}$ C and $r = 1.0 \times 10^{-15}$ m, the electric field is

$$E = \frac{1}{4\pi\varepsilon_0} \frac{q}{r^2}$$

$$= \frac{9.0 \times 10^9 \text{ N} \cdot \text{m}^2 \cdot \text{C}^{-2} \times 1.6 \times 10^{-19} \text{ C}}{(1.0 \times 10^{-15} \text{ m})^2}$$

$$= 1.4 \times 10^{21} \text{ N/C} \qquad (20)$$

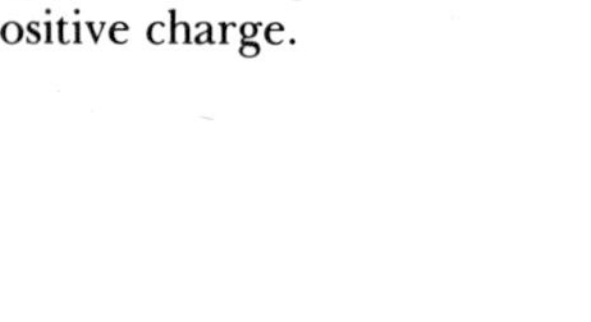

If the protons are touching (Figure 24.16), the center-to-center distance is $2 \times 1.0 \times 10^{-15}$ m. The electric field generated by one proton at this distance is one-quarter of the above value, that is, $E = 3.6 \times 10^{20}$ N/C. The force exerted by one proton on the other is then

$$F = eE = 1.6 \times 10^{-19} \text{ C} \times 3.6 \times 10^{20} \text{ N/C}$$

$$= 58 \text{ N} \qquad (21)$$

Fig. 24.16 Two protons in contact. Each proton may be regarded as a spherical ball of positive charge.

COMMENTS AND SUGGESTIONS: This is the weight of 6 kg; acting on a particle of a mass of only 10^{-27} kg, this represents a gigantic force! In the nucleus of an atom, the large repulsive electric force between the tightly packed protons is more than canceled by an even larger binding force (the "strong" force) that holds the nucleus together.

24.4 Conductors in Electric Fields

As we pointed out in Section 22.4, in metallic conductors — such as copper, silver, aluminum — some of the electrons are free, that is, they can move without restraint within the volume of the metal. If such a conductor is immersed in an electric field, the free electrons accelerate in response to the electric force. The electrons move in a direction opposite to the direction of the electric field, and they continue moving until they reach the surface of the metal. As an excess of electrons accumulates on one part of the conductor, a deficit of electrons will appear on another part of the conductor: negative and positive charges are induced on the conductor. Within the volume of the conductor, the electric field of the induced charges tends to cancel the original electric field in which the conductor was immersed (Figure 24.17). The accumulation of negative and positive charges on the surface of the conductor continues until the electric field generated by these charges exactly cancels the original electric field. Consequently, *when the charge distribution on a conductor reaches static equilibrium, the net electric field within the material of the conductor is exactly zero.* The proof of this statement is by contradiction: if the electric field were different from zero, the free electrons would continue to move, and the charge distribution would *not* (yet) be in equilibrium. For a good conductor (such as copper or aluminum), the equilibrium is reached in a fairly short time, a small fraction of a second.[1]

Electrostatic equilibrium

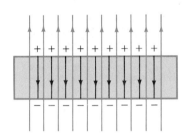

Fig. 24.17 A thick slab of conductor immersed in an electric field. The charges that have accumulated on the surfaces generate an electric field (black) opposite to the original electric field (color).

In a conductor in static equilibrium, any (extra) electric charge deposited on the conductor resides on the surface of the conductor. We can prove this by means of Gauss' Law. Consider a small closed surface inside the conducting material (Figure 24.18). Since $\mathbf{E} = 0$ everywhere in this material, the left side of Eq. (4) vanishes, and therefore the right side must also vanish — which means that the charge enclosed by *any* arbitrary small surface is zero, that is, the charge in *any* small volume of the conductor is zero. Obviously, if the charges are not in the volume of the conductor, they must be on the surface.

Finally, we can say something about the electric field just outside a conductor: *The electric field at the surface of a conductor in static equilibrium is perpendicular to the surface.* The proof is again by contradiction: if the electric field had a component tangential to the surface of the conductor, the free electrons would move along the surface, and the charge distribution would *not* be in equilibrium.

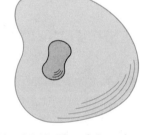

Fig. 24.18 Closed Gaussian surface inside a volume of conducting material.

Note that this argument does not exclude an electric field perpendicular to the surface of the conductor; such an electric field merely pushes the free electrons against the surface, where they are held in equilibrium by the combination of the force exerted by the electric

---

[1] The time for reaching equilibrium depends on such details as the shape and the size of the conductor and the characteristics of the conducting material. Paradoxically, a good conductor takes somewhat longer to reach equilibrium than a poor conductor. In the good conductor, the charges move very easily and they tend to overshoot their equilibrium positions — the charges slosh around on the conductor and take a while to settle in their equilibrium positions.

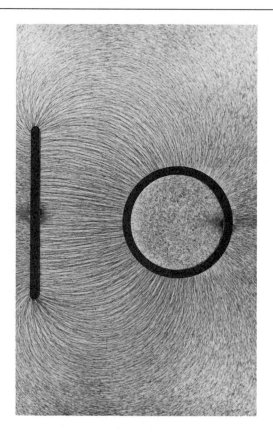

Fig. 24.19 Field lines in the space surrounding a charged flat plate and cylinder, made visible by small bits of thread suspended in oil. (Courtesy H. Waage, Princeton University.)

field and the restraining force exerted by the surface of the conductor. Figure 24.19 displays an experimental demonstration of electric fields perpendicular to the surfaces of conductors. The flat plate and the cylinder shown in this figure are conductors, and we see that the field lines meet the surfaces of these conductors at right angles.

In the preceding paragraphs, we implicitly assumed that the material of the conductor is homogeneous; that is, the material has uniform density and uniform chemical composition. If this is not the case, then our conclusions are not quite valid. For instance, suppose that the body of the conductor consists of two dissimilar metals joined together. Figure 24.20 shows a conductor made of a block of lead joined to a block of silver. In this joined conductor, there will exist some electric field at and near the interface of the metals; and, what is more, such an electric field will exist even if the conductor carries no net charge. This electric field is created by the contact of the two dissimilar metals. To understand how the electric field comes about, imagine that at first the blocks of lead and of silver are separated. Each metal is neutral and contains within it a gas of free electrons. If the blocks are now brought into contact, some of the free electrons from the lead block will flow into the silver block. This is so because silver exerts a slightly stronger hold on free electrons than lead (the energy required to remove a free electron from silver is larger than that from lead). As the extra free electrons accumulate in the silver, their electric repulsions gradually bring the flow to a halt. At equilibrium, the silver will have an excess of electrons (negative charge) and the lead a deficit of electrons (positive charge). These charges will then create an electric field at and near the interface of the joined blocks.

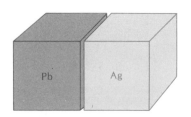

Fig. 24.20 A conductor made of two blocks of lead and silver joined together.

The charges and the electric fields that are created by the contact of dissimilar metals are usually quite small. In the following chapters we will ignore these "contact" charges and fields. However, it is well to keep in mind that these charges and fields play a crucial role in the operation of transistors, solar cells, and other solid-state devices that consist of pieces of dissimilar conductors or semiconductors joined together.

EXAMPLE 7. Find the electric field outside a very large flat conducting surface, such as the upper surface of the slab shown in Figure 24.17, on which there is a uniform surface-charge density of σ coulombs per square meter.

SOLUTION: In view of the symmetry of the charge configuration, the electric field will be perpendicular to the conducting surface, and it will have constant magnitude over any plane parallel to the conducting surface. As Gaussian surface, take the cylinder shown in Figure 24.21. The base area of the cylinder is A; the upper base is outside the conductor, and the lower base is in the conductor. The upper base contributes an amount

$$\int E \cos\theta \, dS = \int_{\substack{\text{upper}\\ \text{base}}} E \, dS = E \int dS = EA \tag{22}$$

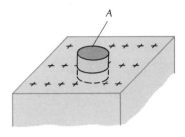

Fig. 24.21 A very large, thick slab of conductor, with a uniform distribution of charge on its surface. The cylindrical Gaussian surface intersects the slab.

to the flux integral. The lower base, in the conductor, does not contribute to the flux integral, since $E = 0$ in this region. Finally, the curved lateral surface does not contribute, since $\cos\theta = 0$. The charge within the Gaussian surface is σA. Hence,

$$EA = \sigma A/\varepsilon_0$$

and

$$E = \sigma/\varepsilon_0 \tag{23}$$

COMMENTS AND SUGGESTIONS: This is a constant electric field, independent of distance from the conducting surface. Of course, if the surface is of finite size, then the electric field will have the constant value (23) only in a region very near the surface.

Note that, for a given surface-charge density, the electric field generated by a conducting surface is twice as strong as the electric field generated by a sheet of charge [see Eq. (13)]. The reason is obvious: in the former case, all field lines that originate on the charges go to the same side of the surface; in the latter case, half go to each side.

Over a small region, any curved conducting surface can be approximated by a flat surface. Hence the expression (23) can be used to find the electric field in a region very near any smooth curved conducting surface [the expression (23) is *not* a good approximation near sharp edges]; it is not even necessary that σ be constant — it can be some smooth function of position.

EXAMPLE 8. At the ground directly below a thundercloud, the electric field is 2×10^4 N/C and points upward. What is the surface-charge density on the ground?

SOLUTION: For the purpose of this problem, we treat the ground as a good conductor. Equation (23) gives the relation between electric field and surface-charge density for points at or near the charged surface,

$$\sigma = \varepsilon_0 E = 8.85 \times 10^{-12} \frac{\text{C}^2}{\text{N} \cdot \text{m}^2} \times 2 \times 10^4 \text{ N/C}$$

$$= +1.8 \times 10^{-7} \text{ C/m}^2$$

SUMMARY

Electric flux through surface:
$$\Phi = \int E_n \, dS$$
$$= \int \mathbf{E} \cdot d\mathbf{S}$$

Gauss' Law (for closed surface): $\oint \mathbf{E} \cdot d\mathbf{S} = Q/\varepsilon_0$

Conductor in electrostatic equilibrium:
  The electric field within the conductor is zero.
  The charge resides on the surface.
  The electric field at the surface is perpendicular and of magnitude $E = \sigma/\varepsilon_0$.

QUESTIONS

1. A point charge Q is inside a spherical Gaussian surface, which is enclosed in a larger, cubical Gaussian surface. Compare the fluxes through these two surfaces.

2. Figure 24.22 shows a Gaussian surface and lines of electric field entering and leaving this surface. Assuming that the number of lines has been normalized according to the convention of Section 23.3, what can you say about the magnitude and the sign of the electric charge within the surface?

3. Suppose that instead of $1/\varepsilon_0$ lines of force per unit charge, we had adopted some other normalization, say, k/ε_0 lines per unit charge. Would this change Gauss' Law?

4. A Gaussian surface contains an electric dipole, and no other charge. What is the electric flux through this surface?

5. Defining the gravitational field as the gravitational force per unit mass, formulate a Gauss' Law for gravity. Check that your law implies Newton's Law of universal gravitation.

6. Suppose that the electric field of a point charge were not exactly proportional to $1/r^2$, but rather proportional to $1/r^{2+a}$ where a is a small number, $a \ll 1$. Would Gauss' Law still be valid? (Hint: Consider Gauss' Law for the case of a point charge.)

7. Prove that if the electric field is uniform in some region, then the charge density must be zero in that region.

8. Problems soluble by Gauss' Law fall into three categories, according to their symmetry: spherical, cylindrical, and planar. Give some examples in each category.

9. A hemisphere of radius R has a charge Q uniformly distributed over its volume. Can we use Gauss' Law to find the electric field?

10. A spherical rubber balloon of radius R has a charge Q uniformly distributed over its surface. The balloon is placed in a uniform electric field of 120 N/C. What is the net electric field inside the rubber balloon?

11. The electric field at the surface of a conductor in static equilibrium is perpendicular to the surface. Is the gravitational field at the surface of a mass in static equilibrium necessarily perpendicular to the surface? What if the surface is that of a fluid, such as water?

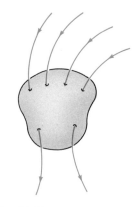

Fig. 24.22

12. Suppose we drop a charged ping-pong ball into a cookie tin and quickly close the lid. What happens to the portions of the electric field lines that are outside of the cookie tin when we close the lid?

13. When an electric current is flowing through a wire connected between a source and a sink of electric charge — such as the terminals of a battery — there is an electric field inside the wire, even though the wire is a conductor. Why does our conclusion about zero electric field inside a conductor not apply to this case?

14. The free electrons belonging to a metal are uniformly distributed over the entire volume of the metal. Does this contradict the result we derived in Section 24.4, according to which the charges are supposed to reside on the surface of a conductor?

PROBLEMS

Sections 24.1 and 24.2

1. Consider the thundercloud described in Problem 23.16 and Figure 23.28. What is the total electric flux coming out of the surface of the cloud?

2. A point charge of 1.0×10^{-8} C is placed inside an uncharged metallic can (say, a closed beer can) insulated from the ground. How many flux lines will emerge from the surface of the can when the point charge is inside?

3. On a clear day, the Earth's atmospheric electric field near the ground has a magnitude of 100 N/C and points vertically down. Inside the ground, the electric field is zero, since the ground is a conductor. Consider a mathematical box of 1 m $\times$ 1 m $\times$ 1 m, half below the ground and half above. What is the electric flux through the sides of this box? What is the charge enclosed by the box?

4. Suppose we suspend a small ball carrying a charge of 1.0×10^{-6} C in the center of a safe and lock the door. The safe is made of solid steel; it has inside dimensions 0.3 m $\times$ 0.3 m $\times$ 0.3 m and outside dimensions 0.4 m $\times$ 0.4 m $\times$ 0.4 m. What is the electric flux through a cubical surface measuring 0.2 m $\times$ 0.2 m $\times$ 0.2 m centered on the ball? A cubical surface measuring 0.35 m $\times$ 0.35 m $\times$ 0.35 m? A cubical surface measuring 0.5 m $\times$ 0.5 m $\times$ 0.5 m?

5. Consider a mathematical surface having the shape of a cube of edge 5 cm. You do not know the electric charge or the electric field inside the cube, but you do know the electric field at the surface: at the top of the cube the electric field has a magnitude of 5×10^5 N/C and points perpendicularly out of the cube; at the bottom of the cube, the electric field has a magnitude of 2×10^5 N/C and points perpendicularly into the cube; on all other faces, the electric field is tangential to the surface of the cube.
   (a) How much electric charge is inside the cube?
   (b) Can you guess what charge distribution inside (and outside) the cube would generate this kind of electric field?

6. A point charge of 2×10^{-8} C is located at the center of a mathematical surface that has the shape of a cube of edge 8 cm. What is the average value of E_n over one face of the cube?

7. A point charge of 6.0×10^{-8} C sits at a distance of 0.30 m above the x–y plane. What is the electric flux that this charge generates through the (infinite) x–y plane?

8. The electric field in a region has the following form as a function of x, y, z:

$$E_x = 5.0x \quad E_y = 0 \quad E_z = 0$$

where E is in newtons per coulomb and x in meters. This represents an electric field in the x direction with a magnitude that increases in direct proportion to x. Show that such an electric field can exist only if the region is filled with some electric charge density. Find the value of the required charge density as a function of x, y, z.

Section 24.3

9. A uranium nucleus is a spherical ball of radius 7.4×10^{-15} m with a charge of $92e$ uniformly distributed over its volume. Find the electric field produced by this charge distribution at $r = 3 \times 10^{-15}$ m, 6×10^{-15} m, and 9×10^{-15} m.

10. Charge is placed on a small metallic sphere which is surrounded by air. If the radius of the sphere is 0.5 cm, how much charge can be placed on the sphere before the air near the sphere suffers electric breakdown? The critical electric field strength that leads to breakdown in air is 3×10^6 N/C.

11. Electrons can penetrate the nucleus because the nuclear material exerts no forces on them other than electric forces. Suppose that an electron penetrates a nucleus of lead. What is the electric force on the electron when it is at a distance of 3.0×10^{-15} m from the center of the nucleus? The nucleus of lead is a uniformly charged ball of radius 7.1×10^{-15} m with a total charge of $82e$.

12. In symmetric fission, a uranium nucleus splits into two equal pieces each of which is a palladium nucleus. The palladium nucleus is spherical with a radius of 5.9×10^{-15} m and a charge of $46e$ uniformly distributed over its volume. Suppose that, immediately after fission, the two palladium nuclei are barely touching (Figure 24.23). What is the value of the total electric field at the center of each? What is the repulsive force between them? What is the acceleration of each? The mass of a palladium nucleus is 1.99×10^{-25} kg.

Fig. 24.23 Two palladium nuclei in contact. Each nucleus is a sphere.

13. Charge is distributed uniformly over the volume of a very long cylindrical plastic[2] rod of radius R. The amount of charge per meter of length of the rod is λ. Find a formula for the electric field at a distance r from the axis of the rod. Assume $r < R$.

14. What is the maximum amount of electric charge per unit length that one can place on a long and straight human hair of diameter 8×10^{-3} cm if the surrounding air is not to suffer electrical breakdown? The air will suffer breakdown if the electric field exceeds 3×10^6 N/C.

*15. A long plastic[2] pipe has inner radius a and outer radius b. Electric charge is uniformly distributed over the region $a < r < b$. The amount of charge is λ coulombs per meter of length of the pipe. Find the electric field in the regions $r < a$, $a < r < b$, and $r > b$.

*16. Charge is uniformly distributed over the volume of a large plane slab of plastic[2] of thickness d. The charge density is ρ coulombs per cubic meter. The midplane of the slab is the y–z plane (Figure 24.24). What is the electric field at a distance x from the midplane? Consider both the case $|x| < d/2$ and the case $|x| > d/2$.

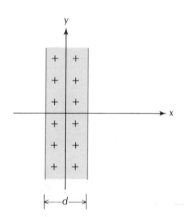

Fig. 24.24 A slab of plastic with electric charge uniformly distributed over its volume.

*17. The tube of a Geiger counter consists of a thin conducting wire of radius 1.3×10^{-3} cm stretched along the axis of a conducting cylindrical shell of radius 1.3 cm (Figure 24.25). The wire and the cylinder have equal and opposite charges of 7.2×10^{-10} C distributed along their length of 9.0 cm. Find a formula for the electric field in the space between the wire and the cylinder; pretend that the electric field is that of an infinitely long wire and cylinder. What is the magnitude of the electric field at the surface of the wire?

Fig. 24.25

*18. A thick spherical shell of inner radius a and outer radius b has a charge Q uniformly distributed over its volume. Find the electric field in the regions $r < a$, $a < r < b$, and $r > b$.

---

[2] Assume that the plastic has no effect on the electric field.

*19. According to a (crude) model, the neutron consists of an inner core of positive charge surrounded by an outer shell of negative charge. Suppose that the positive charge has a magnitude $+e$ and is uniformly distributed over a sphere of radius 0.50×10^{-15} m; suppose that the negative charge has a magnitude $-e$ and is uniformly distributed over a concentric shell of inner radius 0.50×10^{-15} m and outer radius 1.0×10^{-15} m (Figure 24.26). Find the magnitude and direction of the electric field at 1.0×10^{-15}, 0.75×10^{-15}, 0.50×10^{-15}, and 0.25×10^{-15} m from the center.

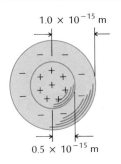

Fig. 24.26 Hypothetical charge distribution inside a neutron.

*20. Positive charge Q is uniformly distributed over the volume of a solid sphere of radius R. Suppose that a spherical cavity of radius $R/2$ is cut out of the solid sphere, the center of the cavity being at a distance of $R/2$ from the center of the original solid sphere (Figure 24.27); the cut-out material and its charge are discarded. What new electric field does the sphere with the cavity produce at the point P at a distance r from the original center? Assume $r > R$.

*21. According to an old (and erroneous) model due to J. J. Thomson, an atom consists of a cloud of positive charge within which electrons sit like plums in a pudding. The electrons are supposed to emit light when they vibrate about their equilibrium positions in this cloud. Assume that in the case of the hydrogen atom the positive cloud is a sphere of radius $R = 0.5$ Å with a charge of e uniformly distributed over the volume of this sphere. The (pointlike) electron is held at the center of this charge distribution by the electrostatic attraction.
  (a) Show that the restoring force on the electron is $e^2 r/(4\pi\varepsilon_0 R^3)$ when the electron is at a distance r from the center ($r \leq R$).
  (b) What is the frequency of small oscillations of the electron moving back and forth along a diameter? Give a *numerical* answer.

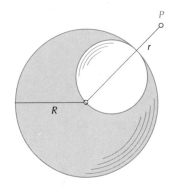

Fig. 24.27 Sphere with spherical cavity.

*22. According to the Thomson model (see also Problem 21), the atom of helium consists of a uniform spherical cloud of positive charge within which sit two electrons. Assume that the positive cloud is a sphere of radius 0.5 Å with a charge $2e$ uniformly distributed over the volume. The two electrons are symmetrically placed with respect to the center (Figure 24.28). What is the equilibrium separation of the electrons?

*23. A charge distribution with spherical symmetry has a charge density ρ coulombs per cubic meter described by the formula $\rho = kr^n$, where k is a constant and $n > -3$.
  (a) What is the amount of charge $Q(r)$ inside a sphere of radius r?
  (b) What is the magnitude of the electric field as a function of r?
  (c) For what value of n is the magnitude of the electric field constant?
  (d) Why is it necessary to assume that $n > -3$?

*24. The tau particle is a negatively charged particle similar to the electron, but of much larger mass — its mass is 3.18×10^{-27} kg, about 3490 times the mass of an electron. Nuclear material is transparent to the tau; thus, the tau can orbit around inside a nucleus, under the influence of the electric attraction of the nuclear charge. Suppose that a tau is in a circular orbit of radius 2.9×10^{-15} m inside a uranium nucleus. Treat the nucleus as a sphere of radius 7.4×10^{-15} m with a charge $92e$ uniformly distributed over its volume. Find the speed, the kinetic energy, the angular momentum, and the frequency of the orbital motion of the tau.

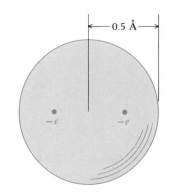

Fig. 24.28 Model of the helium atom.

*25. A very long cylinder of radius R has positive charge uniformly distributed over its volume. The amount of charge is λ coulombs per meter of length of the cylinder. A spherical cavity of radius $R' \leq R$, centered on the axis of the cylinder, has been cut out of this cylinder, and the charge in this cavity has been discarded.
  (a) Find the electric field as a function of distance from the center of the sphere along the axis of the cylinder.
  (b) Find the electric field as a function of distance along a line passing through the center of the sphere perpendicular to the axis of the cylinder. Consider distances both smaller and larger than R'.

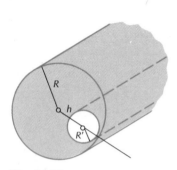

Fig. 24.29

**26. The electric charge of the proton is not concentrated in a point but, rather, distributed over a volume. According to experimental investigations at the Stanford Linear Accelerator, the charge distribution of the proton can be approximately described by a charge density which is an exponential function of the radial distance:

$$\rho = \frac{e}{8\pi b^3} e^{-r/b}$$

where b is a constant, $b = 0.23 \times 10^{-15}$ m. Find the electric field as a function of the radial distance. What is the magnitude of the electric field at $r = 1.0 \times 10^{-15}$ m? [Hint: The following integral may be useful: $\int x^2 e^{-x} dx = -x^2 e^{-x} - 2e^{-x}(x+1)$]

**27. A long cylinder of radius R has a uniform charge density ρ distributed over its volume. A cylindrical hole of radius R' has been drilled parallel to the axis of the cylinder along its full length, and the charge in this hole has been discarded. The axis of the hole is at a distance $h > R'$ from the axis of the cylinder (see Figure 24.29). Find the electric field as a function of the radial distance r on the radial line shown in Figure 24.29. Consider the cases $r < h - R'$, $h - R' < r < h + R'$, $h + R' < r < R$, and $r > R$ separately.

**28. A charge Q is uniformly distributed over the volume of a solid sphere of radius R. A spherical cavity is cut out of this solid sphere (Figure 24.30), and the material and its charge are discarded. Show that the electric field in the cavity will then be uniform, of magnitude $(1/4\pi\varepsilon_0)Qd/R^3$, where d is the distance between the centers of the spheres (Figure 24.30). Make a drawing of the lines of electric field in the cavity.

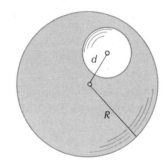

Fig. 24.30 Sphere with spherical cavity.

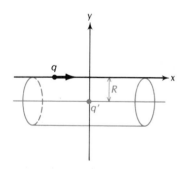

Fig. 24.31 The charge q moves along the x axis. The charge q' sits at a distance R below the x axis. The cylinder, which is supposed to extend to infinity, is the Gaussian surface.

**29. When a point charge q moving at high speed passes by another stationary point charge q', the main effect of the electric forces is to give each charge a transverse impulse. Figure 24.31 shows the charge q moving at (almost) constant velocity v along the x axis in an almost straight line and shows the charge q' sitting at a distance R below the origin. The transverse impulse on q is

$$\int_{-\infty}^{\infty} F_y \, dt = \frac{q}{v} \int_{-\infty}^{\infty} E_y \, dx$$

Evaluate the integral $\int E_y \, dx$ by means of Gauss' Law, and prove that

$$\int_{-\infty}^{\infty} F_y \, dt = \frac{q}{v} \frac{q'}{2\pi\varepsilon_0 R}$$

(Hint: Consider $2\pi R \int E_y \, dx$; show that this is the flux that q' produces through the infinite cylindrical surface indicated in Figure 24.31.)

*30. The formula derived in Problem 29 gives the transverse momentum that a high-speed charged particle acquires as it passes by a stationary charged particle.

(a) Calculate the transverse momentum that an electron of speed 4.0×10^7 m/s acquires as it passes by a stationary electron at a distance of 0.60×10^{-10} m.
(b) What transverse velocity corresponds to this transverse momentum?
(c) What will be the recoil velocity of the stationary electron (if it is free to move)?

Section 24.4

31. The surface of a long cylindrical copper pipe has a charge of λ coulombs per meter (Figure 24.32). What is the electric field outside the pipe? Inside the pipe?

Fig. 24.32 A long pipe with positive charge on its surface.

32. A solid copper sphere of radius 3 cm carries a charge of 10^{-6} C. This sphere is placed concentrically within a spherical, thin copper shell of radius 15 cm carrying a charge of 3×10^{-6} C. Find a formula for the electric field in the space between the sphere and the shell. Find a formula for the electric field outside the shell. Plot these electric fields as a function of radius.

33. A thick spherical shell made of solid metal has an inner radius a, an outer radius b, and is initially uncharged. A point charge q is placed at the center of the shell. Find the electric field in the regions $r < a$, $a < r < b$, and $r > b$. Find the induced surface-charge densities at $r = a$ and $r = b$.

34. On days of fair weather, the atmospheric electric field of the Earth is about 100 N/C; this field points vertically downward (compare Problem 23.2). What is the surface-charge density on the ground? Treat the ground as a flat conductor.

35. You wish to generate a uniform electric field of 2.0×10^5 N/C in the space between two flat parallel plates of metal placed face to face. The plates measure 0.30 cm × 0.30 cm. How much electric charge must you put on each plate? Assume that the gap between the plates is small so that the charge distribution and the electric field are approximately uniform, as for infinite plates.

CHAPTER 25

The Electrostatic Potential

In Chapter 8 we saw that to formulate a law of conservation of energy for a particle moving under the influence of a conservative force, we had to construct a potential energy. In this chapter we will construct the electrostatic potential energy for a charged particle moving under the influence of the electric force generated by a static charge distribution. This potential energy will help us in the calculation of the motion of the particle. Furthermore, this potential energy will give us yet another method for the calculation of the electric force and the electric field generated by a given charge distribution. The method involves two steps: first, find the potential energy of a point charge placed somewhere near the charge distribution; second, find the electric force and the electric field by differentiating this potential energy. Thus, we will have available three alternative methods for the calculation of the electric field: via Coulomb's Law, via Gauss' Law, and via the potential energy. If a problem does not yield to the simple and elegant method based on Gauss' Law, it is usually best to to use the method based on the potential energy, because the mathematics is likely to be less cumbersome than with the method based on Coulomb's Law.

25.1 The Electrostatic Potential

According to the definition given in Chapter 8, a force is conservative if the work it does on a particle depends on the initial and final positions of the particle, but not on the shape of the path connecting these positions. The electric force exerted by a fixed point charge q' on a moving point charge q is a conservative force. The proof is the same as in the case of the gravitational force. Figure 25.1 shows a fixed point charge q' and a second point charge q that moves from P_1 to P_2; the figure also shows two alternative paths I and II from P_1 to P_2. The work done by the electric force along a path is $\int \mathbf{F} \cdot d\mathbf{l}$.[1] To evaluate this integral, we approximate each path by a sequence of infinitesimal radial segments and circular arcs (Figure 25.1). Along the circular arcs the work is zero because the force is perpendicular to the displacement. Along the radial segments the work is not zero. Since for each radial segment belonging to the first path, there is a corresponding radial segment (at the same radius) belonging to the second path, and since the magnitudes of the forces are the same at both these radial segments, the contributions to the work along the first and second paths are equal. This establishes that the work is independent of the shape of the path and that the force is conservative.

More generally, the net electric force exerted by any number of fixed point charges on another given point charge is conservative, since the net force is a sum of the individual conservative forces of all the fixed point charges.

For any conservative force we can construct a corresponding potential energy by means of the recipe given in Section 8.2: Take a reference position P_0 and at this position assign to the potential energy the value $U(P_0)$. At any other position, assign to the potential energy the value [see Eq. (8.4)]

$$U(P) = -\int_{P_0}^{P} \mathbf{F} \cdot d\mathbf{l} + U(P_0) \tag{1}$$

The **electrostatic potential** is defined as the potential energy of a point charge divided by the magnitude of q of this charge; thus, it is defined as the potential energy per unit charge,

$$V(P) = \frac{U(P)}{q} \tag{2}$$

$$= -\int_{P_0}^{P} \frac{\mathbf{F}}{q} \cdot d\mathbf{l} + \frac{U(P_0)}{q} \tag{3}$$

Since the force per unit charge is the electric field ($\mathbf{E} = \mathbf{F}/q$), we can also write this as

$$\boxed{V(P) = -\int_{P_0}^{P} \mathbf{E} \cdot d\mathbf{l} + V(P_0)} \tag{4}$$

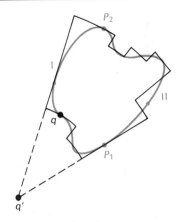

Fig. 25.1 A charge q moves from the position P_1 to the position P_2 while under the influence of the electric field of a charge q'. The path can be approximated by radial segments and circular arcs.

Electrostatic potential

In this equation, $V(P_0)$ plays the role of an additive constant in the potential. In the calculations of potential differences between positions (see below), the additive constant in the potential cancels. Hence, this constant is of no physical significance. We can make any convenient choice for this constant, but we must remain faithful to our choice throughout any subsequent calculation. We will see some examples of convenient choices of $V(P_0)$ in what follows.

---

[1] Note that we are now using the symbol $d\mathbf{l}$ for an infinitesimal displacement, whereas in Chapters 7 and 8 we used the symbol $d\mathbf{r}$; we now want to reserve the letter $\mathbf{r}$ for the radial distance in Coulomb's Law.

Potential difference

The **potential difference** between two arbitrary positions P_1 and P_2 is [compare Eq. (8.5)]:

$$V(P_2) - V(P_1) = -\int_{P_0}^{P_1} \mathbf{E} \cdot d\mathbf{l} + V(P_0) + \int_{P_0}^{P_2} \mathbf{E} \cdot d\mathbf{l} - V(P_0)$$

or

$$V(P_2) - V(P_1) = -\int_{P_1}^{P_2} \mathbf{E} \cdot d\mathbf{l} \tag{5}$$

Alessandro, Conte Volta, *1745–1827, Italian physicist, professor at Pavia. Volta established that the "animal electricity" observed by Luigi Galvani, 1737–1798, in experiments with frog muscle tissue placed in contact with dissimilar metals, was not due to any exceptional property of animal tissues but was also generated whenever any wet body was sandwiched between dissimilar metals. This led him to develop the first "voltaic pile," or battery, consisting of a large stack of moist disks of cardboard (electrolyte) sandwiched between disks of metal (electrodes).*

Hence the potential difference is numerically equal to the work that *you* must do (against the electric force) in order to push one coulomb of charge from the position P_1 to the position P_2.

The unit of electrostatic potential is the **volt** (V),[2]

$$1 \text{ volt} = 1 \text{ V} = 1 \text{ joule/coulomb} = 1 \text{ J/C} \tag{6}$$

The unit of electric field we have employed in the preceding chapters is the N/C. This can be expressed in terms of volts as follows:

$$1\,\frac{\text{N}}{\text{C}} = 1\,\frac{\text{N} \cdot \text{m}}{\text{C} \cdot \text{m}} = 1\,\frac{\text{J}}{\text{C}}\,\frac{1}{\text{m}} = 1\,\frac{\text{V}}{\text{m}} \tag{7}$$

Thus, N/C and V/m are equal units; in practice, volts per meter is the preferred unit for the electric field.

EXAMPLE 1. A very large flat conducting surface carries a uniform surface-charge density which generates a constant electric field **E** (see Example 24.7). What is the electrostatic potential at some distance z above this surface? Assume that the potential is zero at the surface.

SOLUTION: In Figure 25.2, the charged surface coincides with the x–y plane. According to Example 24.7, the electric field generated by the charge on this surface is vertical and is independent of distance:

$$E_z = E = [\text{constant}] \tag{8}$$

If P_0 is a position on the surface, then $V(P_0) = 0$ and

$$V(P) = V(z) = -\int \mathbf{E} \cdot d\mathbf{l} = -\int E_x\,dx - \int E_y\,dy - \int E_z\,dz$$

$$= -\int_0^z E\,dz = -Ez \tag{9}$$

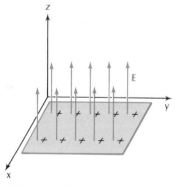

Fig. 25.2 Electric field of a very large charged surface lying in the x–y plane.

Hence the potential is directly proportional to the distance from the surface.

---

[2] Note that the same letter V is used in physics both as a symbol for potential and as an abbreviation for *volt*. This leads to confusing equations such as $V = 3.0$ V (which means $V = 3.0$ volts). If there is a possibility of confusion, it is best not to abbreviate *volt*.

EXAMPLE 2. Near the ground directly below a thundercloud, the electric field is constant, of magnitude 2×10^4 V/m, and is directed upward. What is the potential difference between the ground and a point in the air, 50 m above ground?

SOLUTION: From Eq. (9)

$$V(z) = -Ez = -2 \times 10^4 \text{ volt/m} \times 50 \text{ m} = -1 \times 10^6 \text{ volt}$$

Since the electric field in a conducting body in electrostatic equilibrium is zero, Eq. (5) implies that the potential difference between any two points on or in a conducting body is zero. Thus, *all points on the surface or within the material of a conducting body are at the same potential.* For instance, since soil is a conductor, all points on the surface of the Earth or in the Earth are at the same electrostatic potential. In experiments with electric circuits, it is usually convenient to adopt the convention that the potential of the surface of the Earth is zero, $V = 0$. The surface of the Earth is said to be the **electric ground,** and any conductors connected to it are said to be grounded.

Electric ground

25.2 Calculation of the Electrostatic Potential

To find an explicit expression for the electrostatic potential of a fixed point charge q', we must substitute the radial electric field of such a point charge into the general formula for the potential,

$$V(P) = -\int_{P_0}^{P} \mathbf{E} \cdot d\mathbf{l} + V(P_0) \qquad (10)$$

Since the integral does not depend on the path between P_0 and P, we may take any path that is convenient. Figure 25.3 shows a path consisting of one radial segment and one circular arc. Along the circular arc, the integral receives no contribution. Along the radial segment, the displacement is $d\mathbf{l} = \hat{\mathbf{r}} \, dr$, and the dot product of this with the electric field $\mathbf{E} = (1/4\pi\varepsilon_0)q'\hat{\mathbf{r}}/r^2$ is $\mathbf{E} \cdot d\mathbf{l} = (1/4\pi\varepsilon_0)(q'/r^2) \, dr$. Hence[3]

$$V(P) = V(r) = -\int_{r_0}^{r} \frac{1}{4\pi\varepsilon_0} \frac{q'}{r'^2} dr' + V(P_0) \qquad (11)$$

$$= \frac{q'}{4\pi\varepsilon_0} \left(\frac{1}{r} - \frac{1}{r_0} \right) + V(P_0) \qquad (12)$$

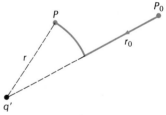

Fig. 25.3 A path from P_0 to P consisting of a radial segment and a circular arc.

For the reference position P_0 we choose a point at infinite distance from the fixed charge; for the value of the potential at this position we take $V(P_0) = 0$. The potential at the position P can then be written

$$\boxed{V(r) = \frac{1}{4\pi\varepsilon_0} \frac{q'}{r}} \qquad (13)$$

Potential of point charge

---

[3] The variable of integration in this integral has been written r' to distinguish it from the limit of integration r.

Thus, the potential in the space surrounding a point charge is inversely proportional to the distance.

The potential energy of a charge q under the influence of the electric field of the charge q' is then

Potential energy of two point charges

$$U(r) = qV(r) = \frac{1}{4\pi\varepsilon_0}\frac{qq'}{r} \qquad (14)$$

EXAMPLE 3. The electron in a hydrogen atom is at a distance of 0.53×10^{-10} m from the nucleus. What is the electrostatic potential generated by the nucleus at this distance? What is the potential energy of the electron?

SOLUTION: The nucleus is (approximately) a point charge with $q' = e = 1.6 \times 10^{-19}$ C. The electrostatic potential generated by the nucleus is

$$V = \frac{1}{4\pi\varepsilon_0}\frac{q'}{r} = \frac{1}{4\pi\varepsilon_0} \times \frac{1.6 \times 10^{-19} \text{ C}}{0.53 \times 10^{-10} \text{ m}}$$

$$= 27 \text{ volts} \qquad (15)$$

The charge of the electron is $q = -e = -1.6 \times 10^{-19}$ C. The potential energy of the electron is

$$U = qV = -e \times 27 \text{ volts} \qquad (16)$$

$$= -1.6 \times 10^{-19} \text{ C} \times 27 \text{ volts} = -4.3 \times 10^{-18} \text{ J} \qquad (17)$$

For the purposes of atomic physics, the joule is a rather large unit of energy, and it is more convenient to leave the answer as in Eq. (16),

$$U = -27e \cdot \text{volt}$$

The product of the fundamental unit of atomic charge and the unit of potential, $e \cdot \text{volt}$, or eV, is a unit of energy. This unit of energy is called an **electron-volt.** It can be converted to joules by substituting the numerical value for e,

Electron-volt, eV

$$1 \text{ eV} = 1 \times 1.60 \times 10^{-19} \text{ C} \times 1 \text{ V} = 1.60 \times 10^{-19} \text{ J} \qquad (18)$$

In chemical reactions between atoms or molecules, the energy released or absorbed by each atom or molecule is typically 1 or 2 eV. Such reactions involve a change in the arrangement of the exterior electrons of the atoms, and the energy of 1 or 2 eV represents the typical amount of energy needed for this rearrangement.

The total mechanical energy of a point charge moving in an electric field is the sum of the electric potential energy and the kinetic energy. The law of conservation of energy for the motion of a point charge q in the electric field of a fixed point charge q' takes the form

Energy of a point charge

$$E = K + U = \tfrac{1}{2}mv^2 + \frac{1}{4\pi\varepsilon_0}\frac{qq'}{r} = [\text{constant}] \qquad (19)$$

This total energy remains constant during the motion. As we saw in Chapter 8, the examination of the energy reveals some general features of the motion. Obviously, if qq' is negative (opposite charges, attractive Coulomb force), then Eq. (19) implies that whenever r increases, v must decrease, and conversely.

EXAMPLE 4. An electron is initially at rest at a very large distance from a proton. Under the influence of the electric attraction, the electron falls toward the proton, which remains (approximately) at rest. What is the speed of the electron when it has fallen to within 0.53×10^{-10} m of the proton?

SOLUTION: The potential energy of the electron is $qV(r) = -eV(r)$. The total energy is the sum of the potential and kinetic energies,

$$E = \tfrac{1}{2}m_e v^2 - eV(r) \tag{20}$$

This total energy is conserved. The initial value of the energy is zero; hence the final value of the energy must also be zero:

$$\tfrac{1}{2}m_e v^2 - eV(r) = 0$$

and

$$v = \sqrt{\frac{2eV(r)}{m_e}}$$

According to Example 3, $V(r) = 27$ volts for $r = 0.53 \times 10^{-10}$ m, so

$$v = \sqrt{\frac{2 \times 1.6 \times 10^{-19} \text{ C} \times 27 \text{ V}}{9.1 \times 10^{-31} \text{ kg}}} = 3.1 \times 10^6 \text{ m/s}$$

EXAMPLE 5. A sphere of radius R carries a total charge Q uniformly distributed over its volume. Find the electrostatic potential inside and outside the sphere.

SOLUTION: Outside the sphere ($r \geq R$), the electrostatic potential is the same as for a point charge,

$$V(r) = \frac{1}{4\pi\varepsilon_0} \frac{Q}{r} \quad \text{for } r \geq R \tag{21}$$

Inside the sphere ($r \leq R$), the electric field is [see Eq. (24.17)]

$$E(r) = \frac{1}{4\pi\varepsilon_0} \frac{Qr}{R^3}$$

The potential difference between R and r can be calculated from this field [see Eq. (5)]:

$$V(r) - V(R) = -\int_R^r \frac{1}{4\pi\varepsilon_0} \frac{Qr'}{R^3} dr'$$

$$= -\frac{Q}{4\pi\varepsilon_0}\left(\frac{r^2}{2R^3} - \frac{1}{2R}\right)$$

Consequently,

$$V(r) = -\frac{Q}{4\pi\varepsilon_0}\left(\frac{r^2}{2R^3} - \frac{1}{2R}\right) + V(R)$$

Since $V(R) = (1/4\pi\varepsilon_0)(Q/R)$, this gives

$$V(r) = -\frac{1}{4\pi\varepsilon_0}\frac{Qr^2}{2R^3} + \frac{1}{4\pi\varepsilon_0}\frac{3Q}{2R} \quad \text{for } r \leq R \tag{22}$$

COMMENTS AND SUGGESTIONS: Figure 25.4 is a plot of the potential as a function of radius. The potential has a maximum at the center. The value of this maximum is

$$V(0) = \frac{1}{4\pi\varepsilon_0}\frac{3Q}{2R} \tag{23}$$

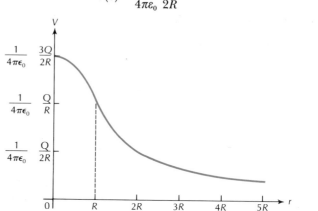

Fig. 25.4 Electrostatic potential of a uniformly charged sphere.

In all the above examples of calculations of electrostatic potentials, we started with a specified electric field. But it is also possible to start with a specified charge distribution and to calculate the electrostatic potential directly from this charge distribution, without bothering with the electric field. For such a calculation, we employ Eq. (13), which tells us how much a point charge contributes to the potential. Any charge distribution can be regarded as consisting of small charge elements, each of which can be treated as a point charge. The net electrostatic potential of the charge distribution is then the sum or integral of all the contributions from all these pointlike charge elements.

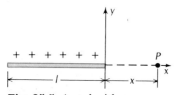

Fig. 25.5 A rod with a uniform distribution of charge.

EXAMPLE 6. A rod of length l has a charge Q uniformly distributed along its length (Figure 25.5). Find the electrostatic potential at a distance x from one end of the rod.

SOLUTION: Consider a small segment dx' of the rod. Since the charge per unit length is Q/l, the charge in this segment is $(Q/l)dx'$. The distance of the segment from the position P is $x - x'$ (here, x' is a negative quantity; see Figure 25.5). According to Eq. (13), the segment then makes the following contribution to the potential at the position P:

$$dV = \frac{1}{4\pi\varepsilon_0}\frac{(Q/l)dx'}{x - x'} \tag{24}$$

The integral of this over the length of the rod gives us the potential at the position P:

$$V(P) = V(x) = \int_{-l}^{0} \frac{1}{4\pi\varepsilon_0}\frac{Q/l}{x - x'}dx'$$

$$= \frac{1}{4\pi\varepsilon_0}\frac{Q}{l}\left[-\ln(x - x')\right]_{-l}^{0} = \frac{1}{4\pi\varepsilon_0}\frac{Q}{l}\ln\left[\frac{x+l}{x}\right] \tag{25}$$

COMMENTS AND SUGGESTIONS: When performing a summation or an integration of the contributions that the charge elements make to the potential, we do not have to worry about the directions of these contributions. The potential is a scalar quantity, with a magnitude but no direction. However, keep in mind that positive charges make a positive contribution to the potential, and negative charges a negative contribution.

25.3 The Electrostatic Field as a Conservative Field

The electric field produced by a static charge distribution is said to be a **conservative field** because the integral $\int \mathbf{E} \cdot d\mathbf{l}$ between any two positions P_1 and P_2 is independent of the path between these positions. As we saw in Section 8.1, this statement about the path independence of the integral is equivalent to the statement that the integral vanishes for any closed path, that is,

$$\oint \mathbf{E} \cdot d\mathbf{l} = 0 \qquad (26)$$

Conservative field

where the circle on the integral sign indicates that the integration is around a closed path. Note that this equation implies that a field line can never form a closed loop — if it did, then the integral $\oint \mathbf{E} \cdot d\mathbf{l}$ around such a closed loop would not be zero. Consequently, any field line must start and end somewhere. Since we derived the conservative property of the electrostatic field by examining the electric field produced by a point charge, it is not surprising that Eq. (26) should imply the existence of starting points and ending points for field lines. These sources and sinks of field lines are, of course, the positive and negative electric charges, and Gauss' Law tells us just how many field lines emerge from each electric charge. It can be shown that Eq. (26) together with Gauss' Law is exactly equivalent to Coulomb's Law, that is, any electric field that satisfies both Eq. (26) and Gauss' Law is necessarily the electric field of a static distribution of point charges.

With Eq. (26) we can easily prove an interesting theorem about static electric fields: *Within a closed, empty cavity inside a homogeneous conductor, the electric field is exactly zero.* Figure 25.6 shows such a cavity. If there were an electric field in this cavity, then there would have to be field lines in the cavity. Consider one of these field lines. Since the cavity is empty (contains no charge), the field line cannot end or begin within the space of the cavity — it must therefore begin and end on the surface of the cavity (see Figure 25.6). The field line cannot penetrate the conducting material, since the electric field is zero in this material. Now take a closed path consisting of one portion that follows the field line and a second portion that lies entirely in the conducting material. Along the first portion the integral $\int \mathbf{E} \cdot d\mathbf{l}$ is positive, and along the second portion it is zero. Hence $\oint \mathbf{E} \cdot d\mathbf{l} \neq 0$ for this closed path. This is in contradiction to Eq. (26) and establishes the impossibility of such an electric field.

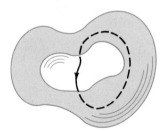

Fig. 25.6 Empty cavity in a volume of conducting material. The solid line is a hypothetical field line. The dashed line completes a closed path.

The absence of electrostatic fields in closed conducting cavities has important practical applications. Delicate electric instruments can be shielded from atmospheric electric fields, and other stray electric fields, by placing them in a box made of sheet metal. Such a box is called a **Faraday cage**. Often, the box is made of fine wire mesh rather than sheet metal; although such wire mesh does not have the perfect

Faraday cage

shielding properties of solid sheet metal, it provides good enough shielding for most purposes.

In the same way, the sheet metal of an automobile provides fairly good shielding against atmospheric electric fields, such as the electric fields of lightning. The windows of the automobile leave some gaps in this shield and permit the penetration of some field lines, but the strength of the external atmospheric fields is attenuated to such an extent that the occupants are quite safe from atmospheric electrical discharges.

25.4 The Gradient of the Potential

The electrostatic potential can be calculated from the electric field by integration,

$$V(P) = -\int_{P_0}^{P} \mathbf{E} \cdot d\mathbf{l} + V(P_0) \tag{27}$$

Conversely, the electric field can be calculated from the potential by differentiation. Consider two nearby positions separated by an infinitesimal displacement $d\mathbf{l}$. The change in potential between these positions is then

$$dV = -\mathbf{E} \cdot d\mathbf{l} = -E \cos \theta \, dl \tag{28}$$

where θ is the angle between the displacement $d\mathbf{l}$ and the electric field. Hence

$$\frac{dV}{dl} = -E \cos \theta \tag{29}$$

This shows that the derivative dV/dl equals the negative of the component of $\mathbf{E}$ in the direction of dl.

If the displacement dl is in the x direction, $dl = dx$ and Eq. (29) becomes

Derivatives of the potential

$$\boxed{\frac{\partial V}{\partial x} = -E_x} \tag{30}$$

Likewise

$$\boxed{\frac{\partial V}{\partial y} = -E_y} \tag{31}$$

and

$$\boxed{\frac{\partial V}{\partial z} = -E_z} \tag{32}$$

Thus, if the potential is a known function of position, the components of the electric field can be calculated by taking derivatives (compare Section 8.3).

In vector notation, we can express Eqs. (30)–(32) as

$$\frac{\partial V}{\partial x}\hat{\mathbf{x}} + \frac{\partial V}{\partial y}\hat{\mathbf{y}} + \frac{\partial V}{\partial z}\hat{\mathbf{z}} = -\mathbf{E} \tag{33}$$

The quantity on the left side of this equation is called the **gradient** of the potential. The gradient is a vector with components $\partial V/\partial x$, $\partial V/\partial y$, and $\partial V/\partial z$. Thus, Eq. (33) asserts that the electric field is the negative of the gradient of the potential.

EXAMPLE 7. The electric potential generated by a large uniformly charged flat conducting plate is

$$V = -Ez \tag{34}$$

where E is a constant [see Eq. (9)]. Calculate the electric field by differentiating this potential.

SOLUTION: Equations (30)–(32) immediately give

$$E_x = -\frac{\partial V}{\partial x} = 0 \tag{35}$$

$$E_y = -\frac{\partial V}{\partial y} = 0 \tag{36}$$

$$E_z = -\frac{\partial V}{\partial z} = +E_0 \tag{37}$$

EXAMPLE 8. The electrostatic potential due to a point charge q' is

$$V = \frac{1}{4\pi\varepsilon_0}\frac{q'}{r} \tag{38}$$

Calculate the electric field by differentiating this potential.

SOLUTION: In order to use Eqs. (30)–(32), we would first have to express r in terms of x, y, z (that is, $r = \sqrt{x^2 + y^2 + z^2}$). It is simpler to return to Eq. (29), which, with $dl = dr$, directly gives the component of **E** in the *radial direction*,

$$E_r = -\frac{dV}{dr} = -\frac{d}{dr}\left(\frac{1}{4\pi\varepsilon_0}\frac{q'}{r}\right) = \frac{1}{4\pi\varepsilon_0}\frac{q'}{r^2} \tag{39}$$

According to Eq. (29), the electric field has components only in those directions in which the potential changes. Since the potential (38) does not change in the tangential direction, the radial electric field given by Eq. (39) is the total field.

The results obtained in the preceding examples do not tell us anything new — since we already knew the electric fields in these examples, the calculations merely verify the consistency of our formulas. However, if we do not know the electric field in advance, then Eqs. (30)–(32) can be useful in a calculation of the electric field, provided we have some means of finding the potential. If the charge distribution is specified, we can find the potential by a summation or an integration over the charge elements in the distribution, as in Example 6. The advantage of this procedure is that it is much easier to sum the potentials of a distribution of point charges than it is to sum their electric fields; in the latter sum it is necessary to take into account the directions of the electric fields and to take into account the three electric field components, which is rather cumbersome. Although the calculation of the electric field via the potential involves two separate steps — evaluation of the potential followed by evaluation of the gradient of the potential — it often is the quickest route to the answer. The next example illustrates such a calculation of the electric field via the potential.

EXAMPLE 9. An amount of charge Q is uniformly distributed along the circumference of a ring of radius R. Find the potential and the electric field on the axis of the ring.

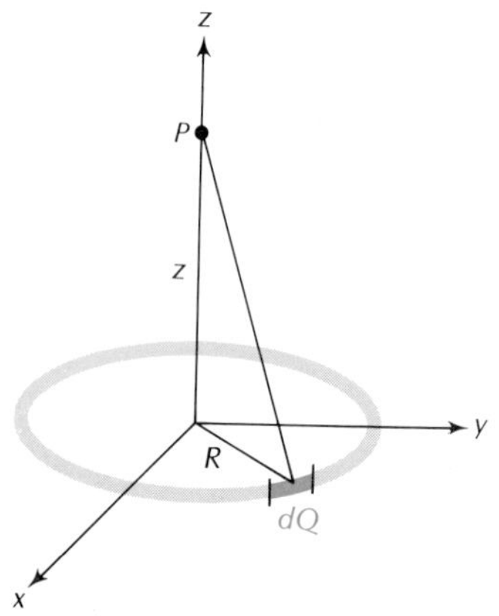

SOLUTION: We already calculated this electric field in Chapter 23 (see Example 23.3). Now we will calculate it again by beginning with the electrostatic potential. At a height z above the plane of the ring (Figure 25.7) the potential contributed by a small charge element dQ is

$$dV = \frac{1}{4\pi\varepsilon_0}\frac{dQ}{r} = \frac{1}{4\pi\varepsilon_0}\frac{dQ}{(z^2+R^2)^{1/2}} \quad (40)$$

Since all charge elements around the ring are at the same distance from the point z, they all contribute equally, and the total potential is simply

$$V = \frac{1}{4\pi\varepsilon_0}\frac{Q}{(z^2+R^2)^{1/2}} \quad (41)$$

Consequently, the z component of the electric field is

$$E_z = -\frac{\partial V}{\partial z} = \frac{1}{4\pi\varepsilon_0}\frac{Qz}{(z^2+R^2)^{3/2}} \quad (42)$$

and the x and y components are zero. This result agrees with Eq. (23.20), but we now obtained it a bit more quickly.

Fig. 25.7 A uniformly charged ring.

Equipotential surface

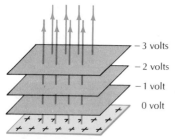

Fig. 25.8 Equipotential surfaces (gray) for a very large sheet with a uniform charge distribution.

A mathematical surface on which the potential function has a fixed, constant value is called an **equipotential surface**. Figure 25.8 shows the equipotential surfaces belonging to the potential of a uniformly charged flat sheet — the equipotential surfaces are parallel planes. Figure 25.9 shows the equipotential surfaces belonging to the potential of a point charge — the equipotential surfaces are concentric spheres.

Note that the electric field is everywhere perpendicular to the equipotentials. This is an immediate consequence of Eq. (28): along any equipotential surface, the potential is constant, that is, $dV = 0$; consequently, the electric field in the direction parallel to the surface is zero, and the electric field must be entirely perpendicular to the surface.

Conversely, if the electric field is everywhere perpendicular to a given surface, then this surface must be an equipotential surface. This, also, is a consequence of Eq. (28): if the electric field is zero in the direction parallel to the surface, then for a displacement $d\mathbf{l}$ parallel to the surface, $dV = 0$, and the potential is constant. Furthermore, since we already know (from Section 24.4) that along the surface of any con-

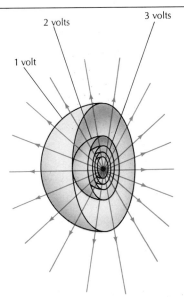

1 volt
2 volts
3 volts

Fig. 25.9 Equipotential surfaces (gray) for a positive point charge. The equipotential surfaces are concentric spheres.

ductor in electrostatic equilibrium the electric field is perpendicular to the surface, we conclude that any conducting surface is an equipotential surface. (This conclusion agrees with the general statement made at the end of Section 25.1: the potential is constant throughout any conductor.)

25.5 The Potential and Field of a Dipole

We will now calculate the electric field generated by an electric dipole. As in the preceding example, we will do this calculation via the potential.

Figure 25.10 shows an electric dipole consisting of point charges $\pm Q$ on the z axis at $z = \pm l/2$. The net potential generated by this pair of charges is just the sum of the individual potentials,

$$V = \frac{1}{4\pi\varepsilon_0}\frac{Q}{r_1} - \frac{1}{4\pi\varepsilon_0}\frac{Q}{r_2} \tag{43}$$

$$= \frac{Q}{4\pi\varepsilon_0}\frac{r_2 - r_1}{r_1 r_2} \tag{44}$$

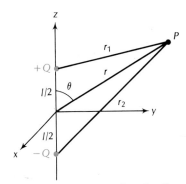

Fig. 25.10 Charges Q and $-Q$ on the z axis at $z = l/2$ and $z = -l/2$.

where r_1 and r_2 are the lengths of the lines QP and $-QP$. From this potential function we can calculate the electric field by taking derivatives. Before we do this, we will make the simplifying assumption that r_1 and r_2 are much larger than l, that is, we will make the assumption that the field point P is at a large distance from the electric dipole. Figure 25.11 shows that under these conditions, r_1 and r_2 are approximately parallel and approximately equal,

$$r_1 \cong r_2 \cong r \tag{45}$$

and their difference is a small quantity,

$$r_2 - r_1 \cong l \cos\theta \tag{46}$$

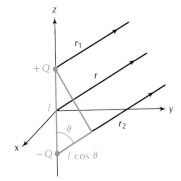

Fig. 25.11 If the point P is at a large distance (beyond the diagram), then the lines QP and $-QP$ are nearly parallel and the difference between their lengths r_1 and r_2 is $l \cos\theta$.

With this, Eq. (44) yields the following approximation for V:

$$V \cong \frac{Q}{4\pi\varepsilon_0} \frac{l \cos\theta}{r^2} \qquad (47)$$

The product of Q and l is the dipole moment

$$p = lQ$$

and hence our approximation for V has the form

Potential of dipole
$$V = \frac{p}{4\pi\varepsilon_0} \frac{\cos\theta}{r^2} \qquad (48)$$

To calculate the components of the electric field, it is convenient to express everything in rectangular coordinates,

$$r = \sqrt{x^2 + y^2 + z^2} \qquad (49)$$

$$\cos\theta = \frac{z}{\sqrt{x^2 + y^2 + z^2}} \qquad (50)$$

so that

$$V = \frac{p}{4\pi\varepsilon_0} \frac{z}{(x^2 + y^2 + z^2)^{3/2}} \qquad (51)$$

The components of the electric field are then

$$E_x = -\frac{\partial V}{\partial x} = \frac{p}{4\pi\varepsilon_0} \frac{3zx}{(x^2 + y^2 + z^2)^{5/2}} \qquad (52)$$

$$E_y = -\frac{\partial V}{\partial y} = \frac{p}{4\pi\varepsilon_0} \frac{3zy}{(x^2 + y^2 + z^2)^{5/2}} \qquad (53)$$

$$E_z = -\frac{\partial V}{\partial z}$$

$$= -\frac{p}{4\pi\varepsilon_0} \left(\frac{1}{(x^2 + y^2 + z^2)^{3/2}} - \frac{3z^2}{(x^2 + y^2 + z^2)^{5/2}} \right) \qquad (54)$$

These expressions for the electric field are only approximations, but they are very good at large distances from the dipole. For instance, if the dipole is within a molecule, then l is very small — 10^{-10} m or less — and our approximation is valid whenever the distance is large compared with 10^{-10} m. Figure 25.12 shows the field lines for the electric field of a dipole.

Equations (52)–(54) describe the electric field surrounding a water molecule, with a dipole moment $p = 6.1 \times 10^{-30}$ C·m (see Section 23.4). By means of this electric field, the water molecule can act on other molecules in its vicinity — it can exert forces on the electric charges in other molecules. The force exerted by the electric dipole field of the water molecule is what makes water such a good solvent.

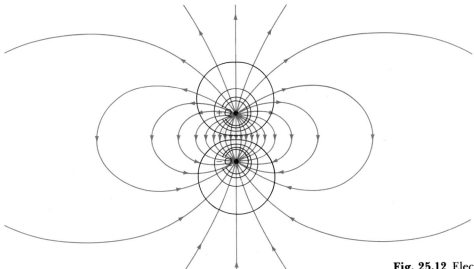

Fig. 25.12 Electric field lines (color) and equipotential surfaces (black) of an electric dipole.

25.6* The Mean-Value Theorem

There exists a beautiful theorem concerning the electrostatic potential in a region free of electric charge. This theorem, called the **mean-value theorem,** states the following:

Mean-value theorem

> *If S is the surface of a (mathematical) sphere whose interior is free of charge, then the potential at the center of the sphere equals the mean value of the potential over the surface.*

The proof of the theorem is simple. Suppose the sphere has a radius R. Then the mean value of the potential over the surface of the sphere is obtained by integrating the potential over the surface and dividing by the area of the surface,

$$\overline{V} = \frac{1}{4\pi R^2} \int V \, dS \tag{55}$$

The essential step of the proof is demonstrating that $\overline{V}$ is independent of R, i.e., $\overline{V}$ is independent of the size of the sphere. Consider the derivative of the integral (55),

$$\frac{d\overline{V}}{dR} = \frac{d}{dR}\left(\frac{1}{4\pi R^2} \int V \, dS\right) \tag{56}$$

Figure 25.13 shows an area element dS. This area is approximately the base of an infinitesimal cone of half-angle $\Delta\theta$; the base has an area

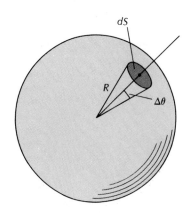

Fig. 25.13 Small circular area on surface of a sphere. The radius of this circular area is approximately $R \, \Delta\theta$.

---

* This section is optional.

$\pi(R\,\Delta\theta)^2$. The contribution to the right side of Eq. (56) from this cone is

$$\frac{d}{dR}\left(\frac{1}{4\pi R^2}\,V\,dS\right) = \frac{d}{dR}\left[\frac{1}{4\pi R^2}\,V\pi(R\,\Delta\theta)^2\right]$$

$$= \frac{d}{dR}\left[\frac{(\Delta\theta)^2}{4}\,V\right] = \frac{(\Delta\theta)^2}{4}\,\frac{dV}{dR}$$

$$= \frac{1}{4\pi R^2}\,\frac{dV}{dR}\,\pi(R\,\Delta\theta)^2$$

$$= \frac{1}{4\pi R^2}\,\frac{dV}{dR}\,dS \tag{57}$$

The entire sphere can be regarded as a collection of such infinitesimal cones. Hence

$$\frac{d\overline{V}}{dR} = \frac{1}{4\pi R^2}\int \frac{dV}{dR}\,dS \tag{58}$$

But, according to Eq. (29), dV/dR is the negative of the component of **E** in the radial direction; i.e., it is the component of **E** in a direction perpendicular to the surface of integration. By Gauss' Law, the surface integral of this component is zero, since the sphere S is free of charge, that is,

$$\frac{d\overline{V}}{dR} = \frac{1}{4\pi R^2}\int \frac{dV}{dR}\,dS = \frac{-1}{4\pi R^2}\int E_n\,dS = 0 \tag{59}$$

This establishes that $\overline{V}$ is independent of R. But if so, then the sphere of radius R and any smaller sphere must have the same value for the corresponding mean potential. In particular, since the potential at the center of the sphere can be regarded as the mean potential for a sphere of infinitesimal radius, it follows that the potential at the center must have the same value as the mean potential over the sphere of radius R. This concludes the proof.

Earnshaw's theorem

The mean-value theorem has an important corollary, called **Earnshaw's theorem**: in a region free of charge, the electrostatic potential cannot have a minimum or a maximum. For suppose it had a minimum at a point P. This would mean that at *all* points in the immediate vicinity the potential is higher. Hence, if we draw a small sphere centered on P, the mean value of the potential over the surface would have to be larger than the potential at the center. This would contradict the mean-value theorem and is therefore impossible. A similar argument shows that a maximum is also impossible. The absence of minima and maxima implies that if the potential increases in some directions, it must decrease in others in such a way that the average of the changes in all directions is zero.

Finally, we will describe a very useful numerical method for the approximate calculation of the electrostatic potential. It will be easiest to describe this method in the context of a special example. Figure 25.14

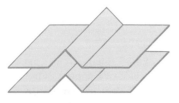

Fig. 25.14 Very large parallel plates with a kink.

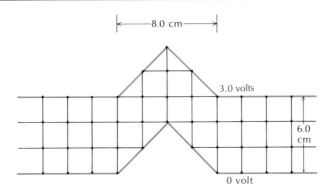

Fig. 25.15 Plates viewed edge on. A coordinate grid has been superimposed on the diagram; the squares of this grid are 2 cm × 2 cm.

shows a pair of large conducting plates with a kink. Figure 25.15 shows the plates edge on and gives the relevant dimensions. Suppose that the potential of the lower plate is 0 volt and that of the upper plate 3.0 volts. What is the potential in the space between the plates?

In order to apply numerical methods to this problem, we must represent the space between the plates by a discrete, and finite, set of points. Figure 25.15 shows a coordinate grid superimposed on the picture of the plates. We will try to find the potential at the intersection points of this grid. This will give us only a rather rough description of the potential; for a more precise description we would have to take a finer grid. Note that the grid is two dimensional. The third dimension (out of the plane of the page in Figure 25.15) can be ignored, because the potential at all points above or below the plane of the page is the same as that at the points shown in Figure 25.15.

To obtain the potential at the grid points, we proceed by the following method of successive approximations: We begin by making some reasonable first guess for the potential. Since for flat plates the potential in the space between would increase regularly from 0 volt to 3.0 volts, the values given in Figure 25.16a are a reasonable first approximation. To find the second approximation to the potential, we rely on the mean-value theorem. According to this theorem, the potential at any point should equal the average of the potentials of all the points that are at a distance of, say, 2 cm from the given point. In our grid, this average is approximated by an average over the four nearest neighbor points. We therefore obtain a second approximation by replacing the potential at each point by this average over the potentials of the four nearest neighbor points. This averaging procedure yields the values in Figure 25.16b. Next we obtain a third approximation by again replacing each of the potentials of Figure 25.16b with the average over the potentials of the four nearest neighbor points. This yields the values in Figure 25.16c, and so on.

After a few successive steps, this iteration procedure yields self-consistent values for the potential, that is, it yields values that do not change appreciably from one step to the next. These values are an approximate answer to the problem. The components of the electric field can then be obtained from the values of the potential by numerical calculation of the derivatives of the potential in the horizontal and vertical directions.

As a check on the approximation, it is a good idea to repeat the calculation with the finer grid size. This tends to get tedious, but the method lends itself very well to programming on a digital computer.

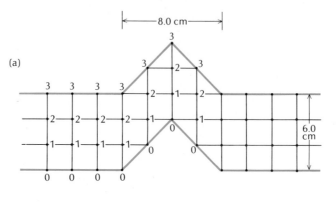

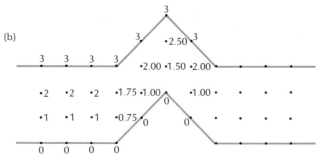

Fig. 25.16 The numbers at the grid points give the potential in volts. The numbers for the right half of the diagram have been omitted; they are the same as those for the left half.

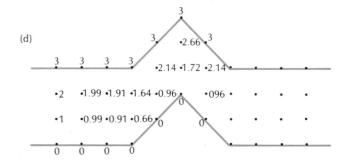

SUMMARY

Definition of electrostatic potential:

$$V(P) = \frac{U}{q} = -\int_{P_0}^{P} \mathbf{E} \cdot d\mathbf{l} + V(P_0)$$

Potential of point charge: $V = \dfrac{1}{4\pi\varepsilon_0} \dfrac{q'}{r}$

Potential energy of two point charges: $U = \dfrac{1}{4\pi\varepsilon_0}\dfrac{qq'}{r}$

Derivatives of the potential: $\dfrac{dV}{dl} = -E\cos\theta$

$$\dfrac{\partial V}{\partial x} = -E_x, \quad \dfrac{\partial V}{\partial y} = -E_y, \quad \dfrac{\partial V}{\partial z} = -E_z$$

Mean-value theorem (for charge-free region): potential at center of any sphere equals mean potential on surface.

QUESTIONS

1. The potential difference between the poles of an automobile battery is 12 volts. Explain what this means in terms of the definition of potential as work per unit charge.

2. An old-fashioned word for electrostatic potential is electrostatic *tension*. Is it reasonable to think of the potential as analogous to mechanical tension?

3. If the electric field is zero in some region, must the potential also be zero? Give an example.

4. A bird sits on a high-voltage power line which is at a potential of 345,000 volts. Does this harm the bird?

5. How would you define the gravitational potential? Are the units of gravitational potential the same as the units of electric potential? According to your definition, what is the gravitational potential difference between the ground and a point 50 m above the ground?

6. Consider an electron moving in the vicinity of a proton. Where is the electrostatic potential produced by the proton highest? Where is the potential energy of the electron highest?

7. Suppose that the electrostatic potential has a minimum at some point. Is this an equilibrium point for a positive charge? For a negative charge? Is the equilibrium stable?

8. Consider a sphere of radius R with a charge Q uniformly distributed over its volume. Where does the potential have a maximum? Where does the magnitude of the electric field have a maximum?

9. Give an example of a conductor that is not an equipotential. Is this conductor in electrostatic equilibrium?

10. If the potential in a three-dimensional region of space is known to be constant, what can you conclude about the electric field in this region? If the potential on a two-dimensional surface is known to be constant, what can you conclude about the electric field on this surface?

11. In many calculations it is convenient to assign a potential of 0 volt to the ground. If so, what is the potential at the top of the Eiffel Tower? What is the potential at the top of your head? (Hint: Your body is a conductor.)

12. Is it true that the surface of a mass in static mechanical equilibrium is a gravitational equipotential surface? What if the surface is that of a fluid, such as water?

13. If a high-voltage power cable falls on top of your automobile, you will probably be safest if you remain inside the automobile. Why?

14. Suppose that several separate solid metallic bodies have been placed near a charge distribution. Is it necessarily true that all of these bodies will have the same potential?

15. If we surround some region with a conducting surface, we shield it from external electric fields. Why can we not shield a region from gravitational fields by a similar method?

16. A cavity is completely surrounded by conducting material. Can you create an electric field in this cavity?

17. Show that different equipotential surfaces cannot intersect.

18. Consider the patterns of field lines shown in Figures 23.13 and 23.14. Roughly, sketch some of the equipotential surfaces for each case.

19. Sketch the equipotential surfaces for the potential described numerically in Figure 25.16d.

PROBLEMS

Section 25.1

1. In order to charge a typical 12-volt automobile battery fully, the charging device must force $+2.0 \times 10^5$ coulombs from the negative terminal of the battery to the positive terminal. How much work must the charging device do during this process?

2. An ordinary flashlight battery has a potential difference of 1.5 V between its positive and negative terminals. How much work must you do to transport an electron from the positive terminal to the negative terminal?

3. On days of fair weather, the atmospheric electric field of the Earth is about 100 V/m; this field points vertically downward (compare Problem 23.2). What is the electric potential difference between the ground and an airplane flying at 600 m? What is the potential difference between the ground and the tip of the Eiffel Tower? Treat the ground as a flat conductor.

4. Consider the arrangement of parallel sheets of charge described in Problem 23.24. Find the potential difference between the upper sheet and the lower sheet.

5. Suppose that, as a function of x, y, z, an electric field has components

$$E_x = 6x^2y \qquad E_y = 2x^3 + 2y \qquad E_z = 0$$

where E is measured in volts per meter and the distances are measured in meters.
   (a) Find the potential difference between the origin and the point $x = 3$, $y = 0$, $z = 0$.
   (b) Find the potential difference between the origin and the point $x = 0$, $y = 2$, $z = 0$.

6. At the Stanford Linear Accelerator (SLAC), electrons are accelerated from an energy of 0 eV to 20×10^9 eV as they travel in a straight evacuated tube 1600 m in length. The acceleration is due to a strong electric field pushing the electrons along. Assume that the electric field is uniform. What must be its strength?

7. The potential difference between the two poles of an automobile battery is 12.0 V. Suppose that you place such a battery in empty space and that you release an electron at a point next to the negative pole of the battery. The electron will then be pushed away by the electric force and move off in some direction.
   (a) If the electron strikes the positive pole of the battery, what will be its impact speed?
   (b) If, instead, the electron moves away toward infinity, what will be its ultimate speed?

8. The gap between the electrodes of a spark plug in an automobile is 0.025 in. In order to produce an electric field of 3×10^6 V/m (required to initiate an electric spark), what minimum potential difference must you apply to the spark plug?

9. Prove that the plane midway between a positive and a negative point charge of equal magnitudes is an equipotential surface. Is this also true if both charges are positive?

Section 25.2

10. A charge of 2×10^{-12} C is placed on a small (pointlike) cork ball. What is the electrostatic potential at a distance of 30 cm from the ball? At a distance of 60 cm?

11. A proton sits at the origin of coordinates. How much work (in electron-volts) must you do to push an electron from the point $x = 1.0$ Å, $y = 0$, $z = 0$ to the point $x = 0.5$ Å, $y = 0.5$ Å, $z = 0$?

12. The nucleus of lead is a uniformly charged sphere with a charge of $82e$ and a radius of 7.1×10^{-15} m. What is the electrostatic potential at the nuclear surface? At the nuclear center?

13. The nucleus of platinum is a uniformly charged sphere with a charge of $78e$ and a radius 7.0×10^{-15} m. What is the electric potential energy of an incident proton arriving at the nuclear surface? At the nuclear center?

*14. The tau particle is similar to an electron and has the same electric charge, but its mass is 3490 times as large as the electron mass. Like the electron, the tau can penetrate nuclear material, experiencing no forces except the electric force. Suppose that a tau is initially at rest a large distance from a lead nucleus. Under the influence of the electric attraction, the tau accelerates toward the nucleus. What is its speed when it crosses the nuclear surface? When it reaches the center of the nucleus? The nucleus of lead is a uniformly charged sphere of radius 7.1×10^{-15} m and charge $82e$.

*15. An alpha particle of kinetic energy 1.7×10^{-12} J is shot directly toward a platinum nucleus from a very large distance. What will be the distance of closest approach? The electric charge of the alpha particle is $2e$ and that of the platinum nucleus is $78e$. Treat the alpha particle and the nucleus as spherical charge distributions, and disregard the motion of the nucleus.

*16. An alpha particle is initially at a very large distance from a plutonium nucleus. What is the minimum kinetic energy with which the alpha particle must be launched toward the nucleus if it is to make contact with the nuclear surface? The plutonium nucleus is a sphere of radius 7.5×10^{-15} m with a charge of $94e$ uniformly distributed over the volume. For the purpose of this problem, the alpha particle may be regarded as a particle (of negligible radius) with a charge of $2e$.

*17. A thorium nucleus emits an alpha particle according to the reaction

$$\text{thorium} \rightarrow \text{radium} + \text{alpha}$$

Assume that the alpha particle is pointlike and that the residual radium nucleus is spherical with a radius of 7.4×10^{-15} m. The charge on the alpha particle is $2e$, and that on the radium nucleus is $88e$.
  (a) At the instant the alpha particle emerges from the nuclear surface, what is its electrostatic potential energy?
  (b) If the alpha particle has no initial kinetic energy, what will be its final kinetic energy and speed when far away from the nucleus? Assume that the radium nucleus does not move. The mass of the alpha particle is 6.7×10^{-27} kg.

*18. Consider again the arrangement of charges within the thundercloud of Figure 23.28. Find the electric potential due to these charges at a point which

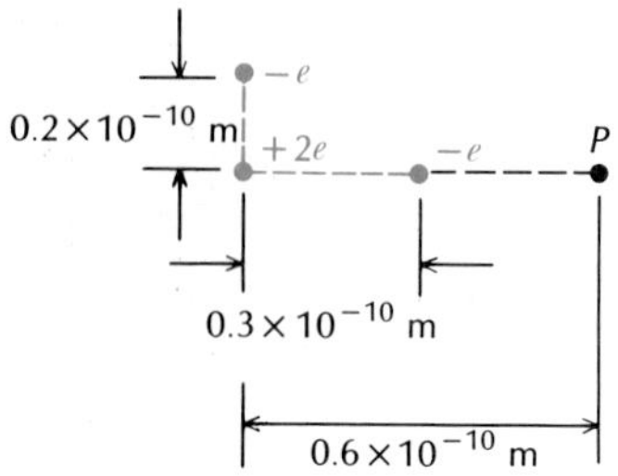

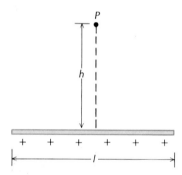

Fig. 25.17 Nucleus (charge $+2e$) and electrons (charge $-e$) of helium atom at one instant of time.

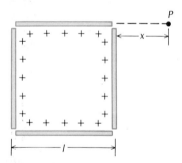

Fig. 25.18 A charged rod of length l.

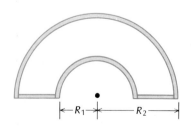

Fig. 25.19

is at a height of 8 km and on the vertical line passing through the charges. Find the electric potential at a second point which is at the same height and has a horizontal distance of 5 km from the first point.

*19. In a helium atom, at some instant one of the electrons is at a distance of 0.3×10^{-10} m from the nucleus and the other electron is at a distance of 0.2×10^{-10} m, 90° away from the first (Figure 25.17). Find the electric potential produced jointly by the two electrons and the nucleus at a point P beyond the first electron and at a distance of 0.6×10^{-10} m from the nucleus.

**20. According to Bohr's theory of the atom (see also Problem 22.24), the electron in a hydrogen atom orbits around the nucleus in a circular orbit. The force that holds the electron in this orbit is the Coulomb force. The size of the orbit depends on the angular momentum — the smallest possible orbit has an angular momentum $\hbar = 1.05 \times 10^{-34}$ J·s; the next possible orbit has angular momentum $2\hbar$; the next $3\hbar$; etc.

(a) Show that if a circular orbit has angular momentum $n\hbar$ (where $n = 1, 2, 3, \ldots$), then its radius is

$$r = \frac{4\pi\varepsilon_0}{m_e e^2} n^2 \hbar^2$$

(b) Show that the orbital energy (kinetic and potential) of the electron in such an orbit is

$$E = -\frac{m_e e^4}{2(4\pi\varepsilon_0)^2 \hbar^2} \frac{1}{n^2}$$

(c) Evaluate this energy for $n = 1$; express your answer in electron-volts.

*21. A total charge Q is distributed uniformly along a straight rod of length l. Find the potential at a point P at a distance h from the midpoint of the rod (Figure 25.18).

*22. Three thin rods of glass of length l carry charges uniformly distributed along their lengths. The charges on the three rods are $+Q$, $+Q$, and $-Q$, respectively. The rods are arranged along the sides of an equilateral triangle. What is the electrostatic potential at the midpoint of this triangle?

*23. A uniformly charged sphere of radius a is surrounded by a uniformly charged concentric spherical shell of inner radius b and outer radius c. The total charge on the sphere is Q, and that on the shell $-Q$. Find the potential at $r = b$, at $r = a$, and at $r = 0$.

*24. Four rods of length l are arranged along the edges of a square. The rods carry charges $+Q$ uniformly distributed along their lengths (Figure 25.19). Find the potential at the point P at a distance x from one corner of the square.

*25. Two semicircular rods and two short straight rods are joined in the configuration shown in Figure 25.20. The rods carry a charge of λ coulombs per meter. Calculate the potential at the center of this configuration.

*26. A long straight wire of radius 0.80 mm is surrounded by an evacuated concentric conducting shell of radius 1.2 cm. The wire carries a charge of -5.5×10^{-8} coulomb per meter of length. Suppose that you release an electron at the surface of the wire. With what speed will this electron hit the conducting shell?

*27. A long plastic pipe has an inner radius a and an outer radius b. Charge is uniformly distributed over the volume $a < r < b$. The amount of charge is λ coulombs per meter of length of the tube. Find the potential difference between $r = b$ and $r = 0$. Assume that the plastic has no effect on the electric field.

*28. A flat disk of radius R has charge Q uniformly distributed over its surface. Find a formula for the potential along the axis of the disk.

Fig. 25.20

*29. The tube of a Geiger counter consists of a thin straight wire surrounded by a coaxial conducting shell. The diameter of the wire is 0.0025 cm and that of the shell is 2.5 cm. The length of the tube is 10 cm however, in your calculation use the formula for the electric field of an infinitely long line of charge. If the potential difference between the wire and the shell is 1.0×10^3 volts, what is the electric field at the surface of the wire? At the cylinder?

**30. An infinite charge distribution with spherical symmetry has a charge density ρ coulombs per cubic meter given by the formula $\rho = k r^{-5/2}$ where k is a constant. Find the potential as a function of the radius. Assume $V = 0$ at $r = \infty$.

**31. A point charge Q is on the positive z axis at the point $z = h$. A point charge $-Q \times R/h$ (where R is a positive length, $0 < R < h$) is on the z axis at the point $z = R^2/h$. Show that the surface of the sphere of radius R about the origin is an equipotential surface.

Sections 25.4 and 25.5

32. In some region of space, the electrostatic potential is the following function of x, y, and z:

$$V = x^2 + 2xy$$

where the potential is measured in volts and the distances in meters. Find the electric field at the point $x = 2$, $y = 2$.

33. In terms of x, y, and z, the potential of a point charge is

$$V(x, y, z) = \frac{1}{4\pi\varepsilon_0} \frac{q'}{\sqrt{x^2 + y^2 + z^2}}$$

(a) By differentiating this potential function, calculate the components E_x, E_y, and E_z of the electric field.
(b) Show that the magnitude $\sqrt{E_x^2 + E_y^2 + E_z^2}$ agrees with the usual expression for the electric field of a point charge.

34. A rod of length l has a charge Q uniformly distributed along its length. Equation (25) gives the potential at the point P at a distance x from one end of the rod. Find the electric field at this point.

*35. Two rods of equal length l form a symmetric cross. The rods carry charges $\pm Q$ uniformly distributed along their lengths. Calculate the potential at the point P at a distance x from one end of the cross (Figure 25.21). Calculate the electric field at this point.

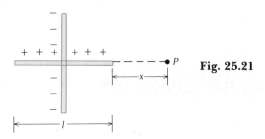

Fig. 25.21

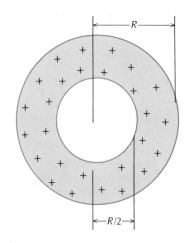

Fig. 25.22

*36. An annulus (a disk with a hole) made of paper has an outer radius R and an inner radius $R/2$ (Figure 25.22). An amount Q of electric charge is uniformly distributed over the paper.
(a) Find the potential as a function of distance on the axis of the annulus.
(b) Find the electric field on the axis of the annulus.

*37. A nucleus of carbon (charge $6e$) and one of helium (charge $2e$) are separated by a distance of 1.2×10^{-13} m and instantaneously at rest. The center of mass of this system is at a distance of 0.4×10^{-13} m from the carbon nucleus.

Take this point as origin and take the x axis along the line joining the nuclei, with the carbon nucleus on the negative x axis.
(a) Find the potential V as a function of x, y, and z.
(b) Find E_x and E_y as a function of x, y, and z.

*38. The water molecule has a dipole moment of 6.1×10^{-30} C·m.
(a) Find the magnitude and direction of the electric field at a point on the axis of the dipole at a distance of 12.0 Å from the molecule.
(b) Find the magnitude and direction of the electric field at a point on a line transverse to the axis of the dipole at the same distance.

**39. A thin cylindrical cardboard tube has a charge Q uniformly distributed over its surface. The radius of the tube is R and the length l.
(a) Find the potential at a point on the axis of the tube at a distance x from the midpoint. Assume $x > l$.
(b) Find the electric field at this point.

**40. A point charge $-2Q$ is at the origin of coordinates; two point charges $+Q$ are on the z axis, at $z = \pm l$, respectively (Figure 25.23).
(a) Show that, for $r \gg l$, the net potential of these charges is approximately

$$V = \frac{2Ql^2}{4\pi\varepsilon_0 r^3} \frac{(3\cos^2\theta - 1)}{2}$$

(b) Calculate E_x, E_y, and E_z, expressing these in terms of the coordinates x, y, and z.

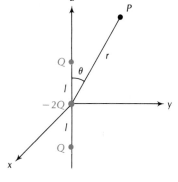

Fig. 25.23 Charges $-2Q$, Q, and Q on the z axis.

Section 25.6

*41. Figure 25.24 shows two large parallel conducting plates seen in cross section. One of the plates has a kink. The potential difference between them is 2.0 V. Use the mean-value theorem to find the potential at the points of the grid in Figure 25.24 to at least two significant figures.

*42. Figure 25.25 shows two large parallel conducting plates seen in cross section. Both plates have rectangular kinks. The potential difference between the plates is 6.0 V. Use the mean-value theorem to find the potential at the grid points to within at least two significant figures.

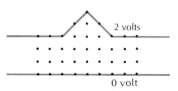

Fig. 25.24 Very large parallel plates, viewed edge on. One of the plates has a kink.

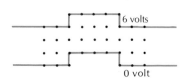

Fig. 25.25 Very large parallel plates, viewed edge on. Both plates have rectangular ridges.

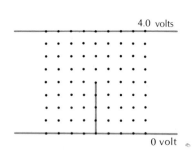

Fig. 25.26 Two very large parallel plates, seen edge on. The lower plate has a protruding thin ridge; the potential is $V = 0$ on this plate and on the ridge.

*43. Two large parallel conducting plates have a potential difference of 4.0 V between them. One of the plates carries a thin vertical ridge (Figure 25.26). Use the mean-value theorem to find the potential at the grid points to within two significant figures or better.

*44. Given that each box of the grid of Figure 25.16d is 2 cm × 2 cm, evaluate numerically the derivative $\Delta V/\Delta z$ at each of the grid points and thereby find E_z. The z direction is the vertical direction (perpendicular to the flat portion of the plate).

*45. Two long concentric tubes of sheet metal have a square cross section; Figure 25.27 shows their cross section. The outer tube is at a potential of 10 V; the inner tube is at a potential of 0 V. Use the mean-value theorem to find the potential at all the grid points shown to within two significant figures.

*46. When calculating the average potential in the method of successive approximations described in Section 25.6, we took into account only the nearest four points surrounding the given point in a *two-dimensional* grid. Since the mean-value theorem applies to a sphere surrounding the given point, we should actually take into account the six nearest points surrounding the given point in a *three-dimensional* grid; that is, we should take into account an extra point in front of and an extra point behind the plane shown in Figure 25.15. Prove that for the problem discussed in Section 25.6, in which the potential in front of the given point and the potential behind the given point are the same as the potential at the given point, the average over the four nearest points in the two-dimensional grid coincides with the average over the six nearest points in the three-dimensional grid.

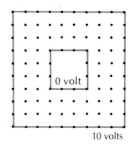

Fig. 25.27 Two concentric tubes (ducts) of sheet metal, seen in cross section.

CHAPTER 26

Electric Energy

In the preceding chapter we dealt with the electric potential energy of a point charge — a test charge — placed in the electric field of a given charge distribution. Now we will calculate the potential energy of the charge distribution by itself, without the test charge. The charge distribution can be regarded as a collection of point charges; since all of these point charges exert forces on one another, it requires a certain amount of work to bring them together into their final configuration if they are initially separated by large distances. This amount of work is the potential energy of the charge distribution.

We will see that this potential energy is stored in the electric field. The distribution of energy in the electric field can be described by an energy density — the energy is concentrated in those regions of space where the electric field is strong. Since the electric field is endowed with energy, we must regard the field as a material object, a fifth state of matter.

26.1 Energy of a System of Point Charges

The electric potential energy of two point charges q_1, q_2 separated by a distance r is [see Eq. (25.14)]

$$U = \frac{1}{4\pi\varepsilon_0} \frac{q_1 q_2}{r} \tag{1}$$

This potential energy can be regarded as the work required to move q_1 from infinity to within a distance r of q_2 or, alternatively, the work required to move q_2 to within a distance r of q_1. It is a *mutual* potential energy which belongs to both q_1 and q_2. This mutual potential energy is associated with the relative configuration of the pair q_1, q_2.

For configurations consisting of more than two charges, the net potential energy can be calculated by writing down a term similar to that in Eq. (1) for *each pair* of charges. For instance, if we are dealing with three charges (Figure 26.1), we have three possible pairs, (q_1, q_2), (q_2, q_3), and (q_1, q_3), so that the net potential energy is

$$U = \frac{1}{4\pi\varepsilon_0}\frac{q_1 q_2}{r_{12}} + \frac{1}{4\pi\varepsilon_0}\frac{q_2 q_3}{r_{23}} + \frac{1}{4\pi\varepsilon_0}\frac{q_1 q_3}{r_{13}} \qquad (2)$$

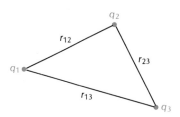

Fig. 26.1 Three point charges.

where r_{12}, r_{23}, and r_{13} are the distances between the pairs. This is the work required to assemble the charges into the final configuration shown in Figure 26.1, starting from an initial condition of infinite separation.

Note that Eq. (2) is identically equal to

$$U = \frac{1}{2}\left(\frac{1}{4\pi\varepsilon_0}\frac{q_2}{r_{12}} + \frac{1}{4\pi\varepsilon_0}\frac{q_3}{r_{13}}\right)q_1 + \frac{1}{2}\left(\frac{1}{4\pi\varepsilon_0}\frac{q_1}{r_{12}} + \frac{1}{4\pi\varepsilon_0}\frac{q_3}{r_{23}}\right)q_2$$

$$+ \frac{1}{2}\left(\frac{1}{4\pi\varepsilon_0}\frac{q_1}{r_{13}} + \frac{1}{4\pi\varepsilon_0}\frac{q_2}{r_{23}}\right)q_3 \qquad (3)$$

Here, the potential energy of each pair of charges appears twice, each time with a factor of $\frac{1}{2}$. The terms in parentheses are the electric potentials produced by the individual charges; for instance, the two terms in the first parenthesis on the right side of Eq. (3) are the sum of the potentials produced by charge 2 and charge 3 at the position of charge 1. Equation (3) therefore leads to the following neat expression for the energy in terms of potentials:

$$U = \tfrac{1}{2}V_{\text{other}}(1)q_1 + \tfrac{1}{2}V_{\text{other}}(2)q_2 + \tfrac{1}{2}V_{\text{other}}(3)q_3 \qquad (4)$$

where $V_{\text{other}}(1)$ is the electric potential produced at the position of charge 1 by the *other* charges (that is, charges 2 and 3),

$$V_{\text{other}}(1) = \frac{1}{4\pi\varepsilon_0}\frac{q_2}{r_{12}} + \frac{1}{4\pi\varepsilon_0}\frac{q_3}{r_{13}} \qquad (5)$$

and similarly for $V_{\text{other}}(2)$ and $V_{\text{other}}(3)$.

By means of a generalization of this argument, we can easily demonstrate that for a configuration consisting of any number of point charges, the electric potential energy is the sum

$$U = \tfrac{1}{2}V_{\text{other}}(1)q_1 + \tfrac{1}{2}V_{\text{other}}(2)q_2 + \tfrac{1}{2}V_{\text{other}}(3)q_3 + \tfrac{1}{2}V_{\text{other}}(4)q_4 + \cdots \qquad (6)$$

Energy of system of point charges

This expression gives the work that must be done to bring the point charges to their final positions starting from initial positions at very large distances from each other. However, this expression is not the total potential energy, because it does not take into account the energy that a point charge has when it is by itself, at a large distance from all other point charges. Such an isolated point charge has potential energy because it takes work to assemble the point charge out of infinitesimal pieces of charge. The energy needed to assemble the point charge is called the **self-energy** of the point charge.

The calculation of the self-energy of point charges, such as electrons, is one of the unsolved problems of physics. A straightforward calculation of the self-energy of an electron yields the absurd result that this energy is infinite. [Roughly stated: The calculated energy is infinite because the potential at the position of the point charge is infinite, that is, $(1/4\pi\varepsilon_0)(q/r) \to \infty$ as $r \to 0$.] Although up to now physicists have found no satisfactory way to calculate the self-energy, they have invented several rules for bypassing this problem. These rules, called renormalization rules, give prescriptions on how to extract experimentally meaningful numbers from the theoretical calculations. In essence, the rules assert that the self-energy is a constant quantity that never has any effect on energy conservation and can therefore be ignored. The theoretical calculations based on this scheme have been extremely successful. For instance, by combining electromagnetic theory, relativity theory, and quantum theory, the electric energy of an electron in a hydrogen atom has been calculated to nine significant figures. But from the mathematical point of view this scheme has some shady aspects.

In the following, we will always ignore the electric self-energy of point charges and pretend that Eq. (6) is the total electric energy.

26.2 Energy of a System of Conductors

Equation (6) permits us to evaluate the electric energy of a system of charged conductors. Figure 26.2 shows several conductors of arbitrary shapes with charges distributed over their surfaces. Suppose that the charges on these conductors are $Q_1, Q_2, Q_3, \ldots$, and that their potentials are $V_1, V_2, V_3, \ldots$, respectively. The charge Q_1 consists of many charge elements distributed over conductor 1. Each of these charge elements is at potential V_1. This potential acting on a given charge element dQ_1 can be regarded as due to the *other* charge elements (on conductor 1 and on other conductors), because the given charge element makes only an insignificant contribution to V_1. Hence, according to Eq. (6), the electric potential energy associated with the charge element dQ_1 on conductor 1 is $\frac{1}{2}(dQ_1)V_1$, and the potential energy associated with all the charge elements on conductor 1 is simply $\frac{1}{2}Q_1V_1$. Since similar arguments apply to the other conductors, we conclude that net electric energy is

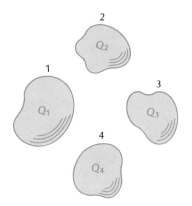

Fig. 26.2 Four conducting bodies carrying electric charges.

Energy of system of conductors

$$U = \tfrac{1}{2}Q_1V_1 + \tfrac{1}{2}Q_2V_2 + \tfrac{1}{2}Q_3V_3 + \cdots \qquad (7)$$

EXAMPLE 1. A metallic sphere of radius R carries a charge Q uniformly distributed over its surface. How much electric energy is stored in this charge distribution?

SOLUTION: In this problem, there is only one conductor, and Eq. (7) reduces to

$$U = \tfrac{1}{2} Q_1 V_1$$

or

$$U = \tfrac{1}{2} QV \tag{8}$$

The potential outside a spherically symmetric charge distribution is given by Eq. (25.21). At $r = R$, this potential is

$$V = \frac{1}{4\pi\varepsilon_0} \frac{Q}{R}$$

According to Eq. (8), the electric energy is then

$$U = \tfrac{1}{2} QV = \tfrac{1}{2} Q \frac{1}{4\pi\varepsilon_0} \frac{Q}{R} = \frac{1}{8\pi\varepsilon_0} \frac{Q^2}{R} \tag{9}$$

EXAMPLE 2. Two large parallel metallic plates of area A are separated by a distance d. Charges $+Q$ and $-Q$ are placed on the plates, respectively (Figure 26.3). What is the electric energy of this charge distribution?

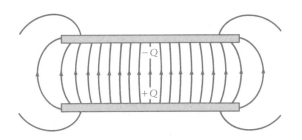

Fig. 26.3 Two very large parallel conducting plates with opposite electric charges. The electric field is approximately constant, except near the edges of the plates.

SOLUTION: Here we are dealing with two conductors, with $Q_1 = Q$ and $Q_2 = -Q$. Thus, Eq. (7) becomes

$$U = \tfrac{1}{2} Q_1 V_1 + \tfrac{1}{2} Q_2 V_2 = \tfrac{1}{2} QV_1 - \tfrac{1}{2} QV_2$$
$$= \tfrac{1}{2} Q(V_1 - V_2)$$

To proceed, we need the potential difference $V_1 - V_2$ between the plates. The electric field in the region between the plates is approximately the field of an infinite charged plate. From Eq. (23.26), the electric field is $E = \sigma/\varepsilon_0$, where σ is the charge per unit area on the plate. Since the total charge on a plate is Q and the area is A, the charge per unit area is $\sigma = Q/A$, and

$$E = \frac{Q}{\varepsilon_0 A} \tag{10}$$

The potential difference between the plates is

$$V_2 - V_1 = -Ed = -\frac{Qd}{\varepsilon_0 A}$$

and the electric potential energy is

$$U = \tfrac{1}{2} Q(V_1 - V_2) = \tfrac{1}{2} \frac{Q^2 d}{\varepsilon_0 A} \tag{11}$$

COMMENTS AND SUGGESTIONS: The expression (10) fails near the edges of the plates, where there is an electric fringing field which is not constant (Figure 26.3). We will investigate the energy in such a variable field in the next section. If the plates are large, then the edge region is only a very small fraction of the total region between the plates, and we can ignore this edge region without introducing excessive errors in our calculation of the energy.

Note that Eq. (11) can be rewritten in the following interesting way:

$$U = \tfrac{1}{2}\varepsilon_0 \left(\frac{Q}{\varepsilon_0 A}\right)^2 Ad$$

$$= \tfrac{1}{2}\varepsilon_0 E^2 \times [\text{volume}] \tag{12}$$

where the "volume" is the volume between the plates, that is, the volume of the region in which there is an electric field. The energy per unit volume of electric field is therefore $\tfrac{1}{2}\varepsilon_0 E^2$. This suggests that the electric energy is distributed over space, being concentrated in those regions where the electric field is strong. In the next section we will confirm that this is indeed true.

26.3 The Energy Density

For the special case of the constant electric field in the region between parallel conducting plates, we found that the energy per unit volume of field is $\tfrac{1}{2}\varepsilon_0 E^2$. We will now prove that this result is also true for an electric field that is not constant, such as the fringing electric field at the edges of the conducting plates. Figure 26.4 shows this field at the edges. Consider a narrow bundle of field lines that start on one conductor and end on the other conductor. The bundle of field lines intercepts two small areas on the surfaces of the two conductors. The electric charges on these small areas are dQ and $-dQ$, respectively. The contribution to the net electrostatic energy from these charges $\pm dQ$ is

$$dU = \tfrac{1}{2}dQ\, V_1 - \tfrac{1}{2}dQ\, V_2 = \tfrac{1}{2}dQ(V_1 - V_2)$$

$$= \tfrac{1}{2}\, dQ \int_1^2 E(l)\, dl \tag{13}$$

where l is the length measured along the bundle of field lines starting at conductor 1, and $E(l)$ is the strength of the electric field at the distance l. [Note that $E(l)\, dl = \mathbf{E} \cdot d\mathbf{l}$ because the path of integration coincides with a field line whose direction is always the same as that of the electric field.] Since dQ is constant, it can be placed inside the integral sign,

$$dU = \tfrac{1}{2} \int dQ\, E(l)\, dl \tag{14}$$

Next we apply Gauss' Law to the tubular volume shown in Figure 26.5. The sides of the tube run along field lines; one cap of the tube is just inside the conductor, the other cap is perpendicular to field lines. The charge in the tube is dQ, and the flux through the surface of the tube is entirely due to the field lines that emerge through the perpendicular cap $dS(l)$ at a distance l. Gauss' Law then tells us

$$\frac{dQ}{\varepsilon_0} = E(l)\, dS(l) \tag{15}$$

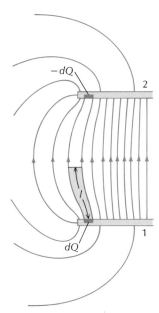

Fig. 26.4 The fringing electric field at the edges of the parallel conducting plates is an example of a field that is not constant. Field lines start on the charge dQ on the lower plate and end on the charge $-dQ$ on the upper plate.

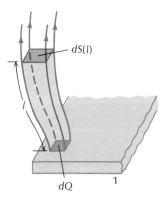

Fig. 26.5 The field lines that start on dQ form a tube of rectangular cross section.

Substituting this expression for dQ into Eq. (14), we obtain

$$dU = \tfrac{1}{2} \int \varepsilon_0 E(l) E(l) \, dS(l) \, dl \tag{16}$$

But $dS(l) \, dl$ is the amount of volume in our tube between l and $l + dl$. Hence Eq. (16) can be written

$$dU = \tfrac{1}{2} \int \varepsilon_0 E^2 \, dv \tag{17}$$

where the integration extends over the volume of the tube from conductor 1 to conductor 2. This shows that the potential energy of the charges $\pm dQ$ on the conductors equals the integral of $\tfrac{1}{2}\varepsilon_0 E^2$ over the volume of the tube that lies within the bundle of field lines originating on these charges. Therefore the potential energy of *all* the charges on the conductors will equal the integral of $\tfrac{1}{2}\varepsilon_0 E^2$ over the volume of *all* the regions in which there is electric field.

Clearly, the above argument in no way depends on the shape of the conductors. Hence the result applies not only to parallel conducting plates, but to any number of conductors of arbitrary shapes. Furthermore, although we obtained this result for the case of field lines that start on one conductor and end on another, it is easy to see that we can obtain the same result for field lines that start at one conductor and go on to infinity. This establishes that for any arbitrary system of charged conductors the electric energy can be expressed as a volume integral of $\tfrac{1}{2}\varepsilon_0 E^2$, that is,

$$\boxed{U = \int \tfrac{1}{2}\varepsilon_0 E^2 \, dv} \tag{18}$$

Energy in electric field

where the integration extends over the volume of all the regions where there is electric field. Incidentally: This equation for the electric energy is also valid for charges placed on nonconductors; we will not deal with the proof of this assertion, but we will take for granted that Eq. (18) is a general result for the energy associated with an electric field in vacuum.

Note that Eq. (7) expresses the energy as a sum over the electric charges, whereas Eq. (18) expresses the energy as an integral over the electric field. The former equation suggests that the energy is in the charges, whereas the latter suggests it is in the field. Thus, these two equations, which are mathematically equivalent, suggest conflicting physical interpretations. To decide which of these alternatives is correct, we need some extra information. The clue is the existence of electric fields that are independent of electric charges. As we will see in Chapter 36, radio waves and light waves consist of electric and magnetic fields traveling through space. Such fields are originally created by electric charges, but they persist even when the charges disappear. For instance, a radio wave or a light beam continues to travel through space long after the radio transmitter has been shut down or the candle has been snuffed out — obviously, the energy of a radio wave resides in the radio wave itself, in its electric and magnetic fields, and not in the electric charges in the antenna of the radio transmitter. We can then argue that if energy is associated with the traveling electric fields of a radio wave, energy should also be associated with the electric fields of a static charge distribution. We will therefore accept the interpretation that the energy (18) is in the electric field. The energy density, or energy per unit volume, in the electric field is

Energy density in electric field

$$u = \tfrac{1}{2}\varepsilon_0 E^2 \qquad (19)$$

EXAMPLE 3. Since energy has mass (see Section 8.6), the electric field should have not only an energy density, but also a mass density. What is the corresponding mass density in a thundercloud, where $E = 2 \times 10^6$ V/m?

SOLUTION:

$$u = \tfrac{1}{2}\varepsilon_0 E^2 = \tfrac{1}{2}\varepsilon_0 \times (2 \times 10^6 \text{ V/m})^2 = 18 \text{ J/m}^3$$

The mass density is the energy density divided by c^2, the square of the speed of light,

$$u/c^2 = \tfrac{1}{2}\varepsilon_0 E^2/c^2 = (18 \text{ J/m}^3)/(3 \times 10^8 \text{ m/s})^2$$

$$= 2.0 \times 10^{-16} \text{ kg/m}^3$$

This is obviously much too small to be detectable.

EXAMPLE 4. To a good approximation, a uranium nucleus can be regarded as a sphere with charge uniformly distributed over its volume. The radius of the nucleus is 7.4×10^{-15} m, and the electric charge is $92e$. What is the electric energy of the nucleus?

SOLUTION: According to Example 24.5, the electric fields outside of the nucleus and inside the nucleus are, respectively,

$$E = \frac{1}{4\pi\varepsilon_0} \frac{q}{r^2} \qquad \text{for } r \geq R \qquad (20)$$

and

$$E = \frac{1}{4\pi\varepsilon_0} \frac{qr}{R^3} \qquad \text{for } r \leq R \qquad (21)$$

The energy in the volume outside the nucleus is then

$$U_{\text{ext}} = \tfrac{1}{2}\varepsilon_0 \int_R^\infty E^2 \, dv = \tfrac{1}{2}\varepsilon_0 \int_R^\infty \left(\frac{1}{4\pi\varepsilon_0} \frac{q}{r^2}\right)^2 dv \qquad (22)$$

The amount of volume in a radial interval dr is $dv = 4\pi r^2 \, dr$, and hence

$$U_{\text{ext}} = \tfrac{1}{2}\varepsilon_0 \int_R^\infty \left(\frac{1}{4\pi\varepsilon_0} \frac{q}{r^2}\right)^2 4\pi r^2 \, dr$$

$$= \frac{q^2}{8\pi\varepsilon_0} \int_R^\infty \frac{1}{r^2} dr = \frac{q^2}{8\pi\varepsilon_0}\left[-\frac{1}{r}\right]_R^\infty = \frac{1}{8\pi\varepsilon_0} \frac{q^2}{R} \qquad (23)$$

The energy in the volume inside the nucleus is

$$U_{\text{int}} = \tfrac{1}{2}\varepsilon_0 \int_0^R E^2 \, dv = \tfrac{1}{2}\varepsilon_0 \int_0^R \left(\frac{1}{4\pi\varepsilon_0} \frac{qr}{R^3}\right)^2 4\pi r^2 \, dr$$

$$= \frac{q^2}{8\pi\varepsilon_0} \frac{1}{R^6} \int_0^R r^4 \, dr = \frac{q^2}{8\pi\varepsilon_0} \frac{1}{R^6} \left[\frac{r^5}{5} \right]_0^R$$

$$= \frac{1}{8\pi\varepsilon_0} \frac{q^2}{5R} \tag{24}$$

The total electrostatic energy is the sum of Eqs. (23) and (24),

$$U = U_{\text{ext}} + U_{\text{int}} = \frac{1}{8\pi\varepsilon_0} \frac{q^2}{R} + \frac{1}{8\pi\varepsilon_0} \frac{q^2}{5R} = \frac{1}{4\pi\varepsilon_0} \frac{3q^2}{5R} \tag{25}$$

Inserting numerical values, we obtain

$$U = \frac{1}{4\pi\varepsilon_0} \frac{3 \times (92 \times 1.6 \times 10^{-19} \text{ C})^2}{5 \times 7.4 \times 10^{-15} \text{ m}}$$

$$= 1.6 \times 10^{-10} \text{ J} \tag{26}$$

COMMENTS AND SUGGESTIONS: Expressed in electron-volts, this energy amounts to about 9.8×10^8 eV. Compared with the typical electric energy of an electron in an atom (about -27 eV for a hydrogen atom; see Example 25.3), this is a very large amount of energy. The energy released in nuclear fission arises from this large electric energy of the nucleus (as described in Interlude XI).

SUMMARY

Energy of a system of point charges:
$$U = \tfrac{1}{2}q_1 V_{\text{other}}(1) + \tfrac{1}{2}q_2 V_{\text{other}}(2) + \tfrac{1}{2}q_3 V_{\text{other}}(3) + \cdots$$

Energy of a system of conductors:
$$U = \tfrac{1}{2}Q_1 V_1 + \tfrac{1}{2}Q_2 V_2 + \tfrac{1}{2}Q_3 V_3 + \cdots$$

Energy density in electric field:
$$u = \tfrac{1}{2}\varepsilon_0 E^2$$

QUESTIONS

1. Suppose we have a system of electric point charges with the electrical potential energy given by Eq. (6). By what factor will this energy change if we increase the values of all the electric charges by a factor of 2?

2. Consider a metallic sphere carrying a given amount of charge. Explain why the electric energy is large if the radius of the sphere is small. Would you expect a similar inverse proportion between the electric energy and the size of a conductor of arbitrary shape?

3. Suppose that we increase the separation between the metallic plates described in Example 2. Does this change the electric energy density? The net electric energy?

4. Consider the electric fields shown in Figures 23.13 and 23.14. In what regions of the latter figure is the electric energy density larger than in the former? Can you guess which of these electric fields has a larger energy density on the average?

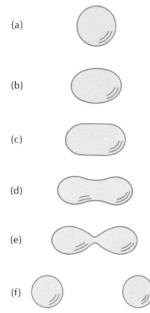

(a)
(b)
(c)
(d)
(e)
(f)

Fig. 26.6

5. Equation (6) suggests that the electric energy is located at the charges, whereas Eq. (18) suggests it is located in the field. How could we perform an experiment to test where the energy is located? (Hint: Energy gravitates.)

6. Figure 26.6 shows a sequence of deformations of a nucleus about to undergo fission. The volume of the nucleus and the electric charge remain constant during these deformations. Which configuration has the highest electric energy? The lowest?

7. Consider a sphere with a uniform distribution of charge over its volume. Where is the energy density within this sphere highest? Lowest?

8. What fraction of the electric energy of a sphere with a uniform distribution of charge is inside the sphere? Outside the sphere?

9. Suppose that a nucleus of charge Q, radius R, and electric energy $(1/4\pi\varepsilon_0)(3q^2/5R)$ fissions into two equal parts of charge $q/2$ each. The nuclear material in the original nucleus and in the final two nuclei has the same density. What is the radius of each of the two final nuclei? How does the sum of the individual electric energies of the two final, separated nuclei compare with the initial electric energy?

10. Since the electric energy density is never negative, how can the mutual electric potential energy of a pair of opposite charges be negative?

PROBLEMS

Section 26.1

1. Consider once more the distribution of charges within the thundercloud shown in Figure 23.28. What is the electric potential energy of this charge distribution?

2. In the water molecule, the hydrogen atoms tend to give up their electrons to the oxygen atom. Crudely, the molecule may be regarded as consisting of a uniformly charged ball of charge $-2e$ and two smaller uniformly charged balls of charge $+e$ each. The dimensions of the molecule are given in Figure 23.21. Calculate the electrostatic energy of this arrangement of three charges; ignore the internal electrostatic energy of the individual balls of charge. Note that in this calculation each of the uniformly charged spherical balls can be treated as though it were a point charge; explain why.

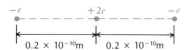

Fig. 26.7 The nucleus (2e) and the electrons (−e) of an atom of helium at an instant of time.

3. Suppose that at one instant the electrons and the nucleus of a helium atom occupy the positions shown in Figure 26.7; at this instant, the electrons are at a distance of 0.20×10^{-10} m from the nucleus. What is the electric potential energy of this arrangement? Treat the electrons and the nucleus as point charges.

*4. Four equal positive charges of magnitude Q are placed on the four corners of a square of side d. What is the electric energy of this system of charges?

*5. Four positive and four negative point charges of equal magnitudes $\pm Q$ are arranged alternately on the corners of a cube of edge d (see Figure 26.8). What is the electric energy of this arrangement?

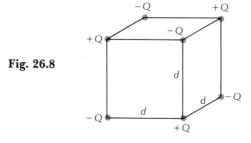

Fig. 26.8

*6. According to the alpha-particle model of the nucleus, some nuclei consist of a regular geometric arrangement of alpha particles. For instance, the nucleus of ^{12}C consists of three alpha particles arranged on an equilateral triangle (Figure 26.9). Assuming that the distance between pairs of alpha particles is 3.0×10^{-15} m, what is the electric energy (in eV) of this arrangement of alpha particles? Treat the alpha particles as pointlike.

*7. According to the alpha-particle model (see also the preceding problem), the nucleus of ^{16}O consists of four alpha particles arranged on the vertices of a tetrahedron (Figure 26.10). If the distance between pairs of alpha particles is 3.0×10^{-15} m, what is the electric energy (in eV) of this configuration of alpha particles? Treat the alpha particles as pointlike.

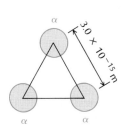
Fig. 26.9

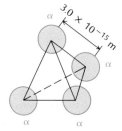
Fig. 26.10

*8. Problem 24.22 describes the Thomson model of the helium atom. The equilibrium separation of the electrons is 0.50 Å. Calculate the electric energy of this configuration. Take into account both the electric energy between the electrons and the positive charge, and the electric energy between the electrons; ignore the energy of the cloud and of the electrons by themselves.

*9. Four equal particles of positive charges q and masses m are initially held at the four corners of a square of side L. If these particles are released simultaneously, what will be their speeds when they have separated by a very large distance?

*10. Two thin rods of length l carry equal charges Q uniformly distributed over their lengths. The rods are aligned, and their nearest ends are separated by a distance x (Figure 26.11). Calculate the mutual electric potential energy. Ignore the self-energy of each rod.

Fig. 26.11

Section 26.2

11. A pair of parallel conducting plates, each measuring 30 cm × 30 cm, are separated by a gap of 1.0 mm. How much work must you do against the electric forces to charge these plates with $+1.0 \times 10^{-6}$ C and -1.0×10^{-6} C, respectively?

12. Two large parallel conducting plates of area 0.2 m² are separated by a distance of 0.5 mm. The plates carry opposite charges, and the electric field in the space between them is 5×10^5 V/m. What is the electric energy?

13. A penny coin is hung from a silk thread inside a closed tin can placed on the ground. Given that the penny coin carries a charge of 2.0×10^{-6} C, and that the potential difference between the tin can and the penny coin is 3.0×10^4 V, find the electric potential energy of this system of two conductors.

*14. A charge of 7.5×10^{-6} C can be placed on a metallic sphere of radius 15 cm before the surrounding air suffers electrical breakdown. What is the electric energy of the sphere with this charge?

*15. A sphere of radius R has a charge Q uniformly distributed over its volume. A thin conducting shell of radius $2R$ surrounds the sphere concentrically. The shell carries a charge $-Q$ on its interior surface. What is the electric energy of this system?

*16. Pretend that an electron is a conducting sphere of radius R with a charge e distributed uniformly over its surface. In terms of e and the mass m_e of the electron, what must be the radius R if the electric energy is to equal the rest-mass energy $m_e c^2$ of the electron? Numerically, what is the value of R?

**17. Consider the Geiger-counter tube described in Problem 25.29. If the tube is initially uncharged, how much work must be done to bring the tube to its operating voltage of 1.0×10^3 V?

Section 26.3

18. Near the surface of the nucleus of a lead atom, the electric field has a strength of 3.4×10^{21} V/m. What is the energy density in this field?

19. The atmospheric electric field near the surface of the Earth has a strength of 100 V/m.
    (a) What is its energy density?
    (b) Assuming that the field has the same magnitude everywhere in the atmosphere up to a height of 10 km, what is the corresponding total energy?

20. Calculate the energy density in each of the electric fields, listed in the first four entries of Table 23.1. Calculate the mass densities that correspond to these energy densities.

21. The nuclei of ^{235}Pu, ^{235}Np, ^{235}U, and ^{235}Pa all have the same radii, about 7.4×10^{-15} m, but their electric charges are $94e$, $93e$, $92e$, and $91e$, respectively. Treating these nuclei as uniformly charged spheres, calculate their electric energies; express your answers in electron-volts.

22. According to a crude model, a proton can be regarded as a uniformly charged sphere of charge e and radius 1.0×10^{-15} m. Find the electric self-energy of the proton. Express your answer in eV.

23. A solid sphere of copper of radius 10 cm with a charge of 1.0×10^{-6} C is placed at the center of a thin, spherical copper shell of radius 20 cm with a charge of -1.0×10^{-6} C. Find a formula for the energy density in the space between the solid sphere and the shell. Find the total electric energy.

*24. One method for the determination of the radii of nuclei makes use of the known difference of electric energy between two nuclei of the same size but different electric charges. For instance, the nuclei ^{15}O and ^{15}N have the same size but their charges are $8e$ and $7e$, respectively. Given that the difference in electric energy is 3.7×10^6 eV, what is the nuclear radius?

*25. A spherical shell of inner radius a, outer radius b carries a charge Q uniformly distributed over its volume. What is the electric energy of this charge distribution?

*26. In analogy to the electric field $\mathbf{E}$ (electric force per unit mass), we can define a gravitational field $\mathbf{g}$ (gravitational force per unit mass).
    (a) The energy density in the electric field is $\varepsilon_0 E^2/2$. By analogy, show that the energy density in the gravitational field is $g^2/(8\pi G)$.
    (b) Calculate the gravitational field energy of the Moon due to its own gravity; treat this body as a sphere of uniform density. What is the ratio of the gravitational field energy to the rest-mass energy of the Moon?[1]

*27. In symmetric fission, the nucleus of uranium (^{238}U) splits into two nuclei of palladium (^{119}Pd). The uranium nucleus is spherical with a radius of

---

[1] Warning: This problem does not take the gravitational *interaction* energy into account. The gravitational interaction energy density is [mass density] × [gravitational potential]. The total gravitational energy is the sum of the field energy and the interaction energy; this total gravitational energy is always negative.

7.4×10^{-15} m. Assume that the two palladium nuclei adopt a spherical shape immediately after fission; at this instant, the configuration is as shown in Figure 26.12. The size of the nuclei in Figure 26.12 can be calculated from the size of the uranium nucleus because the nuclear material maintains a constant density (the initial nuclear volume equals the final nuclear volume).

Fig. 26.12

(a) Calculate the electric energy of the uranium nucleus before fission.
(b) Calculate the total electric energy of the palladium nuclei in the configuration shown in Figure 26.12, immediately after fission. Take into account the mutual electric potential energy of the two nuclei and also the individual electric energies of the two palladium nuclei by themselves.
(c) Calculate the total electric energy a long time after fission when the two palladium nuclei have moved apart by a very large distance.
(d) Ultimately, how much electric energy is released into other forms of energy in the complete fission process (a) through (c)?
(e) If 1 kg of uranium undergoes fission, how much electric energy is released?

**28. Using the model described in Problem 24.19 for the charge distribution of a neutron, calculate the electric self-energy of a neutron. Express your answer in eV.

**29. A long rod of plastic of radius a has a charge of λ coulombs per unit length uniformly distributed over its volume. The rod is surrounded by a concentric cylinder of sheet metal of radius b with a charge of $-\lambda$ coulombs per unit length on its interior surface.
(a) What is the energy density (as a function of radius) in the space between the rod and the cylinder?
(b) What is the energy density in the volume of the rod?
(c) What is the total electric energy per unit length?

CHAPTER 27

Capacitors and Dielectrics

Capacitor   Any arrangement of conductors that is used to store electric charge is called a **capacitor**, or condenser. Since work must be done during the charging process, the capacitor will also store electric potential energy. In our electronic technology, capacitors find widespread application — they are part of the circuitry of radios, electronic calculators, automobile ignition systems, and so on.

The first part of this chapter deals with the properties of capacitors.

Dielectric   The second part deals with the properties of electric fields in regions of space filled with an insulating material, or **dielectric.** Since capacitors are often filled with such a dielectric material, the study of the mutual effects between the electric field and the dielectric material is closely linked to the study of capacitors. But the effects of electric fields and dielectric materials upon one another are also interesting in their own right. For instance, air is a dielectric material, and we ought to inquire how the electric field in air differs from that in vacuum.

27.1 Capacitance

Fig. 27.1 Charge is stored on a conducting sphere.

As a first example of a capacitor, consider an isolated metallic sphere of radius R (Figure 27.1). Obviously, charge can be stored on this sphere. If the amount of charge placed on the sphere is Q, then the potential of the sphere will be

$$V = \frac{1}{4\pi\varepsilon_0} \frac{Q}{R} \tag{1}$$

Thus, the amount of charge stored on the sphere is directly proportional to the potential.

This proportionality holds in general for any conductor of arbitrary shape. The charge on the conductor produces an electric field whose strength is directly proportional to the amount of charge (twice the charge produces twice the field), and the electric field yields a potential which is directly proportional to the field strength (twice the field strength yields twice the potential); hence charge and potential are proportional. We write this relationship as

$$Q = CV \qquad (2)$$

Capacitance of a single conductor

where C is the constant of proportionality. This constant is called the **capacitance** of the conductor. The capacitance is large if the conductor is capable of storing a large amount of charge at a low potential. For instance, the capacitance of a conducting sphere is

$$C = \frac{Q}{V} = \frac{Q}{(1/4\pi\varepsilon_0)(Q/R)} = 4\pi\varepsilon_0 R \qquad (3)$$

Thus, the capacitance of the sphere increases with its radius.

The unit of capacitance is the **farad** (F),

$$1 \text{ farad} = 1 \text{ F} = 1 \text{ coulomb/volt} \qquad (4)$$

Farad, F

This unit of capacitance is rather large; in practice, electrical engineers prefer the **microfarad** and the **picofarad**. A microfarad equals 10^{-6} farad (1 μF = 10^{-6} F), and a picofarad equals 10^{-12} farad (1 pF = 10^{-12} F).

Note that 1 F = 1 C/V = 1 C^2/N·m, so the constant ε_0 can be written

$$\varepsilon_0 = 8.85 \times 10^{-12} \frac{\text{C}^2}{\text{N} \cdot \text{m}^2} = 8.85 \times 10^{-12} \text{ F/m} \qquad (5)$$

The latter expression is the one usually listed in tables of physical constants.

EXAMPLE 1. What is the capacitance of an isolated metallic sphere of radius 20 cm?

SOLUTION: According to Eq. (3),

$$C = 4\pi\varepsilon_0 R = 4\pi \times 8.85 \times 10^{-12} \text{ F/m} \times 0.20 \text{ m}$$

$$= 2.2 \times 10^{-11} \text{ F} = 22 \text{ pF}$$

EXAMPLE 2. What is the capacitance of the Earth, regarded as a conducting sphere?

SOLUTION: The radius of the Earth is 6.4×10^6 m, so

$$C = 4\pi\varepsilon_0 R = 4\pi \times 8.85 \times 10^{-12} \text{ F/m} \times 6.4 \times 10^6 \text{ m}$$

$$= 7.1 \times 10^{-4} \text{ F}$$

COMMENTS AND SUGGESTIONS: As capacitances go, this is a rather large capacitance. However, note that it takes a charge of only about 10^{-3} coulomb to alter the potential of the Earth by 1 volt.

In electrostatic experiments, the Earth is often used as a dump for unwanted positive or negative charge, which alters the potential of the Earth relative to infinity. This alteration conflicts with the convention adopted in Section 25.1, where we treated the Earth as a body at a fixed potential, $V = 0$. But this conflict need not trouble us: for most terrestrial electric experiments, only the potential difference between the apparatus and the ground is relevant, and the alteration of the potential difference between the apparatus and infinity has no immediate effect.

The most common variety of capacitor consists of *two* metallic conductors, insulated from one another and carrying opposite amounts of electric charge $\pm Q$. The capacitance of such a pair of conductors is defined in terms of the *difference* of potential between the two:

Capacitance of a pair of conductors

$$Q = C \, \Delta V \tag{6}$$

In this expression, both Q and ΔV are taken as positive quantities. Thus, the quantity Q is *not* the net charge in the capacitor, but the magnitude of the charge on each plate. The net charge in the two-conductor capacitor is zero.

Figure 27.2 shows such a two-conductor capacitor consisting of two large parallel metallic plates, each of area A, separated by a distance d. The plates carry charges $+Q$ and $-Q$ on their surfaces, respectively. The electric field in the region between the plates is (neglecting edge effects)

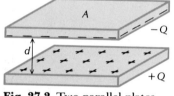

Fig. 27.2 Two parallel plates, with charges $+Q$ and $-Q$.

$$E = \frac{\sigma}{\varepsilon_0} = \frac{Q}{\varepsilon_0 A}$$

and the potential difference is

$$\Delta V = Ed = \frac{Qd}{\varepsilon_0 A} \tag{7}$$

Hence the capacitance of this configuration is

Capacitance of parallel plates

$$C = \frac{Q}{\Delta V} = \frac{Q}{Qd/\varepsilon_0 A} = \frac{\varepsilon_0 A}{d} \tag{8}$$

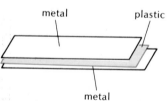

Fig. 27.3 Sheets of aluminum foil separated by a sheet of plastic.

From this formula we see that in order to store a large amount of charge at a low potential, we want a large plate area A, but a small plate separation d. Parallel-plate capacitors are usually manufactured out of two parallel sheets of aluminum foil, a few centimeters wide and several meters long. The sheets are placed very close together, but kept from contact by a thin sheet of plastic sandwiched between (we will deal in Section 27.3 with the effects of the plastic on the capacitance). For convenience, the entire sandwich is covered with another sheet of plastic and rolled up like a roll of toilet paper (Figure 27.3).

EXAMPLE 3. A parallel-plate capacitor consists of two strips of aluminum foil, each with an area of 0.20 m², separated by a distance of 0.10 mm. The space between the foils is empty. By means of a battery, potential difference of

200 V is applied to this capacitor. What is the capacitance of this capacitor? What is the electric charge on each plate? What is the strength of the electric field between the plates?

SOLUTION: According to Eq. (8), the capacitance is

$$C = \frac{\varepsilon_0 A}{d} = \frac{8.85 \times 10^{-12} \text{ F/m} \times 0.20 \text{ m}^2}{1.0 \times 10^{-4} \text{ m}} = 1.8 \times 10^{-8} \text{ F}$$

$$= 0.018 \text{ } \mu\text{F}$$

The magnitude of the charge on each plate is

$$Q = C \, \Delta V = 1.8 \times 10^{-8} \text{ F} \times 200 \text{ volts} = 3.5 \times 10^{-6} \text{ coulomb}$$

and the electric field between the plates is

$$E = \Delta V/d = 200 \text{ volts}/1.0 \times 10^{-4} \text{ m} = 2.0 \times 10^6 \text{ volts/m}$$

27.2 Capacitors in Combination

The capacitors used in practical applications in the electric circuits of radios, amplifiers, television receivers, and so on, are commonly of the two-conductor variety. Schematically, in circuit diagrams, such capacitors are represented as two short parallel lines with terminals emerging from their middles (Figure 27.4). In electric circuits, several capacitors are often wired together, and the combination is wired to other circuit elements. For instance, Figure 27.5 shows a portion of the circuit diagram for a radio receiver, with diverse combinations of several capacitors.

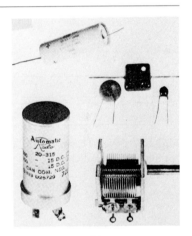

(a)

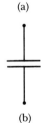

(b)

Fig. 27.4 (a) Different kinds of capacitors used in electric circuits. (b) Symbol for a capacitor in a circuit diagram.

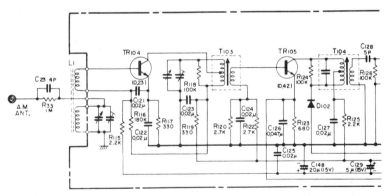

Fig. 27.5 A portion of the circuit diagram of a radio with several capacitors (C), resistors (R), inductors (L), transformers (T), transistors (TR), and diodes (D). We will become acquainted with all these circuit elements in later chapters.

When several capacitors are connected together, the net capacitance of the combination can be calculated from the individual capacitances. The simplest ways of connecting capacitors together are in **parallel** and in **series**.

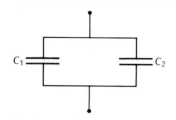

Fig. 27.6 Two capacitors connected in parallel.

Figure 27.6 shows two capacitors connected in *parallel*. If charge is fed into this combination via the two terminals, some of the charge will be stored on the first capacitor and some on the second. The net capacitance of the combination can be found as follows. Since the top plates and the bottom plates of the capacitors are joined by conductors, the potentials of these plates must be equal, and the potential differences across both capacitors must also be equal,

$$\Delta V = \frac{Q_1}{C_1} \quad \text{and} \quad \Delta V = \frac{Q_2}{C_2} \tag{9}$$

Therefore the net charge for the capacitor combination can be expressed as

$$Q = Q_1 + Q_2 = C_1 \Delta V + C_2 \Delta V \tag{10}$$

that is,

$$Q = (C_1 + C_2) \Delta V \tag{11}$$

Comparing this with the definition for capacitance given in Eq. (6), we see that the combination is equivalent to a single capacitor of capacitance

$$C = C_1 + C_2 \tag{12}$$

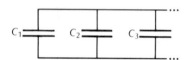

Fig. 27.7 Several capacitors connected in parallel.

Thus, the net capacitance of the parallel combination is simply the sum of the individual capacitances.

It is easy to obtain a similar result for any number of capacitors connected in parallel (Figure 27.7). The net capacitance for such a parallel combination is

Parallel combination of capacitors

$$\boxed{C = C_1 + C_2 + C_3 + \cdots} \tag{13}$$

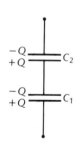

Fig. 27.8 Two capacitors connected in series.

Next, let us consider the alternative way of wiring capacitors together. Figure 27.8 shows two capacitors connected in *series*. Any charge fed into this combination via the two outside terminals will have to remain on the outside plates [the bottom plate of the first capacitor (C_1) and the top plate of the second (C_2); see Figure 27.8]. Thus, the bottom plate will have a charge Q and the top plate a charge $-Q$. But these charges on the outside plates will induce charges on the inside plates (the top plate of the first capacitor and the bottom plate of the second). The charge Q on the bottom plate will attract electrons to the facing plate, and a charge $-Q$ will accumulate on this plate. Corresponding to the excess electrons on the top plate of the first capacitor, there will be a deficit of electrons on the bottom plate of the second capacitor, and a charge $+Q$ will accumulate there. The capacitance of the combination can then be found as follows. The potential differences across the two capacitors are

$$\Delta V_1 = \frac{Q}{C_1} \quad \text{and} \quad \Delta V_2 = \frac{Q}{C_2} \tag{14}$$

The net potential difference between the external terminals is the sum of these,

$$\Delta V = \Delta V_1 + \Delta V_2 = \frac{Q}{C_1} + \frac{Q}{C_2} \tag{15}$$

that is,

$$\Delta V = Q\left(\frac{1}{C_1} + \frac{1}{C_2}\right) \tag{16}$$

Comparing this, again, with the definition of capacitance given by Eq. (6), we see that the combination is equivalent to a single capacitor with

$$\frac{1}{C} = \frac{1}{C_1} + \frac{1}{C_2} \tag{17}$$

Thus, the inverse net capacitance of the series combination is obtained by taking a sum of inverses. Note that the net capacitance is *less* than the individual capacitances. For example, if $C_1 = C_2$, then $C = \tfrac{1}{2}C_1 = \tfrac{1}{2}C_2$.

A similar result applies to any number of capacitors connected in series (Figure 27.9). The net capacitance for such a series combination is given by

$$\boxed{\frac{1}{C} = \frac{1}{C_1} + \frac{1}{C_2} + \frac{1}{C_3} + \cdots} \tag{18}$$

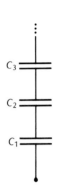

Fig. 27.9 Several capacitors connected in series.

Series combination of capacitors

27.3 Dielectrics

So far, in dealing with problems of electrostatics we have assumed that the space surrounding the electric charge consisted of a vacuum, which has no effect on the electric field, or of air, which has only an insignificant effect on the electric field. However, in dealing with capacitors, we must take the effects of the medium into account. The space between the plates of a capacitor is usually filled with an insulator, or **dielectric**, which drastically changes the electric field from what it would be in a vacuum: the dielectric reduces the strength of the electric field.

To understand this, consider a parallel-plate capacitor whose plates carry some amount of charge per unit area. Suppose that a slab of dielectric, such as glass or polyethylene, fills most of the space between the plates (Figure 27.10). This dielectric contains a large number of atomic nuclei and electrons, but, of course, these positive and negative charges balance each other, so the material is electrically neutral. In an insulator, all the charges are **bound** — the electrons are confined within their atoms or molecules, and they cannot wander about as in a conductor. Nevertheless, in response to the force exerted by the electric field, the charges will move very slightly without leaving their atoms. The positive and negative charges are pulled in opposite direc-

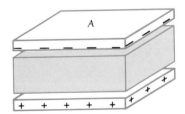

Fig. 27.10 A slab of dielectric between the plates of a capacitor.

Bound charges

tions in the electric field; these opposite pulls tend to separate the positive and the negative charges and thereby create electric dipoles within the dielectric. In most dielectrics, the magnitudes of the charge separations and the magnitudes of the corresponding dipole moments are directly proportional to the strength of the electric field; such dielectrics are said to be **linear.**

Linear dielectric

The details of the mechanism of displacement and separation of charge depend on the dielectric. In some dielectrics — such as glass, polyethylene, or other solids — the creation of dipole moments involves a distortion of the molecules or atoms. By tugging on the electrons and nuclei in opposite directions, the electric field stretches the molecule while producing a charge separation within it (see Figure 27.11).

In other dielectrics — such as distilled water[1] or carbon dioxide — the creation of dipole moments results mainly from a realignment of existing dipoles. In such dielectrics, the molecules have permanent dipole moments, which are randomly oriented when the dielectric is left to itself. The randomness of the orientation of the dipoles means that, on the average, there is no charge separation in the dielectric. But when the dielectric is placed in an electric field, the permanent dipoles experience a torque that tends to align them with the electric field (see Figure 27.12). Random thermal motions oppose this alignment, and the molecules achieve an average equilibrium state in which the average amount of alignment is approximately proportional to the strength of the electric field. This average alignment is equivalent to an average charge separation.

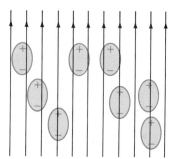

Fig. 27.11 The electric field produces a distortion of molecules.

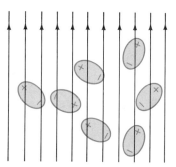

Fig. 27.12 The electric field produces a (partial) alignment of already distorted molecules.

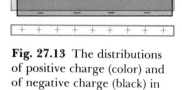

Fig. 27.13 The distributions of positive charge (color) and of negative charge (black) in the slab of dielectric do not overlap precisely.

Polarization

In any case, the microscopic charge separations lead to a relative displacement between the overall, macroscopic distributions of positive and negative charges. When the dielectric is subjected to an electric field, the distributions of positive and negative charges cease to overlap precisely (see Figure 27.13). Consequently, there will be an excess of positive charge on one surface of the slab of dielectric and an excess of negative charge on the opposite surface — the slab of dielectric acquires layers of surface charge. The slab of dielectric is then said to be **polarized.** These surface charges act just like a pair of parallel sheets of positive and negative charge; between the sheets, these charges generate an electric field that is *opposite* to the original applied electric field. The total electric field, consisting of the sum of the field of the

---

[1] Remember that distilled water is an insulator.

free charges on the conducting plates plus the field of the bound charges on the dielectric surfaces, is therefore *smaller* than the field of the free charges alone (Figure 27.14).

In a linear dielectric, the amount by which the dielectric reduces the strength of the electric field can be characterized by the **dielectric constant** κ. This constant is merely the factor by which the electric field in the dielectric between the parallel plates is reduced, that is, if $\mathbf{E}_{\text{free}}$ is the electric field that the free charges produce by themselves and $\mathbf{E}$ the electric field that the free charges and the bound charges produce together, then

$$\mathbf{E} = \frac{1}{\kappa} \mathbf{E}_{\text{free}} \qquad (19)$$

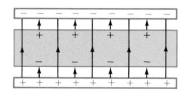

Fig. 27.14 Some electric field lines stop on the negative charges at the bottom of the slab of dielectric. The density of field lines is smaller in the dielectric than in the empty gaps adjacent to the plates.

Electric field in dielectric

where $\kappa \geq 1$.

Table 27.1 lists the values of the dielectric constants of some materials. Note that air has a value very near $\kappa = 1$, that is, the dielectric properties of air are not very different from those of a vacuum, and the electric fields produced by (free) charges placed in air are almost the same as those produced in vacuum. This justifies our policy of neglect of the effects of air in many of the problems in preceding chapters, and we will continue this policy.

Table 27.1 Dielectric Constants of Some Materials[a]

Material	κ
Vacuum	1.
Helium	1.000064
Air	1.00055
Carbon dioxide	1.00092
Carbon tetrachloride	2.2
Paraffin	~2
Polyethylene	2.3
Rubber, hard	2.8
Transformer oil	~3
Plexiglas	3.4
Nylon	3.5
Epoxy resin	3.6
Paper	~4
Bakelite	~5
Pyrex glass	~5
Glass	~6
Porcelain	~7
Water, distilled	80

[a] At room temperature (20° C) and 1 atm.

Incidentally: A metal can be regarded as a dielectric with an infinite dielectric constant — if we substitute $\kappa = \infty$ into Eq. (19), we find that $E = 0$, as it should be inside the metal. This large value of the dielectric constant simply indicates that the material becomes very strongly polarized, that is, the charges in the metal respond very strongly to an electric field.

If the slab of dielectric fills the space between the plates entirely, then the formula (19) for the reduction of the strength of the electric field applies throughout all of this space. Since the potential difference between the capacitor plates is directly proportional to the strength of the electric field, it follows that, for a given amount of free charge on the plates, the presence of the dielectric also reduces the potential difference by the factor κ,

$$\Delta V = \frac{1}{\kappa} \Delta V_0 \qquad (20)$$

where ΔV_0 is the potential difference in the absence of the dielectric. Consequently, the presence of the dielectric increases the capacitance by a factor κ,

$$C = \frac{Q}{\Delta V} = \kappa \frac{Q}{\Delta V_0} = \kappa C_0 \qquad (21)$$

where C_0 is the capacitance in the absence of dielectric. For example, the capacitance of a parallel-plate capacitor filled with dielectric is

$$C = \kappa C_0 = \kappa \frac{\varepsilon_0 A}{d} \qquad (22)$$

By filling the space between the capacitor plates with dielectric, we can therefore obtain a substantial gain in capacitance. Furthermore, the dielectric can prevent electric breakdown in the space between the plates. If this space contains air, sparking will occur between the plates when the electric field reaches a value of about 3×10^6 V/m, and the capacitor will discharge spontaneously. Some dielectrics are better insulators than air, and they will tolerate an electric field that is appreciably larger than 3×10^6 V/m. For instance, Plexiglas will tolerate an electric field of up to 40×10^6 V/m before it suffers electric breakdown (Figure 27.15).

The magnitude of the surface-charge density on the slab of dielectric is given by the simple formula

Fig. 27.15 Electric breakdown of a Plexiglas block in a very strong electric field caused minute perforations in the block and created this beautiful arboreal pattern.

$$\boxed{\sigma_{\text{bound}} = -\frac{\kappa - 1}{\kappa} \sigma_{\text{free}}} \qquad (23)$$

Here the negative sign on the right side of the formula indicates that the negatively charged surface of the dielectric adjoins the positively charged plate of the capacitor (see Figure 27.14). This formula is a direct consequence of Eq. (19). The total electric field E in the dielectric is the sum of the fields of the free charges and the bound charges,

$$E = E_{\text{free}} + E_{\text{bound}} \qquad (24)$$

Hence

$$\frac{E_{\text{free}}}{\kappa} = E_{\text{free}} + E_{\text{bound}} \qquad (25)$$

which, expressed in terms of charge densities, becomes

$$\frac{\sigma_{\text{free}}}{\varepsilon_0 \kappa} = \frac{\sigma_{\text{free}}}{\varepsilon_0} + \frac{\sigma_{\text{bound}}}{\varepsilon_0} \qquad (26)$$

This is equivalent to Eq. (23).

Although we derived the results of this section for the special case of a parallel-plate capacitor, Eqs. (19), (20), (21), and (23) also hold for cylindrical or spherical conductors and concentric shells of dielectric filling the space between the conductors (Figure 27.16). If the dielectric does not entirely fill the space between the conductors, then our results must be modified.

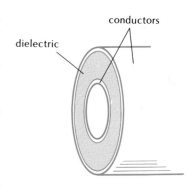

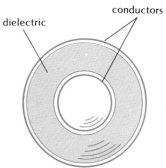

Fig. 27.16 A cylindrical capacitor and a spherical capacitor.

EXAMPLE 4. A parallel-plate capacitor, such as used in a radio, is made of two strips of aluminum foil with a plate area of 0.75 m². The plates are separated by a layer of polyethylene 2.0×10^{-5} m in thickness. Suppose that a potential difference of 30 V is applied to this capacitor. What is the magnitude of the free charge on each plate? What is the magnitude of the bound charge on the surface of the dielectric? What is the electric field in the dielectric?

SOLUTION: With $\kappa = 2.3$ (see Table 27.1), the capacitance is

$$C = \kappa \varepsilon_0 \frac{A}{d} = \frac{2.3 \times 8.85 \times 10^{-12} \text{ F/m} \times 0.75 \text{ m}^2}{2.0 \times 10^{-5} \text{ m}}$$

$$= 7.6 \times 10^{-7} \text{ F}$$

The magnitude of the free charge on each plate is then

$$Q_{\text{free}} = C \, \Delta V = 7.6 \times 10^{-7} \text{ F} \times 30 \text{ V} = 2.3 \times 10^{-5} \text{ coulomb}$$

The bound charge on the surfaces of the dielectric can be found from Eq. (23):

$$Q_{\text{bound}} = A\sigma_{\text{bound}} = -\frac{\kappa - 1}{\kappa} A\sigma_{\text{free}} = -\frac{\kappa - 1}{\kappa} Q_{\text{free}}$$

$$= -\frac{2.3 - 1}{2.3} \times 2.3 \times 10^{-5} \text{ coulomb}$$

$$= -1.3 \times 10^{-5} \text{ coulomb}$$

Here, as in Eq. (23), the negative sign indicates that the bound charge on the dielectric is negative on the surface adjacent to the positive plate of the capacitor.

The electric field in the dielectric is

$$E = \frac{1}{\kappa} E_{\text{free}} = \frac{1}{\kappa} \frac{\sigma_{\text{free}}}{\varepsilon_0} = \frac{1}{\kappa} \frac{Q_{\text{free}}}{\varepsilon_0 A}$$

$$= \frac{1}{2.3} \times \frac{2.3 \times 10^{-5} \text{ coulomb}}{8.85 \times 10^{-12} \text{ F/m} \times 0.75 \text{ m}^2} = 1.5 \times 10^6 \text{ volts/m}$$

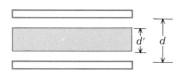

Fig. 27.17 A slab of dielectric between the plates of a capacitor.

EXAMPLE 5. The space between two large parallel conducting plates is partially filled with a parallel slab of dielectric (Figure 27.17). The area of the plates is A, their separation is d, and the thickness of the slab is d'. The dielectric constant of the slab is κ. What is the capacitance of this arrangement?

SOLUTION: Consider a straight path from one plate to the other. A length $d - d'$ of this path is in a vacuum, and a length d' is in the dielectric; the electric field has the value E_{free} in vacuum and E_{free}/κ in the dielectric. Hence, the potential difference between the plates is

$$\Delta V = E_{\text{free}}(d - d') + \frac{1}{\kappa} E_{\text{free}} d'$$

$$= \left[1 - \left(1 - \frac{1}{\kappa}\right) \frac{d'}{d}\right] E_{\text{free}} d$$

$$= \left[1 - \left(1 - \frac{1}{\kappa}\right) \frac{d'}{d}\right] \Delta V_0 \qquad (27)$$

The factor in brackets is less than one; hence ΔV is smaller than ΔV_0 and, correspondingly, the capacitance C is larger than C_0:

$$C = \frac{C_0}{1 - (1 - 1/\kappa) d'/d}$$

$$= \frac{\varepsilon_0 A/d}{1 - (1 - 1/\kappa) d'/d} \qquad (28)$$

COMMENTS AND SUGGESTIONS: Note that for $d' = d$, this formula reduces to $C = \kappa \varepsilon_0 A/d$; and for $d' = 0$, it reduces to $C = \varepsilon_0 A/d$. Both of these extremes are in agreement with our expectations.

27.4 Gauss' Law in Dielectrics

The electric field produced by the bound charges in a dielectric does, of course, obey Gauss' Law. Hence the *total* electric field **E** will obey Gauss' Law:

$$\varepsilon_0 \int \mathbf{E} \cdot d\mathbf{S} = Q_{\text{total}} = Q_{\text{free}} + Q_{\text{bound}} \qquad (29)$$

Here the total charge is the sum of the free charge plus the bound charge.

Unfortunately, Q_{bound} is usually not known beforehand, because the amount of polarization in the dielectric depends on the (unknown) strength of the electric field. Thus, the above form of Gauss' Law is not very helpful.

Since the free charge Q_{free} is usually known, it is better to devise a modified form of Gauss' Law that depends only on this free charge. For simplicity's sake, we will assume throughout the following discussion that the conductors, the distribution of free charge on them, and the dielectrics between them have enough symmetry so that Gauss' Law suffices for the calculation of the electric field. We can then proceed as follows.

Imagine that at first the dielectrics are absent, so that only the free charge produces an electric field. For this situation, Gauss' Law for the electric field $\mathbf{E}_{\text{free}}$ produced by the charge Q_{free} is

$$\varepsilon_0 \oint \mathbf{E}_{\text{free}} \cdot d\mathbf{S} = Q_{\text{free}} \tag{30}$$

Next, imagine that the dielectrics are inserted into their proper places while the amount of free charge on the conductors is held constant. In general, we would expect that this will disturb the electric field, because the bound charges on the dielectrics act back on the free charges and affect the distribution of these charges on the conductors. But if the arrangement of conductors has a high symmetry and the dielectrics have the *same* symmetry (parallel flat conductors with parallel flat slabs of dielectric; concentric cylindrical conductors with concentric cylindrical dielectrics; concentric spherical conductors with concentric spherical dielectrics), then the distribution of free charge is determined by the symmetry, and the dielectric cannot disturb this distribution of free charge. Neither can it disturb the geometrical arrangement of the field lines. Thus, the direction of the electric field at each point is unchanged. Only the strength of the electric field is altered: the new electric field will be smaller than the original one by a factor of κ,

$$\mathbf{E} = \mathbf{E}_{\text{free}}/\kappa \tag{31}$$

Substituting this into Eq. (30), we obtain

$$\boxed{\varepsilon_0 \oint \kappa \mathbf{E} \cdot d\mathbf{S} = Q_{\text{free}}} \tag{32}$$

Gauss' Law in dielectrics

This is **Gauss' Law in dielectrics.** It relates the total electric field $\mathbf{E}$ to the *free* charge Q_{free}. The effect of the bound charge is implicitly contained in the factor κ appearing on the left side of the equation. Although the preceding discussion has focused on arrangements of conductors and dielectrics with high symmetry, the above modified version of Gauss' Law turns out to be valid for conductors and dielectrics of any shape whatsoever.

EXAMPLE 6. Two concentric spheres of sheet metal have radii r_1 and r_2, respectively. The space between these is filled with gas of dielectric constant κ (Figure 27.18). What is the capacitance of this contraption?

SOLUTION: Suppose that the free charge on the inner sphere is Q_{free} and on the outer $-Q_{\text{free}}$. As Gaussian surface, take a sphere of radius r, with $r_2 > r > r_1$ (see Figure 27.18). Equation (32) then becomes

$$\varepsilon_0 \kappa E \times 4\pi r^2 = Q_{\text{free}}$$

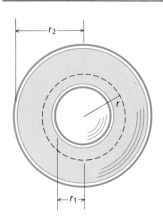

Fig. 27.18 Concentric conducting spheres.

or

$$E = \frac{1}{4\pi\kappa\varepsilon_0}\frac{Q_{\text{free}}}{r^2} \tag{33}$$

The potential difference between r_1 and r_2 is then

$$\Delta V = \int_{r_1}^{r_2}\frac{1}{4\pi\kappa\varepsilon_0}\frac{q_{\text{free}}}{r^2}\,dr = \frac{Q_{\text{free}}}{4\pi\kappa\varepsilon_0}\left(\frac{1}{r_1}-\frac{1}{r_2}\right)$$

and the capacitance

$$C = \frac{Q_{\text{free}}}{\Delta V} = \frac{4\pi\kappa\varepsilon_0}{1/r_1 - 1/r_2} \tag{34}$$

COMMENTS AND SUGGESTIONS: Equation (33) shows that the electric field of a spherical charge placed at the center of a spherical dielectric differs from the electric field of this charge in vacuum by a factor of $1/\kappa$. This result for the electric field is also approximately valid for the electric field near a small spherical charge surrounded by a large volume of dielectric of any shape whatsoever; if the dielectric is very large, then all points of its surface are far away from the central charge, and the shape of the surface has little effect on the electric field near the central charge.

In a more advanced study of dielectrics, it is useful to introduce a quantity **D** called the **electric displacement field**,

$$\mathbf{D} = \varepsilon_0 \kappa \mathbf{E}$$

In terms of this quantity, Gauss' Law becomes [see Eq. (32)]

Electric displacement field

$$\oint \mathbf{D}\cdot d\mathbf{S} = Q_{\text{free}} \tag{35}$$

This version of Gauss' Law is of more general validity than Eq. (32); for instance, it remains valid even in nonlinear dielectrics, where Eq. (32) fails.

27.5 Energy in Capacitors

Capacitors store not only electric charge, but also electric energy. Consider a two-conductor capacitor with charges $\pm Q$ on its plates. If the capacitor contains no dielectric, then the electric potential energy can be calculated directly from Eq. (26.7):

$$U = \tfrac{1}{2}QV_2 + \tfrac{1}{2}(-Q)V_1 = \tfrac{1}{2}Q(V_2 - V_1)$$

where V_1 and V_2 are the potentials of the plates. Thus, the potential energy can be expressed in terms of the charge and the potential difference,

Energy in capacitor

$$U = \tfrac{1}{2}Q\,\Delta V \tag{36}$$

By means of the definition of capacitance, $Q = C\,\Delta V$, this can be put in the alternative forms

$$U = \tfrac{1}{2}C(\Delta V)^2 \qquad (37)$$

or

$$U = \tfrac{1}{2}Q^2/C \qquad (38)$$

If the capacitor contains a dielectric, then the calculation of the energy is a bit more involved. The trouble is that the dielectric, with its bound charges, contributes to the electric potential energy. However, in practice we are usually interested not in the total potential energy, but only in that part of the potential energy that changes as we charge (or discharge) the capacitor, that is, we are interested only in the amount of work required to charge (or discharge) the capacitor. It turns out that this amount of work is correctly given by Eqs. (36)–(38), regardless of whether the capacitor contains a dielectric or not. The quantity Q in these equations is the charge on the plates, i.e., it is the *free* charge. To see this, let us derive Eq. (38) from a different starting point. Imagine that we charge the capacitor gradually, starting with an initial charge $q = 0$ and ending with a final charge $q = Q$. When the plates carry charges $\pm q$, the potential difference between them is q/C, and the work that we must perform to increase the charge on the plates by $\pm dq$ is

$$dU = \frac{q}{C}\,dq \qquad (39)$$

The total work that we must perform to charge the capacitor is then

$$U = \int_0^Q \frac{q}{C}\,dq = \frac{1}{C}\int_0^Q q\,dq = \tfrac{1}{2}Q^2/C \qquad (40)$$

This agrees with Eq. (38) and establishes its general validity. Note that, as the free charges on the capacitor plate increase, the bound charges within the dielectric rearrange themselves, becoming more strongly polarized; the energy required for this rearrangement is already included in Eq. (40) (the properties of the dielectric enter into this formula via the capacitance).

In Section 26.3 we obtained a formula for the energy density in an electric field in a vacuum. That calculation was based on an examination of the electric field in the space between the plates of a parallel-plate capacitor. By repeating this calculation for a parallel-plate capacitor with dielectric, we readily find that the **energy density in the dielectric** is

$$\boxed{u = \tfrac{1}{2}\kappa\varepsilon_0 E^2} \qquad (41)$$

Energy density in dielectric

Formally, this differs from Eq. (26.19) only by an extra factor of κ. Note, however, that the electric field **E** in Eq. (41) is the actual electric field in the dielectric, which already contains an implicit dependence on κ.

EXAMPLE 7. Consider the parallel-plate capacitor of Example 4. What is the stored potential energy? What is the energy density in the dielectric?

SOLUTION: By Eq. (36),

$$U = \tfrac{1}{2} Q\, \Delta V = \tfrac{1}{2} \times 2.3 \times 10^{-5} \text{ coulomb} \times 30 \text{ volts} = 3.4 \times 10^{-4} \text{ J}$$

The energy density can be calculated either from Eq. (41),

$$u = \tfrac{1}{2}\kappa \varepsilon_0 E^2 = \tfrac{1}{2} \times 2.3 \times 8.85 \times 10^{-12} \text{ F/m} \times (1.5 \times 10^6 \text{ volts/m})^2$$

$$= 23 \text{ J/m}^3$$

or else by taking the ratio of energy to volume,

$$u = \frac{U}{Ad} = \frac{3.4 \times 10^{-4} \text{ J}}{0.75 \text{ m}^2 \times 2.0 \times 10^{-5} \text{ m}} = 23 \text{ J/m}^3$$

SUMMARY

Capacitance of a pair of conductors: $C = Q/\Delta V$

Capacitance of parallel plates: $C = \varepsilon_0 A/d$

Parallel combination: $C = C_1 + C_2 + C_3 + \cdots$

Series combination: $1/C = 1/C_1 + 1/C_2 + 1/C_3 + \cdots$

Electric field in dielectric between parallel plates: $E = \dfrac{1}{\kappa} E_{\text{free}}$

Gauss' Law in dielectrics: $\oint \kappa \mathbf{E} \cdot d\mathbf{S} = Q_{\text{free}}/\varepsilon_0$

Energy in capacitor: $U = \tfrac{1}{2} Q\, \Delta V$

Energy density in dielectric: $u = \tfrac{1}{2}\kappa \varepsilon_0 E^2$

QUESTIONS

1. Commercially available large capacitors have a capacitance of 1000 μF. How is it possible that the capacitance of such a device is larger than the capacitance of the Earth?

2. A single-conductor capacitor may be regarded as a two-conductor capacitor with the second plate consisting of a very large conducting shell of infinite radius. Show that for $r_2 \to \infty$, Eq. (34), with $\kappa = 1$, reduces to Eq. (3).

3. Suppose we enclose the entire Earth in a conducting shell of a radius slightly larger than the Earth's radius. Explain why this would make the capacitance of the Earth much larger than the value calculated in Example 2.

4. Equation (8) shows that $C \to \infty$ as $d \to 0$. In practice, why can we not construct a capacitor of arbitrarily large C by making d sufficiently small? (Hint: What happens to E as $d \to 0$ while ΔV is held constant?)

5. If you put more charge on one plate of a parallel-plate capacitor than on the other, what happens to the extra charge?

6. Taking the fringing field into account, would you expect the capacitance of a parallel-plate capacitor to be larger or smaller than the value given by Eq.

(8)? (Hint: How does the fringing affect the density of field lines between the plates?)

7. Explain why there must be a fringing field in the region near the edges of a pair of parallel plates. (Hint: Suppose there were no fringing field, so that the field lines look as in Figure 27.19. Is $\oint \mathbf{E} \cdot d\mathbf{l} = 0$ for the path shown in this figure?)

8. Figure 27.20 shows a capacitor with a **guard rings.** These rings fit snugly around the edges of the capacitor plates, but they are not in electrical contact with the plates. In use, the potential on the rings is adjusted to the same value as the potential on the plates. Explain how the rings keep the field of the plates from fringing.

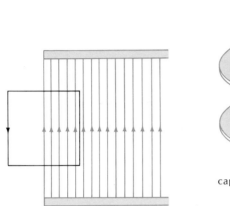

Fig. 27.19   **Fig. 27.20**

9. Figure 27.21 shows the design of an adjustable capacitor used in the tuning circuit of a radio. This capacitor can be regarded as several connected capacitors. Are these several capacitors connected in series or in parallel? If we turn the tuning knob (and the attached colored plates) counterclockwise, does the capacitance increase or decrease?

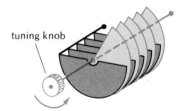

Fig. 27.21 The black plates are connected together, and the colored plates are connected together.

10. Consider a parallel-plate capacitor. Does the capacitance change if we insert a thin conducting sheet between the two plates, parallel to them?

11. Suppose we insert a thick slab of metal between the plates of a parallel-plate capacitor, parallel to the plates. Does the capacitance increase or decrease?

12. Consider a fluid dielectric that consists of molecules with permanent dipole moments. Will the dielectric constant increase or decrease if the temperature increases?

13. Figure 27.22 shows a dielectric slab partially inserted between the plates of a capacitor. Will the electric forces between the slab and the plates pull the slab into the region between the plates or push it out? (Hint: Consider the fringing field.)

14. If we increase the separation between the plates of a parallel-plate capacitor by a factor of 2, while holding the electric charge constant, by what factors will we change the electric field, the potential difference, the capacitance, and the electric energy?

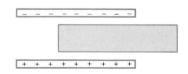

Fig. 27.22

15. Consider the parallel-plate capacitor with the slab of dielectric shown in Figure 27.17. How does the capacitance change if we move the slab up? If we move the slab to the right? If we tilt the slab?

16. Spell out the steps in the derivation of Eq. (41).

PROBLEMS

Section 27.1

1. Consider an isolated metallic sphere of radius R and another isolated metallic sphere of radius $3R$. If both spheres are at the same potential, what is the ratio of their charges? If both spheres carry the same charge, what is the ratio of their potentials?

2. The collector of an electrostatic machine is a metal sphere of radius 18 cm.
   (a) What is the capacitance of this sphere?
   (b) How many coulombs of charge must you place on this sphere to raise its potential to 2.0×10^5 V?

3. Your head is (approximately) a conducting sphere of radius 10 cm. What is the capacitance of your head? What will be the charge on your head if, by means of an electrostatic machine, you raise your head (and your body) to a potential of 100,000 V?

4. A capacitor consists of a metal sphere of radius 5 cm placed at the center of a thin metal shell of radius 12 cm. The space between is empty. What is the capacitance?

5. A capacitor consists of two parallel conducting disks of radius 20 cm separated by a distance of 1.0 mm. What is the capacitance? How much charge will this capacitor store if connected to a battery of 12 V?

6. A parallel-plate capacitor consists of two square conducting plates of area 0.04 m² separated by a distance of 0.2 mm. The plates are connected to the terminals of a 12-V battery.
   (a) What is the charge on each plate? What is the electric field between the plates?
   (b) If you move the plates apart, so their separation becomes 0.3 mm, how much charge will flow from each plate to the terminals of the battery? What will be the new electric field?

*7. What is the capacitance of the Geiger-counter tube described in Problem 25.29? Pretend that the space between the conductors is empty.

Section 27.2

8. What is the combined capacitance if three capacitors of 3.0, 5.0, and 7.5 μF are connected in parallel? What is the combined capacitance if they are connected in series?

*9. Three capacitors, with capacitances $C_1 = 5.0$ μF, $C_2 = 3.0$ μF, and $C_3 = 8.0$ μF, are connected as shown in Figure 27.23. Find the combined capacitance.

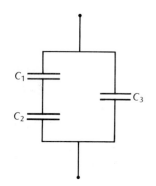

Fig. 27.23

*10. A multiplate capacitor, such as used in radios, consists of four parallel plates arranged one above the other as shown in Figure 27.24. The area of each plate is A, and the distance between adjacent plates is d. What is the capacitance of this arrangement?

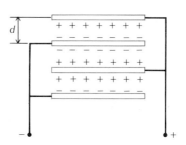

Fig. 27.24

*11. Three capacitors, of capacitances $C_1 = 4.0$ μF, $C_2 = 6.0$ μF, and $C_3 = 8.0$ μF, are connected as shown in Figure 27.25. What is the capacitance of this combination?

*12. Two capacitors, of 2.0 and 6.0 μF, respectively, are initially charged to 24 V by connecting each, for a few instants, to a 24-V battery. The battery is then removed and the charged capacitors are connected in a closed series circuit, the positive terminal of each capacitor being connected to the negative terminal of the other (Figure 27.26). What will be the final charge on each capacitor?

**13. Three capacitors, of capacitances $C_1 = 2.0$ μF, $C_2 = 5.0$ μF, and $C_3 = 7.0$ μF, are initially charged to 36 V by connecting each, for a few instants, to a 36-V battery. The battery is then removed and the charged capacitors are connected in a closed series circuit, with the positive and negative terminals joined as shown in Figure 27.27. What will be the final charge on each capacitor? What will be the voltage across the points PP' in Figure 27.27?

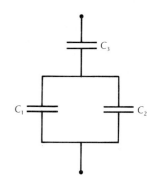

Fig. 27.25

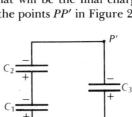

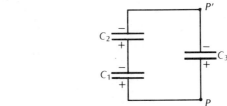

Fig. 27.26 Capacitors connected after they have been charged.

Fig. 27.27 Capacitors connected after they have been charged.

Section 27.3

14. You wish to construct a capacitor out of a sheet of polyethylene of thickness 5×10^{-2} mm and $\kappa = 2.3$ sandwiched between two aluminum sheets. If the capacitance is to be 3.0 μF, what must be the area of the sheets?

15. In order to measure the dielectric constant of a dielectric material, a slab of this material 1.5 cm thick is slowly inserted between a pair of parallel conducting plates separated by a distance of 2.0 cm. Before insertion of the dielectric, the potential difference across these capacitor plates is 3.0×10^5 V. During insertion, the charge on the plates remains constant. After insertion, the potential difference is 1.8×10^5 V. What is the value of the dielectric constant?

*16. A parallel-plate capacitor of plate area A and spacing d is filled with two parallel slabs of dielectric of equal thickness with dielectric constants κ_1 and κ_2, respectively (Figure 27.28). What is the capacitance? (Hint: Check that the configuration of Figure 27.28 is equivalent to two capacitors in series.)

*17. A capacitor with two large parallel plates of area A separated by a distance d is filled with two equal slabs of dielectric side by side (Figure 27.29). The dielectric constants are κ_1 and κ_2. What is the capacitance?

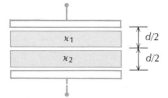

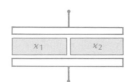

Fig. 27.28 A parallel-plate capacitor with two slabs of dielectric.

Fig. 27.29 A parallel-plate capacitor with two slabs of dielectric.

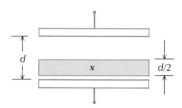

Fig. 27.30 A parallel-plate capacitor, partially filled with dielectric.

*18. Show that the result of Example 5 can also be derived by regarding the capacitor partially filled with dielectric as two capacitors in series, one completely filled with dielectric, one empty.

*19. A parallel-plate capacitor of plate area A and separation d contains a slab of dielectric of thickness $d/2$ (Figure 27.30) and dielectric constant κ. The potential difference between the plates is ΔV.
   (a) In terms of the given quantities, find the electric field in the empty region of space between the plates.
   (b) Find the electric field inside the dielectric.
   (c) Find the density of bound charge on the surface of the dielectric.

*20. Within some limits, the difference between the dielectric constants of air and of vacuum is proportional to the pressure of the air, i.e., $\kappa - 1 \propto p$. Suppose that a parallel-plate capacitor is held at a constant potential difference by means of a battery. What will be the percentage change in the amount of charge on the plates as we increase the air pressure between the plates from 1.0 atm to 3.0 atm?

*21. A parallel-plate capacitor is filled with carbon dioxide at 1 atm pressure. Under these conditions the capacitance is 0.5 μF. We charge the capacitor by means of a 48-V battery and then disconnect the battery so that the electric charge remains constant thereafter. What will be the change in the potential difference if we now pump the carbon dioxide out of the capacitor, leaving it empty?

**22. A parallel-plate capacitor is filled with a layer of distilled water 0.30 cm thick. The dipole moment of a water molecule is 6.1×10^{-30} C·m. Assume that the dipole moments of the water molecules are all perfectly aligned with the electric field. What is the surface-charge density of bound charges on the surface of the layer of water?

Section 27.4

23. A spherical capacitor consists of a metallic sphere of radius R_1 surrounded by a concentric metallic shell of radius R_2. The space between R_1 and R_2 is filled with dielectric having a constant κ. Suppose that the free surface-charge density on R_1 is $\sigma_{\text{free}(1)}$.
   (a) What is the free surface-charge density on the metallic sphere at R_2?
   (b) What is the bound surface-charge density on the dielectric at R_1?
   (c) What is the bound surface-charge density on the dielectric at R_2?

*24. A long cylindrical copper wire of radius 0.20 cm is surrounded by a cylindrical sheath of rubber of inner radius 0.20 cm and outer radius 0.30 cm. The rubber has $\kappa = 2.8$. Suppose that the surface of the copper has a free charge density of 4.0×10^{-6} C/m^2.
   (a) What will be the bound charge density on the inside surface of the rubber sheath? On the outside surface?
   (b) What will be the electric field in the rubber near its inner surface? Near its outer surface?
   (c) What will be the electric field just outside the rubber sheath?

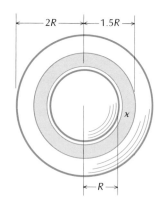

Fig. 27.31 A spherical capacitor, partially filled with dielectric.

*25. A metallic sphere of radius R is surrounded by a concentric dielectric shell of inner radius R, outer radius $3R/2$. This is surrounded by a concentric, thin, metallic shell of radius $2R$ (Figure 27.31). The dielectric constant of the shell is κ. What is the capacitance of this contraption?

*26. Two small metallic spheres are submerged in a large volume of transformer oil of dielectric constant $\kappa = 3.0$. The spheres carry electric charges of 2.0×10^{-6} C and 3.0×10^{-6} C, respectively, and the distance between them is 0.60 m. What is the force on each?

Fig. 27.32 A brass sphere afloat in a lake of oil.

**27. A sphere of brass floats in a large lake of oil of dielectric constant $\kappa = 3.0$. The sphere is exactly halfway immersed in the oil (Figure 27.32). The sphere has a net charge of 2.0×10^{-6} C. What fraction of this electric charge will be on the upper hemisphere? On the lower? (Hint: For the path shown in Figure 27.32, $\oint \mathbf{E} \cdot d\mathbf{l} = 0$; from this prove that the electric fields in the oil and in the air above the oil will be exactly the same.)

**28. A spherical capacitor consists of two concentric spheres of metal of radii R_1 and R_2. The space between these spheres is filled with two kinds of dielectric (Figure 27.33); the dielectric in the upper hemisphere has a constant κ_1, and the dielectric in the lower hemisphere has a constant κ_2. What is the capacitance of this device? (Hint: For the path shown in Figure 27.33 $\oint \mathbf{E} \cdot d\mathbf{l} = 0$; from this prove that the electric fields in both dielectrics are exactly the same.)

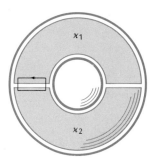

Fig. 27.33 A spherical capacitor with two kinds of dielectric.

**29. In a semiconductor with impurity ions, electrons will orbit around these ions. The sizes of the orbits are considerably larger than the spacings between the semiconductor atoms, and hence an electron may be regarded as moving through a more or less uniform medium of a given dielectric constant.
  (a) Show that the electric force of attraction between an electron and an ion of charge e immersed in a medium of dielectric constant κ is

$$F = \frac{1}{4\pi\varepsilon_0 \kappa} \frac{e^2}{r^2}$$

  (b) Calculate the orbital energy of an electron in a circular orbit according to Bohr's theory described in Problem 25.20.
  (c) The dielectric constant of germanium is $\kappa = 15.8$. Evaluate the orbital energy of an electron moving around an ion embedded in germanium; assume that the electron is in the smallest Bohr orbit. By what factor does your result differ from what you would obtain for an electron moving around this ion in a vacuum?

Section 27.5

30. A parallel-plate capacitor has a plate area of 900 cm² and a plate separation of 0.50 cm. The space between the plates is empty.
  (a) What is the capacitance?
  (b) What is the potential difference if the charges on the plates are $\pm 6.0 \times 10^{-8}$ C?
  (c) What is the electric field between the plates?
  (d) The energy density?
  (e) The total energy?

31. Repeat Problem 30, under the assumption that the space between the plates is filled with Plexiglas.

32. A TV receiver contains a capacitor of 10 μF charged to a potential difference of 2×10^4 V. What is the amount of charge stored in this capacitor? The amount of energy?

33. Two parallel conducting plates of area 0.5 m² placed in a vacuum have a potential difference of 2.0×10^5 V when charges of $\pm 4.0 \times 10^{-3}$ C are placed on them, respectively.
  (a) What is the capacitance of the pair of plates?
  (b) What is the distance between them?
  (c) What is the electric field between them?
  (d) What is the electric energy?

34. A large capacitor has a capacitance of 20 μF. If you want to store an electric energy of 40 J in this capacitor, what potential difference do you need?

*35. A parallel-plate capacitor has plates of area 0.04 m² separated by a distance of 0.5 mm. Initially, the space between the plates is empty and the capacitor is uncharged. A 12-volt battery is available for connection to the capacitor. For each of the following cases, find the resulting capacitance C, po-

tential difference V, and charge Q on the plates; also find the energy delivered by the battery:
   (a) The battery is connected to the capacitor.
   (b) While the battery remains connected, a slab of dielectric with $\kappa = 3$ is inserted between the plates, completely filling the space between them.
   (c) The battery is disconnected, and the dielectric is pulled out from the capacitor.
   (d) The capacitor is discharged, and the battery is again connected to the empty capacitor for some moments. The battery is then disconnected, and the dielectric is inserted between the plates of the capacitor.

*36. Two capacitors of 5.0 μF and 8.0 μF are connected in series to a 24-V battery. What is the energy stored in the capacitors?

*37. A parallel-plate capacitor without dielectric has an area A and a charge $\pm Q$ on each plate.
   (a) What is the electric force F of attraction between the plates?
   (b) How much work must you do against this force in order to increase the plate separation by an amount Δl?
   (c) By means of Eq. (38), calculate the change of ΔU in potential energy during this change.
   (d) By comparing (a) and (c), check that $F = -\Delta U/\Delta l$.
   (Hint: One-half of the electric field between the plates is due to one plate, and one-half is due to the other. Consequently, in the calculation of the electric force on a plate from the product of field times charge, only one-half of the field must be used; the other half gives the electric force of the plate on *itself* and is of no interest.)

*38. Power companies are interested in the storage of surplus electric energy. Suppose we wanted to store 10^6 kW·h of electric energy (half a day's output for a large power plant) in a large parallel-plate capacitor filled with a plastic dielectric with $\kappa = 3.0$. If the dielectric can tolerate a maximum electric field of 5×10^7 V/m, what is the minimum total volume of dielectric needed to store this energy?

*39. Three capacitors are connected as shown in Figure 27.34. Their capacitances are $C_1 = 2.0$ μF, $C_2 = 6.0$ μF, and $C_3 = 8.0$ μF. If a voltage of 200 V is applied to the two free terminals, what will be the charge on each capacitor? What will be the energy in each?

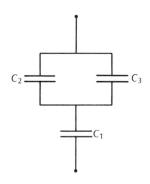

Fig. 27.34

*40. Ten identical 5-μF capacitors are connected in parallel to a 240-V battery. The charged capacitors are then disconnected from the battery and reconnected in series, the positive terminal of each capacitor being connected to the negative terminal of the next. What is the potential difference between the negative terminal of the first capacitor and the positive terminal of the last capacitor? If these two terminals are connected via an external circuit, how much charge will flow around this circuit as the series arrangement discharges? How much energy will be released in the discharge? Compare this charge and this energy with the charge and energy stored in the original, parallel arrangement, and explain any discrepancies.

CHAPTER 28

Currents and Ohm's Law

Under static conditions, there can exist no electric field inside a conductor. But suppose that we suddenly deposit opposite amounts of electric charge on the opposite ends of a long metallic conductor, such as a wire. The conductor will then not be in electrostatic equilibrium, and the charges at the ends will generate an electric field along and inside the conductor (Figure 28.1). This electric field propels the charges toward each other. When the charges meet, they cancel. The electric field then disappears — the conductor reaches equilibrium.

For a good conductor, such as copper, the approach to equilibrium is fairly rapid; typically, the time required to achieve equilibrium is a small fraction of a second. However, we can keep a conductor in a permanent state of disequilibrium if we continually supply more electric charge to its ends. For example, we can connect the two ends of a copper wire to the terminals of a battery or an electric generator. The terminals of such a device act as source and sink of electric charge, just as the outlet and the intake of a pump act as source and sink of water. Under these conditions, electric charge will continually flow from one terminal to the other, forming an electric current.[1]

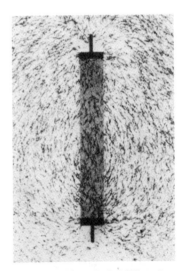

Fig. 28.1 Electric field lines in and near a straight conductor not in equilibrium. The field lines have been made visible by sprinkling grass seeds on the surface of paper on which the conductor has been painted with conducting paint. (From O. Jefimenko, *Am. J. Phys.* **30**, 19, 1962.)

28.1 Electric Current

When a wire is connected between the two terminals of a battery or generator, the electric charges are propelled from one end of the wire to the other by the electric field that exists along and within the wire.

---

[1] We will discuss the inner workings of batteries, generators, and other "pumps" for electric charge in the next chapter.

Most of the field lines originate at the terminals of the battery or generator, but some field lines originate at charges that have accumulated on the wire itself. As Figures 28.1 and 28.2 show, the field lines tend to concentrate within the conductor, and they tend to follow the conductor. If the conductor has no sharp kinks, the field lines are uniformly distributed over the cross-sectional area of the conductor. For instance, if the conductor is a more or less straight wire of constant thickness, then the electric field inside the wire will be of constant magnitude and of a direction parallel to the wire. If the length of the wire is l and if the battery or generator maintains a difference of potential ΔV across its ends, then this constant electric field in the wire has a magnitude

Electric field in uniform wire

$$E = \Delta V / l \qquad (1)$$

This electric field causes the flow of charge, or the **electric current,** from one end of the wire to the other. Before we can explore the dependence of the current on the electric field, we need a precise definition of the current. Suppose that an amount of charge dq flows past some given point of the wire (for instance, the end of the wire) in a time dt; then the electric current is defined as charge divided by time,

Electric current

$$I = \frac{dq}{dt} \qquad (2)$$

Note that if the sides of the wire do not leak (good insulation), then the conservation of electric charge requires that the current be the same everywhere along the wire; thus, the current is simply the rate at which charge enters the wire at one end or the rate at which charge leaves at the other end.

The SI unit of current is the **ampere** (A); this is a flow of charge of one coulomb per second,

$$1 \text{ ampere} = 1 \text{ A} = 1 \text{ C/s} \qquad (3)$$

In metallic conductors the charge carriers are electrons — a current in a metal is nothing but a flow of electrons. In electrolytes the charge carriers are positive ions, negative ions, or both — a current in such a conductor is a flow of ions. For the sake of mathematical uniformity, whenever we need to indicate the direction of the current along a conductor, we will follow the convention that the current has the direction of a hypothetical positive flow of charge. This means that we pretend that the moving charges are always positive charges. Of course, in metals the moving charges are actually negative charges (electrons), and hence the above convention assigns to the current a direction opposite to that of the true motion of the charges. However, as regards the transfer of charge, the transport of negative charge in one direction is equivalent to the transport of positive charge in the opposite direction. Our convention for labeling the direction of the current takes advantage of this equivalence.

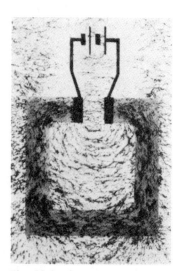

Fig. 28.2 Electric field lines in and near a rectangular conductor connected to a battery carrying an electric current. (From O. Jefimenko, *Am. J. Phys.* **30,** 19, 1962.)

If we divide the current in a conductor by the cross-sectional area of the conductor (Figure 28.3), we obtain the **current density**,[2]

$$j = I/A \qquad (4)$$

Current density

This is really the *average* current density over the area A. We can also define a local current density in terms of the current dI that flows across an infinitesimal portion of dA of cross-sectional area (Figure 28.3),

$$j = \frac{dI}{dA} \qquad (5)$$

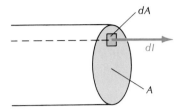

Fig. 28.3 A conductor carries a current from left to right. The conductor has been cut off on the right, so as to give a clear view of its cross section.

Under normal conditions, the current in a wire is uniformly distributed over the entire cross-sectional area of the wire. In terms of the local current density, this means that j is constant over the entire cross-sectional area.

28.2 Resistance and Ohm's Law

We will now examine in detail the behavior of a current in a metallic conductor. Such conductors contain a vast number of free electrons; for example, copper has about 8×10^{22} free electrons per cubic centimeter. These electrons form a gas which fills the entire volume of the metal. Of course, in an electrically neutral conductor, the negative charge of the free electrons is exactly balanced by the positive charge of the ions that make up the crystal lattice of the metal. A current in a metallic conductor is simply a flow of the gas of electrons, while the ions remain at rest.

The flow of the gas of electrons along a metallic wire is analogous to the flow of water along a canal leading down a gentle slope. In such a canal, the force of gravity acting on the water has a component along the canal; this component pushes the water along. But the water does not accelerate — friction between the water and the walls of the canal opposes the motion, and the water moves at a constant speed because the friction exactly matches the push of gravity.

Likewise, the electric field in the wire pushes the gas of electrons along. But the gas of electrons does not accelerate — friction between the gas and the body of the wire opposes the motion, and the gas moves at a constant speed because the friction exactly matches the push of the electric field.

The analogy between the motion of water and the motion of the electron gas extends to the motion of individual water molecules and individual electrons. Although the water in a canal usually has a fairly low speed, perhaps a few meters per second, the individual molecules within the water have a rather high speed — the typical speed of the random thermal motion of water molecules is about 500 m/s at ordi-

---

[2] Do not confuse the symbol A for area with the abbreviation A for ampere.

Fig. 28.4 Path of a water molecule in a canal. The molecule gradually drifts from left to right.

nary temperatures. But since this thermal motion consists of rapid, random zigzags which are just as likely to move the molecule backward as forward, this high speed does not contribute to the net downhill motion of the water. Figure 28.4 shows the motion of a water molecule in the canal; on a microscopic scale, this motion consists of rapid zigzags on which is superimposed a much slower "drift" along the canal.

Likewise, the electron gas moves along the wire at a rather low speed, perhaps 10^{-2} m/s, but the individual electrons have a much higher speed — the typical speed of the random motion of electrons in a metal is about 10^6 m/s (this very high speed is due to quantum-mechanical effects, which we cannot discuss here). Thus, the net motion of an electron also consists of rapid zigzags on which is superimposed a much slower "drift" motion along the wire. Qualitatively, the motion resembles the path of a water molecule shown in Figure 28.4, but the amount of drift per zigzag is even less than that shown in this figure.

The friction between the electron gas and the body of the wire is caused by collisions between the electrons and the ions of the crystal lattice of the wire. An electron moving through a piece of copper will suffer about 10^{14} collisions with ions per second. Each collision slows the electron down, brings it to a stop, or reverses its motion. Because of the disturbing effects of these collisions, the electron never gains much velocity from the electric field that is attempting to accelerate it. The collisions dissipate the kinetic energy that the electron receives from the electric field. The dissipated kinetic energy of the electrons remains in the crystal lattice in the form of random kinetic and potential energy of the ions, that is, it remains as heat.

The average velocity, or drift velocity, that an electron attains in the electric field is proportional to the strength of the electric field,

$$v_d \propto E \tag{6}$$

This proportionality merely reflects the fact that if the electric field is strong, the electron gains more velocity between one collision and the next, and therefore attains a larger average velocity. The electric current carried by the wire is proportional to the average velocity of the electrons,

$$I \propto v_d \propto E \tag{7}$$

The current is also proportional to the cross-sectional area of the wire, because a large cross-sectional area means that more electrons participate in the transport of charge. Hence

$$I \propto AE \tag{8}$$

With $E = \Delta V/l$, this proportionality becomes

$$I \propto \frac{A}{l} \Delta V \tag{9}$$

To transform this into an equality, we rewrite it as

$$I = \frac{1}{\rho} \frac{A}{l} \Delta V \tag{10}$$

where ρ is a constant of proportionality that depends on the characteristics of the material of the wire. This constant is called the **resistivity** of the material.

Resistivity

It is customary to define the **resistance** of the wire as

$$R = \rho \frac{l}{A} \qquad (11)$$

Resistance

Equation (10) can then be expressed in the convenient form

$$I = \frac{\Delta V}{R} \qquad (12)$$

Ohm's Law

This equation is called **Ohm's Law.** It asserts that *the current is proportional to the potential difference between the ends of the conductor.* Note that in Eq. (12) the resistance plays the role of a constant of proportionality. For a wire of uniform cross section, the resistance can be calculated from the simple formula (11). But Ohm's Law is also valid for conductors of arbitrary shape — such as wires of nonuniform cross section — for which the resistance must be calculated from a more complicated formula tailored to the shape and the size of the conductor.

Ohm's Law is valid for metallic conductors and also for nonmetallic conductors (for example, carbon) in which the current is carried by a flow of electrons (the basic mechanism for conduction in metals and nonmetals is the same; the only difference is that the nonmetals have very few free electrons). It is even valid for plasmas and for electrolytes, in which the current is carried by a flow of both electrons and ions. However, we ought to keep in mind that in spite of its wide range of applicability, Ohm's Law is not a general law of nature — such as Gauss' Law — but only an assertion about the electrical properties of some conducting materials.

Georg Simon Ohm, *1787–1854, German physicist, professor at Munich. Ohm was led to his law by an analogy between the conduction of electricity and the conduction of heat: the electric field is analogous to the temperature gradient, and the electric current is analogous to the heat flow. Equation (28.12) is then analogous to Eq. (20.5).*

28.3* The Flow of Free Electrons

Before we deal with some applications of Ohm's Law, let us reexamine the derivation of this law and fill in the constants of proportionality that we left out in Eqs. (6)–(9). To do this, we need to calculate the average motion of the free electrons in a metal in some detail. The high-speed thermal motion of the electrons does not enter directly into the calculation of the average motion, because it is random; however, it enters indirectly, because it determines the collision rate. The large collision rate of a free electron in a metal — for instance, 10^{14} collisions per second for an electron in copper — results from the high speed of the random motion: the electron moves a large distance per second and therefore encounters many ions with which to collide. Since the extra speed that the electron gains from the electric field is

---

* This section is optional.

very small compared with the random speed, the collision rate is nearly unaffected by the electric field. The average time interval τ between collisions is therefore a constant that depends only on the characteristics of the metal. (The average time interval between collisions is related to the mean free path discussed in Section 19.5; it roughly equals the mean free path divided by the average speed. However, for our present purposes, we find it more convenient to use the average time interval than the mean free path.)

We can find the average motion of an electron by examining the losses and gains of momentum of this electron. If the average velocity, or drift velocity, of an electron is v_d, then the average momentum is $m_e v_d$. We expect that, on the average, a collision will absorb all of this momentum, that is, a collision will destroy the forward drift velocity and leave the electron with only the random thermal motion. This means that, in a time interval τ, the electron loses a momentum $m_e v_d$; and the average rate at which the electron loses momentum in collisions is therefore

$$\left(\frac{\Delta p}{\Delta t}\right)_{\text{loss}} = \frac{m_e v_d}{\tau} \qquad (13)$$

On the other hand, the rate at which the electron gains momentum by the action of the electric force is

$$\left(\frac{\Delta p}{\Delta t}\right)_{\text{gain}} = -eE \qquad (14)$$

Under steady-state conditions, the rate of loss of momentum must match the rate of gain. By setting the right sides of Eqs. (13) and (14) equal we immediately obtain

Drift velocity

$$\boxed{v_d = -eE\tau/m_e} \qquad (15)$$

This is the average velocity with which the electron gas flows along the wire. As expected, this velocity is proportional to the strength of the electric field. The negative sign in Eq. (15) indicates that the direction of flow is opposite to the direction of the electric field.

EXAMPLE 1. A potential difference of 3.0 V is applied to the ends of a copper wire 0.5 m long. What is the drift velocity of the free electrons in the wire? In copper at room temperature, the average time interval between collisions is $\tau = 2.7 \times 10^{-14}$ s.

SOLUTION: The electric field in the wire is $E = 3.0 \text{ V}/0.5 \text{ m} = 6.0 \text{ V/m}$. Hence Eq. (15) gives

$$v_d = -\frac{1.6 \times 10^{-19} \text{ C} \times 6.0 \text{ V/m} \times 2.7 \times 10^{-14} \text{ s}}{9.1 \times 10^{-31} \text{ kg}}$$

$$= -2.8 \times 10^{-2} \text{ m/s}$$

COMMENTS AND SUGGESTIONS: Note that this speed is rather low — it takes an electron about a third of a minute to wander from one end of this wire to the other. Nevertheless, if a pulse of current (a signal) is suddenly injected into the wire at one end, a similar pulse of current will emerge from the far end *almost instantaneously.* What happens in this case is that, by means of their electric fields, electrons push on neighboring electrons, and a compressional wave travels through the electron gas; this ejects electrons out of the far end almost instantaneously. The phenomenon is analogous to the propagation of a wave on the water of a canal; the wave travels much faster than the flow of water.

To find the electric current that corresponds to the flow of the electron gas, we need to take into account the number of electrons. Suppose that the metal of the wire has n free electrons per unit volume. The quantity n depends on the metal; for copper $n = 8.5 \times 10^{28}/\text{m}^3$. If the wire has a cross-sectional area A and a length l, then its total number of free electrons is $n \times [\text{volume}] = nAl$ and the total charge associated with these electrons is

$$\Delta q = -enAl \tag{16}$$

It takes a time $\Delta t = l/|v_d|$ for all of these electrons to emerge at one end of the wire. Hence the current in the wire is

$$I = \frac{|\Delta q|}{\Delta t} = \frac{enAl}{l/|v_d|} = \frac{e^2 n \tau}{m_e} AE \tag{17}$$

In this equation, we have ignored the negative sign on the electric charge because the direction of the current is to be reckoned according to the convention described in Section 28.1 and not according to the motion of the electrons. The current is in the same direction as the electric field.

According to Eq. (17), the current is directly proportional to the magnitude of the electric field. It is also directly proportional to the cross-sectional area of the wire. All of this agrees with our earlier, rough argument.

In terms of the potential difference $\Delta V = El$ across the ends of the wire, we can rewrite Eq. (17) as follows:

$$I = \frac{e^2 n \tau}{m_e} AE = \frac{e^2 n \tau}{m_e} \frac{A}{l} (El) \tag{18}$$

$$= \left(\frac{e^2 n \tau}{m_e} \frac{A}{l} \right) \Delta V \tag{19}$$

If we write the factor in parentheses as

$$R = \frac{m_e}{e^2 n \tau} \frac{l}{A} \tag{20}$$

then Eq. (19) becomes

$$I = \frac{\Delta V}{R} \tag{21}$$

which is Ohm's Law. Thus, Eq. (20) gives us a theoretical expression for the resistance in terms of the physical parameters associated with the free-electron gas. In practice, Eq. (20) is not very useful, because there is no direct method for measuring the time per collision τ.

28.4 The Resistivity of Materials

As we saw in Section 28.2, the resistance of a wire of uniform cross section is related to the resistivity by the formula

$$R = \rho \frac{l}{A} \tag{22}$$

We can use this formula to calculate the resistance if the resistivity of the material is known, and we can also use it to calculate the resistivity if the resistance is known. The latter calculation is important in the experimental determination of the resistivity of a material, which is done by measuring the potential difference and current in a wire of given length and cross section made of a sample of the material. This means that Ohm's Law is used both as a definition of resistance and as a law relating current, potential difference, and resistance. In the definitions of force (Section 5.2) and of temperature (Section 19.2) we have already encountered similar instances of such dual uses of laws.

As is obvious from Ohm's Law, the unit of resistance is 1 volt/ampere; this unit is called **ohm (Ω),**

Ohm, Ω

$$1 \text{ ohm} = 1 \text{ } \Omega = 1 \text{ volt/ampere} \tag{23}$$

The unit of resistivity is the ohm-meter ($\Omega \cdot$ m). Table 28.1 gives the resistivities of some conducting materials.

Conductivity

The inverse of the resistivity is called the **conductivity.** The unit of conductivity is 1/ohm-meter, or $\Omega^{-1} \cdot$ m^{-1}. Some tables of properties of materials list the conductivities of conductors rather than the resistivities.

Table 28.1 RESISTIVITIES OF METALS[a]

Material	ρ	Increase in ρ per °C
Silver	1.6×10^{-8} $\Omega \cdot$ m	0.38%
Copper	1.7×10^{-8}	0.39%
Aluminum	2.8×10^{-8}	0.39%
Brass	$\sim 7 \times 10^{-8}$	0.2%
Nickel	7.8×10^{-8}	0.6%
Iron	10×10^{-8}	0.5%
Steel	$\sim 11 \times 10^{-8}$	0.4%
Constantan	49×10^{-8}	0.001%
Nichrome	100×10^{-8}	0.04%

[a] At a temperature of 20°C.

EXAMPLE 2. A wire commonly used for electrical installations in homes is No. 10 copper wire, which has a radius of 0.129 cm. What is the resistance of a piece of this wire 30 m long? What is the potential drop along this wire if it carries a current of 10 A?

SOLUTION: The cross-sectional area of the wire is

$$A = \pi r^2 = \pi (0.129 \times 10^{-2} \text{ m})^2 = 5.2 \times 10^{-6} \text{ m}^2$$

By Eq. (22), the resistance will be

$$R = \rho l/A = 1.7 \times 10^{-8} \, \Omega \cdot m \times 30 \, m/5.2 \times 10^{-6} \, m^2 = 0.098 \, \Omega \quad (24)$$

For a current of 10 A, Ohm's Law then gives a potential drop

$$\Delta V = IR = 10 \, A \times 0.098 \, \Omega = 0.98 \, \text{volt} \quad (25)$$

By combining Eq. (12) and (22), we readily find

$$\frac{I}{A} = \frac{1}{\rho} \frac{\Delta V}{l} \quad (26)$$

Since I/A is the current density and $\Delta V/l$ is the electric field in the conductor, Eq. (26) can be written

$$j = \frac{1}{\rho} E \quad (27)$$

Thus, the current density is directly proportional to the electric field. This is an alternative expression for Ohm's Law. Although our derivation of Ohm's Law began with a wire of uniform cross section, Eq. (27) is valid for conductors of arbitrary shape. This equation relates two *local* quantities: the current density at one point within the conductor and the electric field at that point.

The resistivity of materials depends somewhat on temperature. In ordinary metals, the resistivity increases slightly with temperature. This is due to an increase in the rate of collision between the moving electrons and atoms of the lattice — at high temperature the atoms jump violently around their positions in the lattice, and they are then more likely to disturb the motion of the electrons. The numbers in the first column of Table 28.1 give the resistivity at room temperature (20°C). The numbers in the second column give the percentage increase in the resistivity per degree Celsius.

Dependence of resistivity on temperature

EXAMPLE 3. Suppose that because of a current overload, the temperature of the copper wire of Example 2 increases from 20°C to 50°C. How much does the resistance increase?

SOLUTION: The temperature increase is 30°C. According to Table 28.1, the resistance of copper increases by 0.39% for a temperature increase of 1°C. Hence the resistance increases by $30 \times 0.39\% = 12\%$ for a temperature increase of 30°C. The change of resistance is therefore

$$\Delta R = 0.098 \, \Omega \times 0.12 = 0.012 \, \Omega$$

and the new resistance of the wire will be

$$0.098 \, \Omega + 0.012 \, \Omega = 0.110 \, \Omega$$

At very low temperatures, the resistivity of a metal is substantially less than at room temperature. Some metals, such as lead, tin, zinc, and niobium, exhibit the phenomenon of **superconductivity**: their resistance vanishes completely as the temperature approaches absolute

Superconductivity

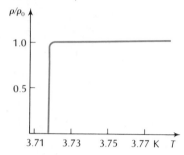

Fig. 28.5 Resistivity of tin as a function of temperature. Below 3.72 K, the resistivity is zero. The resistivity has been expressed as a fraction of the resistivity at 4.2 K, the temperature of liquefaction of helium.

zero. For example, Figure 28.5 shows a plot of resistivity vs. temperature for tin; at an absolute temperature of 3.72 K, the resistivity abruptly vanishes. The resistance of such a superconductor is *exactly* zero. In one experiment, a current of several hundred amperes was started in a superconducting ring; the current continued on its own with undiminished strength for over a year, without any battery or generator to maintain it. In Interlude VIII we will present a detailed discussion of the properties of superconductors.

According to the definition that we gave in Section 22.4, an ideal insulator is a material that does not permit any motion of electric charge. Real insulators, such as porcelain or glass, do permit some very slight motion of charge. What distinguishes them from conductors is their enormously large resistivity. Typically, the resistivity of insulators is more than 10^{20} times as large as that of conductors (see Table 28.2). This means that even when we apply a high voltage to a piece of glass, the flow of current will be insignificant (provided, of course, that the material does not suffer electrical breakdown). In fact, on a humid day it is likely that more current will flow along the microscopic film of water that tends to form on the surface of the insulator than through the insulator itself.

A semiconductor is a material with a resistivity between that of conductors and insulators. The resistivities of semiconductors vary over a wide range; the resistivities may be 10^4 to 10^{15} times as large as the resistivities of conductors (see Table 28.3).

n-type and p-type semiconductors

Semiconductors fall into two categories: *n* type and *p* type. In an *n*-type semiconductor, the carriers of current are free electrons, as in a metal. However, the resistance is higher because the semiconductor has fewer free electrons than a metal. In a *p*-type semiconductor, the carriers of current are "holes" of positive charge. Such a hole is merely the *absence* of an electron in the uniform background distribution of electrons. Since the negative charge of the uniform distribution of electrons balances the positive charge of the ions of the semiconductor material, the absence of an electron at some location implies a preponderance of positive charge at this location. Thus, a hole in the electron distribution effectively acts as a positive charge; and if the hole moves, it acts as a carrier of current.

Table 28.2 Resistivities of Insulators

Material	ρ
Polyethylene	$2 \times 10^{11} \; \Omega \cdot m$
Glass	$\sim 10^{12}$
Porcelain, unglazed	$\sim 10^{12}$
Rubber, hard	$\sim 10^{13}$
Epoxy	$\sim 10^{15}$

Table 28.3 Resistivities of (Pure) Semiconductors

Material	ρ
Silicon	$2.6 \times 10^3 \; \Omega \cdot m$
Germanium	4.2×10^{-1}
Carbon (graphite)	3.5×10^{-5}

Semiconductors usually contain both free electrons and free holes. Whether a semiconductor is n type or p type depends on which kind of charge carrier dominates. The concentration of free electrons and of free holes is largely determined by the impurities that are present in the material. It is a characteristic feature of semiconductors that the addition of impurities to the material has a drastic effect on the resistivity. For instance, the silicon used in electronic devices is often "doped" with small amounts of arsenic or boron; the addition of just one part per million of arsenic will decrease the resistivity of silicon by a factor of more than 10^5. Pure semiconductor materials are hardly ever used in practical applications. It is usually the presence of impurities that gives the semiconductor materials their interesting electric properties. (Further details on semiconductors will be presented in Chapter 44.)

28.5 Resistances in Combination

The metallic wires of any electric circuit have some resistance. But in electronic devices — radios, televisions, amplifiers — the main contribution to the resistance is usually due to devices that are specifically designed to have a high resistance. These devices are called **resistors,** and they are used to control and modify the currents. Resistors are commonly made out of a short piece of pure carbon (graphite) connected between two terminals. Carbon has a high resistivity, and hence a small piece of carbon can have a higher resistance than a long piece of metallic wire. Such resistors obey Ohm's Law (current proportional to potential difference) for a wide range of values of the current; of course, if the resistor is overloaded with current, it will heat up, possibly even burn, and Ohm's Law will fail.

In circuit diagrams, the symbol for a resistor is a zigzag line, reminiscent of the path of an electron inside a wire (Figure 28.6).

For resistors, as for capacitors, the two simplest ways of connecting several resistors are in series and in parallel. Figure 28.7 shows two resistors connected in *series*. It is intuitively obvious that the net resistance of this combination is the sum of the individual resistances,

$$R = R_1 + R_2 \tag{28}$$

The formal derivation of this result begins with the observation that if the potential differences across the individual resistors are ΔV_1 and ΔV_2, then the net potential difference across the combination is

$$\Delta V = \Delta V_1 + \Delta V_2$$

Furthermore, the currents in both resistors must be exactly the same. Hence, by Ohm's Law,

$$\Delta V = IR_1 + IR_2 = I(R_1 + R_2)$$

From this it is clear that the resistance of the combination is equivalent to a single resistance given by Eq. (28).

Fig. 28.6 (a) Different kinds of resistors used in electric circuits. The bands pointed on the resistors specify the resistance in a color code. (b) Symbol for a resistor in a circuit diagram.

Fig. 28.7 Two resistors connected in series.

Series combination of resistors

The generalization of this result to any number of resistors in series (Figure 28.8) is obvious. The net resistance of the series combination is

$$R = R_1 + R_2 + R_3 + \cdots \quad (29)$$

Figure 28.9 shows two resistors in *parallel*. The potential difference across each resistor is the same as the potential difference across the combination. Hence the currents are

$$I_1 = \frac{\Delta V}{R_1} \quad \text{and} \quad I_2 = \frac{\Delta V}{R_2} \quad (30)$$

The total current through the combination is the sum of the individual parallel currents,

$$I = I_1 + I_2 = \frac{\Delta V}{R_1} + \frac{\Delta V}{R_2} = \left(\frac{1}{R_1} + \frac{1}{R_2}\right)\Delta V \quad (31)$$

The resistance of the combination is therefore equivalent to a single resistance given by

$$\frac{1}{R} = \frac{1}{R_1} + \frac{1}{R_2} \quad (32)$$

Fig. 28.8 Several resistors connected in series.

Note that the resistance of the parallel combination is less than each of the individual resistances. For example, if $R_1 = R_2 = 1.0\ \Omega$, then $R = 0.5\ \Omega$.

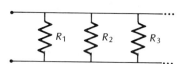

Fig. 28.10 Several resistors connected in parallel.

Fig. 28.9 Two resistors connected in parallel.

The generalization of this to any number of resistors in parallel (Figure 28.10) is again obvious. The net resistance of the parallel combination is given by

Parallel combination of resistors

$$\frac{1}{R} = \frac{1}{R_1} + \frac{1}{R_2} + \frac{1}{R_3} + \cdots \quad (33)$$

EXAMPLE 4. Two resistors, with $R_1 = 10\ \Omega$ and $R_2 = 20\ \Omega$, are connected in parallel (see Figure 28.9). A total current of 1.8 A flows through this combination. What is the potential difference across the combination? What is the current in each resistor?

SOLUTION: According to Eq. (32), the net resistance is given by

$$\frac{1}{R} = \frac{1}{R_1} + \frac{1}{R_2} = \frac{1}{10\ \Omega} + \frac{1}{20\ \Omega} = \frac{30}{200\ \Omega}$$

that is, $R = 6.67\ \Omega$. Hence, the potential difference is

$$\Delta V = IR = 1.8 \text{ A} \times 6.67 \text{ }\Omega = 12.0 \text{ volts}$$

and the individual currents are

$$I_1 = \Delta V/R_1 = 12.0 \text{ volts}/10 \text{ }\Omega = 1.20 \text{ A}$$

$$I_2 = \Delta V/R_2 = 12.0 \text{ volts}/20 \text{ }\Omega = 0.60 \text{ A}$$

COMMENTS AND SUGGESTIONS: Note that the sum of the currents I_1 and I_2 is $1.20 \text{ A} + 0.60 \text{ A} = 1.80 \text{ A}$, as it should be.

In the calculations of the preceding example, we neglected the resistances of the wires connecting the resistors. This is a good approximation if the resistances of these wires are small compared with the resistances of the resistors. If this is not so, then we must take the resistances of the wires into account in our calculation. In circuit diagrams, it is customary to represent the resistance of a wire schematically by an equivalent resistor of suitable magnitude; this means we pretend that all of the resistance of a wire is concentrated, or lumped, in one place. The line segments connecting these equivalent resistors to the rest of the circuit can then be assumed to have zero resistance exactly. For instance, if the resistance of each of the two wires connecting the resistors in Figure 28.9 is 2 Ω, then the schematic circuit diagram that includes the resistances of the wires is as shown in Figure 28.11. Note that in this diagram we have placed the resistors of 2 Ω that represent the resistances of the wires immediately above the resistors R_1 and R_2. We could equally well have placed these equivalent resistors immediately below R_1 and R_2, or one-half of each above and one-half below — all these arrangements are effectively equivalent, and they lead to the same distribution of current between the two branches of the circuit. Also note that we have neglected the resistances of the short leads sticking out of the circuit above and below; if these have significant resistances, then we will also need to take these into account.

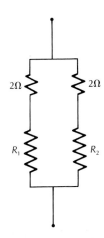

Fig. 28.11 The resistances of the wires in the loop are represented by resistors of 2Ω.

EXAMPLE 5. If, as in the preceding example, a current of 1.8 A flows through the combination of resistors and resistive wires shown schematically in Figure 28.11, what is the potential difference across the combination? What is the current in each resistor?

SOLUTION: The resistance of the left branch of the circuit is $R'_1 = R_1 + 2 \text{ }\Omega = 12 \text{ }\Omega$, and that of the right branch is $R'_2 = R_2 + 2 \text{ }\Omega = 22 \text{ }\Omega$. The net resistance is then given by

$$\frac{1}{R} = \frac{1}{R'_1} + \frac{1}{R'_2} = \frac{1}{12 \text{ }\Omega} + \frac{1}{22 \text{ }\Omega} = \frac{34}{264 \text{ }\Omega}$$

from which $R = 7.76 \text{ }\Omega$. Hence the potential difference across the combination is

$$\Delta V = IR = 1.8 \text{ A} \times 7.76 \text{ }\Omega = 14.0 \text{ volts}$$

and the individual currents are

$$I_1 = \frac{\Delta V}{R'_1} = \frac{14.0 \text{ volts}}{12 \text{ }\Omega} = 1.16 \text{ A}$$

$$I_2 = \frac{\Delta V}{R'_2} = \frac{14.0 \text{ volts}}{22 \text{ }\Omega} = 0.64 \text{ A}$$

SUMMARY

Electric field in uniform wire: $E = \Delta V/l$

Electric current: $I = dq/dt$

Current density: $j = I/A$

Resistance in terms of resistivity: $R = \rho \dfrac{l}{A}$

Ohm's Law: $I = \dfrac{\Delta V}{R}$

Resistivities:
- conductors: $\sim 10^{-8}\ \Omega \cdot m$
- semiconductors: $\sim 10^{-4}$ to $10^{7}\ \Omega \cdot m$
- insulators: $\sim 10^{11}$ to $10^{17}\ \Omega \cdot m$

Series combination of resistors: $R = R_1 + R_2 + R_3 + \cdots$

Parallel combination of resistors: $1/R = 1/R_1 + 1/R_2 + 1/R_3 + \cdots$

QUESTIONS

1. A wire is carrying a current of 15 A. Is this wire in electrostatic equilibrium?

2. Can a current flow in a conductor when there is no electric field? Can an electric field exist in a conductor when there is no current?

3. Figure 28.12 shows a putative mechanical analog of an electric conductor with resistance. A marble rolling down the inclined board is stopped every so often by a collision with a pin and therefore maintains a constant average velocity v_d. Is this a good analog, i.e., is the average velocity v_d proportional to g?

4. By what factor must we increase the diameter of a wire to decrease its resistance by a factor of 2?

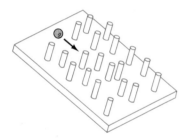

Fig. 28.12 An inclined board with pins nailed on at random.

5. Show that, for a wire of given length made of a given material, the resistance is inversely proportional to the mass of the wire.

6. What deviations from Ohm's Law do you expect if the current is very large?

7. Ohm's Law is an approximate statement about the electrical properties of a conducting body, just as Hooke's Law is an approximate statement about the mechanical properties of an elastic body. Is there an electric analog of the elastic limit?

8. An automobile battery has a potential difference of 12 V between its terminals even when there is no current flowing through the battery. Does this violate Ohm's Law?

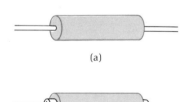

Fig. 28.13

9. Figure 28.13 shows two alternative experimental arrangements for determining the resistance of a carbon cylinder. A known potential difference is applied via the copper terminals, the current is measured, and the resistance is calculated from $R = \Delta V/I$. The arrangement in Figure 28.13a yields a higher resistance than that in Figure 28.13b. Why?

10. Why is it bad practice to operate a high-current appliance off an extension cord?

11. The installation instructions for connecting an outlet to the wiring of a house recommend that the wire be wrapped at least three-quarters of the way around the terminal post (Figure 28.14). Explain.

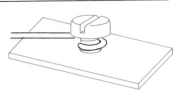

Fig. 28.14

12. The **temperature coefficient of resistivity** α is defined as the fractional increase of resistivity per degree Celsius, $\alpha = (1/\rho)\, d\rho/dT$. According to Table 28.1, what is the value of α for copper?

13. The resistivity of semiconductors and insulators *decreases* with temperature. Can you think of a likely explanation?

14. Figure 20.9 shows a platinum **resistance thermometer** consisting of a coil of fine platinum wire inside a glass tube that can be put in thermal contact with a body. How can this wire be used to determine the temperature of the body?

15. Aluminum wire should never be connected to terminals designed for copper wire. Why not? (Hint: Aluminum has a considerably higher coefficient of thermal expansion than copper.)

16. A wire of copper and a wire of silver are connected in parallel. The two wires have the same length and the same diameter. Which carries more current?

17. Two copper wires are connected in parallel. The wires have the same length, but one has twice the diameter of the other. What fraction of the total current flows in each wire?

PROBLEMS

Sections 28.1–28.3

1. A conducting wire of length 2.0 m is connected between the terminals of a 12-V battery. The resistance of the wire is 3.0 Ω. What is the electric current in the wire? What is the electric field in the wire?

2. When a thin copper wire is connected between the poles of a 1.5-V battery, the current in the wire is 0.50 A. What is the resistance of this wire? What will be the current in the wire if it is connected between the terminals of a 7.5-V battery?

3. The resistance of a 150-W, 110-V light bulb is 0.73 Ω when the light bulb is at its operating temperature. What current passes through this light bulb when in operation? How many electrons per second does this amount to?

4. The resistance of the wire in the windings of an electric starter motor for an automobile is 3.0×10^{-2} Ω. The motor is connected to a 12-V battery. What current will flow through the motor when it is stalled (does not turn)?

*5. An aluminum wire has a resistance of 0.10 Ω. If you draw this wire through a die, making it thinner and twice as long, what will be its new resistance?

*6. A circular loop of superconducting material has a radius of 2.0 cm. It carries a current of 4.0 A. What is the orbital angular momentum of the moving electrons in the wire? Take the center of the loop as origin.

*7. A table lamp is connected to an electric outlet by a copper wire of diameter 0.20 cm and length 2.0 m. Assume that the current through the lamp is 1.5 A, and that this current is steady. How long does it take an electron to travel from the outlet to the lamp?

*8. A high-voltage transmission line consists of a copper cable of diameter

3.0 cm and length 250 km. Assume that the cable carries a steady current of 1500 A.
  (a) What is the electric field inside the cable?
  (b) What is the drift velocity of the free electrons?
  (c) How long does it take one electron to travel the full length of the cable?

Section 28.4

9. The electromagnet of a bell is constructed by winding copper wire around a cylindrical core, like thread on a spool. The diameter of the copper wire is 0.45 mm, the number of turns in the winding is 260, and the average radius of a turn is 5.0 mm. What is the resistance of the wire?

10. The following is a list of some types of copper wire manufactured in the United States:

Gauge No.	Diameter
8	0.3264 cm
9	0.2906
10	0.2588
11	0.2305
12	0.2053

For each type of wire, calculate the resistance for a 100-m segment.

11. To measure the resistivity of a metal, an experimenter takes a wire of this metal of diameter of 0.500 mm and length 1.10 m and applies a potential difference of 12.0 V to the ends. He finds that the resulting current is 3.75 A. What is the resistivity?

12. A high-voltage transmission line has an aluminum cable of diameter 3.0 cm, 200 km long. What is the resistance of this cable?

*13. In silver, the number of free electrons is $n = 5.8 \times 10^{28}$ per cubic meter.
  (a) Using the value of the resistivity given in Table 28.1, calculate the average time interval between collisions of one of these electrons.
  (b) Assuming that the electric field in a current-carrying silver wire is 8.0 V/m, calculate the average drift velocity of an electron.

*14. You want to make a resistor of 1.0 Ω out of a carbon rod of diameter 1.0 mm. How long a piece of carbon do you need?

*15. The resistance of a square centimeter of dry human epidermis is about 10^5 Ω. Suppose that a (foolish) man firmly grasps two wires in his fists. The wires have a radius of 0.13 cm, and the skin of each hand is in full contact with the surface of the wire over a length of 8 cm.
  (a) Calculate the resistance the man offers to a current flowing through his body from one wire to the other. In this calculation you can neglect the resistance of the internal tissues of the human body, because the body fluids are reasonably good conductors, and their resistance is small compared with the skin resistance.
  (b) What current will flow through the body of the man if the potential difference between the wires is 12 V? If it is 110 V? If it is 240 kV? What is your prognosis in each case? See Section 29.8 for the effects of electric currents on the human body.

*16. A lightning rod of iron has a diameter of 0.80 cm and a length of 0.50 m. During a lightning stroke, it carries a current of 1.0×10^4 A. What is the potential drop along the rod?

*17. The air conditioner in a home draws a current of 12 A.
  (a) Suppose that the pair of wires connecting the air conditioner to the fuse box are No. 10 copper wire with a diameter of 0.259 cm and a length

of 25 m each. What is the potential drop along each wire? Suppose that the voltage delivered to the home is exactly 110 V at the fuse box. What is the voltage delivered to the air conditioner?

(b) Some older homes are wired with No. 12 copper wire with a diameter of 0.205 cm. Repeat the calculation of part (a) for this wire.

*18. When the starter motor of an automobile is in operation, the cable connecting it to the battery carries a current of 80 A. This cable is made of copper and is 0.50 cm in diameter. What is the current density in the cable? What is the electric field in the cable?

*19. A copper cable in a high-voltage transmission line has a diameter of 3.0 cm and carries a current of 750 A. What is the current density in the wire? What is the electric field in the wire?

*20. The electromagnet of a bell is wound with 8.2 m of copper wire of diameter 0.45 mm. What is the resistance of the wire? What is the current through the wire if the electromagnet is connected to a 12-V source?

*21. The copper cable connecting the positive pole of a 12-V automobile battery to the starter motor is 0.60 m long and 0.50 cm in diameter.
  (a) What is the resistance of this cable?
  (b) When the starter motor is stalled, the current in the cable may be as much as 600 A. What is the potential drop along the cable under these conditions?

*22. Although aluminum has a somewhat higher resistivity than copper, it has the advantage of having a considerably lower density. Find the mass of a 100-m segment of aluminum cable 3.0 cm in diameter. Compare this with that of a copper cable of the same length and the same resistance. The densities of aluminum and of copper are 2.7×10^3 kg/m^3 and 8.9×10^3 kg/m^3, respectively.

*23. According to the National Electrical Code, the maximum permissible current in a No. 12 copper wire (diameter 0.21 cm) with rubber insulation is 20 A.
  (a) What is the current density in a wire carrying this current?
  (b) What is the potential drop along a 1-m segment of the wire?

*24. An aluminum wire of length 15 m is to carry a current of 25 A with a potential drop of no more than 5 V along its length. What is the minimum acceptable diameter of this cable?

*25. According to safety standards set by the American Boat and Yacht Council, the potential drop along a copper wire connecting a 12-V battery to an item of electrical equipment should not exceed 10%, i.e., it should not exceed 1.2 V. Suppose that a 9-m wire (length measured around the circuit) carries a current of 25 A; what gauge of wire is required for compliance with the above standard? Use the table of wire gauges given in Problem 10. Repeat the calculation for currents of 35 A and 45 A.

*26. A parallel-plate capacitor with a plate area of 8.0×10^{-2} m^2 and a plate separation of 1.0×10^{-4} m is filled with polyethylene. If the potential difference between the plates is 2.0×10^4 V, what will be the current flowing through the polyethylene from one plate to the other?

27. Consider the aluminum cable described in Problem 12. If the temperature of this cable increases from 20°C to 50°C, how much will its resistance increase?

28. What increase of temperature will increase the resistance of a nickel wire from 0.5 Ω to 0.6 Ω?

**29. A solid truncated cone is made of a material of resistivity ρ (Figure 28.15). The cone has a height h, a radius a at one end, and a radius b at the other end. Derive a formula for the resistance of this cone.

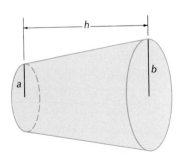

Fig. 28.15

Section 28.5

30. A brass wire and an iron wire of equal diameters and of equal lengths are connected in parallel. Together they carry a current of 6.0 A. What is the current in each?

31. Consider the brass and iron wires described in Problem 30. What is the current in each if they are at a temperature of 90°C instead of a temperature of 20°C?

*32. An electric cable of length 12.0 m consists of a copper wire of diameter 0.30 cm surrounded by a cylindrical layer of rubber insulation of thickness 0.10 cm. A potential difference of 6.0 V is applied to the ends of the cable.
  (a) What will be the current in the copper?
  (b) Taking into account the conductivity of the rubber (see Table 28.2), what will be the current in the rubber?

*33. A water pipe is made of iron with an outside diameter of 2.5 cm and an inside diameter of 2.0 cm. The pipe is used to ground an electric appliance. If a current of 20 A flows from the appliance into the water pipe, what fraction of this current will flow in the iron? What fraction in the water? Assume that water has a resistivity of 0.01 $\Omega \cdot$ m.

*34. A copper wire, of length 0.50 m and diameter 0.259 cm, has been accidentally cut by a saw. The region of the cut is 0.4 cm long, and in this region the remaining wire has a cross-sectional area of only one-quarter of the original area. What is the percentage increase of the resistance of the wire caused by this cut?

*35. The windings of high-current electromagnets are often made of copper pipe. The current flows in the walls of the pipe, and cooling water flows in the interior of the pipe. Suppose the copper pipe has an outside diameter of 1.20 cm and an inside diameter of 0.80 cm. What is the resistance of 30 m of this copper pipe? What voltage must be applied to it if the current is to be 600 A?

*36. An underground telephone cable, consisting of a pair of wires, has suffered a short somewhere along its length (Figure 28.16). The telephone cable is 5 km long, and in order to discover where the short is, a technician first measures the resistance across the terminals AB; then he measures the resistance across the terminals CD. The first measurement yields 30 Ω; the second, 70 Ω. Where is the short?

Fig. 28.16 A pair of wires with a short at the point P (the wires are joined at P).

*37. The air of the atmosphere has a slight conductivity due to the presence of a few free electrons and positive ions.
  (a) Near the surface of the Earth, the atmospheric electric field has a strength of about 100 V/m and the atmospheric current density is 4×10^{-12} A/m². What is the resistivity?
  (b) The potential difference between the ionosphere (upper layer of atmosphere) and the surface of the Earth is 4×10^5 V. What is the total resistance of the atmosphere? (Hint: For the purposes of this problem you may assume that the Earth is flat.)

*38. Three resistors, with resistances of 3.0 Ω, 5.0 Ω, and 8.0 Ω, are connected in parallel. If this combination is connected to a 12.0-V battery, what is the current through each resistor? What is the current through the combination?

*39. A flexible wire for an extension cord for electric appliances is made of 24 strands of fine copper wire, each of diameter 0.053 cm, tightly twisted together. What is the resistance of a length of 1.0 m of this kind of wire?

*40. Two copper wires of diameters 0.26 cm and 0.21 cm, respectively, are connected in parallel. What is the current in each if the combined current is 18 A?

*41. Commercially manufactured superconducting cables consist of filaments of superconducting wire embedded in a matrix of copper (see Figure VIII.17). As long as the filaments are superconducting, all the current flows in them, and no current flows in the copper. But if superconductivity suddenly fails because of a temperature increase, the current can spill into the copper; this prevents damage to the filaments of the superconductor. Calculate the resistance per meter of length of the copper matrix shown in Figure VIII.17. The diameter of the copper matrix is 0.7 mm, and each of the 2100 filaments has a diameter of 0.01 mm.

*42. What is the net resistance of the combination of four resistors shown in Figure 28.17? Each of the resistors has a resistance of 3 Ω.

*43. Three resistors with $R_1 = 2.0$ Ω, $R_2 = 4.0$ Ω, and $R_3 = 6.0$ Ω are connected as shown in Figure 28.18.
  (a) Find the net resistance of the combination.
  (b) Find the current that passes through the combination if a potential difference of 8.0 V is applied to the terminals.
  (c) Find the potential drop and the current for each individual resistor.

*44. Three resistors with $R_1 = 4.0$ Ω, $R_2 = 6.0$ Ω, and $R_3 = 8.0$ Ω are connected as shown in Figure 28.19.
  (a) Find the net resistance of the combination.
  (b) Find the current that passes through the combination if a potential difference of 12.0 V is applied to the terminals.
  (c) Find the potential drop and the current for each individual resistor.

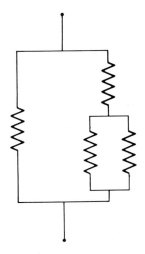

Fig. 28.17

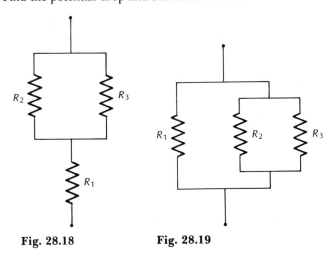

Fig. 28.18   Fig. 28.19

*45. Consider the combination of three resistors described in the preceding problem. If we want a current of 6.0 A to flow through resistor R_2, what potential difference must we apply to the external terminals?

*46. Three resistors, with $R_1 = 4.0$ Ω, $R_2 = 6.0$ Ω, and $R_3 = 2.0$ Ω, are connected as shown in Figure 28.20. This combination is connected to a 1.5-V battery.
  (a) What is the net resistance of this combination?
  (b) What is the current through the combination?
  (c) What is the potential drop and the current for each resistor?

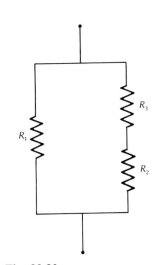

Fig. 28.20

**47. Twelve resistors, each of resistance R, are connected along the edges of a cube (Figure 28.21). What is the resistance between diagonally opposite corners of this cube?

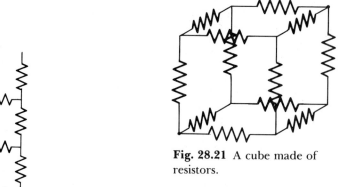

Fig. 28.21 A cube made of resistors.

**48. What is the resistance of an infinite ladder of 1.0-Ω resistors connected as shown in Figure 28.22a? (Hint: The ladder can be regarded as made of two pieces connected in parallel; see Figure 28.22b.)

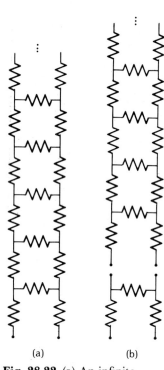

Fig. 28.22 (a) An infinite "ladder" of resistors. The terminals are marked by a pair of dots. (b) One rung has been cut off from the ladder.

CHAPTER 29

DC Circuits

If we want to keep a current flowing in a wire, we must connect its ends to a "pump of electricity," such as a battery or a generator, that can continuously supply electric charges to one end of the wire and remove them from the other. Figure 29.1 shows a circuit consisting of a wire connected to the terminals of a battery. A current will flow around this circuit; in Figure 29.1 we have indicated the direction of the current according to our convention that it is in the direction of flow of (hypothetical) positive charges. As long as the "strength" of the battery and the resistance of the wire remain constant, the current will also remain constant. Such a steady, time-independent current is called a **direct current**, or DC.

Direct current, DC

29.1 Electromotive Force

The battery must do work on the charges in order to keep them flowing around the circuit shown in Figure 29.1. Suppose that a hypothetical positive charge is originally at the point P, at one terminal of the battery. Pushed along by the electric field, the charge moves along the wire. On the average, the kinetic energy of the charge does not change — any kinetic energy that the charge gains from the electric field is dissipated by friction within the wire, and the charge reaches the point P', at the other terminal of the battery, with its original kinetic energy.

Thus, only the potential energy changes. Since the electric field is directed along the wire, the electric potential steadily decreases with distance along the wire, and the charge reaches the point P' with a potential energy lower than its original potential energy. In order to keep the current flowing, the battery must "pump" the charge from the

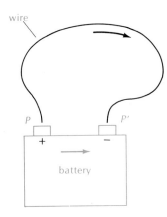

Fig. 29.1 A wire connected between the terminals of a battery.

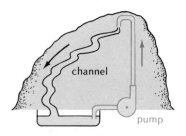

Fig. 29.2 Mechanical analog of the battery–wire circuit of Figure 29.1.

Electromotive force, emf

low-potential terminal to the high-potential terminal, that is, the battery must supply electric potential energy to the charge. The role of the battery is analogous to that of a hydraulic pump system that lifts water from the bottom to the top of a hill (Figure 29.2). The wire is analogous to a channel by means of which the water runs down the hill returning to the pump. The water then flows in a closed hydraulic circuit, just as charge flows in a closed electric circuit. The hydraulic pump of Figure 29.2 can be regarded as a source of gravitational potential energy — it produces this energy from an external supply of chemical or mechanical energy. Likewise, the "pump of electricity" of Figure 29.1 can be regarded as a source of electric potential energy — it produces this energy from a supply of chemical energy.

To measure the "strength" of a source of electric potential energy, we introduce the concept of **electromotive force**, or **emf**. The emf of a source of electric potential energy is defined as *the amount of electric energy delivered by the source per coulomb of positive charge as this charge passes through the source from the low-potential terminal to the high-potential terminal*. Since the emf is energy per unit charge, its units are volts. Keep in mind that the electromotive "force" is not a force, but an energy — the confusing name became attached to it a long time ago, when physicists were not yet making a sharp distinction between force and energy.

If a steady, time-independent current carries one coulomb of charge around the circuit of Figure 29.1 from P to P' along the wire and from P' to P through the source of emf, then the energy that this charge receives from the source of emf must exactly match the energy it loses within the wire. If so, the charge returns to its starting point with exactly the same energy it had originally, and it can repeat this round-trip again and again in exactly the same manner. We can write this energy balance as

$$\mathscr{E} + \Delta V = 0 \qquad (1)$$

where $\mathscr{E}$ represents the emf, or the increase of potential energy, due to the source and ΔV represents the change of potential energy along the wire ($\mathscr{E}$ is positive and ΔV is negative).

According to Eq. (1), the emf $\mathscr{E}$ has the same magnitude as the potential drop in the external circuit connected between the terminals of the source of emf. For example, a battery with an emf of 1.5 V connected to an external circuit will do 1.5 J of work on a coulomb of positive charge that passes through the battery in the forward direction, from the − terminal to the + terminal; and the resistors and other devices in the external circuit will do −1.5 J of work on the charge as it flows around this circuit, from the + terminal to the − terminal. Because of the equality between the emf $\mathscr{E}$ and the potential drop ΔV, the emf is often simply called the **voltage** of the source.

Voltage

EXAMPLE 1. A fresh flashlight battery with a voltage of 1.5 V will deliver a current of 1 A for about 1 h before running down. How much work does the battery do in this time interval?

SOLUTION: The battery does 1.5 J of work on each coulomb that passes through. If the current is 1 A, the charge that passes through in 1 h is 1 A × 3600 s = 3600 C, and the total work is 1.5 J/C × 3600 C = 5400 J.

Note that if one coulomb of positive charge is forced through the battery in the reverse direction (from the + terminal to the − terminal), then the battery will take electric potential energy away from the charge. The charge will then emerge from the battery at a potential which is 1.5 volts lower than the potential with which it entered. The energy taken away from the charge will either be stored within the battery (if it is a reversible battery) or else it will merely be wasted as heat within the battery (if it is an irreversible battery).

29.2 Sources of Electromotive Force

The most important kinds of sources of emf are batteries, electric generators, fuel cells, and solar cells. We will now briefly discuss each of these.

BATTERIES These sources of emf convert chemical energy into electric energy. A very common type of battery is the **lead–acid battery,** which finds widespread use in automobiles. In its simplest form, this battery consists of two plates of lead — the positive electrode and the negative electrode — immersed in a solution of sulfuric acid (Figure 29.3). The positive electrode is covered with a layer of lead dioxide, PbO_2. When the external circuit is closed, the following reactions, already mentioned in Section 22.3, take place at the immersed surfaces of the negative and positive electrodes, respectively:

$$Pb + SO_4^{--} \rightarrow PbSO_4 + 2e^- \tag{2}$$

$$PbO_2 + SO_4^{--} + 4H^+ + 2e^- \rightarrow PbSO_4 + 2H_2O \tag{3}$$

Lead–acid battery

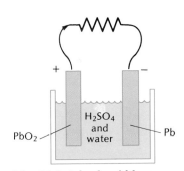

Fig. 29.3 A lead–acid battery connected to an external circuit.

These reactions deposit electrons on the negative electrode and absorb electrons from the positive electrode. Thus, the battery acts as a pump for electrons — the negative electrode is the outlet, the positive electrode is the intake, and the electrons flow from one to the other via the external circuit.

The reactions (2) and (3) deplete the sulfuric acid in the solution, and they deposit lead sulfate on the electrodes. The depletion of sulfuric acid finally halts the reaction — the battery is then "discharged."

The lead–acid battery can be "charged" by simply passing a current through it in the backward direction. This reverses the reactions (2) and (3) and restores the sulfuric acid solution. Note that what is stored in the battery during the "charging" process is not electric charge, but chemical energy. The number of positive and negative electric charges (protons and electrons) in the battery remains constant; what changes is the concentration of chemical compounds. A "charged" battery contains chemical compounds (lead, lead dioxide, sulfuric acid) of relatively high internal energy; a "discharged" battery contains chemical compounds (lead sulfate, water) of lower internal energy.

The single-cell battery shown in Figure 29.3 has an emf of 2.0 V. In an automobile battery, six such cells are stacked together and connected in series to give an emf of 12.0 V. The energy stored in such a battery is typically about 0.5 kW·h.[1] Large banks of batteries, weigh-

---

[1] Recall that 1 kW·h = 1000 W × 3600 s = 3.6×10^6 J.

Dry cell

ing several hundred tons, for the propulsion of submarines store more than 5×10^3 kW·h.

Another familiar type of battery is the **dry cell**, or flashlight battery. The positive electrode consists of manganese dioxide and the negative electrode of zinc. The electrolyte in which these electrodes are "immersed" is a moist paste of ammonium chloride and zinc chloride. The chemical reactions at the electrodes convert chemical energy into electric energy and pump electrons from one electrode to the other via the external circuit. The emf of such a dry cell is 1.5 V. Since there is no liquid to slosh around, these batteries are particularly suitable for portable devices. The energy stored in an ordinary flashlight battery is typically of the order of 2×10^{-3} kW·h.

ELECTRIC GENERATORS Generators convert mechanical energy (kinetic energy) into electric energy. Their operation involves magnetic fields and the phenomenon of induction. We will leave the description of electric generators for Section 32.3.

Fuel cell

FUEL CELLS These resemble batteries in that they convert chemical energy into electric energy. However, in contrast to a battery, neither the high-energy chemicals nor the low-energy reaction products are stored inside the fuel cell. The former are supplied to the fuel cell from external tanks, and the latter are ejected. Essentially, the fuel cell acts as a combustion chamber in which controlled combustion takes place. The fuel cell "burns" a high-energy fuel, but produces electric energy rather than heat energy.

Figure 29.4a shows a fuel cell that "burns" a hydrogen–oxygen fuel. The electrodes of the fuel cell are hollow cylinders of porous carbon; oxygen at high pressure is pumped into the positive electrode and hydrogen into the negative electrode. The electrodes are immersed in a potassium-hydroxide electrolyte. The reactions at the negative and positive electrode are, respectively,

$$2H_2 + 4OH^- \rightarrow 4H_2O + 4e^- \tag{4}$$

$$O_2 + 2H_2O + 4e^- \rightarrow 4OH^- \tag{5}$$

These reactions deposit electrons on the negative electrode and remove electrons from the positive electrode. This pumps electrons from one electrode to the other via the external circuit.

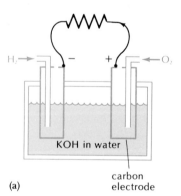

Fig. 29.4 (a) Schematic diagram of a fuel cell connected to an external circuit. (b) Fuel cell used on Skylab.

The net result of the sequence of reactions (4) and (5) is the conversion of oxygen and hydrogen into water. This reaction is the reverse of the electrolysis of water (decomposition of water by an electric current). The excess water is removed from the cell in the form of water vapor.

All fuel cells produce a certain amount of waste heat. The best available fuel cells convert about 45% of the chemical energy of the fuel into electric energy, and they waste the remainder. Fuel cells are still at an experimental stage; however, they have already been put to use as practical power sources aboard the Apollo spacecraft and on Skylab (Figure 29.4b). They are compact and clean; on Skylab, the waste water eliminated from the fuel cell was used both for drinking and for showers.

SOLAR CELLS Solar cells convert the energy of sunlight directly into electric energy. They are made of thin wafers of a semiconductor, such as silicon. Figure 29.5 shows a cross section through a solar cell. It consists of a central core of n-type silicon surrounded by an outer layer of p-type silicon. The outer layer is very thin, only about 10^{-4} cm, and when sunlight penetrates this layer and reaches the boundary between the n-type silicon and the p-type silicon, it pumps electrons from the latter into the former. We will deal with the details of this process in Chapter 44.

The emf of a silicon solar cell is about 0.6 V. However, the current that can be extracted is rather small. Even in full sunlight, a single solar cell of surface area 5 cm² will deliver only 0.1 ampere. Large panels, containing very many solar cells, are needed for the generation of appreciable amounts of electric power (see Figures 29.6 and V.13).

Solar cell

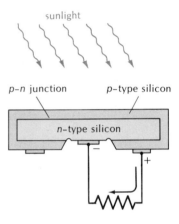

Fig. 29.5 Schematic diagram of a solar cell connected to an external circuit.

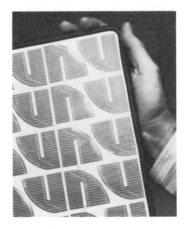

Fig. 29.6 An array of solar cells in a panel designed for charging automobile batteries.

29.3 Single-Loop Circuits

In schematic diagrams of electric circuits, a source with a time-independent emf is represented by a stack of parallel thick and thin lines suggesting the plates of a lead–acid battery. The high-potential terminal is marked with a plus sign and the low-potential terminal with a minus sign. If the terminals of such a source are connected to a network of resistances, a steady current, also called a direct current or DC, will flow through the network.

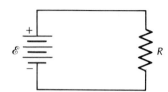

Fig. 29.7 A simple circuit with a source of emf and a resistor.

Figure 29.7 shows a schematic circuit diagram for a very simple circuit consisting of a source of emf, such as a battery, connected to a resistor. The emf of source is $\mathcal{E}$, and the resistance of the resistor is R. The wires from the resistor to the battery are assumed to have negligible resistance (if higher accuracy is required, the resistance of the wires must be included in the circuit diagram). The resistance R in Figure 29.7 could equally well represent a carbon resistor or some other device, such as a light bulb, endowed with an electrical resistance; the resistance R could even represent the resistance of a wire by itself connected between the poles of the battery.

To find the current that will flow through the circuit, we note that according to Ohm's Law the potential drop across the resistor must be

$$\Delta V = -IR \qquad (6)$$

Kirchhoff's (second) rule

But, from Eq. (1), we know that the emf plus the potential drop must equal zero; hence

$$\mathcal{E} - IR = 0 \qquad (7)$$

or

$$I = \frac{\mathcal{E}}{R} \qquad (8)$$

Equation (7) is an instance of **Kirchhoff's rule,** which states:

Around any closed loop in a circuit, the sum of all the emfs and all the potential drops across resistors and other circuit elements must equal zero.[2]

In this sum, the emf of a source is reckoned as positive if the current flows through the source in the forward direction and negative if in the backward direction.

The proof of this general rule is similar to the proof of Eq. (1). If one coulomb of positive charge flows once around a closed loop with several sources of emf and several resistors, it will gain potential energy while passing through each emf and lose potential energy while passing through each resistor. Under steady conditions, the sum of gains and losses must equal zero, since the charge must return to its starting point with no change of energy. In this sum, the emf must be reckoned as positive (gain of potential energy) whenever the charge flows through the source in the forward direction, from the − terminal to the + terminal; and it must be reckoned as negative (loss of potential energy) whenever the charge flows through the source in the backward direction, from the + terminal to the − terminal.

Gustav Robert Kirchhoff (keerkh-hoff), *1824–1887, German physicist, professor at Heidelberg and at Berlin. Mainly known for his development of spectroscopy, he also made many important contributions to mathematical physics, among them, his first and second rules for circuits.*

EXAMPLE 2. Figure 29.8 shows a circuit with two batteries and two resistors. The emfs of the batteries are $\mathcal{E}_1 = 12$ V and $\mathcal{E}_2 = 15$ V; the resistances are $R_1 = 4$ Ω and $R_2 = 2$ Ω. What is the current in the circuit?

SOLUTION: To apply Kirchhoff's rule, we must decide in which direction the current flows around the loop. We will arbitrarily assume that the current

---

[2] This is often called Kirchhoff's second rule. We will become acquainted with his first rule in Section 29.4.

flows in the clockwise direction. If this hypothesis is wrong, our calculation of the current will yield a negative value, and this will indicate that the direction of the actual current is opposite to the hypothetical current.

The sum of all emfs and all potential drops across resistors is

$$\mathcal{E}_1 - IR_1 - \mathcal{E}_2 - IR_2 = 0 \tag{9}$$

Note that $\mathcal{E}_2$ enters with a negative sign into this equation since the hypothetical current passes through this source of emf in the backward direction. The solution of Eq. (9) leads to

$$I = \frac{\mathcal{E}_1 - \mathcal{E}_2}{R_1 + R_2} \tag{10}$$

or

$$I = \frac{12 \text{ V} - 15 \text{ V}}{4 \, \Omega + 2 \, \Omega} = -0.5 \text{ A} \tag{11}$$

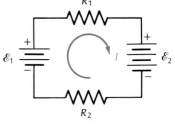

Fig. 29.8 Two sources of emf and two resistors.

The negative sign indicates that the current is *not* clockwise but counterclockwise.

COMMENTS AND SUGGESTIONS: We could have guessed the direction of the current by inspecting the circuit diagram. Obviously, the stronger battery on the right will force the current backward through the weaker battery on the left. However, in the more complicated multiloop circuits of the next section, the directions of the currents will usually not be obvious, and we will have to determine these directions by carefully keeping track of the signs in our calculations.

Incidentally: In Example 2, we neglected the internal resistance of the batteries. The electrolyte in a battery always has some resistance, and this causes the current to suffer a voltage drop even before it leaves the external terminal of the battery. The nominal emf $\mathcal{E}$ quoted on the labels of batteries refers to the potential difference between the terminals when no current is flowing; this is often called the "open-circuit voltage." The internal resistance R_i of the battery may be regarded as connected in series with the emf $\mathcal{E}$ (Figure 29.9). When a current I is flowing, the voltage drops by $\Delta V = -IR_i$ across the internal resistance, and hence the remaining voltage at the external terminals of the battery will be $\mathcal{E} - IR_i$. The internal resistance of a good battery is small, and can often be neglected. But if we need to take it into account, we can do so by simply placing the appropriate internal resistance in series with each battery in the circuit diagram; then we can proceed with the usual calculation of the currents.

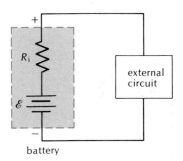

Fig. 29.9 The internal resistance is in series with the nominal emf $\mathcal{E}$.

EXAMPLE 3. A flashlight battery of a nominal emf 1.5 V has an internal resistance of 0.05 Ω. What will be the potential difference across its terminals if the battery is delivering a current of 2 A? A current of 10 A?

SOLUTION: For a current of 2 A, the voltage across the terminals will be

$$\mathcal{E} - IR_i = 1.5 \text{ V} - 2 \text{ A} \times 0.05 \, \Omega = 1.4 \text{ V}$$

For a current of 10 A, the voltage across the terminals will be

$$\mathcal{E} - IR_i = 1.5 \text{ V} - 10 \text{ A} \times 0.05 \, \Omega = 1.0 \text{ V}$$

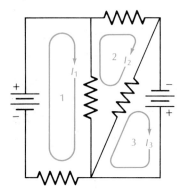

Fig. 29.10 A multiloop circuit. The currents I_1, I_2, and I_3 are regarded as flowing in closed loops. Note that loops 1 and 2 share the middle resistor; and loops 2 and 3 share the diagonal resistor. The net current in the middle resistor is $I_1 - I_2$, and the net current in the diagonal resistor is $I_2 - I_3$.

29.4 Multiloop Circuits

If several sources of emf and several resistors are connected in some complicated circuit with branches and loops, then the currents will flow along several alternative paths. Figure 29.10 shows an example of such a complicated circuit. To find the currents in the circuit, we must solve a simultaneous set of several equations with the currents as unknowns.

The procedure for obtaining the necessary equations is as follows:

a. Regard the given circuit as a collection of several closed current loops. The loops may overlap, but each loop must have at least one portion that does not overlap with other loops (Figure 29.10).
b. Label the currents in the loops $I_1, I_2, I_3, \ldots$, and arbitrarily assign a direction to each of these currents.
c. Apply Kirchhoff's rule to each loop: the sum of all the emfs and all the potential drops across resistors must add to zero around each loop. Note that when calculating the potential drop across a resistor, we must take the product of the resistance and the *net* current through the resistor; if the resistor belongs to two adjacent loops, then the *net* current is the algebraic sum of the two loop currents (Figure 29.10).

Loop method

This procedure, called the **loop method**, will result in the right number of equations for the unknown currents $I_1, I_2, I_3, \ldots$. We can then solve the equations for these unknowns by the standard mathematical methods for the solution of a system of equations with several unknowns. If a current turns out to be negative, its direction is opposite to the direction assigned in step (b).

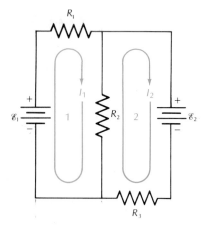

Fig. 29.11 The loops 1 and 2 share the resistor R_2. The currents I_1 and I_2 flow through R_2 in opposite directions; the net current through R_2 is therefore the difference between I_1 and I_2.

EXAMPLE 4. Figure 29.11 shows a circuit with several batteries and resistors. Find the current in each of the resistors.

SOLUTION: Obviously, this circuit can be regarded as consisting of the two loops indicated by the arrows. We label the currents in the loops I_1 and I_2; these symbols have been written next to the arrows that indicate the directions. When calculating the potential drop in the resistor R_2 (which is included in both loops), we must take into account that both loop currents flow through R_2 simultaneously, in opposite directions. The net current through R_2 in the direction of the arrow of loop 1 is therefore $I_1 - I_2$. With this, Kirchhoff's rule for loop 1 yields

$$\mathscr{E}_1 - I_1 R_1 - (I_1 - I_2) R_2 = 0 \tag{12}$$

Likewise, the net current through R_2 in the direction of the arrow of loop 2 is $I_2 - I_1$; and Kirchhoff's rule for loop 2 yields

$$-\mathscr{E}_2 - I_2 R_3 - (I_2 - I_1) R_2 = 0 \tag{13}$$

This gives us two equations for the two unknowns I_1 and I_2.

Before we proceed to the solution of these equations, it is convenient to substitute the numerical values for the known quantities $\mathscr{E}$ and R. For instance, if $\mathscr{E}_1 = 12.0$ V, $\mathscr{E}_2 = 8.0$ V, $R_1 = 4.0\ \Omega$, $R_2 = 4.0\ \Omega$, and $R_3 = 2.0\ \Omega$, then Eqs. (12) and (13) become

$$12 - 8I_1 + 4I_2 = 0 \tag{14}$$

$$-8 - 6I_2 + 4I_1 = 0 \tag{15}$$

with the solution

$$I_1 = 1.25 \text{ A} \qquad I_2 = -0.50 \text{ A} \qquad (16)$$

The negative sign on I_2 indicates that the direction of the current in the second loop is opposite to the direction shown in Figure 29.11.

The current in the resistor R_1 is then $I_1 = 1.25$ A; the current in the resistor R_2 is $I_1 - I_2 = 1.75$ A; and the current in the resistor R_3 is $I_2 = -0.50$ A.

COMMENTS AND SUGGESTIONS: In setting up the equations for the several loops in the circuit, you must handle the directions of the currents and the sign conventions for currents and potentials consistently. After you have assigned directions to the loops by drawing arrows, go around each loop in the direction of the arrow and apply Kirchhoff's rule, with meticulous attention to signs. Each emf in the loop contributes a positive term if the direction of the arrow is the forward direction for this emf, and it contributes a negative term if the direction of the arrow is the backward direction. Each resistor contributes a negative term of the form −[resistance] × [net current], where the net current is the algebraic sum of the current in the loop under consideration and the current in any other loop that flows through this resistor simultaneously. In this algebraic sum, the second current is to be included with a positive sign if its arrow is parallel to that of the first current, and with a negative sign if antiparallel.

Keep in mind that the currents you obtain by setting up and solving the loop equations are the loop currents. If you need to know the net currents through a given emf or a given resistor, you must calculate these afterward by forming the algebraic sum of the loop currents that simultaneously flow through this emf or resistor.

And if you need to know the potential difference between two given points in the circuit, simply sum all the emfs and voltage drops in resistors along a path from one point to the other, complying with the same sign conventions as for a closed loop.

An alternative procedure for obtaining the equations for a multiloop circuit makes use of **Kirchhoff's first rule,** which states:

Kirchhoff's first rule

The sum of all the currents entering any branch point of the circuit (where three or more wires merge) must equal the sum of currents leaving.

This rule expresses charge conservation — the amount of charge entering any branch point in some time interval equals the amount of charge leaving. Our previous procedure for circuits contained Kirchhoff's first rule implicitly, and hence we had no occasion to use this rule explicitly. The alternative procedure uses this Kirchhoff rule explicitly, as follows:

a. Regard the given circuit as a collection of several branches, which begin and end at the points where wires merge loops (see Figure 29.12).
b. Label the currents in the distinct branches $I_1, I_2, I_3, \ldots$, and arbitrarily assign a direction to each of these currents.
c. Apply Kirchhoff's rule, now called Kirchhoff's second rule, to each loop: the sum of all the emfs and all the potential drops across resistors must add to zero around each loop.
d. Apply Kirchhoff's first rule to the branch points: the sum of all the currents entering a branch point must equal the sum of the currents leaving.

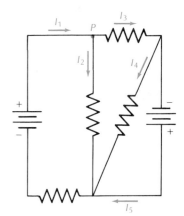

Fig. 29.12 A multiloop circuit. The currents in the distinct branches of the circuit are $I_1, I_2, I_3, I_4,$ and I_5. The point P is a branch point, where three wires merge.

This procedure, called the **branch method**, will result in the right number of equations for the unknowns $I_1, I_2, I_3, \ldots$.

Branch method

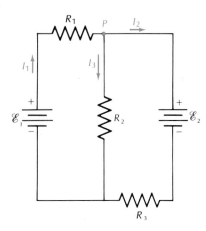

Fig. 29.13 The currents in the distinct branches of this circuit are I_1, I_2, and I_3.

EXAMPLE 5. Again consider the circuit of Figure 29.11. Repeat the calculation of the current in the resistors of this circuit, using the branch method.

SOLUTION: The circuit has three distinct branches; we label the currents in these branches I_1, I_2, and I_3 (Figure 29.13). Kirchhoff's second rule for loop 1 yields

$$\mathcal{E}_1 - I_1 R_1 - I_3 R_2 = 0 \tag{17}$$

Likewise, for loop 2

$$-\mathcal{E}_2 - I_2 R_3 + I_3 R_2 = 0 \tag{18}$$

The last term in Eq. (18) has been given a positive sign because the direction of the current I_3 is *opposite* to the direction of the arrow of loop 2.

Kirchhoff's first rule applied to the branch point P (Figure 29.13) yields

$$I_1 = I_2 + I_3 \tag{19}$$

Equations (17), (18), and (19) are three equations for the three unknowns I_1, I_2, and I_3. Note that in this example, the branch method gives us three equations with three unknowns, whereas the loop method gave us two equations with two unknowns. In general, the branch method suffers from the disadvantage of giving us more unknowns and more equations. Solving Eq. (19) for I_3, we find

$$I_3 = I_1 - I_2 \tag{20}$$

When we substitute this into Eqs. (17) and (18), we obtain Eqs. (12) and (13) — the same equations for I_1 and I_2 as before. Hence the final answers obtained by the branch method are the same as those obtained by the loop method.

COMMENTS AND SUGGESTIONS: Since the loop method involves fewer unknowns and fewer equations, it is usually quicker than the branch method. However, the branch method has the advantage that it leads directly to the currents in the emfs and resistors, whereas the loop method leads to loop currents, which have to be combined into net currents afterward. Feel free to choose whichever method strikes your fancy.

29.5 Energy in Circuits; Joule Heat

As we saw in Section 29.1, to keep a current flowing in a circuit, the batteries or other sources of emf must do work. If an amount of charge dq passes through a source of an emf $\mathcal{E}$, the amount of work done will be

$$dW = \mathcal{E}\, dq \tag{21}$$

Hence the rate at which the source does work is

$$\frac{dW}{dt} = \mathcal{E}\frac{dq}{dt} \tag{22}$$

The rate of work is the power; the rate of flow of the charge is the current. Equation (22) therefore asserts that the electric power delivered by the source of emf to the current is

$$P = \mathcal{E}I \qquad (23)$$

Power delivered by source of emf

Note that in Eq. (23) we have not yet taken into account the algebraic sign of the power. Obviously, we will have to attach a positive sign to the power if the current passes through the source in the forward direction and a negative sign if the current passes through in the backward direction. In the former case the source delivers energy to the current, and in the latter case the source receives energy from the current.

EXAMPLE 6. What power do the two batteries described in Example 4 deliver?

SOLUTION: The current through the first battery is 1.25 A. This current passes through the battery in the forward direction; hence, the power delivered by the battery is positive,

$$P = 1.25 \text{ A} \times 12.0 \text{ V} = 15 \text{ W}$$

Likewise, the power delivered by the other battery is

$$P = 0.50 \text{ A} \times 8.0 \text{ V} = 4.0 \text{ W}$$

The net power delivered by both batteries is 19 W.

The electric potential energy acquired by the charges is carried along the circuit to the resistors, and it is continually dissipated in the resistors. If within a given resistor the charge dq suffers a potential drop ΔV (regarded as a positive quantity), then the loss of potential energy is $dU = \Delta V \, dq$, and the rate at which energy is dissipated is

$$\frac{dU}{dt} = \Delta V \frac{dq}{dt} \qquad (24)$$

Power dissipated in resistor

that is, the power dissipated in the resistor is

$$P = (\Delta V) I \qquad (25)$$

By means of Ohm's Law, $\Delta V = IR$, we can also write this power as

$$P = I^2 R \qquad (26)$$

or as

$$P = (\Delta V)^2 / R \qquad (27)$$

The energy lost by the charges during their passage through a resistor generates heat, that is, it generates random microscopic kinetic and potential energy of the atoms. This conversion of electric energy into thermal energy in a resistor is called **Joule heating.**

Joule heat

In any circuit consisting of several sources of emf and several resistors with steady currents, the total power delivered by the sources of emf must, of course, equal the total power dissipated in the resistors. This equality is a direct consequence of Kirchhoff's rule. The net result of the flow of current in such a circuit is therefore a conversion of energy of the sources of emf into an equal amount of energy of heat.

EXAMPLE 7. What is the rate at which Joule heat is produced in the resistors of Example 4?

SOLUTION: The resistances are $R_1 = 4.0\ \Omega$, $R_2 = 4.0\ \Omega$, and $R_3 = 2.0\ \Omega$; the corresponding currents are 1.25 A, 1.75 A, and 0.50 A. Equation (26) then gives the power dissipated in each resistor,

$$P_1 = (1.25\ \text{A})^2 \times 4.0\ \Omega = 6.25\ \text{W}$$

$$P_2 = (1.75\ \text{A})^2 \times 4.0\ \Omega = 12.25\ \text{W}$$

$$P_3 = (0.50\ \text{A})^2 \times 2.0\ \Omega = 0.50\ \text{W}$$

COMMENTS AND SUGGESTIONS: Note that the net power dissipated is 19.0 W, which agrees with the net power delivered by the batteries (see Example 6).

EXAMPLE 8. A high-voltage transmission line that connects a city to a power plant consists of a pair of copper cables, each with a resistance of 4 Ω. The current flows to the city along one cable, and back along the other.
(a) The transmission line delivers to the city 1.7×10^5 kW of power at 2.3×10^5 V. What is the current in the transmission line? How much power is lost as Joule heat in the transmission line?
(b) If the transmission line were to deliver the same 1.7×10^5 kW of power at 110 V, how much power would be lost as Joule heat? Is it more efficient to transmit power at high voltage or at low voltage?

SOLUTION: (a) Figure 29.14 shows the circuit consisting of power plant, transmission line, and city. In terms of the power and the voltage delivered to the city, the current through the city is

$$I = \frac{P_{\text{delivered}}}{\Delta V_{\text{delivered}}} = \frac{1.7 \times 10^8\ \text{W}}{2.3 \times 10^5\ \text{V}} = 7.4 \times 10^2\ \text{A}$$

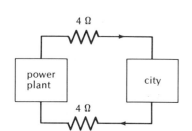

Fig. 29.14 Circuit diagram for a high-voltage transmission line connecting a city to a power plant.

The currents in both portions of the transmission line must, of course, be the same. The combined resistance of both cables is $4\ \Omega + 4\ \Omega = 8\ \Omega$, and hence the power lost in the transmission line is

$$P_{\text{lost}} = I^2 R = (7.4 \times 10^2\text{A})^2 \times 8\ \Omega = 4.4 \times 10^6\ \text{W} \tag{28}$$

Thus, the power lost is 3% of the power delivered.

(b) For $\Delta V_{\text{delivered}} = 110$ V, the current is

$$I = \frac{1.7 \times 10^8\ \text{W}}{110\ \text{V}} = 1.6 \times 10^6\ \text{A}$$

and the power lost is

$$P_{\text{lost}} = I^2 R = (1.6 \times 10^6\ \text{A})^2 \times 8\ \Omega = 1.9 \times 10^{13}\ \text{W} \tag{29}$$

Thus, the power lost is much larger than the power delivered! Obviously, transmission at high voltage is more efficient than transmission at low voltage.

29.6* Electrical Measurements

Measurements of currents, potentials, and resistances in electrical circuits require various specialized instruments. Here we will briefly discuss some of the instruments used in DC circuits.

AMMETER AND VOLTMETER Most electrical measurements are performed with ammeters and voltmeters. The ammeter measures the electric current flowing into its terminals, and the voltmeter measures the potential difference applied to its terminals.

Ammeter and voltmeter

The internal mechanisms of the ammeter and the voltmeter are similar. In the traditional moving-coil instruments, the sensitive element is a small coil of wire, delicately suspended between the poles of a magnet, which suffers a deflection when a current passes through it (see Figure 31.34 for a view of the internal mechanism of these instruments). Thus, ammeters and voltmeters both respond to an electric current passing through the instrument. The difference is that the ammeter has a low internal resistance and permits the passage of whatever current enters its terminals without hindrance, whereas the voltmeter has a very large internal resistance and draws only an extremely small current, even when the potential difference applied to its terminals is large.

Figure 29.15a shows an example of an electric circuit. To measure the electric current at, say, the point *P* in this circuit, the experimenter must cut the wire apart and insert the ammeter. Figure 29.15b shows the correct way of connecting the ammeter (for comparison, Figure 29.15c shows a wrong way). Since the ammeter has a very low internal resistance, its insertion in the circuit usually has an insignificant effect on the current. However, if the resistances in the circuit are very small, then the insertion of an ammeter can have a significant inhibiting effect on the current. For the accurate measurement of the current, the experimenter must select an ammeter whose internal resistance is much lower than the resistances in the circuit.

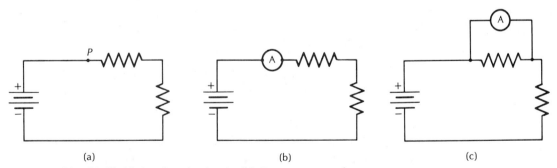

Fig. 29.15 (a) An electric circuit. (b) Correct connection of the ammeter. (c) Incorrect connection of the ammeter.

To measure the potential difference between, say, the points *P* and *P'* in the circuit, the experimenter must connect the terminals of the voltmeter to these points. Figure 29.16a shows the correct way of connecting the voltmeter (for comparison, Figure 29.16b shows a wrong way). Since the voltmeter has a very large internal resistance, it draws

---

* This section is optional.

Fig. 29.16 (a) Correct connection of the voltmeter. (b) Incorrect connection of the voltmeter.

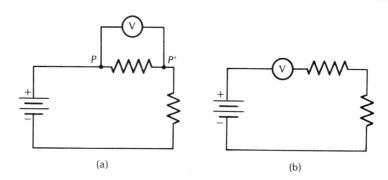

only a very small current, and the alteration in the flow of current through the resistance PP', and the consequent alteration of the potential difference between P and P', are usually insignificant. However, if the resistance PP' across which the voltmeter is connected is large, then the voltmeter can draw a significant fraction of the current, with a consequent decrease of potential across the resistance PP'. For an accurate measurement of the potential, the experimenter must select a voltmeter of an internal resistance much larger than the resistance PP'.

POTENTIOMETER A voltmeter can be used for a direct measurement of the emf of a battery, or other source of emf. However, when the voltmeter is connected across the terminals of the battery, it draws a small amount of current, and the potential difference between the terminals of the battery will then be reduced by the drop in the internal resistance of the battery (see the discussion at the end of Section 29.3). To avoid this difficulty, it is desirable to measure the emf when no current is flowing. The potentiometer is a precision instrument used for such measurements of the emf at zero current, or the "open-circuit voltage."

The potentiometer is an electrical balance — it compares an unknown emf with a known reference emf. In order to attain the condition of zero current for both the unknown emf and the known emf, the comparison is made indirectly, in two steps, by means of a third, auxiliary emf. A schematic diagram of the potentiometer is shown in Figure 29.17. The auxiliary emf $\mathscr{E}_0$ is connected to a long uniform wire of a metal of fairly high resistivity; in Figure 29.17, this wire is represented by the long resistance PN. A steady current will then flow in the wire, and the potential along the wire will be a linearly decreasing function of the length of wire measured from the point P. In the first step of the measurement, the experimenter connects the unknown source of emf $\mathscr{E}_x$ to the end P of the wire and to a sensitive ammeter A. The experimenter then moves the sliding contact O until the ammeter A reads zero current. Under these conditions, the emf of the unknown source exactly matches the potential drop along the length PO of wire. The potentiometer is then said to be balanced. In the second step of the measurement, the experimenter replaces the unknown source of emf $\mathscr{E}_x$ by the known reference source $\mathscr{E}_s$ and again balances the potentiometer. The emf of the reference source will match the potential drop along some length PO' of the wire. The ratio of the lengths PO and PO' is equal to the ratio of the unknown emf $\mathscr{E}_x$ and the known emf $\mathscr{E}_s$,

$$\frac{\mathscr{E}_x}{\mathscr{E}_s} = \frac{PO}{PO'} \tag{30}$$

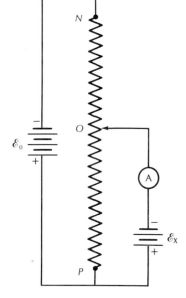

Fig. 29.17 Schematic diagram for a potentiometer.

From this, $\mathscr{E}_x$ can be readily calculated. Note that the result is completely independent of the auxiliary emf; the only relevant property of this emf is that it must remain constant during the steps of the operation of the potentiometer.

The advantage of the potentiometer over the voltmeter is not only that the unknown emf is measured at zero current, but also that the standard cell, used as reference emf, maintains a precisely fixed value of its emf, which provides a more reliable calibration of the potential than is to be had with a voltmeter.

WHEATSTONE BRIDGE The Wheatstone bridge is used for precise comparisons of an unknown resistance with a known reference resistance. Its operation is based on much the same principles as the potentiometer. Figure 29.18 shows a schematic diagram of a Wheatstone bridge. Again, a long uniform wire is connected across an auxiliary source of emf $\mathscr{E}_0$. One end of the unknown resistance is connected to the point P, and one end of a known reference resistance is connected to the point N. The other ends of these two resistances are joined at the ammeter A. To balance the bridge, the experimenter moves the sliding contact O until the ammeter reads zero current. Thus, steady currents I_1 and I_2 flow along the parallel paths of the circuit, and no current crosses the bridge AO. Under these conditions, the potential drop $I_2 R_X$ must equal the potential drop across the segment PO of the wire, and the potential drop $I_2 R_S$ must equal the potential drop across the segment ON. Since the potential drops in the wire are in proportion to the lengths,

$$I_2 R_X \propto PO \text{ and } I_2 R_S \propto ON \tag{31}$$

Taking the ratio of these proportions, we obtain

$$\frac{R_X}{R_S} = \frac{PO}{ON} \tag{32}$$

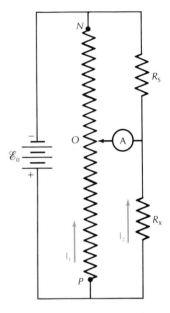

Fig. 29.18 Schematic diagram for a Wheatstone bridge with a slide wire.

This equation permits the calculation of the unknown resistance. Again, the result is independent of the auxiliary emf.

In essence, the operation of the Wheatstone bridge hinges on an adjustment of the ratio of resistances in the circuit. In some variants of the Wheatstone bridge, the lengths (and resistances) PO and ON are held fixed, and the ratio R_X/R_S is varied by using an adjustable, calibrated resistance R_S. If $PO = ON$, then the bridge will balance when $R_X = R_S$; in this case, the dial on the adjustable, calibrated resistance directly tells us the value of the unknown resistance.

29.7* The RC Circuit

Throughout this chapter we have dealt only with steady, time-independent currents. However, Kirchhoff's rule and the methods for solving circuits we have developed in this chapter also apply to time-dependent currents. The only restriction is that the emfs and the currents in the circuit must not vary too quickly. We recall from Section

---

*This section is optional.

29.3 that Kirchhoff's rule amounts to an energy balance for a hypothetical charge moving around the circuit. To permit the formulation of such an energy balance, the currents and emfs in the circuit must not change significantly in the time it takes the charge to complete its trip around the circuit. The relevant travel time is not the (very long) time for a free electron to wander around the circuit; rather, it is the time taken by a hypothetical charge moving with the maximum conceivable speed, that is, the speed of light. A rough criterion for the applicability of Kirchhoff's rule is then that the currents and emfs must not change significantly in an interval equal to the travel time for a light signal around the circuit.

In Chapter 34 we will deal with a variety of circuits with time-dependent currents. Here we will deal with the simple case of a time-dependent current in a circuit consisting of a resistor and a capacitor connected in series and being charged by a battery. Figure 29.19 shows a schematic diagram for such an RC circuit. We assume that the capacitor is initially uncharged and that the battery is suddenly connected at time $t = 0$. Initially, the potential difference across the capacitor is then zero. When the battery is connected, charge flows from the terminals of the battery to the plates of the capacitor. As the charge accumulates on the plates, the potential difference across the plates gradually increases. Obviously, the flow of charge will come to a halt when the potential difference across the plates matches the emf of the battery. This qualitative discussion of the charging process indicates that the current in the circuit is initially large, but gradually tapers off, and ultimately approaches zero.

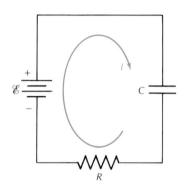

Fig. 29.19 An RC circuit with a battery.

For a precise mathematical treatment of the time dependence of the current in the circuit, we turn to Kirchhoff's rule: the sum of all the emfs and voltage drops around the circuit must be zero. The emf of the battery is $\mathscr{E}$. If the current is I, the voltage drop across the resistor is IR. And if the charge on the capacitor plates has a magnitude Q, the voltage difference across the plates is Q/C. Hence

$$\mathscr{E} - IR - \frac{Q}{C} = 0 \tag{33}$$

Since the current is the rate of change of the charge, we can write this equation as

$$\mathscr{E} - R\frac{dQ}{dt} - \frac{Q}{C} = 0 \tag{34}$$

This is a differential equation — it involves the charge and the first derivative of the charge. In later chapters we will encounter many other examples of circuits with time-dependent currents, for which Kirchhoff's rule leads to differential equations.

To integrate this differential equation, we collect all terms involving Q on one side of the equation, and all terms involving t on the other:

$$\frac{dQ}{C\mathscr{E} - Q} = \frac{dt}{RC} \tag{35}$$

We can now integrate both sides of this equation:[3]

---

[3] The variables of integration have been written with primes to distinguish them from the limits of integration.

$$\int_0^Q \frac{dQ'}{C\mathcal{E} - Q'} = \frac{1}{RC} \int_0^t dt' \tag{36}$$

which gives

$$-\ln(C\mathcal{E} - Q) + \ln(C\mathcal{E}) = \frac{t}{RC}$$

or

$$\ln\left(1 - \frac{Q}{C\mathcal{E}}\right) = -\frac{t}{RC} \tag{37}$$

If we take the exponential function[4] of both sides and recall that, for any variable x, $e^{\ln x} = x$, we obtain the final result

$$1 - \frac{Q}{C\mathcal{E}} = e^{-t/RC}$$

or

$$\boxed{Q = C\mathcal{E}(1 - e^{-t/RC})} \tag{38}$$

Figure 29.20a is a plot of Q/C, or the potential across the capacitor, as a function of time. In agreement with our qualitative discussion, the potential asymptotically approaches a final value $\mathcal{E}$.

The current in the circuit can be evaluated by differentiating Eq. (38):

$$I = \frac{dQ}{dt} = \frac{\mathcal{E}}{R} e^{-t/RC} \tag{39}$$

Figure 29.20b is the plot of this current as a function of time. This current has an initial (maximum) value of $\mathcal{E}/R$, and it gradually decreases and asymptotically tends to zero.

Fig. 29.20 (a) Potential across capacitor vs. time during charging by a battery. (b) Current vs. time.

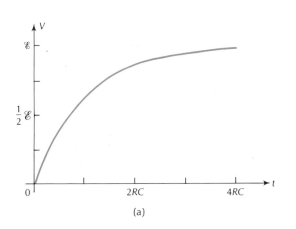
(a)

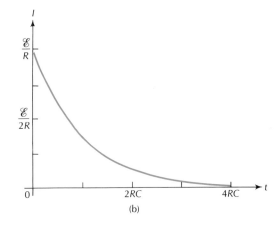
(b)

---

[4] See Appendix 2 for a summary of the properties of logarithms and exponentials.

The quantity RC that appears in the arguments of the exponential functions in Eqs. (38) and (39) is called the **time constant,** or the relaxation time, of the RC circuit. As is obvious from Eq. (39), at the time $t = RC$, the current is smaller than its maximum value by a factor of $1/e$, or about 0.37.

EXAMPLE 9. Suppose that in the circuit shown in Figure 29.19, the resistance is $R = 8.0 \times 10^3$ Ω, the capacitance is $C = 2.0$ μF, and the emf of the battery is $\mathscr{E} = 1.5$ V. What is the initial value of the current, at the instant after the battery has been connected? At what time is the current smaller by a factor of $1/e$? At what time is it smaller by a factor of $1/2$?

SOLUTION: The initial value of the current is

$$I = \frac{\mathscr{E}}{R} = \frac{1.5 \text{ V}}{8.0 \times 10^3 \text{ Ω}} = 1.9 \times 10^{-4} \text{ A}$$

The time at which the current is smaller by a factor of $1/e$ is the time constant, RC:

$$t = RC = 8.0 \times 10^3 \text{ Ω} \times 2.0 \times 10^{-6} \text{ F} = 1.6 \times 10^{-2} \text{ s} = 0.016 \text{ s}$$

According to Eq. (39), the time at which the current is $1/2$ of its initial value is given by

$$1/2 = e^{-t/RC}$$

or

$$\ln(1/2) = -t/RC$$

From which

$$t = RC \ln 2 = 0.016 \text{ s} \times 0.69 = 0.011 \text{ s}$$

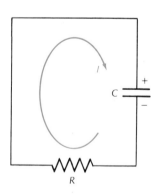

Fig. 29.21 An RC circuit, with the capacitor discharging through the resistor.

Once the charging process has been completed and the current in the circuit has stopped, we can disconnect the battery. The charge will then remain on the capacitor plates (except for a slow leakage through the capacitor or into the air). However, suppose we now connect the free terminals of the capacitor and the resistor, as shown in Figure 29.21. The capacitor will then discharge through the resistor. Obviously, the current will be large at the initial time, and gradually taper off and approach zero as the potential difference across the capacitor decreases. Kirchhoff's rule for the circuit of Figure 29.21 yields

$$-IR - \frac{Q}{C} = 0 \qquad (40)$$

or

$$-R\frac{dQ}{dt} - \frac{Q}{C} = 0 \qquad (41)$$

This equation can be integrated by the same method as Eq. (34), with the result

$$Q = \mathcal{E}Ce^{-t/RC} \tag{42}$$

Here $\mathcal{E}C$ is the charge on the capacitor at the initial time $t = 0$, when the discharging process begins. The current in the circuit is

$$I = \frac{dQ}{dt} = -\frac{\mathcal{E}}{R}e^{-t/RC} \tag{43}$$

This current has a negative sign because it flows in a direction opposite to that of the original current that charged the capacitor. Figures 29.22a and 29.22b give plots of the potential difference Q/C across the discharging capacitor and the current in the circuit. The time constant for the discharging capacitor is, again, RC. At this time, the current will have decreased by a factor of $1/e$ from its initial value.

Fig. 29.22 (a) Potential vs. time during discharging. (b) Current vs. time.

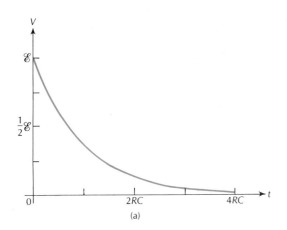

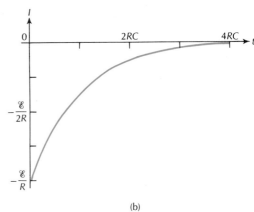

29.8 The Hazards of Electric Currents

As a side effect of the widespread use of electric machinery and devices in factories and homes, each year in the United States about 1000 people die by accidental electrocution. A much larger number suffer nonfatal electric shocks. Fortunately, the human skin is a fairly good insulator, which provides a protective barrier against injurious electric currents. The resistance of a square centimeter of dry human epidermis in contact with a conductor can be as much as 10^5 Ω. However, the resistance varies in a rather sensitive way with the thickness, moisture, and temperature of the skin, and with the magnitude of the potential difference.[5]

The electric power supplied to factories and homes in the United States is usually in the form of alternating currents, or AC. These are oscillating currents, which periodically reverse direction (the standard period for the alternating current supplied by power companies is $\frac{1}{60}$ s). Since most accidental electric shocks involve alternating currents, the following discussion of the effects of currents on the human body will emphasize alternating currents.

---

[5] The variation of resistance with potential difference implies that skin does not obey Ohm's Law.

In the typical accidental electric shock, the current enters the body through the hands (in contact with one terminal of the source of emf) and exits through the feet (in contact with the ground, which constitutes the other terminal of the source of emf in most AC circuits). Thus the body plays the role of a resistor, closing an electric circuit.

The damage to the body depends on the magnitude of the current passing through it. An alternating current of about 0.001 A produces only a barely detectable tingling sensation. Higher currents produce pain and strong muscular contractions. If the victim has grasped an electric conductor, such as an exposed power cable, with the hand, the muscular contraction may prevent the victim from releasing the hold on the conductor. The magnitude of the "let-go" current, at which the victim can barely release the hold on the conductor, is about 0.01 A. Higher currents lock the victim's hand to the conductor. Unless the circuit is broken within a few seconds, the skin in contact with the conductor will then suffer burns and blisters. Such damage to the skin drastically reduces its resistance, which can lead to a fatal increase of the current.

An alternating current of about 0.02 A flowing through the body from the hands to the feet produces a contraction of the chest muscles that halts breathing; this leads to death by asphyxiation if it lasts for a few minutes. A current of about 0.1 A lasting just a few seconds induces fibrillation of the heart. This is a rapid, uncoordinated flutter of the heart muscles, with cessation of the natural rhythm of the heartbeat and cessation of the pumping of blood. Fibrillation usually continues even when the victim is removed from the electric circuit; the consequences are fatal unless immediate medical assistance is available. The treatment for fibrillation involves the deliberate application of a severe electric shock to the heart by means of electrodes placed against the chest; this arrests the motion of the heart completely. When the shock ends, the heart usually resumes beating with its natural rhythm.

A current of a few amperes produces a block of the nervous system and paralysis of the respiratory muscles. Victims of such currents can sometimes be saved by prompt recourse to artificial respiration. At these high values of the current, the effects of AC and DC are not very different. But at lower values, a DC current poses less of a hazard than a comparable AC current, because the former does not trigger the strong muscular contractions triggered by the latter.

In the above we assumed that the path of the current through the body is from the hands to the feet. If the current enters and exits through the same arm or leg, or enters through one leg and exits through the other, no vital organs lie in its path, and the threat to life is lessened. However, an intense current through a limb tends to kill the tissue through which it passes, and may ultimately require the surgical excision of large amounts of dead tissue, and even the amputation of the limb.

Other things being equal, a higher voltage will result in a higher current. The hazard posed by contact with high-voltage sources is therefore obvious. But under exceptional circumstances, even sources of low voltage can be hazardous. Several cases of electrocution by contact with sources of a voltage as low as 12 V have been reported. It seems that in these cases death resulted from an unusually sensitive response of the nervous system; it is also conceivable that an unusually small skin resistance was a contributing factor. Thus, it is advisable to treat even sources of low voltage with respect!

Aid to victims of electric shock should begin with switching off the current. When no switch, plug, or fuse for cutting off the current is accessible, the victim must be pushed or pulled away from the electric conductor by means of a piece of insulating material, such as a piece of *dry* wood or a rope. The rescuer must be careful to avoid electric contact. If the victim is not breathing, artificial respiration must be started at once. If there is no heartbeat, cardiac massage must be applied by trained personnel until the victim can be treated with a defibrillation apparatus.

First aid for electric shock

SUMMARY

Kirchhoff's (second) rule: The sum of emfs and voltage drops around any closed loop in a circuit equals zero.

Power delivered by source of emf: $P = I\mathscr{E}$

Power dissipated by resistor: $P = I\,\Delta V$

Time constant of RC circuit: RC

QUESTIONS

1. Can we use a capacitor as a pump of electricity in a circuit? In what way would such a pump differ from a battery?

2. Does a fully charged battery have the same emf as a partially charged battery?

3. The emf of a battery is often called the "open-circuit voltage." Explain.

4. How would you measure the internal resistance of a battery?

5. Kirchhoff's second rule is equivalent to energy conservation in the electric circuit. Explain.

6. What would happen if we were to connect the ammeter incorrectly, as shown in Figure 29.15c?

7. What would happen if we were to connect the voltmeter incorrectly, as shown in Figure 29.16b?

8. A mechanic determines the internal resistance of an automobile battery by connecting a rugged, high-current ammeter (of nearly zero resistance) directly across the poles of the battery. The internal resistance is inversely proportional to the ammeter reading, $R_i \propto 1/I$. Explain.

9. An **ohmmeter** consists of a reference source of emf (a battery), connected in series with a reference resistance, and an ammeter. When the terminals of the ohmmeter are connected to an unknown resistor (Figure 29.23), the current registered by the ammeter permits the evaluation of the unknown resistance. Explain.

Ohmmeter

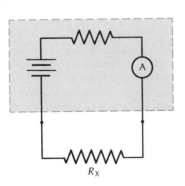

Fig. 29.23

10. If a voltmeter and an ammeter are connected to a resistor simultaneously, their readings can be used to evaluate the resistance, according to $R = \Delta V/I$. Figure 29.24 shows two ways of connecting the ammeter and the voltmeter to the resistor. Explain why both of these methods yield a value of R slightly different from the actual value.

Fig. 29.24

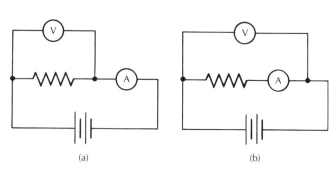

(a) (b)

Shunt

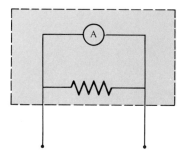

Fig. 29.25 An ammeter with a parallel resistor, or shunt resistor.

11. In order to reduce the effective internal resistance of an ammeter, a physicist connects a resistor in parallel across its terminal (see Figure 29.25). If the resistance of this parallel resistor, or **shunt** resistor, is 1/10 of the resistance of the ammeter, by what factor does the shunt reduce the effective resistance of the modified ammeter? How must the physicist recalibrate the dial of the ammeter? What happens to the sensitivity of the ammeter?

12. A homeowner argues that he should not pay his electric bill, since he is not keeping any of the electrons that the power company delivers to his home — any electron that enters the wiring of his home sooner or later leaves and returns to the power station. How would you answer?

13. The spiral heating elements commonly used in electric ranges *appear* to be made of metal. Why do they not short-circuit when you place an iron pot on them?

14. What are the advantages and what are the disadvantages of high-voltage power lines?

15. In many European countries, electric power is delivered to homes at 220 V, instead of the 110 V customary in the United States. What are the advantages and what are the disadvantages of 220 V?

16. How much does it cost you to operate a 100-watt light bulb for 24 hours? The price of electric energy is 8¢ per kilowatt-hour.

PROBLEMS

Section 29.1

1. The smallest batteries have a mass of 1.5 grams and store an electric energy of about 5×10^{-6} kW·h. The largest batteries (used aboard submarines) have a mass of 270 metric tons and store an electric energy of 5×10^3 kW·h. What is the amount of energy stored per kilogram of battery in each case?

2. A size D flashlight battery will deliver 1.2 A·h at 1.5 V (i.e., it will deliver 1.2 A for 1 h or a larger or smaller current for a correspondingly shorter or longer time). An automobile battery will deliver 55 A·h at 12 V. The flashlight battery is cylindrical with a diameter of 3.3 cm, a length of 5.6 cm, and a mass of 0.086 kg. The automobile battery is rectangular with dimensions 30 cm × 17 cm × 23 cm and a mass of 23 kg.
   (a) What is the available electric energy stored in each battery?

(b) What is the amount of energy stored per cubic centimeter of each battery?

(c) What is the energy stored per kilogram of each battery?

3. A heavy-duty 12-V battery for a truck is rated at 160 A·h, that is, this battery will deliver 1 A for 160 h (or a larger current for a correspondingly shorter time). What is the amount of electric energy that this battery will deliver?

*4. The electric starter motor in an automobile equipped with a 12-V battery draws a current of 80 A when in operation.
   (a) Suppose it takes the starter motor 3.0 s to start the engine. What amount of electric energy has been withdrawn from the battery?
   (b) The automobile is equipped with a generator that delivers 5.0 A to the battery when the engine is running. How long must the engine run so that the generator can restore the energy in the battery to its original level? Assume that all the power delivered to the battery is stored.

Section 29.3

5. In a flashlight, two 1.5-V batteries are connected in series. An 8-Ω light bulb closes the circuit. What is the current in the circuit? What is the electric energy delivered to the light bulb in 1 hour?

6. Find the currents in the circuit shown in Figure 29.11 if the battery on the right is reversed.

7. A voltmeter of internal resistance 5.0×10^4 Ω is connected across the poles of a 12-V battery of internal resistance 0.020 Ω.
   (a) What is the current flowing through the battery?
   (b) What is the voltage drop across the internal resistance of the battery?

*8. To measure the emf and the internal resistance of a battery, an experimenter connects the battery in series to a (resistanceless) ammeter and a variable resistor. She finds that when the variable resistor is set at 1.0 Ω, the current in the circuit 0.40 A; and when the variable resistor is set at 2.0 Ω, the current is 0.22 A. What emf and what internal resistance for the battery can she deduce from this?

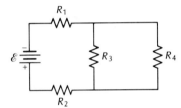

Fig. 29.26

Section 29.4

*9. Four resistors, with $R_1 = 25$ Ω, $R_2 = 15$ Ω, $R_3 = 40$ Ω, and $R_4 = 20$ Ω, are connected to a 12-V battery as shown in Figure 29.26.
   (a) Find the combined resistance of the four resistors.
   (b) Find the current in each resistor.

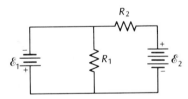

Fig. 29.27

*10. Consider the circuit shown in Figure 29.27. Given that $\mathscr{E}_1 = 6.0$ V, $\mathscr{E}_2 = 10$ V, and $R_1 = 2.0$ Ω, what must be the value of the resistance R_2 if the current through this resistance is to be 2.0 A?

*11. Find the current in the two resistors shown in Figure 29.28. Find the power delivered by the 12-V battery. The resistances are $R_1 = 10$ Ω, $R_2 = 8$ Ω; the emfs are $\mathscr{E}_1 = 10$ V and $\mathscr{E}_2 = 12$ V.

*12. Two batteries of emf $\mathscr{E}$ and of internal resistances R_i and R_i', respectively, are combined in parallel. The combination is connected to an external resistance R. Find the current through each battery.

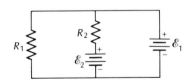

Fig. 29.28

*13. Consider the circuit shown in Figure 29.11 with the resistances and emfs given in Example 4. Suppose we replace the emf $\mathscr{E}_1 = 12.0$ V by a larger emf. How large must we make $\mathscr{E}_1$ if the current I_2 is to charge the battery $\mathscr{E}_2$?

*14. Two batteries with internal resistances are connected as shown in Figure 29.29. Given that $R_1 = 0.50$ Ω, $R_2 = 0.20$ Ω, $\mathscr{E} = 12.0$ V, $\mathscr{E}' = 6.0$ V, $R_i = 0.025$ Ω, and $R_i' = 0.020$ Ω, find the currents in the resistances R_1 and R_2.

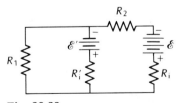

Fig. 29.29

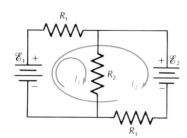

Fig. 29.30

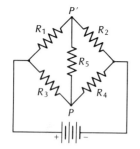

Fig. 29.31

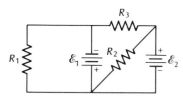

Fig. 29.32

*15. When using the loop method to obtain the equations for a circuit, we can make several choices for the loops. In Figure 29.11, we made one possible choice and obtained Eqs. (29.12) and (29.13). Suppose that instead we make the choice shown in Figure 29.30. What are the two loop equations in this case? Show that the new loop equations are mathematically equivalent to Eqs. (29.12) and (29.13).

**16. Five resistors, of resistances $R_1 = 2.0\ \Omega$, $R_2 = 4.0\ \Omega$, $R_3 = 6.0\ \Omega$, $R_4 = 2.0\ \Omega$, and $R_5 = 3.0\ \Omega$, are connected to a 12-V battery as shown in Figure 29.31.
  (a) What is the current in each resistor?
  (b) What is the potential difference between the points P and P'?

**17. Two batteries, with $\mathscr{E}_1 = 6.0$ V and $\mathscr{E}_2 = 3.0$ V, are connected to three resistors, with $R_1 = 6.0\ \Omega$, $R_2 = 4.0\ \Omega$, and $R_3 = 2.0\ \Omega$, as shown in Figure 29.32. Find the current in each resistor and the current in each battery.

Section 29.5

18. A small electric motor operating on 110 V delivers 0.75 hp of mechanical power. Ignoring friction losses within the motor, what current does this motor require?

19. The rate of flow of water over the Niagara Falls is 2800 m³/s; this water falls a vertical distance of 51 m. At night, one-half of the water is diverted to a power plant. If the plant converts all of the potential energy of this diverted water into electric power, what is the electric power in kilowatts? If this power is fed into a power line at 240 kV, what is the current?

20. A cyclotron accelerator produces a beam of protons of an energy of 700 million eV. The average current of this beam is 1.0×10^{-6} A. What is the number of protons per second delivered by the accelerator? What is the corresponding power delivered by the accelerator?

21. An electric toaster uses 1200 W at 110 V. What is the current through the toaster? What is the resistance of its heating coils?

*22. A hair dryer intended for travelers operates at 110 V and also at 220 V. A switch on the dryer adjusts the dryer for the voltage. At each voltage, the dryer delivers 1000 W of heat. What must be the resistance of the heating coils for each voltage? For such a dryer, design a circuit consisting of two identical heating coils connected to a switch and to the power outlet.

23. While cranking the engine, the starter motor of an automobile draws 80 A at 12.0 V for a time interval of 2.5 s. What is the electric power used by the starter motor? How many horsepower does this amount to? What is the electric energy used up in the given time interval?

*24. A 12-V battery of internal resistance 0.20 Ω is being charged by an external source of emf delivering 6.0 A.
  (a) What must be the minimum emf of the external source?
  (b) What is the rate at which heat is developed in the internal resistance of the battery?

*25. An electric automobile is equipped with an electric motor supplied by a bank of sixteen 12-V batteries connected in series. When fully charged, each battery stores an energy of 2.2×10^6 J.
  (a) What current is required by the motor when it is delivering 12 hp? Ignore friction losses.
  (b) With the motor delivering 12 hp, the car has a speed of 65 km/h (on a level road). How far can the automobile travel before its batteries run down?

*26. The banks of batteries in a submarine store an electric energy of 5×10^3 kW·h. If the submarine has an electric motor developing 1000 hp, how long can it run on these batteries?

*27. An electric toothbrush draws 7 watts. If you use it 4 minutes per day and if electric energy costs you 8¢/kW·h, what do you have to pay to use your toothbrush for one year?

*28. In a small electrostatic generator, a rubber belt transports charge from the ground to a spherical collector at 2.0×10^5 V. The rate at which the belt transports charge is 2.5×10^{-6} C/s. What is the rate at which the belt does work against the electrostatic forces?

*29. A solar panel (an assemblage of solar cells) measures 58 cm × 53 cm. When facing the sun, this panel generates 2.7 A at 14 V. Sunlight delivers an energy of 1.0×10^3 W/m² to an area facing it. What is the efficiency of this panel, that is, what fraction of the energy in sunlight is converted into electric energy?

*30. A battery of emf $\mathscr{E}$ and internal resistance R_i is connected to an external circuit of resistance R. In terms of $\mathscr{E}$, R_i, and R, what is the power delivered by the battery to the external circuit? Show that this power is maximum if $R = R_i$.

*31. A 3.0-V battery with an internal resistance of 2.5 Ω is connected to a light bulb of a resistance of 6.0 Ω. What is the voltage delivered to the light bulb? How much electric power is delivered to the light bulb? How much electric power is wasted in the internal resistance?

*32. Suppose that a 12-V battery has an internal resistance of 0.40 Ω.
  (a) If this battery delivers a steady current of 1.0 A into an external circuit until it is completely discharged, what fraction of the initial stored energy is wasted in the internal resistance?
  (b) What if the battery delivers a steady current of 10.0 A? Is it more efficient to use the battery at low current or at high current?

*33. Two heating coils have resistances of 12.0 Ω and 6.0 Ω, respectively.
  (a) What is the Joule heat generated in each if they are connected in parallel to a source of emf of 110 V?
  (b) What if they are connected in series?

*34. A 40-m cable connecting a lightning rod on a tower to the ground is made of copper and has a diameter of 7 mm. Suppose that during a stroke of lightning the cable carries a current of 1×10^4 A.
  (a) What is the potential drop along the cable?
  (b) What is the rate at which Joule heat is produced?

*35. The maximum current recommended for a No. 10 copper wire, of diameter 0.259 cm, is 25 A. For such a wire with this current, what is the rate of production of Joule heat per meter of wire? What is the potential drop per meter of wire?

*36. The cable connecting the electric starter motor of an automobile with the 12.0-V battery is made of copper and has a diameter of 0.50 cm and a length of 0.60 m. If the starter motor draws 500 A (while stalled), what is the rate at which Joule heat is produced in the cable? What fraction of the power delivered by the battery does this Joule heat represent?

*37. An electric clothes dryer operates on a voltage of 220 V and draws a current of 20 A. How long does the dryer take to dry a full load of clothes? The clothes weigh 6.0 kg when wet and 3.7 kg when dry. Assume that all the electric energy going into the dryer is used to evaporate water (the heat of evaporation is 539 kcal/kg).

*38. A large electromagnet draws a current of 200 A at 400 V. The coils of the electromagnet are cooled by a flow of water passing over them. The water enters the electromagnet at a temperature of 20°C, absorbs the Joule heat, and leaves at a higher temperature. If the water is to leave with a temperature no higher than 80°C, what must be the minimum rate of flow of water (in liters per minute) through the electromagnet?

Section 29.6

39. A battery of unknown emf is being measured with a potentiometer. When this battery is inserted in the potentiometer, the balance is achieved with $PO = 30.2$ cm. When a standard cell of emf 1.50 V is inserted in the potentiometer, the balance is achieved at $PO = 44.5$ cm. What is the emf of the battery?

40. An unknown resistance is being measured with a Wheatstone bridge equipped with a slide wire of length 1.00 m and a reference resistance of 200 Ω. The bridge balances at $PO = 68.4$ cm. What is the value of the unknown resistance?

*41. To measure the internal resistance of a battery, a physicist places the battery in a potentiometer and finds that it balances at a length $PO = 55.2$ cm. Without removing the battery from the potentiometer, he then connects a 2.0-Ω resistor across the terminals of the battery and, while a current is flowing through the resistor-battery circuit, he again balances the potentiometer. The new balance is at 53.4 cm. What is the internal resistance of the battery?

*42. A voltmeter of internal resistance 5000 Ω is connected across the poles of a battery of internal resistance 0.2 Ω. The voltmeter reads 1.4993 volts. What is the actual zero-current emf of the battery?

*43. A circuit consists of a resistor of 3.0 Ω connected to a (resistanceless) battery. To measure the current in this circuit, you insert an ammeter of internal resistance 2×10^{-3} Ω. This ammeter then reads 3.955 A. What was the current in the circuit before you inserted the ammeter?

*44. A voltmeter reads 11.9 V when connected across the poles of a battery. The internal resistance of the battery is 0.020 Ω. What must be the minimum value of the internal resistance of the voltmeter if the reading of the instrument is to coincide with the emf of the battery to within better than 1%?

Section 29.7

45. For a laboratory demonstration, you want to construct an RC circuit with a time constant of 15 s. You have available a capacitor of 20 μF. What resistance do you need?

46. A capacitor with $C = 20$ μF and a resistor with $R = 100$ Ω are suddenly connected in series to a battery with $\mathscr{E} = 6.0$ V.
  (a) What is the charge on the capacitor at $t = 0$? At $t = 0.001$ s? At $t = 0.002$ s?
  (b) What is the final value of the charge?
  (c) What is the rate of increase of the charge at $t = 0$?

47. Ideal capacitors have an infinite internal resistance between their plates (that is, the material between the plates is a perfect insulator). However, real capacitors have a finite internal resistance, and consequently the charge will gradually leak from one plate to the opposite, and the capacitor will gradually discharge when left to itself. If a capacitor of 8.0 μF has an internal resistance of 5.0×10^8 Ω, how long does it take for one-half of its initial charge to leak away?

48. Two capacitors of 2.0 μF and 4.0 μF and a resistor of 8×10^3 Ω are connected to a battery as shown in Figure 29.33. What is the time constant of this circuit?

49. An RC circuit consists of a resistance R, a capacitance C, and a battery connected in series. At what time is the current 1/10 of its initial (maximum) value? At what time is the current 1/100 of its initial value?

*50. A capacitor with $C = 0.25$ μF is initially charged to a potential of 6.0 V. The capacitor is then connected across a resistor and allowed to discharge.

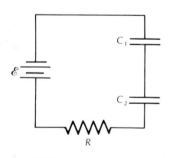

Fig. 29.33

After a time of 5.0×10^{-3} s, the potential across the capacitor has dropped to 1.2 V. What value of the resistance can you deduce from this?

*51. Consider the RC circuit described in Example 9. The final (asymptotic) value of the charge in the capacitor is $\mathscr{E}C$. At what time is the charge in the capacitor one-half of this final value? At what time is the electric energy in the capacitor one-half of its final value?

*52. Integrate Eq. (41) and obtain Eq. (42).

*53. Equation (34) is a differential equation for the charge. Substitute the expression (38) into this equation and verify explicitly that it is a solution.

**54. An RC circuit, with $R = 6 \times 10^5$ Ω and $C = 9$ μF, is connected to a 5-V battery until the capacitor is fully charged. Then, the battery is suddenly replaced with a new 3-V battery of opposite polarity. At what time after this replacement will the voltage across the capacitor be zero? What will be the current in the resistor at this time?

**55. Equations (38) and (39) describe the charge and current in an RC circuit with a battery. From these equations, deduce the total energy dissipated as Joule heat in the resistor in the time interval $t = 0$ to $t = \infty$, and deduce the total energy that the battery has to deliver in this time interval.

INTERLUDE VI

ATMOSPHERIC ELECTRICITY*

The atmosphere is a great electric machine. Thunderstorms are the most spectacular manifestation of electric activity in the atmosphere (Figure VI.1), but even in fair weather the atmosphere is endowed with electric fields and electric currents. The thunderstorms act as giant electrostatic generators, delivering negative charge to the ground, and positive charge to the upper level of the atmosphere. This upper level of the atmosphere, or ionosphere, is a good conductor and the current reaching it quickly spreads laterally, over the entire globe. In fair-weather regions, this current gradually leaks down to the ground, completing the **atmospheric electric circuit** (Figure VI.2).

There are roughly 2000 thunderstorms in action all over the Earth at any given time. The time-average current generated by a thunderstorm is about 1 A (the instantaneous current can of course be much larger — up to 20,000 A in a stroke of lightning). Thus, all the thunderstorms together contribute an average of 2000 A to the current in the atmospheric circuit. Bursting air bubbles at the ocean surface also contribute to the current, by spraying small positively charged droplets into the atmosphere; but this contribution is believed to be appreciably smaller than that of thunderstorms.

Figure VI.3 is a schematic diagram of the atmospheric electric circuit. Both the ionosphere and the ground are good conductors, and each is an equipotential surface; the potential difference between them is about 300,000 V. The resistance of the entire fair-weather atmosphere between ionosphere and ground is about 200 Ω. Most of this resistance is concentrated in the dense, low regions of the atmosphere; correspondingly, most of the 300,000-V potential drop occurs in the low regions of the atmosphere, within a few kilometers from the ground. On the average, the total electric power delivered to the global circuit by thunderstorm activity is about 2000 A × 300,000 V = 6 × 10^8 W, that is, nearly a million kW.

VI.1 THE FAIR-WEATHER ELECTRIC FIELD

The atmospheric current that, in the fair-weather regions of the Earth, flows from the ionosphere to the

* This chapter is optional.

Fig. VI.1 Thunderstorm with lightning.

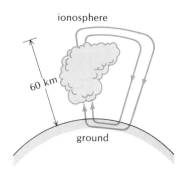

Fig. VI.2 Flow of current in the Earth's atmosphere.

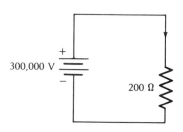

Fig. VI.3 Schematic diagram of the atmospheric electric circuit.

ground is carried by ions — atoms and molecules that have a positive or negative charge due to a deficit or excess in the number of their electrons. Such ions are always present in the atmosphere. They are continually produced by the impacts of cosmic rays on air molecules throughout the atmosphere, by the ultraviolet irradiation of the upper atmosphere, and by the natural and man-made radioactivity near the ground. Thus the current consists of a downward motion of positive ions and an upward motion of negative ions. The driving force that maintains this motion is an **atmospheric electric field** that points vertically downward; this electric field is analogous to the electric field that exists in a current-carrying wire and drives the electrons along the wire. Near the surface of the Earth, over open ground, in fair weather, the strength of this atmospheric electric field is between 100 and 200 V/m. The exact value of the strength of the field depends on local conditions, such as dust in the atmosphere, topography, time of day; the worldwide average value is 130 V/m.

The potential difference between the ground and a point 2 m above the ground (the height of a man) is therefore typically a few hundred volts. Does this mean that we can operate an appliance by plugging one terminal into the ground and the other terminal into air, 2 m above ground? Unfortunately, we cannot: the atmosphere will deliver only an infinitesimal current to the exposed terminal, and therefore that terminal (and the entire appliance) will remain at the potential of the ground. The only consequence of sticking a wire, or any conductor, into the air is to deform the atmospheric equipotentials. The 0-V equipotential will follow the shape of the conductor; successively higher equipotentials will exhibit successively less deformation (Figure VI.4). Buildings, trees, or human bodies are reasonably good conductors; they will deform the equipotentials, and the electric field, in a similar manner.

The lines of the atmospheric electric field end on the surface of the Earth; thus, there must be negative charge on the surface. The amount of charge per unit area is given by Eq. (24.23):

$$\sigma = \varepsilon_0 E = -8.85 \times 10^{-12} \text{ F/m} \times 100 \text{ C/m} \cong -10^{-9} \text{ C/m}^2$$

The total negative charge on the surface of the Earth (including the oceans) is about half a million coulombs.

The electric field near the ground can be measured with a **field meter**. A simple kind of field meter is constructed with a horizontal metallic plate placed near the ground and connected to the ground by means of a wire. The lines of electric field end on the surface of the metallic plate; this surface must therefore carry an electric charge. If a second metallic plate, also con-

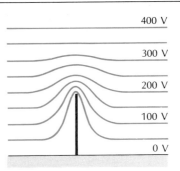

Fig. VI.4 Equipotentials in air in the vicinity of a thin vertical conductor connected to the ground.

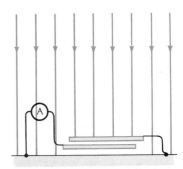

Fig. VI.5 Two horizontal metallic plates, connected to the ground.

nected to the ground, is suddenly placed directly above the first plate, the lines of field will end on the second plate, that is, the second plate shields the first plate from the electric field (Figure VI.5). The charges on the first plate, released from the grip of the electric field, will then quickly flow through the wire into the ground. A sensitive ammeter connected to this wire will detect this flow charge; the total amount of charge that runs out of the plate is a measure of the strength of the electric field.

In practice, the two plates are often given the shape of a Maltese cross (Figure VI.6). The upper plate is per-

Fig. VI.6 A field mill. When in use, the cross is made to rotate at a high speed by a motor within the base.

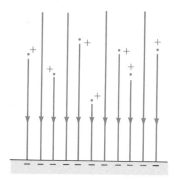

Fig. VI.7 Field lines of the atmospheric electric field start on positive charges in the air and end on negative charges on the ground.

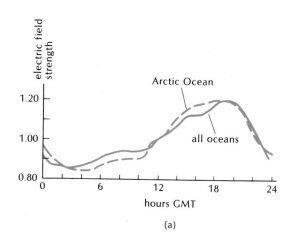

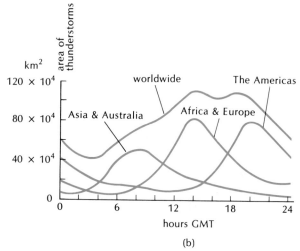

Fig. VI.8 (a) Strength of electric field vs. Greenwich time. The strength of the electric field is expressed as a multiple of the average strength. (b) Thunderstorm activity vs. Greenwich time.

manently kept above the lower plate and made to rotate in the horizontal plane. The arms of the upper plate then successively cover and uncover the arms of the lower plate; each covering and uncovering sends a pulse of current through the wire connecting the lower plate to the ground; the strength of this alternating current is a measure of the electric field. An instrument of this kind is called a **field mill**.

The strength of the atmospheric electric field decreases with altitude. This decrease is related to the decrease of the resistance of the atmosphere: at low altitude the resistance of the atmosphere is large, and at high altitude the resistance is small. Thus, at low altitude a large electric field is needed to maintain a given current, while at high altitude only a small electric field is needed. The decrease of the electric field is engendered by a positive charge density in the atmosphere (Figure VI.7). Note that most of the electric field lines start on positive charges in the lower atmosphere, within a few kilometers from the ground; only very few field lines start on positive charges in the ionosphere. The total positive charge in the fair-weather atmosphere is of the same magnitude as the negative charge on the ground.

The strength of the fair-weather field depends somewhat on the time of day. The magnitude of the diurnal variation can be as large as 20%. Except for local effects due to contamination of the atmosphere by smoke or dust, the variation is simultaneous at all places on the Earth; for example, the field strength is maximum at 18:00 local time (Greenwich Mean Time) in London and simultaneously at 13:00 local time (Eastern Standard Time) in New York.

Figure VI.8a shows the time dependence of the strength of the fair-weather field (as a percentage of the mean strength) over the open ocean; these measurements over the open ocean give the best indication of the global time variation because they are free of spurious effects caused by local sources of pollution.

The time dependence of the field strength is due to a time dependence of thunderstorm activity. Worldwide, thunderstorm activity reaches a peak between 14:00 and 20:00 Greenwich Mean Time; this peak is mainly due to an abundance of mid-afternoon thunderstorms in the Amazon basin. Thus, the maximum in the electric field at 18:00 arises from the enhanced rate at which thunderstorms charge up the atmosphere at about this time (Figure VI.8b).

VI.2 THUNDERSTORMS

Thunderstorms obtain the energy for their violent mechanical and electrical activity from humid air. Thunderstorms are heat engines; their heat reservoir is the heat of evaporation stored in water vapor. One cubic kilometer of air at 17°C, at atmospheric pressure, and at 100% relative humidity contains 1.6×10^7 kg of water vapor. Since the heat of evaporation for water is 586 kcal/kg (at 17°C), condensation of

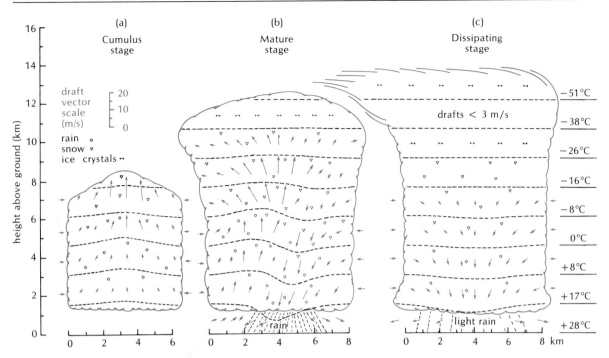

Fig. VI.9 (a) A young thunderstorm cell. The vectors give the velocity of drafts within the cell. (b) A mature thunderstorm cell. (c) An old thunderstorm cell. (From A. J. Chisholm, *Meteorological Monographs,* November 1973, no. 36.)

all the water vapor in 1 km³ of air would release 9.2×10^9 kcal. This is equivalent to the energy released in the explosion of 9.2 kilotons of TNT. A typical thunderstorm involves many cubic kilometers of air, and therefore the amount of stored energy (latent energy) is enormous. Although the efficiency for energy conversion is low, the release of just a small fraction of the energy stored in humid air suffices to account for the activity of a thunderstorm. Tornadoes and hurricanes also obtain their energy from humid air, and their destructive power reflects the large amount of available energy.

A thunderstorm is made of several **cells,** or thunderclouds, within each of which air moves upward or, in the later stages of the cell's development, downward. Each cell is some 8 km across; the base of the cell is at a height of about 1500 m and the top of the cell at a height that initially may be 7500 m, but grows to 12,000 m or even 18,000 m. Within such a cell there are strong updrafts and, at later stages, strong downdrafts, commonly reaching vertical speeds of 12 m/s.

The rising motion of air in a storm cell is powered by the heat released in the condensation of water vapor. Dry air on the surface of the Earth will usually not rise; it is in stable (or neutral) equilibrium. If, because of some disturbance, a mass of dry air is pushed upward to a higher altitude, the reduced pressure of its surroundings leads to expansion of the air; the work done by the air during this expansion reduces its temperature; and this reduced temperature will match the normal reduced temperature of the environment at the higher altitude. Thus, dry air has no tendency to rise.

However, humid air is unstable with respect to vertical displacement. If a mass of humid air initially near the ground is pushed upward to a slightly higher altitude, the expansion and consequent reduction of temperature leads to condensation of a fraction of the water vapor. This supplies the air with extra heat. The air then will be warmer and less dense than the dry air that resides at the higher altitude. Thus, the air continues to rise — it is unstable. Once the upward motion starts, it will continue faster and faster until all the water vapor has condensed, and even then the motion may continue, since the water can supply further energy by freezing into ice.

In a young thunderstorm cell the air rises as in a chimney, drawing in more and more humid air at the base while the top of the cell grows upward at speeds of 10 m/s or more (Figure VI.9a).

A mature thunderstorm cell may extend to a height of 12,000 m or 18,000 m. Inside the cell small water droplets coalesce, forming raindrops; some of the rain freezes into ice and hail; snow may also form. Raindrops and ice particles that are too large and heavy to be supported by the updraft begin to fall. This downward motion drags some air along and starts a downdraft. Once the air begins to descend, it will continue

Fig. VI.10 A thunderstorm with a well-developed anvil cloud.

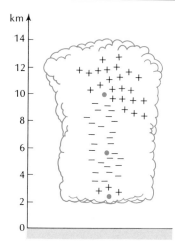

Fig. VI.11 Charge distribution in a typical thundercloud.

descending — humid air is unstable with respect to downward displacement as well as upward displacement. If a mass of humid air (intermixed with water) is pushed downward to a slightly lower altitude and higher pressure, the compression tends to warm the air, but the evaporation of water tends to cool it. The net result is that the air will be somewhat cooler and denser than the dry air surrounding the thunderstorm cell; therefore the mass of humid air continues to descend.

In the mature thunderstorm cell there are both strong updrafts and strong downdrafts. The top of the cell spreads out laterally, forming the characteristic anvil cloud, or **thunderhead** (Figure VI.10). Observations made by looking down on thunderstorms from U-2 aircraft and satellites have revealed that above the anvil cloud, small turrets often sprout upward and reach into the stratosphere. These turrets consist of cloud masses thrown upward by violent updrafts with speeds as great as 100 m/s. During the mature stage, different regions of the thunderstorm cell acquire different electric charges, and lightning discharges ensue. Rain or hail falls out of the bottom of the cell, accompanied by cool, descending air which generates a gusty wind near the ground (Figure VI.9b).

In the late stage of a thunderstorm cell, the updraft ends and the downdraft takes over, covering the entire cell. Then the storm dissipates (Figure VI.9c).

VI.3 GENERATION OF ELECTRIC CHARGE

A mature thunderstorm cell typically has a charge distribution as shown in Figure VI.11. There is a positive charge in the upper part of the cell and a negative charge in the lower part of the cell; furthermore, at the very bottom of the cell, there is often an extra, small, positive charge.

Exactly how these changes are generated remains somewhat of a mystery. There are numerous theories invoking different mechanisms. Most of these theories blame the charge separation on a difference in the size of the charge carriers created in the cloud — the carriers of positive charge are supposed to be smaller and lighter than the carriers of negative charge. The updrafts in the cloud then blow the positive carriers upward, but the negative carriers either remain stationary or fall down. Thus, the upper part of the cloud acquires a positive charge, while the lower part acquires a negative charge.

The charge carriers may be raindrops, hail, ice crystals, or ions; different theories propose different scenarios for how any or all of these particles may become charged. The conjectural charging mechanisms involve some form of electrification by friction, collision, freezing, melting, or thermoelectricity. For instance, according to one theory, the charges are created by collisions between falling hailstones and small drops of water. The vertical downward atmospheric electric field polarizes the hailstones, inducing positive charges at their bottoms and negative charges at their tops. When such a falling hailstone strikes a drop of water, pushing it aside, the drop is likely to pick up a few positive charges during the contact with the hailstone (Figure VI.12); this leaves the hailstone with a net negative charge. According to an alternative theory, the charges are created by collisions between falling hailstones and ions. These ions are present in the air, with equal average concentrations of negative and positive ions. If the hailstone approaches a negative ion, it is likely to attract and capture the ion; if the hailstone approaches a positive ion, it is likely to

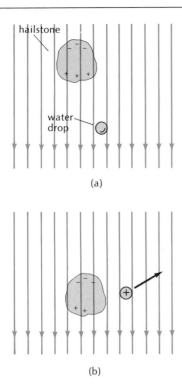

Fig. VI.12 (a) A falling hailstone, polarized by the electric field, approaches an ascending drop of water. (b) In the collision with the lower edge of the hailstone, the water drop picks up some positive charge.

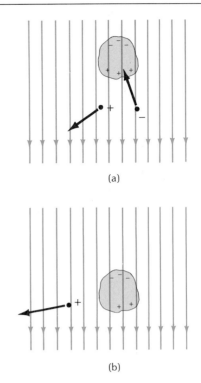

Fig. VI.13 (a) A falling hailstone attracts negative ions, but repels positive ions. (b) Capture of a negative ion gives the hailstone a negative charge.

merely repel the ion (Figure VI.13). Thus the net result of such collisions is that the hailstone again tends to acquire a negative charge. The falling hailstone then carries this negative charge to the bottom of the cloud and updrafts carry the positive charge to the top of the cloud.

Another conjectural charging mechanism relies entirely on updrafts and downdrafts to accumulate charges in the cloud, after a small initial amount of charge has been generated in the cloud by other means. This mechanism can work in two ways. At the top of the cloud, the positive charge (see Figure VI.11) attracts negative ions from the nearby air. If these ions become attached to droplets of water or particles of ice, downdrafts can carry them past the positive charge and deposit them in the lower portion of the cloud, adding to the negative charge already there. In a similar way, at the bottom of the cloud, the negative charge (see Figure VI.11) attracts positive ions released by point discharge from sharp protuberances on the ground (see the next section). If these ions become attached to suitable water droplets or ice particles, updrafts can carry them to the upper portion of the cloud, adding to the positive charge already there.

Although all of these mechanisms, and others as well, presumably contribute to the charging of thunderclouds, it remains unclear which, if any, of the proposed mechanisms is dominant.

VI.4 THE ELECTRIC FIELD OF A THUNDERCLOUD

Figure VI.14a shows a crude model of the charge distributions in a typical thundercloud. For the purposes of this model, it has been assumed that the charge distributions are roughly spherical; the positive and negative charges can then be represented by point charges located at the centers of the charge distributions. The small positive charge at the bottom of the thundercloud has been ignored. The charges of ∓ 40 C are at heights of 5000 m and 10,000 m, respectively. At the ground, directly below these charges, they produce an upward electric field of magnitude

$$E = \frac{1}{4\pi\varepsilon_0}\left(\frac{Q_1}{r_1^2} + \frac{Q_2}{r_2^2}\right)$$

$$= 9.0 \times 10^9 \text{ F/m} \left(\frac{40 \text{ C}}{(5 \times 10^3 \text{ m})^2} - \frac{40 \text{ C}}{(10 \times 10^3 \text{ m})^2}\right)$$

$$= 11 \times 10^3 \text{ V/m} \tag{1}$$

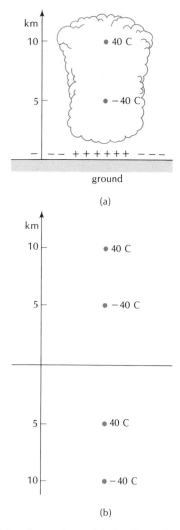

Fig. VI.14 (a) In this crude model, the charge distribution has been approximated as consisting of two point charges. These point charges induce a charge distribution on the ground. (b) The effect of the induced charge distribution on the ground is equivalent to that of two virtual image charges.

But this is not the total electric field. The charges in the cloud induce charges on the ground (a conductor), and these produce an extra electric field. The induced charge distribution on the ground will cover a large area, perhaps some 100 km²; within this area, the charge density will be concentrated directly below the charges in the cloud, and it will gradually decrease with distance.

We can calculate the effect of this induced charge distribution by a clever trick involving three steps: First, replace the ground by a thin, flat conducting plate; this does not change the electric field in the space above the ground, since the thickness of the conducting plate is irrelevant. Second, place two charges of ±40 C in the empty space below the con-

ducting plate (Figure VI.14b). These two new charges are at depths of 5000 m and 10,000 m below the conducting plate; they are called the **image** charges of the original charges in the cloud. The presence of these image charges below the plate does not change the electric field above the plate, since the conductor shields these regions from each other. Third, slide the conducting plate away in a horizontal direction. This, again, does not change the electric field above the plate, since, for the configuration of four charges shown in Figure VI.14b, the midplane is an equipotential surface, and the presence or absence of a coincident conducting surface makes no difference to the electric field.

Having taken these three steps, we recognize that the combination of cloud charges plus charges on the ground produces exactly the same electric field (above the ground) as the combination of cloud charges plus image charges; that is, the induced charges on the ground are virtually equivalent to the image charges. Thus, we can calculate the total electric field at any given point on or above the ground by simply adding the electric fields of the four charges shown in Figure VI.14b. At any given point on the ground, the image charges produce exactly the same electric field as the cloud charges — the total electric field is therefore *twice* the electric field of the cloud charges. Directly below the cloud charges, the electric field is then $2 \times 11 \times 10^3$ V/m $= 22 \times 10^3$ V/m. Incidentally: The above trick for the evaluation of the electric field of some given point charges in the presence of a conducting surface is called the **method of images;** the method only works under rather special conditions, but when it works, it permits a very quick and elegant evaluation.

We can also use this method to calculate the potential difference between the ground and the base of the thundercloud at a height of, say, 2 km (see Figure VI.11). The potential of the ground is zero and the potential of the base of the thundercloud is a sum of four terms, corresponding to the four point charges of Figure VI.14b:

$$V = \frac{1}{4\pi\varepsilon_0}\left(\frac{40\text{ C}}{8 \times 10^3\text{ m}} - \frac{40\text{ C}}{3 \times 10^3\text{ m}}\right.$$
$$\left. + \frac{40\text{ C}}{7 \times 10^3\text{ m}} - \frac{40\text{ C}}{12 \times 10^3\text{ m}}\right)$$
$$= -5.4 \times 10^7 \text{ volts} \tag{2}$$

As these numbers show, the electric fields and the potential differences generated by thunderstorms are quite large.

In the above calculations we assumed a perfectly

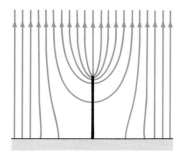

Fig. VI.15 Electric field lines around a thin vertical conductor connected to the ground (lightning rod).

flat Earth. In reality there are protuberances on the surface of the Earth, and near these the electric field strength will be strongly enhanced.[1] It is a general rule that if a sharp conducting point protrudes from an equipotential surface, the electric field will be strong near the point. For example, Figure VI.15 shows the electric field near a lightning rod; the high density of field lines near the tip of the rod indicates a strong field. It is easy to see that some enhancement of the field strength must occur for any kind of protruding point, regardless of the precise shape: the field lines must meet any conductor at right angles; hence they must bend toward the protruding conductor; hence their density increases.

The strong electric fields near any sharp spikes or edges on the ground produce a **point discharge.** For example, experiments with trees have shown that a tree under a thunderstorm will carry a current of about 1 A out of the ground; the current flows into the atmosphere through the tips of the tree's leaves.

The point discharge is due to the electrical breakdown of the air — at a high electric field strength the air becomes a conductor. The sudden increase in the conductivity of air is caused by ionization, that is, the formation of positive ions and free electrons by the disruption of neutral atoms. There are always a few ions and free electrons present in the air, but not enough to conduct any large current. However, if the air is exposed to an intense electric field, then those few electrons are accelerated to high velocity, and when they smash into a neutral atom they can kick extra electrons out of the atom. A chain reaction sets in: each free electron generates an avalanche of free electrons. In dry air under atmospheric pressure, about 3×10^6 V/m are required to start such a discharge.

---

[1] The field strength is also much greater in the vicinity of the charge concentrations in the cloud. On one occasion, a value of 340×10^3 V/m was measured at an airplane flying through a thunderstorm just before the airplane was struck by lightning.

The discharges into the atmosphere can take a variety of forms: point discharge with no visible display, corona discharge (also called St. Elmo's fire) with its characteristic glow of visible light, sparks, or lightning. Although lightning is the most spectacular of these phenomena, in many thunderstorms point discharge actually plays the larger role in the transfer of charge from the thundercloud to the ground — it contributes a current to the atmospheric circuit which is several times larger than the (average) current contributed by lightning.

VI.5 LIGHTNING

An electric field of 3×10^6 V/m is required to induce electric breakdown and produce small sparks in dry air at atmospheric pressure, but the presence of water drops and the lower-than-atmospheric pressure in a thundercloud favor breakdown and, therefore, sparking is likely to begin at somewhat lower electric fields. To initiate a flash of lightning, the electric field has to be very intense only in a small region near the cloud; once an avalanche of electrons starts, it can propagate into regions of less intense electric field. A lightning discharge can take place between the opposite charges inside the thundercloud, or between the thundercloud and clear air, or between the thundercloud and the ground. In the latter case the discharge usually takes place between the negative charge of the cloud and the ground; discharges between the positive charge and the ground are rare.

The sequence of events in a flash of lightning occurs too fast to be resolved by the human eye; however, scientists have used techniques of high-speed photography to analyze the time development of flashes of lightning. A typical flash of lightning between a thundercloud and the ground begins with an electron avalanche in the intense electric field near the negative charge of the thundercloud. As the avalanche moves downward it leaves behind a channel of ionized air, or plasma; electrons from the cloud flow into this channel, giving it a negative charge. The concentration of negative charge near the tip of the channel generates strong electric fields which continue to push the avalanche downward (Figure VI.16a). The channel follows a tortuous path, often with lateral branches whose twists and turns reflect random variations in the density of free electrons in the air ahead of the avalanche. The channel is called a **leader;** it has a radius of several meters (perhaps 5 m) but only its central region is (faintly) luminous. Figure VI.17 shows a time sequence of pictures of the leader. These pictures fail to illustrate one important feature of the leader: the leader does not move downward smoothly — it proceeds in steps of about 50 m with a pause of about 50 μs be-

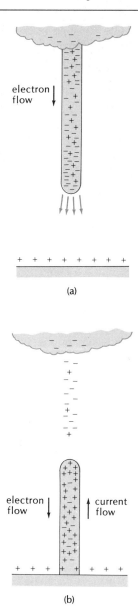

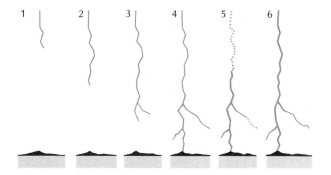

Fig. VI.17 This sequence of diagrams, based on high-speed photographs, shows the downward motion of the leader (first four diagrams) and the upward motion of the return stroke (last two diagrams). The return stroke is much faster than the leader.

Fig. VI.16 (a) The leader. An electron avalanche from the cloud makes a channel of ionized air. Electrons fill this channel. (b) The return stroke. Electrons suddenly drain out of the bottom of the channel, leaving it with a positive charge.

tween steps. Because of this, it is sometimes called the **step leader.**

When the tip of the leader comes near the ground, its intense electric field initiates a discharge from the ground or from a pointed body on the ground. The upward-moving discharge meets the leader 20 m to 100 m above ground. At this instant the circuit between cloud and ground is complete — there is now a continuous conducting path, and the negative charge can flow from cloud to ground with little resistance.

The electrons at the bottom of the channel are the first to move — they drain out into the ground, giving a large current. Then electrons from successively higher positions drain down. Thus, a "drainage front" moves upward toward the thundercloud. The drainage front is the head of a tube of intense current which snakes toward the thundercloud at as much as one-half the speed of light; this tube of current is the **return stroke** (Figures VI.16b and VI.17). The intense current heats the air and produces the bright flash of light that we see as lightning (Figure VI.1). The radius of the tube of current is quite small — it ranges from less than 1 cm to a few centimeters.

The peak current of the return stroke is about 10,000 A to 20,000 A. This strong current lasts less than 100 μs. When it ends, a weaker current (a few hundred amperes) continues to flow for several milliseconds. The strong current can cause severe damage to whatever body lies in its path (Figures VI.18 and VI.19).

The first return stroke is often succeeded by several more return strokes. A typical flash of lightning contains three to five return strokes with intervals of 40 ms between them. Each of the latter return strokes is preceded by its own **dart leader** which travels along the channel of the first return stroke. This channel retains its conductivity for some time and enables the dart leaders to travel downward very fast, without the pauses characteristic of the first step leader.

The total charge delivered to the ground by the complete flash of lightning with its several return strokes amounts to about -25 C (see Table VI.1). Since the potential difference between the base of the cloud and the ground is about 5×10^7 V [see Eq. (2)], the energy dissipated in the lightning channel is about 10^9 J. Most of this energy goes into heat (Joule heating), although a small fraction goes into emission of light and

Fig. VI.18 The grass of this golf course was burnt by the currents produced by a stroke of lightning. Note that the current branches out along the ground in all directions.

Table VI.1 DATA ON LIGHTNING[a]

Step leader	
Length of step	50 m
Time between steps	50 µs
Speed	1.5×10^5 m/s
Charge in channel	−5 C
Dart leader	
Speed	2×10^6 m/s
Charge in channel	−1 C
Return stroke	
Speed	5×10^7 m/s
Peak current	10^4–2×10^4 A
Charge transferred to ground	−2.5 C
Complete lightning flash	
Number of strokes	3–5
Time between strokes	40 ms
Time duration of flash	0.2 s
Charge transferred to ground	
(including continuing current)	−25 C

[a] Based on M. A. Uman, *Lightning*. The numbers given are average values.

Fig. VI.19 Tree exploded by lightning.

of radio waves. Immediately after the current has passed, the plasma in the lightning channel will be at an extremely high temperature (about 30,000 K) and a correspondingly high pressure. The high-pressure plasma expands explosively, generating a shock wave in the surrounding air. The shock wave dissipates within a few meters, gradually changing into a sound wave that propagates outward as a pulse. When this sound pulse reaches our ears, we hear it as thunder. The rumbling noises in thunder are due to the successive arrivals of sound pulses from different portions of the lightning bolt, which are at different distances from our ears.

At some distance from the lightning bolt, we will hear the sound after we see the flash of lightning — because of the high speed of light, the flash arrives almost instantaneously, while the sound takes an appreciable time to cover the intervening distance. Since the speed of sound is about $\frac{1}{3}$ km/s, we can use the following rule of thumb to reckon the distance (in kilometers) to the lightning bolt: Count the seconds between flash and thunder and divide by 3.

The most spectacular displays of lightning occur in tornadoes. In many instances, eyewitnesses reported that the interior of the funnel of a tornado was brightly lit by constant flashes of lightning. Measurements of the radio waves emitted by the lightning in tornadoes[2] confirm these reports; from such measurements scientists have estimated that about 20 flashes of lightning occur each second. Since the energy per flash of lightning is about 10^9 J (see above), the electric power dissipated by a tornado is about 10^9 J $\times$ 20/s = 2×10^{10} W — this is roughly $\frac{1}{20}$ of the combined output of all the electric power plants in the United States!

[2] These radio waves can be picked up on an ordinary TV set; this can give advance warning of the approach of a tornado.

Finally, some brief remarks on the weird phenomenon of ball lightning, or **Kugelblitz.** This consists of a glowing fireball that floats in midair. The sizes of these balls range from 10 cm to 100 cm, although airplane pilots have reported balls as large as 15 m to 30 m inside thunderclouds. The balls sometimes occur after a lightning stroke and sometimes they occur spontaneously, without apparent cause; they usually last only a few seconds. The balls sometimes fall down vertically from the sky, and sometimes drift horizontally near the ground. They have been known to enter buildings through doors, windows, or chimneys. Many of these balls disappear silently without a trace, but some disintegrate explosively with an ear-shattering bang.

It seems likely that the ball is created and maintained by electric effects, but we have no understanding of the mechanism involved in this phenomenon. Several theories have been proposed to explain the ball: according to one theory, it is a ball of plasma held together by magnetic fields; according to another, it is a miniature thundercloud of dust particles acting as a very efficient electrostatic generator. But owing to a lack of precise data and a lack of detailed calculations, the phenomenon remains a mystery.

Further Reading

The following books discuss thunderstorms and lightning at an introductory level:

The Lightning Book by P. E. Viemeister (Doubleday, Garden City, 1961)
The Flight of Thunderbolts by B. F. J. Schonland (Oxford, 1964)
The Nature of Violent Storms by L. J. Battan (Doubleday, Garden City, 1961)

Short but very informative articles at the same level are the following:

"Thundercloud Electricity," B. Vonnegut, *Discovery,* March 1965
"Thunder," A. A. Few, *Scientific American,* July 1975
"Thunder and Lightning," J. Latham, in *Forces of Nature,* edited by V. Fuchs (Thames and Hudson, London, 1977)
"The Tornado," J. T. Snow, *Scientific American,* April 1984.

At a more technical level, the following books and articles provide a wealth of information:

Lightning by R. H. Golde (Academic Press, London, 1977)
Lightning by M. A. Uman (McGraw-Hill, New York, 1969)
Atmospheric Electricity by J. A. Chalmers (Pergamon Press, Oxford, 1967)
Physics of Lightning by D. J. Malan (English University Press, London, 1963)
Ball Lightning and Bead Lightning: Extreme Forms of Atmospheric Electricity by J. D. Barry (Plenum, New York, 1980)
"Atmospheric Electricity" by C. D. Stow, in *Reports on Progress in Physics,* Vol. 32, Part I, 1969

"Some Facts and Speculations Concerning the Origin and Role of Thunderstorm Electricity" by B. Vonnegut, in *Meteorological Monographs,* Vol. 5, No. 27, September 1963

Questions

1. It is possible to measure the potential differences between the atmosphere and the ground by shooting an arrow upward, with a fine wire trailing from its tail. A voltmeter connected between the lower end of the wire and the ground will then register the potential at the arrow. Why does this differ from sticking a stationary conductor into the air?

2. By comparing Figures VI.8a and b, can you conclude that the bursting of bubbles on the ocean surface makes only a small contribution to the current in the atmospheric circuit?

3. In 1752, Benjamin Franklin flew a kite into a thundercloud and demonstrated that lightning is an electric phenomenon. A few months later, a scientist at St. Petersburg was killed by lightning while attempting to repeat this stunt. After that, scientists took the precaution of enclosing themselves in boxes of sheet metal while flying kites into thunderclouds. How does this help?

4. How could you construct a power plant that captures electric energy from a thundercloud?

5. What are the main dangers to an aircraft flying through a thundercloud?

6. A lightning rod serves not only to conduct lightning safely to the ground, but also to inhibit lightning by promoting point discharge. Explain.

7. What parts of an aircraft are most likely to be struck by lightning when flying in or near a thundercloud?

8. During a thunderstorm, would you be safer in an automobile with a sheet metal body or an automobile with a fiber glass body?

9. To protect the crew of a wood or fiber glass boat from the hazards of lightning, all metal parts on the boat should be connected to a thick conducting cable, terminating on a conducting plate on the outside of the hull, below the waterline. Explain.

10. What pattern of current flow do you expect below and on the water surface near the point of entry of lightning? Would you suffer damage if swimming nearby?

11. Is the conductivity of the ground due to a flow of electrons or a flow of ions?

12. The ancient Greeks noticed that when lightning strikes the ground, nearby cattle are killed, but nearby men often survive. The Greeks thought this indicated an affinity between men and gods. Can you think of a better explanation? (Hint: Chicken also survive.)

13. It is dangerous to take showers or baths when a thunderstorm is overhead. Why?

14. According to a familiar saying, lightning never strikes twice. According to Table VI.1, how often does lightning strike, on the average?

15. Suppose you get caught in the open by a thunderstorm with severe lightning. Consider and discuss each of the following options: take shelter beneath a tree standing in isolation; lie down in the open, flat on the ground; sit in a ditch or depression, on your heels, with both feet close together.

16. When lightning strikes a tree, it often explodes the branches or the trunk. Explain. (Hint: The heating of water produces steam.)

17. Suppose that after a flash of lightning, you hear the first thunder in 3 s and the last in 9 s. What can you conclude about the distances of the nearest and the farthest portions of the lightning bolt?

CHAPTER 30

The Magnetic Force and Field

Hans Christian Oersted (örstad), *1777–1851, Danish physicist and chemist, professor at Copenhagen. He observed that a compass needle suffers a deflection when placed near a wire carrying an electric current. This discovery gave the first empirical evidence of a connection between electric and magnetic phenomena.*

The magnetic force most familiar in everyday experience is the force that the magnetic poles of the Earth exert on a compass needle. This magnetic force was known for many centuries, but only during the last century did experimenters discover that electric currents also exert magnetic forces on compass needles, and that electric currents exert magnetic forces on one another. Finally, physicists came to understand that the magnetic force is nothing but an extra electric force acting between charges in motion. This means that between two charges in motion, there acts not only the Coulomb force but also a force that is a function of the velocities of the charges.

In the next section, we will write down an equation for the magnetic force between two moving charges. From this we will derive the equation for the magnetic force between two currents consisting of many moving charges. Our development of magnetic forces in this chapter goes counter to the historical development — the law of magnetic force between currents was discovered long before the law of magnetic force between individual charges. But our development follows the road taken in earlier chapters: we always begin with the fundamental laws that apply to *particles* and from these deduce the laws that apply to systems of particles (such as currents on a wire).

30.1 The Magnetic Force

According to Coulomb's Law, the electric force exerted by a point charge q' on a point charge q is

$$\mathbf{F} = \frac{1}{4\pi\varepsilon_0} \frac{qq'}{r^2} \hat{\mathbf{r}} \qquad (1)$$

where r is the distance between the charges and $\hat{\mathbf{r}}$ a unit vector directed from q' toward q. The force on q' exerted by q is exactly the opposite of that in Eq. (1).

Equation (1) correctly gives the force acting on charges at rest. However, when the charges are *in motion*, there is an extra force acting on these charges. This extra force is the **magnetic force.** The magnitude and the direction of the magnetic force depend not only on the distance between the charges but also on their velocities. In the general case, the dependence on distance and velocities is fairly complicated.

To gain a clear understanding of the magnetic force, it will be helpful to begin with a simple special case and to deal with the general case afterward. Let us consider two point charges q and q' instantaneously moving side by side with parallel velocities $\mathbf{v}$ and $\mathbf{v}'$ (see Figure 30.1). In this special case, the magnetic force on the charge q exerted by the charge q' is

$$\mathbf{F} = -[\text{constant}]\, \frac{qq'}{r^2}\, vv'\hat{\mathbf{r}} \quad \text{for parallel velocities at right angles to } \hat{\mathbf{r}} \quad (2)$$

We must regard this equation for the magnetic force [and the more general Eq. (10) given below] as a fundamental law of physics, which has the same status as Coulomb's Law or Newton's Law of gravitation. The magnetic force varies as the inverse square of the distance, like the electric force and the gravitational force. But, in contrast to these other forces, the magnetic force also depends on the product of the velocities. For charges of equal signs, the magnetic force is attractive if the velocities are parallel, and it is repulsive if they are antiparallel. Note that the magnetic force is zero unless *both* velocities are different from zero; that is, the magnetic force acts only if *both* charges are in motion.

In the SI system of units, the numerical value of the constant in Eq. (2) is

$$[\text{constant}] = 1.00 \times 10^{-7}\, \frac{\text{N} \cdot \text{s}^2}{\text{C}^2} \quad (3)$$

This constant is conventionally written in the form

$$[\text{constant}] = \frac{\mu_0}{4\pi} \quad (4)$$

with

$$\mu_0 = 4\pi \times 10^{-7}\, \frac{\text{N} \cdot \text{s}^2}{\text{C}^2} = 1.26 \times 10^{-6}\, \frac{\text{N} \cdot \text{s}^2}{\text{C}^2} \quad (5)$$

Permeability constant

The quantity μ_0 is called the **permeability constant.**[1]

Our equation (1) for the magnetic force on the charge q exerted by the charge q' then becomes

$$\mathbf{F} = -\frac{\mu_0}{4\pi}\, \frac{qq'}{r^2}\, vv'\hat{\mathbf{r}} \quad \text{for parallel velocities at right angles to } \hat{\mathbf{r}} \quad (6)$$

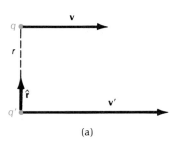

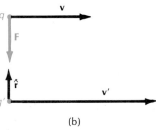

Fig. 30.1 (a) Two point charges q and q' instantaneously moving side by side, with parallel velocities at right angles to $\hat{\mathbf{r}}$. (b) The magnetic force on q is attractive.

---

[1] The value of the constant in Eq. (3) is *exact*. Correspondingly, the value $\mu_0 = 4\pi \times 10^{-7}$ N·s²/C² is also exact. (These values hinge on the definitions of current and of charge in the SI system.)

EXAMPLE 1. Compare the magnitude of the magnetic force for parallel velocities [Eq. (6)] with the magnitude of the electric force.

SOLUTION: The magnitude of the magnetic force is

$$F_{\text{mag}} = \frac{\mu_0}{4\pi} \frac{qq'}{r^2} vv'$$

If the charges have equal signs, the magnetic force is attractive. The magnitude of the electric force is

$$F_{\text{el}} = \frac{1}{4\pi\varepsilon_0} \frac{qq'}{r^2}$$

This force is, of course, repulsive. The net force is the sum of the magnetic and electric forces.

The ratio of the magnitudes of the two forces is

$$\frac{F_{\text{mag}}}{F_{\text{el}}} = \left(\frac{\mu_0}{4\pi} \frac{qq'}{r^2} vv'\right) \bigg/ \left(\frac{1}{4\pi\varepsilon_0} \frac{qq'}{r^2}\right) = \mu_0 \varepsilon_0 vv' \qquad (7)$$

$$= 1.26 \times 10^{-6} \frac{\text{N} \cdot \text{s}^2}{\text{C}^2} \times 8.85 \times 10^{-12} \frac{\text{C}^2}{\text{N} \cdot \text{m}^2} vv'$$

$$= (1.12 \times 10^{-17} \text{ s}^2/\text{m}^2) vv' \qquad (8)$$

The numerical factor appearing in Eq. (8) is equal to $1/(3.0 \times 10^8 \text{ m/s})^2$. Hence we can write Eq. (8) as

$$\frac{F_{\text{mag}}}{F_{\text{el}}} = \frac{v}{3.0 \times 10^8 \text{ m/s}} \frac{v'}{3.0 \times 10^8 \text{ m/s}} \qquad (9)$$

This shows that F_{mag} will be small compared with F_{el} if v and v' are small compared with 3.0×10^8 m/s. Note that 3.0×10^8 m/s is the speed of light; why the speed of light should turn up in a calculation concerning electricity and magnetism is a puzzle which we will solve in Section 35.5.

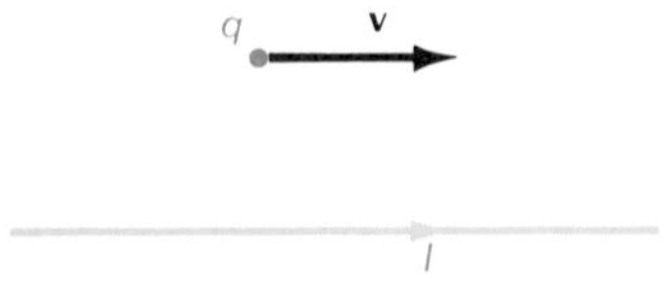

Fig. 30.2 Point charge moving parallel to a current.

According to Eq. (9), the magnetic force between point charges is small compared with the electric force unless the speeds of the charges approach the speed of light. This suggests that the magnetic force becomes significant only when the speeds are so large that Newtonian physics fails, and we enter the realm of the theory of Special Relativity. As we will see in Chapter 41, at such high speeds, a variety of relativistic corrections have to be included in the equation of motion.[2] It would then seem that the magnetic force is no more than just another relativistic correction that has to be included in the equation of motion of charged particles. However, the magnetic force can become important even when the speeds are low. This will happen whenever we are dealing with a charge distribution for which the electric forces cancel. For example, Figure 30.2 shows a moving point charge at some distance from a straight wire carrying a current. Under these conditions, there is no electric force on the point charge — the wire contains equal amounts of positive and negative charges, and the electric attractions

---

[2] If the speeds are large, some extra relativistic corrections will also be needed in Eqs. (1) and (2).

and repulsions on the point charge cancel. The only remaining force on the point charge is the magnetic force produced by the moving charges of the wire. Since the number of moving charges in the wire is very large, the magnetic force can be quite large.

Another charge distribution for which the electric forces cancel and the magnetic forces become important is that of two wires carrying currents. Figure 30.3 shows two straight wires carrying parallel currents. There is no electric force between these wires — the electric forces between the positive and the negative charges on the two wires cancel. However, the magnetic forces do not cancel. The moving charges of one wire exert magnetic forces on the moving charges of the other wire. If the currents are parallel, as in Figure 30.3, these magnetic forces are attractive. If the currents are antiparallel, then the magnetic forces are repulsive. The magnitudes of the net forces on the currents in the wires can be calculated by integrating the magnetic forces of the individual point charges.

Fig. 30.3 Two parallel currents.

The magnetic forces between the currents in wires were discovered by Ampère in 1820, long before it was recognized that these forces are really due to the magnetic forces between individual moving charges in the wires. Experiments with individual moving charges (charged brass balls moving at high speeds) were carried out early in this century. It is very difficult to perform a precise measurement of the magnetic force between individual charges. However, it is fairly easy to perform precise measurements of the force between parallel wires carrying a current (such measurements are actually used for the *definition* of the ampere, which we will give in Section 31.5). This amounts to an indirect experimental verification of the magnetic-force law for individual moving charges.

[As we will see in Section 41.8, the magnetic force can be regarded as generated by a transformation of reference frame applied to an electric force. Note that in a reference frame in which one of the two charges in Eq. (6) is at rest, there is no magnetic force. In this reference frame there is only an electric force. When we now transform to a different reference frame, the magnetic force suddenly makes its appearance — the transformation of reference frame generates an extra magnetic force from an electric force. In Section 41.8 we will use the principles of the theory of Special Relativity to derive the magnetic force from the electric force.]

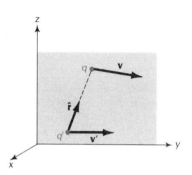

Fig. 30.4 Two point charges with velocities at different angles relative to the unit vector $\hat{\mathbf{r}}$. In this example, the point charges and the vectors $\mathbf{v}$, $\mathbf{v}'$, and $\hat{\mathbf{r}}$ are in the z–y plane, or the plane of the page.

Now we must deal with the general case of two point charges moving with arbitrary velocities. If the instantaneous velocities $\mathbf{v}$ and $\mathbf{v}'$ make some arbitrary angles with the line between the charges (see Figure 30.4), the magnetic force on charge q exerted by charge q' is

$$\mathbf{F} = \frac{\mu_0}{4\pi} \frac{qq'}{r^2} \mathbf{v} \times (\mathbf{v}' \times \hat{\mathbf{r}}) \qquad (10)$$

Magnetic force on moving point charge

This formula involves two vector cross products in succession: we first must cross-multiply $\hat{\mathbf{r}}$ by $\mathbf{v}'$ and then cross-multiply the result by $\mathbf{v}$. The directions of these cross products must be determined by the usual right-hand rule; Figure 30.5 illustrates how the right-hand rule determines the direction of the magnetic force. Note that the magnetic force is always perpendicular to the velocity $\mathbf{v}$ of the charge q.

The general formula (10) includes the simple formula (2) for parallel velocities as a special case. If the charges are instantaneously moving

Fig. 30.5 To find the direction of the cross product $\mathbf{v}' \times \hat{\mathbf{r}}$ from the right-hand rule, place the fingers of the right hand along $\mathbf{v}'$ and curl them toward $\hat{\mathbf{r}}$; the thumb then points out of the plane of the page and indicates the direction of the vector $\mathbf{v}' \times \hat{\mathbf{r}}$. Next, imagine that this vector has been shifted to the position of the charge q. Place the fingers along $\mathbf{v}$ and curl them toward the shifted vector. The thumb then indicates the direction of the cross product $\mathbf{v} \times (\mathbf{v}' \times \hat{\mathbf{r}})$, which is in the plane of the page and perpendicular to $\mathbf{v}$.

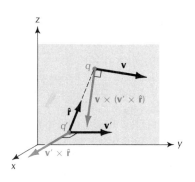

side by side and their velocities are parallel, then the angles in both the cross products in Eq. (10) are right angles, and hence the magnitudes of these cross products are simply the products of the magnitudes of the vectors, in agreement with Eq. (2). From the right-hand rule, we readily find that the direction of the magnetic force is as given in Eq. (2) and Figure 30.1b.

Equation (10) tells us the force exerted by the point charge q' on the point charge q. To find the force exerted by q on q', we must exchange the velocities in Eq. (10), and we must replace $\hat{\mathbf{r}}$ by $\hat{\mathbf{r}}'$ (Figure 30.6; note that $\hat{\mathbf{r}} = -\hat{\mathbf{r}}'$). This gives the magnetic force on q':

$$\mathbf{F}' = \frac{\mu_0}{4\pi} \frac{qq'}{r^2} \mathbf{v}' \times (\mathbf{v} \times \hat{\mathbf{r}}') \qquad (11)$$

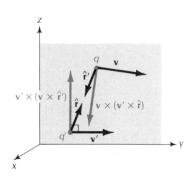

Fig. 30.6 With $\mathbf{v}'$, $\hat{\mathbf{r}}$, and $\mathbf{v}$, as in Figure 30.4, both $\mathbf{v} \times (\mathbf{v}' \times \hat{\mathbf{r}})$ and $\mathbf{v}' \times (\mathbf{v} \times \hat{\mathbf{r}}')$ are in the plane of the page, but they are not of equal magnitudes and opposite directions.

As Figure 30.6 shows, in general the magnetic force $\mathbf{F}'$ exerted by q on q' is *not* opposite and equal in magnitude to the magnetic force $\mathbf{F}$ exerted on q' by q. For magnetic forces, *Newton's Third Law on the equality of action and reaction fails*. Consequently, the law of conservation of momentum will also fail — when magnetic forces act, the momentum of the system of charged particles is not constant.

This is a disaster for Newton, but it is not a disaster for physics. If we take into account the electric fields of the charges, we can see why a violation of momentum conservation is quite reasonable for electric charges in motion. We know that in the space surrounding the electric charges there is an electric field, with an energy density. As the charges move, the field continuously changes, and the energy of the field must continuously redistribute itself, that is, the energy of the field must flow through space. We know that energy has mass; hence, such a changing field involves a flow of field-mass from one part of space to another. Since any moving mass has momentum, we conclude that the moving, changing electric field has momentum. Once we recognize this, we can see why the momentum of the charged particles is not conserved. The charged particles exchange momentum not only with each other but also with the electric field. What is conserved is not the momentum of the particles but the total momentum of *particles and field*. Although Newton's Third Law fails, the law of conservation of momentum remains valid. The explicit calculation of the momentum contained in an electric field is fairly complicated, and we will not attempt it here.

Incidentally: If the system of charged particles consists of wires with steady currents flowing in closed circuits, then we can prove that the net magnetic force of one circuit on another circuit does satisfy Newton's Third Law of the equality of action and reaction. Obviously, if steady currents are flowing in closed circuits, then the fields are constant in time, and their momentum must also remain constant. There is then no possibility of a momentum transfer to the fields, and the conservation of momentum would fail if Newton's Third Law did not hold.

30.2 The Magnetic Field

In Section 23.2 we presented some arguments in favor of the view that the electric force is communicated from one charge to another by action-by-contact, through an electric field. Likewise, the magnetic force is communicated from one charge to another through a **magnetic field**.

The definition of the magnetic field is as follows: To find the magnetic field at some point in the vicinity of moving charges or currents, place a test charge q at that point and give it some velocity $\mathbf{v}$. This charge will then experience a magnetic force depending on its velocity. The magnetic field $\mathbf{B}$ is implicitly defined by the following equation for the magnetic force:

$$\boxed{\mathbf{F} = q\mathbf{v} \times \mathbf{B}} \tag{12}$$

Nikola Tesla *1856–1943, American electrical engineer and inventor. He made many brilliant contributions to high-voltage technology, ranging from new motors and generators to transformers and a system for radio transmission. Tesla designed the power-generating station at Niagara Falls.*

Relation between magnetic force and field

Note that this force is always perpendicular to both $\mathbf{B}$ and $\mathbf{v}$, in the direction specified by the right-hand rule.

To discover the magnitude and direction of the magnetic field at some point, we place a test charge q at this point and launch it with a velocity $\mathbf{v}$ in some direction. By repeating this procedure several times, we discover how the force depends on the direction of the velocity $\mathbf{v}$. In one direction, the force will be zero; this is the direction parallel or antiparallel to $\mathbf{B}$. In the direction perpendicular to this direction, the magnitude of the force will be $F = qvB$, from which we deduce that the magnitude of the magnetic field is $B = F/(qv)$. Hence, in the special case of velocity perpendicular to the magnetic field, the magnitude of the magnetic field is the force per unit charge and unit velocity.

The SI unit of magnetic field is $N/(C \cdot m/s)$; this unit is called the **tesla** (T),

$$1 \text{ tesla} = 1 \text{ T} = 1 \text{ N}/(\text{C} \cdot \text{m/s}) \tag{13}$$

tesla, T

An alternative name for this unit is weber/m². For weak magnetic fields, a smaller unit is often preferred; this is the **gauss**,

$$1 \text{ gauss} \Leftrightarrow 10^{-4} \text{ T} = 10^{-4} \text{ weber/m}^2 \tag{14}$$

gauss

We have written this relationship between gauss and tesla as an equivalence ($\Leftrightarrow$) rather than as an equality ($=$) because the gauss is a cgs unit of magnetic field rather than an SI unit; the definitions of magnetic field in the cgs and SI systems are somewhat different, and the units do not carry the same dimensions and cannot be equated. Table 30.1 gives the values of some typical magnetic fields.

Table 30.1 SOME MAGNETIC FIELDS

At surface of nucleus	$\sim 10^{12}$ T
At surface of pulsar	$\sim 10^{8}$ T
Maximum achieved in laboratory:	
Explosive compression of field lines	1×10^{3} T
Steady	30 T
Large bubble-chamber magnet	2 T
In sunspot	~ 0.3 T
At surface of Sun	$\sim 10^{-2}$ T
Near small ceramic magnet	$\sim 2 \times 10^{-2}$ T
Near household wiring	$\sim 10^{-4}$ T
At surface of Earth	$\sim 5 \times 10^{-5}$ T
In sunlight (rms)	3×10^{-6} T
In Crab nebula	$\sim 10^{-8}$ T
In radio wave (rms)	$\sim 10^{-9}$ T
In interstellar galactic space	$\sim 10^{-10}$ T
Produced by human body	3×10^{-10} T
In shielded antimagnetic chamber	2×10^{-14} T

From Eq. (10) and (12), we see that the magnetic field generated by a point charge q' moving with velocity $\mathbf{v}'$ is

Magnetic field of point charge

$$\mathbf{B} = \frac{\mu_0}{4\pi} \frac{q'}{r^2} (\mathbf{v}' \times \hat{\mathbf{r}}) \tag{15}$$

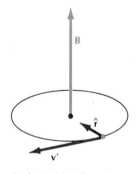

Fig. 30.7 Orbit of an electron around a nucleus, and magnetic field generated by the electron.

EXAMPLE 2. In a hydrogen atom, an electron moves in a circular orbit of radius 5.3×10^{-11} m. The speed of the electron is 2.2×10^{6} m/s. What magnetic field does the electron produce at the center of the orbit?

SOLUTION: The instantaneous velocity $\mathbf{v}'$ and the unit vector $\hat{\mathbf{r}}$ are perpendicular (Figure 30.7). Hence the magnitude of $\mathbf{B}$ is

$$B = \frac{\mu_0}{4\pi} \frac{ev'}{r^2}$$

$$= \frac{1.00 \times 10^{-7} \text{ N} \cdot \text{s/C}^2 \times 1.6 \times 10^{-19} \text{ C} \times 2.2 \times 10^{6} \text{ m/s}}{(5.3 \times 10^{-11} \text{ m})^2}$$

$$= 13 \text{ T} \tag{16}$$

The direction of $\mathbf{B}$ is perpendicular to the plane of the orbit. Since the charge is *negative*, the direction of $\mathbf{B}$ is *opposite* to the direction of $\mathbf{v}' \times \mathbf{r}$ given by the right-hand rule.

The magnetic field can be represented graphically by field lines. As in the case of the electric field, the tangent to the field line indicates the direction of the field, and the density of field lines indicates the strength of the field. Figure 30.8 shows the pattern of magnetic field lines around a moving positive charge. The pattern consists of concentric circles. The density of lines decreases with distance, according to the inverse-square law; and the density of lines becomes infinite at the position of the charge. Ahead and behind the charge, the pattern of field lines also consists of concentric circles, but without the strong concentration near the center. Note that the direction of the field lines

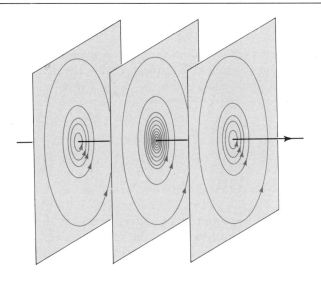

Fig. 30.8 Magnetic field lines of a positive point charge moving toward the right. At the instant shown, the point charge is in the central plane. In this plane, the density of field lines tends toward infinity near the point charge. However, the density of field lines is zero on the line of motion, ahead of and behind the point charge.

follows a simple right-hand rule: Place the thumb along the velocity of the charge; the fingers will then curl in the direction of the field lines. For a negative charge, the direction of the field lines in Figure 30.8 must of course be reversed.

For computations with field lines we will adopt the convention that the number of lines per unit area equals the magnitude of the magnetic field. However, for the purpose of making drawings, this normalization is sometimes unwieldy, and we will alter it in case of need.

The magnetic field lines of a moving charge form closed loops, that is, the magnetic field lines do not begin or end anywhere in the way that the electric field lines begin and end on positive and negative electric charges. In general, the magnetic field lines of any kind of magnetic field must always form closed loops, since there is no known "magnetic charge" that could act as source or sink of magnetic field lines.

Mathematically, we can express these features of the magnetic field lines in terms of a modified version of Gauss' Law. Consider a closed surface of arbitrary shape. The number of magnetic field lines that enter this surface is exactly equal to the number that leave, that is, the **magnetic flux** through the closed surface is zero:

$$\oint \mathbf{B} \cdot d\mathbf{S} = 0 \qquad (17)$$

This equation is true not only for the magnetic field of a single moving charge but also for any arbitrary magnetic field produced by any arbitrary number of moving charges. The net magnetic field produced by many charges acting together is the sum of their individual magnetic fields, and since each of these individual fields satisfies Eq. (17), the net field will also.

The argument of the preceding paragraph hinges on the **principle of superposition** obeyed by the magnetic field: If several moving charges $q_1, q_2, q_3, \ldots$, simultaneously generate magnetic fields $\mathbf{B}_1, \mathbf{B}_2, \mathbf{B}_3, \ldots$, then the net magnetic field generated by all these charges acting together is

Karl Friedrich Gauss, *1777–1855, German mathematician, physicist, and astronomer, director of the Göttingen Observatory. One of the greatest mathematicians of all time, Gauss did his most celebrated work in number theory. Gauss was an indefatigable calculator, and he loved to perform enormously complicated computations, which today would be regarded as impossible without an electronic computer. He developed new methods for calculations in celestial mechanics and successfully predicted the orbit of the asteroid Ceres, briefly seen and then lost by the astronomer G. Piazzi. Later, Gauss became interested in electric and magnetic phenomena, which he researched in collaboration with W. Weber. He also worked on geodetic surveys and invented the electric telegraph.*

Principle of superposition

$$\mathbf{B} = \mathbf{B}_1 + \mathbf{B}_2 + \mathbf{B}_3 + \cdots \qquad (18)$$

We will exploit this superposition principle in the next section when we calculate the magnetic field of a current, that is, the magnetic field of a large number of point charges moving along a wire.

30.3 The Biot–Savart Law

The magnetic fields produced by currents flowing on wires are of much greater practical interest than the magnetic field produced by a single point charge. A current on a wire is a distribution of moving point charges, each of which produces its own individual magnetic field. According to the superposition principle, the magnetic field of the current is then the sum or the integral of all the individual magnetic fields of all the individual point charges.

Figure 30.9 shows a thin wire of arbitrary shape carrying a steady current I. Following the usual convention, we will pretend that the current is due to a flow of positive charge. Consider a small segment dl of this wire. We can regard the moving charge dq' within dl as a point charge which produces a magnetic field

$$d\mathbf{B} = \frac{\mu_0}{4\pi} \frac{dq'}{r^2} (\mathbf{v}' \times \hat{\mathbf{r}}) \qquad (19)$$

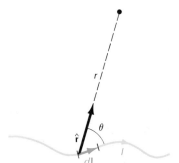

Fig. 30.9 Wire carrying a current I.

with a magnitude

$$dB = \frac{\mu_0}{4\pi} \frac{dq'}{r^2} v' \sin\theta \qquad (20)$$

where θ is the angle between $\mathbf{v}'$ and $\hat{\mathbf{r}}$, or the angle between the wire and $\hat{\mathbf{r}}$ (Figure 30.9). To relate dq' to the current, we begin with the definition of the current

$$dq' = I\, dt \qquad (21)$$

Here dt is the time it takes the charge dq' to flow out of the small segment dl, that is,

$$dt = dl/v' \qquad (22)$$

Hence

$$dq' = I\, dl/v' \qquad (23)$$

With this, Eq. (20) becomes

$$dB = \frac{\mu_0}{4\pi} \frac{I\, dl \sin\theta}{r^2} \qquad (24)$$

To put this in vector form, we treat the segment of wire as a vector $d\mathbf{l}$ tangent to the wire and in the direction of the current (Figure 30.9). The magnetic field (19) can then be expressed as

Jean Baptiste Biot (bio), *1774–1862, French physicist, professor at the Collège de France. His most important work dealt with the refraction and polarization of light, but he was also interested in a broad range of problems in the physical sciences. With Félix Savart, 1791–1841, he confirmed Oersted's discovery of magnetic fields generated by electric currents, and formulated the equation (30.25) for the strength of the magnetic field.*

$$\boxed{d\mathbf{B} = \frac{\mu_0}{4\pi} \frac{I\, d\mathbf{l} \times \hat{\mathbf{r}}}{r^2}} \qquad (25) \qquad \text{Biot–Savart Law}$$

This equation holds true because the right side has the correct magnitude [compare with Eq. (24)], and it also has the correct direction [compare with Eq. (19), noting that $\mathbf{v}'$ and $d\mathbf{l}$ have the same direction].

Equation (25) gives us the magnetic field generated by a short segment of a wire. It is called the **Biot–Savart Law.** The magnetic field generated by a wire of any length and shape can be calculated by integrating this equation along the wire.

EXAMPLE 3. A very long and very thin straight wire carries a steady current I. What is the magnetic field at some distance from the wire?

SOLUTION: Figure 30.10 shows the wire lying along the y axis. The magnetic field lines are concentric circles around the wire; the lines come out of the plane of the page in the region above the wire and go into the plane of the page in the region below. We will calculate the magnetic field at a vertical distance z from the wire.

A small segment dy of the wire contributes a magnetic field

$$dB_x = \frac{\mu_0}{4\pi} \frac{I\, dy\, \sin\theta}{r^2} \qquad (26)$$

The directions of the contributions from all segments dy are parallel (in the region above the wire, the directions of all the contributions to the magnetic field are out of the plane of the page). Hence the magnitude of the net field is

$$B_x = \int_{-\infty}^{\infty} \frac{\mu_0}{4\pi} \frac{I \sin\theta}{r^2}\, dy \qquad (27)$$

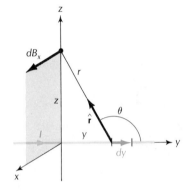

Fig. 30.10 Straight wire carrying a current I along the y axis.

From Figure 30.10,

$$y = z \cot(\pi - \theta) = -z \cot\theta \qquad (28)$$

so

$$dy = z \csc^2 \theta\, d\theta \qquad (29)$$

Furthermore,

$$r = z/\sin(\pi - \theta) = z/\sin\theta \qquad (30)$$

By means of Eqs. (29) and (30) we can change the variable of integration from y to θ; the new limits of integration are then $\theta = 0°$ and $\theta = 180°$, or, in radians, $\theta = 0$ and $\theta = \pi$,

$$B_x = \frac{\mu_0}{4\pi} \frac{I}{z} \int_0^\pi \sin\theta\, d\theta \qquad (31)$$

$$= \frac{\mu_0}{4\pi} \frac{I}{z} \Big[-\cos\theta \Big]_0^\pi = \frac{\mu_0}{4\pi} \frac{I}{z} \times 2$$

Thus,

$$\boxed{B_x = \frac{\mu_0}{2\pi} \frac{I}{z}} \qquad (32) \qquad \textit{Magnetic field of straight wire}$$

COMMENTS AND SUGGESTIONS: The strength of this magnetic field is inversely proportional to the distance from the wire. This dependence on distance is reminiscent of the electric field of a very long straight line of charge; however, the magnetic field forms closed circles around the wire, whereas the electric field of the charged line forms radial lines.

Figure 30.11 shows the pattern of the magnetic field lines for this magnetic field. And Figure 30.12 shows iron filings sprinkled on a sheet of paper placed around a long straight wire carrying a strong current. The iron filings align in the direction of the magnetic field and therefore make the pattern of field lines visible.

The direction of the magnetic field along the circular field lines in Figure 30.11 field is again given by a simple right-hand rule: If the thumb of the right hand is placed along the wire, the fingers will curl along the magnetic field lines in the direction of their arrows.

Right-hand rule for current and magnetic field

Fig. 30.11 (left) Magnetic field lines around a straight wire carrying a current. The wire is perpendicular to the plane of the page, and the current emerges from this plane.

Fig. 30.12 (right) Magnetic field lines around a straight wire, made visible by iron filings sprinkled on a sheet of paper.

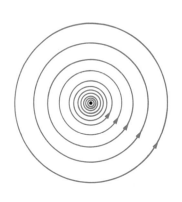
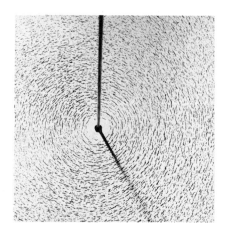

EXAMPLE 4. A square loop of wire of dimension $L \times L$ carries a current I. What is the magnetic field at the center of the loop?

SOLUTION: Figure 30.13a shows the loop. Each side of the loop makes the same contribution to the magnetic field, in magnitude and in direction. Hence the net magnetic field at the center is four times the magnetic field contributed by one side.

Figure 30.13b shows one side of the loop. The calculation of the magnetic field of this side is similar to the calculation of the magnetic field of the infinitely long wire, but the limits for the integration over x are now finite. In terms of the angle θ introduced in Eq. (28), the lower limit is $\theta = 45°$ (or $\theta = \pi/4$) and the upper limit is $\theta = 135°$ (or $\theta = 3\pi/4$). With these new limits, the integral in Eq. (31) becomes

Fig. 30.13 A square loop carrying a current I.

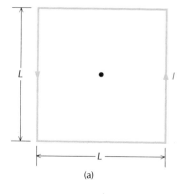
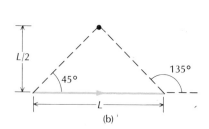

$$B_1 = \frac{\mu_0}{4\pi} \frac{I}{L/2} \int_{\pi/4}^{3\pi/4} \sin\theta \, d\theta$$

$$= \frac{\mu_0}{4\pi} \frac{I}{L/2} \left[-\cos\theta \right]_{\pi/4}^{3\pi/4}$$

$$= \frac{\mu_0}{4\pi} \frac{I}{L/2} \sqrt{2} \tag{33}$$

Multiplying this by 4, we obtain the net magnetic field of the entire loop,

$$B = 4B_1 = \frac{\mu_0}{4\pi} \frac{8\sqrt{2}\, I}{L} \tag{34}$$

EXAMPLE 5. Find the magnetic field on the axis of a circular loop of wire of radius R carrying a current I.

SOLUTION: Figure 30.14 shows the loop and a point on its axis. A small segment dl of the ring produces a magnetic field of magnitude

$$dB = \frac{\mu_0}{4\pi} \frac{I\, dl}{z^2 + R^2} \tag{35}$$

The direction of this magnetic field is perpendicular to the line connecting dl and the point on the axis (Figure 30.14). Upon integration around the circle, the horizontal components of the magnetic field cancel because diametrically opposite segments dl contribute opposite horizontal components. Only the vertical component of **B** survives.

The vertical component of the magnetic field in Eq. (35) is

$$dB_z = dB\cos(90° - \alpha) = dB\sin\alpha = \frac{\mu_0}{4\pi} \frac{I\, dl}{z^2 + R^2} \sin\alpha \tag{36}$$

where, according to Figure 30.14,

$$\sin\alpha = R/\sqrt{z^2 + R^2} \tag{37}$$

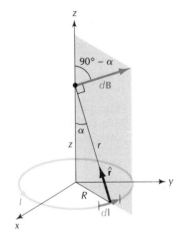

Fig. 30.14 Circular wire loop carrying a current I in the x–y plane.

Hence

$$B_z = \int \frac{\mu_0}{4\pi} \frac{I R\, dl}{(z^2 + R^2)^{3/2}} \tag{38}$$

Since all the terms in the integrand are constant, the integration amounts to multiplication of the integrand by the length of the path of integration (the circumference of the circle). Thus,

$$B_z = \frac{\mu_0}{4\pi} \frac{IR}{(z^2 + R^2)^{3/2}} \times 2\pi R \tag{39}$$

or

$$B_z = \frac{\mu_0}{2\pi} \frac{I\pi R^2}{(z^2 + R^2)^{3/2}} \tag{40}$$

$z = 0$

$\dfrac{\mu_0}{2\pi} \dfrac{I}{r}$

COMMENTS AND SUGGESTIONS: The examples of this section (straight wire, square loop, circular loop) illustrate the typical procedures for the calculation of the magnetic fields of current distributions according to the Biot–Savart Law. To calculate the magnetic field at a given point, always begin by determining the direction of the magnetic field. The direction of the magnetic field contributed by a small current element can be obtained from Eq. (25), by applying the right-hand rule to the cross product $d\mathbf{l} \times \hat{\mathbf{r}}$. If the contributions from all the current elements are in the same direction (as in Examples 3 and 4), you can proceed directly to the integration. If the contributions from different current elements are in different directions, consider the x, y, and z components of the net magnetic field, and decide which of these components are zero, because of cancellations brought about by the symmetry of the current distribution (see Example 5). With a good choice of coordinate axes, the net magnetic field will usually have only one nonzero component, to be evaluated by integration.

Calculations of magnetic fields of complicated current distributions can often be broken up into several individual parts, with simpler individual current distributions whose magnetic fields are easier to calculate (see Example 4). By the superposition principle, the net magnetic field generated jointly by the entire current distribution is then the vector sum of the individual magnetic fields of the individual parts of the current distribution.

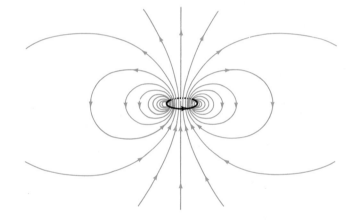

Fig. 30.15 Magnetic field lines of a circular loop of wire carrying a current.

The calculation of Example 5 gives the magnetic field only on the axis of the loop. The calculation of the magnetic field at other points is rather messy. Figure 30.15 shows the general pattern of magnetic field lines throughout space. And Figure 30.16 shows this same pattern of magnetic field lines, as made visible with iron filings.

Fig. 30.16 Iron filings sprinkled on a sheet of a paper placed around a loop carrying a strong current.

30.4 The Magnetic Dipole

At large distances from the loop, the pattern of magnetic field lines in Figure 30.15 is quite similar to the pattern of electric field lines of an electric dipole (see Figure 25.12). Hence, a small loop of current is called a **magnetic dipole**. In this context, *small* means that the size of the loop is small compared with the distance at which we wish to examine the magnetic field; in Eq. (40) it means $R \ll z$. We can then approximate the expression (40) by

$$B_z = \frac{\mu_0}{2\pi} \frac{I\pi R^2}{z^3} \qquad (41)$$

Magnetic dipole

This shows that the magnetic field of a loop of current decreases as the inverse cube of the distance. It is customary to write Eq. (41) as

$$B_z = \frac{\mu_0}{2\pi} \frac{\mu}{z^3} \qquad (42)$$

where

$$\mu = I\pi R^2 \qquad (43)$$

is the **magnetic dipole moment** of the loop.[3]

Note that Eq. (43) has the form

$$\boxed{\mu = [\text{current}] \times [\text{area of loop}]} \qquad (44)$$

Magnetic dipole moment

This turns out to be a general expression for the magnetic dipole moment of a (plane) loop of arbitrary shape. For instance, the magnetic dipole moment of a square or rectangular loop of current can be calculated from Eq. (44), and the magnetic field at a large distance from the loop is then given by Eq. (42).

If we regard a loop of arbitrary shape as made up of many small circular loops, we can understand why Eq. (44) is applicable in general. For instance, Figure 30.17 shows a rectangular loop and an array of small circular loops filling this rectangular loop. Assume that each of the circular loops carries the same current as the rectangular loop. Consider any point within the area of the rectangle where two of the circular loops (almost) touch; since the currents of the touching loops are opposite, they cancel at this point, and they make no contribution to the magnetic field. The only portions of the circular loops that escape this cancellation are the portions lying along the edges of the rectangle. Thus, the effective net current of the array of circular loops is approximately the same as the current around the rectangular loop, and we can regard the magnetic field of the rectangular loop as the sum of the magnetic fields of the circular loops. The magnetic moment of the rectangle is therefore the sum of the magnetic moments of all the loops. Since the sum of the areas of the circular loops equals the

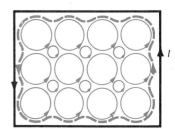

Fig. 30.17 A rectangular loop is approximated by an array of small circular loops. Each circular loop carries the same current as the rectangular loop. The dashed line represents the effective net current of the array of circular loops.

---

[3] The μ in Eqs. (42) and (43) must not be confused with the μ_0 for permeability; they are not related.

area of the rectangle, we obtain Eq. (44) for the magnetic moment of the rectangular loop.

Electrons, protons, and many other elementary particles have magnetic dipole moments. These particles may be regarded crudely as small balls of electric charge spinning about an axis. Such a rotating charge distribution behaves like an array of loops of current. These currents generate a magnetic field which, at a large distance from the particle, is essentially the dipole field of Figure 30.15. The dipole moment of an electron is 9.3×10^{-24} A·m². Thus, the space surrounding an electron is filled not only with electric fields but also with magnetic fields. Figure 30.18 shows the electric and magnetic fields of an electron. Of course, if the electron also has a translational motion, then there will be an extra magnetic field of the type given by Eq. (15) around the electron.

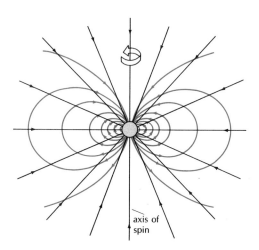

Fig. 30.18 Electric (black) and magnetic (colored) fields of an electron.

The Earth also has a magnetic dipole moment. This is presumably due to currents that flow around in loops deep inside the Earth, in the liquid iron of the core. Geophysicists do not yet know exactly what the energy supply is for the emf that drives these currents, but it is certain that the rotation of the Earth plays a crucial role in the generation of the currents. These currents create a magnetic field which, at large

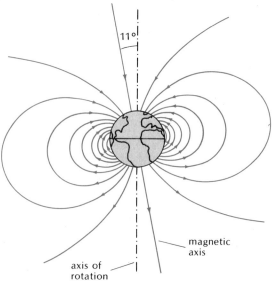

Fig. 30.19 Magnetic field of the Earth. The axis of the magnetic dipole makes an angle of 11° with the axis of rotation of the Earth.

distances from the core, is nearly a dipole field. The magnetic moment of the Earth is 8.0×10^{22} A·m². Figure 30.19 shows the magnetic field lines. Note that the field lines emerge from the surface of the Earth near the geographic South Pole, and they reenter the surface near the geographic North Pole. The force experienced by a compass needle is due to this magnetic field.

SUMMARY

Magnetic force between moving point charges:
$$\mathbf{F} = \frac{\mu_0}{4\pi} \frac{qq'}{r^2} \mathbf{v} \times (\mathbf{v'} \times \hat{\mathbf{r}})$$

Magnetic field of point charge:
$$\mathbf{B} = \frac{\mu_0}{4\pi} \frac{q'}{r^2} (\mathbf{v'} \times \hat{\mathbf{r}})$$

Permeability constant: $\mu_0 = 1.26 \times 10^{-6}$ N·s²/C²

$$\frac{\mu_0}{4\pi} = 1.00 \times 10^{-7} \text{ N·s}^2/\text{C}^2$$

Definition of magnetic field: $\mathbf{F} = q\mathbf{v} \times \mathbf{B}$

Gauss' Law for magnetism: $\oint \mathbf{B} \cdot d\mathbf{S} = 0$

Biot–Savart Law: $d\mathbf{B} = \frac{\mu_0}{4\pi} \frac{I \, d\mathbf{l} \times \hat{\mathbf{r}}}{r^2}$

Magnetic dipole moment: $\mu = $ [current] $\times$ [area of loop]

QUESTIONS

1. List all the physical laws of force you can think of. Which of these laws are fundamental, and which can be derived from others?

2. Give an example of two moving charged particles with velocities such that the mutual magnetic forces do not obey Newton's Third Law. Give an example with velocities such that the mutual magnetic forces do obey Newton's Third Law.

3. How would the magnetic field lines shown in Figure 30.8 differ if the charge of the particle were negative instead of positive?

4. Theoretical physicists have proposed the existence of **magnetic monopoles**, which are sources and sinks of magnetic field lines, just as electric charges are sources and sinks of electric field lines. What would the pattern of magnetic field lines of a positive magnetic monopole look like? Can you guess the pattern of *electric* field lines of a moving magnetic monopole?

Magnetic monopole

5. The Earth's magnetic field at the equator is horizontal, in the northward direction. What is the direction of the magnetic force on an electron moving vertically up?

6. An electron with a vertical velocity passes through a magnetic field without suffering any deflection. What can you conclude about the magnetic field?

7. At an initial time, a charged particle is at some point P in a magnetic field and it has an initial velocity. Under the influence of the magnetic field, the particle moves to a point P'. If you now reverse the velocity of the particle, will it retrace its orbit and return to the point P?

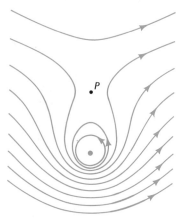

Fig. 30.20 These magnetic field lines result from the superposition of the field of a long straight wire and a uniform horizontal field.

Dip needle

8. An electron moving northward in a magnet is deflected toward the east by the magnetic field. What is the direction of the magnetic field?

9. A Faraday cage shields electric fields. Does it also shield magnetic fields?

10. A long straight wire carrying a current I is placed in a uniform magnetic field B_0 at right angles to the direction of the field. The net magnetic field is the superposition of the field of the wire and the uniform field; the field lines of this net field are shown in Figure 30.20. At the point P, the net magnetic field is zero. What is the distance of this point from the wire?

11. Strong electric fields are hazardous — if you place some part of your body in a strong electric field, you are likely to receive an electric shock. Are strong magnetic fields hazardous? Do they produce any effect on your body?

12. Figure 30.19 shows the magnetic field of the Earth. What must be the direction of the currents flowing in loops inside the Earth to give this magnetic field?

13. The needle of an ordinary magnetic compass indicates the direction of the horizontal component of the Earth's magnetic field. Explain why the magnetic compass is unreliable when used near the poles of the Earth.

14. A **dip needle** is a compass needle that swings about a horizontal axis. If the axis is oriented east–west, then the equilibrium direction of the dip needle is the direction of the Earth's magnetic field. The **dip angle** of the dip needle is the angle that it makes with the horizontal. How does the dip angle vary as you transport a dip needle along the surface of the Earth from the South Pole to the North Pole?

15. Suppose we replace the single loop shown in Figure 30.14 by a coil of N loops. How does this change the formula [Eq. (40)] for the magnetic field?

16. In order to eliminate or reduce the magnetic field generated by the pair of wires that connect a piece of electric equipment to an outlet, a physicist twists these wires tightly about each other. How does this help?

17. Consider a circular loop of wire carrying a current. Describe the direction of the magnetic field at different points in the plane of the loop, both inside and outside of the loop.

18. An infinite flat conducting sheet lies in the x–y plane. The sheet carries a current in the y direction; this current is uniformly distributed over the entire sheet. What is the direction of the magnetic field above the sheet? Below the sheet?

19. Suppose that an infinitely long straight wire lies along the axis of a circular loop. Both the wire and the loop carry currents I. Draw some field lines of the net magnetic field of the wire and the loop.

20. According to Eq. (42), the magnetic field at a large distance from a square loop is approximately the same as the magnetic field of a circular loop of the same area. Why is the shape unimportant? [Hint: The circular loop can be regarded as made of many small (infinitesimal) square loops.]

PROBLEMS

Section 30.1

1. A positron (charge $+e$) and an electron (charge $-e$) are moving side by side with a uniform velocity of 2.0×10^6 m/s along parallel tracks. Is the magnetic force between them attractive or repulsive? By what percentage is the total force (electric and magnetic) acting between these moving charges larger or smaller than that acting between charges at rest at the same distance?

2. Two electrons are (instantaneously) moving at right angles with speeds $v = 3.0 \times 10^6$ m/s and $v' = 1.0 \times 10^6$ m/s in the same plane (Figure 30.21). Their distance is 0.80×10^{-10} m. Calculate the magnetic force of the first electron on the second and that of the second on the first. Show the directions of the forces on a diagram.

3. An electron (charge $-e$) and an antielectron (charge $+e$) are in a circular orbit about each other (Figure 30.22). The orbital radius is 0.53×10^{-10} m and the orbital speed is 1.1×10^6 m/s.
   (a) What is the magnitude of the electric force on each particle? Draw a diagram showing the direction of the electric force on each particle.
   (b) What is the magnitude of the magnetic force on each particle?

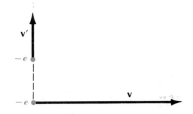

Fig. 30.21 Instantaneous positions and velocities of two electrons.

4. Two positive point charges q and q' are moving along perpendicular paths which intersect. Their speeds v and v' are equal, and they reach the intersection point simultaneously (assume they pass through each other without disruption). What is the magnitude of the magnetic force on each when their distance from the intersection point is d? Show the directions of the magnetic forces on a diagram, before the charges reach the intersection point and afterward.

*5. Prove that the magnetic force (10) can be expressed as

$$\mathbf{F} = \frac{\mu_0}{4\pi} \frac{qq'}{r^2} (\mathbf{v}' \, \hat{\mathbf{r}} \cdot \mathbf{v} - \hat{\mathbf{r}} \, \mathbf{v} \cdot \mathbf{v}')$$

and verify explicitly that this leads to Eq. (2) if the charges are instantaneously moving side by side with parallel velocities. (Hint: Use the result of Problem 3.46.)

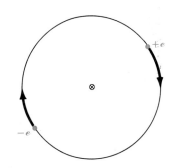

Fig. 30.22 Electron and antielectron in orbit about their common center of mass.

*6. Two protons have equal speeds of 2.0×10^6 m/s. Their directions of motion are in the same plane, and they make an angle of 60°. The distance between the protons is 2.4×10^{-10} m, and this distance is (instantaneously) perpendicular to the velocity $\mathbf{v}$ of one of the protons.
   (a) Calculate the magnetic force on each proton due to the other proton; show the directions of the forces on a diagram.
   (b) Calculate the instantaneous rate of change of the total momentum of the two-proton system.

Section 30.2

7. Show that the magnetic field of a point charge q' moving with a velocity $\mathbf{v}'$ [Eq. (15)] can be written in terms of the electric field of this point charge as $\mathbf{B} = \mu_0 \varepsilon_0 \mathbf{v}' \times \mathbf{E}$.

8. The magnetic field surrounding the Earth typically has a strength of 5×10^{-5} T. Suppose that a cosmic-ray electron of energy 3×10^4 eV is instantaneously moving in a direction perpendicular to the lines of this magnetic field. What is the force on this electron?

9. At a location where the strength of the Earth's magnetic field is 0.60×10^{-4} T, what must be the minimum speed of an electron if the magnetic force on it is to exceed its weight?

10. At the surface of a pulsar, or neutron star, the magnetic field may be as strong as 10^8 T. Consider the electron in a hydrogen atom on the surface of such a neutron star. The electron is at a distance of 0.53×10^{-10} m from the proton and has a speed of 2.2×10^6 m/s. Compare the electric force that the proton exerts on the electron with the magnetic force that the magnetic field of the neutron star exerts on the electron. Is it reasonable to expect that the hydrogen atom will be strongly deformed by the magnetic field?

*11. In New York, the magnetic field of the Earth has a vertical (down) component of 0.60×10^{-4} T and a horizontal (north) component of 0.17×10^{-4} T. What are the magnitude and direction of the magnetic force on an electron of velocity 1.0×10^6 m/s moving (instantaneously) in an east-to-west direction in a television tube?

Section 30.3

12. The current in a lightning bolt may be as much as 2×10^4 A. What is the magnetic field at a distance of 1.0 m from a lightning bolt? The bolt can be regarded as a straight line of current.

13. The cable of a high-voltage power line is 25 m above the ground and carries a current of 1.8×10^3 A.
    (a) What magnetic field does this current produce at the ground?
    (b) The strength of the magnetic field of the Earth is 0.60×10^{-4} T at the location of the power line. By what factor do the fields of the power line and of the Earth differ?

14. 150 turns of insulated wire are wound around the rim of a plywood disk of radius 20 cm. The resistance of this wire is 20 Ω. What voltage must you supply to the terminals of the wire to generate a magnetic field of 8.0×10^{-4} T at the center of the disk?

15. A superconducting circular ring of diameter 3.0 cm carries a current of 12 A. What is the strength of the magnetic field at the center of the ring? Along the axis of the ring at a distance of 3.0 cm from the center?

16. A circular ring of wire of diameter 0.60 m carries a current of 35 A. What acceleration will the magnetic force generated by this ring give to an electron that is passing through the center of the ring with a velocity of 1.2×10^6 m/s in the plane of the ring?

*17. Two very long straight wires carry currents I at right angles. One of the wires lies along the x axis; the other lies along the y axis (see Figure 30.23). Find the magnetic field at a point in the first quadrant.

Fig. 30.23 Two long straight wires.

*18. In a motorboat, the compass is mounted at a distance of 0.80 m from a cable carrying a current of 20 A from an electric generator to a battery.
    (a) What magnetic field does this current produce at the location of the compass? Treat the cable as a long straight wire.
    (b) The horizontal (north) component of the Earth's magnetic field is 0.18×10^{-4} T. Since the compass points in the direction of the net horizontal magnetic field, the current will cause a deviation of the compass. Assume that the magnetic field of the current is horizontal and at right angles to the horizontal component of the Earth's magnetic field. Under these circumstances, by how many degrees will the compass deviate from north?

*19. Two very long, straight, parallel wires separated by distance d carry currents of magnitude I in opposite directions. Find the magnetic field at a point equidistant from the lines, with a distance $2d$ from each line. Draw a diagram showing the direction of the magnetic field.

*20. Three long parallel wires are spaced as shown in Figure 30.24. The wires carry equal currents in the same direction. What is the magnetic field at the point P? Draw a diagram giving the direction of the magnetic field.

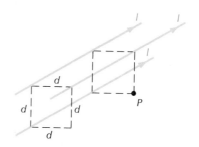

Fig. 30.24 Three long parallel wires. The wires pass through three corners of a square.

*21. Two very long parallel wires separated by a distance of 1.0 cm carry opposite currents of 8.0 A.
  (a) Find the magnetic field at the midpoint between the wires.
  (b) Find the magnetic field in the plane of the wires, at a distance of 2.0 cm from the midline.

*22. A very long wire is bent at a right angle near its midpoint. One branch of it lies along the positive x axis and the other along the positive y axis (Figure 30.25). The wire carries a current I. What is the magnetic field at a point in the first quadrant of the x–y plane?

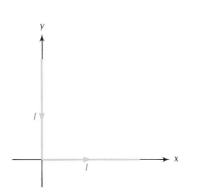

Fig. 30.25 A very long wire bent at a right angle. The wire extends to infinity along the x and y axes.

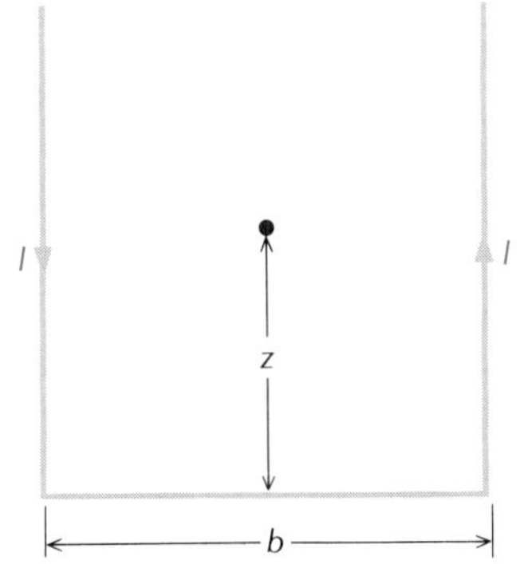

Fig. 30.26 A very long wire bent in the shape of a **U**.

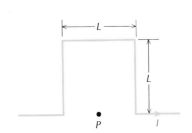

Fig. 30.27 A very long wire with square bends.

*23. A very long wire carrying a current I is bent in the shape of a **U** (Figure 30.26). The two parallel segments are separated by a distance b. Find the magnetic field along the midline at a distance z from the bottom of the **U**. Consider both the case $z > 0$ and $z < 0$.

*24. A very long straight wire with a current I has a square bend near its midpoint (Figure 30.27). Each side of the square has a length L. What is the magnetic field at the point P, halfway between the two lower corners?

*25. A loop of superconducting wire has the shape of a rectangle measuring $L \times 2L$. A current I flows around the wire. What is the strength of the magnetic field at the center of the rectangle?

*26. A square loop of wire, measuring $h \times h$, carries a current I (Figure 30.28). Find the magnetic field at the point P at a distance $h/4$ from the center of the square.

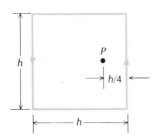

Fig. 30.28 A square loop.

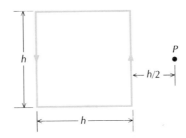

Fig. 30.29 A square loop.

*27. A square loop of wire, measuring $h \times h$, carries a current I (Figure 30.29). Find the magnetic field at the point P at a distance h from the center of the square.

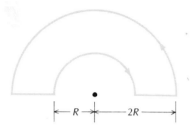

Fig. 30.30 A loop consisting of two semicircles and two straight segments.

*28. A loop of wire has the shape of two concentric semicircles connected by two radial segments (Figure 30.30). The loop carries a current I. Find the magnetic field at the center.

*29. A circular loop of wire is folded along a diameter so as to form two semicircles of radius R intersecting at right angles (Figure 30.31). A current I flows around this loop. What is the magnetic field at the center? Draw a diagram showing the direction of the magnetic field.

*30. A very long strip of copper of width b carries a current I uniformly distributed over the strip. What is the magnetic field at a distance z above the midline of this strip (Figure 30.32).

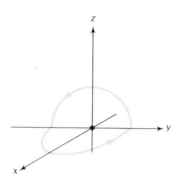

Fig. 30.31 A semicircle in the z–y plane and a semicircle in the x–y plane.

Fig. 30.32 A very long strip carrying a uniformly distributed current I.

*31. An infinite conducting sheet occupies the x–y plane. On this sheet a current flows in the y direction; the current is uniformly distributed over the sheet with σ amperes flowing across each 1-meter segment of the x axis. Find the magnetic field at a distance z from this sheet.

*32. Two semi-infinite wires are in the same plane. The wires make an angle of 45° with each other, and they are joined by an arc of circle of radius R (Figure 30.33). The wires carry a current I. Find the magnetic field at the center of the arc of circle.

*33. Helmholtz coils are often used to make reasonably uniform magnetic fields in laboratories. These coils consist of two thin circular rings of wire parallel to each other and on a common axis, the z axis (Figure 30.34). The rings have radius R and they are separated by a distance which is also R. The rings carry equal currents in the same direction.
  (a) Find the magnetic field at any point of the z axis.
  (b) Show that dB/dz and d^2B/dz^2 are both zero at $z=0$.

Fig. 30.33 Two very long wires joined by an arc of circle.

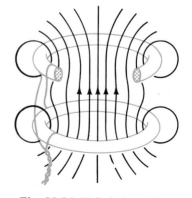

Fig. 30.34 Helmholtz coils.

**34. A current I flows around the surface of a copper pipe of radius R and length L (Figure 30.35). Find a formula for the magnetic field on the axis of the pipe at a distance z from the center. Assume $z > L/2$.

**35. A semi-infinite wire lies along the positive x axis. Another semi-infinite wire lies along the y axis. The wires carry a current I from $y = \infty$ to $x = \infty$ (Figure 30.36).
  (a) Find the magnetic field at a point in the first octant ($x > 0$, $y > 0$, $z > 0$).
  (b) Find the magnetic field at a point in the second octant ($x < 0$, $y > 0$, $z > 0$).

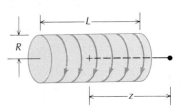

Fig. 30.35 A uniformly distributed current flows around the circumference of a copper pipe.

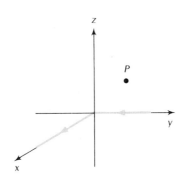

Fig. 30.36 Long straight wires along the x and the y axes.

**36. Two semi-infinite parallel wires lie in the x–y plane. The wires are joined by a semicircle of radius R whose center coincides with the origin of coordinates (Figure 30.37). The wires carry a current I in the direction shown in the figure. Find the x, y, and z components of the magnetic field at a point on the z axis.

**37. An amount of charge Q is uniformly distributed over a disk of paper of radius R. The disk spins about its axis with angular velocity ω. Find the magnetic field produced by this disk at a point on the axis of the disk, at a distance z from the center.

38. The magnetic dipole moment of an electron is 9.3×10^{-24} A·m². What is the magnetic field on the axis of spin of the electron at a distance of 1.0 Å from its center?

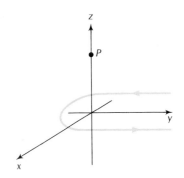

Fig. 30.37 Two long wires in the x–y plane joined by a semicircle.

39. The magnetic field of the Earth is approximately that of a magnetic dipole located at the center of the Earth. The strength of the magnetic field at the surface of the Earth at the magnetic north pole is 6.2×10^{-5} T. What is the strength of the magnetic field at an altitude of 1000 km above the pole? 2000 km? 3000 km? Make a plot of the strength of the magnetic field vs. altitude.

40. The magnetic field of the Earth is that of a magnetic moment of 8.0×10^{22} A·m² located deep within the Earth. Assume that the magnetic moment is due to a circular current flowing along the equator of the liquid core of the Earth; this core has a radius of 3500 km. What must be the current? What must be its direction (eastward or westward)?

41. It can be proved that the dependence on x, y, z of the magnetic field of a magnetic dipole is described by the same function as the dependence on x, y, z of the electric field of an electric dipole. Use this fact to convert the formulas of Eqs. (25.52)–(25.54) into formulas for a magnetic dipole.

*42. A steady current I flows around a wire bent into a square of side L (Figure 30.38). Find the magnetic field at a distance z from the center of the square on a line perpendicular to the face of the square. Show that if $z \gg L$, your answer reduces to $B_z = (\mu_0/2\pi)\mu/z^3$, with a magnetic moment $\mu = IL^2$.

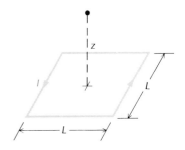

Fig. 30.38 A square loop.

*43. A pipe made of a superconducting material has a length of 0.30 m and a radius of 4.0 cm. A current of 4.0×10^3 A flows around the surface of the pipe; the current is uniformly distributed over the surface. What is the magnetic moment of this current distribution? (Hint: Treat the current distribution as a large number of rings stacked one on top of another.)

*44. An amount of charge Q is uniformly distributed over a disk of paper of radius R. The disk spins about its axis with angular velocity ω. Find the magnetic dipole moment of the disk. Sketch the lines of magnetic field and of electric field in the vicinity of the disk.

45. At a large distance, the pattern of magnetic field lines of a magnetic dipole is the same as that of the electric field lines of an electric dipole. The field of the latter is given by Eqs. (25.52)–(25.54). Use these equations to find the angle at which the magnetic field lines of the magnetic dipole field of the Earth intersect the surface of the Earth, at any given magnetic latitude Θ. This angle is called the **dip angle. Evaluate the dip angle at a magnetic latitude of 45°. [Hint: For $y = 0$, write $x = R_E \cos \Theta$, $z = R_E \sin \Theta$ in Eqs. (25.52)–(25.54).]

CHAPTER 31

Ampère's Law

In Chapter 24 we saw that Gauss' Law gives us a shortcut for calculating the electric field of charge distributions that have a certain amount of symmetry. In this chapter we will see that there exists a corresponding method for calculating the magnetic field of current distributions. The method is based on Ampère's Law, an important relation between the magnetic field and the current.

31.1 Ampère's Law

To obtain this relation, we begin by examining the magnetic field of a very long straight wire carrying a steady current I. According to Eq. (30.32), the magnetic field at a distance r from this wire has a magnitude

$$B = \frac{\mu_0}{2\pi} \frac{I}{r} \tag{1}$$

and the magnetic field lines are concentric circles around the wire. Consider now a closed mathematical path of arbitrary shape around this wire (Figure 31.1) and evaluate the integral $\oint \mathbf{B} \cdot d\mathbf{l}$ around this path. The path can be regarded as consisting of short radial segments and circular arcs. Since $\mathbf{B}$ is in the tangential direction, the radial segments do not contribute to the integral. Since the tangential segments have lengths $r\, d\theta$, we obtain

$$\oint \mathbf{B} \cdot d\mathbf{l} = \oint B r\, d\theta = \oint \frac{\mu_0}{2\pi} \frac{I}{r} r\, d\theta = \frac{\mu_0}{2\pi} I \oint d\theta \tag{2}$$

André Marie Ampère, *1775–1836, French physicist and mathematician, professor at the École Polytechnique and at the Collège de France. After Oersted's discovery of the generation of magnetic fields by electric currents, Ampère demonstrated experimentally that currents exert magnetic forces on each other. He carefully investigated the relationship between currents and magnetic fields, and he established that a magnet is equivalent to a distribution of currents. We are indebted to him not only for Ampère's law but also for the equation (31.35) for the force exerted by a magnetic field on a current element.*

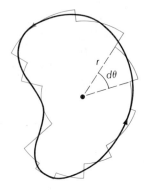

Fig. 31.1 Path of integration around a long straight wire. The wire is perpendicular to the plane of the page. The path can be approximated by radial segments and circular arcs.

Ampère's Law

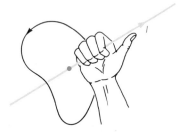

Fig. 31.2 The right-hand rule for Ampère's Law: If the fingers curl in the same direction as the path of integration, the thumb points in the positive direction for the current.

The integral $\oint d\theta$ over the closed path is simply 2π. Hence

$$\oint \mathbf{B} \cdot d\mathbf{l} = \mu_0 I \tag{3}$$

Although the above calculation was based on a planar path, the same result applies to nonplanar paths — any path segment parallel to the wire is perpendicular to the magnetic field, and therefore, as in the above calculation, only the circular arcs of the path contribute to the integral.

It turns out that Eq. (3) is valid not only for the magnetic field produced by a current on a long wire but also for the magnetic field produced by an arbitrary distribution of steady currents. This is **Ampère's Law**:

*The integral of **B** around any closed mathematical path equals μ_0 times the current intercepted by the area spanning the path,*

$$\boxed{\oint \mathbf{B} \cdot d\mathbf{l} = \mu_0 I} \tag{4}$$

Note that the current in Eq. (4) need not flow on thin wires; it may flow on a conductor of any shape, or it may even flow without any conductor (for example, a beam of protons in vacuum). Furthermore, note that the current is to be reckoned as positive if it is related to the direction around the path by the right-hand rule (thumb points along current when fingers curl in the direction of path; Figure 31.2), and negative otherwise. Although Ampère's Law can be derived rigorously from our expression for the magnetic field of a small current element [see Eq. (30.25)], we will not attempt to give this general proof.

For the calculation of magnetic fields, Ampère's Law plays a role similar to that of Gauss' Law for the calculation of electric fields. Provided that a distribution of currents has sufficient symmetry, Ampère's Law completely determines the magnetic field, as we will see in the following examples.

EXAMPLE 1. Deduce the magnetic field of a current on a very long thin straight wire by means of Ampère's Law.

SOLUTION: By considerations of symmetry, the magnetic field lines of such a current will have to be either concentric circles, or radial lines, or parallel lines in the same direction as the wire. Radial lines would require that the field lines start on the wire, which is impossible, since the field lines must form closed loops. Parallel lines in the direction of the wire are likewise inconsistent with closed loops. Thus, the field lines must necessarily be concentric circles. Furthermore, by symmetry, the magnetic field must have a constant magnitude along each circle.

Taking a path that follows one of these circles, of radius r, we see that the integral of **B** around this path is

$$\oint \mathbf{B} \cdot d\mathbf{l} = \oint Br \, d\theta = 2\pi r B \tag{5}$$

and Ampère's Law then gives us

$$2\pi r B = \mu_0 I \tag{6}$$

that is

$$B = \frac{\mu_0}{2\pi} \frac{I}{r} \quad (7)$$

Thus, Ampère's Law gives us back the formula with which we started.

COMMENTS AND SUGGESTIONS: The path of integration in this calculation was chosen so that the magnetic field is everywhere tangential to the path (the path follows a field line) and of constant magnitude. The evaluation of the integral $\oint \mathbf{B} \cdot d\mathbf{l}$ then becomes trivial, and we can "solve" Ampère's Law for the unknown magnetic field. In the other examples of the use of Ampère's Law in this and the next section, we will find it convenient to make similar choices of paths of integration, with the magnetic field either tangential to all portions of the path or tangential to some portions of the path and perpendicular elsewhere.

Incidentally: We were able to deduce the magnetic field of a steady current on a straight wire from Ampère's Law, but we are not able to deduce the magnetic field of an individual moving charge — such a point charge does *not* constitute a steady current, and Ampère's Law is not directly applicable.

EXAMPLE 2. A very long straight copper wire has a circular cross section of radius R. The wire carries a current I_0 uniformly distributed over the cross-sectional area of the wire. What is the magnetic field inside the wire? Outside the wire?

SOLUTION: Consider a circular path of radius r inside the wire (Figure 31.3). Symmetry tells us that the magnetic field lines are circles. Taking a path of integration along one of these circles, we find

$$\oint \mathbf{B} \cdot d\mathbf{l} = 2\pi r B$$

Since the current is uniformly distributed over the volume of the wire, the amount of current I intercepted by the area within the circular path is in direct proportion to this area,

$$I = I_0 \frac{\pi r^2}{\pi R^2} = I_0 \frac{r^2}{R^2} \quad (8)$$

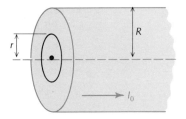

Fig. 31.3 Segment of a wire of radius R with a current I_0 uniformly distributed over the cross-sectional area.

Ampère's Law then leads to

$$2\pi r B = \mu_0 I \quad (9)$$

or

$$2\pi r B = \mu_0 I_0 r^2 / R^2 \quad (10)$$

and

$$B = \frac{\mu_0}{2\pi} \frac{I_0 r}{R^2} \quad (11)$$

Note that the magnetic field is zero at the center of the wire ($r = 0$) and reaches a maximum value at the surface of the wire ($r = R$).

Outside the wire, we can find the magnetic field by repeating the calculation of Example 1. This calculation depends only on cylindrical symmetry and not on the thickness of the wire. The magnetic field is therefore exactly the same as for a thin wire,

$$B = \frac{\mu_0}{2\pi} \frac{I_0}{r} \quad (12)$$

COMMENTS AND SUGGESTIONS: Our calculation relies on the implicit assumption that the material of the wire is nonmagnetic and has no effect on the magnetic field. This is a valid assumption for a copper wire, but it would not be a valid assumption for a wire of a magnetic material, such as iron. Throughout this chapter and the next, we will assume that all the materials in the conductors and in the space surrounding the conductors are nonmagnetic.

31.2 Solenoids

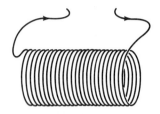

Fig. 31.4 A solenoid.

A solenoid is a thin conducting wire wound in a tight helical coil of many turns (Figure 31.4). A current in this wire will produce a strong magnetic field within the coil. Because of the similarity of the current distributions, a tight coil produces essentially the same magnetic field as a large number of rings stacked next to one another.[1] The magnetic field of such a stack of rings can be calculated by summing the magnetic fields of the individual rings. Figure 31.5 shows the pattern of field lines of the resultant magnetic field. Figure 31.6 shows the pattern of field lines of an actual solenoid with a rather loose coil, made visible with iron filings. The field is strong inside the solenoid but fairly weak outside. The detailed calculation of this magnetic field is rather messy because the individual magnetic fields of the rings of current must be added vectorially. We will not attempt this calculation here; instead, we will calculate the magnetic field of an **ideal solenoid**, that is, a very long (infinitely long) solenoid with very tightly wound coils, so the current distribution on the surface of the solenoid is nearly uniform.

Ideal solenoid

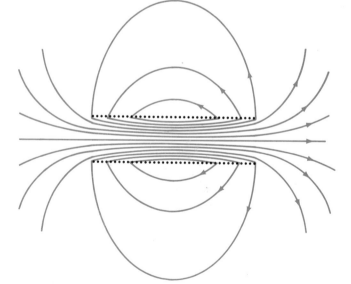

Fig. 31.5 Magnetic field lines of a solenoid.

---

[1] There is, however, one small difference: in a cylindrical stack of rings the current flows in closed circles around the surface of the cylinder, whereas in a cylindrical solenoid the current has an extra component parallel to the axis of the cylinder (the current is helical). If the solenoid is tightly wound with thin wire, this axial component of the current is insignificant compared to the circular component and we can ignore it.

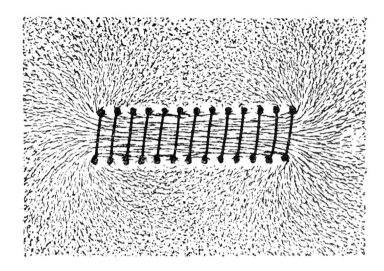

Fig. 31.6 Iron filings sprinkled on a sheet of paper inserted in a solenoid. Note that inside the solenoid the distribution of field lines is nearly uniform.

To find this magnetic field, we begin with an appeal to symmetry, as in Example 1. The ideal solenoid has translational symmetry (along the axis of the solenoid) and rotational symmetry (around the axis). For consistency with these symmetries, the magnetic field lines inside the solenoid will then have to be either concentric circles, or radial lines, or lines parallel to the axis. Concentric circles and radial lines are unacceptable; the former would require the presence of a current along the axis (compare Example 1), and the latter would require that the field lines start on the axis, which is impossible. Thus the field lines inside the solenoid must all be parallel to the axis. These field lines emerge from the (distant) end of the solenoid and return to the other end (Figure 31.5). For an ideal, very long solenoid, these external field lines will spread over a very large region of space; hence, the density of field lines and the magnetic field outside the solenoid are nearly zero.

We can now use Ampère's Law to determine the magnitude of the magnetic field. Consider the rectangular path shown in Figure 31.7 and evaluate the integral of **B** along this path. The horizontal side external to the solenoid does not contribute to the integral (**B** is zero); and the vertical sides do not contribute to the integral (**B** is perpendicular to the path). Hence, only the horizontal side within the solenoid contributes. The magnetic field has some constant magnitude along this side, and if the length of this side is l,

$$\oint \mathbf{B} \cdot d\mathbf{l} = Bl \tag{13}$$

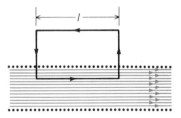

Fig. 31.7 Magnetic field lines of a very long solenoid.

The total current intercepted by the area within the rectangle is the current I_0 in the wire multiplied by the number N of wires passing through this area. Ampère's Law then tells us

$$Bl = \mu_0 N I_0 \tag{14}$$

or

$$B = \mu_0 I_0 (N/l) \tag{15}$$

The ratio N/l is the number of turns of wire per unit length, commonly designated by n. Thus,

Field in solenoid

$$B = \mu_0 I_0 n \qquad (16)$$

Note that this result is independent of the length of the vertical sides of the path of integration; hence **B** has the same magnitude everywhere within the solenoid. This shows that the magnetic field within an ideal solenoid is perfectly uniform.

Electromagnet

An **electromagnet** is essentially a solenoid with a gap, or, what amounts to the same thing, a pair of solenoids with their ends placed close together (Figure 31.8). Magnetic field lines come out of one solenoid and go into the other (of course, field lines will also have to come out of the solenoids at their other ends and close on themselves). The first solenoid is called the north pole of the electromagnet, and the second the south pole. If the gap is small, then the magnetic field in this gap is almost the same as that inside the solenoids. In most electromagnets, the space inside the solenoids is filled with an iron core (Figure 31.9); as we will see in Section 33.3, the iron enhances the magnetic field, making it much stronger than the value given by Eq. (16).

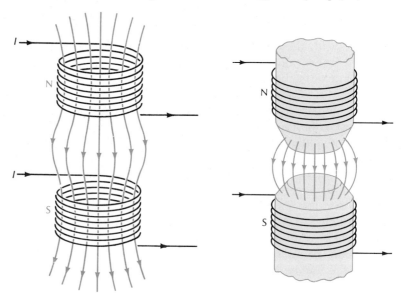

Fig. 31.8 (left) An electromagnet with two coils. The coil from which field lines emerge into the gap is called the north pole; the coil into which field lines enter is called the south pole.

Fig. 31.9 (right) An electromagnet with iron pole pieces.

EXAMPLE 3. A solenoid used for the investigation of the effect of magnetic fields on the propagation of light in a liquid consists of 180 turns of wire wound on a narrow tube 19 cm long. The current in the wire is 5.0 A. What is the strength of the magnetic field within the tube?

SOLUTION: The number of turns per unit length is

$$n = 180/0.19 \text{ m} = 9.5 \times 10^2/\text{m} \qquad (17)$$

A narrow solenoid can be regarded as a long solenoid, and the magnetic field near its middle is approximately the same as that of an ideal, infinite solenoid. Hence, according to Eq. (16),

$$B = \mu_0 I_0 n$$

$$= 1.26 \times 10^{-6} \text{ N} \cdot \text{m}^2/\text{C}^2 \times 5.0 \text{ A} \times 9.5 \times 10^2/\text{m}$$

$$= 6.0 \times 10^{-3} \text{ T}$$

EXAMPLE 4. A **toroid** is a conducting wire wound in a tight coil in the shape of a torus (a doughnut). It may be thought of as a solenoid that has been bent so that its ends meet (Figure 31.10a). What is the magnetic field inside a toroid with N turns of wire carrying a current I_0?

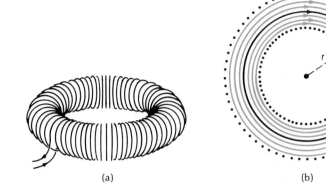 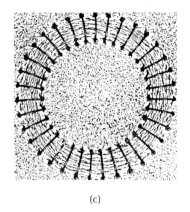

(a) (b) (c)

SOLUTION: By symmetry, the magnetic field lines inside the toroid must be closed circles (Figure 31.10b). If we integrate **B** along one of these circular field lines, we obtain

$$\oint \mathbf{B} \cdot d\mathbf{l} = 2\pi r B \tag{18}$$

Fig. 31.10 (a) A toroid. (b) Magnetic field lines within a toroid. (c) Magnetic field lines made visible with iron filings.

The total current intercepted by the circular area within this path equals the total number N of wires passing through the area, multiplied by the current I_0 in each wire. Hence, from Ampère's Law,

$$2\pi r B = \mu_0 N I_0 \tag{19}$$

or

$$B = \frac{\mu_0}{2\pi} \frac{N I_0}{r} \tag{20}$$

Thus, the magnetic field is inversely proportional to the radial distance from the axis of the torus.

COMMENTS AND SUGGESTIONS: The magnetic fields of a torus [Eq. (20)] and of a long straight wire [Eq. (7)] have the same dependence on distance from the axis of symmetry. However, the magnetic field of the wire extends through all space, whereas the magnetic field of the torus is confined to the interior of the torus.

31.3 Motion of Charges in Electric and Magnetic Fields

The force exerted by a magnetic field on a charged particle is

$$\mathbf{F} = q\mathbf{v} \times \mathbf{B} \tag{21}$$

This force is always at right angles to the velocity of the particle; con-

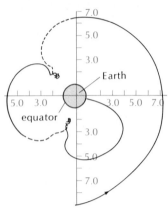

Fig. 31.11 Trajectory of a cosmic ray of high energy in the magnetic field of the Earth. The trajectory is three dimensional; portions above the equatorial plane, or the plane of the page, are shown with a solid line; portions below the equatorial plane are shown with a dashed line. The numbers along the axes give distances in Earth radii.

sequently, the force changes the *direction* of the velocity of the particle, but not the *magnitude* of the velocity. The formal proof of this is easy:

$$\frac{d}{dt}v^2 = \frac{d}{dt}\mathbf{v}\cdot\mathbf{v} = 2\mathbf{v}\cdot\frac{d\mathbf{v}}{dt} = 2\mathbf{v}\cdot\mathbf{F}/m = 0 \qquad (22)$$

where the last equality expresses the fact that force and velocity are perpendicular. Thus, the magnetic force changes only the momentum of the particle, not the kinetic energy — the magnetic force *does no work* on the particle.

In general, the motion of a particle in a magnetic field is quite complex. For example, Figure 31.11 shows the trajectory of a cosmic-ray particle approaching the Earth and being deflected by the magnetic field (the effects of gravity on such a cosmic ray are insignificant). In what follows, we will consider only the relatively simple case of motion in a uniform magnetic field.

Figure 31.12 shows a region with a uniform magnetic field, directed into the plane of the page. Suppose that a positively charged particle has an initial velocity in the plane of the page; this initial velocity is perpendicular to the magnetic field. The magnetic force $q\mathbf{v}\times\mathbf{B}$ is perpendicular to both $\mathbf{v}$ and $\mathbf{B}$; its direction is shown in Figure 31.12. The acceleration then has a constant magnitude

$$a = F/m = qvB/m \qquad (23)$$

and its direction is always perpendicular to the velocity. Such an acceleration is characteristic of uniform circular motion. Thus, the particle will move in a circle of some radius r, and the acceleration given by Eq. (23) will play the role of centripetal acceleration, that is,

$$\frac{qvB}{m} = \frac{v^2}{r} \qquad (24)$$

This leads to the following formula for the radius of the circular motion:

$$r = \frac{mv}{qB} \qquad (25)$$

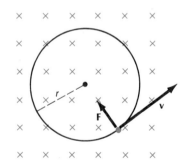

Fig. 31.12 Positively charged particle with uniform circular motion in a uniform magnetic field. The magnetic field points perpendicularly into the plane of the page; the crosses show the tails of the magnetic field vectors.

Figure 31.13 is a photograph of a beam of electrons executing such circular motion in a cathode-ray tube placed in a magnetic field.

The angular velocity of the circular motion is

$$\omega = \frac{v}{r} = \frac{qB}{m} \qquad (26)$$

and the frequency is

Cyclotron frequency

$$\boxed{\nu = \frac{\omega}{2\pi} = \frac{qB}{2\pi m}} \qquad (27)$$

This is called the **cyclotron frequency** because the operation of cyclotrons (described below) involves particles moving with this frequency

Fig. 31.13 Electrons moving in a circle in a cathode-ray tube in a magnetic field. The tube contains a gas at a very low pressure, and the atoms of the gas glow under the impact of the electrons; this makes the electron beam visible.

in a magnetic field. Note that the cyclotron frequency is independent of the speed of the circular motion — in a uniform magnetic field, slow particles and fast particles (of a given charge and mass) move around circles at the same frequency, but the slow particles move along smaller circles than the fast particles.

It is useful to write Eq. (25) in terms of the momentum $p = mv$; this leads to

$$r = \frac{p}{qB} \qquad (28)$$

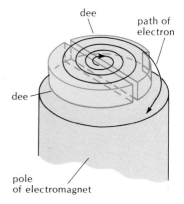

Fig. 31.14 Trajectory of a particle within the dees of a cyclotron. In this diagram, the upper pole of the electromagnet has been omitted for the sake of clarity.

Cyclotron

The advantage of this equation is that it remains valid even when the particle moves with relativistic velocity. Thus, Eq. (28) is more general than our derivation of it.

The **cyclotron** is a device for the acceleration of protons, deuterons, or other charged particles used in high-energy collision experiments. It consists of an evacuated cavity placed between the poles of a large electromagnet; within the cavity there is a flat metallic can cut into two **D**-shaped pieces, or **dees** (Figures 31.14 and 31.15). An oscillating high-voltage generator is connected to the dees; this creates an oscillating electric field in the gap between the dees. The frequency of the voltage generator is adjusted so that it coincides with the cyclotron frequency of Eq. (27). An ion source at the center of the cyclotron releases protons. The electric field in the gap between the dees gives each of these protons a push, and the uniform magnetic field in the cyclotron then makes the proton travel on a semicircle inside of the first dee. When the proton returns to the gap after one-half period, the high-voltage generator will have reversed the electric field in the gap; the proton therefore receives an additional push, which sends it into the second dee. There it travels on a semicircle of slightly larger radius corresponding to its slightly larger energy, and so on. Each time the proton crosses the gap between the dees, it receives an extra push and extra energy. Thus, the protons travel along arcs of circles of stepwise increasing radius. When the protons reach the outer edge of the dees, they leave the cyclotron as a high-energy beam (Figure 31.14).

Fig. 31.15 The dees of one of the early small cyclotrons built by E. O. Lawrence.

EXAMPLE 5. One of the first cyclotrons, built by E. O. Lawrence at Berkeley in 1932, had dees with a diameter of 28 cm, and its magnet was capable of producing a field of 1.4 T. What was the maximum kinetic energy of the protons accelerated by this cyclotron?

SOLUTION: When the proton reaches its maximum energy, its orbit has a radius of 14 cm. Since the magnetic field is 1.4 T, the momentum of such a proton is

$$p = mv = eBr$$

$$= 1.6 \times 10^{-19} \text{ C} \times 1.4 \text{ T} \times 0.14 \text{ m}$$

$$= 3.1 \times 10^{-20} \text{ kg} \cdot \text{m/s}$$

and the kinetic energy is

$$K = p^2/(2m)$$

$$= (3.1 \times 10^{-20} \text{ kg} \cdot \text{m/s})^2 / (2 \times 1.67 \times 10^{-27} \text{ kg})$$

$$= 2.9 \times 10^{-13} \text{ J}$$

COMMENTS AND SUGGESTIONS: Nuclear physicists like to measure energies in MeV, where

$$1 \text{ MeV} = 10^6 \text{ eV} = 1.60 \times 10^{-13} \text{ J}$$

In these units, the above kinetic energy is 1.8 MeV.

Cyclotrons of large size can be used to accelerate protons up to an energy of about 30 MeV. Above this energy, cyclotrons begin to fail because relativistic effects change the frequency of the orbital motion of the protons. At high energies the masses of the protons increase, according to Einstein's mass-energy formula $\Delta m = \Delta E/c^2$; consequently, the orbital frequency decreases. In order to keep the pushes of the electric field in phase with the motion of the proton, we must gradually decrease the frequency of oscillation of the high-voltage generator. A modified cyclotron that automatically performs such an adjustment of frequency is called a **synchrocyclotron**. It accelerates a bunch of protons by matching its frequency of oscillation to the frequency of the orbital motion of the bunch; when it ejects the bunch, it begins to accelerate the next bunch, and so on. Machines of this kind have been used to achieve energies up to several hundred MeV.

Synchrocyclotron

At even higher energies, cyclotrons become impractical because they require excessively large magnets. It is then better to keep the protons moving around a circle of fixed radius in a large magnet of annular shape (Figure 31.16). As the protons gradually acquire more energy, the strength of the magnetic field must be gradually increased [see Eq. (28)]. Accelerator machines that automatically perform this increase of magnetic field are called **synchrotrons**. The most powerful synchrotrons built to date are the machines at Fermilab (Batavia, Illinois) and at CERN (Organisation Européenne pour la Recherche Nucléaire, on the Swiss–French border near Geneva). The Fermilab machine accelerates protons up to an energy of 1,000,000 MeV.

Synchrotron

The relation (28) between the momentum and the radius of the orbit of a charged particle in a magnetic field is widely exploited in

Fig. 31.16 The Bevatron at the Lawrence Berkeley Laboratory. This is a synchrotron.

high-energy physics to determine the momenta of elementary particles produced in collisions. Since the charge on any elementary particle is $\pm e$ (or some multiple of this), a measurement of the radius of the motion in a given magnetic field immediately determines the momentum. For example, Figure 31.17 shows the tracks of several particles in a bubble chamber filled with superheated liquid hydrogen, that is, liquid hydrogen suddenly raised to a temperature slightly above the boiling point. As we described in Section 20.5, in such a bubble chamber the tracks of individual particles become visible, because the passage of a charged particle through the superheated, unstable liquid triggers the formation of a fine trail of bubbles, which can be photographed. Figure 31.17 shows such a photograph of the tracks of several kaons crossing the bubble chamber and the tracks of several new particles produced by the collision of one of the kaons with a nucleus of one of the hydrogen atoms. The bubble chamber is placed within a large electromagnet; and the magnetic field curves the tracks of all the particles — the tracks are arcs of circles. The radii of these arcs can be measured on the photograph, and this immediately yields the momenta of the particles.

In our discussion of the motion of a charged particle in a uniform magnetic field, we assumed that the initial velocity was perpendicular

Ernest Orlando Lawrence, *1901–1958, American experimental physicist, professor at Berkeley, founder and director of the Radiation Laboratory (now Lawrence Berkeley Laboratory). He was awarded the Nobel Prize in 1939 for the invention and development of the cyclotron.*

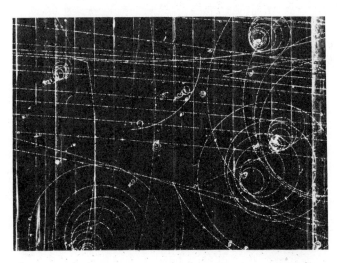

Fig. 31.17 Tracks of charged particles in a bubble chamber. The nearly straight tracks were made by high-energy kaons incident from the left. The spirals were made by electrons of relatively low energy.

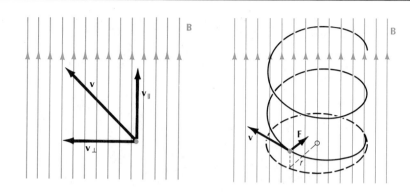

Fig. 31.18 (left) The velocity of a particle has a component parallel to **B** and a component perpendicular to **B**.

Fig. 31.19 (right) Positively charged particle with helical motion in uniform magnetic field.

to the magnetic field (see Figure 31.12). If this is not the case, then we can regard the velocity as consisting of two components: one component $v_\perp$ perpendicular to **B** and one component $v_\parallel$ parallel to **B** (Figure 31.18). Since the magnetic force is perpendicular to **B**, only the component $v_\perp$ changes, while $v_\parallel$ remains constant. The component $v_\perp$ then gives rise to uniform circular motion with radius $r = mv_\perp/qB$ in a direction perpendicular to **B**, while the component $v_\parallel$ gives rise to uniform translational motion in a direction parallel to **B**. The combination of these two motions is a helical motion with axis along the magnetic field — the particle spirals around the magnetic field lines (Figure 31.19). Such spiraling is a general feature of motion in a magnetic field; it will happen even if the magnetic field is a function of position. For example, Figure 31.20 shows a particle spiraling about the magnetic field lines of the field of the Earth. A detailed analysis

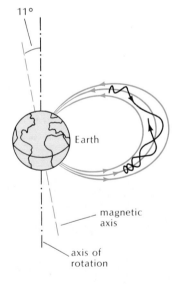

Fig. 31.20 Particle spiraling around the magnetic field lines of the Earth's field.

shows that near the poles, where the field lines converge, the component $v_\parallel$ of the velocity parallel to the field lines will decrease, then vanish, and then reverse (provided $v_\parallel$ is not too large). Hence the particle spirals toward one pole, then halts its approach, and then spirals back toward the other pole, and so on. The **Van Allen belts** surrounding the Earth (Figure 31.21) consist of a large number of electrons and protons spiraling back and forth along the field lines in this manner.

Van Allen belts

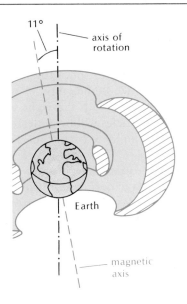

Fig. 31.21 The Van Allen radiation belts. Near the poles of the Earth, where the horns of the belts are close to the upper atmosphere, charged particles sometimes spill into the atmosphere, producing the luminous glow known as the Aurora Borealis.

31.4 Crossed Electric and Magnetic Fields; the Hall Effect

If electric and magnetic fields act on a particle simultaneously, then the force has both an electric and a magnetic part,

$$\boxed{\mathbf{F} = q\mathbf{E} + q\mathbf{v} \times \mathbf{B}} \qquad (29)$$

Lorentz force

This is called the **Lorentz force.**

As an example of the simultaneous action of electric and magnetic fields, let us consider **crossed fields,** that is, uniform electric and magnetic fields at right angles to one another. Figure 31.22 shows electric field lines and magnetic field lines at right angles. The motion of a particle in such fields is usually fairly complicated, but under special circumstances the motion becomes very simple. Suppose that in Figure 31.22, a positively charged particle enters the field region from the left. The magnetic force $q\mathbf{v} \times \mathbf{B}$ is then opposite to the electric force $q\mathbf{E}$, and if the velocity has just the right magnitude, these forces will cancel, and the particle will continue its original motion on a straight line. The condition for this is

Crossed fields

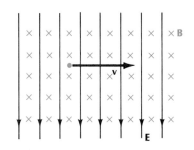

$$\mathbf{E} = -\mathbf{v} \times \mathbf{B} \qquad (30)$$

or, considering magnitudes,

$$E = vB$$

The "right" velocity for cancellation of the forces is then

$$\boxed{v = E/B} \qquad (31)$$

Fig. 31.22 Electric (black) and magnetic (color) fields at right angles. The crosses show the tails of the magnetic field vectors. A positively charged particle moves from left to right.

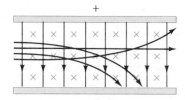

Fig. 31.23 A velocity selector with electric and magnetic fields at right angles. A beam of positively charged particles enters from the left. Particles of excessive velocity are deflected upward; particles of insufficient velocity are deflected downward.

This cancellation is the basic principle behind the **velocity selectors** (or velocity filters) often used in physics laboratories to select particles of some desired velocity from a beam containing particles with a large variety of velocities. It is only necessary to shoot the beam into a region containing crossed **E** and **B** fields whose magnitudes are related to the desired velocity by Eq. (31). Particles with the right velocity will then proceed undeflected; all other particles will be deflected to one side or another, and they are thereby eliminated from the beam (see Figure 31.23).

Such a velocity selector with crossed electric and magnetic fields was first used by J. J. Thomson in his experimental investigations of cathode rays, which led him to the discovery of the electron in 1897. Cathode rays had been observed in electric discharge tubes. These are glass tubes with electrodes at the ends connected to a source of high voltage. When the tube contains gas at low pressure (about 10^{-3} atm), the electric discharge from one electrode to the other is luminous, as in the now ubiquitous neon signs. But if the tube is almost completely evacuated, the electric discharge becomes invisible. Under these circumstances, the cathode (negative electrode) emits rays that propagate in straight lines through the space within the tube. Although the rays by themselves are invisible, their presence can be easily perceived when they strike the wall of the tube or the atoms of residual gas in the tube; the walls or the gas then phosphoresce in a brilliant bluish or greenish color (see Figure 31.13). Thomson investigated the nature of these cathode rays by subjecting them to electric and magnetic fields, and he demonstrated that the deflections experienced by the rays in such fields are as expected for a charged particle of very small mass. Figure 31.24 shows the apparatus used by Thomson to determine the speed of the cathode rays. The electrically charged parallel plates produce an electric field, and the electromagnet produces a magnetic field at right angles to this electric field. By adjusting the magnitudes of these crossed fields, Thomson could achieve the cancellation of the electric and magnetic forces and calculate the speed of his cathode rays, according to Eq. (31). He then measured the deflection of his

Sir Joseph John Thomson, *1856–1940, English experimental physicist, director of the Cavendish Laboratory at Cambridge, and president of the Royal Society. His discovery of the electron marks the beginning of modern experimental physics. Thomson received the Nobel Prize in 1906 for his investigations on electric discharges in gases. His experiments with beams of positive ions led to the first separation of the isotopes of a chemical element.*

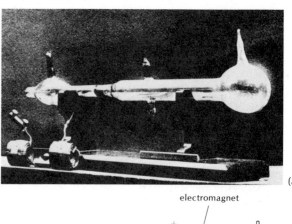

Fig. 31.24 (a) Cathode-ray tube of J. J. Thomson. (b) Schematic diagram. The parallel plates in the tube are electrically charged, with opposite charges. One pole of the electromagnet is behind the tube; the other pole has been omitted for the sake of clarity.

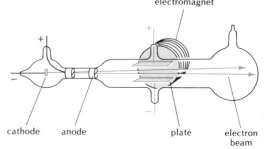

cathode rays when subjected to the electric field alone, and from this deflection and the known speed, he calculated the ratio of charge to mass (e/m_e). With the reasonable assumption — later verified by other experiments — that the charge has the same magnitude as that of a hydrogen ion, he evaluated the mass of the particle in the cathode rays, a particle eventually given the name *electron.*

With some modifications and refinements, the method employed by Thomson for the determination of the mass of the electron is still employed for precise mass determinations of ions. In modern **mass spectrometers,** a beam of ions is subjected to electric and magnetic fields, and the deflections are used to evaluate the mass. The precise values of the masses of atoms listed in charts of isotopes (see, for example, the charts in the Prelude and in Chapter 45) were obtained by these means.

Mass spectrometer

Crossed **E** and **B** fields also play a role whenever a wire or some other conductor carries a current in a magnetic field. Figure 31.25 shows a metallic strip carrying a current; the strip is placed in a magnetic field, perpendicular to the surface of the strip. If the current is toward the right, the motion of the free electrons in the strip must be toward the left with some average velocity **v**. These free electrons experience a magnetic force $-e\mathbf{v} \times \mathbf{B}$, directed upward in Figure 31.25. This magnetic force will tend to push the electrons toward the upper edge of the strip, and some electrons will accumulate there, leaving the lower edge with a deficit of electrons — the upper edge acquires a negative charge and the lower edge a positive charge. There will then exist an electric field in the strip, as shown in Figure 31.25. This electric field is perpendicular to the magnetic field; that is, the **E** and **B** fields inside the strip are crossed fields. Under equilibrium conditions, the transverse magnetic force on an electron will be matched by the transverse electric force. The condition for this balance of forces is Eq. (30). Thus, inside the strip, there exists an electric field of magnitude $E = vB$; consequently, in a strip of width l, there is a potential difference

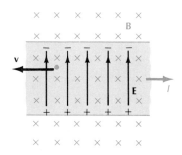

Fig. 31.25 A metallic strip carrying a current has been placed in a uniform magnetic field. The crosses show the tails of the magnetic field vectors. The current flows from left to right; the electrons making up this current move from right to left. The magnetic force on an electron is directed upward; the electric force is directed downward.

$$\Delta V = \int E \, dl = vBl \qquad (32)$$

between the upper and the lower edge of the strip, the lower edge being at a higher potential. This generation of potential difference between opposite edges of a conductor carrying a current in a magnetic field is called the **Hall effect**.

Hall effect

We can express the Hall potential difference in terms of the current by means of Eq. (28.17):

$$v = \frac{I}{enA}$$

where A is the cross-sectional area of the conductor and n the number of free electrons per unit volume. This gives

$$\Delta V = \frac{IBl}{enA} \qquad (33)$$

For example, a copper strip of cross-sectional area $0.5 \text{ cm} \times 0.02 \text{ cm}$ carrying a current of 20 A in a magnetic field of 1.5 T develops a potential difference of

$$\Delta V = \frac{20 \text{ A} \times 1.5 \text{ T} \times 0.005 \text{ m}}{1.6 \times 10^{-19} \text{ C} \times 8.5 \times 10^{28}/\text{m}^3 \times 0.01 \times 10^{-4} \text{ m}^2}$$

$$= 1.1 \times 10^{-5} \text{ volt}$$

In practice, the Hall effect is often used to determine the density of free electrons n in a metal. Also, the Hall effect provides direct empirical evidence that the current carriers in metals are negative charges — if the current carriers were positive charges, then the direction of **v** in Figure 31.25 would be opposite to that shown, and the positive charges would accumulate at the top of the strip rather than on the bottom.

31.5 Force on a Wire

If a wire carrying a current is placed in a magnetic field, the moving charges within the wire will experience a force. Since the motion of the charges in the wire is constrained by the wire (the charges must move along the wire, regardless of the magnetic force they experience), any force acting on these charges is merely transferred to the wire, and therefore the wire as a whole will experience a force equal to that acting on the charges.

According to Eq. (30.23), the amount of moving charge in a segment dl of the wire is

$$dq = I \frac{dl}{v}$$

where, as always, we pretend that the moving charge is positive. If θ is the angle between the velocity and **B**, or, equivalently, the angle between the wire segment and **B** (Figure 31.26), then the force on the wire segment is

$$dF = dq\,vB \sin\theta = I\,dl\,B \sin\theta \qquad (34)$$

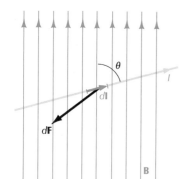

Fig. 31.26 A long straight wire in a uniform magnetic field. The force on a wire segment $d\mathbf{l}$ is perpendicular to $d\mathbf{l}$ and to **B**.

This force can be expressed in vector form by an argument similar to that given in connection with Eq. (30.25):

Force on wire segment

$$\boxed{d\mathbf{F} = I\,d\mathbf{l} \times \mathbf{B}} \qquad (35)$$

where the vector $d\mathbf{l}$ is along the wire in the direction of the current.

To find the net force on the entire wire we must integrate Eq. (35) along the wire.

EXAMPLE 6. Two very long parallel wires separated by a distance r carry currents I and I', respectively. Find the magnetic force acting on each wire.

SOLUTION: Each wire generates a magnetic field, which exerts a force on the other wire. Figure 31.27 shows the magnetic field **B** that the current I' produces in the space surrounding the current I. This magnetic field has a magnitude

$$B = \frac{\mu_0}{2\pi} \frac{I'}{r} \qquad (36)$$

and is perpendicular to the current I. By Eq. (35), the force exerted by the magnetic field **B** on a segment dl of the wire carrying the current I is

$$dF = IB\, dl = \frac{\mu_0}{2\pi} \frac{II'}{r} dl \qquad (37)$$

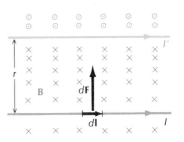

Fig. 31.27 A long straight wire carrying a current I in the magnetic field of a long straight wire carrying a current I'. The magnetic field lines of the current I' are circles perpendicular to the plane of the page. The crosses show the tails of the magnetic field vectors; the dots show the tips of the magnetic field vectors.

The force on a segment dl' of the other wire carrying the current I' is of the same magnitude, but opposite direction.

The forces acting between the wires are attractive if the currents are parallel, and repulsive if they are antiparallel. Of course, the net forces between two infinitely long wires are infinite. It is therefore useful to consider the force per unit length; this is finite and of magnitude

$$\frac{dF}{dl} = \frac{\mu_0}{2\pi} \frac{II'}{r} \qquad (38)$$

The official definition of the **SI unit of current** is based on the force per unit length between two long parallel wires. This force can be measured very precisely by holding one wire stationary and suspending the other from a balance; the wires are connected in series so that the currents are exactly equal ($I = I'$). The force per unit length and the distance can be measured experimentally, and the value of I calculated from Eq. (38) is then the current in amperes. The constant μ_0 appearing in Eq. (38) is assigned the value $\mu_0 = 4\pi \times 10^{-7}$ N · s^2/C^2 by *definition*.[2]

The official definition of the **SI unit of electric charge** is based on the unit of current. The coulomb is defined as the amount of charge that a current of one ampere delivers in one second.

SI unit of current

SI unit of electric charge

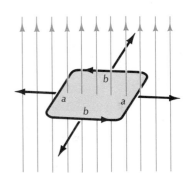

Fig. 31.28 Rectangular loop of current perpendicular to a uniform magnetic field.

31.6 Torque on a Current Loop

The action of a magnetic field on a loop of wire carrying a current will in general result not only in a net force on the loop but also in a torque. The force and torque depend on the shape and the orientation of the loop. We will consider the special case of a rectangular loop placed in a uniform magnetic field.

Figure 31.28 shows the loop oriented perpendicular to the magnetic field. The forces on the four sides of the loop can be calculated from Eq. (35). The forces on opposite sides are opposite, because the currents are opposite. Hence the forces cancel in pairs and the loop will be in equilibrium.

Figure 31.29 shows the loop oriented at an angle with the magnetic field. Again the forces on opposite sides are opposite, but the lines of action of the forces pulling to the left and the right do not coincide — these forces exert a torque which tends to rotate the loop. Since **B** is constant, Eq. (35) tells us that the magnitudes of the forces on the left and on the right sides are

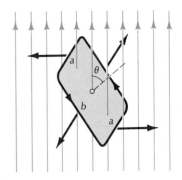

Fig. 31.29 Rectangular loop of current at an angle with a uniform magnetic field.

---

[2] In contrast, the constant ε_0 appearing in Coulomb's Law must be determined by experiment or calculated from μ_0 and the speed of light (see Section 36.1).

$$F = IaB \tag{39}$$

where a is the length of the side (Figure 31.29). The torque due to this pair of forces is

$$\tau = F(b/2)\sin\theta + F(b/2)\sin\theta$$

$$= IabB\sin\theta \tag{40}$$

where b is the length of the other side (see Figure 31.29) and θ is the angle between the directions of **B** and of the normal to the loop.

According to the definition given in Section 30.4, the product of the area of the loop and the current around it is the **magnetic dipole moment** of the loop,

$$\mu = Iab \tag{41}$$

Note that if the loop consists of several turns of wire, then Eq. (41) must be evaluated with the net current flowing around the rim of the loop. This net current equals the number N of turns of wire times the current I_0 in one turn so that

$$\mu = NI_0 ab$$

In any case, the torque is

$$\tau = \mu B \sin\theta \tag{42}$$

This formula turns out to be valid not only for the rectangular loops but also for loops of arbitrary shape. The direction of the torque is such that it tends to twist the loop into an orientation perpendicular to the magnetic field.

To express the torque in vector form, we must regard the magnetic dipole moment as a vector; this vector is perpendicular to the loop, in a direction given by the right-hand rule (fingers curled in same direction as current, thumb along μ; see Figure 31.30). With this vector, we can write

$$\boxed{\tau = \mu \times \mathbf{B}} \tag{43}$$

Fig. 31.30 Right-hand rule for the magnetic moment: If the fingers are curled in the same direction as the current in the loop, the thumb points in the direction of μ.

Torque on current loop

This expression for the torque on a magnetic dipole in a magnetic field is mathematically similar to the expression for the torque on an electric dipole in an electric field [see Eq. (23.30)]. The torque tends to twist the dipole so as to bring μ into alignment with **B**.

We can also write a potential energy such that changes in this potential energy represent the work that *you* must do against the torque (43) when changing the orientation of the loop. The expression for the potential energy is

$$\boxed{U = -\mu \cdot \mathbf{B}} \tag{44}$$

Potential energy of current loop

The derivation of this expression is entirely analogous to the derivation of the expression (23.32) for the potential energy of an electric dipole in an electric field. As expected, the potential energy (44) has a minimum for **μ** parallel to **B**, and a maximum for **μ** antiparallel to **B**.

A loop of current suitably pivoted on an axis acts as a compass needle; the normal of the loop seeks to align itself with the magnetic field. This similarity is no accident. As we will see in Section 33.3, a compass needle is a small permanent magnet, which contains a large number of small current loops. The mechanism underlying the alignment of a compass needle with a magnetic field is therefore the same as for a current loop.

In an electric motor, the torque on a current loop pivoted on an axis and placed in a strong magnetic field brings about the rotational motion of the current loop and thereby provides the means of converting electric energy into mechanical energy of rotational motion. The following example illustrates the operation of a simple electric motor.

EXAMPLE 7. A simple electric motor consists of a rectangular coil of wire that rotates on a longitudinal axle in a magnetic field of 0.50 T (Figure 31.31). The coil measures 10 cm × 20 cm; it has 40 turns of wire, and the current in the wire is 8.0 A.
(a) As a function of the angle θ between the lines of **B** and the normal to the coil, what is the torque that the magnetic field exerts on the coil?
(b) To keep the sign of the torque constant, a slotted sliding contact, or **commutator,** mounted on the axle reverses the current in the coil whenever θ passes through 0 and π radians. Plot this torque as a function of θ.

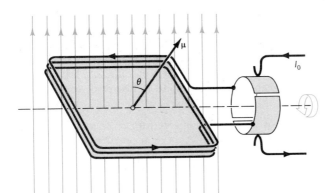

Fig. 31.31 An electric motor.

SOLUTION: (a) According to Eq. (42), the torque on the coil is

$$\tau = NI_0 abB \sin\theta$$

$$= 40 \times 8.0 \text{ A} \times 0.10 \text{ m} \times 0.20 \text{ m} \times 0.50 \text{ T} \times \sin\theta$$

$$= 3.2 \text{ N}\cdot\text{m} \sin\theta$$

(b) If the torque always has the same sign (say, positive), we can write it as

$$\tau = 3.2 \text{ N}\cdot\text{m} |\sin\theta|$$

Figure 31.32 shows a plot of this torque.

COMMENTS AND SUGGESTIONS: Practical electric motors consist of many such coils, each with its commutator, arranged at regular intervals around the axle. The plot for the net torque of this arrangement is a sum of plots such as that shown in Figure 31.32, but with different starting angles. This averages out the ups and downs of Figure 31.32, and yields a torque which is nearly constant as a function of angle.

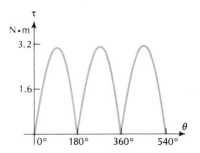

Fig. 31.32 Torque as a function of angle according to Example 7.

SUMMARY

Ampère's Law: $\oint \mathbf{B} \cdot d\mathbf{l} = \mu_0 I$

Field in solenoid: $B = \mu_0 I_0 n$

Circular orbit in magnetic field: $r = \dfrac{p}{qB}$

Cyclotron frequency: $\nu = \dfrac{qB}{2\pi m}$

Lorentz force: $\mathbf{F} = q\mathbf{E} + q\mathbf{v} \times \mathbf{B}$

Force on wire segment: $d\mathbf{F} = I\, d\mathbf{l} \times \mathbf{B}$

Torque on current loop: $\boldsymbol{\tau} = \boldsymbol{\mu} \times \mathbf{B}$

Potential energy of current loop: $U = -\boldsymbol{\mu} \cdot \mathbf{B}$

QUESTIONS

1. Suppose you evaluate the integral $\oint \mathbf{B} \cdot d\mathbf{l}$ for the magnetic field of the Earth along a closed path that coincides with a meridian circle. What is the value of the integral?

2. Is the magnetic field a conservative field?

3. A long solenoid has been placed inside another long solenoid of larger radius. The solenoids are coaxial and both have the same number of turns per unit length and the same current. What is the formula for the magnetic field in the region within the smaller solenoid? Between the smaller and the larger solenoid?

4. Figure 31.33 shows a solenoid of arbitrary cross section; this solenoid is a (noncircular) cylinder. Suppose that there are n turns of wire per unit length and that the current in the wire is I. Use Ampère's Law to show that the magnetic field in the solenoid is $\mu_0 In$, the same as for a circular cylinder.

5. Consider a long solenoid and a long straight wire along its axis, both carrying some current. Describe the field lines within the solenoid.

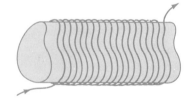

Fig. 31.33 Cylinder of arbitrary, noncircular cross section.

6. The drawing of Figure 31.8 shows the field lines near the gap of an electromagnet. Describe the pattern of field lines beyond the edges of the drawing.

7. Cosmic rays are high-speed charged particles — mostly protons — that crisscross interstellar space and strike the Earth from all directions. Why is it easier for the cosmic rays to penetrate through the magnetic field of the Earth near the poles than anywhere else?

8. Is it possible to define a magnetic potential energy for a charged particle moving in a magnetic field?

9. If we want a proton to orbit all the way around the equator in the Earth's magnetic field, must we send it eastward or westward?

10. Consider the strip of metal placed in a magnetic field, as shown in Figure 31.25. How does the Hall potential difference between the edges of the strip change if we reverse the current? If we reverse the magnetic field? If we reverse both?

11. If we replace the metallic strip in Figure 31.25 by a semiconducting strip of p-type semiconductor, will the lower edge remain at the higher potential?

12. The Earth's magnetic field at the equator is horizontal and in the northward direction. Assume that the atmospheric electric field is downward. The electric and magnetic fields are then crossed fields. In what direction must we launch an electron if it is to move without any deflection?

13. If a strong current flows through a thick wire, it tends to cause a compression of the wire. Explain.

14. If a wire carrying a current is placed in a magnetic field, the field exerts forces on the moving electrons in the wire. How does this cause a push on the wire?

15. A horizontal wire carries a current in the eastward direction. The wire is located in a magnet producing a uniform magnetic field. What must be the direction of this field if it is to compensate for the weight of the wire?

16. In the SI system, we first define the unit of current (by means of the magnetic force between wires), and we then define the unit of charge in terms of the current. Could we proceed in the opposite manner: first define the unit of electric charge (by means of the electric force between charges), and then define the unit of current as a flow of charge? What would be the advantages and the disadvantages?

17. Figure 31.34 is a schematic diagram showing the mechanism of an **ammeter**. The mechanism consists of a moving coil suspended between the poles of a permanent magnet. The coil can rotate about an axis perpendicular to the magnetic field, and is held in an equilibrium position by a spiral spring. To operate the ammeter, the current of the external circuit is made to pass through the coil. Explain how this ammeter registers the current. Explain how a sensitive ammeter of this kind can be used as a voltmeter by connecting the coil in series with a large resistance.

18. The **tangent galvanometer** is an old form of ammeter, consisting of an ordinary magnetic compass mounted at the center of a coil whose axis is horizontal and oriented along the east–west line (Figure 31.35). If there is no current in the coil, the compass needle points north. Explain how the compass needle will deviate from north when there is a current in the coil.

19. A simple indicator of electric current, first used by H. C. Oersted in his early experiments, consists of a compass needle placed below a wire stretched in the northward direction. Explain how the angle of the compass needle indicates the electric current in the wire.

20. Positive charge is uniformly distributed over a sphere. If this sphere spins about a vertical axis in a counterclockwise direction as seen from above, what is the direction of the magnetic moment vector?

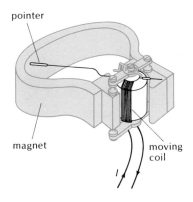

Fig. 31.34 Mechanism of the moving-coil ammeter.

Ammeter

Tangent galvanometer

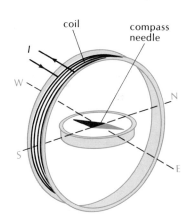

Fig. 31.35 Tangent galvanometer.

PROBLEMS

Section 31.1

1. A wire of superconducting niobium, 0.20 cm in diameter, can carry a current of up to 1900 A. What is the strength of the magnetic field just outside of the wire when it carries this current?

2. A long straight wire of copper with a radius of 1 mm carries a current of 20 A. What is the instantaneous magnetic force and the corresponding acceleration on one of the conduction electrons moving at 10^6 m/s along the surface of the wire in a direction opposite to that of the current? What is the direction of the acceleration?

3. A current I flows in a thin wire bent into a circle of radius R. The axis of the circle coincides with the z axis. What is the value of the integral $\int \mathbf{B} \cdot d\mathbf{l}$ along the z axis from $z = -\infty$ to $z = +\infty$?

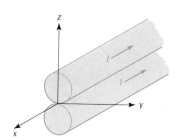

Fig. 31.36

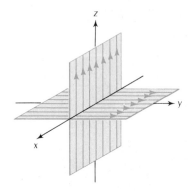

Fig. 31.37 Two large plates intersecting at right angles.

Fig. 31.38

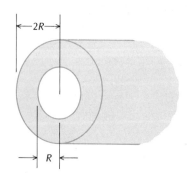

Fig. 31.39

4. In a proton accelerator, protons of velocity 3×10^8 m/s form a beam of current of 2×10^{-3} A. Assume that the beam has a circular cross section of radius 1 cm and that the current is uniformly distributed over the cross section. What is the magnetic field that the beam produces at its edge? What is the magnetic force on a proton at the edge of the beam?

*5. Six parallel aluminum wires of small, but finite, radius lie in the same plane. The wires are separated by equal distances d, and they carry equal currents I in the same direction. Find the magnetic field at the center of the first wire. Assume that the current in each wire is uniformly distributed over its cross section.

*6. Two long parallel wires of copper of radius R are in contact along their full length (see Figure 31.36). The wires carry equal currents I, in the same direction. The currents are uniformly distributed over the volumes of the wires. Find the magnetic field in the midplane (z–x plane) of the wires, as a function of the distance z from the point of contact. Where is this magnetic field maximum and what is the value of the maximum magnetic field?

*7. A very large, thin conducting plate lies in the x–y plane. The plate carries a current in the y direction. The current is uniformly distributed over the plate with σ ampere flowing across each 1-m length along the x axis. Use Ampère's Law to find the magnetic field at some distance from the plate. (Hint: The magnetic field lines are parallel to the plate.)

*8. One very large conducting plate coincides with the x–y plane. Another very large conducting plate coincides with the x–z plane. Each plate carries a uniformly distributed current with σ amperes flowing across each 1-m length perpendicular to the current (Figure 31.37). Find the magnetic field in each of the four quadrants. (Hint: See Problem 7.)

Section 31.2

9. A long solenoid has 15 turns per centimeter. What current must we put through its windings if we wish to achieve a magnetic field of 5.0×10^{-2} T in its interior?

10. The electromagnet of a small electric bell is a solenoid with 260 turns in a length of 2.0 cm. What magnetic field will this solenoid produce if the current is 8.0 A?

11. Figure 31.38 shows a "solenoid" made of one turn of a sheet of copper. The solenoid has a length of 20 cm, and the current flowing through it is 2×10^3 A. What is the magnetic field in this solenoid? Assume that the current is uniformly distributed over the sheet of copper, and treat the solenoid as very long.

12. A toroid used in plasma research has 240 turns of wire carrying a current of 7.2×10^4 A. The inner radius of the toroid is 0.50 m, and the outer radius is 1.50 m. What is the strength of the magnetic field at the inner radius? At the outer radius?

*13. A long copper pipe with thick walls has an inner radius R and an outer radius $2R$ (Figure 31.39). A current I flows along this wall, uniformly distributed over the volume of copper. Find the magnetic fields at the radius $\tfrac{3}{2}R$ and at the radius $3R$.

*14. A long solenoid of n turns per unit length carries a current I, and a long straight wire lying along the axis of this solenoid carries a current I'. Find the net magnetic field within the solenoid, at a distance r from the axis. Describe the shape of the magnetic field lines.

*15. A coaxial cable consists of a long cylindrical copper wire of radius r_1 surrounded by a cylindrical shell of inner radius r_2, outer radius r_3 (Figure 31.40). The wire and the shell carry equal and opposite currents I uniformly distributed over their volumes. Find formulas for the magnetic field in each of the regions $r < r_1$, $r_1 < r < r_2$, $r_2 < r < r_3$, $r > r_3$.

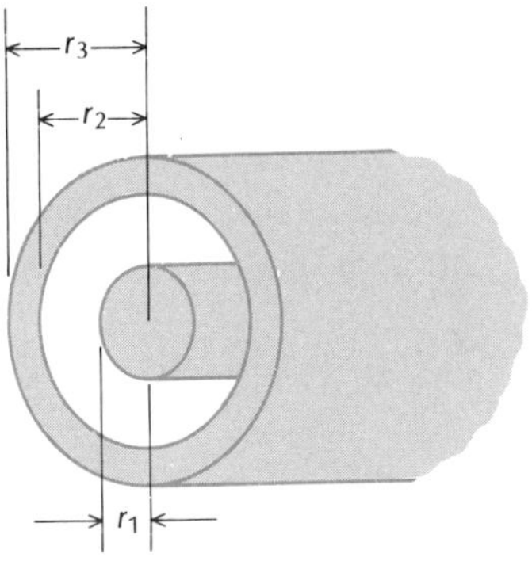

*16. Suppose we place a long straight wire along the axis of symmetry of the toroid shown in Figure 31.10. If a current I_0 flows in the windings of the toroid and, simultaneously, a current I' flows on the straight axial wire, what is the net magnetic field in the toroid? What value of the current I' is required to make the magnetic field in the toroid twice as strong as that without the straight wire? If we reverse this current I', what happens to the net magnetic field?

*17. The coil of an electromagnet consists of a large number of layers of very thin wire wound on a cylindrical core. The inner radius of the windings is r_1 and the outer radius is r_2. The number of turns in each layer is n per unit length, and the number of layers is n' per unit length in the radial direction. The current in the turns of wire is I. Find formulas for the magnetic field in the region $r < r_1$ and in the region $r_1 < r < r_2$.

Fig. 31.40 Coaxial cylinder and cylindrical shell.

Section 31.3

18. A bubble chamber, used to make the tracks of protons and other charged particles visible, is placed between the poles of a large electromagnet that produces a uniform magnetic field of 20 T. A high-energy proton passing through the bubble chamber makes a track that is an arc of a circle of radius 3.5 m. According to Eq. (28), what is the momentum of the proton?

19. A proton of energy 1.0×10^7 eV moves in a circular orbit in the magnetic field near the Earth. The strength of the field is 0.50×10^{-4} T. What is the radius of the orbit?

20. In principle, a proton of the right energy can orbit the Earth in an equatorial orbit under the influence of the Earth's magnetic field. If the orbital radius is to be 6.5×10^3 km and the magnetic field at this radius is 0.33×10^{-4} T, what must be the momentum of the proton?

21. In the Crab nebula (the remnant of a supernova explosion), electrons of a momentum of up to about 10^{-16} kg·m/s orbit in a magnetic field of about 10^{-8} T. What is the orbital radius of such electrons? Note that it is necessary to use Eq. (28) for the calculation of the orbital radius.

22. At the Fermilab accelerator, protons of momentum 5.3×10^{-16} kg·m/s are held in a circular orbit of diameter 2.0 km by a vertical magnetic field. What is the strength of the magnetic field required for this?

*23. Figure 31.41 shows the tracks of an electron (charge $-e$) and an antielectron (charge $+e$) created in a bubble chamber. When the particles made these tracks, they were under the influence of a magnetic field of a magnitude 1.0 T and of a direction perpendicular to and into the plane of the figure. What is the momentum of each particle? Assume that they are moving in the plane of the figure and that this figure is $\frac{1}{10}$ natural size. Which is the track of the electron and which is the track of the antielectron?

24. Some astrophysicists believe that the radio waves of 10^9 Hz reaching us from Jupiter are emitted by electrons of fairly low (nonrelativistic) energies orbiting in Jupiter's magnetic field. What must be the strength of this field if the cyclotron frequency is to be 10^9 Hz?

*25. In the Brookhaven AGS accelerator, protons are made to move around a circle of radius 128 m by a magnetic force exerted by a vertical magnetic field. The maximum field that the magnets of this accelerator can generate is 1.3 T.
  (a) Calculate the maximum permissible momentum of the protons.
  (b) Calculate the maximum kinetic energy [use the relativistic relation between energy and momentum, Eq. (41.52)].
  (c) Calculate the orbital frequency of such protons.

Fig. 31.41 Tracks of an electron and an antielectron (top) in a bubble chamber. The tracks spiral because the particles suffer a loss of energy as they pass through the liquid. For the purposes of Problem 23, concentrate on the initial portions of the tracks.

*26. You want to confine an electron of energy 3.0×10^4 eV by making it circle inside a solenoid of radius 10 cm under the influence of the force exerted by the magnetic field. The solenoid has 120 turns of wire per centimeter. What minimum current must you put through the wire if the electron is not to hit the wall of the solenoid?

Section 31.4

*27. A velocity selector with crossed **E** and **B** fields is to be used to select alpha particles of energy 2.0×10^5 eV from a beam containing particles of several energies. The electric field strength is 1.0×10^6 V/m. What must be the magnetic field strength?

*28. The Earth's magnetic field at the equator has a magnitude of 5×10^{-5} T; its direction is horizontal toward the north. Suppose that the Earth's atmospheric electric field is 100 V/m; its direction is vertically down. With what velocity (magnitude and direction) must we launch an electron if the electric force is to cancel the magnetic force?

*29. In a mass spectrometer, a beam of ions is first made to pass through a velocity selector with crossed **E** and **B** fields. The selected ions then are made to enter a uniform magnetic field **B'** where they move in arcs of circles (Figure 31.42). The radius of these circles depends on the masses of the ions. Assume that an ion has a single charge e.

(a) Show that in terms of E, B, B', and of the impact distance l marked in Figure 31.42, the mass of the ion is

$$m = \frac{eBB'l}{2E}$$

(b) Assume that in such a spectrometer, ions of the isotope ^{16}O impact at a distance of 29.20 cm and ions of a different isotope of oxygen impact at a distance of 32.86 cm. The mass of the ions of ^{16}O is 16.00 u. What is the mass of the other isotope?

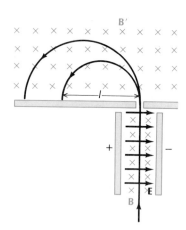

Fig. 31.42 Schematic diagram of a mass spectrometer. The magnetic fields **B** and **B'** are perpendicular to the plane of the page. The crosses show the tails of the magnetic field vectors.

30. A copper wire of diameter 3.0 mm is placed at right angles to a magnetic field of 2.0 T. The wire carries a current of 50 A. What is the Hall potential difference between opposite sides of the wire?

*31. Repeat the preceding problem, with the assumption that the wire is placed at an angle of 60° with respect to the magnetic field.

Section 31.5

32. A straight wire is placed in a uniform magnetic field; the wire makes an angle of 30° with the magnetic field. The wire carries a current of 6.0 A, and the magnetic field has a strength of 0.40 T. Calculate the force on a 10-cm segment of this wire. Show the direction of the force in a diagram.

33. Figure 31.43 shows a balance used for the measurement of a magnetic field. A loop of wire carrying a precisely known current is partially immersed in the magnetic field. The force that the magnetic field exerts on the loop can be measured with the balance, and this permits the calculation of the strength of the magnetic field. Suppose that the short side of the loop measures 10.0 cm, the current in the wire is 0.225 A, and the magnetic force is 5.35×10^{-2} N. What is the strength of the magnetic field?

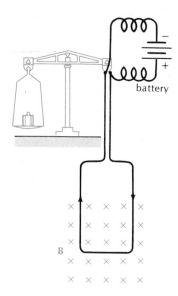

Fig. 31.43 A current balance. The magnetic field **B** is perpendicular to the plane of the page. The crosses show the tails of the magnetic field vectors.

34. The electric cable supplying an electric clothes dryer consists of two long straight wires separated by a distance of 1.2 cm. Opposite currents of 20 A flow on these wires. What is the magnetic force experienced by a 1.0-cm segment of wire due to the entire length of the other wire?

35. Two parallel cables of a high-voltage power line carry opposite currents of 1.8×10^3 A. The distance between the cables is 4.0 m. What is the magnetic

force pushing on a 50-m segment of one of these cables? Treat both cables as very long straight wires.

*36. A rectangular loop of wire of dimensions 12 cm × 18 cm is near a long straight wire. The rectangle and the straight wire lie in the same plane. One of the short sides of the rectangle is parallel to the straight wire and at a distance of 6.0 cm; the long sides are perpendicular to the straight wire (Figure 31.44). A current of 40 A flows on the straight wire and a current of 60 A flows around the loop. What are the magnitude and direction of the net magnetic force that the straight wire exerts on the loop?

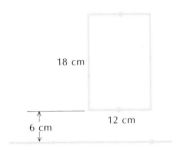

Fig. 31.44 Long straight wire and rectangular loop.

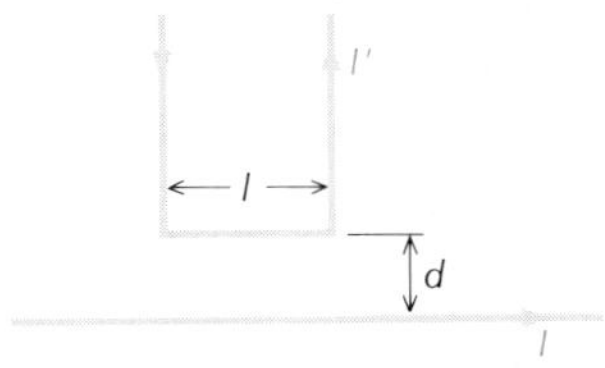

Fig. 31.45

*37. Consider the long straight wire and the **U**-shaped wire shown in Figure 31.45. The wires lie in the same plane. The bottom of the **U** has a length l, and the sides of the **U** are very long. The wires are separated by a distance d, and they carry currents I and I', respectively. What is the force that the straight wire exerts on the **U**-shaped wire?

*38. A closed loop of arbitrary (possibly three-dimensional) shape is placed near a very long straight wire. Currents I and I', respectively, flow around the loop and along the straight wire. Prove that the force exerted by the straight wire on the loop must be perpendicular to the wire. (Hint: Use Newton's Third Law.)

39. A thin flexible wire carrying a current I hangs in a uniform magnetic field **B (Figure 31.46). A weight attached to one end of the wire provides a tension T. Within the magnetic field, the wire will adopt the shape of an arc of circle.
(a) Show that the radius of this arc of circle is $r = T/BI$.
(b) Show that if we remove the wire and launch a particle of charge $-q$ from the point P with a momentum $p = q\,T/I$ in the direction of the wire, it will move along the same arc of circle. (This means that the wire can be used to simulate the orbit of the particle. Experimental physicists sometimes use such wires to check the orbits of particles through systems of magnets.)

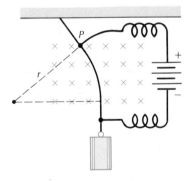

Fig. 31.46 A wire with a weight hanging in a magnetic field.

**40. An electromagnetic launcher, or rail gun, consists of two parallel conducting rails along which slides a conducting projectile (Figure 31.47). To "fire" this gun, a large charged capacitor is suddenly connected to the rear ends of the rails, causing an intense current to flow around the circuit formed

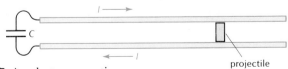

Fig. 31.47 An electromagnetic launcher.

by the rails and the projectile. The magnetic field produced by the current then exerts a strong force on the current flowing across the projectile, accelerating the projectile along the rails. Such rail guns can achieve accelerations of the order of 10^5 m/s^2 and muzzle speeds of a few km/s, much in excess of what can be achieved with ordinary guns employing an explosive charge. Estimate the current needed in the rails to attain an acceleration of 10^5 m/s^2 with a projectile of 0.2 kg. For the sake of simplicity, pretend that the magnetic field of the rails is approximately that of two infinitely long, thick wires and that the projectile can be treated as a transverse wire segment immersed in this magnetic field. Assume that the transverse length of the projectile is four times the diameter of each rail.

Section 31.6

41. The coil in the mechanism of an ammeter is a rectangular loop, measuring 1 cm × 2 cm, with 120 turns of wire. The coil is immersed in a magnetic field of 0.01 T. What is the torque on this coil when it is parallel to the magnetic field and carries a current of 0.001 A?

42. A horizontal circular loop of wire of radius 20 cm carries a current of 25 A. At the location of the loop, the magnetic field of the Earth has a magnitude of 0.39×10^{-4} T and points down at an angle of 16° with the vertical. What is the magnitude of the torque that this magnetic field exerts on the loop?

43. The proton has a magnetic moment of 1.41×10^{-26} A·m^2.
    (a) If this magnetic moment makes an angle of 45° with a uniform magnetic field of 0.80 T, what is the torque on the proton?
    (b) If the magnetic moment is initially oriented antiparallel to the field, how much energy will be released when the proton flips into the parallel orientation?

44. Consider the electric motor of Example 7. What is the average value of the torque over a complete rotation? If this motor rotates at the rate of 50 rev/s what is the average horsepower that it delivers to its axle?

45. A molecule with a magnetic moment $\mu = 9.3 \times 10^{-24}$ A·m^2 is in a uniform magnetic field $B = 0.80$ T. What is the difference in the potential energy between the parallel and antiparallel orientations for μ and **B**? Express your answer in electron-volts.

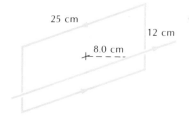

Fig. 31.48 Long straight wire and rectangular loop.

**46. A rectangular loop of wire of dimension 12 cm × 25 cm faces a long straight wire. The two long sides of the loop are parallel to the wire, and the two short sides are perpendicular; the midpoint of the wire is 8.0 cm from the wire (Figure 31.48). Currents of 95 A and 70 A flow in the straight wire and the loop, respectively.
    (a) What translational force does the straight wire exert on the loop?
    (b) What torque about an axis parallel to the straight wire and through the center of the loop does the straight wire exert on the loop?

**47. A compass needle is attached to an axle which permits it to turn freely in a horizontal plane so that only the horizontal component of the magnetic field affects its motion. The magnetic moment of the needle is 9.0×10^{-3} A·m^2, its moment of inertia is 2.0×10^{-8} kg·m^2, and the horizontal component of the magnetic field is 0.19×10^{-4} T.
    (a) What is the torque on the needle as a function of the angle θ between the needle and the north direction?
    (b) What is the frequency of small oscillations of the needle about the north direction? [Hint: Compare the equation for the rotational motion of the needle with Eq. (15.71) for the motion of a pendulum.]

CHAPTER 32

Electromagnetic Induction

In this chapter we will discover that electric fields can be generated not only by charges but also by changing magnetic fields. Whenever the magnetic field lines move or change in any way, they will generate an electric field, called an **induced electric field.** This kind of electric field exerts the usual electric forces on charges — in this regard the induced electric field does not differ from an ordinary electrostatic electric field. However, the induced and electrostatic electric fields differ in that the latter is conservative while the former is not. This means that the path integral of **E** around a closed path is not zero if **E** is an induced electric field. The path integral of the induced electric field is called the **induced emf.** As we will see, a famous law formulated by Michael Faraday asserts that the magnitude of this induced emf is directly proportional to the rate at which the magnetic flux sweeps across the path.

We begin with a discussion of the induced emf produced by the motion of a conducting body in a constant magnetic field. Such a body cuts across field lines, and thereby generates an induced emf, called a motional emf.

Induced emf

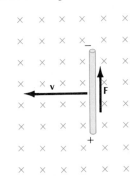

Fig. 32.1 Metallic rod moving with velocity **v** through a uniform magnetic field. A free electron within this rod experiences a force in the upward direction.

32.1 Motional Emf

Suppose that we push a rod of metal with some velocity through a uniform magnetic field, such as the magnetic field of a large electromagnet. If the rod and the velocity **v** of the rod are perpendicular to each other and to the magnetic field (Figure 32.1), then the free electrons in the metal will experience a magnetic force evB directed along the rod. The electrons will therefore flow along the rod, accumulating negative charge on the upper end and leaving positive charge on the

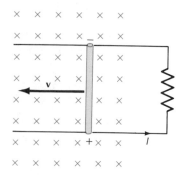

Fig. 32.2 If the ends of the rod are in sliding contact with a pair of wires, a current will flow around the circuit.

Motional emf

Michael Faraday, *1791–1867, English physicist and chemist. Faraday was apprentice to a bookbinder when he became interested in science. He attended lectures by Humphry Davy, the famous chemist, and prepared a set of lecture notes which so impressed Davy that he appointed Faraday his assistant at the laboratory of the Royal Institution. Ultimately, Faraday succeeded Davy as director of the laboratory. Faraday's earliest research lay in chemistry, but he soon turned to research in electricity and magnetism, making contributions of the greatest significance. His discovery of electromagnetic induction was no accident, but arose from a systematic experimental investigation of whether magnetic fields can generate electric currents. Although Faraday was essentially an experimenter, with no formal training in mathematics, he made an important theoretical contribution by introducing the concept of field lines and by recognizing that electric and magnetic fields are physical entities.*

lower end. This flow of charge will stop when the electric repulsion generated by the accumulated charges balances the magnetic force evB. However, if the ends of the rod are in sliding contact with a pair of long wires that provide a (stationary) return path (Figure 32.2), then the electrons will not accumulate on the ends of the rod — they will flow continually around the circuit. Thus, the moving rod acts as a "pump of electricity." The rod is a source of emf. The upper end of the rod in Figure 32.2 is the negative terminal of this source, and the lower end is the positive terminal.

The emf associated with the rod is the work done by the driving force on a hypothetical unit positive charge that passes from the negative end of the rod to the positive. The driving force on a unit positive charge is vB, and if the length of the rod is l, the work done by this force is

$$\mathscr{E} = lvB \tag{1}$$

This is called a **motional emf** because it is generated by the motion of the rod through the magnetic field. Note that we have ignored the component of velocity associated with motion of the charge *along* the rod. This component leads to a magnetic force perpendicular to the length of the rod, and that force is canceled by the constraining force that the sides of the rod exert on the charges, preventing their escape. The cancellation of one part of the magnetic force is crucial for the result stated in Eq. (1) — if it were not for this cancellation, the total magnetic force would do zero work [compare Eq. (31.22)]. Of course, the energy delivered to the current by the emf ultimately comes not from the magnetic field but from the mechanical device that pushes the rod. The magnetic field merely plays the role of mediator, transferring the mechanical energy into electric energy. If we did not apply an external push to the rod of Figure 32.2, it would slow down and stop as its kinetic energy is converted into electric energy and delivered to the current.

The quantity lvB appearing on the right side of Eq. (1) can be given an interesting interpretation in terms of the magnetic flux. As we already mentioned in Section 30.2, the **magnetic flux** is defined in the same way as the electric flux. For an arbitrary open or closed surface, the magnetic flux is the integral of the normal component of the magnetic field over the surface:

$$\Phi_B = \int \mathbf{B} \cdot d\mathbf{S} \tag{2}$$

The unit of magnetic flux is the weber (Wb),

$$1 \text{ weber} = 1 \text{ Wb} = 1 \text{ T} \cdot \text{m}^2 \tag{3}$$

With our usual convention for the density of field lines, the magnetic field is simply equal to the net number of magnetic field lines intercepted by the surface.

Now, let us consider Eq. (1). The quantity lv in this equation is the area swept out by the moving rod per unit time, and the product lvB is therefore the magnetic flux swept across by the rod. We can then express Eq. (1) as

$$\mathscr{E} = [\text{rate of sweeping of magnetic flux}] \tag{4}$$

The advantage of Eq. (4) over Eq. (1) is that the former equation is of general validity — it holds for rods and wires of arbitrary shape moving through arbitrary magnetic fields (this can be proved by treating a wire of arbitrary shape as made of small straight segments chosen so the magnetic field is approximately constant in the vicinity of each segment). Furthermore, as we will see in the next section, the relation between the induced emf and the rate of sweeping of magnetic flux holds even if the magnetic field is time dependent.

Instead of attributing the emf to the rate of sweeping of magnetic flux, we can equivalently attribute it to the rate of intersecting of magnetic field lines by the rod, since the flux in the area swept by the rod is equal to the number of field lines in this area. The concept of intersecting of field lines by a moving rod or wire is due to Faraday; it provided him with a sharp mental picture of the mechanism of induction and helped him formulate the general law of induction. However, we must be careful not to interpret Faraday's vivid phrase *intersecting of field lines* in a literal sense — the field lines are not physical objects, but merely mathematical constructs, and they suffer no damage when a rod passes through them.

Note that Eq. (4) gives us only the magnitude of the emf. To find the sign of the emf, we must remember that within the moving body that sweeps across the field lines, a positive charge will be pushed in the direction of **v** × **B**.

Wilhelm Eduard Weber, *1804–1891, German physicist, professor at Göttingen. He worked on problems in magnetism, and devised a system of units for electric and magnetic quantities.*

EXAMPLE 1. A straight metallic rod is rotating about its midpoint on an axis parallel to a uniform magnetic field (Figure 32.3a). The length of the rod is $2l$, and the angular velocity of rotation is ω. What is the induced emf between the midpoint of the rod and each end? What is the induced emf between the two ends?

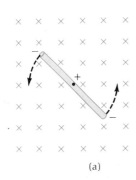 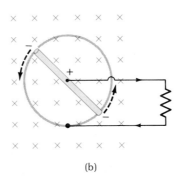

(a) (b)

Fig. 32.3 (a) Rod rotating in a uniform magnetic field. (b) If the ends of the rod are in sliding contact with a circular track, a current will flow as shown.

SOLUTION: Consider one-half of the rod, from the midpoint to one end. This piece takes a time $2\pi/\omega$ to sweep out the circular area πl^2. Hence the rate at which this piece sweeps out area is $\pi l^2 \omega / 2\pi = l^2 \omega / 2$ and the rate at which it sweeps across magnetic flux is $B l^2 \omega / 2$. Equation (4) then leads to an induced emf

$$\mathscr{E} = Bl^2\omega/2 \tag{5}$$

between the midpoint and each end. With the direction of motion as shown in Figure 32.3a, the midpoint will act as the positive terminal of a source of emf, and each moving end as the negative terminal.

COMMENTS AND SUGGESTIONS: The emf between the two ends is zero because both ends act as negative terminals, with the same emf (the two halves of the rod act as two batteries connected in parallel). Note that a careless calculation with Eq. (4) would have led us to the wrong conclusion that the emf for the entire rod is twice that for each half, since the entire rod sweeps across twice as much flux as each half — but we must take into account the polarities of these emfs!

Homopolar generator

If the moving ends of the rod are in sliding contact with a circular track connected to the midpoint via an external circuit (Figure 32.3b), then the rod will generate a current in the circuit. The device shown in Figure 32.3b acts as a generator of DC voltage; this device is called a **homopolar generator.** For practical applications, homopolar generators are constructed with a rotating disk rather than a rotating rod. This does not affect the emf of the generator, but it helps to reduce its internal resistance. Homopolar generators are used in applications requiring a large current, but only a fairly small voltage, such as in electroplating.

Magnetohydrodynamic generator

EXAMPLE 2. A **magnetohydrodynamic (MHD) generator** consists of a rectangular channel within which flows a hot, ionized gas, or plasma. The channel is placed in a strong magnetic field (Fig. 32.4). What emf will be induced between opposite sides of the stream of plasma? The width of the channel is 50 cm, the speed of the gas is 800 m/s, and the strength of the magnetic field is 6.0 T.

SOLUTION: Consider a straight path, fixed in the plasma, from one side of the stream of plasma to the other. This path sweeps across magnetic flux, just as does the straight rod discussed at the beginning of this section; hence Eq. (1) applies,

$$\mathscr{E} = vBl = 800 \text{ m/s} \times 6.0 \text{ T} \times 0.50 \text{ m} = 2.4 \times 10^3 \text{ V} \tag{6}$$

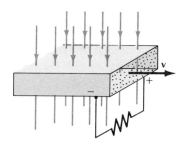

Fig. 32.4 A channel in which plasma flows toward the right. The channel is placed in a vertical downward magnetic field.

If, as shown in Figure 32.4, the magnetic field points downward, then the left side of the stream of plasma acts as the positive terminal of a source of emf and the right side as the negative terminal. This source of emf will drive a current if we connect an external circuit, as shown in Figure 32.4. Note that the two sides of the channel must be made of a conductor, so that they are in good electrical contact with the plasma, but the top and bottom of the channel must be made of an insulator.

Magnetohydrodynamic generators are expected to serve as auxiliary generators in power plants burning fossil fuel (Figure 32.5). The MHD generator

Fig. 32.5 Experimental MHD generator at the Argonne National Laboratory. This generator has a circular channel of diameter ~ 1 m, which is surrounded by a powerful superconducting magnet.

uses the exhaust gas released by the burning of this fuel. For this purpose, the fuel must be burned in a special combustion chamber, similar to that of a jet engine, so as to produce a high-speed stream of very hot exhaust gas. After the hot gas emerges from the MHD generator, it can supply heat to a boiler providing steam for a conventional power generator.

COMMENTS AND SUGGESTIONS: The emf induced across a stream of plasma flowing in a channel is quite analogous to the voltage produced by the Hall effect across the stream of electrons flowing in a conducting strip (see Section 31.4). The Hall voltage can be regarded as an induced, motional emf.

32.2 Faraday's Law

As we saw in the preceding section, the induced emf between the ends of a rod moving through a magnetic field equals the rate at which the rod sweeps across magnetic flux. Of course, the rod will sweep across flux whenever there is *relative motion* between the rod and the magnet producing the magnetic field. Relativity suggests that it should not make any difference whether we move the rod past a fixed magnet or move the magnet past a fixed rod — and experiments do indeed confirm that in both cases the induced emf is exactly the same.

But there is another way in which we can sweep across flux: we can hold the rod fixed and increase or decrease the strength of the magnetic field by increasing or decreasing the current. To understand why flux will be swept under these conditions, we must first take a look at what happens to the field lines of a current when the current increases or decreases. Figure 32.6a shows the field lines produced by a current on a long straight wire. If we increase the current, the magnetic field increases, that is, the number of field lines increases. Figure 32.6b shows the field lines of a stronger current. Where do the extra field lines come from? Obviously, the current has to make them. When the current slowly increases, it makes new, small circles of field lines in its immediate vicinity; meanwhile the circles that already exist in Figure 32.6a expand and move outward, just like the ripples on the surface of a pond that are created when we drop a stone into the pond. Thus, the pattern shown in Figure 32.6a gradually grows into the new pattern shown in Figure 32.6b. Note that the pattern grows from the inside out.

If we decrease the current, field lines must disappear. But this is not quite the reverse of the creation of field lines — the circles of Figure 32.6b do not contract and disappear at the center. The pattern cannot change from outside in; it must first change *near* the current. When the current slowly decreases, it makes new small circles of field lines of *opposite* direction (negative field lines) in its immediate vicinity. These opposite circles expand and move outward, gradually canceling the original field lines.

In any case, a change in the strength of magnetic field involves moving field lines. If a rod is located in the vicinity of the current, the moving field lines will sweep across the rod, that is, flux will sweep across the rod. Hence, an increase or decrease of the strength of magnetic field can cause flux to sweep across a *stationary* rod.

Does such a sweeping of flux induce an emf in the rod? This is a question that can be resolved only by experiment — and the answer is

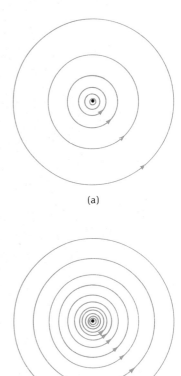

Fig. 32.6 (a) Magnetic field lines of a current on a very long wire. The wire is perpendicular to the plane of the page. (b) Magnetic field lines of a stronger current.

yes. The flux sweeping across the stationary rod induces an emf which is given by the same formula as before. The general statement is known as **Faraday's Law of induction:**

The induced emf along any moving or fixed mathematical path in a constant or changing magnetic field equals the rate at which magnetic flux sweeps across the path,

Faraday's Law

$$\mathscr{E} = [\text{rate of sweeping of magnetic flux}] \qquad (7)$$

Note that we have expressed Faraday's Law as an assertion about a mathematical *path*. The emf will exist between the ends of this path regardless of whether we place a rod or wire along this path; that is, whenever a unit positive charge moves along this path, it will gain an amount $\mathscr{E}$ of energy from the induced electric field regardless of whether the charge moves on a rod or through empty space. The rod or wire merely serves as a convenient conduit for the charge. Of course, for the practical exploitation of the emf, we will have to provide a rod, wire, or some other conductor along which the current can flow, and we will also have to provide a return path for the current.

For a closed path, Faraday's Law can be conveniently stated in terms of the magnetic flux through the area within the path. When we reckon the net rate at which flux sweeps across a closed path, we must take into account that the flux can make a positive or a negative contribution to the rate of sweeping, depending on whether the flux enters the area bounded by the path or leaves it. The net rate of sweeping of magnetic flux is equal to the rate of change of the flux intercepted by the area within the path. Hence, Faraday's Law can be stated as follows:

The induced emf around a closed mathematical path in a magnetic field is equal to the rate of change of the magnetic flux intercepted by the area within the path,

$$\mathscr{E} = -\frac{d\Phi_B}{dt} \qquad (8)$$

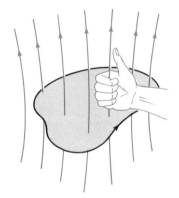

Fig. 32.7 A closed path in a changing magnetic field. The sign of the magnetic flux intercepted by the area within the path is given by a right-hand rule: If the fingers are curled in the direction in which we are reckoning the emf around the path (indicated by the black arrow), the thumb points in the direction that the magnetic field must have in order to give a positive flux.

The minus sign that we have inserted in this equation indicates how the polarity of the induced emf is related to the sign of the flux and of the rate of change of flux. The sign of the magnetic flux is fixed by the following **right-hand rule:** Curl the fingers in the direction in which we are reckoning the emf around the path; the magnetic flux is then positive if the magnetic field points in the direction of the thumb and negative otherwise. For example, in Figure 32.7, this right-hand rule indicates that the field lines shown make a positive contribution to the magnetic flux; consequently, if the strength of the magnetic field is increasing, $d\Phi_B/dt$ will be positive and, according to Eq. (8), the induced emf around the closed path will be negative. If the path in Figure 32.7 coincides with a conducting wire, the induced emf will therefore drive a current around this wire in a direction opposite to that indicated by the arrow.

32.3 Some Examples; Lenz' Law

In this section, we will look at some examples of the calculation of induced emfs. We will also lay down a simple rule for finding the polarity of the emf.

EXAMPLE 3. A rectangular coil of wire of 150 turns measuring 0.20 m × 0.10 m forms a closed circuit. The resistance of the coil is 5.0 Ω. The coil is placed in an electromagnet, face on to the magnetic field (Figure 32.8). Suppose that when the electromagnet is suddenly switched off, the strength of the magnetic field decreases at the rate of 20 T per second. What is the induced emf in the coil? The induced current? What is the direction of the induced current?

SOLUTION: The flux intercepted by the area within the coil is $\Phi_B = NAB$, where N is the number of turns, A the area, and B the magnetic field. Hence the induced emf is

$$\mathscr{E} = -\frac{d\Phi_B}{dt} = -NA\frac{dB}{dt}$$

$$= 150 \times (0.20 \text{ m} \times 0.10 \text{ m}) \times 20 \text{ T/s}$$

$$= 60 \text{ V}$$

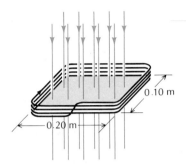

Fig. 32.8 A rectangular coil in a uniform, decreasing magnetic field.

and the induced current is

$$I = \mathscr{E}/R = 60 \text{ V}/5.0 \text{ Ω} = 12 \text{ amperes}$$

To determine the polarity of the induced current, let us reckon the emf around the loop in the direction indicated by the arrow in Figure 32.8. By the right-hand rule, the field lines shown make a positive contribution to the magnetic flux. Since the magnetic field is *decreasing*, $d\Phi_B/dt$ will then be *negative* and, according to Eq. (8), the induced emf will be positive. This induced emf will therefore drive a current in the direction of the arrow.

COMMENTS AND SUGGESTIONS: Note that here each turn of the coil makes its own contribution to the flux — the net flux through the coil is N times the flux through one turn. Always remember to include this factor of N when dealing with a coil of several turns.

EXAMPLE 4. A long solenoid has a circular cross section of radius R. The solenoid is connected to a source of alternating current so that the magnetic field inside the solenoid is

$$B = B_0 \sin \omega t \tag{9}$$

where B_0 and ω are constants. What is the induced emf around a concentric circular path of radius r inside the solenoid or outside the solenoid?

SOLUTION: Inside the solenoid ($r < R$), the magnetic field is uniform. Figure 32.9 shows a circle of radius r. The area of this circle intercepts a magnetic flux $-\pi r^2 B$. [If the positive direction for **B** is into the page and the direction in which we are reckoning the emf is counterclockwise (Figure 32.9), then, by the right-hand rule discussed in connection with Eq. (8), the flux will have a sign opposite to that of **B** — at the instant shown in Figure 32.9, the flux is negative.] The rate of change of the flux is $-\pi r^2 dB/dt$. Hence Faraday's Law gives us the induced emf

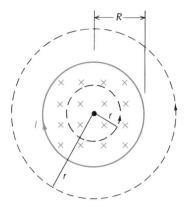

Fig. 32.9 Cross-sectional view of solenoid showing uniform magnetic field. The dashed circles are two alternative paths for reckoning the emf.

$$\mathcal{E} = -\frac{d\Phi_B}{dt} = \pi r^2 \frac{dB}{dt} = \pi r^2 \omega B_0 \cos \omega t \qquad (10)$$

The emf increases in proportion to the area of the circle, reaching a maximum value if the radius of the circle coincides with the radius of the solenoid ($r = R$).

Note that as regards the time dependence, the emf is 90° out of phase with the magnetic field — when the magnetic field (9) has maximum magnitude ($\sin \omega t = \sin \pi/2 = 1$), the emf (10) has minimum magnitude ($\cos \omega t = \cos \pi/2 = 0$).

Outside the solenoid ($r > R$), there is no magnetic field. The magnetic flux intercepted by a circle of radius r (Figure 32.9) is therefore simply the flux $-\pi R^2 B$ inside the solenoid, and the rate of change in this flux is $-\pi R^2 \, dB/dt$.[1] Instead of Eq. (10) we now obtain

$$\mathcal{E} = -\frac{d\Phi_B}{dt} = \pi R^2 \frac{dB}{dt} = \pi R^2 \omega B_0 \cos \omega t \qquad (11)$$

Thus the emf remains constant as we go away from the solenoid. Figure 32.10 shows a plot of the value of the induced emf as a function of the radius of the circular path.

Fig. 32.10 Induced emf as a function of radius of the circular path (at time $t = 0$).

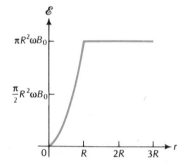

COMMENTS AND SUGGESTIONS: The solenoid and its current did not directly enter into our calculation — the induced emf is entirely determined by the magnetic field. This implies that we will get the same kind of induced emf if the magnetic field is that in the gap of an electromagnet, such as that shown in Figure 31.9, rather than that in a solenoid.

Betatron

Induced emfs in electromagnets find an important application in the design of electron accelerators. In the **betatron** (Figure 32.11), electrons travel in a circular orbit between the poles of a large electromagnet. The electrons are held in their circular orbit by the magnetic field. The acceleration of the electron is accomplished by quickly increasing the strength of the magnetic field. This change of the magnetic field induces an emf around the circular orbits of the electrons,

---

[1] This leads to an intriguing question: The rate of change of flux ought to equal the rate of sweeping of flux by the circle, but if there is no magnetic field outside of the solenoid, how can the circle sweep across flux? The answer is that when we increase the current, a new field line will be created inside the solenoid and move toward the center; whenever this happens, the return portion of this field line must move from the current outward to infinity (the magnetic field lines form *closed* loops around the current). This outward-moving field line sweeps across the outer circle of Figure 32.9.

Fig. 32.11 The small machine on the cart in the center is the first betatron, built in 1940. The large machine in the background is the largest betatron. This machine, at the University of Illinois, attained an electron energy of 340 MeV.

which accelerates the electrons, increasing their energy. This acceleration continues as long as the magnetic field is increasing; at the same time, the increasing magnetic field provides just the right amount of extra centripetal force to keep the electrons in a circular orbit of constant (or nearly constant) radius. In a typical betatron, the magnetic field increases over a time interval of a few milliseconds, and in this time the electrons move around their circular orbit a few hundred thousand times, acquiring a final energy of 100 MeV or so. They are then suddenly guided out of the machine (by means of deflecting electric or magnetic fields), and the magnetic field is then permitted to decrease, readying the machine for the next cycle of acceleration.

The polarity of any induced emf can always be found out by an argument similar to that given in Examples 3 and 4. But there is a simple rule for finding the direction, a rule known as **Lenz' Law:**

The induced emfs are always of such a polarity as to oppose the change that generated them.

Lenz' Law

The exact meaning of *change* depends on the problem at hand and is best made clear by some examples. For instance, the moving rod of Figure 32.12 generates a counterclockwise current. Consider now the magnetic force on this current in the rod; this force has the direction of $I\,d\mathbf{l} \times \mathbf{B}$, which points to the right in Figure 32.12. Thus, the force associated with the induced current *opposes* the motion of the rod, that is, it opposes the change (motion) that generates the current. This is in agreement with Lenz' Law.

Likewise, if there were a circular wire loop within the solenoid of Figure 32.9, a counterclockwise current would flow in it [assuming that B in Eq. (9) is increasing, $0 < \omega t < \pi/2$]. Such a current flowing around the loop would produce a magnetic field within the loop opposite to the field B of Eq. (9). Thus, the induced current around the loop *opposes* the change of flux within the loop, that is, it opposes the change that generates the current.

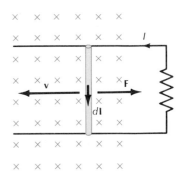

Fig. 32.12 The magnetic field exerts a force on the current in the rod.

Note that Lenz' Law is consistent with energy conservation but that the contrary of Lenz' Law is not. If the contrary of Lenz' Law were true, then the force on the rod in Figure 32.12 would be toward the left so as to speed up the rod. This would develop into a runaway situation: as the rod acquires higher velocity, the force would increase and accelerate it more and more. There would then be no need for us to push the rod through the magnetic field . . . it would accelerate spontaneously! Clearly, this would be a violation of energy conservation.

EXAMPLE 5. An electromagnetic generator consists of a coil of N turns of wire that rotates about an axis perpendicular to a constant magnetic field. Sliding contacts connect the coil to an external circuit (Figure 32.13). What emf does the coil deliver to the external circuit? The coil has an area A and rotates with an angular frequency ω.

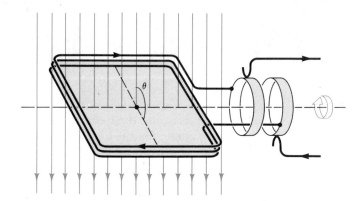

Fig. 32.13 Electromagnetic generator.

SOLUTION: The coil makes an angle $\theta = \omega t$ with the magnetic field. The magnetic flux through the coil is $\Phi = NAB \sin \theta = NAB \sin \omega t$. By Faraday's Law [see Eq. (8)], the induced emf is

$$\mathcal{E} = -\frac{d\Phi_B}{dt} = -\frac{d}{dt}(NAB \sin \omega t)$$

$$= -NAB\omega \cos \omega t \tag{12}$$

At the instant shown in Figure 32.13, the magnetic flux through the coil is decreasing. According to Lenz' Law, the induced current must therefore flow around the coil clockwise, so as to contribute an extra magnetic flux that tends to oppose the decrease of flux.

COMMENTS AND SUGGESTIONS: The function given by Eq. (12) is an **alternating emf,** or **AC voltage,** that oscillates between positive and negative values. A plot of this emf is shown in Figure 32.14. To obtain a constant emf, or DC

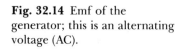

Fig. 32.14 Emf of the generator; this is an alternating voltage (AC).

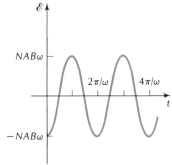

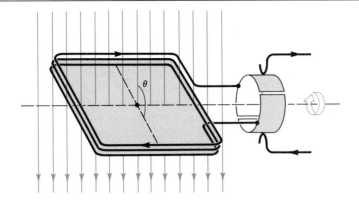

Fig. 32.15 Electromagnetic generator with commutator.

voltage, from a generator, we can use a slotted sliding contact, or **commutator,** arranged so that the emf at the external terminals always has the same sign (Figure 32.15). A plot of the emf produced by such a generator is shown in Figure 32.16. This emf oscillates between zero and a maximum value but always remains positive. To eliminate or reduce the oscillations of the emf, we can connect several generators in series, each with its coil at a slightly different initial angle to the magnetic field. In the sum of the emfs, the oscillations then tend to average out. For example, the plot of Figure 32.17 shows the combined emf of three coils whose positions are always 120° apart as they rotate. The emf delivered by this generator is fairly constant, with only a small ripple.

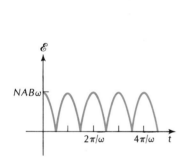
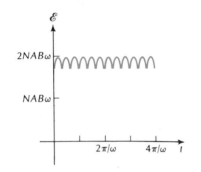

Fig. 32.16 (left) Emf of generator with commutator.

Fig. 32.17 (right) Emf of generator with three coils 120° apart connected in series; each loop has a commutator. The ripple in the voltage is about 14% of the maximum voltage.

32.4 The Induced Electric Field

It is instructive to reexamine the generation of the induced emf in a rod moving through a magnetic field from the point of view of a reference frame in which the rod is at rest. Consider again the rod of Figure 32.1 moving at constant velocity **v.** In a reference frame moving with the rod at the same velocity **v,** the free charges within the rod have no average velocity and there is *no magnetic force.* However, the free charges must experience some other kind of force that pushes them along the rod. The question is then: What is this new kind of force in the moving reference frame that produces the same effect as the magnetic force in the stationary reference frame? Since the only kinds of force that act on electric charges are the magnetic force and the electric force, the "new" kind of force must be due to a "new" kind of electric field, an electric field that exists in the moving reference frame, but not in the stationary reference frame (Figure 32.18). This "new" electric field in the moving reference frame arises indirectly from the currents that are responsible for the magnetic field in

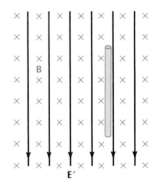

Fig. 32.18 In the moving reference frame of the rod, there is a magnetic field **B** (colored crosses) and also an electric field **E'** (black).

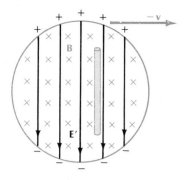

Fig. 32.19 In the moving reference frame, the solenoid has a magnetic field **B** (colored crosses) and an electric field **E′** (black). The edge of the solenoid has acquired an electric charge.

the stationary reference frame — it turns out that although the conductors that carry these currents are electrically neutral relative to the stationary reference frame, they acquire an excess of positive or negative charge relative to the moving reference frame (for details of the mechanism involved in this, see Section 41.8). For instance, if the magnetic field through which the rod moves is that of a very long solenoid (Figure 32.19), then the upper portion of the solenoid acquires a positive charge and the lower portion a negative charge, as seen in the moving reference frame.

We can determine the magnitude of the induced electric field from a consistency requirement: the "new" electric force qE' in the moving reference frame must coincide with the magnetic force qvB in the stationary reference frame. This tells us that the electric field in the moving reference frame must have a magnitude

$$E' = vB \tag{13}$$

This electric field does work on the free charges and therefore represents a source of emf.[2] The work done on a hypothetical unit positive charge that passes through the rod is

$$\mathscr{E} = El' \tag{14}$$

In view of Eq. (13), this value of the emf coincides with the value that we obtained in Eq. (1). Thus, the motional emf can be calculated either in a stationary reference frame or else in a moving reference frame; in the former case it arises from a magnetic field and in the latter from an electric field.

In our simple example, the electric field **E′** is constant along the rod. More complicated problems may involve an electric field **E′** that depends on position. Equation (13) indicates that such a dependence can come about in two ways: if the magnetic field varies along the rod (a nonuniform magnetic field) or if the velocity varies along the rod (a rigid rod with rotational motion or perhaps a flexible rod undergoing a deformation). If **E′** depends on position, then we must of course replace Eq. (14) by an integral,

$$\mathscr{E} = \int \mathbf{E}' \cdot d\mathbf{l} \tag{15}$$

where the integration extends from one end of the rod to the other.

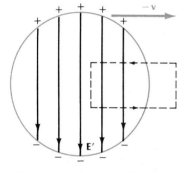

Fig. 32.20 Electric field of a long solenoid as seen in the moving reference frame. For the dashed path shown, which has one portion outside of the electric field **E′**, the integral $\oint \mathbf{E}' \cdot d\mathbf{l}$ is *not* zero.

Induced electric field

The electric field **E′** that exists in the reference frame of the moving rod is called an **induced electric field.** An important feature of this kind of field is that it is *not conservative*. Figure 32.20 shows the electric field **E′** of a solenoid as seen in the reference frame of the moving rod. If we integrate **E′** around the closed path drawn in this figure, we will obviously get a result different from zero — only one side of the rectangle gives a contribution to the integral. This establishes that **E′** is not a conservative field since, for a conservative field, the integral has to be zero for *every* arbitrary closed path.[3]

---

[2] Note that this is an external electric field (generated by charges *not* on the rod). Besides, there may be an internal electric field (generated by charges that accumulate on the ends of the rod); for now, we will not concern ourselves with this latter field.

[3] We might ask, Why can we not use a similar argument to establish that the electric field of a capacitor is nonconservative? The answer is that when we take a path that has one portion inside the capacitor and one portion outside, *both* portions contribute to the

In general, an induced electric field will exist in the reference frame of any path, or segment of a path, moving through a magnetic field. An induced electric field will also exist in the reference frame of a stationary path placed in a changing, time-dependent magnetic field. The induced electric field is always related to the induced emf by Eq. (15). Hence, in terms of the induced electric field, we can write Faraday's Law as

$$\int \mathbf{E}' \cdot d\mathbf{l} = [\text{rate of sweeping of magnetic flux}] \tag{16}$$

for an arbitrary path or as

$$\oint \mathbf{E}' \cdot d\mathbf{l} = -\frac{d\Phi_B}{dt} \tag{17}$$

for a closed path. Here the induced electric field $\mathbf{E}'$ associated with some segment $d\mathbf{l}$ of the path must be reckoned in a reference frame in which the segment $d\mathbf{l}$ is (instantaneously) at rest.

EXAMPLE 6. Consider again the long solenoid with the time-dependent magnetic field described in Example 4. What is the induced electric field inside and outside the solenoid?

SOLUTION: We have already evaluated the induced emf in Example 4. Inside the solenoid ($r < R$), its value is given by Eq. (10); and outside the solenoid ($r > R$), its value is given by Eq. (11). The electric field is related to this emf by

$$\oint \mathbf{E} \cdot d\mathbf{l} = \mathscr{E} \tag{18}$$

where the path of integration is a closed concentric circle of radius r. Note that the electric field $\mathbf{E}$ in this equation is an electric field in the original reference frame (there is no moving reference frame involved in this problem). We therefore are entitled to denote the electric field by $\mathbf{E}$, rather than by $\mathbf{E}'$. To evaluate the left side of Eq. (18), we observe that the field lines of the electric field must be closed concentric circles, because this is the only pattern of electric field lines consistent with the rotational symmetry of the solenoid. Hence the path of integration coincides with a field line and

$$\oint \mathbf{E} \cdot d\mathbf{l} = 2\pi r E \tag{19}$$

Combining Eqs. (10) and (19), we then obtain, inside the solenoid,

$$2\pi r E = \pi r^2 \omega B_0 \cos \omega t$$

or

$$E = \tfrac{1}{2} r \omega B_0 \cos \omega t \tag{20}$$

The electric field is zero on the axis of the solenoid ($r = 0$) and increases ($\propto r$) as we move away from the axis.

---

path integral because the capacitor has a fringing field that extends beyond the edges of the plates. In earlier chapters we ignored this fringing field because it was unimportant, but for the evaluation of the path integral it is very important (see Question 27.7). In contrast, the ideal, infinitely long solenoid has no fringing field — the magnetic and the electric fields outside the solenoid are exactly zero.

Combining Eqs. (11) and (19), we obtain, outside the solenoid,

$$2\pi r E = \pi R^2 \omega B_0 \cos \omega t$$

or

$$E = \frac{1}{2} \frac{R^2}{r} \omega B_0 \cos \omega t \qquad (21)$$

The electric field decreases in strength ($\propto 1/r$) as we go away from the solenoid.

COMMENTS AND SUGGESTIONS: The electric field attains its maximum strength at the edge of the solenoid ($r = R$). Figure 32.21 is a plot of the strength of the electric field as a function of r. Figure 32.22 shows a picture of the field lines. According to Eqs. (20) and (21), the electric field at the initial time ($t = 0$) is positive, that is, it has the same direction as the direction of integration around the circular path in Figure 32.9 (counterclockwise).

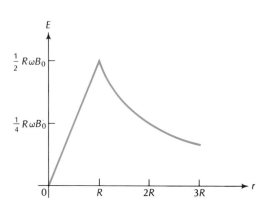

Fig. 32.21 Magnitude of the induced electric field as a function of radius (at $t = 0$).

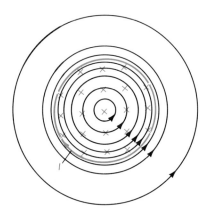

Fig. 32.22 Electric field lines (black) within solenoid and outside of solenoid.

The preceding example shows how a time-dependent magnetic field generates an electric field. This induced electric field differs from the electrostatic fields of the preceding chapters in that the field lines form closed loops — the field lines do not start on electric charges. Furthermore, the induced field is nonconservative; this becomes immediately obvious when we integrate the electric field along a closed field line [see Eq. (19)]. As we will see in Chapter 35, the electric fields of a light wave or a radio wave also are instances of induced electric fields, with closed field lines.

32.5 Inductance

If a conductor carrying a time-dependent current is near some other conductor, then the changing magnetic field of the former can induce an emf in the latter. Thus, a time-dependent current in one conductor can induce a current in another nearby conductor. For instance, consider the two coils of Figure 32.23. The first coil carries a time-de-

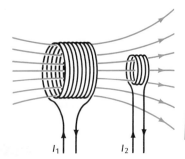

Fig. 32.23 Coil 1 creates a magnetic field. Some of the field lines pass through coil 2.

pendent current, which generates a magnetic field. The changing magnetic flux Φ_{B_1} through the second coil induces an emf in this coil. The emf in the second coil is

$$\mathcal{E}_2 = -\frac{d\Phi_{B_1}}{dt} \tag{22}$$

The flux Φ_{B_1} depends on the strength of the magnetic field $\mathbf{B}_1$ in the second coil produced by the current I_1 in the first coil. This field strength is directly proportional to I_1 [see Eq. (30.25) for the dependence of magnetic field on current]; hence the flux Φ_{B_1} is also proportional to I_1. We can write this relationship between Φ_{B_1} and I_1 as

$$\boxed{\Phi_{B_1} = L_{21} I_1} \tag{23}$$

and then Eq. (22) becomes

$$\boxed{\mathcal{E}_2 = -L_{21} \frac{dI_1}{dt}} \tag{24}$$

Here L_{21} is a constant of proportionality which depends on the size of the coils, their distance, and the number of turns in each, that is, L_{21} depends on the geometry of Figure 32.23. This constant is called the **mutual inductance** of the coils.

Mutual inductance

Equation (24) states that the emf induced in coil 2 is proportional to the rate of change of current in coil 1. But the converse is also true: if coil 2 carries a current, then the emf induced in coil 1 is proportional to the rate of change of the current in coil 2,

$$\mathcal{E}_1 = -L_{12} \frac{dI_2}{dt} \tag{25}$$

The constants of proportionality L_{21} and L_{12} appearing in Eqs. (24) and (25) are exactly equal, that is, $L_{21} = L_{12}$. Although we will accept this assertion without proof, we note that the result is quite reasonable: the mutual inductance reflects the geometry of the *relative* arrangement of the coils, and that is of course the same in both cases.

The SI unit of inductance is called the **henry** (H),

$$1 \text{ henry} = 1 \text{ H} = 1 \text{ V} \cdot \text{s/A} \tag{26}$$

Incidentally: The permeability constant is commonly expressed in terms of this unit of inductance,

$$\mu_0 = 1.26 \times 10^{-6} \text{ H/m} \tag{27}$$

Joseph Henry, *1797–1878, American experimental physicist, professor at Princeton and first director of the Smithsonian Institution. He made important improvements in electromagnets by winding coils of insulated wire around iron pole pieces, and invented an electromagnetic motor and a new, efficient telegraph. He discovered self-induction and investigated how currents in one circuit induce currents in another.*

EXAMPLE 7. A long solenoid has n turns per unit length. A ring of wire of radius r is placed within the solenoid, perpendicular to the axis (Figure 32.24). What is the mutual inductance of ring and solenoid?

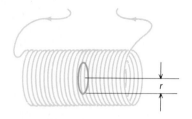

Fig. 32.24 A ring of wire inside a solenoid.

SOLUTION: If the current in the solenoid windings is I_1, the magnetic field is $B_1 = \mu_0 n I_1$ [see Eq. (31.16)], and the flux through the ring is

$$\Phi_{B_1} = \pi r^2 \mu_0 n I_1$$

Comparing this with Eq. (23), we recognize that

$$L_{12} = \pi r^2 \mu_0 n \tag{28}$$

Self-inductance

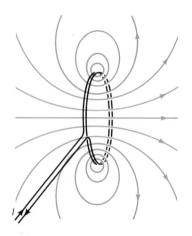

Fig. 32.25 A coil and its magnetic field.

A conductor by itself has a **self-inductance.** Consider a coil with a time-dependent current (Figure 32.25). The magnetic field of this coil will then produce a time-dependent magnetic flux and, by Faraday's Law, an induced emf. The net emf acting on the coil is then the sum of the external emf (applied to the terminals in Figure 32.25) and the self-induced emf. This means that whenever the current is time dependent, the coil will act back on the current and modify it (we will see how to calculate the net resulting current in Section 34.3). For this reason, the self-induced emf is often called a **back emf.** From Lenz' Law we immediately recognize that the self-induced emf always acts in such a direction to *oppose* the change in the current, that is, it attempts to maintain the current constant.

In terms of the flux through the circuit, the definition of the self-inductance is of the same form as Eq. (23),

$$\boxed{\Phi_B = LI} \tag{29}$$

and therefore the induced emf is

$$\boxed{\mathcal{E} = -L \frac{dI}{dt}} \tag{30}$$

Inductors used in radios and other electronic devices consist of small solenoids. In circuit diagrams, the symbol for an inductor is a coiled line (see Figure 32.26, on page 797).

EXAMPLE 8. A long solenoid has n turns per unit length and a radius R. What is its self-inductance per unit length?

SOLUTION: The magnetic field inside the solenoid is $B = \mu_0 n I$. The number of loops in a length l is nl; each of these loops has a flux $\pi R^2 B$. Hence the flux through all the loops in a length l is

$$\Phi_B = \pi R^2 B n l = \pi R^2 \mu_0 n^2 I l \tag{31}$$

and the self-inductance of the long solenoid is

$$L = \Phi_B / I = \mu_0 n^2 \pi R^2 l \tag{32}$$

The inductance per unit length is therefore $\mu_0 n^2 \pi R^2$.

32.6 Magnetic Energy

Inductors store magnetic energy, just as capacitors store electric energy. When we connect an external source of emf to an inductor and start a current through the inductor, the back emf will oppose the increase of the current, and the external emf must do work in order to overcome this opposition and establish the flow of current. This work is stored in the inductor, and it can be recovered by removing the external source of emf from the circuit. As the current begins to decrease, the inductor will supply a back emf which tends to keep the current flowing for a while (at a decreasing rate). Thus, the inductor delivers energy to the current.

To calculate the amount of stored energy, we note that when the current increases at the rate of dI/dt, the back emf is

$$\mathcal{E} = -L\, dI/dt \tag{33}$$

The inductor does work on the current at the rate

$$I\mathcal{E} = -LI\, dI/dt \tag{34}$$

Here the negative sign implies that the energy is delivered by the current to the inductor rather than vice versa. In a time dt, the energy stored in the inductor is therefore

$$dU = -I\mathcal{E}\, dt = LI\, dI \tag{35}$$

If we integrate this from the initial value of the current ($I' = 0$) to the final value ($I' = I$), we obtain

$$U = \int_0^I LI'\, dI' = L \int_0^I I'\, dI' \tag{36}$$

or

$$\boxed{U = \tfrac{1}{2}LI^2} \tag{37}$$

Magnetic energy in inductor

This equation for the magnetic energy in an inductor is analogous to Eq. (27.40) for the electric energy in a capacitor.

EXAMPLE 9. A solenoid has a radius of 2.0 cm; its winding has one turn of wire per millimeter. A current of 10 A flows through the winding. What is the amount of energy stored per unit length of the solenoid?

SOLUTION: The inductance per unit length is [see Eq. (32)]

$$L/l = \mu_0 n^2 \pi R^2$$

and the energy per unit length is

$$U/l = \tfrac{1}{2}(L/l)I^2 = \tfrac{1}{2}\mu_0 n^2 \pi R^2 I^2$$

$$= \tfrac{1}{2} \times 1.26 \times 10^{-6}\ \text{H/m} \times (10^3/\text{m})^2 \times \pi \times (0.020\ \text{m})^2 \times (10\ \text{A})^2$$

$$= 7.9 \times 10^{-2}\ \text{J/m}$$

The energy stored in a solenoid can be expressed in terms of the magnetic field. Consider a portion of length l of the solenoid. The inductance of this portion is [see Eq. (32)]

$$L = \mu_0 n^2 \pi R^2 l \tag{38}$$

so that

$$U = \tfrac{1}{2}\mu_0 n^2 \pi R^2 l I^2 \tag{39}$$

Since for a solenoid $B = \mu_0 nI$, we can write this as

$$U = \frac{1}{2\mu_0} B^2 \pi R^2 l \tag{40}$$

or

$$U = \frac{1}{2\mu_0} B^2 \times [\text{volume}] \tag{41}$$

where the volume $\pi R^2 l$ is the volume of magnetic field in the portion of length l of the solenoid.

According to Eq. (41), the quantity

Energy density in magnetic field

$$u = \frac{1}{2\mu_0} B^2 \tag{42}$$

can be regarded as the magnetic energy per unit volume. Although we have derived this equation only for the special case of a long solenoid, it turns out to be generally valid (in vacuum and in nonmagnetic materials). The magnetic field, just like the electric field, stores energy. The magnetic energy density is proportional to B^2 just as the electric density is proportional to E^2 [compare Eq. (26.19)].

EXAMPLE 10. Near the surface, the Earth's magnetic field typically has a strength of 0.3×10^{-4} T and the Earth's atmospheric electric field typically has a strength of 100 V/m. What is the energy density in each field?

SOLUTION: The magnetic energy density is

$$u_B = \frac{1}{2\mu_0} B^2 = \frac{1}{2 \times 1.26 \times 10^{-6} \text{ H/m}} \times (0.3 \times 10^{-4} \text{ T})^2$$

$$= 3.6 \times 10^{-4} \text{ J/m}^3$$

and the electric energy density is

$$u_E = \frac{\varepsilon_0}{2} E^2 = \frac{(8.85 \times 10^{-12} \text{ F/m})}{2} \times (100 \text{ V/m})^2$$

$$= 4.4 \times 10^{-8} \text{ J/m}^3$$

32.7* The RL Circuit

The RL circuit consists of a resistor and an inductor connected in series to a battery or some other source of emf. This circuit provides a

---
* This section is optional.

good illustration of the effects of self-inductance on a current. Figure 32.26 is the schematic diagram for this circuit. In the diagram, the resistance R is shown connected to an ideal, resistanceless inductor; if the wire in the coils of the inductor has some resistance, then this resistance must be included in R. We assume that the current in the inductor is initially zero, and that the battery is suddenly connected at time $t = 0$. The current then starts to increase. But the self-inductance will generate an emf across the inductor, which, by Lenz' Law, opposes the increase of the current. Because of this self-induced emf, the current in the circuit cannot increase suddenly; it can increase only gradually. The self-induced emf continues to oppose the current and to restrain its growth until this current attains its steady, final value, $I = \mathscr{E}/R$. Qualitatively, the gradual growth of current in an RL circuit is analogous to the increase of charge in an RC circuit. As we will see, the equations governing the approach to steady state in the RL circuit are mathematically similar to the equations governing the RC circuit.

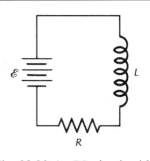

Fig. 32.26 An RL circuit with a battery.

For the calculation of the time dependence of the current in the RL circuit, we appeal to Kirchhoff's rule. If the instantaneous rate of change of the current is dI/dt, the emf contributed by the inductor is $-L dI/dt$. Hence

$$\mathscr{E} - L\frac{dI}{dt} - IR = 0 \qquad (43)$$

This equation has the same mathematical form as Eq. (29.34) for the charge in the RC circuit,

$$\mathscr{E} - R\frac{dQ}{dt} - \frac{Q}{C} = 0 \qquad (44)$$

Comparing Eq. (43) with Eq. (44), we see that I is mathematically analogous to Q, L is analogous to R, and R is analogous to $1/C$. By making corresponding replacements in Eq. (29.38), we can then immediately write down the solution of Eq. (43):

$$I = \frac{\mathscr{E}}{R}(1 - e^{-tR/L}) \qquad (45)$$

Figure 32.27 is a plot of this current as a function of time. The current grows quickly at first and more slowly later, and it asymptotically approaches the final value $\mathscr{E}/R$. The **time constant** of the RL circuit is

Time constant

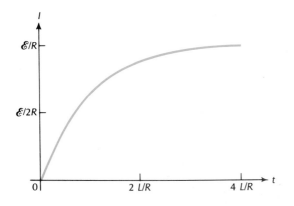

Fig. 32.27 Growing current in the RL circuit vs. time.

L/R. At this time, the current will have grown to within a factor of $1/e$ of its final value.

When the current has reached its final, steady value, we can suddenly switch the battery out of the circuit and "discharge" the current of the inductor directly into the resistor. The current will then gradually decay as a function of time and asymptotically approach zero. It is easy to show that the current in such a short-circuited RL circuit is

$$I = \frac{\mathscr{E}}{R} e^{-tR/L} \tag{46}$$

This decaying current is plotted in Figure 32.28. The time constant for the decaying current is, again, L/R. At this time, the current will have decreased by a factor of $1/e$ from its initial, maximum value.

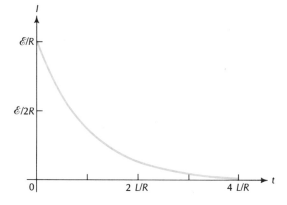

Fig. 32.28 Decaying current in the RL circuit vs. time.

SUMMARY

Motional emf in a rod: $\mathscr{E} = vBl$

Faraday's Law: $\mathscr{E} =$ [rate of sweeping of magnetic flux]

Magnetic flux: $\Phi_B = \int \mathbf{B} \cdot d\mathbf{S}$

Faraday's Law for closed loop: $\mathscr{E} = -\dfrac{d\Phi_B}{dt}$

Lenz' Law: Induced emf opposes change.

Mutual inductance: $\Phi_{B_1} = L_{21} I_1$

$$\mathscr{E}_2 = -L_{21} \frac{dI_1}{dt}$$

Self-inductance: $\Phi_B = LI$

$$\mathscr{E} = -L \frac{dI}{dt}$$

Self-inductance of solenoid: $\mu_0 n^2 \pi R^2$ per unit length

Magnetic energy in inductor: $U = \tfrac{1}{2} LI^2$

Energy density in magnetic field: $u = \dfrac{1}{2\mu_0} B^2$

Time constant of RL circuit: L/R

QUESTIONS

1. At the latitude of the United States, the magnetic field of the Earth has a downward component, larger than the northward component. Suppose that an airplane is flying due west in this magnetic field. Will there be an emf between its wingtips? Which wingtip will be positive? Will there be a flow of current?

2. What is the magnetic flux that the magnetic field of the Earth produces through the surface of the Earth?

3. Is Eq. (8) valid for an open path? Is Eq. (7) valid for a closed path?

4. A long straight wire carries a steady current. A square conducting loop is in the same plane as the wire. If we push the loop toward the wire, how is the direction of the current induced in the loop related to the direction of the current in the wire?

5. A **flip coil** serves to measure the strength of a magnetic field. It consists of a small coil of many turns connected to a sensitive ammeter. The coil is placed face on in the magnetic field and then suddenly flipped over. How does this indicate the presence of the magnetic field?

Flip coil

6. The **magneto** used in the ignition system of old automobile engines consists of a permanent magnet mounted on the flywheel of the engine. As the flywheel turns, the magnet passes by a stationary coil, which is connected to the spark plug. Explain how this device produces a spark.

Magneto

7. A long straight wire carries a current that is increasing as a function of time. A rectangular loop is near the wire, in the same plane as the wire. How is the direction of the current induced in the loop related to the direction of the current in the wire?

8. Consider two adjacent rectangular circuits, in the same plane. If the current in one circuit is suddenly switched off, what is the direction of the current induced in the other circuit?

9. Figure 32.29 shows two coils of wire wound around a plastic cylinder. If the current in the left coil is made to increase, what is the direction of the current induced in the coil on the right?

Fig. 32.29

10. A circular conducting ring is being pushed toward the north pole of a bar magnet. Describe the direction of the current induced in the ring.

11. Ganot's *Éléments de Physique*, a classic nineteenth-century textbook, states the following rules for the current induced in one loop face to face with another loop:
    I. The distance remaining the same, a continuous and constant current does not induce any current in an adjacent conductor.
    II. A current, at the moment of being closed, produces in an adjacent conductor an inverse current.
    III. A current, at the moment it ceases, produces a direct current.
    IV. A current which is removed, or whose strength diminishes, gives rise to a direct induced current.
    V. A current which is approached, or whose strength increases, gives rise to an inverse induced current.

Explain these rules on the basis of Lenz' Law.

12. A sheet of aluminum is being pushed between the poles of a horseshoe magnet (Figure 32.30). Describe the direction of flow of the induced currents, or **eddy currents**, in the sheet. Explain why there is a strong friction force that opposes the motion of the sheet.

Fig. 32.30

Eddy currents

13. Some beam balances use a magnetic damping mechanism to stop excessive swinging of the beam. This mechanism consists of a small conducting plate at-

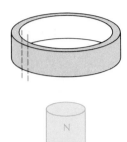

Fig. 32.31 A conducting ring falling toward a bar magnet. The dashed lines indicate where a slot will be cut through the ring.

tached to the beam and a small magnet mounted on a fixed support near this plate. How does this damp the motion of the beam? (Hint: See the preceding question.)

14. A bar magnet, oriented vertically, is dropped toward a flat horizontal copper plate. Explain why there will be a repulsive force between the bar magnet and the copper plate. Is this an elastic force, that is, will the bar magnet bounce if the magnet is very strong?

15. A conducting ring is falling toward a bar magnet (Figure 32.31). Explain why there will be a repulsive force between the ring and the magnet. Explain why there will be no such force if the ring has a slot cut through it (see dashed lines in Figure 32.31).

16. You have two coils, one of slightly smaller radius than the other. To achieve maximum mutual inductance, should you place these coils face to face or one inside the other?

17. Two circular coils are separated by some distance. Qualitatively, describe how the mutual inductance varies as a function of the orientation of the coils.

18. If a strong current flows in a circuit and you suddenly break the circuit (by opening a switch), a large spark is likely to jump across the switch. Explain.

19. What arguments can you give in favor of the view that the magnetic energy of an inductor is stored in the magnetic field, rather than in the current?

PROBLEMS

Section 32.1

1. An automobile travels at 88 km/h along a level road. The vertical downward component of the Earth's magnetic field is 0.58×10^{-4} T. What is the induced emf between the right and the left door handles separated by a distance of 2.1 m? Which side is positive and which negative?

2. The DC-10 jet aircraft has a wingspan of 47 m. If such an aircraft is flying horizontally at 960 km/h at a place where the vertical component of the Earth's magnetic field is 0.60×10^{-4} T, what is the induced emf between its wingtips?

3. In order to detect the movement of water in the ocean, oceanographers sometimes rely on the motional emf generated by this movement of the water through the magnetic field of the Earth. Suppose that, at a place where the vertical magnetic field is 0.70×10^{-4} T, two electrodes are immersed in the water separated by a distance of 200 m measured perpendicularly to the movement of the water. If a sensitive voltmeter connected to the electrodes indicates a potential difference of 7.0×10^{-3} V, what is the speed of the water?

4. A homopolar generator consists of a metal disk rotating about a horizontal axis in a uniform horizontal magnetic field. The external circuit is connected to contact brushes touching the disk at the rim and at the axis. If the radius of the disk is 1.2 m and the strength of the magnetic field is 6.0×10^{-2} T, at what rate (rev/s) must you rotate the disk to obtain an emf of 6.0 V?

5. The rate of flow of a conducting liquid, such as detergent, tomato pulp, beer, liquid sodium, or sewage, can be measured with an electromagnetic **flowmeter** that detects the emf induced by the motion of the liquid in a magnetic field. Suppose that a plastic pipe of diameter 10 cm carries beer with a speed of 1.5 m/s. The pipe is in a transverse magnetic field of 1.5×10^{-2} T. What emf will be induced between the opposite sides of the column of liquid?

6. Pulsars, or neutron stars, rotate at a fairly high speed, and they are surrounded by strong magnetic fields. The material in neutron stars is a good

conductor; and hence a motional emf is induced between the center of the neutron star and the rim (this is similar to the emf induced in a rotating metallic rod; see Example 1). Suppose that a neutron star of radius 10 km rotates at the rate of 30 rev/s and that the magnetic field has a strength of 10^8 T. What is the emf induced between the center of the star and a point on its equator?

7. A helicopter has blades of length 4.0 m rotating at 3 rev/s in a horizontal plane. If the vertical component of the Earth's magnetic field is 0.65×10^{-4} T, what is the induced emf between the tip of the blade and the hub?

8. Sharks have delicate sensors on their bodies that permit them to sense small differences of potential. They can sense electrical disturbances created by other fish, and they can also sense the Earth's magnetic field and use this for navigation. Suppose that a shark is swimming horizontally at 25 km/h at a place where the magnetic field has a strength of 4.7×10^{-5} T and points down at an angle of 40° with the vertical. Treat the shark as a cylinder of diameter 30 cm. What is the largest induced emf between diametrically opposite points on the sides of the shark when heading east? North? Northeast? Make a rough plot of the induced emf vs. the direction of travel of the shark around all points of the compass. Indicate which side of the shark is positive, which negative. (Hint: The component of **B perpendicular to the velocity **v** is proportional to $|\mathbf{v} \times \mathbf{B}|$.)

Sections 32.2 and 32.3

9. In Idaho, the magnetic field of the Earth points downward at an angle of 69° below the horizontal. The strength of the magnetic field is 0.59×10^{-4} T. What is the magnetic flux through 1 m² of ground in Idaho?

10. An electric generator consists of a rectangular coil of wire rotating about its longitudinal axis which is perpendicular to a magnetic field of 2.0×10^{-2} T. The coil measures 10.0 cm × 20.0 cm and has 120 turns of wire. The ends of the wire are connected to an external circuit. At what speed (in rev/s) must you rotate this coil in order to induce an alternating emf of amplitude 12.0 V between the ends of the wire?

*11. A very long train whose metal wheels are separated by a distance of 4 ft 9 in. is traveling at 80 mi/h on a level track. The vertical component of the magnetic field of the Earth is 0.62×10^{-4} T.
 (a) What is the induced emf between the right wheels and the left wheels?
 (b) The wheels are in contact with the rails, and the rails are connected by metal cross ties which close the circuit and permit a current to flow from one rail to the other. Since the number of cross ties is very large, their combined resistance is nearly zero; most of the resistance of the circuit is within the train. What is the current that flows in each axle of the train? The axles are cylindrical rods of iron of diameter 3 in. and length 4 ft 9 in.
 (c) Calculate the power dissipated and the effective friction force on the train.

Fig. 32.32 A long solenoid surrounded by a circular coil.

*12. (a) A long solenoid has 300 turns of wire per meter and has a radius of 3.0 cm. If the current in the wire is increasing at the rate of 50 amperes per second, at what rate does the strength of the magnetic field in the solenoid increase?
 (b) The solenoid is surrounded by a coil of wire with 120 turns (Figure 32.32). The radius of this coil is 6.0 cm. What induced emf will be generated in this coil while the current in the solenoid is increasing?

*13. A metal rod of length l and mass m is free to slide, without friction, on two parallel metal tracks. The tracks are connected at one end so that they and the rod form a closed circuit (Figure 32.33). The rod has a resistance R, and the tracks have negligible resistance. A uniform magnetic field is perpendicular to the plane of this circuit. The magnetic field is increasing at a con-

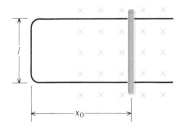

Fig. 32.33 Rod sliding on parallel tracks. The crosses show the tails of the magnetic field vectors.

stant rate dB/dt. Initially the magnetic field has a strength B_0 and the rod is at rest at a distance x_0 from the connected end of the rails. Express the acceleration of the rod at this instant in terms of the given quantities.

*14. A square loop of dimension 8.0 cm × 8.0 cm is made of copper wire of radius 1.0 mm. The loop is placed face on in a magnetic field which is increasing at the constant rate of 80 T/s. What induced current will flow around the loop? Draw a diagram showing the direction of the field and the induced current.

*15. A very long solenoid with 20 turns per centimeter of radius 5.0 cm is surrounded by a rectangular loop of copper wire. The rectangular loop measures 10 cm × 30 cm, and its wire has a radius of 0.05 cm. The resistivity of copper is 1.7×10^{-8} Ω·m. Find the induced current in the rectangular loop if the current in the solenoid is increasing at the rate of 5×10^4 A/s.

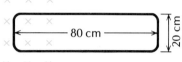

Fig. 32.34

*16. A rectangular loop measuring 20 cm × 80 cm is made of heavy copper wire of radius 0.13 cm. Suppose you shove this loop, short side first, at a speed of 0.40 m/s into a magnetic field of 5.0×10^{-2} T. The rectangle is face on to the magnetic field, and the trailing short side remains outside of the magnetic field (see Figure 32.34). What induced current will flow around the loop?

*17. A washer (annulus) of aluminum is lying on top of a vertical solenoid (see Figure 32.35). When the current in the solenoid is suddenly switched on, the washer flies upward. Carefully explain why the end of the solenoid exerts a repulsive force on the washer under these conditions. (Hint: Take into account that, at the end of the solenoid, the magnetic field lines spread out.)

Fig. 32.35

**18. A square loop of dimension $l \times l$ is moving at speed v toward a straight wire carrying a current I. The wire and the loop are in the same plane, and two of the sides of the loop are parallel to the wire. What is the induced emf of the loop as a function of the distance d between the wire and the nearest side of the loop?

**19. A circular coil of insulated wire has a radius of 9.0 cm and contains 60 turns of wire. The ends of the wire are connected in series with a 15-Ω resistor closing the circuit. The normal to the loop is initially parallel to a constant magnetic field of 5.0×10^{-2} T. If the loop is flipped over, so that the direction of the normal is reversed, a pulse of current will flow through the resistor. What amount of charge will flow through the resistor? Assume that the resistance of the wire is negligible compared with that of the resistor.

20. A flux meter used to measure magnetic fields consists of a small coil of radius 0.80 cm wound with 200 turns of fine wire. The coil is connected by means of a pair of tightly twisted trailing wires to a galvanometer that measures the electric charge flowing through the coil. The resistance of the coil–wire–galvanometer circuit is 6.0 Ω. Suppose that when the coil is quickly moved from a place outside the magnetic field to a place inside the magnetic field the galvanometer registers a flow of charge of 0.30 C. What is the magnetic flux through the coil? What is the component of **B perpendicular to the face of the coil? Assume that **B** is uniform over the area of the coil.

**21. A long straight wire carries a current that increases at a steady rate dI/dt.
(a) What is the rate of increase of the magnetic field at a radial distance r?
(b) What is the induced emf around the loop shown in Figure 32.36?

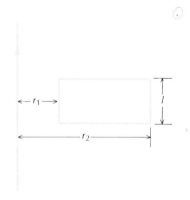

Fig. 32.36 Long straight wire and rectangular loop.

Section 32.4

22. A long solenoid of radius 3.0 cm has 2×10^3 turns of wire per meter. The wire carries a current of 6.0 A. Suppose that this solenoid is moving relative to you at a velocity of 400 m/s in a direction perpendicular to its axis.
(a) What is the induced electric field inside the solenoid, as seen in your reference frame?
(b) What is the electric field in the reference frame of the solenoid?

(c) What is the induced emf between two points on the edge of the solenoid at opposite ends of a diameter perpendicular to the velocity?

23. The magnetic field of a betatron has an amplitude of oscillation of 0.90 T and a frequency of 60 Hz. What is the amplitude of oscillation of the induced electric field at a radius of 0.80 m? What is the amplitude of oscillation of the induced emf around a circular path of this radius?

*24. A boxcar of a train is 2.5 m wide, 9.5 m long, and 3.5 m high; it is made of sheet metal and is empty. The boxcar travels at 60 km/h on a level track at a place where the vertical component of the Earth's magnetic field is 0.62×10^{-4} T. Assume that the boxcar is not in electrical contact with the ground.
  (a) What is the induced emf between the sides of the boxcar?
  (b) Taking into account the electric field contributed by the charges that accumulate on the sides, what is the net electric field inside the boxcar (in the reference frame of the boxcar)?
  (c) What is the surface-charge density on each side? Treat the sides as two very large parallel plates.

*25. A disk of metal of radius R rotates about its axis with a frequency ν. The disk is in a uniform magnetic field B, parallel to the axis of the disk.
  (a) Find the induced emf between the center of the disk and a point on the disk at a radial distance r (where $r < R$).
  (b) Find the strength of the induced electric field at this point.

Section 32.5

26. Two coils are arranged face to face, as in Figure 32.23. Their mutual inductance is 2.0×10^{-2} H. The current in coil 1 oscillates sinusoidally with a frequency of 60 Hz and an amplitude of 12 A,

$$I_1 = 12 \sin(120\pi t)$$

where the current is measured in amperes and the time in seconds.
  (a) What is the magnetic flux that this current generates in coil 2 at time $t = 0$?
  (b) What is the induced emf that this current induces in coil 2 at time $t = 0$?
  (c) What is the direction of the induced current in coil 2 at time $t = 0$, according to Lenz's Law? Assume that the positive direction for the current I_1 is as shown by the arrows in the figure.

*27. A current of 15 A in a coil produces a magnetic flux of 0.10 Wb through each of the turns of an adjacent coil of 60 turns. What is the mutual inductance?

*28. A long solenoid has 400 turns per meter. A coil of wire of radius 1.0 cm with 30 turns of insulated wire is placed inside the solenoid, its axis parallel to the axis of the solenoid. What is the mutual inductance? What emf will be induced around the coil if the current in the solenoid windings changes at the rate of 200 A/s?

29. A loop of wire carrying a current of 100 A generates a magnetic flux of 50 Wb through the area bounded by the loop.
  (a) What is the self-inductance of the loop?
  (b) If the current is decreased at the rate $dI/dt = 20$ A/s, what is the induced emf?

30. A long solenoid has 2000 turns per meter and a radius of 2.0 cm.
  (a) What is the self-inductance for a 1.0-m segment of this solenoid?
  (b) What back emf will this segment generate if the current in the solenoid is changing at the rate of 3.0×10^2 A/s?

31. A solenoid of self-inductance 2.2×10^{-3} H in which there is initially no current is suddenly connected in series with the poles of 24-V battery. What is the instantaneous rate of increase of the current in the solenoid?

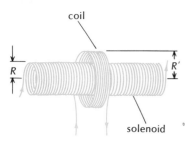

Fig. 32.37 A long solenoid and a circular coil.

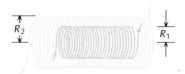

Fig. 32.38 Two long concentric solenoids.

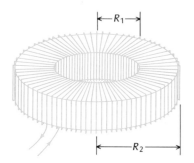

Fig. 32.39 A toroid with a square cross section.

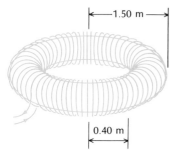

Fig. 32.40 A toroid.

*32. A long solenoid of radius R has n turns per unit length. A circular coil of wire of radius R' with 200 turns surrounds the solenoid (Figure 32.37). What is the mutual inductance? Does the shape of the coil of wire matter?

*33. Two long concentric solenoids of n_1 and n_2 turns per unit length have radii R_1 and R_2, respectively (Figure 32.38). What is the mutual inductance per unit length of the solenoids? Assume $R_1 < R_2$.

**34. A toroid with a square cross section (Figure 32.39) has an inner radius R_1, an outer radius R_2. The toroid has N turns of wire carrying a current I; assume that N is very large, so that the current can be regarded as uniformly distributed over the surface of the toroid.
  (a) Find the magnetic field as a function of radius.
  (b) Find the flux passing through one turn, and find the self-inductance of the toroid.

**35. Two inductors of self-inductances L_1 and L_2 are connected in series. The inductors are magnetically shielded from one another so that neither produces flux in the other. Show that the self-inductance of the combination is $L = L_1 + L_2$.

**36. Two inductors of self-inductance L_1 and L_2 are connected in parallel. The inductors are magnetically shielded from one another so that neither produces flux in the other. Show that the self-inductance of the combination is given by $1/L = 1/L_1 + 1/L_2$.

Section 32.6

37. A ring of thick wire has a self-inductance of 4.0×10^{-8} H. How much work must you do to establish a current of 25 A in this ring?

38. The strongest magnetic field achieved in a laboratory is 10^3 T. This field can be produced only for a short instant by compressing the magnetic field lines with an explosive device. What is the energy density in this field?

39. For each of the first six entries in Table 30.1, calculate the energy density in the magnetic field.

40. For a crude estimate of the energy in the Earth's magnetic field, pretend that this field has a strength of 0.5×10^{-4} T from the ground up to an altitude of 6×10^6 m above ground. What is the total magnetic energy in this region?

41. A current of 5.0 A flows through a cylindrical solenoid of 1500 turns. The solenoid is 40 cm long and has a diameter of 3.0 cm.
  (a) Find the magnetic field in the solenoid. Treat the solenoid as very long.
  (b) Find the energy density in the magnetic field, and find the magnetic energy stored in the space within the solenoid.

*42. A large toroid built for plasma research in the Soviet Union has a major radius of 1.50 m and a minor radius of 0.40 m (Figure 32.40). The average magnetic field within the toroid is 4.0 T. What is the magnetic energy?

*43. Consider a point charge q moving at velocity $\mathbf{v}$ through empty space. What is the energy density in the magnetic field at a distance r ahead of the point charge? At a distance r to one side of the point charge?

*44. According to one proposal, the surplus energy from a power plant could be temporarily stored in the magnetic field within a very large toroid. If the strength of the magnetic field is 10 T, what volume of the magnetic field would we need to store 1.0×10^5 kW·h of energy? If the toroid has roughly the proportions of a doughnut, roughly what size would it have to be?

**45. Two long, straight, concentric tubes made of sheet metal carry equal currents in opposite directions. The inner tube has a radius of 1.5 cm and the outer tube a radius of 3.0 cm. The current on the surface of each is 120 A. What is the magnetic energy in a 1.0-m segment of these tubes?

**46. A toroid of square cross section (Figure 32.39) has an inner radius R_1

and an outer radius R_2. The toroid has N turns of wire carrying a current I; assume that N is very large.
   (a) Find the magnetic energy density as a function of radius.
   (b) By integrating the energy density, find the total magnetic energy stored in the solenoid.
   (c) Deduce the self-inductance from the formula $U = \tfrac{1}{2}LI^2$.

*47. A transmission line consists of two concentric tubes of thin copper sheet metal with radii R_1 and R_2 (see Figure 32.41). The current flows to the left along one of the tubes and back along the other, completing a circuit; the current is uniformly distributed over the surface of each tube. Show that the self-inductance per unit length of this transmission line is

$$\frac{\mu_0}{2\pi} \ln \frac{R_2}{R_1}$$

(Hint: Calculate the magnetic flux through the 1-m long rectangle shown in Figure 32.41.)

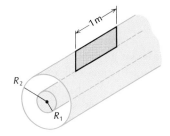

Fig. 32.41

Section 32.7

48. An inductor with $L = 2$ H and a resistor with $R = 100\ \Omega$ are suddenly connected in series to a battery with $\mathcal{E} = 6.0$ V.
   (a) What is the current at $t = 0$? At $t = 0.01$ s? At $t = 0.02$ s?
   (b) What is the final, steady value of the current?
   (c) What is the rate of increase of the current at $t = 0$?

49. Design an RL circuit with an arrangement of switches so that the battery can be suddenly switched out of the circuit and the current in the inductor can be suddenly fed into the resistor. If the self-inductance of the inductor is 0.2 H, what resistance do you need to obtain a time constant of 10 s in your circuit?

*50. Apply Kirchhoff's rule to an RL circuit from which the battery has been suddenly removed, and derive Eq. (46) for the decay of the current.

*51. An inductor is suddenly connected in series to a resistor with $R = 10\ \Omega$. The initial current in the inductor is 3.4 A. After a time of 6.0×10^{-2} s, the current has dropped to 1.7 A. What is the self-inductance?

*52. An RL circuit with $L = 0.50$ H and $R = 0.025\ \Omega$ is initially connected to a battery of 1.2 V. When the current reaches its maximum, steady value, the battery is suddenly switched out of the circuit and the current is switched into the resistor. What is the maximum value of the current? At what time will the current in the inductor drop to one-half of its maximum value? At what time will the energy in the inductor drop to one-half of its maximum value?

*53. An RL circuit consists of two inductors of self-inductances $L_1 = 4$ H and $L_2 = 2$ H connected in parallel to each other, and connected in series to a resistor of $6\ \Omega$ and a battery of 3 V (Figure 32.42). Assume that the inductors have no mutual inductance.
   (a) When the battery is suddenly connected, what is the initial rate of change of the current in each inductor?
   (b) What is the final, steady current in the resistor? What are the final, steady currents in each inductor?

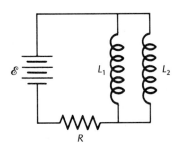

Fig. 32.42

*54. What is the Joule heat dissipated by the current (46) in the resistor in the time interval $t = 0$ to $t = \infty$? Compare with the initial magnetic energy in the inductor.

*55. To measure the self-inductance and the internal resistance of an inductor, a physicist first connects the inductor across a 3.0-V battery. Under these conditions, the final, steady current in the inductor is 24 A. The physicist then suddenly short-circuits the inductor with a thick (resistanceless) wire placed across its terminals. The current then decreases from 24 A to 12 A in 0.22 s. What are the self-inductance and the internal resistance of the inductor?

INTERLUDE VII

PLASMA*

Plasma is a very hot, ionized gas, a gas so hot that the violent thermal collisons dissociate all or many of its atoms into positive ions and electrons. If we reckon solid, liquid, and gas as the first three states of matter, then plasma is the fourth state of matter. We might call it pure fire, since an ordinary flame is part plasma and part hot gas.

Most of the matter in the universe is plasma. The Sun and all the stars are giant balls of plasma — about 99% of the total visible mass in the universe is found in these balls of plasma. Only in planets, in pulsars, and in some clouds of interstellar gas and dust do we find solids, liquids, and gases — these bodies make up only a small fraction of all the matter in the universe. In our immediate environment, naturally occurring plasma is quite rare, so much so that it was not recognized as a separate state of matter until late in the nineteenth century. Lightning bolts, St. Elmo's fire, the Aurora Borealis, and the ionosphere are plasmas — on the Earth, these are the only forms of naturally occurring plasmas. However, modern technology makes use of many forms of artificially produced plasmas. The gas in fluorescent tubes and in neon signs is plasma; the luminous arc of an electric welder is plasma; the exhaust fire of a rocket is plasma; and the fireball of a nuclear bomb is plasma.

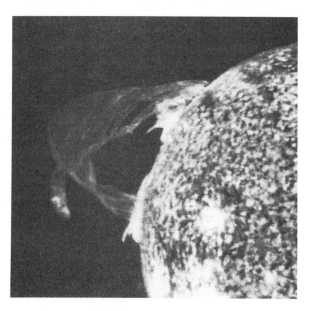

Fig. VII.1 Giant prominence on the Sun photographed December 19, 1973, by an ultraviolet camera on Skylab. This prominence spans more than 588,000 km across the solar surface.

VII.1 THE FOURTH STATE OF MATTER

A plasma contains a mixture of positive ions, electrons, and neutral atoms. The extent of ionization depends on the temperature: if the temperature is low, the plasma will contain a substantial number of neutral atoms; if the temperature is high, almost all the atoms will be ionized.

An ordinary gas also contains some ions and electrons, but not enough to make it into a plasma. If we heat a gas to higher and higher temperatures, we will gradually transform it into a plasma; however, the transition from gas to plasma is not sharply defined — there is no sudden change of phase, such as in the melting of a solid or the evaporation of a liquid. For example, the flame of a candle is on the borderline between hot gas and plasma; if it had more ions and electrons it would be a plasma, if it had less it would be an ordinary gas. The crucial distinction between an ordinary hot gas and a plasma lies in their electromagnetic properties. Plasma is an electric conductor, and macroscopic electric and magnetic fields dominate its behavior, whereas an ordinary gas is an insulator and it does not respond in any drastic manner to electric and magnetic fields. Figure VII.1 shows a prominence of plasma on the surface of the Sun. This prominence has a complicated shape; it arcs up and down because its behavior is dominated by the magnetic field of the Sun rather than by gravity. In contrast, the flame of a candle has a simple shape; it streams upward because of the buoyancy of the hot gas. If we place the flame between the poles of a magnet, we find that its behavior is not changed in any noticeable way.

The criterion for a plasma is that the abundance of positive and negative charges must be so large that any local imbalance between the concentrations of these charges is impossible. Any incipient accumulation of, say, positive charge rapidly attracts negative charges which restore the balance of charge. On a

---

* This chapter is optional.

small scale, the positive and negative charges in a plasma move randomly, but on a large scale they move collectively — the positive and negative charges are strongly linked by their electric interaction, and they move in unison so as to prevent any imbalance of charge. Thus, although the plasma contains a large number of free charges, it remains electrically neutral because, on the average, each unit volume contains equal amounts of positive and negative charge.

Matter in the plasma state has less order than matter in the solid, liquid, or gas state. However, the average electric neutrality of plasma represents some vestigial order. If we heat plasma to an extreme temperature, even this remaining order will be lost, and the matter will turn into a chaotic medley of individual particles. This will happen when the thermal motion of the particles is so energetic that the electric forces cannot restrain them. Note that whether or not neutrality can be preserved depends on the scale of distance. On a small scale, neutrality always fails, whereas on a very large scale it hardly ever fails, because it takes enormous amounts of energy to separate a large number of electric charges by a large distance. To gain some insight into what determines the critical scale of distance, let us examine the interaction between a plasma and a static electric field.

We know that, under static conditions, a good conductor will prevent an electric field from penetrating its interior — the conductor shields the electric field by accumulating some charge on its surface. This also happens in a plasma. If we insert a pair of electrodes into a plasma (Figure VII.2), negative charges (electrons) will be drawn toward the positive electrode, and positive charges (ions) toward the negative electrode. Provided the electrodes are kept from direct contact with the plasma by a layer of dielectric insulator, the electrons and ions will accumulate, forming more or less static charge distributions around the electrodes and shielding their electric fields.

Because of thermal agitation, the charges cannot remain exactly static at the electrodes; instead, they form a restless cloud in the vicinity of each electrode. Only inside the cloud does the electrode disturb the plasma; outside of the cloud, the plasma hardly feels the presence of the electrode. The size of the cloud depends on the temperature — the cloud is large if the temperature is high, because the random thermal back-and-forth motions of the charges will then tend to disperse them over larger distances.

The thickness of the cloud is called the shielding distance, or the **Debye distance.** It is given by the following formula:

$$D = \sqrt{\frac{\varepsilon_0 kT}{ne^2}} \qquad (1)$$

where n is the number of electrons (or ions) per unit volume, T the temperature, and k Boltzmann's constant (see Chapter 19). This distance determines how far an electric field penetrates within the plasma, and it also determines the magnitude of the deviations from neutrality within a plasma. On a scale large compared to the Debye distance, the plasma remains electrically neutral because any concentrations of charge are immediately hidden in a cloud of opposite charge. Consequently, an ionized gas will exhibit plasma behavior on a scale of distance that is large compared to the Debye distance. For example, the ionized gas in an ordinary neon tube has an electron density of about $10^9/\text{cm}^3$ and these electrons have a temperature of about 20,000 K. Equation (1) then gives

$$D = \sqrt{\frac{8.85 \times 10^{-12}\ \text{F/m} \times 1.38 \times 10^{-23}\ \text{J/K} \times 20{,}000\ \text{K}}{10^{15}/\text{m}^3 \times (1.6 \times 10^{-19}\ \text{C})^2}}$$

$$= 3 \times 10^{-4}\ \text{m} = 0.3\ \text{mm}$$

Hence a glob of this gas will behave as a plasma whenever it is larger than a few millimeters.

Table VII.1 gives some examples of plasmas. Note that although most of the plasmas in Table VII.1 are luminous, some are not. The reason is that some

Table VII.1 SOME PLASMAS

Plasma	Electron temperature	Electron density
Sun: center	2×10^7 K	$10^{26}/\text{cm}^3$
surface	5×10^3	10^6
corona	10^6	10^5
Fusion experiments (tokamak)	2×10^8	10^{14}
Fireball of atomic bomb	$\sim 10^7$	10^{20}
Solar wind	1×10^5	5
Lightning bolt	3×10^4	10^{18}
Glow discharge (neon tube)	2×10^4	10^9
Ionosphere	$\sim 2 \times 10^3$	10^5
Ordinary flame	2×10^3	10^8

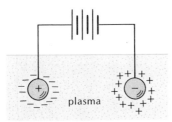

Fig. VII.2 Two spherical electrodes immersed in a plasma. A cloud of electrons accumulates around the positive electrode and a cloud of ions accumulates around the negative electrode.

Fig. VII.3 Corona of Sun visible during the eclipse of June 8, 1918.

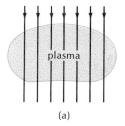

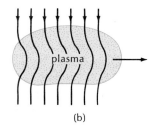

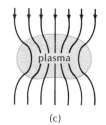

Fig. VII.4 (a) Magnetic field lines in a ball of plasma. (b) If we move the ball of plasma, it drags the magnetic field lines along. (c) If we compress the ball of plasma, it compresses the field lines.

plasmas are so tenuous that they do not give off an appreciable amount of light even though their temperature is very high. The ionosphere of the Earth and the corona of the Sun are examples of such very faint plasmas. Only during a total eclipse does the faint light from the solar corona become visible (Figure VII.3).

VII.2 PLASMAS AND THE MAGNETIC FIELD

Many plasmas — both in nature and in our laboratories — are immersed in magnetic fields. The magnetic fields may originate from external sources or else from currents flowing within the plasma itself. In contrast to static electric fields which cannot exist inside a plasma, static magnetic fields can exist. However, such magnetic fields are subject to a severe restriction: the magnetic field lines are *frozen* in the plasma. If the plasma is stationary, then the field lines within the plasma are also stationary; if the plasma flows, then the field lines flow with it, each segment of field line following the motion of the small volume element of fluid in which it is embedded. For example, Figure VII.4 shows the magnetic field lines passing through a ball of plasma initially placed between the poles of an electromagnet, and what happens to the field lines as we move the ball about or compress it.

This intimate attachment of the field lines to the plasma is an immediate consequence of its high conductivity. In an ideal plasma, the conductivity is infinite and therefore the electric field in the rest frame of any given volume element of plasma must vanish. But we know, from Faraday's Law, that changes in the magnetic field induce electric fields. Since the electric field is forbidden, changes in the magnetic field are also forbidden. This means that in the rest frame of any given element of plasma, the magnetic field lines must remain fixed — they must move with the element of plasma. This freezing of the magnetic field lines within the plasma may be regarded as an extreme instance of Lenz' Law — any incipient change in the magnetic field immediately induces a current whose magnetic field combines with the original magnetic field in such a way that the net magnetic field remains constant. Note that the plasma not only keeps any interior field lines locked inside (as in Figure VII.4), but it also keeps any exterior field lines outside. For example, if a ball of plasma with no magnetic field lines within it approaches a region with field lines, it will deform the latter and push them aside (Figure VII.5).

This effect plays an important role in shaping the magnetic field of the Earth. If left to itself, the magnetic field of the Earth would be essentially a dipole field (Figure 30.19). But this dipole field is affected by the solar wind, a stream of plasma blowing radially outward from the Sun. The solar wind consists of a neutral mixture of electrons and protons moving outward from the Sun at a speed of about 400 km/s. The push of this wind against the magnetic field of the Earth causes a deformation — it compresses the field on the side facing the Sun and elongates the field on the side

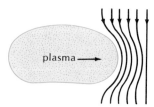

Fig. VII.5 If we move a ball of plasma into a magnetic field, it pushes the field lines aside.

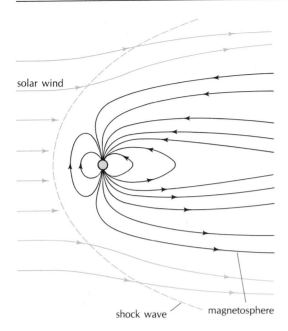

Fig. VII.6 Magnetic field of Earth as distorted by the push of the solar wind. The speed of the wind is 400 m/s. At the leading edge of the magnetic field a shock wave forms.

strength of the magnetic field is B, then this extra magnetic pressure within the plasma is $B^2/2\mu_0$. The net pressure within any region of the plasma is then the sum of the ordinary kinetic pressure of the plasma particles plus the extra magnetic pressure. Note that numerically the magnetic pressure equals the energy density of the magnetic field [see Eq. (32.42)]. This is no accident — the magnetic pressure arises precisely from the changes of magnetic energy, in the following way: If we compress some volume of plasma, we will also compress the magnetic field lines frozen in this volume; this increase of density of the field lines involves an increase of the energy stored in the magnetic field. By the argument given in the preceding paragraph, the magnetic field will then exert a force that opposes the compression, that is, a pressure.

Carefully designed arrangements of magnetic fields are used for the confinement of extremely hot plasmas in experiments on thermonuclear fusion. Such arrangements of magnetic fields are called **magnetic bottles**; they can hold a plasma that is too hot to be held by an ordinary vessel of metal or glass. For instance, Figure VII.7 shows a magnetic bottle that confines a

away from the Sun (Figure VII.6). The region of space occupied by the magnetic field of the Earth is called the **magnetosphere**. Because of the action of the solar wind, the magnetosphere acquires the shape of an elongated raindrop with a tail extending to a distance of at least several hundred thousand kilometers.

Incidentally: The speed of the solar wind past the Earth is supersonic, that is, the speed is larger than the speed of sound waves (compressional waves) in the plasma. Consequently, the supersonic motion of the Earth relative to the solar wind generates a shock wave in the plasma, just like the motion of a supersonic aircraft generates a shock wave in air (Figure VII.6).

A deformation of the magnetic field lines, such as the deformations shown in Figures VII.4 and VII.5, involves an increase of the density of the magnetic field lines in some regions of space; thus it involves an increase of the energy stored in the magnetic field. In order to acquire this energy from the plasma, the magnetic field must exert a force on the plasma, a force that opposes the deformation. If the magnetic field is very strong, then the force with which it opposes the motion of the plasma may be so large that it prevents this motion altogether — the magnetic field lines then confine the plasma and hold it as though it were in a cage.

The force that the magnetic field exerts on the plasma can be described (in part) as a **magnetic pressure** acting within the plasma. It turns out that if the

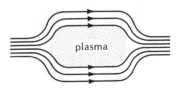

Fig. VII.7 A magnetic bottle.

plasma by means of the magnetic field produced by a solenoid. In the central region, the coils of the solenoid are uniformly spaced, but at each end an extralarge coil intensifies the magnetic field and brings the field lines together. We can understand the confinement of the plasma by this bottle in terms of the magnetic pressure $B^2/2\mu_0$. This pressure is large in the strong magnetic field surrounding the plasma; hence this pressure tends to push the plasma inward, balancing the kinetic pressure that tends to push the plasma outward.

Alternatively, we can understand the confinement of the plasma in terms of the microscopic motions of the charged particles in the plasma. As we know from Section 31.3, a charged particle in a magnetic field moves in a helix around the magnetic field lines (Figure VII.8). Thus, on the average, the particle gradually drifts along the direction of the field lines but does not wander off in a transverse direction. Near the ends of the bottle, where the field lines converge, the particle will be reflected. Figure VII.9 shows how the converging field lines give rise to a component of force that

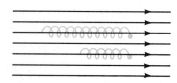

Fig. VII.8 Charged particles with helical motion around the magnetic field lines.

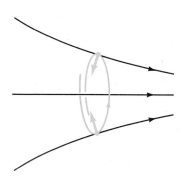

Fig. VII.9 Charged particle and converging magnetic field lines. The magnetic force has a component along the axis of the helix.

halts the drift of the particle along the direction of the field lines and pushes it back (note that the circular motion of the particle does *not* stop or change direction; only the drift motion does). A magnetic field with convergent field lines is called a **magnetic mirror**.

A different method of confinement uses the magnetic field produced by a current flowing through the plasma, that is, it uses the plasma's own magnetic field. Figure VII.10 shows a cylindrical column of plasma, with a longitudinal current. The magnetic field produced by this current encircles the plasma and exerts a pressure on it. The magnetic field has its maximum value at the surface of the plasma; the magnetic pressure is therefore large at the surface and it pushes the plasma inward, balancing the kinetic pressure that pushes the plasma outward. The current used to hold together the very hot column of plasma in a thermonuclear fusion experiment may be as much as a million amperes.

We can also understand the confinement of such a column of plasma in terms of the attractive magnetic force between parallel current elements; obviously,

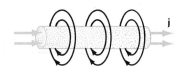

Fig. VII.10 A cylindrical plasma column with an electric current passing through it.

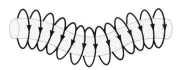

Fig. VII.11 Kink instability of a plasma column.

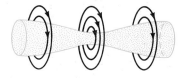

Fig. VII.12 Sausage instability of a plasma column.

this attractive force tends to compress the plasma and hold it together. The compression of a plasma by its own magnetic forces is called the **pinch effect**.

Unfortunately, the confinement provided by magnetic bottles is not perfect. The plasma tends to leak out at the ends and, what is worse, the plasma interacts with the magnetic field, giving rise to a variety of instabilities that destroy the delicate balance between the magnetic pressure and the kinetic pressure. For instance, Figures VII.11 and VII.12 show a kink instability and a sausage instability in a pinched plasma column. The cause of these instabilities can be qualitatively understood by examination of the field lines. In Figure VII.11 the field lines are bunched together above the kink and spread apart below — hence the magnetic pressure pushes the plasma downward, further increasing the kinking of the column. In Figure VII.12 the field lines are concentrated at the neck of the sausage; this squeezes the neck, further decreasing its diameter.

Both of these kinds of instability can be eliminated, or reduced, by means of an axial magnetic field in the plasma column (Figures VII.13 and VII.14). Such a field

Fig. VII.13 A plasma column with an extra axial magnetic field resists the kink instability.

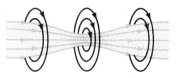

Fig. VII.14 A plasma column with an extra axial magnetic field also resists the sausage instability.

can be generated by placing the plasma in the core of a solenoid or by inducing an additional flow of current around the plasma column, effectively making it act as a solenoid. This magnetic field opposes the kink, because deforming the axial magnetic field lines requires energy. Likewise, the magnetic field opposes the sausage constriction, because compressing the magnetic field lines again requires energy.

Dozens of other, more complicated, instabilities have been identified and studied by plasma physicists. The development of countermeasures to all these instabilities remains one of the central problems in plasma physics.

VII.3 WAVES IN A PLASMA

The dynamical behavior of plasma is much more intricate than that of a gas. In the latter, the pressure completely determines the macroscopic motion, whereas in the former, electric and magnetic forces play a large role. For instance, consider the propagation of a sound wave in a plasma. This is a longitudinal wave with alternating zones of compression and rarefaction (Figure VII.15). In an ordinary gas, the restoring force in the sound wave is simply the excess pressure of the compressed zones. In a plasma, there is an additional restoring force: the concentrated positive electric charge of the ions of the plasma in the compressed zones gives rise to a repulsive electric force. Of course, the electrons of the plasma attempt to shield the ions and cancel their electric repulsion, but because the electrons have large random thermal motions (larger than those of the ions), the shielding is not quite perfect, and a residual electric repulsive force remains.

Even more complicated dynamical effects can occur in a plasma immersed in a magnetic field. Suppose that the plasma is in an originally uniform magnetic field. A sound wave propagating in a direction perpendicular to this magnetic field will have alternating zones of compression and rarefaction and, since the magnetic field lines are frozen in the plasma, they must also have corresponding zones of compression

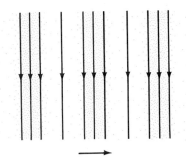

Fig. VII.16 A sound wave in a plasma in the presence of a magnetic field. The compression of the plasma produces a compression of the magnetic field.

and rarefaction (Figure VII.16). The compression of the magnetic field lines generates an increase of the magnetic pressure. Thus, the restoring force that governs the propagation of this wave arises from a combination of kinetic pressure and magnetic pressure. A wave of this kind is called a **magnetosonic wave.**

The magnetosonic wave is a longitudinal wave — it is merely a sound wave modified by magnetic effects. Surprisingly, it turns out that a plasma immersed in a magnetic field can also support a *transverse* wave. Consider a plasma in a uniform magnetic field. If one layer of this plasma is given a transverse displacement, the magnetic field lines will suffer a transverse deformation. As is obvious from Figure VII.17, such a defor-

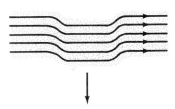

Fig. VII.17 The displacement of a layer of plasma produces a deformation of the magnetic field.

mation leads to an increase of the density of field lines in the layer, and hence to an increase of magnetic energy. Consequently, the embedded magnetic field exerts a restoring force on the plasma, opposing the deformation. Under the influence of the restoring force, the deformation propagates along the magnetic field lines. This is a transverse wave, similar to a wave on a string (see Chapter 16); it propagates along the magnetic field lines just as though these lines were strings under tension. A wave of this kind is called an **Alfvén wave.**[1]

Fig. VII.15 Sound wave in a gas or a plasma.

[1] **Hannes Alfvén,** 1908–, Swedish physicist. He received the Nobel Prize in 1970 for his work in the theory of plasmas.

The propagation of radio waves in a plasma also has many intricate features. Since a plasma is a good conductor and shields electric fields, we expect it will shield the electric fields of a radio wave, that is, it will reflect radio waves just as a metal does. This is true for radio waves of low frequency, but for radio waves of high frequency the shielding fails because the electrons do not have enough time to respond to the electric fields. Thus, high-frequency radio waves can penetrate into a plasma and propagate through it. The minimum frequency that can pass through a plasma is called the **cutoff frequency;** the value of this frequency depends on the electron density — it increases with the square root of the electron density. For instance, the ionosphere of the Earth will pass radio waves of frequency above about 30 MHz (radar and TV) but it will reflect waves of lower frequency (short waves, medium waves, and long waves).

The reflection of radio waves by the ionosphere is of crucial importance for radio communications. We can send radio waves from one end of the globe to the other by successively bouncing them back and forth between the ionosphere and the ground (Figure VII.18). The ionosphere actually consists of several re-

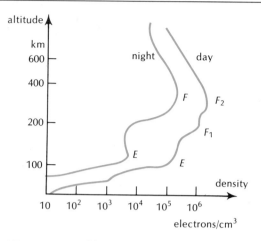

Fig. VII.19 Density of free electrons in the atmosphere as a function of altitude (at a time of sunspot maximum). The peaks in the electron density at altitudes of ~120 km and ~300 km constitute the E and the F layers, respectively.

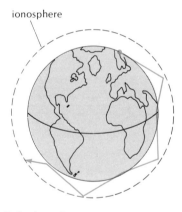

Fig. VII.18 Path of a radio wave around the Earth.

flecting layers with characteristics that vary with day and night and also with the sunspot cycle (Figure VII.19). Radio waves of different wavelength are reflected by different layers; for this reason, the choice of radio frequency band is critical in long-range radio communication. Incidentally: The radio blackout suffered by a space capsule during reentry of the atmosphere is also due to a plasma effect. The friction between the space capsule and the air gradually burns off the heat shield of the capsule, and the flames surround the capsule with a layer of plasma; this layer stops radio waves and prevents communication.

If the plasma is immersed in a magnetic field, then the behavior of radio waves depends in a complicated way on the frequency of the wave, the magnitude of the magnetic field, and the direction of propagation of the wave relative to the direction of the magnetic field. High-frequency radio waves can propagate in any direction (although their speed depends on direction), but low-frequency radio waves can propagate only if their direction lies within a limited range of angles near the direction of the magnetic field. Hence magnetic field lines can act as guides for radio waves. This causes the weird phenomenon of whistlers that can be picked up with an audio amplifier at a few kilohertz. These are radio signals of low frequency emitted by flashes of lightning. The signals propagate in the tenuous plasma surrounding the Earth; they travel from one hemisphere to the other guided by the magnetic field lines of the Earth (Figure VII.20). The high-frequency components of the signal travel faster than the low-frequency components. At the receiver, the result is a whistle consisting of a descending glide tone

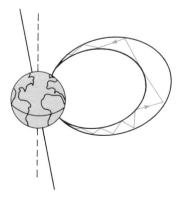

Fig. VII.20 The magnetic field of the Earth acts as a waveguide for low-frequency radio waves.

(glissando) that begins at a high pitch and ends at low pitch.

VII.4 THERMONUCLEAR FUSION

The Sun generates its heat by a nuclear fusion reaction burning hydrogen into helium. But for us on Earth, this reaction is not a viable energy source because it is a very slow reaction. A fusion reaction burning deuterium or tritium into helium is much more practical. In fact, the H-bomb derives its energy not from the fusion of hydrogen, but from the fusion of deuterium and tritium (see Section XI.3). Essentially, a fusion reactor brings the power of an H-bomb under control just like a fission reactor brings the power of an A-bomb under control.

Fusion is a very promising energy source because fuel for fusion is available in great abundance. Deuterium occurs naturally in molecules of heavy water (HDO) — about 0.03% of the water in the oceans is heavy water. Tritium is radioactive and does not occur naturally.[2] However, it can be easily produced by bombarding lithium with neutrons in a nuclear reactor. The available supply of deuterium is sufficient to satisfy all our energy requirements for a billion years; the supply of lithium is sufficient for a million years.

The most suitable reactions for the generation of fusion power are the **deuterium–deuterium reaction** and the **deuterium–tritium reaction**. The first of these involves the following sequence of four steps:

$$\begin{aligned} D + D &\to {}^3He + n \\ D + D &\to T + p \\ D + T &\to {}^4He + n \\ D + {}^3He &\to {}^4He + p \\ \hline 6D &\to 2\,{}^4He + 2p + 2n + 43.1 \text{ MeV} \end{aligned} \quad (1)$$

The net result is the fusion of six deuterium nuclei into two helium nuclei with the release of two protons, two neutrons, and 43.1 MeV of energy. The energy released for each fused deuterium nucleus is smaller than the energy released for each fissioned uranium nucleus — about 7 MeV per fused deuterium nucleus as compared to about 200 MeV per fissioned uranium nucleus. But weight for weight, the energy released in the fusion of deuterium is four times as large as in the fission of uranium. One further advantage of fusion over fission is that the reaction products from the former are harmless, whereas the reaction products from the latter are radioactive. In the reaction (1) the only potentially hazardous products are the neutrons; but these can be captured in a suitable absorber placed around the reactor vessel.

[2] Deuterium (D, or ^{2}H) and tritium (T, or ^{3}H) are two isotopes of hydrogen.

The deuterium–tritium reaction requires the presence of lithium and proceeds in two steps:

$$\begin{aligned} D + T &\to {}^4He + n \\ n + {}^6Li &\to {}^4He + T \\ \hline D + {}^6Li &\to 2\,{}^4He + 22.4 \text{ MeV} \end{aligned} \quad (2)$$

Thus, the net result is the fusion of deuterium and lithium nuclei into helium nuclei. The tritium only exists in an intermediate step in this reaction, being produced from the lithium. The technology required for the D–T reaction is much more complicated than that for the D–D reaction. The reactor chamber containing a mixture of deuterium and tritium would have to be surrounded by a blanket of lithium where neutrons can be absorbed and tritium can be generated; the tritium must afterward be extracted from the blanket and fed into the chamber. In spite of these complications, the D–T reaction is regarded as more promising than the D–D reaction because its ignition temperature is somewhat lower.

Both the D–D and D–T reactions will proceed only at extremely high temperature. The nuclei will fuse only if they are brought into contact by violent collisions — since the nuclei are positively charged, they experience an electric repulsion, and to overcome this, the collision must be initiated with a large kinetic energy. The only feasible way to give the nuclei the necessary kinetic energy is by thermal motion. To attain a sufficiently large kinetic energy, a very high temperature is required — about 5×10^9 K for D–D fusion and about 1×10^8 K for D–T fusion. These temperatures are higher than at the center of the Sun. Nuclear reactions at such extreme temperatures are called **thermonuclear.** At these temperatures the deuterium or tritium atoms will be completely ionized, and the nuclei and electrons will form a plasma.

To ignite the nuclear fire it is sufficient to mix the reactants and heat them to high temperature. But if the nuclear fire is to yield a profitable amount of energy, it is also necessary to keep the reactants together for some minimum length of time. Obviously, to break even, we must gain an amount of energy equal to the amount that we invested to heat the plasma. A calculation shows that for the D–T reaction this requires a minimum time (in seconds)

$$\tau = \frac{2 \times 10^{14}}{n} \quad (3)$$

and for the D–D reaction a minimum time

$$\tau = \frac{5 \times 10^{15}}{n} \quad (4)$$

where n is the number of ions per cubic centimeter. For instance, according to Eq. (3), a D–T plasma of density $10^{14}/cm^3$ must be confined for at least 2 s if it is to yield a profitable amount of energy. And herein lies the great problem of plasma physics — the plasma tends to escape and disperse. The challenge is to invent some means of confining the plasma long enough to extract significant amounts of energy.

VII.5 PLASMA CONFINEMENT

Since the plasma in a thermonuclear fusion reactor must be kept very hot, we must prevent thermal contact with the walls of the reactor vessel. This can be achieved by suspending the plasma in the middle of the reactor vessel, in a magnetic field. But a magnetic field cannot confine the plasma completely. The minimum rate of leakage of the plasma is determined by diffusion caused by the random thermal motion. In practice, the leakage is much worse: the plasma tends to develop a variety of instabilities which make it squirt out of its magnetic confinement at a much faster rate than the minimum rate of leakage.

Hoping to conquer these instabilities, plasma physicists have constructed a variety of machines with fanciful names: Stellarator, Scylla, Scyllac, Tokamak, Alcator, etc. These machines are supposed to hold the plasma by means of intricate magnetic fields and they are supposed to heat the plasma to the point where fusion begins. Several of these machines have succeeded in triggering some fusion reactions, but none has generated any useful amount of power. Of these devices, the tokamak, originally developed in the Soviet Union and later copied in the United States, has so far proven the most successful.

In the **tokamak,** the plasma is confined to a toroidal region, that is, a column bent into a ring. This avoids the loss of plasma from the ends of the column, since there are no ends. The magnetic fields confining the plasma are in part generated by external solenoidal windings and in part by a current induced in the plasma itself. Thus, both of the simple confinement mechanisms mentioned in Section VII.2 come into play. Figure VII.21 shows the arrangement of magnet coils in a typical tokamak. The current along the plasma column is generated by induction, by means of a changing magnetic field provided by central induction coils. The current induced by this magnetic field is usually in excess of a million amperes.

The Princeton Tokamak Fusion Test Reactor (TFTR; see Figure VII.22) has achieved a temperature of 2×10^8 K, ten times hotter than the temperature in the center of the Sun. The plasma in the torus had a density of about $6 \times 10^{13}/cm^3$ and remained confined for about ⅙ of a second. Thus, the plasma was near its

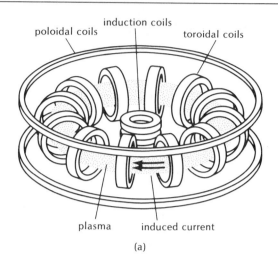

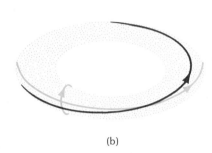

Fig. VII.21 (a) Magnet coils in a tokamak. (b) The magnetic field of the toroidal coils lies along the plasma; the field lines are large horizontal circles (color). The magnetic field of the induced current flowing along the plasma column encircles the column; the field lines are small vertical circles (color). The net magnetic field consists of helical lines (black) wrapped around the plasma column. (The magnetic field of the poloidal coil has been neglected.)

Fig. VII.22 The torus of the Princeton Tokamak Fusion Test Reactor (TFTR) during construction.

Fig. VII.23 Neutral-beam injectors (left) attached to the Princeton TFTR (right). The torus is almost completely hidden within the support frame.

Fig. VII.25 Ying-yang magnet with superconducting coils for the tandem mirror machine.

thermonuclear ignition temperature, but its density and confinement time did not reach the break-even point set by Eq. (3). In tokamaks, the Joule heat of the current induced along the toroidal plasma provides some of the heating needed to start the thermonuclear reaction. In the Princeton tokamak, extra heating is provided by injectors that shoot intense beams of high-speed neutral atoms into the torus (Figure VII.23).

Another promising machine is the **tandem mirror** under development at the Lawrence Livermore National Laboratory. This machine is a straight solenoid closed off at each end by a magnetic mirror consisting of two magnets arranged in tandem (Figure VII.24). One of these magnets has an ordinary circular coil, but the other magnet has several interlocking "ying-yang" coils, shaped somewhat like the seams on a baseball. In the magnetic field of such a tandem magnet, the electrons and the positive ions of the plasma tend to separate to some extent. This charge separation gener-

ates an electric field which aids in the confinement of the plasma. Hence the action of the tandem mirror is part magnetic and part electric. Figure VII.25 shows the magnet coils for Livermore under construction.

Fusion reactors based on the tokamak design or the mirror design aim to reach the break-even point [see Eqs. (3) and (4)] by confining a low-density plasma for a fairly long time. As the fuel burns, fresh fuel would be injected into the reactor chamber, either in the form of puffs of gas or in the form of frozen pellets of a D–T mixture. In contrast, fusion reactors based on the Scylla design aim to reach the break-even point with a much higher-density plasma and a much shorter confinement time. Such a reactor would burn its fuel in a burst lasting perhaps 10^{-3} s and then it would receive a fresh load of fuel for the next spurt of power. The pulsed release of heat in such a reactor is analogous to the pulsed release of heat in one of the cylinders of an internal combustion engine, which also burns its fuel in bursts and delivers spurts of power.

Instead of attempting to confine the plasma with magnetic fields, some experimenters are now exploring the possibility of a fusion reactor using the uncontrolled, explosive burning of small solid pellets of fuel. In such an **inertial confinement** reactor, one pellet at a time is dropped into a combustion chamber and suddenly vaporized into a very hot and very dense plasma by an intense pulse of light from a powerful laser or else by an intense pulse of beamed electrons or ions from an electron or ion gun. This sudden ignition will burn the deuterium and tritium before the plasma has a chance to expand and disperse. This is an extreme case of pulsed burning, lasting only about 10^{-9} s. The pellet explodes, like a miniature H-bomb. During the explosion, the plasma is confined to some extent by its own inertia — the outer layers of plasma do not directly participate in the fusion, but their inertia holds

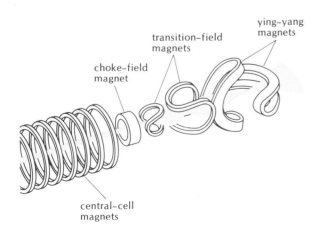

Fig. VII.24 Magnet coils of the tandem mirror machine.

the core of the pellet together for a long enough time for the reactions to proceed. The success of such a fusion scheme depends on the development of very powerful and very efficient lasers (see Interlude X). Part of the output of the reactor will have to be used to energize the lasers.

Like the heat from an ordinary fission reactor, the heat from a fusion reactor can be removed from the reactor core and used to make steam to drive turbines generating mechanical and electric power. There is, however, an ingenious alternative method for the direct generation of electric power. The plasma in the combustion chamber of the fusion reactor is a mixture of electrons and positive ions. If this plasma is allowed to stream out of the chamber into a magnetic field, the negative and positive charges will separate. These opposite charges can be collected on two sets of plates; this results in a direct conversion of the plasma energy into electric energy. In principle, such a plasma generator is somewhat similar to a magnetohydrodynamic generator.

The fusion reactors of the future will probably be very large machines delivering more than 1000 MW of power (Figure VII.26). But besides the confinement problem, there are several other problems that remain to be solved before we can exploit fusion power commercially. For instance, the design of the walls of the combustion chamber and the design of the magnets will require advances in technology. The walls are exposed to an abundant flux of neutrons emitted by the fusion reaction and also to the impact of charged particles escaping from the plasma. Metals exposed to such a flux of neutrons suffer from swelling and embrittlement. New, resistant alloys will have to be developed for the walls. Furthermore, the magnets surrounding the reactor vessel will probably have to be constructed out of a superconductor; otherwise their power consumption would be prohibitive. These magnets will have to be much larger than any magnets built to date, and they will have to stand up to enormous magnetic forces between the currents in their windings.

A fusion reactor is much cleaner than a fission reactor in that it does not produce an abundance of radioactive residues. However, the flux of neutrons released by the fusion reaction induces radioactivity in the walls of the reactor chamber, in the magnets, and wherever else it is absorbed — the impact of neutrons transmutes ordinary stable nuclei into radioactive nuclei. The possible leakage of tritium from the reactor is an additional environmental hazard. Tritium is radioactive and, since it is chemically identical to hydrogen, it could easily contaminate our water supplies. If the confinement problem can be solved, these secondary problems will undoubtedly also be solved. We can then look forward to an abundant supply of relatively clean energy.

Further Reading

The Fourth State of Matter by B. Bova (New American Library, New York, 1974) is an elementary introduction to the properties and applications of plasmas. *A Physicist's ABC on Plasma* by L. A. Artsimovich (MIR Publishers, Moscow, 1978) is a very concise, slightly mathematical introduction to plasma physics, with emphasis on the magnetic confinement problem. The article "World Energy Reserves & Some Speculations on the Future of Nuclear Fusion Energy" by J. L. Tuck in *Cosmology, Fusion & Other Matters*, edited by F. Reines (Colorado Associated University Press, Boulder, 1972), gives a neat survey of available energy reserves and explains the advantages and principles of fusion reactors.

The following magazine articles deal with applications of plasma physics:

"VLF Emissions from the Magnetosphere," R. N. Sudan and J. Denavit, *Physics Today*, December 1973

"Fusion Reactors," B. B. Kadomtsev and T. K. Fowler, *Physics Today*, November 1975

"The Earth's Magnetosphere," S. Akasofu and L. J. Lanzerotti, *Physics Today*, December 1975

"Fusion Power with Particle Beams," G. Yonas, *Scientific American*, November 1978

"Recent Progress in Tokamak Experiments," M. Murakami and H. P. Eubank, *Physics Today*, May 1979

"Alternate Concepts in Magnetic Fusion," F. F. Chen, *Physics Today*, May 1979

"Progress toward a Tokamak Fusion Reactor," H. P. Furth, *Scientific American*, August 1979

"The Active Solar Corona," R. Wolfson, *Scientific American*, February 1983

"The Engineering of Magnetic Fusion Reactors," R. W. Conn, *Scientific American*, October 1983

"Reaching Ignition in the Tokamak," H. P. Furth, *Physics Today*, March 1985

"Comparative Magnetospheres," L. J. Lanzerotti and S. M. Krimigis, *Physics Today*, November 1985

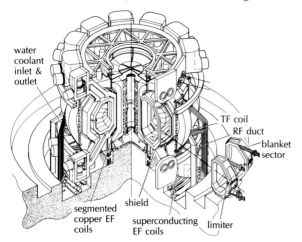

Fig. VII.26 The design for the 1200-MW STARFIRE fusion reactor proposed by the Argonne National Laboratory.

"The Structure of Comet Tails," J. C. Brandt and M. B. Niedner, *Scientific American*, January 1986

"The Earth's Magnetotail," E. W. Hones, *Scientific American*, March 1986

"The Plasma Universe," H. Alfvén, *Physics Today*, September 1986

Questions

1. Is there any plasma in the room you are in?

2. The ancient Greeks thought that there were four elements: earth, water, air, and fire. How does this list compare with the list of the four states of matter?

3. Can matter exist as a mixture of two or more of the four states? Can matter consisting of a single chemical element exist as a mixture of two or more states? Give examples.

4. The effect of magnetic fields on flames, sometimes called the diamagnetism of flames, was first discovered by P. Bancalari and thoroughly investigated by Faraday, who described the following experiment with an electromagnet producing a strong horizontal magnetic field:

> ... a ball of cotton ... soaked in ether ... will give a flame six or seven inches high. This large flame rises freely and naturally between the poles; but as soon as the magnet is rendered active, it divides and passes off in two flames, the one on one side, and the other on the other side of the axial line.

Explain this repulsive effect of the magnetic field on the flame.

5. According to the table of masses of the Sun and the planets printed on the endpapers of this book, roughly what fraction of the Solar System is plasma?

6. The chemical composition of the Sun is usually said to be 70% hydrogen and 30% helium. Are there actually any hydrogen or helium atoms on the Sun? What would be a more accurate description of the composition of the Sun?

7. How can the solar corona be hotter than the solar surface?

8. A neon tube contains plasma at very high temperature. Why does the tube not feel hot to the touch?

9. According to Table VII.1, the temperature of the ionosphere is 2×10^3 K. Would a piece of paper burn if exposed to the ionosphere?

10. The free-electron gas in a metal has some of the features of a plasma. Does this mean we should regard a metal as plasma?

11. If you compress a gas (at fixed temperature), it becomes a liquid. If you compress a plasma, does it become a gas?

12. In the science fiction story *The Black Cloud* by the astronomer Fred Hoyle, a large cloud of plasma behaves as an intelligent being. How could the brain of such a creature store information in currents and magnetic fields? How could such a creature think?

13. Would you expect a kink or a sausage instability to develop in a plasma confined in a solenoidal magnetic field as shown in Figure VII.7?

14. Explain why the magnetic mirrors shown in Figure VII.7 permit some leakage along the center line.

15. Figure VII.19 shows that the electron density in the ionosphere during the day is much larger than during the night (by up to a factor of 100). Can you guess why?

16. Consider a radio wave propagating around the Earth by bouncing back and forth between the ionosphere and the ground. Would you expect that a receiver at some given distance from the transmitter can be reached by two waves with different numbers of bounces?

17. What method could be used to separate the heavy water needed as fuel for a fusion reactor from ordinary water?

18. The Sun is a gigantic fusion reactor. What confines the plasma in this reactor? Why can we not use this method of confinement on Earth?

19. What would happen if the plasma in a fusion reactor were to come in contact with the wall of the reactor chamber?

20. Why is a runaway reaction possible in a fission reactor, but not in a fusion reactor?

21. Why does the reaction (1) require a higher temperature than the reaction (2)? (Hint: Which step in these reactions involves more electric repulsion?)

CHAPTER 33

Magnetic Materials

Within the atom and the nucleus, charged particles are continually moving about — electrons orbit around the nucleus and protons orbit around each other inside the nucleus. The orbital motions may be regarded as flowing electric currents within the atom, and these currents generate magnetic fields. Besides their orbital motions, the charged particles within atoms have rotational motions — electrons, protons, and neutrons all spin about their axes. The rotational motions may be regarded as flowing electric currents inside the particles, and these currents also generate magnetic fields.

The magnetic fields arising from currents flowing in loops inside atoms, nuclei, and particles can be described in terms of the corresponding magnetic dipole moments. If many of these small dipole moments within a sample of material are aligned, they will produce a strong magnetic field. Such an alignment can be achieved by placing the sample of material in the magnetic field produced by some external currents of, say, an electromagnet. The magnetic field produced by the internal atomic and subatomic currents will then be strong, and it will modify the original magnetic field produced by the external currents of the electromagnet. For instance, if a piece of iron is placed between the poles of an electromagnet, the magnetic field is strengthened drastically — it may become several thousand times stronger than the original magnetic field. Exactly how much the original magnetic field is increased or decreased depends on the response of the atomic and subatomic dipoles to the original magnetic field. According to the nature of their magnetic response, we can classify materials as paramagnetic, ferromagnetic, or diamagnetic. Before we discuss these types of magnetic materials, we will take a brief look at the magnitudes of the magnetic moments contributed by electrons, protons, and neutrons.

33.1 Atomic and Nuclear Magnetic Moments

An electron moving in an orbit around a nucleus produces an average current along its orbit. Strictly, the calculation of atomic orbits and currents requires quantum mechanics; but for the sake of simplicity, let us do this calculation with classical mechanics. If the electron has a circular orbit with radius r and speed v (Figure 33.1), then the time for one complete circular motion is $2\pi r/v$. The charge moved in this time is e, and hence the average current along the orbit is

$$I = \frac{e}{2\pi r/v} = \frac{ev}{2\pi r} \tag{1}$$

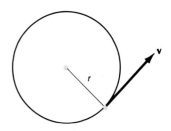

Fig. 33.1 Electron in a circular orbit around a nucleus.

Such a circulating current will give rise to a magnetic moment [see Eq. (30.44)]

$$\mu = I \times [\text{area}] = \frac{ev}{2\pi r} \times \pi r^2 \tag{2}$$

$$= \frac{evr}{2} \tag{3}$$

In terms of the angular momentum $L = m_e vr$, the magnetic moment can be expressed as[1]

$$\boxed{\mu = \frac{e}{2m_e} L} \tag{4}$$

Orbital magnetic moment

This says that the magnetic moment of an orbiting electric charge is proportional to its angular momentum.

It turns out that the relation (4) is valid not only for circular orbits but also for any other periodic orbit. Even more important: the relation (4) remains valid when we repeat the calculation using quantum mechanics. The net magnetic moment of the atom is the sum of the magnetic moments of all its electrons. Hence Eq. (4) can also be regarded as a relation between the net orbital angular momentum and the net magnetic moment of the entire atom.

It is a fundamental tenet of quantum mechanics that the magnitude of the orbital angular momentum is always some integer multiple of the constant value $\hbar = 1.06 \times 10^{-34}$ J·s.[2] Thus, the possible values of the orbital angular momentum are[3]

$$L = 0, \hbar, 2\hbar, 3\hbar, \ldots \tag{5}$$

Quantization of angular momentum

Because angular momentum exists only in discrete packets, it is said to

---

[1] In this equation, the magnetic moment μ must not be confused with the permeability μ_0.

[2] This is Planck's constant divided by 2π, that is, $\hbar = h/2\pi = 6.63 \times 10^{-34}$ J·s/2π. See Appendix 8 for a more precise value of the constant $\hbar$.

[3] Strictly, these numbers are the possible values for the maximum component of the angular momentum in a chosen direction, say, the z direction.

be **quantized.** The constant $\hbar$ is the fundamental quantum of angular momentum, just as e is the quantum of electric charge. For instance, the oxygen atom has a net orbital angular momentum $L = 1\hbar$, and hence, according to Eq. (4), the orbital magnetic moment is

$$\mu = \frac{e}{2m_e}\hbar = 9.27 \times 10^{-24} \text{ A} \cdot \text{m}^2 \tag{6}$$

Besides the magnetic moment generated by the orbital motion of the electrons, we must also take into account that generated by the rotational motion of the electrons. Crudely, an electron may be thought of as a small ball of negative charge rotating about an axis at a fixed rate. The spin, or intrinsic angular momentum, of the electron has a value of $\hbar/2 = 0.53 \times 10^{-34}$ J·s. This kind of rotational motion again involves circulating charge and gives the electron a magnetic moment. This intrinsic magnetic moment has a fixed magnitude

Spin magnetic moment

$$\boxed{\mu_{\text{spin}} = \frac{e}{2m_e}\hbar = 9.27 \times 10^{-24} \text{ A} \cdot \text{m}^2} \tag{7}$$

Bohr magneton

which is called a **Bohr magneton.**[4] The direction of this magnetic moment is opposite to the direction of the spin angular momentum (Figure 33.2). Note that Eq. (4) is not valid for the spin magnetic moment. If we were to insert a spin angular momentum of $\hbar/2$ into Eq. (4), we would obtain a magnetic moment $e\hbar/(4m_e)$, one-half the value given by Eq. (7).

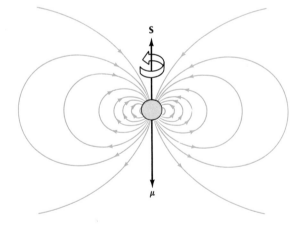

Fig. 33.2 Magnetic field lines of an electron. The magnetic moment μ is opposite to the spin angular momentum **S**.

The net magnetic moment of the atom is obtained by combining both the orbital and spin moments of all the electrons, taking into account the directions of these moments. In the oxygen atom, the net resultant magnetic moment is 13.9×10^{-24} A·m². In some atoms — such as helium and argon — the magnetic moments cancel. But in most atoms the net magnetic moment is different from zero. Thus, most atoms behave like small magnetic dipoles. Table 33.1 gives the magnetic moments of some atoms and ions.

---

[4] See Appendix 8 for more precise values of the magnetic moments of the elementary particles.

Table 33.1 MAGNETIC MOMENTS OF SOME ATOMS AND IONS

Atom	Magnetic moment
H	9.27×10^{-24} A·m²
He	0
Li	9.27×10^{-24}
O	13.9×10^{-24}
Ne	0
Na	9.27×10^{-24}
Ce^{+++}	19.8×10^{-24}
Yb^{+++}	37.1×10^{-24}

The nucleus of the atom also has a magnetic moment. This is due in part to the orbital motion of the protons inside the nucleus and in part to the rotational motion of individual protons and neutrons. The spin of each of these particles is $\hbar/2$, the same as that of an electron. The spin magnetic moment of a proton is 1.41×10^{-26} A·m², and that of a neutron is 0.97×10^{-26} A·m² (the former magnetic moment is parallel to the axis of spin, and the latter is antiparallel).[5] The magnetic moment of a proton or a neutron is small compared with that of an electron, and in reckoning the total magnetic moment of an atom, the nucleus can usually be neglected.

33.2 Paramagnetism

The behavior of the magnetic dipoles of the atoms or ions determines whether the material will be paramagnetic, ferromagnetic, or diamagnetic.

In a **paramagnetic material,** the atoms or ions have permanent magnetic dipole moments. When the material is left to itself, these dipoles are randomly oriented and their magnetic fields average to zero. However, if the material is immersed in an external magnetic field of, say, an electromagnet, the torque on the dipoles tends to align them with the field (see Section 31.6). This alignment will not be perfect, because of the disturbances caused by random thermal motions. But even a partial alignment of the dipoles will have an effect on the magnetic field. The material becomes **magnetized** and contributes an extra magnetic field that *enhances* the original magnetic field.

Paramagnetic material

Magnetization

To see how such an increase of magnetic field comes about, consider a piece of paramagnetic material placed between the poles of an electromagnet. Figure 33.3 shows the alignment of magnetic dipoles in this material; for the sake of simplicity, Figure 33.3b shows a case of perfect alignment. The magnetic dipoles are due to small current loops within the atoms. Figure 33.4a shows the aligned current loops. Now look at any point inside the magnetic material where two of these current loops (almost) touch. Since the currents at this point are opposite, they cancel. Thus, everywhere inside the material, the current is effectively zero. However, at the surface of the material, the current does not cancel. The net result of the alignment of current loops is therefore a current running along the surface of the magnetized mate-

---

[5] See Appendix 8 for more precise values of the magnetic moments of the elementary particles.

Fig. 33.3 (a) A piece of paramagnetic material. In the absence of an external magnetic field, the magnetic dipoles are oriented at random. (b) In the presence of an external magnetic field, the magnetic dipoles align with the magnetic field. The figure shows an ideal case of perfect alignment; in practice, the alignment will be only partial.

Fig. 33.4 (a) The magnetic dipoles of Figure 33.3b can be regarded as small current loops. At the point P, the currents of adjacent loops are opposite, and they cancel. (b) The small aligned current loops of (a) are equivalent to a current along the surface of the piece of material. The extra magnetic field produced by this current is parallel to the original magnetic field.

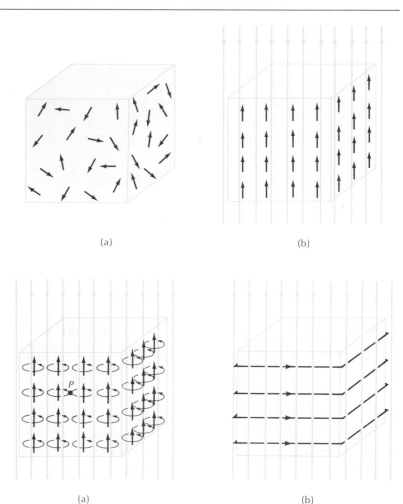

(a)   (b)

(a)   (b)

rial (Figure 33.4b). The material consequently behaves like a solenoid — it produces an extra magnetic field in its interior. This extra magnetic field has the *same* direction as the original, external magnetic field. Hence the total magnetic field in a paramagnetic material is larger than the original magnetic field produced by the currents of the electromagnet.

The alignment of the magnetic dipoles in a magnetized paramagnetic material is analogous to the alignment of the electric dipoles in a dielectric material. However, there is a crucial difference: the alignment of magnetic dipoles *increases* the original magnetic field, whereas the alignment of electric dipoles *decreases* the original electric field.

The increase of the strength of the magnetic field by the paramagnetic material can be described by the **relative permeability constant** κ_m. This constant is simply the factor by which the original magnetic field is increased. By analogy with the notation we adopted in Chapter 27 for the free charges and their electric field, we will adopt the notations I_{free} and B_{free}, respectively, for the external currents in the electromagnet and for the original, external field that these currents produce. These free currents in the electromagnet and the internal currents in the paramagnetic material acting together produce a total magnetic field

Relative permeability constant

$$\mathbf{B} = \kappa_m \mathbf{B}_{\text{free}} \tag{8}$$

Table 33.2 PERMEABILITIES OF SOME PARAMAGNETIC MATERIALS[a]

Material	κ_m
Air	1.000304
Oxygen	1.00133
Oxygen (−190°C, liquid)	1.00327
Manganese chloride	1.00134
Nickel monoxide	1.000675
Manganese	1.000124
Platinum	$1 + 13.8 \times 10^{-6}$
Aluminum	$1 + 8.17 \times 10^{-6}$

[a] At room temperature (20°C) and 1 atm unless otherwise noted.

where $\kappa_m > 1$. Table 33.2 lists the values of κ_m for some paramagnetic materials.

Just as Gauss' Law in the presence of a dielectric material contains an extra factor κ [see Eq. (27.32)], Ampère's Law in the presence of a paramagnetic material contains an extra factor κ_m. A simple argument shows that the revised form of Ampère's Law is

$$\oint \frac{1}{\kappa_m} \mathbf{B} \cdot d\mathbf{l} = \mu_0 I_{\text{free}} \qquad (9)$$

Ampère's Law in magnetic materials

If the arrangement of currents and paramagnetic materials is sufficiently symmetric, then we can use Eq. (9) to calculate the magnetic field in the usual way.

EXAMPLE 1. A long solenoid filled with air has a magnetic field of 1.20 T in its core. By how much will the magnetic field decrease if the air is pumped out of the core while the current is held constant?

SOLUTION: For air, $\kappa_m = 1.00030$. Hence the magnetic field with air is larger than that without by a factor 1.00030, that is,

$$B_{\text{free}} = B_{\text{air}}/1.00030 \cong B_{\text{air}} (1 - 0.00030)$$

where we have used the approximation $1/(1 + x) \cong 1 - x$ for small x. The decrease of the magnetic field is then

$$\Delta B = B_{\text{free}} - B_{\text{air}} = -0.00030 \times 1.20 \text{ T}$$

$$= -3.6 \times 10^{-4} \text{ T}$$

33.3 Ferromagnetism

As is obvious from Table 33.2, the increase of magnetic field produced by a paramagnetic material is quite small. By contrast, the increase produced by a **ferromagnetic material** can be enormous. And what is more, such a material will remain magnetized even if it is not immersed in an external magnetic field. A material that retains magnetization will make a **permanent magnet.**

Ferromagnetic material

Permanent magnet

The intense magnetization in ferromagnetic materials is due to a strong alignment of the spin magnetic moments of electrons. In these materials, there exists a special force that couples the spins of the electrons in adjacent atoms in the crystal, a force created by some subtle quantum-mechanical effects (which we cannot discuss here). This spin–spin force tends to lock the spins of the electrons in a parallel configuration. This force acts in the crystals of only five chemical elements: iron, cobalt, nickel, dysprosium, and gadolinium. However, it also acts in the crystals of alloys and of oxides of a large number of other elements.

Since this special spin–spin force is fairly strong, we must ask, Why is it that ferromagnetic materials are ever found in a nonmagnetized state? Why is it that not every piece of iron is a permanent magnet? The answer is that, on a small scale, ferromagnetic materials are *always* magnetized. A crystal of ordinary iron consists of a large number of small **domains** within which all the magnetic dipoles are perfectly aligned. But the direction of alignment varies from one domain to the next (Figure 33.5). Hence, on a large scale, there is no discernible alignment, because the domains are oriented at random. The sizes and shapes of the domains depend on the crystal. Typically, the sizes of domains range from a tenth of a millimeter to a few millimeters, although in a large uniform crystal the length of a domain may be several centimeters. Figure 33.6 shows domains in a crystal of iron.

The formation of domains results from the tendency of the material to settle into the state of least energy (equilibrium state). The state of least energy for the spins would be a state of complete alignment. But such a complete alignment would generate a large magnetic field around the material; that is, it would make the material into a (very strong) permanent magnet. Energetically, this is an unstable configuration, because there is a very large amount of energy in the magnetic field. The domain arrangement is a compromise. The spins align within the domains, but the domains do not align — the magnetic energy is then small because there is little magnetic field, and the spin–spin energy is then also reasonably small because *most* adjacent spins are aligned.

However, if the material is immersed in an external magnetic field, all dipoles tend to align along this field. The domains will then change in two ways: those domains that already are more or less aligned with

Domain

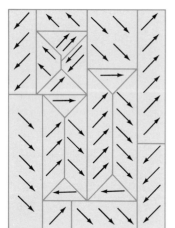

Fig. 33.5 Magnetic domains. Within each domain the magnetic dipoles have perfect alignment.

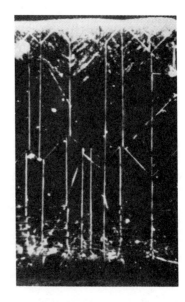

Fig. 33.6 Magnetic domains in a piece of iron. These domains are 0.1–0.3 mm across.

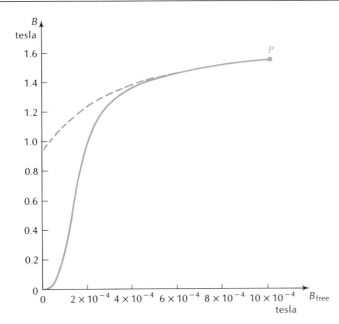

Fig. 33.7 Magnetic field B in annealed iron as a function of B_{free}. The solid curve shows B if the iron is immersed in an external magnetic field B_{free} that increases from an initial value of zero to a final value of 10×10^{-4} T. The dashed curve shows B if the external magnetic field is subsequently reduced to zero.

the field tend to grow in size at the expense of their neighbors, and, furthermore, some domains will rotate their dipoles in the direction of the field.

If *all* the magnetic dipoles in a piece of ferromagnetic material align, their contribution to the magnetic field will be very large. For example, within a piece of magnetized iron, this contribution can be as much as 2.1 T, a rather strong magnetic field. Figure 33.7 is a plot of the actual magnetic field B in a piece of iron in a solenoid as a function of the magnetic field B_{free} contributed by the free current in the solenoid [$B_{\text{free}} = \mu_0 I_0 n$; see Eq. (31.16)]. If the value of B_{free} is 2.0×10^{-4} T, the value of B is about 1.0 T — the iron increases the magnetic field by a factor of about 5000!

The solid portion of the plot in Figure 33.7 was obtained by starting with a piece of unmagnetized iron (annealed iron), and subjecting it to an increasing magnetic field B_{free}. The dashed portion of the plot was obtained by gradually reducing the magnetic field B_{free} to zero, after it had reached the value indicated by the point P. The dashed plot shows that even when B_{free} is reduced to zero, the iron retains some magnetization — the iron has become a permanent magnet. Thus, the iron retains some memory of the magnetic field to which it was exposed earlier. Such a dependence of the state of a system on its past history is called **hysteresis.**

Hysteresis

The hysteresis of a ferromagnetic material is due to a sluggishness in the rearrangement of the domains. Once the domains have become aligned in response to a strong external magnetic field, they tend to stay that way. If we remove the ferromagnetic material from the external magnetic field, the domains will suffer some rearrangement, but they will not lose their alignment completely. The remaining alignment gives the material a permanent distribution of magnetic dipole moments over its volume. The magnetic field of a permanent magnet is produced by these remaining aligned dipole moments.

As we saw in Section 33.2, the aligned dipoles effectively amount to a current running around the surface of the magnetized material. In a strong bar magnet, this surface current may amount to several

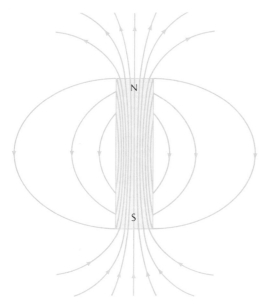
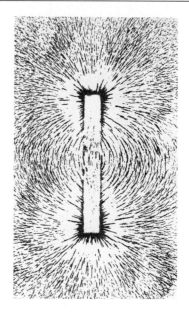

Fig. 33.8 Calculated magnetic field lines of a bar magnet, treated as a solenoid of finite length.

Fig. 33.9 Observed magnetic field lines of a bar magnet, made visible by sprinkling iron filings on a sheet of paper.

hundred amperes per centimeter of length of the magnet. The magnetic field produced by such a current distribution is obviously similar to that of a solenoid of finite length — the field lines emerge from the magnet at one end (the north pole) and reenter the magnet at the other end (the south pole). Figures 33.8 and 33.9 compare the calculated and observed field lines of a bar magnet.

Incidentally: The maximum magnetization that a ferromagnetic material will retain after it has been removed from the external magnetic field depends on the temperature. The higher the temperature, the less the remaining magnetization. Above a certain critical temperature, called the **Curie temperature**, the magnetization disappears completely. For example, iron will not retain any magnetization if the temperature is in excess of 1043°C.

Curie temperature

33.4 Diamagnetism

In both paramagnetic and ferromagnetic materials the important effect is the alignment of *permanent* magnetic dipoles. This is quite analogous to the alignment of permanent electric dipoles in a dielectric. But we know that in some dielectrics the polarization is due to induced electric dipoles rather than permanent dipoles. Is it likewise possible for a material to acquire *induced* magnetic dipoles?

Diamagnetic material

In **diamagnetic materials** the magnetization arises from such induced magnetic dipoles. To see how the magnetic field can induce dipole moments in the atoms, imagine that a sample of some material is placed between the poles of an electromagnet which is initially switched off. If the electromagnet is now switched on, the external magnetic field must increase from its initial (zero) value to its final value. Thus, for a short while, the magnetic field will be time dependent, and it will therefore induce an electric field within the sample.

Let us consider the effect of this electric field on the motion of an electron. We will again pretend that classical mechanics applies to this problem.

Figure 33.10 shows the orbit of an electron within an atom and also shows the increasing magnetic field. If the electron moves clockwise, as seen from above, the induced electric field (compare Example 32.4) will speed the electron up, giving it a larger orbital angular momentum and magnetic moment. The increment of angular momentum produces a magnetic field (within the orbit) that *opposes* the original magnetic field. This is obvious from Lenz' Law: the change in the motion of the electron amounts to an induced current, and the magnetic field associated with this current must be such as to oppose the original increasing magnetic field.

This argument establishes that induced magnetic moments reduce the strength of the magnetic field. The argument is quite general and applies to any kind of electron orbit and any kind of material. However, in paramagnetic and ferromagnetic materials the reduction of magnetic field due to induced magnetic moments is more than compensated by the increase of magnetic field due to alignment of permanent magnetic moments. If the material has no permanent magnetic moments, then the effect of the induced magnetic moments becomes noticeable, and the material will be diamagnetic.

Diamagnetism is a very small effect. The following calculation gives some idea of the size of this effect. When the material is placed in a magnetic field **B**, the electrons will experience a force $-e\mathbf{v} \times \mathbf{B}$ in addition to the usual electric force acting within the atom. Figure 33.11 shows the simple case of an electron in a circular orbit of radius r around a nucleus; the orbit is perpendicular to the magnetic field **B**. If the nucleus produces an electric field **E**, then the net force on the electron is $-e\mathbf{E} - e\mathbf{v} \times \mathbf{B}$. This force must match the product of mass and centripetal acceleration,

$$eE + evB = m_e v^2/r \qquad (10)$$

In terms of the angular frequency of the motion, the equation of motion becomes

$$eE + e\omega r B = m_e \omega^2 r \qquad (11)$$

Let us compare this with the equation of motion in the absence of the magnetic field (undisturbed atom, $B = 0$),

$$eE = m_e \omega_0^2 r \qquad (12)$$

If we subtract Eq. (12) from Eq. (11), we obtain a relation between the frequencies ω and ω_0:

$$e\omega B = m_e (\omega^2 - \omega_0^2) \qquad (13)$$

It is convenient to express this in terms of the increment of frequency,

$$\Delta\omega = \omega - \omega_0 \qquad (14)$$

If the magnetic field is not excessively strong, $\Delta\omega$ will be small compared with ω_0 and hence

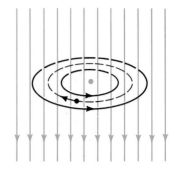

Fig. 33.10 Electron in a circular orbit around a nucleus immersed in an increasing magnetic field. The black circles indicate the induced electric field.

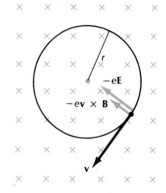

Fig. 33.11 Electron in a circular orbit around a nucleus immersed in a magnetic field. The centripetal force on the electron is the sum of the electric attraction of the nucleus and the magnetic force.

$$\omega^2 - \omega_0^2 = (\omega_0 + \Delta\omega)^2 - \omega_0^2$$

$$= 2\omega_0 \Delta\omega + (\Delta\omega)^2 \cong 2\omega_0 \Delta\omega \quad (15)$$

so that Eq. (13) becomes

$$e\omega B \cong 2m_e \omega_0 \Delta\omega \quad (16)$$

Since ω and ω_0 are nearly equal, we can cancel them on both sides of this equation without introducing any additional errors. This leads to

$$\boxed{\Delta\omega = \frac{eB}{2m_e}} \quad (17)$$

Larmor frequency This frequency is called the **Larmor frequency.** It tells us how much faster the electron will move around its orbit because of the presence of the magnetic field (of course, if the electron is initially moving in a direction opposite to that shown in Figure 33.11, then the electron will move *slower* by the same amount). Note that our calculation implicitly assumes that the orbital radius does not change as the magnetic field **B** is switched on. This assumption can be justified with an extra calculation that verifies that the work done on the electron by the induced emf changes the kinetic energy by just the amount required for the change of angular frequency at a fixed radius.

Corresponding to the change $\Delta\omega$ in the orbital frequency, there will be a change in the orbital magnetic moment. From Eq. (3),

$$\mu = \frac{evr}{2} = \frac{er^2\omega_0}{2} \quad (18)$$

Hence

$$\Delta\mu = \frac{er^2}{2} \Delta\omega \quad (19)$$

Dividing these equations into one another, we obtain an expression for the fractional change in the magnetic moment:

$$\frac{\Delta\mu}{\mu} = \frac{\Delta\omega}{\omega_0} \quad (20)$$

Let us insert some numbers. Typically the frequency of motion of an electron in an atom is $\omega_0 \cong 10^{16}$/s. If the magnetic field is fairly strong, say, $B = 1.0$ T, then

$$\Delta\omega = \frac{eB}{2m_e} = \frac{1.6 \times 10^{-19}\text{ C} \times 1.0\text{ T}}{2 \times 9.1 \times 10^{-31}\text{ kg}}$$

$$= 8.8 \times 10^{10}/\text{s} \quad (21)$$

Consequently,

$$\frac{\Delta\mu}{\mu} = \frac{\Delta\omega}{\omega_0} = \frac{8.8 \times 10^{10}/\text{s}}{10^{16}/\text{s}} \cong 10^{-5} \quad (22)$$

Table 33.3 PERMEABILITIES OF SOME DIAMAGNETIC MATERIALS[a]

Material	κ_m
Bismuth	$1 - 1.9 \times 10^{-5}$
Beryllium	$1 - 1.3 \times 10^{-5}$
Methane	$1 - 3.1 \times 10^{-5}$
Ethylene	$1 - 2.0 \times 10^{-5}$
Ammonia	$1 - 1.4 \times 10^{-5}$
Carbon dioxide	$1 - 0.53 \times 10^{-5}$
Glass (heavy flint)	$1 - 1.5 \times 10^{-5}$

[a] At room temperature (20°C) and 1 atm.

that is, the magnetic moment changes by only about 1 part in 10^5. This gives an indication of the small size of the diamagnetic effect.

The diamagnetic characteristics of a material can be described by a relative permeability κ_m that indicates by what factor the magnetic field is changed [compare Eq. (8)]. In the paramagnetic case $\kappa_m > 1$, but in the diamagnetic case $\kappa_m < 1$.

Table 33.3 lists some diamagnetic materials and the corresponding values of κ_m. In all cases, the value of κ_m is very near to 1.

SUMMARY

Magnetic moment of orbiting electron: $\mu = \dfrac{e}{2m_e} L$

Permeability constant: $\mathbf{B} = \kappa_m \mathbf{B}_{\text{free}}$

Magnetic materials: paramagnetic: $\kappa_m \gtrsim 1$
ferromagnetic: $\kappa_m \gg 1$
diamagnetic: $\kappa_m < 1$

Ampère's Law in paramagnetic and diamagnetic materials:

$$\oint \frac{1}{\kappa_m} \mathbf{B} \cdot d\mathbf{l} = \mu_0 I_{\text{free}}$$

Larmor frequency: $\Delta\omega = \dfrac{eB}{2m_e}$

QUESTIONS

1. Show that the SI unit of magnetic moment ($A \cdot m^2$) is the same as joule per tesla (J/T).

2. In the preceding chapters we ignored the paramagnetism of air in calculations of the magnetic field of wires, solenoids, etc. What percentage error in the magnetic field do we commit if we pretend that air has the magnetic properties of vacuum?

3. How would you measure the magnetic moment of a compass needle?

4. A bar magnet has a north pole and a south pole. If you break the bar magnet into two halves, do you obtain isolated north and south poles?

5. If we regard the Earth as a bar magnet, where is the magnetic north pole?

6. Consider a closed mathematical surface enclosing one of the poles of a bar magnet. What is the magnetic flux through this surface?

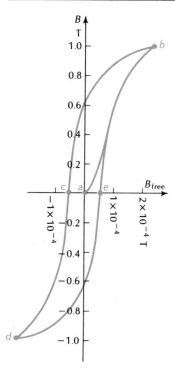

Fig. 33.12 Hysteresis loop for a piece of transformer steel. The configuration of the material initially corresponds to the point a. It is then successively brought to the points b, c, d, and e by suitable alterations of the external magnetic field B_{free}.

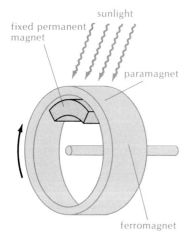

Fig. 33.13 Presnyakov wheel.

7. It is possible to magnetize an iron needle by pointing it north and giving it a few blows with a hammer. Explain.

8. If you drop a permanent magnet on a hard floor, it can become partially demagnetized. Explain.

9. Why does a magnet attract an (unmagnetized) piece of iron?

10. If you sprinkle iron filings on a sheet of paper placed in a magnetic field, the filings orient themselves along magnetic field lines (see Figure 33.9). Explain.

11. Other things being equal, a horseshoe magnet produces a stronger magnetic field than a bar magnet. Why?

12. You can make a chain of paper clips by touching one end of a clip to the pole of a magnet, then touching the free end to another paper clip, and so on. Explain.

13. When a mariner's compass is installed aboard an iron ship, it must be "adjusted" by placing several small permanent "correcting" magnets in its vicinity. What is the purpose of these magnets?

14. In the 1850s, Lord Kelvin redesigned the mariner's compass, partly by lengthy trials in his own yacht and partly by theoretical analysis. He made the compass card (a circular disk, free to rotate, which carries the compass needles) lighter and concentrated most of its weight at the rim, and he used much smaller compass needles than had been customary before. What are the advantages of Kelvin's design?

15. The "keeper" for a horseshoe magnet is a small bar of iron that is placed across the poles when the magnet is not in use. What purpose does this serve?

16. Figure 33.12 shows a plot of B vs. B_{free} for a piece of transformer steel. This plot includes negative values of B and B_{free}. Qualitatively, explain the shape of this curve, called the **hysteresis loop**.

17. Why does the magnetization of a ferromagnet decrease with temperature?

18. Figure 33.13 shows a magnetic solar motor, called the Presnyakov wheel. The wheel is made of a ferromagnetic material. A fixed permanent magnet is installed near the top, where sunlight strikes the wheel. The heat of the sunlight converts the top portion of the ferromagnet temporarily into a paramagnet, with a much smaller value of the permeability. Explain why this wheel keeps turning after it has been given an initial push.

19. Does iron have diamagnetism?

20. The magnetic permeability of paramagnets and ferromagnets is strongly dependent on temperature, but the permeability of diamagnets is nearly independent of temperature. Why?

21. The practice of dowsing for water is still prevalent in some backward communities. The dowser holds a rod of wood, brass, or plastic in his hands in a peculiar way, and the rod supposedly dips wherever water is beneath the ground. The effect on the rod cannot be gravitational (a small body of water does not produce a perceptible disturbance of the Earth's average gravitational field). Hence, if the effect exists, it must be electric or magnetic. Discuss, keeping in mind that the ground is a conductor, and that the dowsing rod is nonmagnetic material.

PROBLEMS

Section 33.1

1. Problem 22.24 gives the radii of the possible circular orbits of an electron in a hydrogen atom. For each such orbit, calculate the orbital magnetic moment.

*2. Two electrons are separated by a distance of 1.0×10^{-10} m. The first electron is on the axis of spin of the second.
   (a) What is the magnetic field that the magnetic moment of the second electron produces at the position of the first?
   (b) The potential energy of the magnetic moment of the first electron in this magnetic field depends on the orientation of the electrons. What is the potential energy (in electron-volts) if the spins of the two electrons are parallel? If antiparallel? Which orientation has the least energy?

*3. The electron in a hydrogen atom is 0.53×10^{-10} m away from the proton. What magnetic field does the magnetic moment of the proton produce at the position of the electron? Assume that the electron is instantaneously located on the axis of spin of the proton.

*4. The field of a fixed magnetic dipole μ located at the origin and oriented parallel to the z axis has the following components as a function of z and x in the plane $y = 0$:

$$B_x = \frac{\mu_0 \mu}{4\pi} \frac{3zx}{(x^2 + z^2)^{5/2}} \qquad B_y = 0,$$

$$B_z = -\frac{\mu_0 \mu}{4\pi} \left(\frac{1}{(x^2 + z^2)^{3/2}} - \frac{3z^2}{(x^2 + z^2)^{5/2}} \right)$$

Suppose that a second magnetic dipole of moment μ' is located at the point x, z (with $y = 0$); this dipole is free to rotate in the x–z plane.
   (a) What is the orientation of least magnetic energy of this second dipole in the field of the first dipole, i.e., what is the angle that the second dipole will make with the z axis?
   (b) What is the numerical value of the angle in the case $x = 0$, $z = r$? In the case $x = r$, $z = 0$? In the case $x = z = r/\sqrt{2}$?

*5. Two magnetic dipoles are separated by a fixed distance r in the horizontal plane. The dipoles are free to rotate about a vertical axis (you may imagine that the dipoles are compass needles, but the magnetic field of the Earth is absent). If the dipoles settle into the configuration of least magnetic energy, what will be their orientation? Draw a diagram of the dipoles in this orientation. Prove your answer. (Hint: You may want to use the expression for the magnetic field given in Problem 4).

**6. A compass needle has the shape of a thin rod of length 2.0 cm pivoted at its center so that it can swing freely in the horizontal plane. The compass needle has a mass of 0.12 g and a magnetic moment of 3.2×10^{-4} A·m². The compass needle is in Hawaii, where the horizontal, northerly component of the magnetic field is 2.9×10^{-5} T. What is the frequency of small rotational oscillations of the compass needle about the northerly direction?

*7. Assume that the proton is a spherical ball within which the positive charge and the mass are uniformly distributed. Assume that the proton rotates rigidly with a spin angular momentum of $\hbar/2 = 0.53 \times 10^{-34}$ J·s. From this information, calculate the magnetic moment of a proton. [Hint: Use Eq. (4) with the mass of the proton. The result of this classical calculation differs by a factor of 5.6 from the actual value; this is due to a failure of classical physics when applied to subatomic particles.]

*8. Assume that the charge distribution of a neutron is as indicated by the model described in Problem 24.19 and that the mass distribution is uniform. Assume that the neutron rotates rigidly with a spin angular momentum of $\hbar/2 = 0.53 \times 10^{-34}$ J·s. According to this crude model, what is the magnetic moment of a neutron? What is the direction of the magnetic moment relative to the direction of the spin angular momentum? [Hint: Apply Eq. (4) to the positive and the negative charge distributions separately.]

9. As described in Section 33.1, the proton has a magnetic moment of $\mu = 1.41 \times 10^{-26}$ A·m² parallel to the axis of its spin angular momentum. If the proton is in a magnetic field **B and the magnetic moment makes an angle θ with this field, the torque exerted by this field on the magnetic moment will be $\tau = \mu B \sin\theta$ and the direction of this torque will be perpendicular to μ. Since the proton has a spin angular momentum **S** parallel to the magnetic moment, the torque will cause a precession of the spin about the direction of the magnetic field (see Section 13.6 for a discussion of the precession of a top under the influence of a torque).

(a) Show that the precession frequency of the proton is

$$\omega = \mu B / S$$

or, since $S = \hbar/2 = 0.53 \times 10^{-34}$ J·s,

$$\omega = 2\mu B/\hbar$$

(b) What is the precession frequency of a proton in a magnetic field of 0.20 T?

*10. In a hydrogen atom, the electron orbits around the proton on a circular orbit of radius 0.53×10^{-10} m. As in Section 33.1, this orbiting electron can be regarded as a ring of current.
 (a) Calculate the magnetic field that the ring of current produces at its center.
 (b) Using the result of Problem 9, calculate the precession frequency of the proton in this magnetic field.

Section 33.2

11. A current of 25 A flows in a long solenoid of 1500 turns per meter.
 (a) If the interior of this solenoid is empty (a vacuum), what is the strength of the magnetic field?
 (b) If the interior is filled with liquid oxygen while the current stays constant, what will be the percentage change in the magnetic field?

12. Suppose that the dipole moments of all the atoms in a 20-g sample of lithium are perfectly aligned. What is the strength of the magnetic field on the axis of the dipoles at a distance of 1.0 m?

13. The space within a solenoid is to be filled with a mixture of air (paramagnetic) and methane (diamagnetic) so that the net permeability constant is exactly $\kappa_m = 1$. What percentage of air and methane should we use?

14. At a temperature of 20°C and a pressure of 1 atm the relative permeability of air is 1.000304. Calculate the relative permeability of air at the same temperature but at a pressure of 3.0 atm. Assume that $\kappa_m - 1$ is proportional to the pressure.

*15. Initially, the space within a long solenoid is empty. It is then filled with liquid oxygen, a paramagnetic material. What is the percentage change of the self-inductance of the solenoid? Does the self-inductance increase or decrease?

*16. Show that the self-inductance per unit length of a very long solenoid filled with a paramagnetic material is $\kappa_m \mu_0 n^2 \pi R^2$, where n is the number of turns of wire per unit length and R is the radius of the solenoid.

*17. Show that the energy density in the magnetic field in a very long solenoid filled with paramagnetic material is $u = B^2/(2\kappa_m \mu_0)$.

Section 33.3

18. A permanent bar magnet of iron has a magnetic field of 0.03 T in its interior. The magnet is 15 cm long. What is the effective current running around its surface? Treat the magnet as though it were a very long cylindrical solenoid.

19. A long solenoid has 1200 turns per meter with a current of 6.0 A. The solenoid is filled with a ferromagnetic material. The value of the magnetic field B in this material is 2.0 T. What is the value of κ_m under these conditions?

20. In an iron crystal, two of the electrons of each atom participate in the alignment of spins; the magnetic field of a permanent magnet is caused by the magnetic moment of these electrons. Suppose that all of these electrons in a compass needle of mass 0.60 g are perfectly aligned. The compass needle is at a place where the horizontal, northerly component of the Earth's magnetic field is 2.4×10^{-5} T; the compass needle is free to swing in the horizontal plane. What is the torque on the compass needle when it is at an angle of 45° with the northerly component of the magnetic field? Does your answer depend on the shape of the compass needle?

21. Figure 33.7 is a plot of the magnetic field B in an iron-filled solenoid as a function of the magnetic field B_{free} that the solenoid would produce without the iron. This plot was prepared under the assumption that initially, when the current is zero, the iron is not magnetized. The value of the relative permeability depends on B_{free}. What is the value of κ_m when $B = 0.4$ T? When $B = 0.8$ T? When $B = 1.2$ T? Make a rough plot of κ_m vs. B_{free}. At what value of B_{free} is κ_m maximum?

22. Under conditions of maximum magnetization, the dipole moment per unit volume in cobalt is 1.5×10^5 A·m²/m³. Assuming that this magnetization is due to completely aligned electrons, how many such electrons are there per unit volume? How many aligned electrons per atom? The density of cobalt is 8.9×10^3 kg/m³ and the atomic mass is 58.9 g/mole.

*23. You want to generate a magnetic field of 1.2 T in an iron-core magnet consisting of a long solenoid of 600 turns per meter filled with annealed iron. Use Figure 33.7 to calculate the current required in the solenoid.

*24. Suppose that a long solenoid with a core of annealed iron generates a magnetic field of 1.0 T when the current in its windings is 8.0 A. If we remove the iron core, what current will be needed to generate the same magnetic field? Use Figure 33.7 for the magnetic properties of annealed iron.

*25. In iron, two of the electrons of each atom participate in the alignment of spins that causes magnetization. Suppose that a cylindrical piece of iron, of radius 1.0 cm and length 8.0 cm, is completely magnetized along its axis, all the available electrons being in perfect alignment.
   (a) What is the number of aligned electrons?
   (b) The dipole moment of each electron is 9.27×10^{-24} A·m². What is the total dipole moment of all the aligned electrons?
   (c) What surface current running around the surface of the cylinder will give the same total dipole moment?
   (d) What magnetic field does this surface current produce in the interior of the iron?

*26. The alignment of electron spins in a ferromagnetic material implies that the magnetized material has angular momentum. Suppose that a rod of iron 2.0 cm in diameter and 30 cm long is totally magnetized so that two of the electrons of each atom have their spins parallel to the axis of the rod. Suppose that the magnetization is suddenly reversed so that the spins become antiparallel to the axis. What is the change of angular momentum? The density of iron is 7.9 g/cm³ and the atomic mass is 55.8 g/mole.

*27. A solenoid is 20 cm long and has 40 turns of wire. A current of 2.0 A flows in the wire. Assume that the solenoid is very thin, and can therefore be treated as very long.
   (a) If the solenoid is empty, what is the magnetic flux linked by (all) the turns of the solenoid? What is the self-inductance?
   (b) If we fill the solenoid with a core of annealed iron (keeping the current constant at 20 A), what will be the new values of the flux and the self-in-

ductance? By what factor is the self-inductance of the filled solenoid larger than that of the empty solenoid? Use Figure 33.7 for the magnetic properties of annealed iron.

**28. A long solenoid is filled with iron. The solenoid has 1800 turns per meter, and the current in each turn is 50 A. Calculate the magnetic field inside the solenoid, assuming that two of the electrons of each atom are completely aligned. (Hint: The magnetic field consists of two contributions: the field of the solenoid wire plus the field of the aligned electrons. The latter field can be calculated by replacing the electrons by a surface current as in Problem 25.)

Section 33.4

29. To determine the diamagnetic permeability of a sample of liquid, an experimenter takes a long solenoid with a steady current and measures the change of the magnetic field when the air in the solenoid is replaced by the sample of liquid. He finds that the magnetic field decreases by 0.033%. What is the value of κ_m for the liquid?

*30. An electron in a hydrogen atom moves around a circular orbit of radius 0.53×10^{-10} m at a speed of 2.2×10^6 m/s. Suppose that the hydrogen atom is transported into a magnetic field of 0.50 T. The magnetic field is parallel to the orbital angular momentum.
  (a) What is the change of the frequency of the motion of the electron? Does the frequency increase or decrease?
  (b) What is the change of the speed of the electron? Assume the radius of the motion remains constant.
  (c) What is the change of the energy of the electron?
  (d) Check that the electric field induced by the change in magnetic flux through the orbit has the right direction to produce the change of speed and energy.

*31. A long cylindrical bar magnet of diameter 1.0 cm has a magnetic field of 0.060 T in its interior. If you take a fine saw and cut the magnet in two pieces, the magnetic force will hold the pieces together. Estimate the magnitude of this magnetic force. [Hint: Suppose that you pull the pieces apart by a distance dx. The magnetic field in the gap between the pieces will then still be (nearly) 0.060 T. The magnetic energy in the gap equals the work that you must have done against the magnetic force while pulling the pieces apart.]

CHAPTER 34

AC Circuits

The current delivered by power companies to homes and factories is an oscillating function of time. This is called alternating current, or AC. Power companies prefer alternating currents to direct currents because of the ease with which alternating voltages can be stepped up or down by means of transformers. This makes it possible to step up the output of a power plant to several hundred thousand volts, transmit the power along a high-voltage line that minimizes the Joule losses, and finally step down the power to 230-volt AC or 115-volt AC just before delivery to the consumer.

All the appliances connected to ordinary outlets in homes therefore involve circuits with oscillating currents. Furthermore, electronic devices — such as radio transmitters and receivers — involve a variety of circuits with oscillating currents of high frequency. Many of these circuits have natural frequencies of oscillation. Such circuits exhibit the phenomenon of resonance when the natural frequency matches the frequency of a signal applied to the circuit. For instance, the tuning of a radio relies on an oscillating circuit whose frequency of oscillation is adjusted by means of a variable capacitor (attached to the tuning knob) so that it matches the frequency of the radio signal.

34.1 Simple AC Circuits with an External Electromotive Force

The simplest conceivable AC circuit consists of a pure resistor connected to an oscillating source of emf. In the circuit diagram (Figure 34.1), the source of emf is symbolized by a wavy line enclosed in a circle. This circuit might represent an electric heater or an incandescent

Fig. 34.1 Resistor connected to a source of alternating emf.

lamp plugged into an ordinary wall outlet. The oscillating emf is of the form

$$\mathscr{E} = \mathscr{E}_{max} \sin \omega t \tag{1}$$

where $\mathscr{E}_{max}$ is the amplitude of the oscillating emf and ω the angular frequency. In the United States and in Canada, the oscillating voltage available at the outlets of private homes has an amplitude $\mathscr{E}_{max} = 163$ V and a frequency of 60 Hz, that is, an angular frequency of $\omega = 2\pi \times 60/\text{s}$. This kind of voltage is usually called "115-volt AC" for reasons that will become clear shortly.

If we apply Kirchhoff's rule to the circuit of Figure 34.1, we obtain, at an instant of time,

$$\mathscr{E} - IR = 0 \tag{2}$$

or

Current in resistor circuit

$$I = \frac{\mathscr{E}}{R} = \frac{\mathscr{E}_{max} \sin \omega t}{R} \tag{3}$$

Thus, the current oscillates in phase with the emf, and its maxima and minima coincide with those of the emf (Figure 34.2).

The instantaneous electric power dissipated in the resistor is

$$P = I\mathscr{E} = \frac{\mathscr{E}_{max}^2 \sin^2 \omega t}{R} \tag{4}$$

This power oscillates between zero and a maximum value $\mathscr{E}_{max}^2/R$ (Figure 34.3). The time-average power can be obtained by averaging

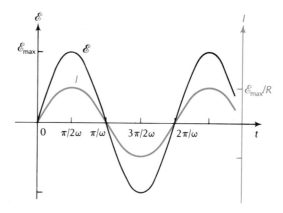

Fig. 34.2 The emf (black) and current (color) in the resistor circuit as a function of time.

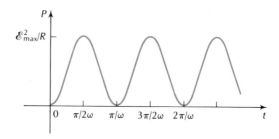

Fig. 34.3 Instantaneous power dissipated in the resistor as a function of time. The power is always positive or zero.

$\sin^2 \omega t$ over one cycle. We can easily verify that the average value of $\sin^2 \omega t$ is $\frac{1}{2}$,

$$\frac{1}{2\pi/\omega}\int_0^{2\pi/\omega} \sin^2 \omega t\, dt = \frac{1}{2\pi/\omega}\int_0^{2\pi/\omega} \tfrac{1}{2}(1-\cos 2\omega t)\, dt = \tfrac{1}{2} \quad (5)$$

Hence the average power is

$$\boxed{\overline{P} = \frac{\mathscr{E}_{max}^2}{2R}} \quad (6)$$

Average power absorbed by resistor

This is often written in the form

$$\boxed{\overline{P} = \frac{\mathscr{E}_{rms}^2}{R}} \quad (7)$$

where the quantity $\mathscr{E}_{rms}$, called the **root-mean-square voltage,** is the square root of the time average of the square of the voltage,

$$\mathscr{E}_{rms} = \sqrt{\overline{\mathscr{E}^2}} = \sqrt{\frac{\mathscr{E}_{max}^2}{2}} = \frac{\mathscr{E}_{max}}{\sqrt{2}} \quad (8)$$

Root-mean-square voltage

In engineering practice, an AC voltage is usually described in terms of $\mathscr{E}_{rms}$. For example, if $\mathscr{E}_{max} = 163$ V, then $\mathscr{E}_{rms} = 163/\sqrt{2}$ V $= 115$ V. An oscillating voltage with this value of $\mathscr{E}_{max}$ is described as "115-volt AC."

Comparison of Eqs. (7) and (29.23) shows that the average AC power dissipated in the resistor is equal to the DC power dissipated in the same resistor when connected to a steady DC voltage of magnitude $\mathscr{E}_{rms}$. Thus, a voltage of 115-volt AC (with $\mathscr{E}_{max} = 163$ V) delivers the same average power to the resistor as 115-volt DC.

EXAMPLE 1. A 115-V AC incandescent light bulb is rated at 150 W. What is the resistance of this light bulb (when at its operating temperature)?

SOLUTION: We have $\mathscr{E}_{rms} = 115$ V and $\overline{P} = 150$ W. Hence

$$R = \mathscr{E}_{rms}^2/\overline{P} = 88.2\ \Omega \quad (9)$$

Next, let us examine a circuit consisting of a capacitor connected to our oscillating source of emf (Figure 34.4). The voltage across the capacitor is Q/C, and therefore Kirchhoff's rule gives

$$\mathscr{E} - Q/C = 0 \quad (10)$$

With Eq. (1) for $\mathscr{E}$, this yields

$$Q = C\mathscr{E} = C\mathscr{E}_{max} \sin \omega t \quad (11)$$

The current in the circuit is $I = dQ/dt$, or

$$I = \omega C \mathscr{E}_{max} \cos \omega t \quad (12)$$

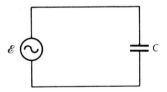

Fig. 34.4 Capacitor connected to a source of alternating emf.

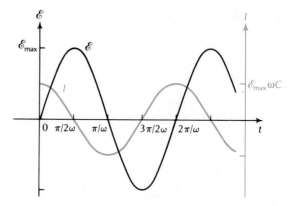

Fig. 34.5 The emf (black) and current (color) in the capacitor circuit as a function of time.

Comparison of Eqs. (12) and (1) shows that the current is a quarter cycle (90°) out of phase with the emf; for instance, at $t = 0$ the magnitude of the emf is minimum, and the current is maximum (Figure 34.5). Since the maxima in the current occur a quarter cycle *before* the maxima in the emf, we say that the current **leads** the emf.

It is customary to write Eq. (12) as

Current in capacitor circuit

$$I = \frac{\mathcal{E}_{max} \cos \omega t}{X_C} \quad (13)$$

where

Capacitive reactance

$$X_C = \frac{1}{\omega C} \quad (14)$$

is called the **capacitive reactance.** The quantity X_C plays roughly the same role for a capacitor in an AC circuit as does the resistance for a resistor [compare Eqs. (3) and (13)]. Note, however, that the reactance depends not only on the characteristics of the capacitor but also on the frequency at which we are operating the circuit. The unit of reactance is the ohm, as it is for resistance.

The instantaneous power delivered to the capacitor is

$$P = I\mathcal{E} = \omega C \mathcal{E}_{max}^2 \cos \omega t \sin \omega t \quad (15)$$

The time dependence of this expression is contained in the factor $\cos \omega t \sin \omega t$. Since this equals $\frac{1}{2} \sin 2\omega t$, we recognize that the power oscillates at a frequency 2ω. But the important point is that the average power delivered is zero — as Figure 34.6 shows, within one cycle,

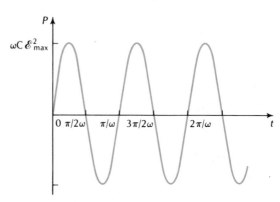

Fig. 34.6 Instantaneous power delivered to the capacitor as a function of time. The power oscillates between positive and negative values.

there is as much positive power as negative power. The source of emf does work on the capacitor during part of the cycle, but the capacitor does work on the source during other parts of the cycle, so that, on the average, the *power is zero*. The ideal capacitor does not consume electric power, because it has no means of dissipating electric energy.

Finally, we will examine a circuit consisting of an inductor connected to an oscillating source of emf. In the circuit diagram (Figure 34.7), the inductor is represented by a coiled line. The induced emf in the inductor (back emf) is $L\,dI/dt$, and, by Kirchhoff's rule, this must balance the applied emf,

$$\mathcal{E} - L\frac{dI}{dt} = 0 \tag{16}$$

Fig. 34.7 Inductor connected to a source of alternating emf.

which gives

$$\frac{dI}{dt} = \frac{\mathcal{E}}{L} = \frac{\mathcal{E}_{max}\sin\omega t}{L} \tag{17}$$

By integrating this we obtain[1]

$$I = -\frac{\mathcal{E}_{max}\cos\omega t}{\omega L} \tag{18}$$

Again, comparison of Eqs. (18) and (1) shows that the current is a quarter cycle (90°) out of phase with the emf (Figure 34.8). However, because of the minus sign in Eq. (18), the maxima in the current occur a quarter cycle *after* the maxima in the emf — the current **lags** the emf.

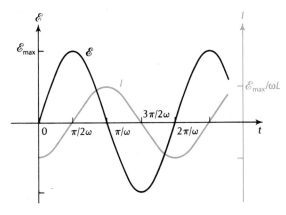

Fig. 34.8 The emf (black) and current (color) in the inductor circuit as a function of time.

We can write Eq. (18) as

$$\boxed{I = -\frac{\mathcal{E}_{max}\cos\omega t}{X_L}} \tag{19}$$

Current in inductor circuit

where

$$X_L = \omega L \tag{20}$$

Inductive reactance

---

[1] Here we assume that the constant of integration is zero. If this constant were not zero, the circuit would carry an additional time-independent current.

is the **inductive reactance.** The unit of this reactance is, again, the ohm.

The instantaneous power delivered to the inductor is

$$P = I\mathscr{E} = -\frac{1}{\omega L}\mathscr{E}_{\max}^2 \cos \omega t \sin \omega t \qquad (21)$$

As in the case of the capacitor, the average power is zero.

34.2 The Freely Oscillating LC Circuit

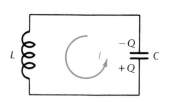

Fig. 34.9 Inductor and capacitor connected in series.

An LC circuit consists of an inductor and a capacitor connected in series (Figure 34.9). The circuit has no source of emf; nevertheless, a current will flow in this circuit provided that the capacitor is initially charged. The potential on one plate is then initially high and that on the other plate is low. A current will begin to flow around the circuit from the positive plate to the negative. If the circuit had no inductance, the current would merely neutralize the charge on the plates, that is, the capacitor would discharge, and that would be the end of the current. But the inductor makes a difference: the inductor initially opposes the buildup of the current, but once the current has become established, the inductor will keep it going for some extra time. Hence, *more* charge flows from one capacitor plate to the other than required for neutrality, and reversed charges accumulate on the capacitor plates. When the current finally does stop, the capacitor will again be fully charged, with reversed charges. And then a reversed current will begin to flow, and so on. Thus, the charge sloshes back and forth around the circuit.

The LC system is analogous to a mass–spring system. The inductor is analogous to the mass — it tends to keep the current constant and provides "inertia." The charged capacitor is analogous to the stretched spring — it tends to accelerate the current and provides a "restoring force."

The equation of motion for the LC system follows from Kirchhoff's rule: the sum of emfs and voltage drops around the circuit must add to zero. Going around the circuit in the direction of the arrow shown in Figure 34.9, we find that the induced emf in the inductor (back emf) is

$$-L\frac{dI}{dt}$$

and the voltage across the capacitor is

$$-\frac{Q}{C}$$

Hence

$$-L\frac{dI}{dt} - \frac{Q}{C} = 0 \qquad (22)$$

Note that here Q is reckoned as positive when the charge on the lower plate is positive, and I is reckoned as positive when the charge on the lower plate is increasing.

Since $I = dQ/dt$, we can also write Eq. (22) as

$$L\frac{d^2Q}{dt^2} + \frac{1}{C}Q = 0 \tag{23}$$

This equation has exactly the same mathematical form as the equation for the simple harmonic oscillator [Eq. (15.22)]:

$$m\frac{d^2x}{dt^2} + kx = 0$$

Obviously, Q plays the role of x, whereas L replaces m, and $1/C$ replaces k. Hence the solution of Eq. (23) can be immediately written down by recalling the solution for the simple harmonic oscillator [Eq. (15.25)]:

$$Q = Q_0 \cos\left(\frac{1}{\sqrt{LC}}t + \phi\right) \tag{24}$$

Here, the phase constant ϕ must be adjusted to fit the initial conditions of the problem. Since we are assuming that at $t = 0$ the capacitor is fully charged ($Q = Q_0$) and the current is zero ($dQ/dt = 0$), the correct choice for the phase constant is $\phi = 0$. Thus, the charge is

$$Q = Q_0 \cos\left(\frac{1}{\sqrt{LC}}t\right)$$

and the current is

$$I = \frac{dQ}{dt} = -\frac{Q_0}{\sqrt{LC}}\sin\left(\frac{1}{\sqrt{LC}}t\right) \tag{25}$$

According to these equations, the charge and the current oscillate with a natural frequency

$$\boxed{\omega_0 = \frac{1}{\sqrt{LC}}} \tag{26}$$

Natural frequency of LC circuit

Figure 34.10 is a plot of the charge and the current in an LC circuit oscillating according to Eq. (25).

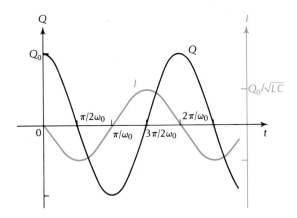

Fig. 34.10 Charge (black) on the capacitor and current (color) in the inductor as a function of time.

EXAMPLE 2. A primitive radio transmitter, such as those built in the early days of "wireless telegraphy," consists of an LC circuit oscillating at high frequency (Figure 34.11a). The circuit is inductively coupled to an antenna (Figure 34.11b) so that the oscillating current in the circuit induces an oscillating current on the antenna; the latter current then radiates electromagnetic waves. Suppose that the inductance in the circuit of Figure 34.11a is 20 μH. What capacitance do we need if we want to produce oscillations of a frequency of 1500 kHz?

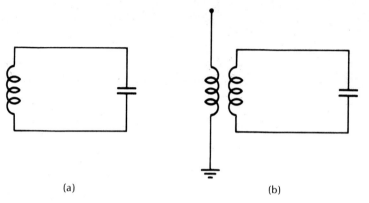

Fig. 34.11 (a) An LC circuit in a radio. (b) The LC circuit is coupled to the antenna by the mutual inductance of the two inductors.

SOLUTION: The angular frequency is $2\pi \times 1500 \times 10^3$ radians/s. Hence, from Eq. (26),

$$C = \frac{1}{\omega_0^2 L}$$

$$= \frac{1}{(2\pi \times 1500 \times 10^3/\text{s})^2 \times 20 \times 10^{-6} \text{ H}}$$

$$= 5.6 \times 10^{-10} \text{ F} = 560 \text{ pF}$$

The energy of the LC system is the sum of the energies stored in the capacitor and in the inductor [see Eqs. (27.40) and (32.37)], that is,

$$U = \tfrac{1}{2}\frac{Q^2}{C} + \tfrac{1}{2}LI^2 \qquad (27)$$

or

$$U = \tfrac{1}{2}\frac{Q^2}{C} + \tfrac{1}{2}L\left(\frac{dQ}{dt}\right)^2 \qquad (28)$$

On physical grounds we expect that the energy remains constant during the oscillations. To prove this conservation theorem for the energy, we need only differentiate Eq. (28) with respect to time:

$$\frac{dU}{dt} = \frac{Q}{C}\frac{dQ}{dt} + L\frac{dQ}{dt}\frac{d^2Q}{dt^2} = \frac{dQ}{dt}\left(\frac{Q}{C} + L\frac{d^2Q}{dt^2}\right) \qquad (29)$$

The expression on the right side is zero because of the relation between Q and d^2Q/dt^2 [see Eq. (23)].

The quantity $\tfrac{1}{2}Q^2/C$ is a potential energy (electrostatic potential energy). The quantity $\tfrac{1}{2}L(dQ/dt)^2$ may be regarded as a "kinetic" energy.

Thus, the expression (28) is analogous to the expression

$$\tfrac{1}{2}kx^2 + \tfrac{1}{2}m\left(\frac{dx}{dt}\right)^2$$

for the energy of a harmonic oscillator. If the capacitor is initially charged but no current is flowing, then the energy is initially purely potential (it is stored in the electric fields in the capacitor). As the current begins to flow, the potential energy decreases and the "kinetic" energy increases. At the instant when the capacitor is completely discharged, the current reaches its maximum value. The potential energy is then zero and the energy is purely "kinetic" (it is stored in the magnetic fields in the inductor). Beyond this instant, the potential energy increases at the expense of the "kinetic" energy. When the capacitor is completely charged, with reversed charge, the current stops flowing. At this instant the energy is again purely potential. The process now repeats with the current flowing in the opposite direction. Figure 34.12 is a plot of potential energy and "kinetic" energy as a function of time for an LC circuit oscillating according to Eqs. (24) and (25).

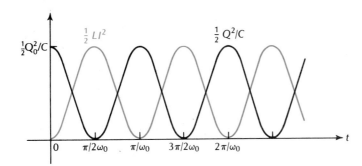

Fig. 34.12 Energy in the capacitor ($\tfrac{1}{2}Q^2/C$; black) and energy in the inductor ($\tfrac{1}{2}LI^2$; color) as a function of time.

So far we have assumed that our LC circuit contains no resistance. This is somewhat unrealistic since, at the very least, the wires connecting the circuit elements will have some resistance. The remainder of this section gives a description of the effects of resistance.

Figure 34.13 shows an LCR circuit, that is, an LC circuit with resistance. The resistance plays a role analogous to that of the friction force in the harmonic oscillator (see Section 15.6). The resistance gradually converts electric energy into heat; hence the electric energy decreases with time. This leads to damped oscillations of gradually decreasing amplitude.

Figure 34.14 shows the charge on the capacitor as a function of time for an LCR circuit. The charge can be described by the following equation:

$$Q = Q_0 e^{-\gamma t/2} \cos \omega_0 t \qquad (30)$$

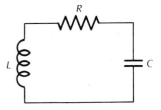

Fig. 34.13 Inductor, capacitor, and resistor connected in series.

This, of course, is based on the assumption that the capacitor is fully charged at the initial time $t=0$. The damping constant γ represents the frictional effects. It can be shown that γ is proportional to the resistance,

$$\gamma = \frac{R}{L} \qquad (31)$$

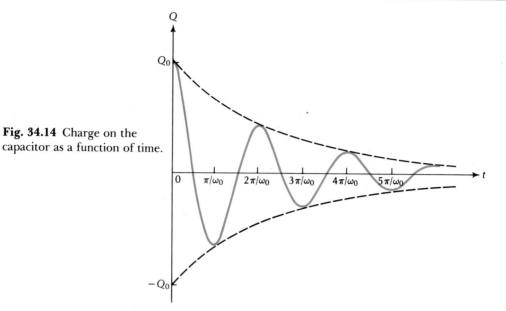

Fig. 34.14 Charge on the capacitor as a function of time.

As we saw in Section 15.6, the damping constant is directly related to the energy ΔU lost per period,

$$\frac{\Delta U}{U} = -2\pi \frac{\gamma}{\omega_0} = -\frac{2\pi R}{\omega_0 L} \qquad (32)$$

Electrical engineers usually express this energy loss in terms of the "*Q*" or the **quality factor** of the circuit,[2]

"Q" of freely oscillating LCR circuit

$$\boxed{\text{``}Q\text{''} = 2\pi |U/\Delta U| = \omega_0 L/R} \qquad (33)$$

Typical circuits of low resistance in radio transmitters and receivers have a "*Q*" of up to 100.

Besides damping the amplitude of the oscillations, the resistance also produces another effect: it reduces the frequency of oscillation. Intuitively, we expect such a reduction of frequency because the friction in the resistance will slow the oscillations. Mathematically, one can show that the angular frequency for the natural oscillations of the LCR circuit is

$$\omega_0 = \frac{1}{\sqrt{LC}} \sqrt{1 - \frac{CR^2}{4L}} \qquad (34)$$

Here $\sqrt{1 - CR^2/4L}$ represents the factor by which the frequency is decreased as compared with an LC circuit without resistance.

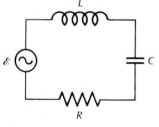

Fig. 34.15 Inductor, capacitor, and resistor connected in series to a source of alternating emf.

34.3 The LCR Circuit with an External Emf

Figure 34.15 shows a series LCR circuit with an oscillating source of emf. This emf acts as a driving force that pushes a current around the

---

[2] This "*Q*" must not be confused with electric charge.

circuit. Besides the current directly attributable to the driving emf, there may also be a current attributable to the natural oscillations of the LCR circuit, described in the preceding section. This kind of current is a transient current; its initial amplitude of oscillation depends on the initial conditions in the circuit, and this amplitude gradually damps away. In the following discussion, we will assume that the driven LCR circuit has been oscillating for a long time, so the transient current has damped to a negligible level. The current in the circuit then depends only on the driving emf, and the circuit is said to have reached **steady state.** As in the case of the driven harmonic oscillator (see Section 15.6), we expect that if the driving frequency coincides with the natural frequency ω_0 of the circuit, then the oscillations will build up to a very large value. This is the condition for **resonance.**

Steady state

Resonance

Suppose that the oscillating emf driving the circuit has an angular frequency ω:

$$\mathscr{E} = \mathscr{E}_{max} \sin \omega t \tag{35}$$

Under steady-state conditions, the current will then oscillate with the same angular frequency. This implies that the current will be of the form[3]

$$I = I_{max} \sin(\omega t - \phi) \tag{36}$$

We are now faced with the task of finding how the amplitude I_{max} and the phase angle ϕ of the current are related to the known parameters of the circuit.

We begin with Kirchhoff's rule: the external emf $\mathscr{E}$ must match the sum of the voltages across the resistor, the capacitor, and the inductor,

$$\mathscr{E} = \Delta V_R + \Delta V_C + \Delta V_L \tag{37}$$

To evaluate the three terms on the right side of this equation, we make use of the results of Section 34.1. The voltage across the resistor is in phase with the current:

$$\Delta V_R = IR = RI_{max} \sin(\omega t - \phi) \tag{38}$$

The voltage across the capacitor is one-quarter cycle behind the current,

$$\Delta V_C = -X_C I_{max} \cos(\omega t - \phi) \tag{39}$$

and the voltage across the inductor is one-quarter cycle ahead of the current,

$$\Delta V_L = X_L I_{max} \cos(\omega t - \phi) \tag{40}$$

Substitution of these into Eq. (37) yields

$$\mathscr{E}_{max} \sin \omega t = RI_{max} \sin(\omega t - \phi) + (X_L - X_C)I_{max} \cos(\omega t - \phi) \tag{41}$$

---

[3] The extra minus sign inserted in front of the phase angle ϕ conforms with standard usage in electrical engineering.

With the trigonometric identities for the sine and the cosine of the difference of two angles this becomes

$$\mathcal{E}_{max} \sin \omega t = RI_{max}(\sin \omega t \cos \phi - \cos \omega t \sin \phi)$$

$$+ (X_L - X_C)I_{max}(\cos \omega t \cos \phi + \sin \omega t \sin \phi)$$

$$= [R \cos \phi + (X_L - X_C)\sin \phi]I_{max} \sin \omega t$$

$$+ [-R \sin \phi + (X_L - X_C)\cos \phi]I_{max} \cos \omega t \qquad (42)$$

If we examine this equation at time $t = 0$, we find that

$$0 = -R \sin \phi + (X_L - X_C)\cos \phi \qquad (43)$$

or

Phase angle for series LCR circuit

$$\boxed{\tan \phi = \frac{X_L - X_C}{R}} \qquad (44)$$

From this we obtain

$$\sin \phi = \frac{1}{\sqrt{1 + 1/\tan^2 \phi}} = \frac{X_L - X_C}{\sqrt{R^2 + (X_L - X_C)^2}} \qquad (45)$$

$$\cos \phi = \frac{1}{\sqrt{1 + \tan^2 \phi}} = \frac{R}{\sqrt{R^2 + (X_L - X_C)^2}} \qquad (46)$$

With these expressions for $\sin \phi$ and $\cos \phi$, Eq. (42) reduces to

$$\mathcal{E}_{max} \sin \omega t = \frac{R^2 + (X_L - X_C)^2}{\sqrt{R^2 + (X_L - X_C)^2}} I_{max} \sin \omega t \qquad (47)$$

from which

$$I_{max} = \frac{\mathcal{E}_{max}}{\sqrt{R^2 + (X_L - X_C)^2}} \qquad (48)$$

Equations (44) and (48) are the desired expressions for I_{max} and ϕ in terms of the known parameters of the circuit.

The quantity

$$Z = \sqrt{R^2 + (X_L - X_C)^2}$$

or

Impedance for series LCR circuit

$$\boxed{Z = \sqrt{R^2 + (\omega L - 1/\omega C)^2}} \qquad (49)$$

is called the **impedance** of the series LCR circuit. In terms of this quantity the current is

$$I = \frac{\mathscr{E}_{max} \sin(\omega t + \phi)}{Z} \tag{50}$$

Current in series LCR circuit

The relations among the voltages and the currents in the circuit elements can be represented graphically by a **phasor diagram.** In such a diagram the amplitude of a sinusoidal function is represented by a line segment of length equal to the amplitude of oscillation, and the phase is represented by the angle between this line segment and the horizontal axis. The line segment is called a **phasor.** For instance, the function $I = I_{max} \sin(\omega t - \phi)$ is represented by the phasor shown in Figure 34.16. Obviously, the projection of this phasor on the vertical axis is $I_{max} \sin(\omega t - \phi)$, i.e., the projection is exactly I. Likewise, we can represent each of the voltages on the right side of Eq. (37) by phasors (Figure 34.17). The vector sum of these phasors then represents the sum of these voltages. Kirchhoff's rule demands that this sum be equal to the external emf (Figure 34.18). We can then derive the expressions for the amplitude and for the phase angle of the current by direct inspection of Figure 34.18.

Phasor

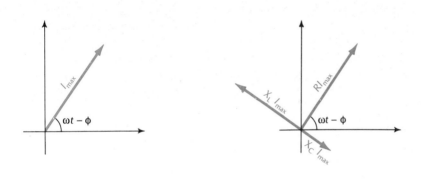

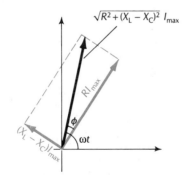

Fig. 34.16 (left) Phasor representing the function $I = I_{max} \sin(\omega t - \phi)$. The phasor makes an angle of $\omega t - \phi$ with the horizontal axis. The projection of the phasor on the vertical axis is $I_{max} \sin(\omega t - \phi)$.

Fig. 34.17 (right) The three phasors representing ΔV_R, ΔV_C, and ΔV_L. Their phase angles are $\omega t - \phi$, $\omega t - \phi - 90°$, and $\omega t - \phi + 90°$, respectively.

Fig. 34.18 In the phasor diagram, the sum of the voltages ΔV_R, ΔV_L, and ΔV_C is represented by the vector sum (black) of the three vectors (color) shown in Figure 34.17. From the diagram we see that this vector sum has a magnitude $\mathscr{E}_{max} = \sqrt{R^2 + (X_L - X_C)^2} \, I_{max}$. The vector sum makes an angle ϕ with the phasor ΔV_R (or with the phasor representing I); from the diagram we see that $\tan \phi = (X_L - X_C)/R$.

The amplitude of the oscillations of the current in the circuit depends critically on the frequency. With $X_C = 1/\omega C$ and $X_L = \omega L$, Eq. (48) exhibits the following dependence on frequency:

$$I_{max} = \frac{\mathscr{E}_{max}}{\sqrt{R^2 + (\omega L - 1/\omega C)^2}} \tag{51}$$

Figure 34.19 is a plot of this maximum current as a function of fre-

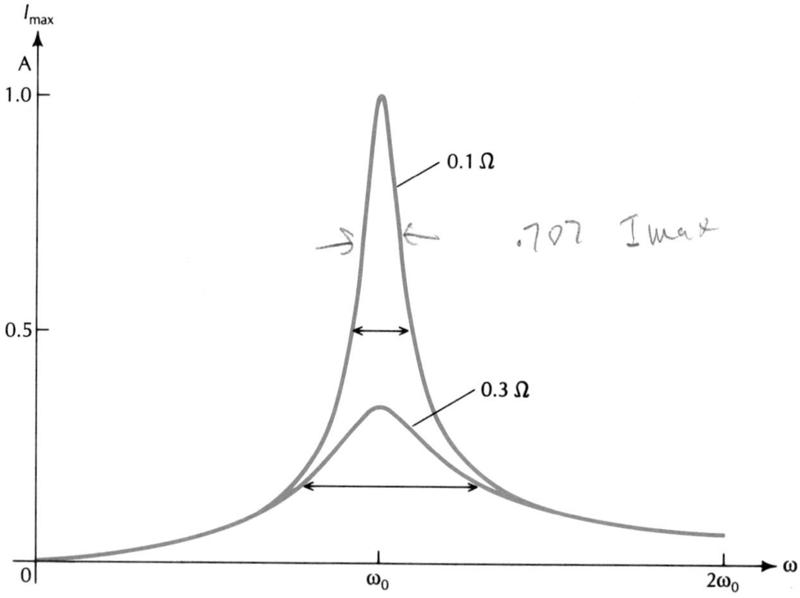

Fig. 34.19 Maximum current in an LCR circuit as a function of angular frequency. The two curves give the currents for two different values of the resistance ($R = 0.1\ \Omega$ and $R = 0.3\ \Omega$), but for the same values of the inductance, the capacitance, and the driving emf ($L = 10^{-6}$ H, $C = 10^{-6}$ F, $\mathscr{E} = 0.1$ V). The arrows indicate the bandwidths.

quency. The current is weak when ω is near zero, and it is also weak when ω is very large. However, when ω is near $\omega_0 = 1/\sqrt{LC}$, the current becomes very strong. This is a resonance phenomenon: the oscillations become large when the driving force pushes the circuit with the same frequency as that of the natural oscillations [see Eq. (26)]. When the frequency ω exactly matches the natural frequency ($\omega = \omega_0 = 1/\sqrt{LC}$), the amplitude of the oscillations of the current reaches the value

$$I_{max} = \frac{\mathscr{E}_{max}}{R} \tag{52}$$

This shows that the height of the resonance peak in Figure 34.19 decreases in inverse proportion to R. For comparison, the second curve in Figure 34.19 is a plot of the maximum current in an LCR circuit with a larger value of the resistance. This other resonance peak is lower and broader. The width of the peak, measured between the points below and above resonance at which the curve has one-half of its maximum height, is called the **bandwidth** of the LCR circuit. This terminology is adopted from radio engineering, where frequency intervals are commonly called bands. LCR circuits are widely used in radio receivers to selectively enhance incoming radio signals of a frequency close to the resonant frequency of the circuit (compare Example 2). The radio signal arriving at the antenna constitutes a source of emf, and it will generate a large current if its frequency matches the resonant frequency of the circuit.

Let us now examine the phase of the current given by Eq. (50). If the driving frequency is below resonance ($\omega < \omega_0$), then $X_L - X_C$ is negative, and ϕ is also negative [see Eq. (44)]. The current then leads the external emf; that is, the maxima in the current occur earlier than those in the emf (see Figure 34.20a). If the driving frequency is above resonance, ($\omega > \omega_0$), then $X_L - X_C$ is positive, and ϕ is also positive [see Eq. (44)]. The current then lags the external emf; that is, the maxima in the current occur later than those in the emf (Figure 34.20b).

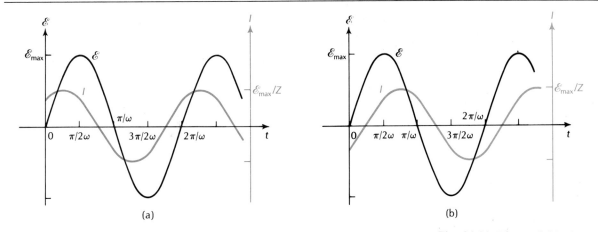

Fig. 34.20 The emf (black) and current (color) in the LCR circuit as a function of time. (a) $\omega < \omega_0$; the current leads the emf. (b) $\omega > \omega_0$; the current lags the emf.

Finally, what is the power delivered by the source of emf to the LCR circuit? The instantaneous power is

$$P = I\mathcal{E} = \frac{1}{Z}\mathcal{E}_{max}^2 \sin(\omega t - \phi)\sin \omega t \tag{53}$$

or

$$P = \frac{1}{Z}\mathcal{E}_{max}^2 (\sin^2 \omega t \cos \phi - \cos \omega t \sin \omega t \sin \phi) \tag{54}$$

To find the time-average power, we note that the time-average of $\sin^2 \omega t$ is $\frac{1}{2}$ [see Eq. (5)], and the time-average of $\cos \omega t \sin \omega t$ is zero. Hence

$$\overline{P} = \frac{\mathcal{E}_{max}^2}{2Z} \cos \phi \tag{55}$$

which we can also write as

$$\boxed{\overline{P} = \frac{\mathcal{E}_{rms}^2}{Z} \cos \phi} \tag{56}$$

Average power absorbed by LCR circuit

This equation is analogous to the familiar equation $P = \mathcal{E}_{rms}^2/R$ for the power delivered to a resistor. However, Eq. (56) includes an extra factor $\cos \phi$, called the **power factor**. This factor takes into account the phase between the current and the emf. For a circuit with zero resistance, the phase angle is $\pm 90°$ [see Eq. (44) with $R = 0$], and the power factor is $\cos(\pm 90°) = 0$. Thus, as expected, such a circuit consumes no power.

Power factor

EXAMPLE 3. An LC circuit with $L = 3.0 \times 10^{-4}$ H and $C = 2.0 \times 10^{-6}$ F is being driven by an oscillating source delivering an AC emf of amplitude 0.40 V at a frequency of 5.0×10^4 radians/s. What is the peak instantaneous voltage across the inductor? Across the capacitor?

SOLUTION: The natural frequency of this circuit, which we will need later in our calculation, is

$$\omega_0 = \frac{1}{\sqrt{LC}}$$

$$= \frac{1}{\sqrt{3.0 \times 10^{-4} \text{ H} \times 2.0 \times 10^{-6} \text{ F}}}$$

$$= 4.1 \times 10^4 \text{ radians/s} \tag{57}$$

The instantaneous voltage across the inductor is

$$\Delta V_L = X_L I_{max} \cos(\omega t - \phi) = \omega L I_{max} \cos(\omega t - \phi)$$

With Eq. (51) and with $R = 0$ this gives

$$\Delta V_L = \frac{\omega L \mathscr{E}_{max} \cos(\omega t - \phi)}{\omega L - 1/\omega C} \tag{58}$$

This has a peak value of

$$\frac{\omega L \mathscr{E}_{max}}{\omega L - 1/\omega C} = \frac{\mathscr{E}_{max}}{1 - 1/(\omega^2 LC)} = \frac{\mathscr{E}_{max}}{1 - \omega_0^2/\omega^2}$$

$$= \frac{0.40 \text{ V}}{1 - \frac{(4.1 \times 10^4/\text{s})^2}{(5.0 \times 10^4/\text{s})^2}} = 1.20 \text{ V}$$

The instantaneous voltage across the capacitor is

$$\Delta V_C = -X_C I_{max} \cos(\omega t - \phi)$$

$$= -\frac{1}{\omega C} \frac{\mathscr{E}_{max} \cos(\omega t - \phi)}{\omega L - 1/\omega C} \tag{59}$$

This has a peak value of

$$-\frac{1}{\omega C} \frac{\mathscr{E}_{max}}{\omega L - 1/\omega C} = -\frac{\mathscr{E}_{max}}{\omega^2/\omega_0^2 - 1}$$

$$= -\frac{0.40 \text{ V}}{\frac{(5.0 \times 10^4/\text{s})^2}{(4.1 \times 10^4/\text{s})^2} - 1} = -0.80 \text{ V}$$

COMMENTS AND SUGGESTIONS: After a calculation of the voltages across the elements of an LCR circuit, it is a good idea to check explicitly whether the results satisfy Kirchhoff's rule, Eq. (37). For such a check, we need the value of the phase angle. Since, in the present example, the resistance is zero and $X_L - X_C > 0$, Eq. (44) yields a phase angle of $\phi = 90°$. According to Eqs. (58) and (59), the sum of the instantaneous voltages across the inductor and the capacitor is then

$$\Delta V_L + \Delta V_C = 1.20 \text{ V} \times \cos(\omega t - 90°) - 0.80 \text{ V} \times \cos(\omega t - 90°)$$

$$= 0.40 \text{ V} \times \cos(\omega t - 90°) = 0.40 \text{ V} \times \sin \omega t$$

This is, indeed, equal to the driving emf.

EXAMPLE 4. Figure 34.21 shows a parallel LCR circuit with an oscillating source of emf. What is the net current delivered by the source of emf?

SOLUTION: If the emf of the source is $\mathscr{E} = \mathscr{E}_{max} \sin \omega t$, then this is also the emf acting on the inductor, the capacitor, and the resistor individually. The currents in the inductor, capacitor, and resistor are then, respectively [see Eqs. (19), (13), and (3)]:

$$I_L = -\frac{\mathscr{E}_{max} \cos \omega t}{X_L}$$

$$I_C = \frac{\mathscr{E}_{max} \cos \omega t}{X_C}$$

$$I_R = \frac{\mathscr{E}_{max} \sin \omega t}{R}$$

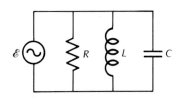

Fig. 34.21 Inductor, capacitor, and resistor connected in parallel to a source of alternating emf.

The net current delivered by the source of emf is the sum of these currents. To evaluate this sum, we can use a phasor diagram. Figure 34.22 shows the phasors representing the currents I_L, I_C, and I_R. The net current is represented by the vector sum of these phasors. By inspection of Figure 34.22, we see that the vector sum has a magnitude

$$I_{max} = \mathscr{E}_{max} \sqrt{\frac{1}{R^2} + \left(\frac{1}{X_L} - \frac{1}{X_C}\right)^2} \tag{60}$$

or

$$I_{max} = \mathscr{E}_{max} \sqrt{\frac{1}{R^2} + \left(\frac{1}{\omega L} - \omega C\right)^2} \tag{61}$$

The angle between the phasor representing the net current and the phasor representing the current in the resistor (or the phasor representing the emf) is given by

$$\boxed{\tan \phi = \frac{1/X_L - 1/X_C}{1/R}} \tag{62}$$

Phase angle for parallel LCR circuit

We can then write the current as

$$I = I_{max} \sin(\omega t - \phi) = \mathscr{E}_{max} \sqrt{\frac{1}{R^2} + \left(\frac{1}{\omega L} - \omega C\right)^2} \sin(\omega t - \phi)$$

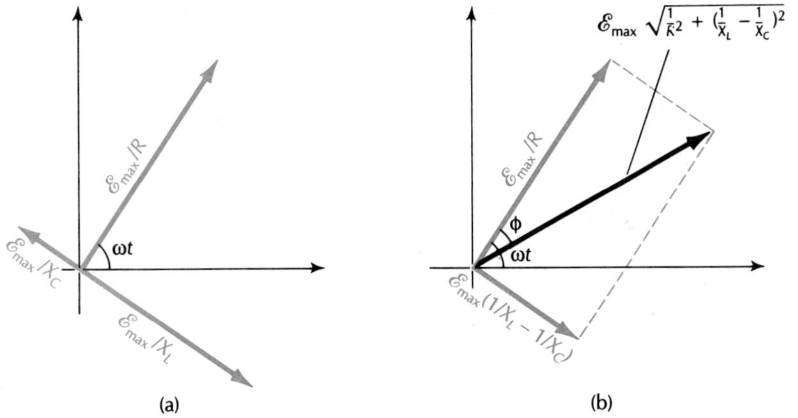

Fig. 34.22 (a) The three phasors (color) representing I_L, I_C, and I_R. (b) The vector sum (black) of these phasors represents the net current.

Current in parallel LCR circuit

or as

$$I = \frac{\mathcal{E}_{max} \sin(\omega t - \phi)}{Z} \qquad (63)$$

where

Impedance for parallel LCR circuit

$$Z = 1 \Big/ \sqrt{\frac{1}{R^2} + \left(\frac{1}{\omega L} - \omega C\right)^2} \qquad (64)$$

is the impedance of the parallel LCR circuit.

COMMENTS AND SUGGESTIONS: If the frequency coincides with the resonant frequency $\omega = \omega_0 = 1/\sqrt{LC}$, the current I specified by Eq. (63) is *minimum*. This is so because, at this frequency, the currents in the inductor and the capacitor are of equal magnitudes; since these currents are always in opposite directions, the equality of their magnitudes implies their cancellation.

34.4 The Transformer

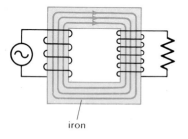

Fig. 34.23 A transformer.

A transformer consists of two coils arranged in such a way that (almost) all the magnetic field lines generated by one of them pass through the other. This can be achieved by winding both the coils on a common iron core (Figure 34.23). As we saw in Chapter 33, the iron increases the strength of the magnetic field in its interior by a large factor. Since the field is much stronger inside the iron than outside, the field lines must concentrate inside the iron; thus the iron tends to keep the field lines together and acts as a conduit for the field lines from one coil to the other.

Each coil is part of a separate electric circuit (Figure 34.24). The **primary** circuit has a source of alternating emf, and the **secondary** circuit has a resistance or some other load that consumes electric power. The alternating current in the primary circuit induces an alternating emf in the secondary circuit. We will show that the induced emf $\mathcal{E}_2$ in the secondary circuit is related as follows to the emf $\mathcal{E}_1$ in the primary circuit:

Emf in primary and secondary circuits of transformer

$$\mathcal{E}_2 = \mathcal{E}_1 \frac{N_2}{N_1} \qquad (65)$$

where N_1 and N_2 are, respectively, the numbers of turns in the primary and secondary coils.

Fig. 34.24 Circuit diagram for the transformer. The parallel lines represent the mutual inductance.

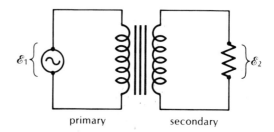

To prove Eq. (65), we begin with Kirchhoff's rule as it applies to the primary circuit: the emf $\mathscr{E}_1$ of the source must equal the induced emf $\mathscr{E}_{1,\text{ind}}$ across the primary coil. But by Faraday's Law, the induced emf equals the rate of change of flux,

$$\mathscr{E}_1 = \mathscr{E}_{1,\text{ind}} = -\frac{d\Phi_1}{dt} \qquad (66)$$

Likewise, the emf $\mathscr{E}_2$ delivered to the load must equal the induced emf $\mathscr{E}_{2,\text{ind}}$ in the secondary coil, which, in turn, equals the rate of change of flux in that coil,

$$\mathscr{E}_2 = \mathscr{E}_{2,\text{ind}} = -\frac{d\Phi_2}{dt} \qquad (67)$$

Since the same numbers of magnetic field lines pass through both coils, the fluxes and their rates of change are necessarily in the ratio N_2/N_1,

$$\frac{d\Phi_2}{dt} = \frac{N_2}{N_1}\frac{d\Phi_1}{dt} \qquad (68)$$

From Eqs. (66), (67), and (68) we obtain

$$\mathscr{E}_2 = -\frac{N_2}{N_1}\frac{d\Phi_1}{dt} = \frac{N_2}{N_1}\mathscr{E}_1 \qquad (69)$$

which we wanted to prove.

If $N_2 > N_1$, we have a step-up transformer; and if $N_2 < N_1$, a step-down transformer.

EXAMPLE 5. Door bells and buzzers usually are designed for 12-volt AC, and they are powered by small transformers which step down 115-volt AC to 12-volt AC. Suppose that such a transformer has a primary winding with 1500 turns. How many turns must there be in the secondary winding?

SOLUTION: Equation (65) applies to the instantaneous voltages. It is therefore also valid for the rms voltages. With our numerical values

$$N_2 = N_1 \frac{\mathscr{E}_2}{\mathscr{E}_1} = 1500 \times \frac{12\text{ V}}{115\text{ V}} = 157 \text{ turns}$$

As long as the secondary circuit is open and carries no current ($I_2 = 0$), an ideal transformer does not consume electric power. Under these conditions, the primary circuit consists of nothing but the source of emf and an inductor, that is, it is a pure L circuit. In such a circuit the power delivered by the source of emf to the inductor averages to zero (see Section 34.1).

If the secondary circuit is closed, a current will flow ($I_2 \neq 0$). This current contributes to the magnetic flux in the transformer and induces a current in the primary circuit. The current in the latter is then different from that in a pure L circuit, and the power will *not* average to zero over a cycle. In an ideal transformer, the electric power that the primary circuit takes from the source of emf exactly matches the power that the secondary circuit delivers to the load, and thus

$$I_1 \mathscr{E}_1 = I_2 \mathscr{E}_2 \qquad (70)$$

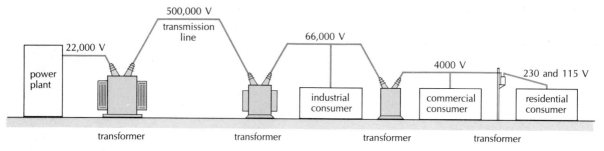

Fig. 34.25 Typical voltage transformations during the transmission of electric power from power plant to consumer.

Good transformers approach this ideal condition fairly closely: about 99% of the power supplied to the input terminals emerges at the output terminals; the difference is lost as heat in the iron core and in the windings.

Transformers play a large role in our electric technology. As we saw in Section 29.5, transmission lines for electric power operate much more efficiently at high voltage since this reduces the Joule losses. To take advantage of this high efficiency, power lines are made to operate at several hundred kilovolts (see Figure 34.25). The voltage must be stepped up to this value at the power plant, and, for safety's sake, it must be stepped down just before it reaches the consumer. For these transformations, banks of large transformers are used at both ends of the transmission line (see Figure 34.26). Transformers are also widely used in TV receivers, computers, X-ray machines, and so on, where high voltages are required.

Fig. 34.26 A large transformer used for stepping 500,000 volts down to 66,000 volts. This transformer is capable of handling 10^6 kW. The iron core and the primary and secondary coils are immersed in oil, which circulates through the external cooling tubes (left and right) and eliminates the waste heat produced in the core.

SUMMARY

Average AC power absorbed by resistor: $\overline{P} = \dfrac{\mathscr{E}_{\max}^2}{2R}$

$$= \dfrac{\mathscr{E}_{\text{rms}}^2}{R}$$

Natural frequency of LC circuit: $\omega_0 = \dfrac{1}{\sqrt{LC}}$

Energy loss in freely oscillating series LCR circuit: $"Q" = 2\pi |U/\Delta U|$
$$= \omega_0 L/R$$

Impedance of series LCR circuit: $Z = \sqrt{R^2 + (\omega L - 1/\omega C)^2}$

Impedance of parallel LCR circuit: $Z = 1 \bigg/ \sqrt{\dfrac{1}{R^2} + \left(\dfrac{1}{\omega L} - \omega C\right)^2}$

emf in transformer: $\mathscr{E}_2 = \mathscr{E}_1 \dfrac{N_2}{N_1}$

QUESTIONS

1. In most European countries, the voltage available at outlets in homes is 240-V AC. What is the actual amplitude of oscillation of this voltage?

2. You can perceive the 120-Hz flicker (two peaks of intensity per AC cycle) in a fluorescent light tube (by sweeping your eye quickly across the tube), but you cannot perceive any such flicker in an incandescent light bulb. Explain.

3. Do the electrons from the power station ever reach the wiring of your house?

4. Some electric motors operate only on DC, others only on AC. What is the difference between these motors?

5. If you connect a capacitor across a 115-V outlet, does any current flow through the connecting wires? Through the space between the capacitor plates? Does the outlet deliver instantaneous electric power? Average electric power?

6. It is sometimes said that a capacitor becomes a short circuit at high frequencies, and that an inductor becomes an open circuit at high frequencies. Explain.

7. Can you blow a fuse by connecting a very large capacitor across an ordinary 115-V outlet?

8. How could you use an LC circuit to measure the capacitance of a capacitor?

9. In Section 34.3 we asserted that under steady-state conditions the current in an LCR circuit has the same frequency as the driving emf. Is this also true if the circuit is not in steady state? Give an example.

10. Consider a series LCR circuit. Can the voltage across the capacitor ever be larger than $\mathscr{E}_{max}$? Across the inductor? Across the resistor? (Hint: Inspect the phasor diagram.)

11. Roughly plot the phase angle φ given by Eq. (44) as a function of ω. What is the phase angle at resonance?

12. If you use an AC voltmeter to measure the driving emf and the voltages across the inductor, the capacitor, and the resistor in a series LCR circuit, you will find that the emf is larger than the voltage across the resistor, but smaller than the sum of the voltages across the inductor, capacitor, and resistor. Explain.

13. What is the impedance of a series LCR circuit at resonance?

14. If we substitute $R = 0$ in Eq. (50), we obtain the equation for the current in a driven LC circuit. What must we substitute to obtain the equation for a driven CR circuit? A driven LR circuit?

15. Show that the equation for the current in a series LCR circuit [Eq. (50)] includes Eqs. (3), (13), and (19) as special cases. What are the individual impedances of a resistor, a capacitor, and an inductor?

16. Show that the equation for the current in a parallel LCR circuit [Eq. (63)] includes Eqs. (3), (13), and (19) as special cases.

17. Consider the parallel LCR circuit described in Example 4. If the driving frequency is greater than $1/\sqrt{LC}$, does the net current lead or lag the emf? If the driving frequency is smaller than $1/\sqrt{LC}$?

18. Show that the time-average power absorbed by a series LCR circuit [Eq. (56)] is maximum at resonance.

19. In a parallel LCR circuit, the time-average power absorbed is $\mathscr{E}_{max}^2/2R$. Explain.

20. Why can we not use a transformer to step up the voltage supplied by a battery?

21. Does an electric motor absorb more electric power when pulling a mechanical load than when running freely?

PROBLEMS

Section 34.1

1. An electric heater plugged into a 115-V AC outlet uses an average electric power of 1200 W.
   (a) What is the rms current and the maximum instantaneous current through the heater?
   (b) What is the maximum instantaneous power and the minimum instantaneous power?

2. An immersible heating element used to boil water consumes an (average) electric power of 400 W when connected to a source of 115 volts AC. Suppose that you connect this heating element to a source of 115 volts DC. What power will it consume?

3. A high-voltage power line operates on an rms voltage of 230,000 volts AC and delivers an rms current of 740 A.
   (a) What are the maximum instantaneous voltage and current?
   (b) What are the maximum instantaneous power and the average power delivered?

4. The GG-1 electric locomotive develops 4600 hp; it runs on an AC voltage of 1100 V.
   (a) What rms current does this locomotive draw?
   (b) Why is it advantageous to supply the electric power for locomotives at high voltage (and fairly low current)?

5. An electric heater operating with a 115-V AC power supply delivers 1200 W of heat.
   (a) What is the rms current through this heater?
   (b) What is the maximum instantaneous current?
   (c) What is the resistance of this heater?

*6. An AC current of 20 A flows in a copper wire of diameter 0.30 cm connected to an electric outlet. The drift velocity of the free electrons in the copper will then oscillate at 60 Hz. What is the maximum value of the instantaneous drift velocity? What is the maximum value of the acceleration of the drift velocity?

*7. A circuit consists of a resistor connected in series to a battery; the resistance is 5 Ω and the emf of the battery is 12 V. The wires (of negligible resistance) connecting these circuit elements are laid out along a square of 20 cm × 20 cm (Figure 34.27). The entire circuit is placed face on in an oscillat-

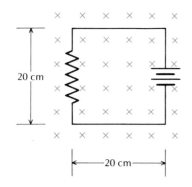

Fig. 34.27

ing magnetic field. The instantaneous value of the magnetic field is

$$B = B_0 \sin \omega t$$

with $B_0 = 0.15$ T and $\omega = 360$ radians/s.
  (a) Find the instantaneous current in the resistor.
  (b) Find the average power dissipated in the resistor.

*8. A circuit consists of two capacitors of 6.0×10^{-8} F and 9.0×10^{-8} F connected in series to an oscillating source of emf (Figure 34.28). This source delivers a sinusoidal emf $\mathcal{E} = 1.8 \sin(120\pi t)$, where $\mathcal{E}$ is in volts and t in seconds.
  (a) Find the charge on each capacitor as a function of time.
  (b) At what time is the charge on the capacitors maximum? At what time minimum?
  (c) What is the maximum energy in the capacitors? What is the time-average energy?

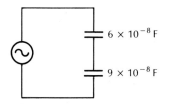

Fig. 34.28

*9. An inductor of 1.6×10^{-3} H is connected to a source of alternating emf. The current in the inductor is $I = I_0 \cos \omega t$, with $I_0 = 180$ A and $\omega = 120\pi$ radians/s.
  (a) What is the potential difference across the inductor at time $t = 0$? At time $t = 1/240$ s?
  (b) What is the energy in the inductor at time $t = 0$? At time $t = 1/240$ s?
  (c) What is the instantaneous power delivered by the source of emf to the inductor at time $t = 0$? At time $t = 1/240$ s?

*10. Consider the circuit shown in Figure 34.29. The emf is of the form $\mathcal{E}_0 \sin \omega t$. In terms of this emf and the capacitance C and inductance L, find the instantaneous currents through the capacitor and the inductor. Find the instantaneous current and the instantaneous power delivered by the source of emf.

*11. A capacitor with $C = 8.0 \times 10^{-7}$ F is connected to an oscillating source of emf. This source provides an emf $\mathcal{E} = \mathcal{E}_{max} \sin \omega t$, with $\mathcal{E}_{max} = 0.20$ V and $\omega = 6.0 \times 10^3$ radians/s.
  (a) What is the reactance of a capacitor?
  (b) What is the maximum current in the circuit?
  (c) What is the current at time $t = 0$? At time $t = \pi/4\omega$?

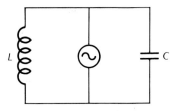

Fig. 34.29 Inductor and capacitor connected to a source of alternating emf.

*12. An inductor with $L = 4.0 \times 10^{-2}$ H is connected to an oscillating source of emf. This source provides an emf $\mathcal{E} = \mathcal{E}_{max} \sin \omega t$, with $\mathcal{E}_{max} = 0.20$ V and $\omega = 6.0 \times 10^3$ radians/s.
  (a) What is the reactance of the inductor?
  (b) What is the maximum current in the circuit?
  (c) What is the current at time $t = 0$? At time $t = \pi/4\omega$?

Section 34.2

13. What is the natural frequency for an LC circuit consisting of a 2.2×10^{-6} F capacitor and a 8.0×10^{-2} H inductor?

14. A radio receiver contains an LC circuit whose natural frequency of oscillation can be adjusted, or tuned, to match the frequency of incoming radio waves. The adjustment is made by means of a variable capacitor. Suppose that the inductance of the circuit is 15 μH. Over what range of capacitances must the capacitor be adjustable if the frequencies of oscillation of the circuit are to span the range from 530 kHz to 1600 kHz?

*15. What is the natural frequency of oscillation of the circuit shown in Figure 34.30? The capacitances are 2.4×10^{-5} F each, and the inductance is 1.2×10^{-3} H.

*16. An LC circuit has an inductance of 5.0×10^{-2} H and a capacitance of 5.0×10^{-6} F. At $t = 0$, the capacitor is fully charged so that $Q_0 = 1.2 \times 10^{-4}$ C. What is the energy in this circuit? At what time, after $t = 0$, will the energy be purely magnetic? At what *later* time will it be purely electric?

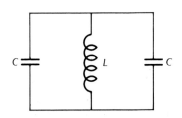

Fig. 34.30 Two equal capacitors connected to an inductor.

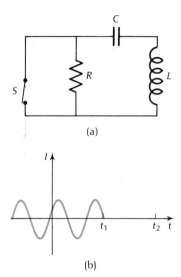

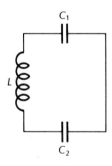

Fig. 34.31 (a) Inductor, capacitor, and resistor connected in a circuit. The switch S is initially closed. (b) Current in the circuit as a function of time. At $t = t_1$, the switch S is opened.

Fig. 34.32

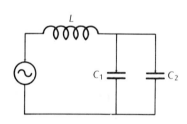

Fig. 34.33 Two capacitors and an inductor connected to a source of alternating emf.

*17. Consider the LC circuit described in Problem 16. If we add a resistance of 120 Ω to this circuit (in series), what will be the frequency of the natural oscillations? By what factor is this smaller than the frequency of the circuit without resistance?

*18. The circuit of Figure 34.31a is oscillating with the switch S closed. The graph of current vs. time is shown in Figure 34.29b.
  (a) At time t_1 the switch S is suddenly opened. Is the frequency of oscillation increased, decreased, or unchanged? In the space on the left in Figure 34.29b sketch the graph of current for times after t_1.
  (b) At time t_2 the switch S is closed. Sketch the graph of current after this time.

*19. What is the resonant frequency of the circuit shown in Figure 34.32? The inductance is $L = 1.5$ H, and the capacitances are $C_1 = 20$ μF and $C_2 = 10$ μF.

*20. An LCR series circuit has $L = 0.5$ H, $C = 2.0 \times 10^{-5}$ F, and $R = 10$ Ω. At time $t = 0$, the capacitor is fully charged with a voltage of 24 V across its plates.
  (a) What is the initial energy in the circuit?
  (b) What is the percentage loss of energy per period?
  (c) At what time after $t = 0$ will the energy in the circuit have fallen to one-half of its initial value? One-tenth of its initial value?
  (d) Plot the energy as a function of time for the time interval $0 \text{ s} \leq t \leq 0.1$ s.

Section 34.3

21. A driven LC circuit consists of an inductor of 3.0×10^{-3} H, a capacitor of 2.0×10^{-8} F, and an oscillating source of emf operating at a frequency of 1.0×10^5 radians/s, all connected in series. If the amplitude of the current in the circuit is to be 6.0×10^{-3} A, what must be the amplitude of the alternating emf?

*22. A series LC circuit is being driven by an audio generator that delivers an emf of amplitude 0.50 V. When the generator delivers the emf at a frequency of 2.0×10^3 radians/s, the maximum current is 1.0×10^{-1} A; when the generator delivers the same emf at a frequency of 1.5×10^3 radians/s, the maximum current is 2.7×10^{-2} A. From this information deduce the values of the inductance and the capacitance.

*23. A capacitor of $C = 24.0$ μF and an inductor of $L = 0.180$ H are connected in series with an oscillating source that delivers an emf $\mathscr{E} = 12.0 \sin 377t$, where t is measured in seconds and $\mathscr{E}$ in volts.
  (a) What is the maximum instantaneous current?
  (b) What is the maximum instantaneous voltage across the capacitor?
  (c) Across the inductor?

*24. A series LC circuit with $C = 1.5 \times 10^{-7}$ F and $L = 2.5 \times 10^{-4}$ H is being driven by an audio generator that delivers a sinusoidal emf with an amplitude of 0.80 V and a frequency 2.2×10^4 Hz.
  (a) Plot the emf as a function of time.
  (b) Plot the current as a function of time.

*25. A series LC circuit with $L = 3.0 \times 10^{-4}$ H and $C = 2.0 \times 10^{-5}$ F is being driven by an oscillating source of emf delivering an AC voltage of amplitude 0.40 V and frequency of 1.6×10^4 radians/s. What is the maximum instantaneous value of the current? What is the maximum instantaneous voltage across the inductor? Across the capacitor?

*26. Consider the circuit shown in Figure 34.33. The oscillating source delivers a sinusoidal emf of amplitude 0.80 V and frequency of 400 Hz. The inductance is 5.0×10^{-2} H, and the capacitances are 8.0×10^{-7} F and 16.0×10^{-7} F. Find the maximum instantaneous current in each capacitor.

*27. Consider the circuit described in Example 3. If we add a resistor with $R = 10\ \Omega$ in series with the other elements of this circuit, what will be the peak instantaneous voltage across the inductor? Across the capacitor? Across the resistor? What will be the average power dissipated in the resistor?

*28. (a) Consider the driven LCR circuit described by Eqs. (35)–(51). If $L = 6.0 \times 10^{-2}$ H, $C = 3.0 \times 10^{-6}$ F, $R = 1.2 \times 10^2\ \Omega$, $\mathscr{E}_{max} = 24.0$ V, and $\omega = 2.5 \times 10^3$ radians/s, calculate the maximum value of the current in this circuit under steady-state conditions.
(b) At what time after $t = 0$ does the alternating emf of the source reach its maximum value? At what time does the current reach its maximum value?
(c) Make a plot of $\mathscr{E}$ as a function of time. On top of this plot, make a plot of I as a function of time.

*29. An LCR circuit consists of an inductor of 1.5×10^{-2} H, a capacitor of 2.8×10^{-6} F, and a resistor of $5.0\ \Omega$ connected in series to a source of alternating emf. The source delivers a voltage of $\mathscr{E} = \mathscr{E}_{max} \sin \omega t$, with $\mathscr{E}_{max} = 0.60$ V and $\omega = 6.0 \times 10^4$ radians/s.
(a) What is the impedance of this circuit?
(b) What is the maximum current in this circuit?
(c) What is the phase angle of the current?
(d) Draw a phasor diagram, with the appropriate lengths and angles for the phasors.

*30. An LCR circuit consists of an inductor of 1.2×10^{-2} H, a capacitor of 2.4×10^{-6} F, and a resistor of $2.0\ \Omega$ connected in series to a source of alternating emf with an amplitude of 0.80 V.
(a) At what frequency will this circuit be in resonance with the driving voltage?
(b) What is the maximum current in the circuit at resonance?
(c) What is the average dissipation of power in the circuit at resonance?

*31. An LR circuit consists of an inductor with $L = 2.0 \times 10^{-4}$ H and a resistor with $R = 1.2\ \Omega$ connected in series to an oscillating source of emf. This source generates a voltage $\mathscr{E} = \mathscr{E}_{max} \sin \omega t$, with $\mathscr{E}_{max} = 0.50$ V and $\omega = 3 \times 10^3$ radians/s. Find the maximum current in the circuit. Find the phase angle of the current and draw a phasor diagram, with the correct lengths and angles for the phasors. Find the average dissipation of power in the resistor.

*32. An RC circuit consists of a resistor with $R = 0.80\ \Omega$ and a capacitor with $C = 1.5 \times 10^{-4}$ F connected in series to an oscillating source of emf. The source generates a voltage $\mathscr{E} = \mathscr{E}_{max} \sin \omega t$, with $\mathscr{E}_{max} = 0.40$ V and $\omega = 9 \times 10^3$ radians/s. Find the maximum current in the circuit. Find the phase angle of the current and draw a phasor diagram, with the correct lengths and angles for the phasors. Find the average dissipation of power in the resistor.

*33. Show that the time-average power absorbed from the external emf by a series LCR circuit can be expressed in the form

$$\overline{P} = \frac{\mathscr{E}_{rms}^2 R}{R^2 + \left(\omega L - 1/\omega C\right)^2}$$

Show that this power is maximum at resonance. What is the value of the maximum power?

*34. An empty solenoid has an inductance of 1.0×10^{-3} H, and its windings have a resistance of $4.0\ \Omega$. This internal resistance can be regarded as connected in series with the inductance, so the solenoid connected to a source of emf is equivalent to an LR circuit. Suppose that the source of emf delivers an oscillating voltage of amplitude 2.0 V and a frequency of 6000 radians/s. What is the average power consumed by the solenoid? If we insert an iron core into this solenoid and thereby increase its inductance by a factor of 4000, what will be the average power consumed?

*35. Consider the circuit of Example 3. What is the maximum energy in this circuit at one instant of time? The minimum energy?

*36. Derive an expression for the instantaneous power absorbed by the parallel LCR circuit of Example 4. Show that the time-average power is $\mathcal{E}_{max}^2/(2R)$.

**37. Consider an LC circuit (Figure 34.34) with a driving emf $\mathcal{E} = \mathcal{E}_{max} \sin \omega t$.
 (a) Show that if $\omega \gg 1/\sqrt{LC}$, the amplitude of the potential difference across the terminals in Figure 34.34 is $\Delta V \cong (X_C/X_L) \mathcal{E}_{max}$.
 (b) Show that ΔV is much smaller than $\mathcal{E}_{max}$. Thus, this circuit can be used as a **filter** that strongly attenuates high-frequency components in the driving emf.

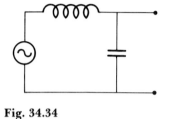

Fig. 34.34

Section 34.4

38. A transformer used to step up 110 V to 5000 V has a primary coil of 100 turns. What must be the number of turns in the secondary coil?

39. A transformer operating on a primary voltage of 110 volts AC delivers a secondary voltage of 6.0 volts AC to a small electric buzzer. If the current in the secondary circuit is 3.0 A, what is the rms current in the primary circuit? Assume that no electric power is lost in the transformer.

40. The generators of a large power plant deliver an electric power of 2000 MW at 22 kilovolts AC. For transmission, this voltage is stepped up to 400 kV by a transformer. What is the rms current delivered by the generators? What is the rms current in the transmission line? Assume that the transformer does not waste any power.

41. A power station feeds 1.0×10^8 W of electric power at 760 kV into a transmission line. Suppose that 10% of this power is lost in Joule heat in the transmission line. What percentage of the power would be lost if the power station were to feed 340 kV into the transmission line instead of 760 kV, other things being equal?

42. The largest transformer ever built handles a power of 1.50×10^9 W. This transformer is used to step down 765 kV to 345 kV. What is the rms current in the primary? What is the current in the secondary? Assume that no electric power is lost by the transformer.

43. A transformer consists of two concentric coils of thin insulated wire of low resistance. One coil has 800 turns and the other has 200 turns. The second coil is wound tightly around the first, so the two have nearly the same radius. If the first coil is connected to a source of oscillating emf supplying a voltage of amplitude 60 V and frequency 1500 Hz, what will be the voltage across the terminals of the second coil?

INTERLUDE VIII

SUPERCONDUCTIVITY*

Of all gases, helium has the lowest liquefaction temperature. All other gases had been liquefied during the nineteenth century, but helium resisted the best efforts of low-temperature physicists until 1908 when it was finally liquefied by H. Kamerlingh Onnes[1] at the University of Leiden. The temperature of boiling helium is 4.2 K (at atmospheric pressure).

The availability of liquid helium laid the realm of low-temperature physics open to exploration. Any material can be cooled to 4.2 K merely by immersing it in liquid helium. Furthermore, lower temperatures can be attained by pumping on the helium; this means that with a vacuum pump connected to a closed vessel containing the liquid helium, the space above the liquid is partially evacuated to a low pressure. The liquid then cools by evaporation until it reaches a lower temperature corresponding to the lower pressure (recall that the boiling point of a liquid is lowered by a reduction of pressure; see Section 20.5). By this method, temperatures slightly below 1 K can be attained.

At such extremely low temperatures, materials develop very unusual properties. Many metals and alloys become **superconductors,** that is, their resistance to electric currents vanishes completely. Liquid helium, at a temperature of 2.2 K, becomes a **superfluid,** that is, its internal friction disappears and it can flow without drag through very fine capillary holes. These strange properties are a manifestation of a high degree of order within the material. As we know from the Third Law of Thermodynamics, the entropy vanishes at absolute zero, that is, the disorder caused by random thermal disturbances disappears. The material then settles into a definite microscopic state. This microscopic state is a quantum state that cannot be adequately described by classical mechanics. Thus, superconductivity and superfluidity are macroscopic manifestations of the underlying quantum behavior of matter.

---

* This chapter is optional.
[1] **Heike Kamerlingh Onnes,** 1853–1926, Dutch physicist and professor at Leiden. He was awarded the Nobel Prize in 1913 for his investigations of the properties of matter at low temperatures.

VIII.1 ZERO RESISTANCE

Superconductivity was discovered by Onnes in 1911 during electrical experiments with a sample of frozen mercury. Figure VIII.1 shows a plot of the measured values of the resistivity of mercury as a function of temperature. At a temperature of 4.15 K, the resistivity drops sharply, and below this critical temperature, the resistivity is zero.

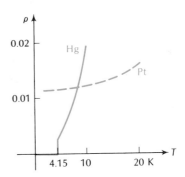

Fig. VIII.1 Resistivity of a sample of mercury as a function of temperature. The resistivity has been expressed as a fraction of the resistivity at 273 K. Below 4.15 K, mercury is a superconductor. For comparison, the dashed curve shows the resistivity of platinum, which is a normal conductor.

The sudden change of resistivity indicates that the material has suffered some drastic alteration of state. Experimentation reveals that in superconducting mercury, as in any metal, the carriers of electric current are free electrons. Hence, it must be that these free electrons suffer some alteration in their state. We will discuss what happens to the electrons in a later section.

Besides mercury, several other metals exhibit superconductivity at low temperatures. Table VIII.1 lists some superconducting metals and their transition temperatures. Furthermore, a large variety of compounds and alloys exhibit superconductivity. For many years, the highest known transition temperature was 23 K, found in a niobium–germanium alloy. But recently several new ceramic compounds containing copper oxide were discovered, with transition temperatures ranging

Table VIII.1 SOME SUPERCONDUCTORS

Material	T_c
Zinc	0.87 K
Aluminum	1.20
Indium	3.40
Tin	3.72
Mercury	4.15
Lead	7.19
Niobium-titanium	8.9–9.3
Niobium	9.26
Niobium-tin	17–18
Niobium-germanium	23
Lanthanum-barium copper oxide	30
Yttrium-barium copper oxide	93
Thallium-calcium-bismuth copper oxide	125

from 30 K to well over 100 K.[2] These surprisingly high transition temperatures have caused much excitement in scientific and engineering circles, because they are above the temperature of boiling nitrogen (77 K); hence, adequate cooling can be achieved with cheap liquid nitrogen, instead of expensive liquid helium. The discovery of the new ceramic superconductors is expected to lead to a vast increase in the practical applications of superconductivity.

A superconductor is a perfect conductor — its resistance is truly zero. Even the most precise experiments have not been able to detect any residual resistance in a superconductor. If a closed loop of superconducting wire initially has a current, then this current will keep flowing around the loop on its own accord as long as the wire is kept cold. Such a steady current that flows without any resistive loss is called a **persistent current**. In one case, a current that was started in a superconducting loop kept on flowing for two and a half years with undiminished strength, and it would probably still be flowing today if the experimenters had not run out of liquid helium for cooling their apparatus. Of course, in view of measurement errors, we cannot prove that the resistivity in superconductors is exactly zero; but if it is not zero, it is certainly extremely small — no more than 10^{-15} times the resistivity of the best normal conductors.

Persistent currents induced by changing magnetic fields bring about some spectacular levitation effects. If a small bar magnet is dropped toward a superconducting lead dish (Figure VIII.2), the magnetic field induces persistent currents along the surface of the dish; by Lenz' Law, the direction of these currents is such that their magnetic force on the magnet is repulsive. When the magnet is close to the dish, this magnetic

[2] **J. Georg Bednorz,** 1950–, and **K. Alex Müller,** 1927–, physicists at the IBM Zürich Research Laboratory, received the Nobel Prize for their discovery of new ceramic superconductors with exceptionally high transition temperatures.

Fig. VIII.2 A small bar magnet levitated above a superconducting dish. (Courtesy A. Leitner, Rensselaer Polytechnic Institute.)

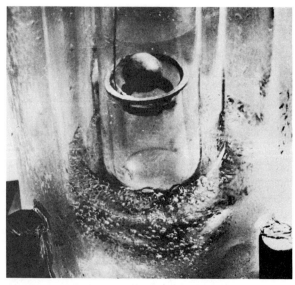

Fig. VIII.3 A small superconducting sphere levitated above a superconducting ring carrying a current.

force is large enough to support the weight of the magnet — the magnet floats forever above the lead dish.

A similar effect can be demonstrated with a superconducting lead ring and sphere (Figure VIII.3). The ring is stationary and it carries an initial current that has been induced by means of an external magnetic field. If now a lead sphere is dropped toward the ring, it will remain floating at some height above the ring. This is analogous to the levitation effect described in the preceding paragraph; the ring with its current and associated magnetic field plays the role of the magnet, and the lead sphere plays the role of the lead dish.

VIII.2 THE CRITICAL MAGNETIC FIELD

Superconducting loops with persistent currents make magnetic fields — they are magnets. Such superconducting magnets are similar to permanent magnets in that they do not require any electric power supply to maintain their current and their magnetic field. Only an initial energy input is needed to get the persistent current started. This suggests that superconductors should permit us to produce extremely intense magnetic fields with little expenditure of energy.

Unfortunately, intense magnetic fields have an adverse effect on superconductors: intense magnetic fields destroy superconductivity. For instance, at a temperature near absolute zero, a magnetic field of 0.041 T will destroy the superconductivity of mercury. At a temperature near the critical temperature (4.15 K), an even smaller magnetic field suffices to destroy the superconductivity. The minimum magnetic field that will quench the superconductivity of a material is called the **critical magnetic field**, B_c. Its strength depends on the temperature; Figure VIII.4 shows a plot of critical field strength for mercury as a function of temperature.

This breakdown of superconductivity imposes serious restrictions on the maximum current that can be carried by a superconductor. The current in, say, a wire will itself generate a magnetic field and, if this magnetic field is intense enough, it will cause a breakdown of the superconductivity of the wire. For example, a superconducting wire of mercury, 0.2 cm in diameter at a temperature near absolute zero, can carry a current of at most 200 A; a larger current will lead to a breakdown of the superconductivity. Such restrictions must be kept in mind in the design of superconducting magnets. We will see in Section VIII.4 that some compounds and alloys can tolerate substantially larger magnetic fields.

VIII.3 THE MEISSNER EFFECT

As we know from our study of ordinary conductors (see Chapter 28), a steady current in such a conductor requires an electric field to overcome the resistance. The electric field within a conductor carrying a given current is directly proportional to the resistance — a conductor of large resistance has a large electric field and a conductor of small resistance has a small electric field (see Ohm's Law in Section 28.2). A superconductor, with zero resistance, always has zero electric field in its interior. Furthermore, it follows from this that the rate of change of magnetic field in a superconductor must always be zero; if it were not, then the changing magnetic flux would induce an electric field, in contradiction with the requirement that the electric field remain zero. For example, if we transport a superconducting cylinder into a magnetic field (Figure VIII.5), it will push the magnetic field lines aside so that none of these penetrate the cylinder. What happens here is that as the cylinder touches the magnetic field, currents are induced on the surface of the cylinder, and the magnetic field of these currents produces just the right deformation of the magnetic field lines to prevent their penetration into the cylinder. This behavior is characteristic of a perfect conductor; for instance, a ball of plasma exhibits the same behavior and for the same reason — a ball of plasma is an (almost) perfect conductor.

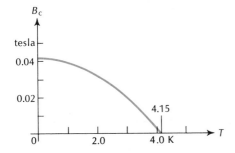

Fig. VIII.4 Critical magnetic field as a function of temperature for mercury.

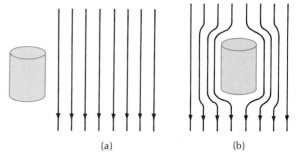

Fig. VIII.5 Superconducting cylinder transported into a magnetic field.

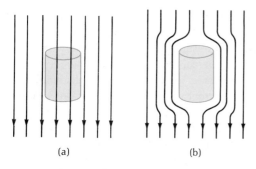

Fig. VIII.6 Conducting cylinder in a magnetic field: (a) normal metal, before T_c is reached, and (b) superconductor, after T_c is reached.

But it turns out that a superconductor is more than just a perfect conductor. A superconductor not only prevents the penetration of magnetic field lines that are initially outside of the superconducting material, but it also *expels* any magnetic field lines that are initially inside the material (before it becomes superconducting). Figure VIII.6 shows a cylinder of lead, above its critical temperature, placed in a magnetic field. This is an ordinary conductor and the magnetic field lines penetrate it without hindrance. However, if we cool the lead to its critical temperature, it will expel these field lines and, when it reaches its superconducting state, the magnetic field within it will be zero. This behavior is to be contrasted with that of a mere perfect conductor. Figure VIII.7 shows a ball of gas, placed in a magnetic field. If we convert this gas into a plasma (by ionizing it), the magnetic field lines are trapped (or frozen) in the plasma — when we switch the magnet off, the magnetic field lines inside the plasma remain unchanged, the induced currents in the plasma generating just enough magnetic field to keep the flux constant.

The expulsion of magnetic flux from a metal during the transition from the normal to the superconducting state is called the **Meissner effect.** It means that a superconductor is not only a perfect conductor, but also a perfect diamagnet. (Recall that a diamagnetic material immersed in a magnetic field tends to reduce the strength of the magnetic field in its interior; see Section 33.4.)

The elimination of the magnetic field from the interior of the superconductor is brought about by currents that flow along the surface of the superconductor; the magnetic field generated by these currents cancels the magnetic field generated by the external sources.

To see that the currents carried by a superconductor must always be surface currents, we note that if any currents were to flow within the volume of the superconductor, then Ampère's Law would demand a nonzero value of **B** within the volume (see Figure VIII.8); such a nonzero value of **B** would be in contradiction to the Meissner effect. From this we can draw the general conclusion that *any* current carried by a superconductor must flow on the surface; this applies to induced currents as well as to currents originating from other sources of emf. Although from a macroscopic point of view these superconductor currents are surface currents, from a microscopic point of view they do not flow exactly on the surface, but rather in a thin layer or skin. The thickness of this layer carrying the current is typically about 10^{-5} cm. Within the surface layer the magnetic field is not quite zero; the Meissner effect is incomplete and some magnetic flux penetrates.

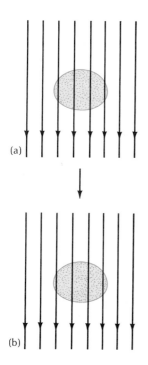

Fig. VIII.7 Ball of gas in a magnetic field: (a) normal gas, before ionization, and (b) plasma, after ionization.

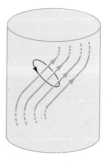

Fig. VIII.8 The lines with arrows show the flow of a hypothetical current within the volume of a superconductor. The closed loop is the path of integration for Ampère's Law; since some current passes through this loop, $\oint \mathbf{B} \cdot d\mathbf{l}$ cannot be zero.

In Figure VIII.6 we illustrated the Meissner effect for a superconducting cylinder. Strictly, the ideal Meissner effect, with complete expulsion of the magnetic flux from the entire volume of the metal, occurs only if the metal has the shape of a very long cylinder (a wire) aligned with the magnetic field. For other shapes, the extent of the expulsion of the magnetic flux depends on the geometry. In general, the volume of the metal splits into domains of superconducting material and normal material. If we increase the strength of the magnetic field, the size of the normal domains increases at the expense of the superconducting domains and, when the field reaches the critical strength, the entire volume of metal becomes normal.

VIII.4 SUPERCONDUCTORS OF THE SECOND KIND

In most pure superconducting metals, the expulsion of magnetic flux from each of the superconducting domains in the metal is an all or nothing affair: if the metal is held at a fixed temperature and immersed in a magnetic field, it will prevent the penetration of the magnetic flux as long as the magnetic field is weaker than the critical value; but the metal will suddenly cease to be a superconductor when the magnetic field becomes stronger than the critical value, and it will then freely permit the penetration of the magnetic flux.

But in niobium, in vanadium, in alloys of other metals, and in the new ceramic superconductors, the expulsion of magnetic flux from a superconducting domain is a much more complicated affair. The material will permit a partial penetration of flux if the magnetic field is of intermediate strength; and it will finally cease to be a superconductor and permit complete penetration of flux when the magnetic field becomes stronger. Thus, the material has two critical values of magnetic field strength: at a value B_{c_1} the flux begins to penetrate, and at a value B_{c_2} the flux penetrates completely and superconductivity breaks down. For example, at a temperature of 4.2 K, the niobium–tin alloy Nb_3Sn has critical values $B_{c_1} = 0.019$ T and $B_{c_2} = 22$ T. The high value of B_{c_2} is of great practical importance — the alloy retains its superconductivity even in a very strong magnetic field where any pure metal would lose its superconductivity.

Materials such as Nb_3Sn that permit a partial penetration of the magnetic flux when immersed in a magnetic field of intermediate strength are called **superconductors of the second kind,** or of type II. When immersed in a magnetic field of intermediate strength, such a superconductor is in a mixed state: the bulk of the material is superconducting, but it is threaded by very thin filaments of normal material; these filaments are oriented parallel to the external

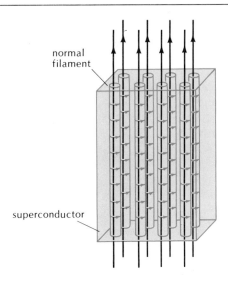

Fig. VIII.9 Filaments of normal material in a superconductor. The filaments serve as conduits for the magnetic field lines.

magnetic field, and they serve as conduits for the penetrating lines of this external magnetic field (Figure VIII.9). A current circulates around the perimeter of each filament; this current shields the bulk of the superconductor from the magnetic field in the filament. The flow of this current has the character of a vortex; because of this, the filaments are usually called **vortex lines.**

The amount of flux associated with each vortex line has a fixed value related to Planck's constant and the electric charge of the electron,[3]

$$\Phi_0 = \frac{h}{2e} = 2.07 \times 10^{-15} \text{ T} \cdot \text{m}^2 \qquad (1)$$

This means that the flux is *quantized* — Φ_0 represents a **quantum of flux,**[3] just as the electric charge e of an electron represents a quantum of electric charge. In a superconductor of the second kind, an increase of the strength of the external magnetic field will not cause an increase of the flux associated with each vortex line; instead it will cause an increase in the number of vortex lines threading the superconductor. The stronger the external magnetic field, the more densely will the vortex lines be packed. Figure VIII.10 shows the vortex lines (viewed end on) in a sample of lead–indium alloy; the vortex lines are packed in a regular triangular pattern.

VIII.5 THE BCS THEORY

We know from Chapter 28 that the resistivity of a metal is caused by collisions between the free electrons and the ions of the crystal lattice of the metal.

[3] See Appendix 8 for a more precise value of the magnetic flux quantum.

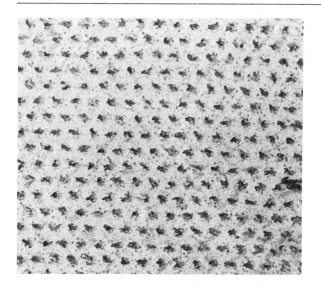

Fig. VIII.10 Ends of vortex lines at the surface of a sample of superconducting lead–indium. The vortex lines have been made visible by dusting with powdered iron. The separation between the vortex lines is about 0.005 cm.

The resistivity depends on temperature because the random thermal motion of the ions of the lattice increases the likelihood that an electron will suffer collisions. In fact, in a perfect lattice with no thermal vibrations, an electron can travel without ever suffering any collisions. If we adopt the naive classical picture of an electron as a pointlike particle, we can see that travel without collisions is possible if the electron moves along a straight line between the rows of atoms. This classical picture of the motion of electrons has been shown to be invalid by modern quantum mechanics, according to which electrons have wave properties (see Section 43.6); however it turns out that the quantum-mechanical picture of the motion of electron waves leads to a similar conclusion. The electron wave can travel through a perfect crystal lattice without suffering any scattering (deviation), because whatever effect is caused by one atom is canceled by the effects of other atoms. The quantum-mechanical picture indicates that in a perfect crystal lattice the electron wave can travel unhindered in *any* direction. Imperfections in the regularity of the position of the atoms of the lattice will hinder the propagation of the electron wave and give the metal a finite resistivity.

Thus, the random thermal vibrations of the ions contribute to the resistivity. At low temperatures, the reduction of these thermal vibrations brings about a reduction of resistivity; at zero temperature, when the thermal vibrations disappear, the resistivity should also disappear (except for a residual contribution due to impurities and dislocations of the lattice).

It therefore came as no surprise to physicists that the experimental values of the resistivity were small at low temperatures — what came as a surprise was that the resistivity of some metals vanished completely at a few degrees above absolute zero.

The details of the mechanism underlying superconductivity were finally spelled out in the **Bardeen–Cooper–Schrieffer** (BCS)[4] **theory** of superconductivity, some 50 years after the discovery of this phenomenon. The key to this mechanism is the formation of electron pairs (Cooper pairs). After many false starts, theoretical physicists recognized that the free electrons in a metal are not quite free, but they interact with one another via the lattice. The negative charge of each free electron exerts an attractive force on the positive charges of the ions of the lattice; consequently, the nearby ions contract slightly toward the electron. This slight concentration of positive charge, in turn, attracts other electrons. The net effect is that a free electron exerts a small attractive force on another free electron. Although this attractive force is too small to be of any consequence at room temperature, it is strong enough to permanently bind two electrons into a pair when the temperature is within a few degrees of absolute zero, where the thermal disturbances nearly disappear. In a superconducting metal in electrostatic equilibrium (no current), each **Cooper pair** consists of two electrons of exactly opposite momenta. Obviously such a configuration makes no sense from a classical point of view — if two particles have opposite and constant momenta, they will travel away from each other in opposite directions; they will then cease to interact and they cannot remain bound. However, the configuration makes sense from a quantum-mechanical point of view, where each particle is described by a wave — if two waves have opposite directions of motion they can continue to overlap for a long time and they can continue to interact. Thus, the BCS theory is intrinsically quantum mechanical, and the language of classical physics can only convey a crude outline of the theory.

In a superconducting metal that is carrying a current, the electron pairs have a net momentum in the direction opposite to that of the current, and they transport electric charge. The electron pairs move through the lattice without resistance, because whenever the lattice scatters one of the electrons and changes its momentum, it will also scatter the other electron of the pair and change its momentum by an opposite amount. Consequently, the lattice cannot change the net momentum of a pair — it can neither slow down nor speed up the average motion of the paired electrons.

The quantum character of superconductivity shows up quite explicitly in the quantization of the current

[4] **John Bardeen,** 1908–, **Leon N. Cooper,** 1930–, and **J. Robert Schrieffer,** 1931–, American physicists. They shared the Nobel Prize in 1972 for their theory of superconductivity.

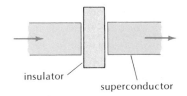

Fig. VIII.11 A Josephson junction. The thickness of the layer of insulator, an oxide, is usually less than 100 Å.

and of the magnetic flux in closed superconducting loops. Delicate experiments with small superconducting loops have established that the current is restricted to a discrete set of values defined by the condition that the flux intercepted by the area within the loop is always a multiple of the basic quantum of flux, $\Phi_0 = 2.07 \times 10^{-15}$ T·m². Even more spectacular quantum effects occur in a **Josephson junction**,[5] consisting of a thin layer of insulator placed between two adjacent pieces of superconductor (Figure VIII.11). Classically, the layer of insulator constitutes an inpenetrable barrier for electrons; but quantum-mechanically the electron waves can tunnel through this barrier. Consequently, the thin layer of insulator permits the passage of a sizable fraction of the superconducting current. The interplay of the electron waves on the two sides of the Josephson junction gives rise to some remarkable interference phenomena. For instance, if we apply a DC potential difference ΔV across the junction, the result is an AC current of frequency $2e\,\Delta V/h$. And if we apply an AC potential difference of frequency $2e\,\Delta V/h$ across the junction, the result is a combination of AC and DC currents. The measurement of the frequency of the oscillating current produced by a steady potential difference has been used for a new determination of the ratio of the fundamental constants e and h, with unprecedented precision. Such measurements also provide the most precise method for the determination of potential differences.

Other remarkable interference phenomena arise when two Josephson junctions are connected in parallel (Figure VIII.12), an arrangement called a **SQUID** (superconductive *q*uantum *i*nterference *d*evice). The net current passing through this device depends on the magnetic flux intercepted by the area spanned by the loop: the current is zero whenever the flux is a half-integer multiple of Φ_0, and the current is maximum whenever the flux is an integer multiple of Φ_0. This sensitive dependence of the current on the magnetic flux can be exploited for extremely precise measurements of magnetic fields.

[5] **Brian D. Josephson**, 1940–, English physicist, and **Ivar Giaever**, 1929–, American physicist, shared the Nobel Prize in 1973 for the prediction of the properties of the Josephson junction and for the discovery of the tunneling of superconducting currents through insulators, respectively.

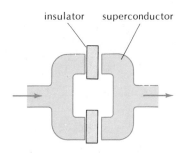

Fig. VIII.12 Two Josephson junctions connected in parallel form a SQUID.

VIII.6 TECHNOLOGICAL APPLICATIONS

Practical applications of superconductivity have focused on the development of electromagnets with superconductive coils capable of generating intense magnetic fields. Conventional electromagnets, with copper windings and iron cores, are among the least energy-efficient devices produced by our technology. They consume enormous amounts of electric power merely to keep a constant current running around in circles. Except for the magnetic energy initially stored in the electromagnet at startup, all of the electric power is wasted as heat. By contrast, a magnet with superconducting coils requires no electric power at all to keep running. Of course, the superconducting magnet must be cooled to a low temperature and maintained at this temperature. The initial cooling and the continual removal of any heat that leaks into the magnet from the environment require refrigeration and a supply of coolant, such as liquid helium.

Fig. VIII.13 A high-field superconducting magnet. This magnet produces a field of 15 T in a bore of 7 cm.

Figure VIII.13 shows a superconducting magnet capable of generating a magnetic field of 15 T; a magnetic field of this strength is very difficult to attain with a conventional magnet. In Figure VIII.14 we see a dramatic display of the strength of the magnetic fields of

Fig. VIII.14 Nails scattered on a board align with the field lines of the intense magnetic field generated by two superconducting magnets. The coils of these magnets are immersed in Dewar flasks full of liquid helium.

superconducting magnets: thousands of iron nails align along the field lines connecting two such magnets.

Figure VIII.15 shows the coils of a large superconducting magnet for a bubble chamber at the Argonne National Laboratory. The coils are 4.8 m across and they produce a magnetic field of 1.8 T. After the initial current has been established in this magnet, the only power required is 190 kW for a liquid-helium refrigerator; in contrast, a conventional magnet of similar size would require about 10,000 kW of electric power. The construction costs for this superconducting magnet are about the same as for a conventional magnet of similar size, but the yearly running costs are lower by a factor of 10.

Superconducting magnets have recently been installed at the Fermilab accelerator. This machine was originally built with conventional magnets, but in order to make the machine capable of handling particles of higher energy, new superconducting magnets producing more intense magnetic fields were required. Fermilab developed 7-m-long supermagnets producing a magnetic field of up to 4.5 T. Figure VIII.16b shows one of these magnets. The complete accelerator ring, 6.2 km in circumference, has 774 of these magnets plus 240 extra magnets used for focusing the beam of high-energy particles. The ring of supermagnets is placed just below the ring of ordinary magnets (Figure VIII.16a). Particles are first accelerated in the ordinary ring and then fed into the super ring, where they reach a final energy of 10^6 MeV.

The construction of superconducting coils must

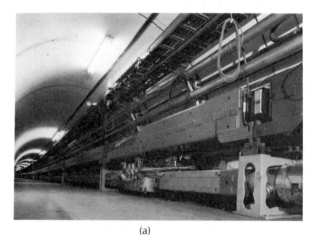

(a)

Fig. VIII.15 The coils of this large superconducting magnet for a bubble chamber at Argonne National Laboratory are made of niobium–titanium alloy embedded in copper.

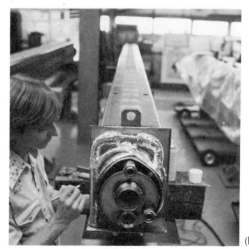

(b)

Fig. VIII.16 (a) The tunnel of the Tevatron accelerator at Fermilab. The upper ring consists of ordinary magnets; the lower ring consists of superconducting magnets. (b) One of the superconducting magnets used at Fermilab.

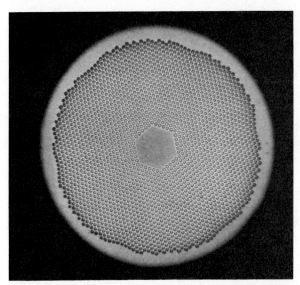

Fig. VIII.17 This strand of a superconducting cable consists of 2100 thin filaments of NbTi alloy embedded in a copper matrix. The diameter of this strand is about 0.7 mm.

overcome some tricky manufacturing problems. The superconducting material commonly employed in these coils is either a niobium–titanium alloy (NbTi) or else a niobium–tin compound (Nb_3Sn). The latter material can withstand the highest current densities and magnetic fields (up to 2×10^5 A/cm² and up to 15 T), but it is extremely brittle and cannot be drawn into wires to be wound around a magnet core. Instead, it is necessary to first wind a coil with separate strands of niobium wrapped in tin and then heat the entire coil to fuse these materials into the Nb_3Sn compound.

Figure VIII.17 shows a cross section through one strand of a superconducting cable of NbTi alloy. The alloy is in the form of very thin filaments embedded in a copper matrix. A bundle of thin filaments has more surface area than a single thick wire of the same cross section as the bundle; since the supercurrents always run on the surface of the superconductor, the large area of the filaments permits the flow of a large current. The complete cable consists of several strands of NbTi alloy and several wires of copper wound around each other. As long as the NbTi alloy is in its superconducting state, the copper plays no role; in fact, relative to the superconductor, the copper behaves as an insulator, and all the current flows in the superconductor. But if the superconductor should ever fail because of inadequate cooling or excessive magnetic fields, the current can continue to flow in the parallel wires of copper. This protects the coils of the magnet from the explosive conversion of magnetic energy into heat that would occur if a large current (10^5 A or more) were suddenly halted by a large resistance.

Whereas electromagnets are among the least efficient of our technological devices, electric generators and electric motors are among the most efficient. Typically, an ordinary electric generator will convert close to 99% of the supplied mechanical power into electric power. Electric generators with superconducting coils could probably be even more efficient, but their main advantage is their smaller size and smaller weight — for a given power output, the weight of a superconducting generator can be 10 times less than that of a conventional generator. Several prototypes of superconducting generators and motors have been built (Figure VIII.18). Superconducting generators may find application in nuclear power plants, because conven-

tional generators capable of handling the very large power output of these plants must be of huge size, and they therefore face serious structural problems because of the weight of the moving parts. Superconducting motors may also find applications in marine propulsion where, again, weight is a problem.

The transmission and storage of electric power are another important application of superconductors. Resistive losses in conventional transmission lines are of the order of 10% or 15% of the total power transmitted. A superconducting cable, with zero resistance, would of course eliminate the loss. However, from a financial point of view, the elimination of the resistive loss makes only a minor contribution to the thriftiness of superconducting cables; what makes a major contribution is their compact size — large amounts of power can be carried by a single cable of fairly small dimensions that can be installed in an underground tunnel not requiring the expensive strips of real estate on which conventional transmission lines are erected. Figure VIII.19 shows an experimental version of a superconducting transmission line; the outside diameter of the pipe enclosing the cable is 40 cm and it can carry a power of 1000 MW, roughly the total power of a nuclear generating station.

Incidentally: An AC current in a superconductor of the second kind is not entirely free of resistive losses. The trouble is that the periodic oscillation of the current produces a similar oscillation of the strength of the magnetic field, and consequently lines of magnetic flux (vortex lines) move periodically in and out of the superconductor. This motion of the vortex lines involves some "friction" and it generates some heat. To avoid this loss, the AC current must be carried by a superconductor of the first kind, without magnetic flux in its interior.

On conventional transmission lines, the electric power is carried at very high voltage to minimize the resistive loss. In a superconducting cable, high voltage is not necessary — one can dispense with the awkward step-up and step-down transformers at the generating plant and at the consumer station. Furthermore, if transformation is not required, it becomes advantageous to transmit DC power rather than AC power. A superconducting cable for DC power is cheaper than a cable for AC power; it can be made of a superconductor of the second kind which will tolerate higher current densities and, furthermore, the design can be kept simpler.

Another possible application of superconductivity is the storage of electric energy. Because of daily fluctuations in the demand for electric power by a city, it is desirable to hold large amounts of electric energy temporarily in some kind of storage system. The power can be fed into the storage system when the demand is low and released when the demand is high. Table VIII.2 compares the energy density in several systems. It is obvious from this table that magnetic fields constitute one of the most concentrated forms of stored energy.

Table VIII.2 Energy Storage

System	Energy density
Magnetic fields, 10 T	11 kW · h/m^3
Electric field, 10^5 kV/m	0.01
Water in reservoir, height 100 m	0.27
Compressed air, 50 atm	5
Hot water, 100°C	18[a]

[a] This number takes into account that the thermodynamic conversion efficiency for this heat into mechanical energy is at most 20%.

In order to store 100 MW · h, the volume of the magnetic field (at 10 T) would have to be about 10^4 m^3. A large toroid about 5 m thick and 100 m across would provide this volume. The windings of the toroid would of course have to be made of superconductor. The toroid would have to withstand enormous forces due to the magnetic interaction of the currents. And it would be exposed to a severe danger: if, by some failure, one part of the winding became resistive, the sudden generation of heat would probably cause the entire toroid to become resistive, and the sudden conversion of all of the magnetic energy into heat would

Fig. VIII.19 Superconducting power transmission line at Brookhaven National Laboratory.

then melt the toroid and blow it apart with great violence.

As a final example of an application of superconductivity, we briefly consider magnetic suspension of trains. In Section VIII.1 we saw that a permanent magnet will remain suspended over a superconducting sheet — the former induces currents in the latter, and the magnetic force on the currents is repulsive. If we replace the permanent magnet by a superconducting coil with a current (a superconducting magnet), we will get the same repulsion effect. Taking advantage of this, we could attach superconducting magnets to the underside of a train and levitate it on superconducting rails. The suspension is frictionless — the train sits on a cushion of magnetic fields and slides without friction. In practice, this ideal arrangement is not feasible because it would be much too expensive to lay down hundreds of kilometers of superconducting rails. However, it is also possible to obtain levitation with rails made of an ordinary conductor (Figure VIII.20). In this case, there is no repulsive force between the superconducting magnets and the rails when they are at rest, but there is a force when they are in motion relative to one another, because the moving magnetic field induces currents in the rails. These currents suffer some resistive loss, but the effective friction would be reasonably small. Engineers estimate that at speeds in excess of 200 km/h, a train riding on superconducting magnets would be much safer than a train on iron wheels.

Fig. VIII.20 Magnetically levitated train during a test run in Japan. The "rails" supporting the train consist of rows of conducting coils. The moving magnetic field of the train induces currents in the coils, which produce a repulsive force.

Further Reading

The Quest for Absolute Zero by K. Mendelssohn (Halsted Press, New York, 1977) is a splendid account of the exciting discoveries in low-temperature physics; it includes a chapter on superconductivity. *Near Zero* by D. K. C. MacDonald (Doubleday, Garden City, 1961) is a much briefer and very elementary introduction. *Cryophysics* by K. Mendelssohn (Interscience Publishers, New York, 1960) is an advanced textbook that gives a concise survey of low-temperature physics; although the book is aimed a senior-level students, many sections are accessible to lower-level students.

Superconductivity by A. W. B. Taylor (Wykeham, London, 1970) is a concise introductory survey with a minimum of mathematics. *Superconductivity* by D. Schoenberg (Cambridge University Press, Cambridge, 1965) is a somewhat more advanced survey; written long before the days of the BCS theory, it is somewhat outdated, but its classic description of the phenomenological properties of superconductors remains useful.

The entire March 1986 issue of *Physics Today* is devoted to superconductivity and its applications.

The following magazine articles deal with applications of superconductivity:

"Large-scale Applications of Superconductivity," B. B. Schwartz and S. Foner, *Physics Today,* July 1977
"Superconductors in Electric-Power Technology," T. H. Geballe and J. K. Hulm, *Scientific American,* November 1980
"Superconducting Electronics," D. G. McDonald, *Physics Today,* February 1981
"The Road to Superconducting Materials," J. K. Hulm, J. E. Kunzler, and B. T. Matthias, *Physics Today,* January 1981
"The Superconducting Computer," J. Matisoo, *Scientific American,* May 1980

Questions

1. Why would you expect helium to be more difficult to liquefy than any other gas?

2. The graph in Figure 28.5 gives the impression that the resistance of tin is constant for $T > 3.72$ K. Is this true?

3. Is Ohm's Law valid for a superconductor?

4. Is it reasonable to regard superconductors as another state of matter? If so, how many states of matter are there?

5. Is a ring of superconductor with a persistent current a perpetual motion machine, i.e., does it violate the law of conservation of energy?

6. Suppose that a closed ring of superconducting wire is initially outside of a magnetic field. If we push the ring into the magnetic field, the magnetic flux through the ring remains zero. Explain.

7. Figure VIII.6b shows a superconducting cylinder that has expelled the magnetic field by the Meissner effect. Describe the currents that must be flowing along the surface of this cylinder.

8. Suppose that a piece of metal immersed in a magnetic field is initially a normal conductor, and is then cooled so that it becomes a superconductor. To expel the magnetic field by the Meissner effect, the metal must acquire electric currents. Where does the energy for these currents come from?

9. If the resistance of a superconductor were small but not zero, could the Meissner effect be sustained for a long time?

10. The levitated magnet shown in Figure VIII.2 was placed near the superconducting dish *after* the dish became superconducting. Consequently, does Figure VIII.2 demonstrate infinite conductivity, or the Meissner effect, or both?

11. (a) If you push a ring of superconductor into a magnetic field, a current will be induced along the ring; if you then remove the ring from the magnetic field, the current disappears. Explain. (b) If you place a ring of normal conductor in a magnetic field, then cool it so it becomes a superconductor, and then remove it from the magnetic field, the ring will be left with a persistent current. Explain.

12. Suppose that a very large flat metallic plate is placed between the poles of an electromagnet. If this plate is cooled so that it becomes superconducting, it will not expel the magnetic field lines. Why not?

13. The equation describing the graph in Figure VIII.4 is $B_c(T) = (1 - T^2/T_c^2)B_c(0)$. At what temperature is $B_c(T) = \frac{1}{2}B_c(0)$?

14. The existence of a critical magnetic field implies the existence of a critical current. For a cylindrical wire of radius r, the critical current is given by the equation $I_c = 2\pi r B_c/\mu_0$, known as the Silsbee rule. Explain why the critical current is proportional to the radius of the wire.

15. Would you expect that a noncrystalline material, such as glass, could exhibit superconductivity?

16. Consider a pair of electrons moving around a circular superconducting ring carrying a current. According to the discussion in Section 43.4, the orbital angular momentum of the electrons is quantized. Explain how this leads to the quantization of the electric current and the magnetic flux associated with this current.

17. A superconducting cable carrying a large current and a copper cable are connected in parallel. Is it really true that the copper cable carries no current at all? (Hint: What is the potential difference from one end of the superconducting cable to the other?)

CHAPTER 35

The Displacement Current and Maxwell's Equations

We already know that a changing magnetic field induces an electric field. In this chapter, we will discover that the converse is also true: a changing electric field induces a magnetic field. The law describing this induction effect of electric fields was formulated by James Clerk Maxwell, who thereby achieved a wide-ranging unification of the laws of electricity and magnetism — these laws became known as Maxwell's equations. The next four chapters of this book are nothing but applications of these basic equations.

The mutual induction of electric and magnetic fields gives rise to the phenomenon of self-supporting electromagnetic oscillation in empty space. If, initially, there exists an oscillating electric field, it will induce a magnetic field, and this will induce a new electric field, and this will induce a new magnetic field, and so on. Thus, these fields can perpetuate each other. Of course, an oscillating charge or current is needed to get the fields started, but after this initiation the fields continue to oscillate on their own. These self-supporting oscillations are **electromagnetic waves,** either traveling waves or standing waves. We will examine such electromagnetic waves in the space within a conducting cavity and in the space surrounding an accelerated electric charge.

James Clerk Maxwell, *1831–1879, Scottish physicist. He was professor at King's College, London, and later professor at Cambridge, where he supervised the construction of the Cavendish Laboratory. Maxwell at first attempted to explain the behavior of electric and magnetic fields in terms of a complicated mechanical model, according to which all of space was filled with an elastic medium, or ether, consisting of a multitude of small rotating vortex cells. But ultimately Maxwell discarded the mechanical model and treated the electric and magnetic fields as physical entities that exist in their own right, without any need for an underlying medium. He published his electromagnetic theory in 1873 in his celebrated* Treatise on Electricity and Magnetism. *In this book he laid down a complete and consistent set of laws for all electromagnetic phenomena and thereby accomplished for electricity and magnetism what Newton had accomplished for mechanics.*

35.1 The Displacement Current

The magnetic fields we became acquainted with in Chapter 30 were produced by currents, according to Ampère's Law. This law asserts that the integral of **B** around any closed path is related to the current

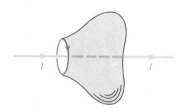

Fig. 35.1 A circular path (color) is spanned by a baglike mathematical surface. The wire carrying the current enters the "mouth" of the bag and is intercepted by the bottom of the bag. Alternatively, the circular path can be spanned by a flat circular surface. The same current will then be intercepted by this flat circular surface.

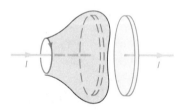

Fig. 35.2 A similar baglike surface that passes between the plates of a capacitor. The baglike surface intercepts no current.

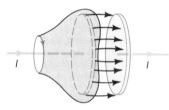

Fig. 35.3 The baglike surface intercepts electric flux.

intercepted by an arbitrary surface that spans this path:

$$\oint \mathbf{B} \cdot d\mathbf{l} = \mu_0 I \tag{1}$$

If all the currents flow along unbroken, continuous circuits, then the shape of the surface used to intercept the current does not affect the validity of Eq. (1). For example, Figure 35.1 shows a straight wire carrying a current and a circular path of integration spanned by a baglike curved surface with a circular mouth. This baglike curved surface intercepts the same current as a flat circular surface spanning the mouth — both surfaces give the same result in Eq. (1).

However, if the wire carrying the current is interrupted by a capacitor, then Ampère's Law develops a flaw. Suppose that a small capacitor is connected to long straight wires as in Figure 35.2. Suppose, further, that the wires carry a constant current I which steadily charges up the capacitor. (This of course requires that the voltage of the source of emf driving the current steadily increases so as to allow for the steadily increasing voltage on the capacitor.) The distribution of current on the interrupted wire of Figure 35.2 differs from that on the continuous wire of Figure 35.1 only in that a short piece of current is missing between the capacitor plates. Except in the immediate vicinity of the capacitor, the magnetic field in Figure 35.2 will therefore be about the same as in Figure 35.1. Let us now apply Ampere's Law to the baglike curved surface shown in Figure 35.2. This surface passes between the capacitor plates, and it therefore intercepts *no current*. Hence the right side of Eq. (1) is zero, but the left side is obviously not zero — Ampère's Law fails for this choice of surface.

The question is then: Can we modify Ampère's Law so that it gives the right answer even when the surface passes between capacitor plates? To discover the required modification, we note that although the surface does not intercept current in the region between the capacitor plates, it does intercept *electric flux* (see Figure 35.3). The positive plate of the capacitor is a source of electric field lines, and the negative plate is a sink; field lines going from one plate to the other cross the baglike surface. This suggests that the required correction to Ampère's Law involves the electric flux Φ.

It is not hard to see that to achieve this correction for our baglike surface, the right side of Eq. (1) should be replaced by $\mu_0 \varepsilon_0 \, d\Phi/dt$,

$$\oint \mathbf{B} \cdot d\mathbf{l} = \mu_0 \varepsilon_0 \frac{d\Phi}{dt} \tag{2}$$

To check that this works, let us calculate the amount of flux. If the electric charge on the positive plate is Q, then Q/ε_0 lines of electric field start on this plate. All these lines cross the baglike surface, and the flux is

$$\Phi = \frac{Q}{\varepsilon_0} \tag{3}$$

The rate of change of the flux is then

$$\frac{d\Phi}{dt} = \frac{1}{\varepsilon_0} \frac{dQ}{dt} = \frac{1}{\varepsilon_0} I \tag{4}$$

Hence the term $\mu_0 \varepsilon_0 \, d\Phi/dt$ on the right side of Eq. (2) equals

$$\mu_0\varepsilon_0 \frac{d\Phi}{dt} = \mu_0\varepsilon_0 \frac{1}{\varepsilon_0} I = \mu_0 I \tag{5}$$

This shows that the calculation with the baglike surface passing between the plates and intercepting the electric flux [Eq. (2)] gives the same result as a calculation with some other surface cutting across the wire and intercepting the current [Eq. (1)].

It can of course also happen that electric flux and current are present at the same place; for example, if the space between the capacitor plates is filled with some slightly conducting material (a leaky capacitor), then a surface passing between the plates will intercept *both* some electric flux and some current. In such a case, this electric flux and this current must be combined, and Ampère's Law becomes

$$\boxed{\oint \mathbf{B}\cdot d\mathbf{l} = \mu_0 I + \mu_0\varepsilon_0 \frac{d\Phi}{dt}} \tag{6}$$

Maxwell's modification of Ampère's Law

where it is to be understood that I is the current intercepted by whatever surface is used in the calculation [thus, the current I in Eq. (6) is not necessarily the same as the current on the wires].

The quantity of $\varepsilon_0\, d\Phi/dt$ is called the **displacement current,**

$$\boxed{I_\mathrm{d} = \varepsilon_0 \frac{d\Phi}{dt}} \tag{7}$$

Displacement current

This quantity is of course not a true current, but in Eq. (6) it has the same effect as though it were:

$$\oint \mathbf{B}\cdot d\mathbf{l} = \mu_0(I + I_\mathrm{d}) \tag{8}$$

Note that the displacement current between the capacitor plates has the same direction as the ordinary current arriving at the plates; the displacement current is the continuation of the ordinary current "by other means." The sign of the electric flux in Eq. (6) is fixed by the same right-hand rule as for the ordinary current: Wrap the fingers around the path of integration, in the direction of integration; the flux is to be reckoned as positive if the electric field is in the direction of the thumb.

Equation (6) is Maxwell's modification of Ampère's Law. The great importance of this equation lies in its general validity — Maxwell boldly proposed that this equation is valid not only for a capacitor connected to wires but also for any arbitrary system of electric fields, currents, and magnetic fields.

Apart from the term $\mu_0 I$, the Maxwell–Ampère Law of Eq. (6) is analogous to the Faraday Law of Eq. (32.17). The latter relates the path integral of the electric field to the rate of change of magnetic flux. The former relates the path integral of the magnetic field to the rate of change of the electric flux. Hence these laws indicate a certain symmetry in the effects of electric and magnetic fields on one another.

Just as a time-dependent magnetic field can induce an electric field, a time-dependent electric field can induce a magnetic field. The following example is quite analogous to Example 32.4.

EXAMPLE 1. A parallel-plate capacitor has circular plates of radius R. The capacitor is connected to a source of alternating emf, so the electric field between the plates oscillates according to

$$E = E_0 \sin \omega t \tag{9}$$

where E_0 and ω are constants. What is the induced magnetic field inside and outside the capacitor? Assume that the electric field is uniform inside the capacitor; neglect the fringing of the field at the edges.

SOLUTION: Inside the capacitor ($r < R$), the electric field is uniform. Figure 35.4 shows this electric field. To find the induced magnetic field, consider a circle of radius r. The flux through this circle is $\pi r^2 E$, and the rate of change of this flux is $\pi r^2 \, dE/dt$. According to Eq. (6), we then have

$$\oint \mathbf{B} \cdot d\mathbf{l} = \mu_0 \varepsilon_0 \pi r^2 \, dE/dt \tag{10}$$

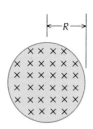

Fig. 35.4 The electric field between the capacitor plates. The crosses show the tails of the electric field vectors.

The field lines of **B** must be closed concentric circles, because this is the only pattern of field lines consistent with the rotational symmetry of the electric field shown in Figure 35.4. Since the path of integration coincides with a field line, we can immediately evaluate the left side of Eq. (10),

$$\oint \mathbf{B} \cdot d\mathbf{l} = 2\pi r B \tag{11}$$

Combining Eq. (10) and (11), we then obtain

$$2\pi r B = \mu_0 \varepsilon_0 \pi r^2 \frac{dE}{dt} \tag{12}$$

or

$$B = \frac{\mu_0 \varepsilon_0}{2} r \frac{dE}{dt} \tag{13}$$

$$= \frac{\mu_0 \varepsilon_0}{2} r \omega E_0 \cos \omega t \tag{14}$$

Outside the capacitor ($r > R$), the electric field is zero. Hence the electric flux through a circle of radius r is $\pi R^2 E$. We can then obtain the magnetic field by going through the same steps as above, with the result

$$B = \frac{\mu_0 \varepsilon_0}{2} \frac{R^2}{r} \frac{dE}{dt} \tag{15}$$

$$= \frac{\mu_0 \varepsilon_0}{2} \frac{R^2}{r} \omega E_0 \cos \omega t \tag{16}$$

The formulas (14) and (16) are obviously analogous to (32.20) and (32.21).

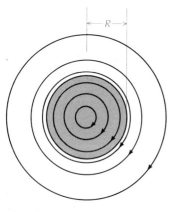

Fig. 35.5 The magnetic field between the capacitor plates.

COMMENTS AND SUGGESTIONS: Figure 35.5 shows the magnetic field lines. The direction of these lines is related to that of the current flowing toward the capacitor plates by the usual right-hand rule. Note that the magnetic field given by Eq. (16) outside the capacitor is exactly that of a long straight wire ($B \propto 1/r$). As far as this magnetic field is concerned, the removal of a short piece of wire and its replacement by a capacitor makes no difference at all.[1]

---

[1] However, a more precise calculation, taking into account the fringing of the electric field at the edges of the capacitor plates, shows that near the plates the magnetic field differs slightly from that of a straight wire.

35.2 Maxwell's Equations

Equation (6) is the last of the fundamental laws that we need to completely describe the behavior of electric and magnetic fields. There are four fundamental laws, as follows:[2]

Gauss' Law for electricity [Eq. (24.6)],

$$\oint \mathbf{E} \cdot d\mathbf{S} = Q/\varepsilon_0 \qquad (17)$$

Gauss' Law for magnetism [Eq. (30.17)],

$$\oint \mathbf{B} \cdot d\mathbf{S} = 0 \qquad (18)$$

Faraday's Law [Eq. (32.12)],

$$\oint \mathbf{E} \cdot d\mathbf{l} = -\frac{d\Phi_B}{dt} \qquad (19)$$

Maxwell–Ampère's Law [Eq. (6)],

$$\oint \mathbf{B} \cdot d\mathbf{l} = \mu_0 I + \mu_0 \varepsilon_0 \frac{d\Phi}{dt} \qquad (20)$$

Taken as a whole, these laws are known as **Maxwell's equations,** because Maxwell supplied the missing link between the magnetic and the electric fields [Eq. (20)] and thereby brought electromagnetic theory to perfection. Maxwell recognized that these equations imply a dynamic interplay between electric and magnetic fields, an interplay that couples and unifies electric and magnetic phenomena.

Maxwell's equations

The physical basis for each of these four equations may be briefly summarized as follows. Gauss' Law for electricity is based on Coulomb's Law describing the forces of attraction and repulsion between stationary charges. Gauss' Law for magnetism asserts that there exist no magnetic monopoles. Faraday's Law describes the induction of an electric field by motion or by a changing magnetic field. And finally, Maxwell–Ampère's Law is based on the law of magnetic force between moving charges, and it also contains the induction of a magnetic field by a changing electric field.

The above equations must be supplemented by the equation for the **Lorentz force** on a point charge:

---

[2] These equations are valid for vacuum. In the presence of dielectric or magnetic materials they must be modified either by implicityly including the bound charges of the dielectric in Q and the currents of the magnetic material in I, or by inserting the dielectric constant and the relative permeability into the appropriate terms of the equations.

Lorentz force

$$\boxed{\mathbf{F} = q\mathbf{E} + q\mathbf{v} \times \mathbf{B}} \qquad (21)$$

This expression is based on the definitions of **E** and **B** [see Eqs. (23.15) and (30.12)].

Maxwell's equations provide a complete description of the interactions among charges, currents, electric fields, and magnetic fields. All the properties of the fields can be deduced by mathematical manipulation of these equations. If the distribution of charges and currents is given, then these equations uniquely determine the corresponding fields. Even more important, Maxwell's equations uniquely determine the time evolution of the fields, starting from a given initial condition of these fields. Thus, these equations accomplish for the dynamics of electromagnetic fields what Newton's equations of motion accomplish for the dynamics of particles.

Calculations with Maxwell's equations are often best done by converting the integral equations (17)–(20) into *differential equations;* however, the latter involve partial derivatives, and for this reason we will not attempt to use them here.

Although the empirical foundations on which we based the development of Maxwell's equations were restricted to charges at rest or charges in uniform motion, these equations also govern the fields of accelerated charges and the fields of light and radio waves. In the remaining sections of this chapter and in the next chapters, we will calculate these fields of electromagnetic waves from our equations, and we will see that the results are in agreement with the observed properties of light and radio waves.

35.3* Cavity Oscillations

As we saw in Example 1, the time-dependent electric field in a parallel-plate capacitor will induce a magnetic field. This implies that the capacitor has some self-inductance, since the changing, time-dependent magnetic field will, in turn, induce an electric field, that is, it will induce an emf. We therefore expect that a parallel-plate capacitor placed in a circuit with a source of alternating emf will behave very much like the driven LC circuit of Section 34.3. The following is a calculation of the oscillations of a parallel-plate capacitor.[3] However, instead of treating this system as an LC circuit with some effective inductance, we will directly examine the electric and magnetic fields between the plates and their dependence on position and time.

Figure 35.6 shows a parallel-plate capacitor with circular plates connected to a source of alternating emf of frequency ω. The electric field between the plates will then oscillate at the same frequency. As a first approximation, we ignore the induction effects within the capacitor. The electric field between the plates is then uniform[4] and parallel to the axis. We use the symbol $E_{(1)}$ for this field, and we write

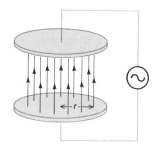

Fig. 35.6 Electric field lines in a parallel-plate capacitor connected to a source of alternating emf. The fringing of the electric field lines has been ignored.

---

* This section is optional.

[3] Based on *The Feynman Lectures on Physics,* by R. P. Feynman, R. B. Leighton, and M. Sands.

[4] As always, we disregard the fringing effects at the edges of the capacitor.

$$E_{(1)} = E_0 \sin \omega t \qquad (22)$$

where E_0 is a constant.

According to Example 1, such a time-dependent electric field induces a magnetic field [see Eq. (14)]

$$B_{(1)} = \frac{\mu_0 \varepsilon_0}{2} r \omega E_0 \cos \omega t \qquad (23)$$

The magnetic field lines are closed circles (Figure 35.7).

This time-dependent magnetic field in turn induces an extra electric field, which will have to be added to the field $E_{(1)}$ of Eq. (22). The direction of the extra field is again parallel to the axis. To find an expression for this field, let us apply Faraday's Law to the path shown in Figure 35.7. If we use the symbol $\mathbf{E}_{(2)}$ for the induced electric field, then

$$\oint \mathbf{E}_{(2)} \cdot d\mathbf{l} = -\frac{d\Phi_{B_{(1)}}}{dt} \qquad (24)$$

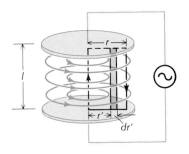

Fig. 35.7 Magnetic field lines. The path of integration is a rectangle of height l and width r. The direction of integration is clockwise.

The integral on the left side does not receive any contribution from the horizontal segments (Figure 35.7), since there the field is perpendicular to the path. Furthermore, we will suppose that $E_{(2)} = 0$ along the axis of the capacitor,[5] so that the integral receives no contribution from the vertical segment along the axis (Figure 35.7). Only the other vertical segment makes a contribution. If the latter is at a distance r from the axis, the left side of Eq. (24) becomes

$$-E_{(2)}(r)\, l \qquad (25)$$

where l is the distance between the plates (the minus sign is needed because we reckon the electric field as positive if it is directed upward, whereas the segment of path is directed downward).

To evaluate the right side of Eq. (24), we begin by calculating the flux $\Phi_{B_{(1)}}$. Consider a strip of height l and width dr'; the flux through this strip is $B_{(1)} l\, dr'$, and the flux through all such strips is an integral,

$$\Phi_{B_{(1)}} = \int_0^r B_{(1)}\, l\, dr' \qquad (26)$$

$$= \int_0^r \left(\frac{\mu_0 \varepsilon_0}{2} r' \omega E_0 \cos \omega t \right) l\, dr'$$

$$= \frac{\mu_0 \varepsilon_0}{2} l \omega E_0 \cos \omega t \int_0^r r'\, dr'$$

$$= \frac{\mu_0 \varepsilon_0}{4} l r^2 \omega E_0 \cos \omega t \qquad (27)$$

---

[5] The choice $E_{(2)} = 0$ along the axis is a matter of definition. It means that we regard $E_{(1)}$ of Eq. (22) as the *exact* electric field along the axis of the capacitor, whereas we regard the deviations from $E_{(1)}$ occurring at all other points as induced corrections which remain to be calculated.

Hence

$$\frac{d\Phi_{B_{(1)}}}{dt} = -\frac{\mu_0\varepsilon_0}{4} lr^2\, \omega^2 E_0 \sin\omega t \qquad (28)$$

Upon inserting Eqs. (25) and (28) into Eq. (24), we immediately find the induced electric field

$$E_{(2)} = -\frac{\mu_0\varepsilon_0}{4} r^2\omega^2 E_0 \sin\omega t \qquad (29)$$

Thus, the electric field $E_{(1)}$ must be corrected by adding to it the electric field $E_{(2)}$:

$$E = E_0 \sin\omega t - \frac{\mu_0\varepsilon_0}{4} r^2\omega^2 E_0 \sin\omega t + \cdots$$

$$= \left(1 - \frac{\mu_0\varepsilon_0}{4}\omega^2 r^2\right) E_0 \sin\omega t + \cdots \qquad (30)$$

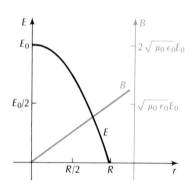

Fig. 35.8 Electric field (black) and magnetic field (color) in a parallel-plate capacitor as a function of radius.

But the correction to the electric field implies a corresponding correction to the induced magnetic field, and this implies a further correction to the electric field, and so on. This calculation therefore leads to an infinite sequence of successive approximations for the fields. The result takes the form of a power series in ascending powers of the radial variable r. The two terms in parentheses in Eq. (30) are merely the first two terms of the power series; the other terms have been indicated by the three dots in Eq. (30). For a start let us ignore all the extra corrections and use Eq. (30) as it stands. One interesting feature of this solution is that the electric field vanishes at a radius R such that

$$1 - \frac{\mu_0\varepsilon_0}{4}\omega^2 R^2 = 0 \qquad (31)$$

that is,

$$R = \frac{2}{\sqrt{\mu_0\varepsilon_0}}\frac{1}{\omega} \qquad (32)$$

A more exact calculation taking into account the extra terms in Eq. (30) shows that the precise radius at which the electric field vanishes is somewhat larger,

$$R = \frac{2.405}{\sqrt{\mu_0\varepsilon_0}}\frac{1}{\omega} \qquad (33)$$

Fig. 35.9 A conducting cylindrical wall of radius R closes off the sides of the parallel-plate capacitor.

Cavity oscillations

Figure 35.8 is a plot of the radial variation of the electric and magnetic fields.

The above solution for the fields between capacitor plates also gives us the solution for another problem, of much greater practical importance: the problem of the electromagnetic oscillations of a closed conducting cavity. We can convert our pair of capacitor plates into a closed cavity by inserting a conducting cylindrical wall of radius R between the plates (Figure 35.9). The cylindrical wall does not interfere with the electric field of Eq. (30), because this electric field is zero anyhow at the radius R. Thus, our solution for the fields between capacitor plates is also the solution for the fields in such a cavity. Note that here the choice of the correct radius is crucial: if the cylindrical wall had a radius different from R, then the electric field would not "fit" this cylinder.

Figure 35.10 shows a closed cylindrical can and the lines of electric and magnetic field within it. The external circuit with its driving emf has been omitted in this figure. The currents can flow from one horizontal end face of the cylinder to the other via the vertical wall; the external wire of Figure 35.6 is therefore not needed. And the external source of emf is not needed either — once the oscillations of the cavity have been started, they are self-sustaining. If charge is initially placed on the end faces,[6] then the cavity begins to oscillate spontaneously just as the LC circuit of Section 34.2 begins to oscillate when charge is initially placed on the capacitor. And, just as in the LC circuit, the energy sloshes back and forth between potential energy (in electric fields) and "kinetic" energy (in magnetic fields).

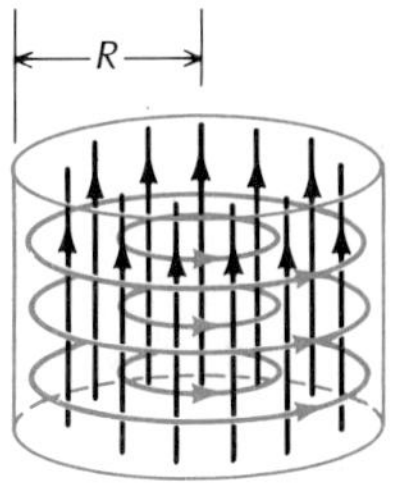

The electromagnetic oscillations in such a conducting cavity are analogous to the acoustic vibrations in a closed organ pipe. Equation (33) determines the resonant frequency of a cavity of given radius. This is analogous to Eq. (17.2), which determines the resonant frequency of an organ pipe. The electromagnetic oscillations in a cavity can be regarded as a standing wave (we will discuss electromagnetic waves in the next chapter), and Eq. (33) can then be regarded as the condition for a proper fit of the wave in the cavity. We know from the study of sound that an organ pipe has several possible harmonic standing waves of different resonant frequencies. It turns out that this is also true for an oscillating electromagnetic cavity, but in order to find the higher harmonics and their frequencies, we would have to calculate the higher-order terms in Eq. (30).

Electromagnetic cavities have many practical applications in the technology of microwaves, that is, high-frequency radio and radar waves. For instance, one of the devices commonly used to generate high-frequency radar waves is the **klystron**, consisting of a cavity in which electromagnetic oscillations are excited by an incident beam of electrons (Figures 35.11 and 35.12). The excitation of electromagnetic

Fig. 35.10 A closed cylindrical can with the same radius R, and the electric (black) and magnetic (color) field lines in its interior. This diagram does not show the time dependence of the fields (both fields reverse direction every half cycle; when the electric field is maximum, the magnetic field is zero and conversely).

Fig. 35.11 One of the 245 large klystrons in the gallery at the Stanford Linear Accelerator Center. These klystrons generate the electromagnetic waves used in the accelerator, which is buried 8 m below the floor.

---

[6] The charge must be placed on the *interior* of the end faces. If the charge were placed on the exterior, it would not produce any field inside the cavity.

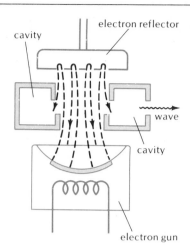

Fig. 35.12 Schematic diagram of a reflex klystron. An electron beam (dashed lines) emerges from an electron gun and moves upward, passing by the open gaps in the walls of the cavity; it is then reflected by an electric field and moves downward, again passing by the gaps. The electric fields of this beam excite oscillations in the cavity.

oscillations in a cavity by a beam of electrons is analogous to the excitation of sound oscillations in an organ pipe by a stream of air. The electrons give up their kinetic energy to the electric and magnetic fields, and thereby increase the strength of the oscillations. As we mentioned, the oscillations can be regarded as standing waves. A small opening on the side of the klystron (see Figure 35.12) permits some of these waves to spill out into space in the form of a traveling radio wave.

In a radar transmitter, the waves are not released into space directly from the generator, but are first guided into a dishlike antenna (a radar antenna) mounted at some suitable height. The waves are carried from the generator to the antenna by a hollow conducting tube or **waveguide** (Figure 35.13). Such a tube is nothing but a cavity with two open ends. The mathematical analysis of the electromagnetic oscillations in a waveguide is quite similar to that in a closed cavity — the electric and magnetic fields induce each other just as in a cavity. However, the oscillations in a waveguide take the form of traveling waves rather than that of standing waves. This can be readily understood in terms of the analogy with sound waves. We know that a standing sound wave in, say, a closed organ pipe consists of two traveling waves of opposite directions (see Section 16.6). If we suddenly open the two ends of such an organ pipe, the standing wave will separate into its traveling parts, each spilling out of one end of the organ pipe.

Waveguides for microwaves are usually copper pipes of rectangular cross section. Figure 35.14 shows the pattern of field lines of a traveling wave in such a waveguide.

Fig. 35.13 A piece of waveguide of rectangular cross section. The flanges at the ends are used for the accurate joining of adjacent pieces.

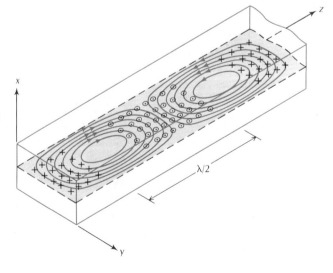

Fig. 35.14 Electric (black) and magnetic (color) field lines in a waveguide. The wave is traveling to the right. The dots and the crosses are the tips and the tails of the electric field lines, respectively. The closed loops are the magnetic field lines.

35.4* The Electric Field of an Accelerated Charge

We started our study of electromagnetic theory with the electric field of a charge at rest — this is the Coulomb field given by Eq. (23.16). Later, we examined the fields of a charge in motion with uniform velocity — in addition to the Coulomb field, such a charge has a magnetic field given by Eq. (30.15). Now we will investigate the fields of a charge with *accelerated motion*. We will find that in this case there are extra electric and magnetic fields, called **radiation fields,** that spread outward from the position of the charge, like ripples on a pond in which a stone has been dropped. These radiation fields constitute an **electromagnetic wave** which travels outward from the charge at the speed of light and carries energy and momentum away from the charge. The electric and magnetic fields in the electromagnetic wave are self-supporting — once these fields have been generated by the accelerated charge and started on their outward journey, they proceed on their own, and they become completely independent of the subsequent motion of the charge.

Radiation fields

We can derive a formula for the electric radiation field emanating from an accelerated charge by the following argument. Suppose that the charge is initially at rest, then is quickly accelerated for some short time interval τ, and then continues to move with a constant final velocity. We can summarize the motion thus:

$t < 0$ charge is at rest

$0 \leq t < \tau$ charge has acceleration a

$t \geq \tau$ charge moves at constant velocity $v = a\tau$ (34)

The initial electric field lines originate at the initial position of the charge. But the field lines at some later time ($t > \tau$) must originate on the new position of the charge. The field lines cannot change from their initial configuration to their final configuration instantaneously; rather, a disturbance must travel outward from the position of the charge and gradually change the field lines. The disturbance takes the form of a kink connecting the old and the new field lines.

The disturbance travels at some speed c (we will prove below that the speed of the disturbance is actually the speed of light; our notation anticipates this result). Figure 35.15 shows the situation at some time after the acceleration ceases. The disturbance begins at the time 0 when the acceleration begins. The leading edge of the disturbance (outer edge of kink) travels outward from the initial position of the charge, and in a time t it reaches out to a distance ct. Beyond the sphere of radius ct, the electric field is still the old field with field lines centered on the initial position of the charge.

The disturbance ceases as soon as the acceleration ceases. The field in the vicinity of the uniformly moving charge then settles into the new radial configuration centered on the new position of the charge. The trailing edge of the disturbance (inner edge of kink) marking the cessation of the acceleration travels outward from the position that the charge has at the time τ, and at some later time t it reaches out to a

---

* This section is optional.

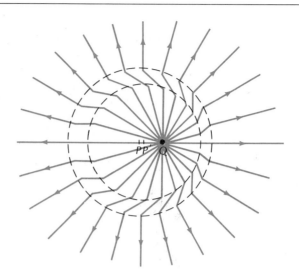

Fig. 35.15 Electric field lines of a charge that has suffered an acceleration. Here Q is the present position of the charge; P is the initial position of the charge. Between P and P' the charge suffered a constant acceleration. Between P' and Q the charge moved at constant velocity. The outer dashed sphere (outer edge of kink) has radius ct and is centered on P; the inner dashed sphere (inner edge of kink) has radius $c(t-\tau)$ and is centered on P'.

distance $c(t-\tau)$. Within the sphere of radius $c(t-\tau)$ the electric field is the new radial field centered on the position of the uniformly moving charge.

The disturbance produced by the accelerated charge is confined to the space between the larger and smaller spheres in Figure 35.15. The field lines in this zone must connect the lines of the new field of the uniformly moving charge with the lines of the old field of the stationary charge. We have drawn the connecting line segments in Figure 35.15 as straight segments. Strictly, this is only an approximation based on the rationale that a curve can be approximated by a straight line. (A careful argument shows that the segments are actually slightly curved, but the curvature is unimportant if the speed of the charge is much smaller than the speed of light, $v \ll c$.)

In the zone of the kink, the electric field has both a radial component and a tangential, or **transverse,** component. This transverse component is characteristic of the field of an accelerated charge.

Figure 35.16 shows one electric field line in detail. This figure relies on the assumption that a long time has elapsed since the acceleration interval, that is, $t \gg \tau$. Consequently, on the scale of this diagram, the distance that the charge has covered while under acceleration is very small compared with the distance that the charge has covered at uniform velocity. The inner and outer circles of Figure 35.16 are then nearly concentric, and the distance from the initial position of the charge to the position at time t is nearly vt.[7] Figure 35.16 also contains the implicit assumption that the final speed of the charge is much less than the speed of light, that is, $v \ll c$. (As we remarked above, if this assumption does not hold, then the field line in the zone of the kink is appreciably curved, and the straight-line approximation of Figure 35.16 is not satisfactory.)

By inspection of Figure 35.16, we see that the ratio of the transverse and radial components of the electric field in the zone of the kink is

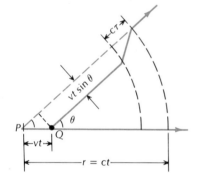

Fig. 35.16 One of the electric field lines (color). This line makes an angle θ with the direction of motion of the charge. Here we have neglected the small distance PP' (see Figure 35.15) that the charge covers while under acceleration. With this approximation, the distance PQ is $\sim vt$, and the distance between the field line and the dashed line is $\sim vt \sin\theta$.

---

[7] The exact distance is $\frac{1}{2}v\tau + v(t-\tau)$ which, for $t \gg \tau$, indeed reduces to $\sim vt$.

$$\frac{E_\theta}{E_r} = \frac{vt \sin\theta}{c\tau} = \frac{at \sin\theta}{c} \tag{35}$$

where, of course, $a = v/\tau$. The radial electric field equals the number of lines passing through a unit area on a sphere; this number is not affected by the angle that the lines make with the radial direction. Thus, E_r is simply the usual Coulomb field:

$$E_r = \frac{1}{4\pi\varepsilon_0} \frac{q}{r^2} \tag{36}$$

Inserting this into Eq. (35), we obtain

$$E_\theta = \frac{1}{4\pi\varepsilon_0} \frac{q}{r^2} \frac{at \sin\theta}{c} \tag{37}$$

Since $t \gg \tau$, the radii of the inner and outer dashed circles in Figure 35.16 are nearly the same; that is, the difference $\Delta r = c\tau$ between them is very small compared with the radius $r = ct$. We can therefore substitute this value of the radius into Eq. (37), so

$$\boxed{E_\theta = \frac{1}{4\pi\varepsilon_0} \frac{qa \sin\theta}{c^2 r}} \tag{38}$$

Radiation field of accelerated charge

This transverse electric field is the **radiation field** of the accelerated charge. Note that it is directly proportional to the acceleration a. Also note that it varies as the inverse distance, not the inverse square of the distance. The radiation field therefore decreases less sharply with distance than the Coulomb field — it remains significant at large distances, where the Coulomb field practically disappears.

The radiation field produced by an acceleration a acting at time 0 reaches the distance r at a time r/c after time 0, that is, the field needs time to propagate from the position of the charge to a remote point. Mathematically, we can incorporate this retardation in our equation by expressing the acceleration as a function $a(t)$ of time,

$$E_\theta(t + r/c, r) = \frac{1}{4\pi\varepsilon_0} \frac{q \sin\theta}{c^2 r} a(t)$$

and consequently

$$E_\theta(t, r) = \frac{1}{4\pi\varepsilon_0} \frac{q \sin\theta}{c^2 r} a\left(t - \frac{r}{c}\right) \tag{39}$$

This formula indicates that the value of the function E_θ at time t is related to the value of the function a at the earlier time $t - r/c$. For example, if $r = 3.8 \times 10^8$ m (the distance from the Earth to the Moon), then $r/c = 3.8 \times 10^8$ m$/(3.0 \times 10^8$ m/s$) = 1.3$ s. Thus, the radiation field that an accelerated charge on the Earth produces at the Moon at time t depends on the acceleration at time $t - 1.3$ s, that is, it depends on the acceleration at a time 1.3 s earlier, the radiation field needing this much time to propagate from the charge to the Moon.

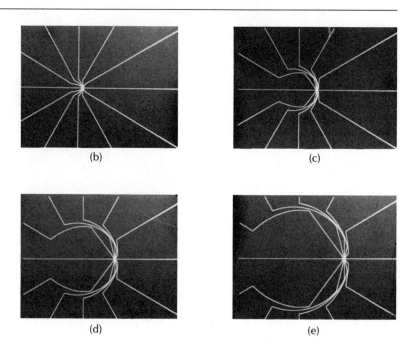

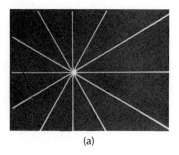

Fig. 35.17 Time-dependent electric field lines for a charge that suddenly accelerates and then moves with uniform velocity. (a) The charge is at rest. (b) The charge has begun to accelerate. (c)–(e) The charge moves with uniform velocity. Note that the field lines in the region of the kink are curved; this is due to the very high speed of the motion.

If the acceleration lasts only a short time, as in the special case shown in Figure 35.15, then the electric radiation field at any point also lasts only a short time. The radiation field described by Eq. (39) then constitutes a single wave pulse that propagates outward from the place at which the acceleration occurred. Figure 35.17 shows a sequence of snapshots from a filmed computer simulation of the electric field lines of an accelerated charge, whose acceleration lasted only a short time. These snapshots give a clear impression of the outward propagation of the kinks in the field lines.

The formula (39) for the radiation field is perfectly general and applies even if the acceleration is not constant and lasts for any length of time. For instance, if the charge oscillates back and forth with simple harmonic motion of frequency ω, so

$$a = a_0 \sin \omega t \tag{40}$$

then

Radiation field of oscillating charge

$$E_\theta(t,r) = \frac{1}{4\pi\varepsilon_0} \frac{q \sin\theta}{c^2 r} a_0 \sin\omega\left(t - \frac{r}{c}\right) \tag{41}$$

This represents a (spherical) harmonic wave of frequency ω traveling in the radial direction with a speed c. The amplitude of the wave gradually decreases ($\propto 1/r$) as the wave spreads out. As we will see in the next chapter, the function (41) describes a radio wave emitted from an antenna or a light wave emitted from an atom.

Figure 35.18 shows a sequence of snapshots from a computer simulation of the electric field lines of a charge oscillating back and forth with simple harmonic motion, as indicated by Eq. (40). The kinks in the field lines can be seen to propagate outward, forming a regular succession of wave peaks and wave troughs.

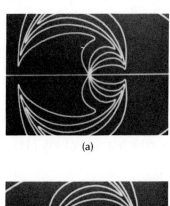
(a)

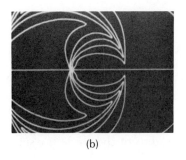
(b)

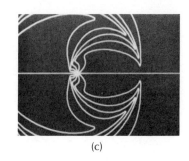
(c)

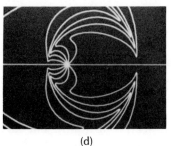
(d)

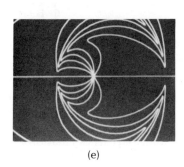
(e)

Fig. 35.18 Time-dependent field lines of a charge moving with simple harmonic motion. The pictures display one-half cycle of the motion.

EXAMPLE 2. In a head-on collision with an atom, a fast-moving electron suffers a deceleration of 4.0×10^{23} m/s². What is the electric radiation field generated by the electron at a distance of 10 cm at right angles to the direction of motion? Pretend that classical mechanics can be applied to this problem.

SOLUTION: With $\theta = 90°$, Eq. (38) gives

$$E_\theta = \frac{1}{4\pi\varepsilon_0} \frac{1.6 \times 10^{-19} \text{ C} \times 4.0 \times 10^{23} \text{ m/s}^2}{(3.0 \times 10^8 \text{ m/s})^2 \times 0.10 \text{ m}}$$

$$= 6.4 \times 10^{-2} \text{ V/m}$$

COMMENTS AND SUGGESTIONS: Radiation emitted by the collision of a beam of fast-moving electrons with the atoms in a target (block of metal) is called **Bremsstrahlung**.[8] This is essentially the mechanism by means of which X rays are produced in an X-ray tube (see Figure IV.1).

35.5* The Magnetic Field of an Accelerated Charge

When a charge accelerates, the disturbance traveling outward consists not only of electric fields but also of magnetic fields. The initial magnetic field of a charge at rest is zero, whereas the final magnetic field of a moving charge is that given by Eq. (30.15). Hence a disturbance must move outward and gradually change the magnetic field from its initial to its final value.

The magnetic field can be regarded as an induced field obeying the Maxwell–Ampère Law

---

[8] German for "braking radiation."
* This section is optional.

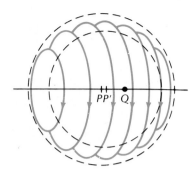

Fig. 35.19 Magnetic field lines in the zone of the kink. These lines are circles around the line of motion. At any point, the magnetic field is perpendicular to the electric field shown in Figure 35.15.

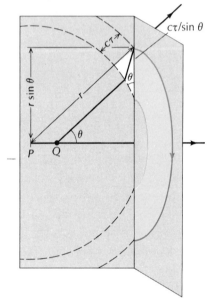

Fig. 35.20 The magnetic field line (color) is a circle of radius $r \sin \theta$ perpendicular to the plane of the page. The area bounded by this field line is a circle of the same radius perpendicular to the plane of the page. This circle intercepts the transverse electric field in the annular region marked with color. The width of this annular region equals the hypotenuse of the small white triangle, which is $\sim c\tau/\sin \theta$.

$$\oint \mathbf{B} \cdot d\mathbf{l} = \mu_0 \varepsilon_0 \frac{d\Phi}{dt} \quad (42)$$

By symmetry, the magnetic field lines are circles about the direction of motion of the charge. Figure 35.19 shows the magnetic field lines in the zone of the kink. Note that the magnetic field is everywhere perpendicular to the electric field. Let us evaluate Eq. (42) for a circular path that follows one of the magnetic field lines (see Figure 35.20). The radius of the circle is $r \sin \theta$, and hence the left side of Eq. (42) is

$$\oint \mathbf{B} \cdot d\mathbf{l} = 2\pi r(\sin \theta) B \quad (43)$$

In the evaluation of the right side of Eq. (42) we need to take into account two contributions to the flux: from the radial electric field E_r and from the transverse electric field E_θ. It turns out that the flux produced by the radial electric field induces the familiar magnetic field dependent on velocity, that is, the magnetic field given by Eq. (30.15). We will not present the detailed calculation establishing this result; instead we leave this calculation as a problem (see Problem 33). The flux produced by the transverse electric field induces a magnetic field dependent on the acceleration. Since in the context of this section we are interested only in the magnetic field characteristic of accelerated motion, we will take into account only the contribution of the flux from this transverse electric field E_θ.

Taking a flat circular area for the evaluation of the flux, we see that this area intercepts the transverse electric field E_θ only in a narrow annular region of radius $r \sin \theta$ and width $c\tau/\sin \theta$ (see Figure 35.20). The area of this region is

$$[\text{area}] \cong [\text{circumference}] \times [\text{width}] = 2\pi r \sin \theta \times \frac{c\tau}{\sin \theta} = 2\pi r c \tau \quad (44)$$

The transverse field E_θ makes an angle of θ with this area, and hence the flux is

$$\Delta \Phi = E_\theta \sin \theta \times [\text{area}] = 2\pi r c \tau E_\theta \sin \theta \quad (45)$$

In a time τ, all of this flux disappears since the kink moves beyond the annular area. Hence

$$\frac{d\Phi}{dt} = \frac{\Delta \Phi}{\tau} = 2\pi r c E_\theta \sin \theta \quad (46)$$

Using Eqs. (43) and (46) in Eq. (42), we then obtain

$$B = c\mu_0 \varepsilon_0 E_\theta \quad (47)$$

This equation expresses the magnetic field of the pulse of radiation in terms of the transverse electric field, the speed of propagation c, and the familiar constants ε_0 and μ_0. As we pointed out above, the speed of propagation of the pulse coincides with the speed of light,

$$c = 3.00 \times 10^8 \text{ m/s} \quad (48)$$

However, we can do better than this. In what follows, we will derive a theoretical expression for the speed of propagation, and we will confirm that the theoretical prediction agrees with the experimental value given by Eq. (48).

The preceding calculation regarded the magnetic field **B** as induced by E_θ. But the converse is also true: the electric field E_θ is induced by the magnetic field **B** according to Faraday's Law

$$\oint \mathbf{E}_\theta \cdot d\mathbf{l} = -\frac{d\Phi_B}{dt} \tag{49}$$

To extract some useful information from this law, it is convenient to evaluate the integral for the path shown in Figure 35.21; this path consists of two semicircles joined by short radial segments. The transverse electric field is different from zero only along the larger semicircle. Hence the left side of Eq. (49) is

$$\oint \mathbf{E}_\theta \cdot d\mathbf{l} = \int_{P_1}^{P_2} E_\theta \, dl \tag{50}$$

where the points P_1 and P_2 are shown in Figure 35.21. The magnetic flux on the right side can be calculated by taking small area elements of the shape shown in Figure 35.21 with dimensions $c\tau \times dl$:

$$\Delta\Phi_B = \int \mathbf{B} \cdot d\mathbf{S} = \int Bc\tau \, dl = c\tau \int_{P_1}^{P_2} B \, dl \tag{51}$$

In a time τ, the two dashed circles of Figure 35.21, with the magnetic field between them, sweep beyond the fixed path shown in this figure. Hence

$$\frac{d\Phi_B}{dt} = \frac{-\Delta\Phi_B}{\tau} = -c \int_{P_1}^{P_2} B \, dl \tag{52}$$

Faraday's Law now becomes

$$\int_{P_1}^{P_2} E_\theta \, dl = c \int_{P_1}^{P_2} B \, dl$$

In view of Eq. (47), this can be written

$$\int_{P_1}^{P_2} E_\theta \, dl = c^2 \mu_0 \varepsilon_0 \int_{P_1}^{P_2} E_\theta \, dl \tag{53}$$

Obviously, the consistency of this equation requires that

$$1 = c^2 \mu_0 \varepsilon_0 \tag{54}$$

or

$$\boxed{c = \frac{1}{\sqrt{\mu_0 \varepsilon_0}}} \tag{55}$$

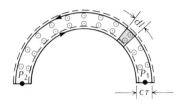

Fig. 35.21 The path of integration for Eq. (49) is marked with a solid line. As in the other figures, the boundaries of the zone of the kink are marked with dashed lines. The larger semicircle of the path is in the zone of the kink, but the smaller semicircle is not. The magnetic field is perpendicular to the plane of the page; the colored dots show the tips of the magnetic field vectors.

Speed of electromagnetic wave

We have therefore derived a theoretical expression for the speed of propagation of an electromagnetic disturbance in vacuum. The numerical value of the right side of Eq. (55) is

$$c = 1/(1.26 \times 10^{-6} \text{ H/m} \times 8.85 \times 10^{-12} \text{ F/m})^{1/2}$$

$$= 3.00 \times 10^8 \text{ m/s} \tag{56}$$

We see that this agrees with the experimental value (48) for the speed of light (in Section 36.1 we will make a more precise comparison of the theoretical and experimental values of the speed of light). The derivation of Eq. (55) was one of the great and early triumphs of Maxwell's electromagnetic theory of light.

By means of Eq. (55), we can somewhat simplify the relation between B and E_θ. We have

$$B = c\mu_0\varepsilon_0 E_\theta = \frac{c}{c^2} E_\theta \qquad (57)$$

or

Relation between electric and magnetic radiation fields

$$\boxed{B = \frac{E_\theta}{c}} \qquad (58)$$

EXAMPLE 3. In Example 2 we found that the sudden deceleration of an electron during a collision produces a transverse electric field of 6.4×10^{-2} V/m at a certain distance. What is the magnetic field produced by the deceleration?

SOLUTION: According to Eq. (58),

$$B = E_\theta/c$$
$$= (6.4 \times 10^{-2} \text{ V/m})/(3.0 \times 10^8 \text{ m/s})$$
$$= 2.1 \times 10^{-10} \text{ T}$$

SUMMARY

Displacement current: $I_d = \varepsilon_0 \dfrac{d\Phi}{dt}$

Maxwell's equations: $\oint \mathbf{E} \cdot d\mathbf{S} = Q/\varepsilon_0$

$\oint \mathbf{B} \cdot d\mathbf{S} = 0$

$\oint \mathbf{E} \cdot d\mathbf{l} = -\dfrac{d\Phi_B}{dt}$

$\oint \mathbf{B} \cdot d\mathbf{l} = \mu_0 I + \mu_0 \varepsilon_0 \dfrac{d\Phi}{dt}$

Transverse electric field of accelerated charge:

$$E_\theta = \frac{1}{4\pi\varepsilon_0} \frac{qa \sin\theta}{c^2 r}$$

Speed of electromagnetic wave: $c = 1/\sqrt{\mu_0\varepsilon_0}$

Transverse magnetic field: $B = \dfrac{E_\theta}{c}$

QUESTIONS

1. The displacement current between the plates of a capacitor has the same magnitude as the conduction current in the wires connected to the capacitor, and yet the magnetic field produced by the former current near and within the capacitor is much weaker than that produced by the latter current near and within the wire. Explain.

2. Consider the electric field of a single positive electric charge moving at constant velocity. What is the direction of the displacement current intercepted by a circular area perpendicular to the velocity in front of the charge? Behind the charge?

3. A capacitor is connected to an alternating source of emf. Does the displacement current between the capacitor plates lead or lag the emf?

4. Which of Maxwell's equations permits us to deduce the electric Coulomb field of a static charge? Which of Maxwell's equations permits us to deduce the magnetic field of a charge moving with uniform velocity?

5. Suppose that there exist magnetic monopoles, that is, positive and negative magnetic charges that act as sources and sinks of magnetic field lines, analogous to positive and negative electric charges. Which of Maxwell's equations would have to be modified to take into account such monopoles? Qualitatively, what are the required modifications?[9]

6. Given Eq. (30) for the electric field between the capacitor plates described in Section 35.3, is there a well-defined potential difference between the plates?

7. Some microwave ovens have rotating turntables that continually turn the meat while cooking. What is the purpose of this arrangement? (Hint: Microwave ovens are cavities with standing electromagnetic waves.)

8. Cavities are capable of oscillating at much higher frequencies than ordinary LC circuits. What limits the maximum frequency attainable with LC circuits?

9. In practice, the electromagnetic oscillations in a cavity are damped by "frictional" losses. How does this "friction" arise?

10. Efficient waveguides are manufactured out of a very good conductor, such as copper, sometimes with a silver lining. Why is high conductivity essential for high efficiency?

11. As we stated in Section 35.5, the magnetic field of a charge moving at uniform velocity can be regarded as an induced magnetic field produced by the changing electric flux in the space surrounding the charge. Check that the direction of the induced magnetic field agrees with the direction given by Eq. (30.15).

12. Since the magnetic field of a charge moving at uniform velocity can be regarded as an induced magnetic field produced by the changing electric flux, we might expect that the electric field (Coulomb field) can be regarded as an induced electric field produced by a changing magnetic flux. Is this the case?

13. Is the transverse electric radiation field E_θ a conservative field?

14. In Figure 35.16 the electric field lines have a sharp discontinuity in slope. Is this discontinuity unphysical? On what is this discontinuity to be blamed?

15. Roughly sketch the electric field lines (at one instant of time) for a positive charge that is initially moving at uniform velocity and suddenly stops.

---

[9] For a quantitative discussion of these modifications, see Problem 12.

16. In a collision with an atom, an electron suddenly stops. Describe the directions of the electric and magnetic radiation fields at some distance from the electron at right angles to the acceleration.

17. Given that the magnitudes of the magnetic and electric radiation fields of an accelerated charge are related by $B = E_\theta/c$, compare the energy densities in these fields.

18. Consider a small angular patch (Figure 35.22) in the radiation field of the accelerated charge described in Section 35.4. The volume of this small patch is [area] × [thickness] = $r^2 \, d\theta \, d\alpha \times c\tau$, and the energy in this volume is $\frac{1}{2}\varepsilon_0 E_\theta^2 \times (r^2 \, d\theta \, d\alpha \times c\tau)$. Show that this energy remains constant as the pulse of radiation propagates outward.

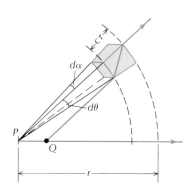

Fig. 35.22

19. A charged particle moving around a circular orbit at uniform speed has a centripetal acceleration and therefore produces a radiation field. However, a uniform current flowing around a circular loop does *not* produce a radiation field. Is this a contradiction? Explain.

PROBLEMS

Section 35.1

1. A parallel-plate capacitor is being charged by a current of 4.0 A.
   (a) What is the displacement current between its plates?
   (b) What is the rate of change of the electric flux intercepted by each plate?

2. A parallel-plate capacitor consists of circular plates of radius 0.30 m separated by a distance of 0.20 cm. The voltage applied to the capacitor is made to increase at a steady rate of 2.0×10^3 V/s. Assume that the electric charge distributes itself uniformly over the plates, and ignore the fringing effects.
   (a) What is the rate of increase of the electric field between the plates?
   (b) What is the magnetic field between the plates at a radius of 0.15 m? At 0.30 m?

3. A parallel-plate capacitor has circular plates of radius 20 cm and a uniform electric field between the plates. The capacitor is being charged at a rate of 0.10 A.
   (a) What is the net displacement current between the plates?
   (b) What is the displacement current between the plates within the radial interval $0 \leq r \leq 5$ cm?

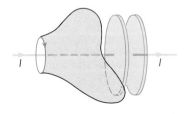

Fig. 35.23

*4. Suppose that instead of arranging the baglike surface as in Figure 35.2, we arrange it as in Figure 35.23. This baglike surface intersects the plates of the capacitor and intercepts both the current in the wire *and* some of the flux between the capacitor plates. Show that, nevertheless, Eq. (6) remains valid for this choice of surface. (Hint: Consider what happens at the intersection of the surface with the capacitor plate.)

*5. An emf $\mathscr{E}_0 \sin \omega t$, with $\mathscr{E}_0 = 0.50$ V and $\omega = 4.0 \times 10^3$ radians/s, is applied to the terminals of a capacitor with $C = 2.0$ pF. What is the displacement current between the capacitor plates?

*6. The space between the plates of a leaky capacitor is filled with a material of resistance 5.0×10^5 Ω. The capacitor has a capacitance of 2.0×10^{-6} F, its plates are circular, with a radius of 30 cm, and its electric field is uniform. At time $t = 0$, the initial voltage across the capacitor is zero.
   (a) What is the displacement current if we increase the voltage at the steady rate of 1.0×10^3 V/s?
   (b) At what time will the real current leaking through the capacitor equal the displacement current?
   (c) What is the magnitude of the magnetic field between the plates at radius $r = 20$ cm at $t = 0$? At $t = 1$ s? At $t = 2.0$ s?

*7. A parallel-plate capacitor with circular plates of radius 25 cm separated by a distance of 0.15 cm is connected to a source of alternating emf. The voltage across the plates oscillates with an amplitude of 5000 V and a frequency of 60 Hz. What is the amplitude of the magnetic field between the plates at a distance of 20 cm from the axis of the capacitor?

*8. A parallel-plate capacitor has circular plates of area A separated by a distance d. A thin straight wire of length d lies along the axis of the capacitor and connects the two plates (Figure 35.24); this wire has a resistance R. The exterior terminals of the plates are connected to a source of alternating emf with a voltage $V = V_0 \sin \omega t$.
  (a) What is the current in the thin wire?
  (b) What is the displacement current through the capacitor?
  (c) What is the current arriving at the outside terminals of the capacitor?
  (d) What is the magnetic field between the capacitor plates at a distance r from the axis? Assume that r is less than the radius of the plates.

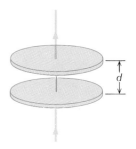

Fig. 35.24 Parallel-plate capacitor with a thin wire connecting the inside faces of the plates.

*9. Suppose that the parallel-plate capacitor discussed in Section 35.1 (see Figure 35.2) is filled with a slab of dielectric with a dielectric constant κ.
  (a) By retracing the arguments of Section 35.1, show that the displacement current must be

$$I_d = \kappa \varepsilon_0 \frac{d\Phi}{dt}$$

and that Maxwell's modification of Ampère's Law must be

$$\oint \mathbf{B} \cdot d\mathbf{l} = \mu_0 I + \kappa \mu_0 \varepsilon_0 \frac{d\Phi}{dt}$$

  (b) If the parallel-plate capacitor described in Problem 1 is filled with a dielectric with $\kappa = 2.0$, how do the answers to that problem change?

Section 35.2

*10. Prove that Maxwell's equations [Eqs. (17)–(20)] mathematically imply the conservation of electric charge; that is, prove that if no electric current flows into or out of a given volume, then the electric charge within this volume remains constant. [Hint: Begin with Eq. (17), $Q = \varepsilon_0 \Phi$, and evaluate dQ/dt by means of Eq. (20); note that if the baglike surface used for the evaluation of Eq. (20) is actually a closed surface (a bag whose mouth has been shrunk to zero), then the path integral of $\mathbf{B}$ is zero.]

*11. Write Maxwell's equations for a medium of dielectric constant κ and relative permeability κ_m.

**12. Suppose that there exist magnetic monopoles, that is, positive and negative magnetic charges that act as sources and sinks of magnetic field lines analogous to positive and negative electric charges. The magnetic field generated by a magnetic charge q_m is an inverse-square field, $B = q_m/4\pi r^2$. Write a new set of Maxwell's equations that take into account the magnetic charge. Be careful with the constants ε_0 and μ_0 and with the signs.

Section 35.3

13. An oscillating emf of amplitude 2.0 V is applied between the centers of the plates of a parallel-plate capacitor with large circular plates. The frequency of the emf is 1.5×10^{10} radians/s and the distance between the plates is 1.0 cm. Using the approximate equations derived in Section 35.3, find the amplitude of oscillation of the electric and magnetic fields at the center of the capacitor and also at a radial distance of 2.0 cm from the center.

14. Consider a closed cylindrical metallic can. You want to establish oscillating electromagnetic fields of the type shown in Figure 35.6 in this can. If the oscil-

lations are to have a frequency of 2.0×10^9 Hz, what must be the radius of the can?

*15. (a) Find the magnetic field $B_{(2)}$ that the electric field $E_{(2)}$ of Eq. (29) will induce.
   (b) By adding $B_{(1)}$ and $B_{(2)}$, find the corrected magnetic field, $B = B_{(1)} + B_{(2)}$.
   (c) What is the value of the corrected magnetic field at the radius R given by Eq. (32)?

**16. If oscillating electric and magnetic fields of the kind described by Eqs. (30) and (23) exist in a closed cylindrical cavity, currents must periodically flow back and forth between the top and the bottom of the cylinder.
   (a) Calculate the instantaneous amount of electric charge on the top and bottom faces of a cylinder of radius $R = 2/(\sqrt{\mu_0 \varepsilon_0}\omega)$ if the electric field in the cylinder is given by Eq. (30).
   (b) Calculate the current that must flow between the top and bottom faces.

**17. Consider the electric and magnetic fields given by Eq. (30) and (23).
   (a) Find the energy density as a function of radius in each of these fields.
   (b) By integrating the energy density, find the electric and magnetic energies between the capacitor plates within a radius $R = 2/(\sqrt{\mu_0 \varepsilon_0}\omega)$.
   (c) Plot the electric energy and the magnetic energy as a function of time. Is the sum of these energies constant?

Section 35.4

18. In the electron gun of a TV tube, an electron is accelerated by a constant electric field and acquires a kinetic energy of 3.2×10^{-15} J within a distance of 2.0 cm. What is the magnitude of the electric radiation field that this accelerated electron generates at a distance of 5.0 cm at right angles to its direction of motion?

19. Consider the electron whose transverse electric and magnetic fields we calculated in Examples 2 and 3.
   (a) For comparison, find the Coulomb field of this electron at the given point.
   (b) Assuming that the electron has a speed of 2.0×10^7 m/s, find the magnetic field, according to Eq. (30.15), of this electron at the given point.

20. In a collision with an atom, an electron suffers a deceleration of 2.0×10^{24} m/s². What is the magnitude of the electric radiation field that this electron generates at a distance of 20 cm at an angle of 45° to the direction of the deceleration? At what time after the instant of collision does this radiation field arrive at that distance?

21. In an X-ray tube, a beam of high-speed electrons is made to strike a block of metal. The sudden deceleration of the electrons brings about the emission of intense electromagnetic radiation (X rays). Suppose that an electron of initial energy 2×10^4 eV decelerates uniformly and comes to rest within a distance of 5×10^{-9} m. What will be the magnitude of the electric radiation field at a distance of 0.3 m from the point of impact in a direction perpendicular to the direction of the acceleration?

*22. An electron of energy 2.0×10^4 eV is in a circular orbit under the influence of a constant magnetic field of 5.0×10^{-2} T.
   (a) Consider a point of the orbit. What are the magnitude and direction of the instantaneous acceleration of the electron at this point?
   (b) Consider a point at a distance of 1.5 m from the electron in a direction perpendicular to that of the instantaneous acceleration. What is the transverse electric field due to the acceleration that reaches this point (after some suitable time delay)?

*23. Repeat Problem 22, substituting a proton for the electron. By what factor does the transverse electric field generated by the proton differ from that generated by the electron?

*24. Two protons of energy 1.5×10^5 eV collide head on.
   (a) What is the distance of closest approach? What is the instantaneous acceleration at this point?
   (b) What is the transverse electric field that each proton produces at a distance of 3.0×10^{-10} m from the point of collision in a direction perpendicular to the line of motion of the protons?
   (c) What is the net transverse electric field of the two protons taken together?

*25. On a radio antenna (a straight piece of wire), electrons move back and forth in unison. Suppose that the velocity of the electrons is $v = v_0 \cos \omega t$, where $v_0 = 8.0 \times 10^{-3}$ m/s and $\omega = 6.0 \times 10^6$ radians/s.
   (a) What is the maximum acceleration of one of these electrons?
   (b) Corresponding to this maximum acceleration, what is the strength of the transverse electric field produced by one electron at a distance of 1.0 km from the antenna in a direction perpendicular to the antenna? What is the time delay (or retardation) between the instant of maximum acceleration and the instant at which the corresponding electric field reaches a distance of 1.0 km?
   (c) There are 2.0×10^{24} electrons on the antenna. What is the collective electric field produced by all the electrons acting together? Assume that the antenna is sufficiently small so that all the electrons contribute just about the same electric field at a distance of 1.0 km.

*26. The accelerated electrons in the antenna of a radio station produce an electric field of 1.0 V/m at a distance of 1.0 km from the radio station. The antenna is a straight vertical wire made of copper, 5.0 m long with a diameter of 2.0×10^{-3} m. Assume that all the free electrons in the copper contribute equally to the field. What must be the average acceleration of each electron?

*27. Consider an electron in a hydrogen atom orbiting the nucleus in a circular orbit of radius 0.53×10^{-10} m under the influence of the electrostatic force of attraction. Calculate the acceleration of this electron, and then calculate the magnitude of the electric radiation field that, according to Eq. (39), the electron produces at a distance of 10 cm from the atom measured along a line tangent to the circular orbit at the position of the electron. (The result of this classical calculation does not agree with the actual behavior of the radiation emitted by the electron; quantum effects play a crucial role in this problem.)

Section 35.5

28. In a Van de Graaff accelerator, a proton is given an acceleration of 1.1×10^{14} m/s².
   (a) Find the strengths of the transverse electric and magnetic fields at a distance of 0.50 m from the proton at an angle of 45° with the acceleration.
   (b) Draw a diagram showing the direction of the acceleration, and the direction of the electric and magnetic fields calculated in part (a).

29. Show that the energy density in the transverse electric field of an accelerated charge equals the energy density in the corresponding magnetic field.

*30. A thorium nucleus emits an alpha particle and thereby transforms itself into a radium nucleus. Assume that the alpha particle is pointlike and that the residual radium nucleus is spherical with a radius of 7.4×10^{-15} m. The charge on the alpha particle is $2e$ and that on the radium nucleus is $88e$.
   (a) Calculate the acceleration of the alpha particle at the instant it leaves the surface of the nucleus.
   (b) Calculate the magnitudes of the electric and magnetic radiation fields that, according to Eqs. (39) and (58), the alpha particle produces at a distance of 0.50×10^{-10} m from the nucleus along a direction perpendicular to the direction of the acceleration. (This classical calculation does not agree with the actual behavior of the radiation emitted by the alpha particle; quantum effects are important in this problem.)

*31. An AC current $I = I_0 \cos 2\pi\nu t$, with $I_0 = 15.0$ A and $\nu = 60$ Hz, flows in a copper wire of diameter 0.26 cm.
  (a) Find formulas for the average velocity (drift velocity) and the corresponding acceleration of the free electrons as a function of time.
  (b) The acceleration reaches a maximum magnitude every $\frac{1}{120}$ s. What is this maximum magnitude?
  (c) For this maximum acceleration, what are the transverse and magnetic electric fields that one of the free electrons produces at a distance of 0.20 m perpendicularly away from the wire?
  (d) The velocity reaches a maximum magnitude every $\frac{1}{120}$ s. What is the maximum magnitude of the velocity? What is the magnetic field that, according to Eq. (30.15), an electron moving with this velocity produces at a distance of 0.20 m perpendicularly away from the wire?

*32. The electric field in a copper wire of diameter 0.26 cm carrying a current of 12.0 A is 3.9×10^{-2} V/m.
  (a) What is the acceleration of one of the free electrons of copper in this electric field?
  (b) What are the transverse electric and magnetic fields that the accelerated electron produces at a distance of 4.0 m perpendicularly away from the wire?
  (c) Suppose that all the free electrons in a segment of this wire 5.0×10^{-2} m long simultaneously produce such transverse electric and magnetic fields. What are the net transverse electric and magnetic fields of all these electrons acting together?
  (d) Compare the magnetic field calculated in part (c) with the static magnetic field associated with the current of 12.0 A in the segment of wire.

**33. Suppose that a charge moves with *uniform* velocity. By a calculation similar to that in Eqs. (44)–(47), show that the *radial* electric field (Coulomb field) of this charge induces a magnetic field. Show that the magnitude of this magnetic field coincides with the magnitude of the familiar magnetic field [see Eq. (30.15)] of a uniformly moving charge. Hence the latter can be regarded as an induced magnetic field.

**34. (a) Integrate the energy density of the electric radiation field over the volume of the zone of the kink, and find the total electric energy in the radiation field of the accelerated charge described in Section 35.4.
  (b) Do the same for the energy of the magnetic radiation field.

CHAPTER 36

Light and Radio Waves

As we saw in the preceding chapter, an accelerated charge creates a propagating electromagnetic wave pulse which spreads outward from the charge. In essence, this wave pulse is a disturbance of the familiar electric and magnetic fields with which we began our study of electricity and magnetism [see Eqs. (23.16) and (30.15)]. As long as the charge moves with uniform velocity, these fields accompany the charge — they move as though they were rigidly attached to the charge. But if the charge is forced to accelerate, then parts of the fields break away; they become independent of the charge, and they travel outward as an electric and magnetic disturbance.

When such a wave pulse is repeated periodically, it forms an **electromagnetic wave.** The fields in the wave are self-supporting — the electric field induces the magnetic field, and the magnetic field induces the electric field. Because the electric and magnetic fields mutually support each other, the wave requires no medium for its propagation, and it readily propagates in a vacuum.

Electromagnetic wave

Radio waves are electromagnetic waves. They are created by the accelerated back-and-forth motion of electrons on an antenna. Light waves are also electromagnetic waves. They are created by the oscillations of electrons within atoms. We cannot observe the creation of light waves as directly as we can that of radio waves, but we can compare the predictions of electromagnetic theory with the results of many experiments testing the interaction of light and matter, and the emission of light by matter. In the next three chapters, we will deal with a few instances of the interaction between light and matter — reflection, refraction, diffraction. We will leave the study of the emission of light by matter to the last chapter, because the atomic processes that lead to the emission of light must be described in terms of quantum theory.

The earliest evidence for the electromagnetic character of light waves emerged from Maxwell's theoretical calculation of the speed of electromagnetic waves. From his equations, Maxwell predicted that electromagnetic waves, consisting of dynamic electric and magnetic fields that mutually induce each other, forming self-supporting electromagnetic oscillations, propagate with a speed $c = 1/\sqrt{\mu_0 \varepsilon_0}$. Numerically, this predicted speed coincides with the measured speed of light waves, a coincidence that led Maxwell to propose that light waves are electromagnetic waves. Maxwell's theory of self-supporting electromagnetic oscillations in empty space received direct experimental confirmation at the hands of Heinrich Hertz, who generated the first artificial radio waves by means of sparks triggered in a gap in a high-frequency LC circuit.

The most precise modern method for the determination of the speed of light relies on separate measurements of the wavelength and of the frequency of the light emitted by a stabilized laser. The speed can then be evaluated as the product of these, that is, $c = \lambda \nu$. This method was first developed by K. M. Evenson et al. at the National Bureau of Standards. The best available results of such determinations of the speed of light (in vacuum) give

Standard value of the speed of light

$$c = 299{,}792{,}458 \text{ m/s}$$

This value has an uncertainty of only ± 1 m/s. Thus, the modern method gives us the speed of light to within nine significant figures!

As we stated in Chapter 1, this value of the speed of light was adopted as a standard of speed in 1983, and it is now used as the basis for the definition of the meter.

Older methods for the determination of the speed of light relied on the direct timing of a light signal traveling back and forth over a precisely measured distance. For instance, in 1926, A. A. Michelson performed many careful measurements of the time required for a light signal to travel between Mt. Wilson and Mt. San Antonio, in California. But uncertainties in the atmospheric conditions seriously limited the accuracy of his determination of the speed of light.

In this chapter we will be concerned mainly with the characteristics of electromagnetic waves after they have emerged from their source and are propagating on their own through empty space. By contrast, in the preceding chapter we were concerned mainly with the process of production of electromagnetic waves by an accelerated charge.

Heinrich Rudolf Hertz *1857–1894, German physicist, and professor at Bonn. He supplied the first experimental evidence for the electromagnetic waves predicted by Maxwell's theory. Hertz generated these waves by means of an electric spark, measured their speed and wavelength, and established their similarity to light waves in the phenomena of reflection, refraction, and polarization.*

36.1 The Plane Wave Pulse

In the preceding chapter, we derived formulas for the electric and magnetic fields in the wave pulse produced by an accelerated charge; we obtained these fields, called radiation fields, from a careful analysis of the behavior of propagating disturbances, or "kinks," in the field lines. The most important features of the electric radiation field of an accelerated charge can be summarized as follows: The magnitude of the electric radiation field is directly proportional to the acceleration, and it is inversely proportional to the radial distance from the charge.

If we look at the electric radiation field in a fixed direction relative to the direction of the acceleration, then the magnitude of this electric field has the following dependence on distance and time:

$$E(r,t) = [\text{constant}] \times \frac{a(t - r/c)}{r} \tag{1}$$

where $a(t)$ represents the acceleration as a function of time, and c is the speed of propagation of the disturbance (speed of light). The time dependence of the terms in Eq. (1) takes into account that the electric field at time t depends on the acceleration at the earlier time $t - r/c$; the time delay r/c is exactly the time required for the disturbance to travel from the charge to the distance r. By comparing Eq. (1) with Eq. (16.2), the standard equation for a traveling wave, we recognize that the former does indeed represent a traveling wave that propagates with a speed c in the radial direction. The factor $1/r$ appearing in Eq. (1) indicates how the electromagnetic wave gradually weakens as it spreads farther and farther out.

Often we will be interested in the behavior of the wave over only a limited range of distances, a range very small compared with r. For example, we might be interested in the electric and magnetic fields that the moving charges on the antenna of a distant radio transmitter produce in the room in which we are sitting. If the length of the room is, say, 5 m and the distance to the transmitter is 50 km, then r changes by only 1 part in 10^4 from one end of the room to the other. Under these conditions the factor $1/r$ in Eq. (1) can be treated as approximately constant. Equation (1) then reduces to

$$E(r,t) = [\text{constant}] \times a(t - r/c) \tag{2}$$

where the constant now includes the factor $1/r$.

The electric radiation field is perpendicular to the direction of propagation. For the sake of clarity, let us assume that the direction of propagation is parallel to the x axis and that the electric field is parallel to the y axis. Then we can write Eq. (2) as

$$E_y(x,t) = [\text{constant}] \times a_y(t - x/c) \tag{3}$$

This is a plane wave. The wave fronts of this wave are flat surfaces parallel to the y–z plane. For instance, Figure 36.1 shows the electric field of such a wave for an acceleration that lasts for some short time interval, and is constant during this time interval. The radiated electric field then consists of a short wave pulse within which the electric field is constant. Note that the flat wave front shown in Figure 36.1 is simply a small patch of the spherical wave front shown in Figure 35.15; such a small patch of a spherical wave front will appear flat if its size is much smaller than the radius of the sphere.

The electric field of the wave pulse induces a magnetic field, and this magnetic field, in turn, induces an electric field. Thus, the electric and magnetic fields in the wave pulse mutually induce each other, and they thereby perpetuate the wave. In Section 35.5 we already examined this process of mutual induction for the fields of the spherical wave pulse created by an accelerating charge. Now, we will reexamine this process for the plane wave pulse, and we will see that the results are independent of how the wave pulse was created.

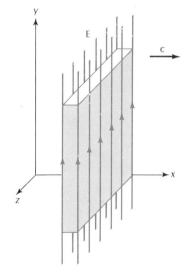

Fig. 36.1 Electric field lines of a plane wave pulse consisting of a region of uniform electric field. The wave front, or front surface of the pulse, is a flat plane. The entire pattern of field lines travels to the right with the wave speed c.

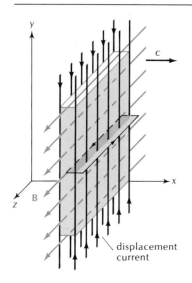

Fig. 36.2 The colored lines indicate the magnetic field of the plane wave pulse. The black lines indicate the displacement current. The displacement current forms a sheet at the front surface of the pulse and another sheet at the rear surface.

Since the wave pulse sweeps through the room at the speed c, it lasts only some short time at any one point of the room. This means the electric field of the wave pulse is time dependent: initially it is zero, then it suddenly increases to a constant value E_y when the front surface of the pulse arrives, then it remains constant for a while, and finally it drops to zero when the rear surface of the pulse passes. This time-dependent electric field induces a magnetic field. Figure 36.2 shows this induced magnetic field. The direction of the magnetic field is perpendicular to the direction of propagation and also perpendicular to the electric field. We can deduce the direction of the magnetic field from the direction of the displacement current that generates it. In Figure 36.2, a small stationary loop (black) perpendicular to the electric field lines registers a rate of change of electric flux at the instant the front surface of the wave pulse sweeps across the loop, and again at the instant the rear surface of the wave pulse sweeps across the loop. Hence, the displacement current of our wave pulse consists of two sheets of current, one with upward current at the front edge of the pulse, and one with downward current at the rear edge of the pulse. A sheet of current can be regarded as an infinite array of parallel straight wires. Such an infinite array of wires produces a uniform magnetic field whose direction is perpendicular to the direction of the current, according to the usual right-hand rule. Since the currents in the two sheets shown in Figure 36.2 have opposite directions, their magnetic fields combine in the region between the sheets and cancel in the region outside of the sheets.

To determine the magnitude of the uniform magnetic field in the region of the wave pulse, we apply the Maxwell–Ampère Law,

$$\oint \mathbf{B} \cdot d\mathbf{l} = \mu_0 \varepsilon_0 \frac{d\Phi}{dt} \tag{4}$$

to the stationary closed loop parallel to the x–z plane shown in Figure 36.3. This loop measures $\Delta x \times \Delta z$; for the sake of convenience, we will assume that the width Δx of the loop coincides with the thickness of the wave pulse. At the instant shown, one edge of this loop is inside the wave pulse, and the opposite edge is still outside. The integral of $\mathbf{B}$ around the loop receives a contribution only from the edge inside the wave pulse:

$$\oint \mathbf{B} \cdot d\mathbf{l} = B_z \, \Delta z \tag{5}$$

Since in a time $\Delta t = \Delta x / c$, the wave front sweeps over the loop and fills the loop with electric flux, the rate of change of electric flux is

$$\frac{\Delta \Phi}{\Delta t} = \frac{E_y \, \Delta x \, \Delta z}{\Delta x / c} = c E_y \, \Delta z$$

Hence

$$B_z \, \Delta z = c \mu_0 \varepsilon_0 E_y \, \Delta z \tag{6}$$

so the magnitude of the magnetic field is

$$B_z = c \mu_0 \varepsilon_0 E_y \tag{7}$$

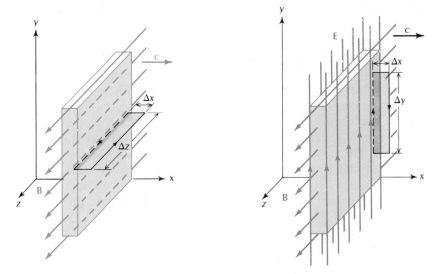

Fig. 36.3 (left) Stationary rectangular loop parallel to the x–z plane. The loop measures $\Delta x \times \Delta z$. At the instant shown, one edge of the loop is inside the front surface of the wave pulse.

Fig. 36.4 (right) Stationary rectangular loop in the x–y plane. The loop measures $\Delta x \times \Delta y$. At the instant shown, one edge of the loop is inside the front surface of the wave pulse.

This equation expresses the magnitude of the magnetic field in terms of the magnitude of the electric field, the speed of propagation of the wave pulse, and the constants ε_0 and μ_0. [Equation (7) coincides with Eq. (35.47), which we derived for the spherical wave pulse.]

Since the magnetic field of the wave pulse is also a time-dependent magnetic field, it induces an electric field. For the sake of consistency, this induced electric field must coincide with the electric field E_y of the wave pulse. Thus, the electric and the magnetic fields of the wave pulse mutually induce each other. The requirement of consistency between these two mutually induced fields leads to the theoretical expression for the speed of light. To derive this expression, we apply Faraday's Law,

$$\oint \mathbf{E} \cdot d\mathbf{l} = -\frac{d\Phi_B}{dt} \tag{8}$$

to the stationary closed loop in the x–y plane shown in Figure 36.4. This loop measures $\Delta x \times \Delta y$; one edge of this loop is, again, inside the wave pulse, and the opposite edge is outside. The integral of the electric field around this loop is $E_y \Delta y$; and the rate of change of magnetic flux through the loop is $\Delta \Phi_B / \Delta t = -B_z \Delta x \Delta y / (\Delta x / c)$. Hence

$$E_y \Delta y = c B_z \Delta y$$

and therefore

$$B_z = (1/c) E_y \tag{9}$$

Comparing this with Eq. (7), we see that consistency demands

$$c \mu_0 \varepsilon_0 = 1/c \tag{10}$$

or

$$\boxed{c = \frac{1}{\sqrt{\mu_0 \varepsilon_0}}} \tag{11}$$

Speed of electromagnetic wave

This is the theoretical expression for the speed of electromagnetic waves in vacuum first obtained by Maxwell. [Not surprisingly, Eqs. (9) and (11) coincide with Eqs. (35.58) and (35.55), derived for the spherical wave pulse.]

When comparing this theoretical value for the speed with the experimental value, we must resist the temptation of blithely inserting the best available modern values into Eq. (11). Such a procedure would be circular because the best available value for ε_0 listed in modern tables of constants — such as the table in Appendix 8 — is actually calculated from Eq. (11). For a meaningful test of Eq. (11), we must insert a value for ε_0 determined directly by electrical measurements. The best determination of this kind available, carried out by scientists at the U.S. Bureau of Standards early in this century, gave $\varepsilon_0 = 8.85433 \times 10^{-12}$ F/m. With this value for ε_0 and with the (exact) value $4\pi \times 10^{-7}$ H/m for μ_0, the theoretical prediction for the speed of electromagnetic waves is

$$c = 1/(4\pi \times 10^{-7} \text{ H/m} \times 8.85433 \times 10^{-12} \text{ F/m})^{1/2}$$

$$= 2.99790 \times 10^8 \text{ m/s} \tag{12}$$

This value is in excellent agreement with the best available experimental value of the speed of light in vacuum,

$$c = 2.99792458 \times 10^8 \text{ m/s} \tag{13}$$

This remarkable agreement constitutes clear evidence for the view that light is an electromagnetic wave. Other evidence for this view is found in the polarization properties of light, which agree with the polarization properties of electromagnetic waves (see the next section).

36.2 Plane Harmonic Waves

Although the above results for the magnitude of the magnetic field and the speed of electromagnetic waves were derived for the special case of a wave pulse consisting of a region of constant electric field, these results are of general validity, because an arbitrary wave can be regarded as a succession of short wave pulses with piecewise constant electric fields. Accordingly, we can write Eq. (9) as a general relation between the instantaneous magnitudes of the (perpendicular) electric and magnetic fields:

Relation between electric and magnetic fields of wave

$$\boxed{B(r,t) = \frac{1}{c} E(r,t)} \tag{14}$$

The direction of the magnetic field is perpendicular to both the electric field and the direction of propagation. The directions of **E** and **B** are related by a right-hand rule: If the fingers are curled from **E** toward **B**, then the thumb lies along the direction of propagation.

A case of great practical importance involves the radiation fields generated by charges moving with simple harmonic motion. For instance, the currents and charges on the antenna of a radio transmitter move up and down the antenna with simple harmonic motion. For this kind of motion, the acceleration is a harmonic function of time,

$$a_y = [\text{constant}] \times \sin \omega t \quad (15)$$

where ω is the angular frequency of the motion. With this, Eq. (3) for the electric field of the plane wave becomes

$$E_y(x,t) = [\text{constant}] \times \sin(\omega t - \omega x/c) \quad (16)$$

Obviously, the function on the right side represents a harmonic wave of angular frequency ω traveling in the x direction with a speed c [compare Eq. (16.14)]. We will write Eq. (16) as

$$\boxed{E_y(x,t) = E_0 \sin\left(\omega t - \frac{\omega x}{c}\right)} \quad (17)$$

Electric field of plane wave polarized in y direction

where the quantity E_0 is the amplitude of the harmonic wave (compare Section 16.2). The magnetic field is perpendicular both to **E** and to the direction of propagation; according to Eq. (14), the magnitude of the magnetic field is E/c. Thus,

$$\boxed{B_z(x,t) = \frac{E_0}{c} \sin\left(\omega t - \frac{\omega x}{c}\right)} \quad (18)$$

Magnetic field of plane wave polarized in y direction

Equations (17) and (18) describe a **plane harmonic wave.** The wave fronts of this wave are flat surfaces parallel to the y–z plane. Note that the directions of **E** and **B** within this wave are related by the right-hand rule given at the beginning of this section. Note also that the electric and magnetic fields are in phase, i.e., **B** is at maximum wherever **E** is at maximum. Figure 36.5 shows the field lines of the plane wave.

Besides the plane wave described above, there is another possible plane wave with the same direction of propagation, but with its fields rotated by 90°. This other wave has an electric field along the z axis and a magnetic field along the y axis:

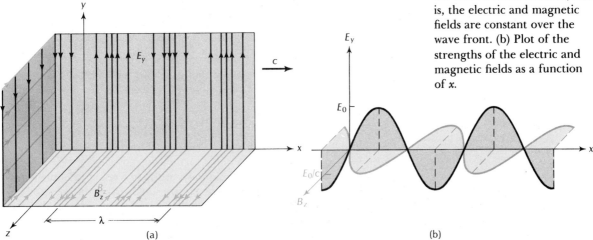

Fig. 36.5 (a) Electric (black) and magnetic (color) field lines of a plane harmonic wave traveling toward the right, shown at one instant of time. The electric field is vertical and the magnetic field is horizontal. The field lines fill all of space, but this diagram shows only the field lines in the x–y plane, in the x–z plane, and on a wave front (dark colored surface on left). The distribution of field lines over the wave front is uniform; that is, the electric and magnetic fields are constant over the wave front. (b) Plot of the strengths of the electric and magnetic fields as a function of x.

Electric and magnetic fields of plane wave polarized in z directon

$$E_z(x,t) = E_0 \sin\left(\omega t - \frac{\omega x}{c}\right) \quad (19)$$

$$B_y(x,t) = -\frac{E_0}{c} \sin\left(\omega t - \frac{\omega x}{c}\right) \quad (20)$$

Figure 36.6 shows the field lines of this wave.

Polarization

The direction of the electric field is called the direction of **polarization** of the wave. The wave described by Eq. (17) is polarized in the y direction, and the wave described by Eq. (19) is polarized in the z direction. It is, of course, also possible to construct waves polarized in some intermediate direction, say, at 45° to the y and z axes; but such waves are superpositions of the standard waves described by Eqs. (17) and (19) and therefore nothing essentially new. Thus, for a given direction of propagation, electromagnetic waves have only *two* independent directions of polarization. We will discuss diverse phenomena involving polarization of light in the next chapter.

Fig. 36.6 (a) Electric (black) and magnetic (color) field lines of another plane wave traveling toward the right, shown at one instant of time. The electric field is horizontal and the magnetic field is vertical. (b) Plot of the strengths of the electric and magnetic fields as a function of x.

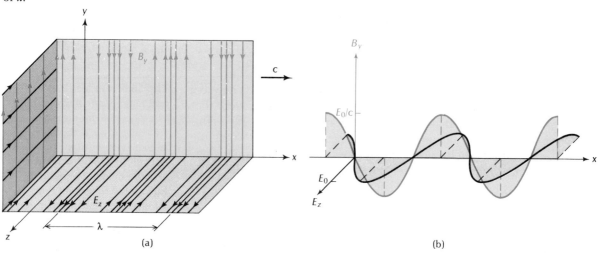

EXAMPLE 1. At a place at some distance from a radio transmitter, the amplitude of the electric field of a harmonic radio wave is 2.0×10^{-3} V/m. What is the amplitude of the magnetic field of this wave?

SOLUTION: Equation (18) relates the instantaneous magnitudes of the electric and magnetic fields of the wave, and it therefore also relates the amplitudes of oscillation of these fields:

$$B_0 = \frac{E_0}{c} = \frac{2.0 \times 10^{-3} \text{ V/m}}{3.0 \times 10^8 \text{ m/s}} = 6.7 \times 10^{-12} \text{ T}$$

EXAMPLE 2. Suppose that the radio wave of the preceding example is propagating in the x direction and that its electric field makes an angle of 20° with the y direction. The wave can be regarded as a superposition of the standard waves (17) and (19) polarized in the y direction and the z direction. What is the amplitude of oscillation of the wave polarized in the y direction? In the z direction?

SOLUTION: Figure 36.7 shows the electric field vector in the y–z plane at an instant when the oscillating electric field of the wave is at its maximum value, 2.0×10^{-3} V/m. The y component of this electric field vector gives us the amplitude of the wave polarized in the y direction:

$$E_0 = 2.0 \times 10^{-3} \text{ V/m} \times \cos 20° = 1.88 \times 10^{-3} \text{ V/m} \quad (21)$$

And the z component gives us the amplitude of the wave polarized in the z direction:

$$E_0 = 2.0 \times 10^{-3} \text{ V/m} \times \cos 70° = 0.68 \times 10^{-3} \text{ V/m} \quad (22)$$

36.3 The Generation of Electromagnetic Waves

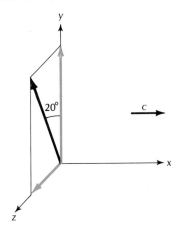

Fig. 36.7 The electric field vector in the y–z plane and its components.

For an electromagnetic wave, as for any other kind of wave, the product of the wavelength and the frequency equals the wave speed

$$\boxed{\lambda v = c} \quad (23)$$

Here the frequency v is related to the angular frequency ω in the usual way,

$$v = \omega/2\pi \quad (24)$$

Equation (23) shows how the wavelength depends on the frequency of the wave, or the frequency of the source that generates the wave. For instance, the electric charges on the antenna of an FM radio station typically oscillate back and forth with a frequency of 10^8 Hz (or 100 MHz); correspondingly, the wavelength of the radiation emitted by these accelerated charges has a wavelength of

$$\lambda = \frac{c}{v} = \frac{3.0 \times 10^8 \text{ m/s}}{10^8/\text{s}} = 3.0 \text{ m} \quad (25)$$

The oscillations of the charges on the antenna are produced by means of a resonating LC circuit coupled to the antenna by a mutual inductance (see Figure 34.11b). This is essentially the method used to generate **long waves**, **medium waves** (AM), and **short waves** (including FM), as well as **TV waves**. Such radio waves span a wavelength range from 10^5 m to a few centimeters.

Kinds of electromagnetic radiation

Waves of shorter wavelength, or **microwaves**, are best generated by an electromagnetic resonating cavity such as a klystron (see Section 35.3); the antenna in this case is merely a horn, connected to the cavity by a waveguide, that permits the waves to spill out into space. This method can be used to generate waves of wavelength as short as about a millimeter. Shorter wavelengths cannot be generated by currents oscillating on macroscopic laboratory equipment; however, short wavelengths are easily generated by electrons oscillating within molecules and atoms subjected to stimulation by heat or by an electric current. Depending on the details of the motion, the electrons in molecules and in atoms will emit **infrared** radiation, **visible** light, **ultraviolet** radia-

Bremsstrahlung

tion, or **X rays;** the corresponding wavelengths range from 10^{-3} m to 10^{-11} m. X rays can also be generated by the acceleration that high-speed electrons suffer during impact on a target; this is **Bremsstrahlung** (see Example 35.2). Radiations of even shorter wavelengths are emitted by protons and neutrons moving within a nucleus; these are **gamma rays**, with wavelengths as short as 10^{-13} m. Of course, the motion of subatomic particles and their emission of radiation cannot be calculated by classical mechanics or classical electromagnetic theory; such calculations require quantum mechanics and quantum electrodynamics.

In an ordinary light source, such as a neon tube, the individual atoms or molecules radiate independently. The emerging light consists of a superposition of many individual light waves with random phase differences, random directions of polarization, and diverging directions of propagation; such light waves with random, unpredictable phase differences are said to be **incoherent**. In a **laser**, the atoms or molecules radiate in unison, by a quantum-mechanical phenomenon called **stimulated emission**. The emerging light is a superposition of light waves with exactly the same phases, the same directions of polarization, and the same directions of propagation; such light waves with no phase differences, or with predictable phase differences, are said to be **coherent**. Since the individual light waves in this kind of light combine constructively, the light beam emerging from the laser is very intense, and it is also very sharply collimated.

Stimulated emission

Another important mechanism for the generation of electromagnetic waves is **cyclotron emission**. This involves high-speed electrons undergoing centripetal acceleration while spiraling in a magnetic field (see Section 31.3). Depending on the speed of the electron and the strength of the magnetic field, the radiation may consist of radio waves, X rays, or anything in between. Most of the radio waves reaching us from stars, pulsars, and radio galaxies are generated by this process.

Fig. 36.8 Wavelength and frequency bands of electromagnetic radiation.

Figure 36.8 displays the wavelength and the frequency bands of

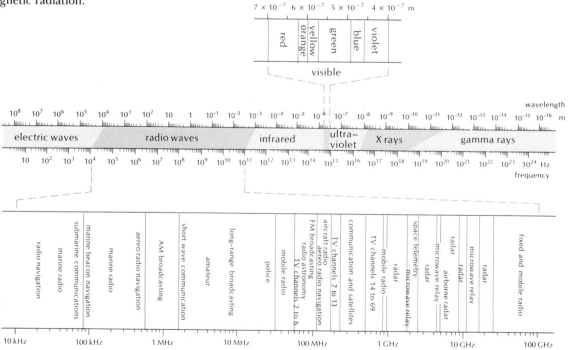

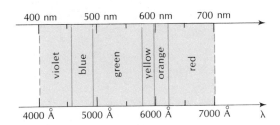

Fig. 36.9 Colors of visible light, with wavelengths expressed in angstroms and in nanometers.

electromagnetic radiation. The bands overlap to some extent because the names assigned to the different ranges of wavelengths depend not only on the values of the wavelength but also on the method used to generate and/or detect the radiation. For example, radiation of a wavelength of a tenth of a millimeter will be called a radio wave (microwave) if detected by a radio receiver, but it will be called infrared radiation if detected by a heat sensor.

Visible light is electromagnetic radiation of a wavelength between about 7×10^{-7} m and 4×10^{-7} m. This is the range of wavelengths over which our eyes are sensitive to light. We perceive different wavelengths within the visible region as having different colors. Figure 36.9 shows how colors are correlated with wavelengths (see also the color plate between pages 898 and 899). The wavelengths in this figure have been expressed in angstroms ($1\text{Å} = 10^{-10}$ m) and in nanometers ($1\text{nm} = 10^{-9}$ m), two units commonly used for the measurement of wavelengths of light.

Incidentally: Our eyes are almost completely insensitive to the polarization of light waves. We can detect the polarization only with special equipment such as Polaroid sunglasses. Radio and TV antennas are, of course, very sensitive to the direction of polarization of radio waves, and they must have the proper orientation to pick up a strong signal.

36.4 Energy of a Wave

The electric and magnetic fields of an electromagnetic wave contain energy. As the wave moves along, so does this energy — the wave transports energy.

Let us calculate the flow of energy in a plane wave or in an approximately plane portion of a spherical wave. Suppose the wave is of the kind described by Eqs. (17) and (18), i.e., it is a wave propagating in the positive x direction. The densities of electric and magnetic energy are, respectively [see Eqs. (26.19) and (32.42)],

$$u_E = \frac{\varepsilon_0}{2} E^2 \quad \text{and} \quad u_B = \frac{1}{2\mu_0} B^2 \tag{26}$$

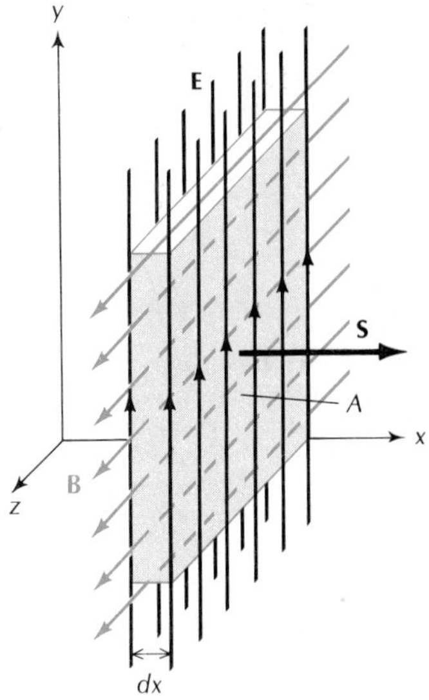

Within a thin rectangular slab of this wave, of thickness dx and frontal area A (Figure 36.10), the fields are nearly constant. The total amount of energy in this slab is

$$dU = (u_E + u_B) \times [\text{volume}] \tag{27}$$

$$= \left(\frac{\varepsilon_0}{2} E^2 + \frac{1}{2\mu_0} B^2 \right) \times A\, dx \tag{28}$$

Fig. 36.10 A slab of electric (black) and magnetic (color) fields in a plane wave propagating toward the right. The slab of thickness dx and frontal area A moves with the wave.

With $E = cB$ and $\varepsilon_0 = 1/\mu_0 c^2$, we can write this as

$$dU = \left[\frac{1}{2\mu_0 c^2} E(cB) + \frac{1}{2\mu_0} B\left(\frac{E}{c}\right)\right] A \, dx \qquad (29)$$

$$= \frac{1}{\mu_0 c} EBA \, dx \qquad (30)$$

Note that the two terms on the right side of Eq. (29) are equal, that is, the electric and magnetic energy densities in an electromagnetic wave are equal.

Since the forward speed of the slab is c, the amount of energy dU moves out of the (stationary) volume $A \, dx$ in a time $dt = dx/c$. Hence the rate of flow of energy is

$$\frac{dU}{dt} = \frac{1}{\mu_0} EBA \qquad (31)$$

and the rate of flow of energy per unit frontal area, or the **energy flux**, is

Energy flux

$$\frac{1}{A}\frac{dU}{dt} = \frac{1}{\mu_0} EB \qquad (32)$$

In Figure 36.10 the energy flows toward the right, along the direction of propagation of the wave. We can describe the energy flux in terms of a vector that has a magnitude equal to the right side of Eq. (32) and a direction along the flow. The vector satisfying these conditions is

Poynting vector

$$\boxed{\mathbf{S} = \frac{1}{\mu_0} \mathbf{E} \times \mathbf{B}} \qquad (33)$$

The vectors $\mathbf{E}$ and $\mathbf{B}$ are perpendicular, and hence the magnitude of the vector $\mathbf{S}$ given by Eq. (33) coincides with the magnitude given by Eq. (32). Furthermore, the direction of $\mathbf{E} \times \mathbf{B}$ in Figure 36.10 is toward the right, as desired. The vector $\mathbf{S}$ is called the **Poynting vector**. It turns out that this vector gives the energy flux not only in a plane wave but in any arbitrary electromagnetic field. The units of energy flux are $J/(m^2 \cdot s)$, or W/m^2.

John Henry Poynting, *1852–1914, British physicist, professor at the University of Birmingham. He introduced the Poynting vector while presenting a proof of a general theorem on energy transfers in the electromagnetic fields.*

Since both the electric and magnetic fields of a wave oscillate in time, so does the energy flux. If the value of x in Eqs. (17) and (18) is fixed, say, $x = 0$, we have

$$E = E_0 \sin \omega t \quad \text{and} \quad B = \frac{E_0}{c} \sin \omega t \qquad (34)$$

so that

$$S = \frac{1}{\mu_0 c} E_0^2 \sin^2 \omega t \qquad (35)$$

Thus, the energy flux oscillates between zero and a maximum value $E_0^2/\mu_0 c$. The time-average energy flux is

$$\boxed{\bar{S} = \frac{1}{2\mu_0 c} E_0^2} \tag{36}$$

Time-average magnitude of Poynting vector

EXAMPLE 3. At a distance of 6.0 km from a radio transmitter, the amplitude of the oscillating electric field of the radio wave is $E_0 = 0.13$ V/m. What is the time-average energy flux? What is the total power radiated by the radio transmitter? Assume that the transmitter radiates uniformly in all directions.

SOLUTION: From Eq. (36) we find

$$\bar{S} = \frac{1}{2\mu_0 c} E_0^2$$

$$= \frac{1}{2 \times 1.26 \times 10^{-6} \text{ H/m} \times 3 \times 10^8 \text{ m/s}} \times (0.13 \text{ V/m})^2$$

$$= 2.2 \times 10^{-5} \text{ W/m}^2$$

To obtain the total power, we must multiply the power per unit area by the area of a sphere of radius 6.0 km:

$$\bar{P} = 4\pi r^2 \bar{S} \tag{37}$$

$$= 4\pi \times (6.0 \times 10^3 \text{ m})^2 \times 2.2 \times 10^{-5} \text{ W/m}^2$$

$$= 1.0 \times 10^4 \text{ W} = 10 \text{ kW}$$

COMMENTS AND SUGGESTIONS: Note that, according to Eq. (37), the energy flux $\bar{S}$ for a spherical wave spreading out in all directions from a source is inversely proportional to the square of the distance,

$$\bar{S} \propto \frac{1}{r^2} \tag{38}$$

This decrease of the flux is consistent with Eq. (1), according to which the amplitude of the electric field is inversely proportional to the distance, $E_0 \propto 1/r$. We must take this dependence of flux and amplitude on distance into account whenever we want to investigate the spreading of the wave over a large range of distances (but we can ignore this dependence over a small range of distances, where the wave can be approximated as a plane wave of constant energy flux).

36.5 Momentum of a Wave

An electromagnetic wave carries not only energy but also momentum. This can be readily understood in terms of Einstein's mass–energy relation, which tells us that the energy in the wave has mass. As the energy moves so does the mass, and such moving mass has momentum.

The following simple calculation permits us to derive a quantitative formula for the amount of momentum in an electromagnetic wave. Imagine that a plane electromagnetic wave polarized in the y direction and traveling in the x direction strikes a charged particle (Figure

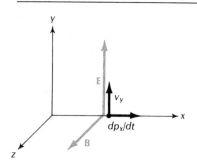

Fig. 36.11 A plane wave traveling toward the right strikes a particle. The electric and magnetic fields of the wave exert forces on the particle.

36.11). The electric and magnetic fields of the wave will then exert forces on the particle, that is, they will change the momentum of the particle. Let us look at the x component of the rate of change of momentum. The electric field is entirely in the y direction — it exerts no force in the x direction. The magnetic field is entirely in the z direction — it exerts the following force in the x direction:

$$\frac{dp_x}{dt} = q(\mathbf{v} \times \mathbf{B})_x = q(v_y B_z - v_z B_y) = q v_y B_z$$

Since $B_z = E_y/c$ [see Eqs. (17) and (18)], we can also write this rate of change of momentum as

$$\frac{dp_x}{dt} = \frac{q}{c} v_y E_y \tag{39}$$

Let us compare this with the rate at which the particle acquires energy from the wave. Only the electric field does work on the particle. The rate of work, or the rate of increase of the energy of the particle, is

$$\frac{dW}{dt} = q\mathbf{E} \cdot \mathbf{v} = q E_y v_y \tag{40}$$

(here we have used the symbol W for the energy of the particle because the symbol E is reserved for the electric field). If we compare the right sides of Eqs. (39) and (40), we recognize that

$$\frac{dp_x}{dt} = \frac{1}{c} \frac{dW}{dt}$$

that is, the rate at which the particle acquires x momentum from the wave is proportional to the rate at which it acquires energy. Whenever the particle absorbs energy from the wave, it will also absorb x momentum,

$$dp_x = \frac{1}{c} dW \tag{41}$$

Any gain of energy and momentum by the particle implies a corresponding loss of energy and momentum by the wave. If we imagine that the wave loses all of its energy and momentum to the particle (and disappears),[1] we recognize that the total amount of energy and the total amount of x momentum stored in the wave must be related by a formula similar to Eq. (41):

Momentum of wave

$$\boxed{P_x = \frac{U}{c}} \tag{42}$$

---

[1] Actually, a single charged particle cannot absorb all of a wave (the particle will scatter the wave); but a cloud of many charged particles can completely absorb a wave.

This is the general relation between the momentum and the energy of a plane wave. This relation is also valid for the amounts of momentum and energy in some portion of the wave.

Since we already know that the energy flows with the wave, we can conclude that the momentum also flows with the wave. By means of Eq. (31), we can then express the momentum flow within a frontal area A of the wave as

$$\frac{dP_x}{dt} = \frac{1}{c}\frac{dU}{dt} = \frac{1}{c\mu_0} EBA \qquad (43)$$

This shows that the momentum flow oscillates in the same way as the energy flow. The time-average momentum flow is

$$\frac{\overline{dP_x}}{dt} = \frac{1}{2\mu_0 c^2} E_0^2 A \qquad (44)$$

Whenever a wave strikes a body and is absorbed by it, the wave will exert a force on the body and transfer momentum to it. If the body has a frontal area A facing the wave, the rate of change of momentum, or the force, is simply given by Eq. (43),

$$F_x = \frac{dP_x}{dt} = \frac{1}{c\mu_0} EBA$$

or

$$F_x = \frac{1}{c} SA$$

where S is the magnitude of the Poynting vector. The force per unit frontal area is then

$$\boxed{\frac{F_x}{A} = \frac{S}{c}} \qquad (45) \qquad \textit{Pressure of radiation}$$

This is called the **pressure of radiation**, or the pressure of light. Note that the formula (45) hinges on the assumption that the wave is completely absorbed. If the wave is partially or totally reflected, then the force is *larger* than given by Eq. (45). For instance, if the wave is totally reflected by a body, then the body must not only absorb all of the initial momentum but also supply the momentum for the reversed motion of the wave. The force is then twice as large as that given by Eq. (45).

The magnitude of the radiation-pressure force is usually insignificant compared with other ordinary forces that act on a body.

EXAMPLE 4. The average energy flux in the sunlight incident on the Earth is $\overline{S} = 1.4 \times 10^3$ W/m². What force does the pressure of light exert on the Earth? How does this compare with the gravitational force that the Sun exerts on the Earth? Pretend that all the light striking the Earth is absorbed.

SOLUTION: The cross-sectional area that the Earth offers to the stream of sunlight is πR_E^2. The total power absorbed by the Earth is then

$$\frac{dU}{dt} = \overline{S}A$$

$$= 1.4 \times 10^3 \text{ W/m}^2 \times \pi \times (6.4 \times 10^6 \text{ m})^2$$

$$= 1.8 \times 10^{17} \text{ W} \tag{46}$$

Hence

$$F = \frac{1}{c}\frac{dU}{dt} = \frac{1.8 \times 10^{17} \text{ W}}{3.0 \times 10^8 \text{ m/s}} = 6.0 \times 10^8 \text{ N}$$

The gravitational force that the Sun exerts on the Earth is much greater:

$$F = \frac{GM_S M_E}{r^2}$$

$$= \frac{6.7 \times 10^{-11} \text{ N} \cdot \text{m}^2/\text{kg}^2 \times 2.0 \times 10^{30} \text{ kg} \times 6.0 \times 10^{24} \text{ kg}}{(1.5 \times 10^{11} \text{ m})^2}$$

$$= 3.6 \times 10^{22} \text{ N}$$

The force of radiation pressure on the Earth is insignificant compared with the gravitational pull of the Sun. However, on a body of very small mass, such as a grain of dust in interplanetary space, the force of radiation pressure can be as large as, or even larger than, the gravitational pull of the Sun. A simple calculation shows that if a grain of dust is smaller than about 10^{-6} m across, then radiation pressure overcomes gravitation and blows the grain away from the Sun.[2] The tails of comets are a spectacular demonstration of this effect. These tails consist of a plume of dust that is blown away from the comet's head by the pressure of sunlight (Figure 36.12). Regardless of the direction of motion of the comet, the tail always points away from the Sun, much as a pennant fluttering from the mast of a ship always points away from the wind.

Fig. 36.12 Comet Mrkos and its tails, photographed in August 1957. This comet has both a dust tail (haze, upper right) and an ion tail (streaks, upper center). The dust tail is produced by the pressure of sunlight; the ion tail is produced by the pressure of the solar wind.

36.6 The Doppler Shift of Light

The frequency of light waves, like that of waves of any kind, suffers a Doppler shift whenever the emitter of waves is in motion relative to the receiver. However, in contrast to sound waves in air or surface waves in water, light waves in a vacuum are not propagating in any material medium that can serve as a preferred reference frame with respect to which we can reckon their velocity. Light waves have the same speed c with respect to *all* inertial reference frames (this well-established observational fact lies at the root of the theory of Special Relativity, as discussed in Chapter 41). Hence, for the calculation of the Doppler shift, we must first decide which reference frame is the most

---

[2] For *very* small grains of dust (less than 10^{-7} m across) and for molecules this will not happen, because such very small objects scatter light instead of absorbing it. This cuts down the force exerted by light.

suitable. Since we want to find the frequency as seen by the *receiver*, it is clear that the relevant reference frame is the rest frame of the receiver.

In this reference frame, the calculation of the Doppler shift can be done exactly as in Section 17.3. Since the receiver is at rest in the "medium" (that is, the reference frame) in which light has the speed c, the formula for the Doppler shift of light must be [see Equation (17.13)]

$$\nu = \frac{1}{1 \pm v/c} \nu_0 \qquad (47)$$

where, as before, ν_0 is the frequency radiated by the emitter, ν is the frequency detected by the receiver, and v is the speed of the emitter. Here, the positive sign is to be used if the emitter is receding, the negative sign if approaching. Since the calculations of Section 17.3 relied on Newtonian physics, which is valid only if the speed is low compared with the speed of light ($v \ll c$), Eq. (47) can be replaced by the approximation

$$\nu \cong (1 \mp v/c)\nu_0 \qquad (48)$$

without any significant loss of accuracy. From this we see that the frequency shift $\Delta\nu = \nu - \nu_0$ is given by

$$\boxed{\frac{\Delta\nu}{\nu_0} \cong \mp \frac{v}{c}} \qquad \begin{array}{l}\text{+ for approaching emitter} \\ \text{- for receding emitter}\end{array} \qquad (49)$$

Approximate Doppler shift

Measurements of the Doppler shift of light play a crucial role in astronomy. The velocities of recession of remote stars relative to us can be determined via the Doppler shift of the starlight reaching us. The rotational velocities of stars can likewise be determined by a careful measurement of the difference between the Doppler shifts of the light from one edge of the star and the opposite edge.

Although Eq. (47) is adequate whenever the speed of the source is small compared with the speed of light, it must be modified for relativistic effects if the speed of the source is large. As we will see in Section 41.4, the emitter is subject to a relativistic time-dilation effect. This, in itself, would *reduce* the frequency of the emitter by a factor $\sqrt{1 - v^2/c^2}$. We must therefore insert this extra factor in Eq. (47),

$$\nu = \frac{\sqrt{1 - v^2/c^2}}{1 \pm v/c} \nu_0 \qquad (50)$$

This is the exact formula for the Doppler shift of light. Since $1 - v^2/c^2 = (1 - v/c)(1 + v/c)$, this formula can be simplified by appropriate cancellations:

$$\boxed{\nu = \sqrt{\frac{1 - v/c}{1 + v/c}} \nu_0} \qquad \text{for receding emitter} \qquad (51)$$

Exact Doppler shift

$$\boxed{\nu = \sqrt{\frac{1 + v/c}{1 - v/c}} \nu_0} \qquad \text{for approaching emitter} \qquad (52)$$

For low speeds there is of course no appreciable difference between these exact formulas and the approximate formula (47).

The velocities of recession of galaxies and quasars partaking of the expansion of the universe (see Interlude II) give the light and radio waves from these objects very large Doppler shifts. For instance, the light from the quasar 3C 147 (Figure 36.13) has a frequency shift amounting to a factor of 1.55. If we insert this factor in Eq. (51),

$$\frac{\nu}{\nu_0} = \frac{1}{1.55} = \sqrt{\frac{1-v/c}{1+v/c}} \tag{53}$$

we find a velocity of recession $v = 0.41c$, that is, 41% of the speed of light! The velocities of recession of some other quasars are even higher.

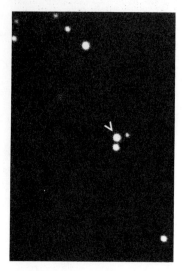

Fig. 36.13 The quasar 3C 147.

SUMMARY

Electric and magnetic fields of plane wave:

$$E_y = E_0 \sin\left(\omega t - \frac{\omega x}{c}\right)$$

$$B_z = \frac{E_0}{c} \sin\left(\omega t - \frac{\omega x}{c}\right)$$

Poynting vector: $\mathbf{S} = \dfrac{1}{\mu_0} \mathbf{E} \times \mathbf{B}$

Time-average energy flux of plane wave: $\bar{S} = \dfrac{1}{2\mu_0 c} E_0^2$

Momentum of wave: $P_x = \dfrac{U}{c}$

Radiation pressure: $\dfrac{F_x}{A} = \dfrac{S}{c}$

Doppler shift: $\dfrac{\Delta \nu}{\nu_0} \cong \mp \dfrac{v}{c}$ at low speed

$$\nu = \sqrt{\frac{1-v/c}{1+v/c}}\, \nu_0 \quad \text{(receding)}$$

$$\nu = \sqrt{\frac{1+v/c}{1-v/c}}\, \nu_0 \quad \text{(approaching)}$$

QUESTIONS

1. Since the speed of light has now been adopted as the standard of speed and has been assigned the value 2.99792458×10^8 m/s *by definition*, why is it still meaningful to compare this number with Maxwell's theoretical prediction for the speed of light?

2. The round-trip travel time for a radio signal to the Moon is 2.6 s. Does this have a noticeable effect on radio communications with astronauts on the Moon?

3. A severe limitation on the speed of computation of large electronic computers is imposed by the speed of light because the electric signals on the connecting wires within the computer are electromagnetic waves ("guided waves"), which travel at a speed roughly equal to the speed of light. If the computer measures about 1 m across, what is the minimum travel time required for a typical signal sent from one part of the computer to another? What is the maximum number of signals that can be sent back and forth (sequentially) per second? Is there any way to avoid the limitation imposed by the travel time of signals?

4. Describe the direction of polarization of the radiation field of an accelerated charge at a few typical points in the space surrounding the charge. Show that the direction of polarization is never perpendicular to the direction of acceleration.

5. Consider transverse waves on a string. How many independent directions of polarization do such waves have? How could you construct a mechanical polarization filter that permits the passage of only a wave polarized in a preferential direction?

6. Figure 36.14 is a plot of the sensitivity of the human eye as a function of the wavelength of light. The sensitivity is maximum at about 5.5×10^{-7} m and drops to about 1% at 6.9×10^{-7} m and at 4.3×10^{-7} m. Suppose that the sensitivity of your eye were constant over the entire interval of wavelengths shown in Figure 36.14. How would this alter your visual perception of some of the things you see in your everyday life?

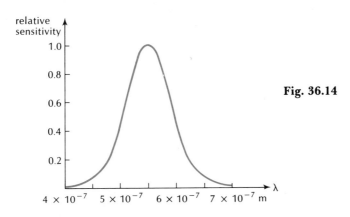

Fig. 36.14

7. An atom radiates visible light of a wavelength several thousand times longer than the size of the atom. How is this possible?

8. Why does the radio reception fade in the receiver of your automobile when you enter a tunnel?

9. Short-wave radio waves are reflected by the ionosphere of the Earth; this makes them very useful for long-range communication. Explain.

10. Why is the effect of the pressure of sunlight on a grain of dust floating in interplanetary space more significant if the grain is very small?

11. Could we propel a spaceship by shining a very intense light out of its back? What would be the advantages of such a propulsion scheme?

12. Astronauts of the future could travel all over the Solar System in a spaceship suspended from a large "sail" coated with a reflecting material. The pressure of sunlight on the sail could propel the spaceship in any direction. How would the astronauts have to orient their (flat) sail to move radially away from the Sun? To move at right angles to the radial direction? To move radially toward the Sun?

13. Suppose that a light wave is totally reflected by a *moving* mirror. Would

you expect the energy of the light wave to change during this reflection? (Hint: Consider the work done by the radiation pressure.)

14. Einstein's first derivation of his famous equation $E = mc^2$ relied on the equation for the momentum of light. Einstein considered a closed boxcar, free to roll on frictionless rails, with a light source at one end and an absorber at the other end (Figure 36.15). If the source at one end emits a light pulse of energy E, the recoil sets the boxcar in motion; this motion stops when the light pulse reaches the other end and is absorbed. Given that the center of mass of the closed system should remain stationary during this process, how could Einstein conclude that the transfer of the energy E from one end to the other involves a transfer of mass?

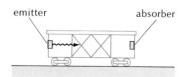

Fig. 36.15

15. In the seventeenth century, the Danish astronomer Ole Roemer noticed that the orbital periods of the moons of Jupiter, as observed from the Earth, exhibit some systematic irregularities: the periods are slightly longer when the Earth is moving away from Jupiter, and slightly shorter when the Earth is moving toward Jupiter. Roemer attributed this apparent irregularity to the finite speed of propagation of light, and exploited it to make the first determination of the speed of light. Explain how one can use the lengthening or the shortening of the observed period to deduce the speed of light. (This shift of period may be regarded as the earliest discovery of a Doppler shift.)

16. Equations (51) and (52) give the Doppler shift for an emitter that is approaching or receding along the line of sight. Taking into account the relativistic time-dilation effect, would you expect a Doppler shift for an emitter moving at right angles to the line of sight? (This is called the transverse Doppler shift.)

PROBLEMS

Section 36.1

1. At a distance of 6.0 km from a radio transmitter, the amplitude of the electric radiation field of the emitted radio wave is $E_0 = 0.13$ V/m. Taking into account the decrease of the amplitude of the wave with distance, what will be the amplitude of the radio wave when it reaches a distance of 12.0 km? A distance of 18.0 km?

2. Radio station WWV at Fort Collins, Colorado, continually transmits time signals at several short-wave frequencies. These short waves are reflected by the ionosphere and also by the ground. The waves bounce back and forth between the ionosphere and the ground, and they can therefore roughly follow the curvature of the Earth's surface and travel all the way around the Earth. Suppose that a navigator on a ship in the south of the Indian Ocean, halfway around the Earth, listens to the WWV time signals. If he sets his chronometer according to what he hears, roughly how late will he be?

3. (a) One type of antenna for a radio receiver consists of a short piece of straight wire; when the electric field of a radio wave strikes this wire it makes currents flow along it, which are detected and amplified by the receiver. Suppose that the electric field of a radio wave is vertical. What must be the orientation of the wire for maximum sensitivity?
   (b) Another type of antenna consists of a circular loop; when the magnetic field of a radio wave strikes this loop it induces currents around it. Suppose that the magnetic field of a radio wave is horizontal. What must be the orientation of the loop for maximum sensitivity?

*4. Consider the plane wave pulse shown in Figures 36.1 and 36.2. If the magnitude of the electric field in this pulse is 4×10^{-3} V/m, what is the magnitude of the displacement current flowing along the front surface of the pulse per meter of length measured perpendicularly to the current?

*5. Two plane-wave pulses of the kind described in Section 36.1 are traveling in opposite directions. Their polarizations are parallel and the magnitudes of their electric fields are 2×10^{-3} V/m.
  (a) What are the electric energy density and the magnetic energy density in each pulse?
  (b) Suppose that at one instant the two pulses overlap. What are the magnitudes of the electric field and the magnetic field in this superposition?
  (c) What are the electric energy density and the magnetic energy density?

Section 36.2

6. A plane electromagnetic wave travels in the eastward direction. At one instant the electric field at a given point has a magnitude of 0.60 V/m and points down. What are the magnitude and direction of the magnetic field at this instant? Draw a diagram showing the electric field, the magnetic field, and the direction of propagation.

*7. An electromagnetic wave traveling along the x axis consists of the following superposition of two waves polarized along the y and z directions, respectively:

$$\mathbf{E} = \hat{\mathbf{y}} E_0 \sin(\omega t - \omega x/c) + \hat{\mathbf{z}} E_0 \cos(\omega t - \omega x/c)$$

This electromagnetic wave is said to be **circularly polarized.**
  (a) Show that the magnitude of the electric field is E_0 at all points of space at all times.
  (b) Consider the point $x = y = z = 0$. What is the angle between **E** and the z axis at time $t = 0$? $t = \pi/2\omega$? $t = \pi/\omega$? $t = 3\pi/2\omega$? Draw a diagram showing the y and z axes and the direction of **E** at these times. In a few words, describe the behavior of **E** as a function of time.

*8. An electromagnetic wave has the form

$$\mathbf{E} = \hat{\mathbf{x}} E_0 \sin(\omega t + \omega z/c) + 2\hat{\mathbf{y}} E_0 \sin(\omega t + \omega z/c)$$

  (a) What is the direction of propagation of this wave?
  (b) What is the direction of polarization, that is, what angle does the direction of polarization make with the x, y, and z axes?
  (c) Write down a formula for the magnetic field of this wave as a function of space and time.

Section 36.3

9. An ordinary radio receiver, such as found in homes across the country, has an AM dial and an FM dial. The AM dial covers a range from 530 to 1600 kHz and the FM dial a range from 88 to 108 MHz. What is the range of wavelengths for AM? For FM?

10. At many coastal locations, radio stations of the National Weather Service transmit continuous weather reports at a frequency of 162.5 MHz. What is the wavelength of these transmissions?

11. For radio communication with submerged submarines, the U.S. Navy uses ELF (extremely low frequency) radio waves of wavelength 4000 km; such waves can penetrate for some distance below water. What is the frequency of such waves?

12. Figure 36.14 gives the sensitivity of the human eye as a function of the wavelength of light. For what color is the sensitivity maximum? For what color is the sensitivity one-half of the maximum? One-quarter of the maximum?

Section 36.4

13. A plane electromagnetic wave travels in the northward direction. At one instant, the electric field at a given point has a magnitude of 0.50 V/m and is

in the eastward direction. What are the magnitude and direction of the magnetic field at the given point? What are the magnitude and direction of the Poynting vector?

14. The average energy flux of sunlight incident on the top of the Earth's atmosphere is 1.4×10^3 W/m². What are the corresponding amplitudes of oscillation of the electric and magnetic fields?

15. At a distance of several kilometers from a radio transmitter, the electric field of the emitted radio wave has a magnitude of 0.12 V/m at one instant of time. What is the energy density in this electric field? What is the energy density in the magnetic field of the radio wave?

16. The quasar 3C 273 is at a distance of 2.8×10^9 light-years from the Earth. The flux of radio waves of 1410 Hz reaching the Earth from this quasar is 4.1×10^{-25} W/m² (this is the flux in a frequency interval of 1 Hz around 1410 Hz; comparable amounts of flux are found at other radio frequencies). What is the power captured by the Arecibo radio telescope, a dish of diameter 300 m? What is the power emitted by the quasar? Assume the quasar radiates uniformly in all directions.

17. In the United States, the accepted standard for the safe maximum level of continuous whole-body exposure to microwave radiation is 10 milliwatts/cm².[3]

   (a) For this energy flux, what are the corresponding amplitudes of oscillation of the electric and magnetic fields?
   (b) Suppose that a man of frontal area 1.0 m² completely absorbs microwaves with an intensity of 10 milliwatts/cm² incident on this area and that the microwave energy is converted to heat within his body. What is the rate (in calories per second) at which his body develops heat?

18. A silicon solar cell of frontal area 13 cm² delivers 0.20 A at 0.45 V when exposed to full sunlight of energy flux 1.0×10^3 W/m². What is the efficiency for conversion of light energy into electric energy?

19. A magnifying glass of diameter 10 cm focuses sunlight into a spot of diameter 0.50 cm. The energy flux in the sunlight incident on the lens is 0.10 W/cm².
   (a) What is the energy flux in the focal spot? Assume that all points in the spot receive the same flux.
   (b) Will newspaper ignite when placed at the focal spot? Assume that the flux required for ignition is 2 W/cm².

20. The beam of a powerful laser has a diameter of 0.2 cm and carries a power of 6 kW. What is the time-average Poynting vector in this beam? What are the amplitudes of the electric and the magnetic fields?

21. Calculate the instantaneous Poynting vector for the electromagnetic wave described in Problem 7.

22. A radio transmitter emits a time-average power of 5 kW in the form of a radio wave with uniform intensity in all directions. What are the amplitudes of the electric and magnetic fields of this radio wave at a distance of 10 km from the transmitter?

23. A radio receiver has a sensitivity of 2×10^{-4} V/m. At what maximum distance from a radio transmitter emitting a time-average power of 10 kW will this radio receiver still be able to detect a signal? Assume that the transmitter radiates uniformly in all directions.

24. A TV transmitter emits a spherical wave, that is, a wave of the form given

---

[3] It is of interest that in the Soviet Union, where many experiments on the effects of microwaves on the human body have been performed, the accepted standard is much lower, 10 microwatts/cm².

by Eq. (1). At a distance of 5 km from the transmitter, the amplitude of the wave is 0.22 V/m. What is the time-average power emitted by the transmitter?

*25. At night, the naked, dark-adapted eye can see a star provided the energy flux reaching the eye is 8.8×10^{-11} W/m².
   (a) Under these conditions, how many watts of power enter the eye? The diameter of the dark-adapted pupil is 7.0 mm.
   (b) Assume that in our neighborhood there are, on the average, 3.5×10^{-3} stars per cubic light-year and that each of these emits the same amount of light as the Sun (3.9×10^{26} W). If so, how far would the faintest visible star be? How many stars could we see in the sky with the naked eye?

*26. Binoculars are usually marked with their magnification and lens size. For instance, 7 × 50 binoculars magnify angles by a factor of 7, and their collecting lenses have an aperture of diameter 50 mm. Your pupil, when dark adapted, has an aperture of diameter 7.0 mm. When you are observing a distant pointlike light source at night, by what factor do these binoculars increase the energy flux penetrating your eye?

*27. A steady current of 12 A flows in a copper wire of radius 0.13 cm.
   (a) What is the longitudinal electric field[4] in the wire?
   (b) What is the magnetic field at the surface of the wire?
   (c) What is the magnitude of the radial Poynting vector at the surface of the wire?
   (d) Consider a 1.0-m segment of this wire. According to the Poynting vector, what amount of power flows into this piece of wire from the surrounding space?
   (e) Show that the power calculated in part (d) coincides with the power of the Joule heat developed in the 1.0-m segment of wire.

*28. A stationary electric point charge is located in a uniform magnetic field. Draw a diagram showing the lines of electric and magnetic field. Sketch the direction of the Poynting vector at a few places. Roughly sketch the flow lines for the energy according to the Poynting vector.

**29. A coaxial cable transmits DC power from a source of emf to a resistive load (Figure 36.16). The cable consists of a long straight conducting wire of radius a surrounded by a conducting shell of radius b; assume that these conductors have zero resistance. The source has an emf $\mathscr{E}$ and delivers a power P.
   (a) Show that the electric field within the coaxial cable is

$$E = \frac{\mathscr{E}}{\ln(b/a)} \frac{1}{r} \quad \text{for } a < r < b$$

   (b) What is the current in each conductor of the cable? Show that the magnetic field within the coaxial cable is

$$B = \frac{\mu_0}{2\pi} \frac{P}{\mathscr{E}} \frac{1}{r} \quad \text{for } a < r < b$$

   (c) What is the Poynting vector within the coaxial cable?
   (d) Integrate the Poynting vector over the annular surface $a < r < b$, and show that the result equals P.

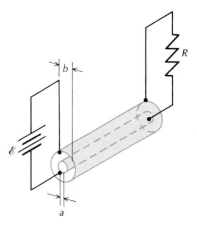

Fig. 36.16 A coaxial cable consisting of a central wire surrounded by a cylindrical shell. The radius of the wire is a; the radius of the shell is b.

Section 36.5

*30. According to a proposed scheme, solar energy is to be collected by a large power station on a satellite orbiting the Earth. The energy is then to be

---

[4] At the surface and outside the wire, the electric field also has a radial component. This contributes a *longitudinal* component to the Poynting vector, corresponding to the transport of energy *along* the wire.

transmitted down to the surface of the Earth as a beam of microwaves. At the surface of the Earth, the beam is to have a width of about 10 km × 10 km and is to carry a power of 5×10^9 W.
(a) What would be the time-average Poynting vector in this beam?
(b) What would be the amplitudes of the electric and magnetic fields in the beam?
(c) What pressure would the beam exert on the surface of the Earth?

*31. According to another proposal for the exploitation of solar energy, a large mirror is to be placed in orbit around the Earth. The mirror reflects sunlight and focuses it on a collector on the surface of the Earth.
(a) One version of this proposal calls for a mirror 370 km in diameter. What force does the pressure of sunlight exert on such a mirror when face on to the Sun?
(b) To prevent the mirror from drifting away under the influence of this force, it will have to be controlled with rocket thrusters. The mass of the mirror and its control system has been estimated as 5.7×10^9 kg. If the rocket thrusters are *not* switched on, how far would the mirror drift in one hour, starting from rest? Ignore the orbital motion of the mirror.

*32. A grain of dust is spherical, with a diameter of 1.0×10^{-6} m. It consists of material with a density 2.0×10^3 kg/m³. The grain is floating in interplanetary space, its distance from the Sun equal to the Earth–Sun distance.
(a) What is the attractive force of the Sun's gravity on the grain?
(b) What is the repulsive force of the Sun's radiation pressure? Which force is larger? Assume that the grain completely absorbs the incident sunlight.
(c) Repeat the above calculations if the diameter of the grain is 0.50×10^{-6} m.
(d) Show that the ratio of the forces of gravity and radiation pressure is independent of the distance of the grain from the Sun.

*33. Astronauts of the future could travel all over the Solar System in a spaceship equipped with a large "sail" coated with a reflecting material. Such a "sail" would act as a mirror; the pressure of sunlight on this mirror could support and propel the spaceship. How large a "sail" do we need to support a spaceship of 70 metric tons (equal to the mass of Skylab) against the gravitational pull of the Sun? Ignore the mass of the "sail."

*34. A beam of light is incident on a mirror at an angle of 30° with the normal and is entirely reflected. The energy incident in some time interval is 10^5 J. How much momentum is transferred to the mirror in this time interval?

*35. A sinusoidal electromagnetic wave travels to the right in a coaxial waveguide consisting of an inner wire surrounded by an outer shell. The radius of the inner conductor is a and the radius of the outer conductor is b. Figure 36.17 shows the electric field lines and the magnetic field lines at one instant of time.

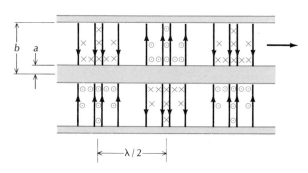

Fig. 36.17 Electric and magnetic field lines of an electromagnetic wave traveling toward the right in a coaxial waveguide. The drawing shows a longitudinal cross section of the waveguide. The electric field lines are radial lines. The magnetic field lines are circles around the central conductor; the dots and the crosses mark the heads and the tails of the magnetic field vectors.

(a) Sketch the direction of the Poynting vector.

(b) The electric and magnetic fields in the region between the conductors have the mathematical forms

$$E = \frac{A}{r} \sin\left(\frac{\omega}{c} x - \omega t\right)$$

$$B = \frac{A}{cr} \sin\left(\frac{\omega}{c} x - \omega t\right)$$

where A is a constant, r is the radial distance from the axis of the wire ($b > r > a$), x is the distance along the wire (direction of propagation), and the other symbols have their usual meaning. Find the instantaneous power that comes out of the coaxial waveguide at its far end at $x = L$. Find the average power.

(c) Suppose that at the far end of the waveguide, the wave is absorbed by a perfect absorber. What is the average force exerted by the wave on the absorber?

*36. Suppose that a magnetic monopole (source of magnetic field lines) and an electric point charge (source of electric field lines) are at some distance from one another.

(a) Draw a diagram showing the overlapping electric and magnetic field lines. Sketch the direction of the Poynting vector at a few places. Describe in words or pictures along what paths the energy flows in the space around the monopole and the charge.

(b) The Poynting vector gives not only the flow of energy but also of momentum. Accordingly, do you expect that the fields sketched in part (a) possess angular momentum?

**37. Question 14 describes how Einstein derived his equation $E = mc^2$, which expresses the equivalence of energy and mass. Formulating the arguments presented in this question mathematically, calculate how much mass the light pulse of energy E transports from one end of the boxcar to the other, and thereby derive Einstein's equation.

Section 36.6

38. In the spectrum of the light reaching us from the quasar PKS 0106+01, astronomers find some distinctive light emitted by hydrogen atoms. The wavelength of this light as received on Earth is 3776 Å. The wavelength as emitted by the atoms on the quasar is 1216 Å. Find the velocity of recession that will produce this shift of wavelength.

39. The light reaching us from the nearby galaxy NGC 221 has a Doppler shift, mainly due to the orbital motion of the Sun around the center of our Galaxy. A wavelength $\lambda_0 = 3968.5$ Å generated by calcium atoms in NGC 221 has shifted to $\lambda = 3965.8$ Å when detected by us on Earth. What is the radial speed of NGC 221 relative to us? Are we moving toward or away from this galaxy?

40. A spent Scout rocket, falling toward the surface of the Earth at 7.0 km/s, emits a radio signal at a frequency of 2203.08 MHz. What is the frequency of the radio signal received on the surface of the Earth?

*41. The light reaching us from all the distant galaxies exhibits a red shift caused by the motion of recession of these galaxies. This is called the **cosmological red shift.** Astronomers usually measure this red shift by means of a wavelength generated by calcium atoms; in the rest frame of the atoms the wavelength is 3968.5 Å.[5] The following table lists several distant galaxies and

---

[5] This is a dark spectral line, that is, an absorption line.

the corresponding Doppler-shifted wavelength detected on the Earth. The table also lists the distances of these galaxies.

Galaxy	Wavelength	Distance
In Virgo cluster	3984 Å	78 × 10⁶ light-years
In Ursa Major cluster	4167	980
In Corona Borealis cluster	4254	1400
In Bootes cluster	4485	2500
In Hydra cluster	4776	4000

(a) Calculate the velocities of recession of these galaxies. Make a plot of the distances vs. the velocities.

(b) Assuming that the galaxies had the same velocities at earlier times, calculate at what time all of the galaxies were at zero distance from our Galaxy.

*42. Doppler radar units, employed by police to measure the speed of automobiles, consist of a transmitter of microwaves and a receiver. The transmitter sends a wave of fixed frequency toward the target and the receiver detects the reflected wave sent back by the target. During this process the wave suffers two Doppler shifts: first when reaching the target and second when returning to the receiver.

(a) Suppose that an automobile approaches a Doppler radar unit at a speed of 96 km/h. Calculate the fractional change of frequency between the transmitted and the received waves.

(b) In the radar unit, the transmitted and received frequencies are allowed to beat against one another. Suppose that the frequency of the transmitted microwaves is 8.0×10^9 Hz. What is the beat frequency?

*43. Astronomers measure the velocity of rotation of stars by observing the Doppler shifts of light emitted by the approaching and the receding edges of the star. Suppose that the surface of a star emits a spectral line of a wavelength of 4101.74 Å. On the Earth, the wavelength received from the opposite edges of the surface are 4101.77 Å and 4101.71 Å.[6] What are the velocities of approach and recession? What is the angular velocity of rotation of the star if its radius is 7.0×10^8 m and its axis of rotation is perpendicular to the line of sight?

*44. Suppose that an astronomer attempting to measure the velocity of rotation of a star by the method described in Problem 43 finds that the wavelengths he receives from opposite edges of the star are 4103.82 Å and 4103.76 Å, respectively. As in Problem 43, the wavelength of this light is 4101.74 Å when emitted by the surface of the star. What can the astronomers conclude from this regarding the rotational and translational velocities of the star? Assume that the axis of rotation is perpendicular to the line of sight.

45. What is the percentage difference between the Doppler-shifted frequencies predicted by the formulas (48) and (51) for a receding emitter with $v/c = 0.10$? For a receding emitter with $v/c = 0.90$?

46. The wavelength of the light reaching us from the quasar Q1208+1011, which is the most distant yet discovered, has a wavelength 4.80 times as large as the wavelength emitted by similar light sources on Earth. According to Eq. (51), what speed of recession can we deduce from this?

---

[6] The light received from other points of the surface will have a Doppler shift somewhere between the Doppler shifts of the light from the edges. Thus, the light received on Earth contains a certain continuous range of wavelengths — the sharp spectral line is *broadened* by the Doppler shift. Thermal motions of the atoms on the surface of the star contribute a further broadening.

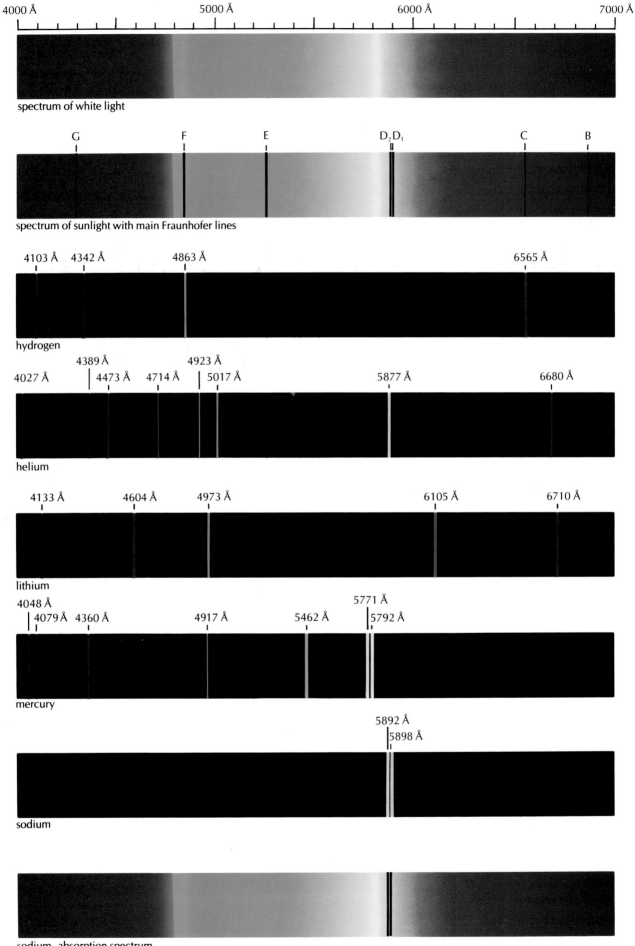

CHAPTER 37

Reflection, Refraction, and Polarization

So far we have examined the propagation of electromagnetic waves only in a vacuum. There, a plane wave will simply propagate in a fixed direction at the constant speed c. But if the wave encounters a region filled with matter — a sheet of metal, a pane of glass, or a layer of water — then the wave will interact with the matter and suffer changes in speed, direction, intensity, and polarization. These changes can, of course, be calculated from Maxwell's equations, taking into account the electric charges and currents induced by the action of the wave in the matter. Sometimes the calculations get extremely complicated and, furthermore, Maxwell's equations often tell us more than we want to know. For instance, if a wave is incident on a water surface, we may wish to compute the angle at which it penetrates the water, but we often do not need to know the changes in intensity and polarization. We will see that a simple rule, called Huygens' Construction, permits us to discover, without complicated calculations, how the change of direction of propagation is related to the change of speed of the wave as it penetrates a region filled with matter. Huygens' Construction is a geometric construction based on the behavior of wave fronts, and it bypasses Maxwell's equations.

Christiaan Huygens, *1629–1695, Dutch mathematician and physicist. Besides making important contributions to mechanics, he refined and extended the wave theory of light first proposed by Hooke, according to which light is an undulation of a pervasive elastic medium, or ether, filling all of space. In his* Traite de la Lumière, *Huygens placed this theory on a firm foundation by showing that reflection and refraction could be explained in terms of wave propagation.*

37.1 Huygens' Construction

The propagation of a wave can be conveniently described by means of the **wave fronts**, or wave crests, that is, the points at which the wave has maximum amplitude[1] at some instant of time. For example, Figure

---

[1] By amplitude of an electromagnetic wave we will hereafter always mean the amplitude of the electric field.

Huygens' Construction

Fig. 37.1 Spherical wave fronts at one instant of time. At a later time, each of these wave fronts will have moved outward by some distance.

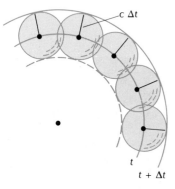

Fig. 37.2 Huygens' Construction for the propagation of a wave front. The inner solid arc shows the wave front at a time t; the outer solid arc shows the propagated wave front at time $t + \Delta t$. The dashed arc shows the propagated wave front for a time-reversed (convergent) wave.

Law of reflection

37.1 shows the instantaneous spherical wave fronts of the radio wave emitted by a radio transmitter. With the passing of time, each of these wave fronts travels in the outward direction.

The rule governing the propagation of wave fronts is **Huygens' Construction:**

> *To find the change of position of a wave front in a small time interval Δt, draw many small spheres of radius [wave speed] $\times \Delta t$ with centers on the old wave front. The new wave front is the surface of tangency to these spheres.*

The small spheres employed in this construction are called **wavelets.** Figure 37.2 shows how Huygens' Construction applies to the propagation of one of the spherical wave fronts of Figure 37.1. The wave speed in this example is simply c, and hence the radius of the wavelets is $c \, \Delta t$. Note that the wavelets in Figure 37.2 have two tangent surfaces, one in the outward direction and one in the inward direction. Obviously, only the former surface is appropriate to our problem (the latter surface would be appropriate if we were dealing with a convergent wave sent toward, say, a radio transmitter and precisely focused on it; this wave would be the time reverse of the wave sent out by the transmitter).

Huygens' Construction applies not only to propagation of electromagnetic waves in a vacuum but also to propagation in any transparent medium. As we will see in the following sections, this construction allows us to derive the laws of reflection and refraction. Although our emphasis will be on the propagation of light, Huygens' Construction is a general feature of wave propagation — it applies just as well to sound waves and water waves. The laws of reflection and refraction for all such waves are essentially the same, and some of our examples will take advantage of this.

37.2 Reflection

When a light wave encounters a material surface — such as the surface of a pane of glass, or the surface of a pond — part of the wave penetrates the surface and part is reflected. The amount of wave reflected depends on the characteristics of the surface. If the surface consists of very smoothly polished metal — such as the silvered surface of a mirror — almost all of the wave is reflected. In this section we will deal with the reflected part of the wave; in the next section we will deal with the part of the wave that penetrates from one medium into the other.

The **law of reflection** for a wave incident on a flat surface at an angle is well known:

> *The angle of incidence equals the angle of reflection.*

(The same law of reflection is also valid for a wave incident on a small portion of a curved surface, since such a portion can be approximated by a flat, tangent surface.) To derive this law of reflection from Huygens' Construction, we begin with Figure 37.3a, which shows wave fronts approaching a reflecting surface; at the instant shown, one edge of the leading wave front is barely touching the surface at the point P. Figure 37.3b shows some Huygens wavelets a short time later; the por-

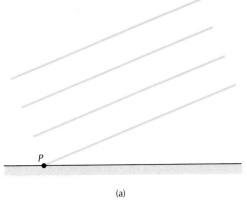

(a)

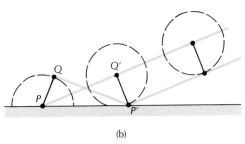

(b)

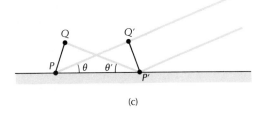

(c)

Fig. 37.3 (a) Wave fronts approaching a reflecting surface. The leading wave front barely touches the reflecting surface. (b) Huygens wavelets erected on the leading wave front of part (a). (c) The incident wave front PQ' makes an angle θ with the reflecting surface; the reflected wave front QP' makes an angle θ' with this surface.

tions of these wavelets below the reflecting surface have been omitted as irrelevant. The new wave front constructed on these wavelets touches the reflecting surface at the point P'. Obviously, to the right of the point P', the new wave front is simply parallel to the old wave front, that is, this part of the wave has not yet been reflected. To find the new wave front to the left of the point P', we draw a straight line that starts at P' and is tangent to the wavelet centered on P. This straight line represents the part of the wave that has already been reflected. To see that the incident wave front and the reflected wave front make the same angle with the reflecting surface, we appeal to Figure 37.3c. The right triangles $PQ'P'$ and $P'QP$ are congruent since they have a common hypothenuse (PP') and their short sides (PQ and $P'Q'$) are equal. Hence the angles θ and θ' are equal.

The direction of propagation of a wave is commonly described by the **rays** of the wave. These are lines perpendicular to the wave fronts. For example, Figure 37.4 shows the rays associated with the incident and reflected wave fronts.

The angle θ (or θ') between the wave front and a reflecting surface is obviously equal to the angle between the ray and the normal to the surface. The angles θ and θ' are called the **angles of incidence and of reflection** (Figure 37.5). Thus, from Huygens' Construction, we have

Rays

Angle of incidence and of reflection

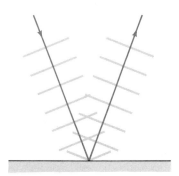
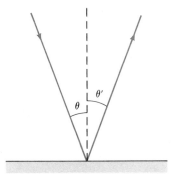

Fig. 37.4 (left) The rays of the wave are perpendicular to the wave fronts.

Fig. 37.5 (right) The angle of incidence θ and the angle of reflection θ'. These angles are the same as in Figure 37.3c.

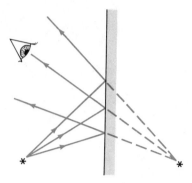

Fig. 37.6 Rays emerging from a point source are reflected by a mirror. The extrapolated rays (dashed) appear to come from a point source beyond the mirror.

Virtual image

deduced that the angle of incidence equals the angle of reflection, which is the law of reflection. Note that, in all of the above, we have implicitly taken for granted that the incident and the reflected rays are coplanar. This is evident from considerations of symmetry, and it can also be deduced from Huygens' Construction by taking into account that, in three dimensions, the wave fronts in Figures 37.3a and b are planes and the wavelets are spheres.

When light from some source strikes a flat mirror, the reflection of the light leads to the formation of an image of the source. Figure 37.6 shows a point source of light and the rays emerging from it; the figure also shows the reflected rays. If we extrapolate the reflected rays to the far side of the mirror, we find that they all appear to come from a point source of light. This apparent point source is the **image.** To an eye looking into the mirror, the image looks like the original source — the eye perceives the mirror image as existing in the space beyond the mirror. But the mirror image is an illusion; the light does not come from beyond the mirror. This kind of illusory image that gives the impression that light rays emerge from where they do not is called a **virtual image.**

If, instead of a single luminous point, our light source consists of an extended object made of many luminous points, then the mirror image will also be an extended object. Note that the mirror image of an object is a mirror-reversed object. For instance, Figure 37.7 shows some written letters and their mirror images. This reversal is commonly referred to as a reversal of left and right. However, it is more accurately described as a reversal of front to back — mirror writing is ordinary writing seen from behind. And the mirror image of, say, a hand facing north is a hand facing south (the reversal is not an ordinary "about face," but involves passing the front of the hand through its back, thereby converting a right hand into a left hand, and vice versa; Figure 37.8).

Fig. 37.7 (left) Some letters and their images in a mirror.

Fig. 37.8 (right) A hand facing north and its image in a mirror.

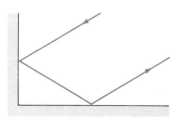

Fig. 37.9 A corner reflector.

Two mirrors at right angles form a **corner reflector.** A ray of light incident on this reflector is sent back in exactly the same direction it came from. Figure 37.9 shows how this works in two dimensions. But the same principle also applies in three dimensions; there it involves three mirrors at right angles to each other (a corner), and the light ray is sent back upon three successive reflections. Reflectors on automo-

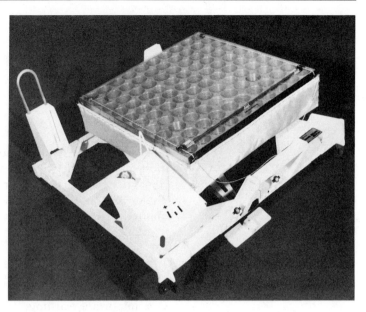

Fig. 37.10 Radar reflector for a sailboat.

Fig. 37.11 Corner reflectors for installation on the surface of the Moon.

biles and bicycles make use of such reflectors arranged in an array of a large number of small corners. Radar reflectors on boats apply the same principle; Figure 37.10 shows such a reflector to be hung on the mast of a sailboat. Finally, Figure 37.11 shows an array of corner reflectors deposited on the Moon by the Apollo 11 astronauts. This was used as a target for a laser beam sent from the Earth to the Moon, in an experiment that determined the Earth–Moon distance with very high precision (± 15 cm) by measuring the travel time of a pulse of laser light sent from the Earth to the Moon and reflected back to the Earth (see also Interlude X.3).

37.3 Refraction

The speed of light in a material medium — such as air, water, or glass — differs from the speed of light in a vacuum. We can recognize this immediately by recalling the theoretical formula for the speed of light derived from Maxwell's equations,

$$c = \frac{1}{\sqrt{\varepsilon_0 \mu_0}} \qquad (1)$$

We know from Chapters 27 and 33 that, in a material medium with given dielectric and magnetic characteristics, the quantities ε_0 and μ_0 in Maxwell's equations get replaced by $\kappa\varepsilon_0$ and $\kappa_m\mu_0$ [compare Eqs. (27.32) and (33.9)]. Hence, the formula (1) for the speed of light likewise gets replaced by

$$v = \frac{1}{\sqrt{\kappa\kappa_m\varepsilon_0\mu_0}} \qquad (2)$$

This is usually written as

Speed of light in a material medium

$$v = \frac{c}{n} \tag{3}$$

where

$$n = \sqrt{\kappa \kappa_m} \tag{4}$$

The quantity n is called the **index of refraction** of the material. In most materials the value of κ_m is very near 1 except, of course, in ferromagnetic materials, where light does not propagate anyhow (see Tables 33.1 and 33.3). Therefore the factor κ_m in Eq. (4) can often be omitted. This permits us to write

Index of refraction

$$n = \sqrt{\kappa} \tag{5}$$

One important warning: The value of the dielectric κ depends on the frequency of the electric field. Hence the values of the dielectric constants from Table 27.1 cannot be inserted in Eq. (5), because the former values apply only to static fields, whereas we are now concerned with the high-frequency fields of a light wave.

Table 37.1 gives the values of the index of refraction for a few materials. The values in this table apply to light waves of medium frequency (yellow-green light). The index of refraction is slightly larger for blue light and slightly smaller for red light; we will deal with this complication later in this section.

With the wave speed $v = c/n$, the relation between frequency and wavelength becomes

$$\lambda \nu = v = \frac{c}{n}$$

or

$$\lambda = \frac{c}{n}\frac{1}{\nu} \tag{6}$$

Table 37.1 INDICES OF REFRACTION OF SOME MATERIALS[a]

Material	n
Air, 1 atm, 0°C	1.00029
1 atm, 15°C	1.00028
1 atm, 30°C	1.00026
Water	1.33
Ethyl alcohol	1.36
Castor oil	1.48
Quartz, fused	1.46
Glass, crown	1.52
light flint	1.58
heavy flint	1.66

[a] For light of wavelength ~5500 Å.

For example, if a wave penetrates from air into water, where $n = 1.33$, its speed is reduced by a factor of 1.33, but its frequency remains constant. Consequently, Eq. (6) shows that its wavelength will be reduced by a factor of 1.33. The fact that the frequency of the wave remains constant can be understood in terms of the atomic mechanism underlying the change of wave velocity. When the wave strikes the water surface, it shakes the electrons of the water molecules; this acceleration of electric charges produces extra waves, which combine with the original wave. The net electromagnetic wave within the water is then a superposition of the original wave plus the extra waves, and it is this superposition that makes up the wave within the water with its changed speed and direction. The superposition has the same frequency as the original wave because the shaking of electrons proceeds at the original frequency and therefore the extra waves generated also will have exactly this frequency.

Note that Eq. (6) does not imply that a light source changes color when immersed in water. The color we perceive depends on the *frequency* of the light reaching our eyes; and this frequency is independent of whether the light source, our eyes, or both are immersed in water or in air.

When a wave strikes the surface of a transparent material, part of it is reflected and part of it penetrates the material. In the preceding section we have investigated the direction of propagation of the reflected wave; now let us investigate the penetrating wave. Again, we will use Huygens' Construction to find out what the wave does when it strikes the material surface. In vacuum the speed of light is c; in the material it is c/n. Figure 37.12a shows the approaching wave fronts at one instant of time; the left edge of one of these wave fronts barely touches the surface. Figure 37.12a also shows the Huygens wavelets that determine the position of the wave front at some later time; only the forward portions of these wavelets are relevant. Above the material

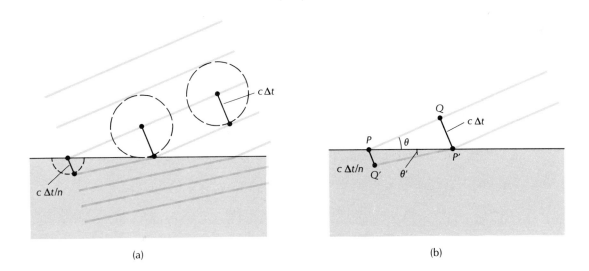

Fig. 37.12 (a) Huygens wavelets erected on a wave front whose edge barely touches the material surface. (b) The incident wave front makes an angle θ with the material surface; the refracted wave front makes an angle θ'.

surface, the wavelet has a radius $c\,\Delta t$; below the material surface, the wavelet has a smaller radius $(c/n)\,\Delta t$. The reduction of the speed of propagation of one side of the wave front causes the wave front to swing around, changing its direction of advance. The right triangles $PP'Q$ and $PP'Q'$ have the hypotenuse PP' in common (Figure 37.12b). In terms of the length PP' of this common hypotenuse, the sines of the angles between the wave fronts and the material surface are

$$\sin\theta = \frac{c\,\Delta t}{PP'} \qquad (7)$$

and

$$\sin\theta' = \frac{(c/n)\,\Delta t}{PP'} \qquad (8)$$

The ratio of this pair of equations is

$$\frac{\sin\theta}{\sin\theta'} = \frac{c}{c/n} \qquad (9)$$

or

Law of refraction

$$\boxed{\sin\theta = n\sin\theta'} \qquad (10)$$

This equation describes the change of direction of a wave upon penetration into a transparent material. This change of direction is called **refraction**, and Eq. (10) is called the **law of refraction**, or **Snell's Law**. The angle θ is the angle of incidence, and θ' is the angle of refraction. It is usually convenient to measure these angles between the rays and the normal to the material surface; Figure 37.13 shows the incident and refracted rays, and the angles of incidence and refraction. Note that the ray in the material is bent toward the normal, that is, $\theta' < \theta$. Also note that, as in the case of reflection, the incident and the refracted rays are coplanar.

Our formula (10) describing refraction at the interface between vacuum and a material is a special case of a general formula describing refraction at the interface of two materials. If the indices of refraction of the materials are n_1, n_2 and the angles between the rays and the normals are θ_1, θ_2, then

$$n_1 \sin\theta_1 = n_2 \sin\theta_2 \qquad (11)$$

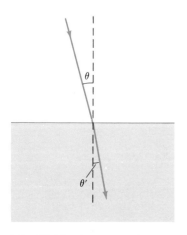

Fig. 37.13 The angle of incidence θ and the angle of refraction θ'.

This equation can be derived by some simple changes in the argument that led to Eq. (10).

EXAMPLE 1. A light ray is incident on one side of a plate of flint glass of index of refraction 1.60 (see Figure 37.14). If the angle of incidence is 60°, what is the angle of refraction? What is the angle at which the ray emerges from the plate of glass at the other side?

SOLUTION: With $n = 1.60$ and $\theta = 60°$, Eq. (10) gives

$$\sin \theta' = \frac{\sin 60°}{1.60} = 0.541$$

from which $\theta' = 32.8°$.

Since the two sides of the plate are parallel, the angle of incidence of the interior ray on the other side of the plate must be equal to 32.8°. If we now regard the angle θ' in Eq. (10) as the angle of incidence, and if we substitute $\theta' = 32.8°$, we immediately find that the ray emerges at an angle of 60° from the plate of glass.

COMMENTS AND SUGGESTIONS: In Eq. (10), we are at liberty to regard either of the angles θ or θ' as the angle of incidence. The equation makes no intrinsic distinction between incidence and refraction — it merely relates the angle θ' of the ray inside the material to the angle θ of the ray outside. The direction of propagation of the ray does not affect the validity of Eq. (10).

Note that we can obtain the angle at which the ray emerges from the plate of glass by appealing to a symmetry argument: if the ray suffers a deflection when entering the plate, it must suffer an opposite deflection when leaving the plate, and therefore the ray emerges at the same angle as the one at which it entered.

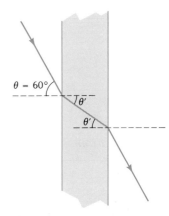

Fig. 37.14 A ray of light enters a plate of glass and emerges on the other side. The angles θ' that the interior ray makes with the normals at the right and the left sides of the plate are equal.

EXAMPLE 2. A small shiny fish is in the water 1.0 m below the surface. Where does a fisherman looking down into the water see the fish; that is, where is the image of the fish?

SOLUTION: Figure 37.15 shows two light rays from the fish to the eye of the fisherman. The first light ray is perpendicular to the surface and is not bent. The second light ray is bent away from the normal. Since the angles are small, Eq. (10) gives

$$\theta = n\theta' = 1.33\theta' \tag{12}$$

Extrapolation of the refracted ray into the water shows that it intersects the vertical ray at the point P', above the point P. Hence the image is above the object. The image distance OP' and the object distance OP are related as follows (Figure 37.15):

$$OP \tan \theta' = OP' \tan \theta \tag{13}$$

For small angles, this yields the approximation

$$\frac{OP'}{OP} = \frac{\theta'}{\theta} = \frac{1}{1.33} \tag{14}$$

Hence the image distance is smaller than the object distance by a factor of 1.33. For instance, if $OP = 1$ m, then $OP' = 0.75$ m. The fish seems to be nearer to the surface than it is.

COMMENTS AND SUGGESTIONS: Our calculation hinges on the assumption that the angles are small (so $\sin \theta \cong \theta$, $\tan \theta \cong \theta$). If the angles are not small, then the image distance depends on the angles in a complicated way.

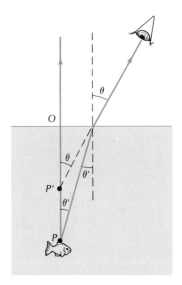

Fig. 37.15 A small shiny fish acts as a source of light. Note that the direction of propagation of the ray is opposite to that shown in Figure 37.13; but this does not affect the validity of Eq. (10). The extrapolated ray (dashed) appears to come from the point P'.

For a ray of light attempting to leave water, there is a critical angle beyond which refraction is impossible. As the light ray emerges into air, it is bent away from the vertical; in the extreme case, it is bent so much that it lies almost along the water surface (Figure 37.16). This extreme case corresponds to $\theta = 90°$ in Eq. (10),

$$n \sin \theta' = \sin 90° = 1 \tag{15}$$

The critical angle for this extreme form of refraction is therefore

Critical angle for total internal reflection

$$\boxed{\theta'_{crit} = \sin^{-1}\left(\frac{1}{n}\right)} \tag{16}$$

which for $n = 1.33$ gives

$$\theta'_{crit} = \sin^{-1}\left(\frac{1}{1.33}\right) = 48.75° \tag{17}$$

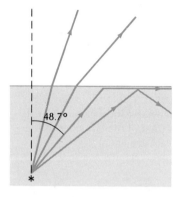

Fig. 37.16 A ray approaching a water surface from below with an angle of incidence $\theta' = 48.7°$ is refracted along the water surface.

If a light ray strikes a water surface from below at an angle larger than this, refraction is impossible. The only alternative is reflection — the water surface behaves like a perfect mirror. This phenomenon is called **total internal reflection.** It can occur whenever the index of refraction of the medium containing the light ray is larger than the index of refraction of the adjacent medium.

Total internal reflection has many important applications in optics. For instance, in periscopes and in binoculars, the light is reflected down the tube by internal reflection in a prism (Figure 37.17); this gives a much better image than reflection in a mirror. In an optical fiber, light moves along a thin rod made of a transparent material; the light zigzags back and forth between the walls of the rod, undergoing a sequence of total internal reflections — the fiber acts as a pipe for light (Figure 37.18). Such optical fibers are being used to replace telephone cables. The electrical impulses normally carried on a cable are converted into pulses of infrared laser light which can be transmitted via an optical fiber. The efficiency of optical fibers is very high because they can carry many conversations simultaneously; in modern telephone systems, single optical fibers are being used to carry 670 telephone conversations simultaneously.

In most materials the index of refraction depends somewhat on the

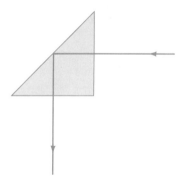

Fig. 37.17 Internal reflection in a prism.

Fig. 37.18 (above) Internal reflection in an optical fiber. (right) Light enters the optical fiber at the top right and emerges at the center.

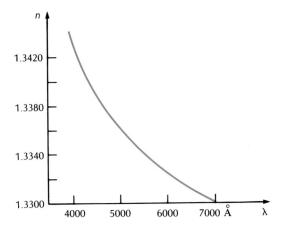

Fig. 37.19 Index of refraction of light in water as a function of wavelength. The index of refraction varies by about 1% over the range of visible wavelengths.

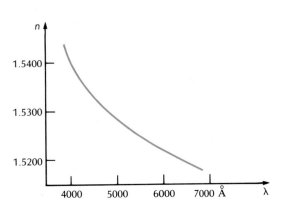

Fig. 37.20 Index of refraction of light in glass (Schott telescope flint glass) as a function of wavelength.

wavelength of light. Usually, the index of refraction increases as the wavelength decreases. For instance, Figures 37.19 and 37.20 are plots of the index of refraction of light in water and in flint glass as a function of wavelength (the wavelengths plotted along the horizontal axis in these figures are measured in air, before the light penetrates the material). When a ray of light containing several wavelengths, or colors, is refracted by a medium with an index of refraction that depends on wavelength, the refracted rays of different colors will emerge at somewhat different angles. The separation of a ray by refraction into distinct rays of different colors is called **dispersion.**

Dispersion

EXAMPLE 3. The index of refraction for red light in water is 1.330 and for violet light it is 1.342. Suppose that a ray of light approaches a water surface with an angle of incidence of 80°. What are the angles of refraction for red light and for violet light?

SOLUTION: For $n = 1.330$, Eq. (10) yields

$$\sin \theta' = \frac{\sin 80°}{1.330} = 0.740 \tag{18}$$

and

$$\theta' = 47.8°$$

For $n = 1.342$, Eq. (10) yields

$$\sin \theta' = \frac{\sin 80°}{1.342} = 0.734 \tag{19}$$

and

$$\theta' = 47.2°$$

COMMENTS AND SUGGESTIONS: The violet light is bent more toward the normal than the red light (Figure 37.21). For $\theta = 80°$, the difference in the angles of refraction is about 0.6°. Thus, refraction in water slightly separates a light ray according to colors. A beautiful demonstration of this effect is found in rainbows, which are produced by the refraction of sunlight in water droplets.

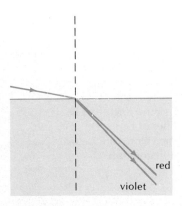

Fig. 37.21 Refraction of red and of violet light in water. The difference between the angles of the refracted rays has been exaggerated for the sake of clarity.

Prism

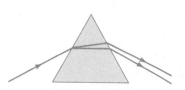

Fig. 37.22 Refraction of red and of violet light by a prism.

Spectrum

Spectral lines

In optical and spectroscopic laboratories, **prisms** are employed to separate light rays into their constituent colors. The basic mechanism is the same as that discussed in Example 3: the glass in the prism has slightly different indices of refraction for light of different wavelengths, and hence it bends rays of different colors by different amounts (Figure 37.22). In passing through a prism, the light is refracted twice: first at the air–glass interface and then at the glass–air interface. Under normal operating conditions, a good prism will introduce a difference of several degrees between the angular directions of the emerging red and violet light rays.

The pattern of colors produced by the analysis of a light ray by means of a prism is called the **spectrum** of the light. The white light emitted by the Sun has a continuous spectrum consisting of a mixture of all the colors (see color plate between pages 898 and 899). The colored light emitted by the atoms of a chemical element in an electric discharge tube, such as a neon tube, has a discrete spectrum consisting of just a few discrete colors. Each of these discrete colors is absolutely pure, i.e., it is light of a single wavelength. For example, hydrogen atoms emit the following discrete colors: red (6563 Å), blue-green (4861 Å), blue-violet (4340 Å), and violet (4102 Å). Such discrete colors are called **spectral lines**.[2] Figure 37.23 shows the spectral lines of hydrogen as displayed by means of a prism illuminated with light from a fine slit. Each of the lines in this figure is a separate image of the slit made by a separate color after refraction by the prism.

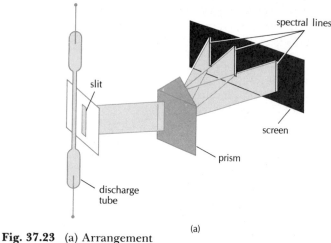

(a)

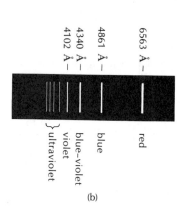

(b)

Fig. 37.23 (a) Arrangement for the analysis of light by means of a prism. Each discrete color produces a spectral line. (b) The spectral lines of hydrogen. (See also the color plate between pages 898 and 899.)

37.4 Polarization

We know from Section 36.2 that electromagnetic waves of a given direction of propagation can have two independent directions of polarization. These two polarizations are analogous to the two possible polarizations of transverse waves on a string. If a string is stretched horizontally, we can shake it up and down or else from side to side at right angles to the direction of the string. This generates two different

---

[2] Hydrogen also emits spectral lines in the ultraviolet and in the infrared; see Chapter 43. Note that the wavelengths listed above are measured in the air. The corresponding wavelengths in vacuum are 6565Å, 4863Å, 4342Å, and 4103Å.

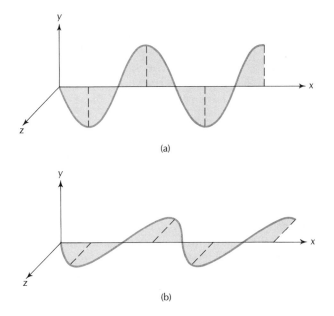

Fig. 37.24 The two independent directions of polarization of the transverse waves on a horizontal string are (a) vertical and (b) horizontal at right angles to the string.

waves, with vertical polarization and horizontal polarization, respectively (Figure 37.24). Any wave of an intermediate direction of polarization can be regarded as a superposition of these two standard polarizations. The two standard polarizations for the waves on a string are analogous to the two standard polarizations of an electromagnetic wave (compare Figures 36.5 and 36.6 with Figure 37.24).

The polarization of electromagnetic waves depends on the characteristics of the source of these waves. In general, the direction of polarization — that is, the direction of the electric field — lies in the same plane as the direction of motion of the accelerating charges or currents that serve as the source of the wave. For instance, the radio waves emitted at right angles from a straight antenna, such as the antenna of a CB radio on an automobile or the antenna of a small microwave generator, are polarized in a direction parallel to the antenna (Figure 37.25).

Although an individual light wave, like any other kind of electromagnetic wave, is always polarized in some definite direction, the light emitted by an ordinary source, such as the Sun or an incandescent light bulb, does not exhibit any noticeable polarization. A light beam from such a source consists of a superposition of a large number of light waves contributed by the large number of atoms in the source. Since the atoms in the source are oriented at random over all directions, the polarizations of their light waves are distributed at random over all directions, and the instantaneous electric field of the superposition of these waves fluctuates rapidly from one direction to another. This kind of light is called **unpolarized light.** To give this kind of light a polarization, we must pass it through a polarizing filter, or a **polarizer**, which selects waves of a preferential direction of polarization and blocks all other waves.

The simplest polarizer is a sheet of Polaroid. Such a sheet contains long chains of organic molecules (polyvinyl alcohol) aligned parallel to each other over the sheet; the preferential direction that permits passage of the electric field of the wave is *perpendicular* to the direction of alignment of the molecules. An analogous polarizing filter for microwaves can be constructed out of a number of thin conducting rods or

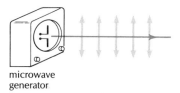

microwave generator

Fig. 37.25 The wave emitted at right angles from the straight antenna of the microwave generator is polarized in the direction parallel to the antenna. Since the electric field oscillates between positive and negative values, we use double-headed arrows to indicate the direction of polarization.

Polarizer

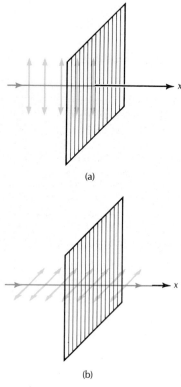

Fig. 37.26 An array of vertical wires (a) blocks the passage of a microwave of vertical polarization but (b) permits the passage of a microwave of horizontal polarization.

wires arranged parallel to each other (Figure 37.26). The preferential direction of polarization that permits the passage of the electric field of the wave is then *perpendicular* to the direction of the wires, because the wires have very little effect on a perpendicular electric field; on the other hand, an electric field parallel to the wires causes strong currents to flow in the wires, which both reflect the wave and dissipate its energy.

Another simple polarizer takes advantage of the phenomenon of **polarization by reflection.** Whenever light is incident on a transparent material, some portion of the light is reflected; that is, the surface of the material acts as an imperfect mirror. The fraction of light reflected depends on the angle of incidence and on the index of refraction of the material (in general, the reflection is quite strong if the angle of incidence is near 90°, as you can easily verify by looking at any smooth surface at a grazing angle). Furthermore, the fraction of light reflected depends on the polarization of the light — the reflection is always somewhat stronger for light polarized at right angles to the plane of incidence than for light polarized in the plane of incidence. Thus, when an unpolarized beam of light is incident on a surface, both the reflected beam and the refracted beam will be partially polarized. In the reflected beam, the light polarized at right angles to the plane of incidence will be somewhat stronger than the light polarized in the plane of incidence; correspondingly, in the refracted beam, the light polarized at right angles to the plane of incidence will be somewhat weaker than the light polarized in the plane of incidence. At a critical angle — called **Brewster's angle** — the surface does not reflect any of the light polarized in the plane of incidence and permits all of this light to pass into the material (total refraction; see Figure 32.27). The value of the critical angle is given by **Brewster's Law:**

$$\boxed{\tan \theta_B = n} \tag{20}$$

Brewster's Law

This expression can be derived from Maxwell's equations, but we will not deal with the derivation, since it is somewhat complicated.

Fig. 37.27 A beam of unpolarized light is incident on the glass surface at the Brewster angle. Since this beam contains equal contributions of horizontally and vertically polarized light, we represent it by two overlapping horizontal (black) and vertical (colored) double-headed arrows. The reflected beam (black ray) contains only light polarized at right angles to the plane of incidence.

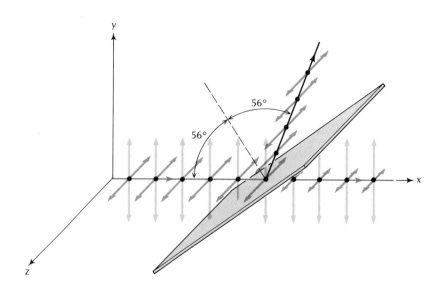

For glass with an index of refraction $n = 1.5$, Brewster's angle is $\theta_B = 56°$. At this angle, the light reflected by the glass is completely polarized at right angles to the plane of incidence, and thus a plate of glass oriented at this angle achieves maximum segregation of light waves of the two different polarizations. The intensity of the reflected beam is low; only about 15% of the light polarized at right angles to the plane of incidence is reflected, and the remaining 85% is refracted (at the Brewster angle, the refraction of the light polarized in the plane of incidence is total, but the reflection of the light polarized at right angles to the plane of incidence is *not* total). For the practical exploitation of polarization by reflection, it is best to use a stack of several parallel plates of glass (see Figure 37.28); in such a stack, at every glass–air surface, some of the light polarized at right angles to the plane of incidence is reflected, and therefore every plate enhances the segregation of light of the two different polarizations.

The most efficient polarizers rely on the phenomenon of double refraction, or **birefringence,** displayed by calcite, quartz, and some other crystals. When a ray of light enters one of these crystals, it forms *two* distinct refracted rays (see Figure 37.29). The difference in refraction between these two rays arises from a difference in the response of the crystal to the two different polarizations of the light. The angles of refraction depend in a complicated way on the orientation of the polarization relative to the atomic planes of the crystal. For a calcite crystal oriented as shown in Figure 37.29, the light polarized in the plane of incidence has an index of refraction of 1.486 and the light polarized at right angles to the plane of incidence has an index of refraction of 1.658. If we cut a prism with cleverly contrived angles out of a calcite crystal, we can achieve total internal reflection for the latter light, but avoid internal reflection for the former. This kind of selective internal reflection means that the calcite can be used to segregate light of the two directions of polarization.

Birefringence

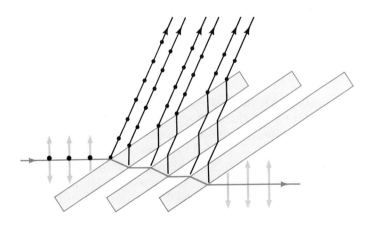

37.28 A stack of parallel plates of glass. At each glass–air surface, some of the light polarized at right angles to the plane of incidence, or the plane of the page, is reflected upward (black rays). If the number of glass plates in the stack is large, the residual light, emerging on the far right, is almost completely polarized in the plane of incidence (colored ray).

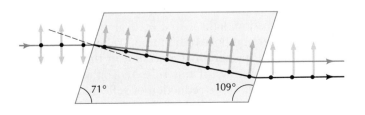

37.29 Refraction of light in a crystal of calcite. The incident ray forms two distinct refracted rays, with different polarizations.

37.30 An arrangement of two polarizers. The polarizers, which may be sheets of Polaroid, stacks of glass plates, or calcite crystals, are marked with lines, which indicate the preferential directions of polarization for which waves will pass freely through the polarizers. The preferential directions of the two polarizers differ by an angle ϕ. The electric field incident on the second polarizer can be regarded as a superposition of two electric fields, respectively parallel and perpendicular to the preferential direction of the second polarizer.

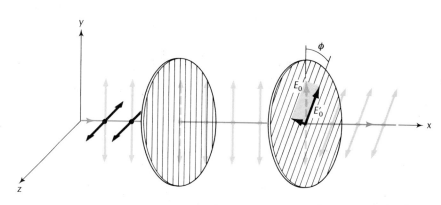

A variety of interesting experiments can be performed with two or more polarizers arranged in tandem. For instance, Figure 37.30 shows a simple arrangement of two polarizers. Unpolarized light is incident on the first polarizer, which selects waves of vertical polarization and allows them to pass; the light emerging from this first polarizer is therefore vertically polarized. When this light is incident on the second polarizer, or analyzer, its electric field vector makes an angle ϕ with the preferential direction. We can regard a light wave with such an electric field as a superposition of two light waves, whose electric fields are, respectively, parallel and perpendicular to the preferential direction of the analyzer. The analyzer permits the passage of the former wave but blocks the latter. If the amplitude of the wave incident on the analyzer is E_0, then the amplitude of the parallel wave is

$$E'_0 = E_0 \cos \phi \qquad (21)$$

Since the intensity of a wave is proportional to the square of the amplitude, the intensity of the wave transmitted by the analyzer is smaller than that of the wave incident on the analyzer by a factor $\cos^2 \phi$:

$$[\text{transmitted intensity}] = [\text{incident intensity}] \times \cos^2\phi \qquad (22)$$

Law of Malus

This relation between the intensity incident on the analyzer and the intensity transmitted by the analyzer is called the **law of Malus**. Note that if $\phi = 90°$, the transmitted intensity is zero; that is, such "crossed" polarizers block the light completely.

This blocking of light by crossed polarizers can be readily demonstrated by means of two Polaroid sunglasses (see Figure 37.31). However, these sunglasses cannot be used for a quantitative test of Eq. (22), because the reduction of intensity of light by the sunglasses is caused not entirely by polarization but also by the color in the glass. The preferential direction of the Polaroid in such sunglasses is arranged vertically, so as to provide a drastic reduction of the intensity of sunlight reflected from any horizontal surfaces within the field of view. According to the above discussion, light with a vertical plane of incidence will acquire a partial horizontal polarization upon reflection, and the vertical arrangement of the preferential direction in the sunglasses favors the reduction of the intensity of this reflected light reaching the eye (see Figure 37.32). Thus, in the reduction of reflected glare, the Polaroid sunglasses act as analyzers for the light polarized by reflection. For direct sunlight, the sunglasses simply act as polarizers.

Étienne Malus (mälüs), *1775–1812, French army engineer and physicist, noted for his mathematical and experimental investigations in optics and in double refraction. He discovered the polarization of light by reflection while looking through a calcite crystal at sunlight reflected by the windows of the Luxembourg Palace in Paris.*

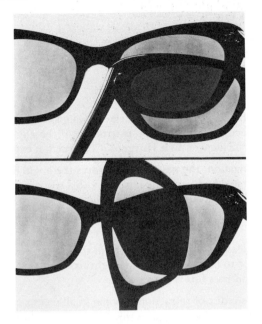

Fig. 37.31 The light that passes through the distant pair of sunglasses becomes polarized in the vertical direction, because this is the preferential direction for the Polaroid in the sunglasses. If the near pair of sunglasses is oriented parallel to the distant pair ($\phi = 0°$), it permits the passage of the polarized light. But if the near pair of sunglasses is oriented perpendicular to the distant pair ($\phi = 90°$), it stops the polarized light completely.

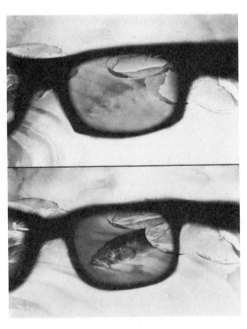

Fig. 37.32 When viewed through ordinary sunglasses (top), the fish is almost completely hidden by the glare reflected by the water. When viewed through Polaroid sunglasses (bottom), the fish becomes visible because the glare is reduced.

SUMMARY

Huygens' Construction: The new wave front is the surface of tangency of the wavelets erected on the old wave front.

Law of reflection: The angle of incidence equals the angle of reflection.

Index of refraction: $v = c/n$
$n = \sqrt{\kappa}$

Law of refraction: $\sin\theta = n \sin\theta'$

Critical angle for total internal reflection:

$$\theta'_{crit} = \sin^{-1}\left(\frac{1}{n}\right)$$

Brewster's Law: $\tan \theta_B = n$

Law of Malus: [transmitted intensity] = [incident intensity] $\times \cos^2 \phi$

QUESTIONS

1. When light is incident on a smooth surface — a glass surface, a painted surface, a water surface — the reflection is strongest if the angle of incidence is near 90° (grazing incidence). Can Huygens' Construction explain this?

2. In celestial navigation, the navigator measures the angle between the Sun, or some other celestial body, and the horizon with a sextant. If the navigator is on dry land, where the horizon is not visible, he can measure instead the angle between the Sun and its reflection in a pan full of water, and divide this angle by two. Explain.

3. Suppose we release a short flash of light in the space between two parallel mirrors placed face to face. Why does this light flash not travel back and forth between the two mirrors forever?

4. To measure the index of refraction of some small transparent chips of plastic of irregular shape, a physicist places the chips in a glass beaker full of water. She then gradually adds sucrose to the water until the chips suddenly become nearly invisible. Explain why a measurement of the index of refraction of the sucrose solution then yields the index of refraction of the chips of plastic.

5. If you immerse one-half of a stick in the water, it will appear bent. Explain.

6. At sunset, the image of the Sun remains visible for some time after the actual position of the Sun has sunk below the horizon. Explain.

7. After a navigator measures the angle between the Sun and the horizon with a sextant, he must make a correction for the refraction of sunlight by the atmosphere of the Earth. Does this refraction increase or decrease the apparent angle between the Sun and the horizon?

8. Figure 37.33 shows water waves refracted in the shallows near a headland. This refraction deflects the waves toward the headland. Explain this, keeping

Fig. 37.33

in mind that the speed of water waves in shallow water is proportional to the square root of the depth ($v = \sqrt{gh}$).

9. If the preferential directions of two adjacent sheets of Polaroid are at right angles, no light will pass through. However, if you now slip a third sheet of Polaroid between the other two and orient its preferential direction so that it lies between the directions of the other two, then some light will pass through the three sheets. Explain.

10. It has been proposed that we could eliminate the glare of the headlights of approaching automobiles by covering the windshields and the headlights with sheets of Polaroid. What orientation should we pick for the sheets of Polaroid installed on windshields and on headlights so that the light of every approaching automobile is blocked out, but our own light is not?

11. Some small-boat sailors like to wear Polaroid sunglasses because these make disturbances of the water surface stand out with exceptional contrast, and thereby make it easier to spot approaching gusts of wind. Why are Polaroid sunglasses better for this purpose than ordinary sunglasses? (Hint: Consider polarization by reflection.)

12. Suppose you are given a sheet of Polaroid that has no markings identifying its preferential direction. You have available a beam of unpolarized light. How can you determine the preferential direction of the sheet? (Hint: Cut the sheet in two and place one sheet behind the other rotated by 90° around the axis of the beam, so that the light is completely blocked. What will happen if you now rotate one sheet about a *transverse axis*, that is, an axis perpendicular to the beam?

13. The scattered light reaching you from the blue sky is (partially) polarized: if you look straight up, the direction of polarization is perpendicular to the direction of the Sun. How does this polarization arise? [Hint: Consider a beam of (unpolarized) sunlight passing overhead. The electric field of the light waves in this beam accelerates electrons in the molecules of air, and the radiation emitted by these electrons constitutes the scattered light of the sky. Since the direction of acceleration is perpendicular to the incident beam, what can you say about the polarization of the radiation emitted downward, toward you?]

PROBLEMS

Section 37.2

1. According to a (questionable) story, Archimedes set fire to the Roman ships besieging Syracuse by focusing the light of the Sun on them with mirrors. Suppose that Archimedes used flat mirrors. How many flat mirrors must simultaneously reflect sunlight at a piece of canvas if it is to catch fire? The energy flux of the sunlight at the surface of the Earth is 0.1 W/cm², and the energy flux required for ignition of canvas is 4 W/cm². Assume that the mirrors reflect the sunlight without loss.

*2. A vertical mirror, oriented toward the Sun, throws a rectangular patch of sunlight on the floor in front of the mirror. The size of the mirror is 0.5 m × 0.5 m, and its bottom rests on the floor. If the Sun is 50° above the horizon, what is the size of the patch of sunlight on the floor?

*3. Two mirrors meeting at a right angle make a corner reflector (see Figure 37.9). Prove that a ray of light reflected successively by both mirrors will emerge on a path antiparallel to its original path.

Section 37.3[3]

4. A ray of sunlight approaches the surface of a smooth pond at an angle of incidence of 40°. What is the angle of refraction?

---

[3] In all problems, assume that the index of refraction of air is 1, unless otherwise stated.

5. The walls of a (filled) aquarium are made of glass with an index of refraction 1.5. If a ray strikes the glass from the inside at an angle of incidence of 45°, what is the angle at which it emerges into the air?

6. The speed of sound in air is 340 m/s, and in water it is 1500 m/s. If a sound wave in air approaches a water surface with an angle of incidence of 10°, what will be the angle of refraction?

7. Make a plot of the angle of incidence vs. the angle of refraction for light rays incident on a water surface. What is the maximum angle of refraction?

8. A ship's navigator observes the position of the Sun with his sextant and measures that the Sun is exactly 39° away from the vertical. Taking into account the refraction of the Sun's light by air, what is the true angular position of the Sun? For the purpose of this problem, assume that the Earth is flat and that the atmosphere can be regarded as a flat, transparent plate of uniform density and an index of refraction 1.0003.

*9. A ray of light strikes a plate of window glass of index of refraction 1.5 and thickness 2.0 mm with an angle of incidence of 50°. Find the transverse displacement between the transmitted ray and the extrapolation of the incident ray.

*10. A point source of light is placed above a thick plate of glass of index of refraction n (Figure 37.34). The distance from the source to the upper surface of the plate is l, and the thickness of the plate is d. A ray of light from the source may suffer either a single reflection at the upper surface, or a single reflection at the lower surface, or multiple alternating reflections at the lower and upper surfaces. Thus, each ray splits into several rays, giving rise to multiple images. In terms of l and d, find the distance of the first, second, and third images below the upper surface of the plate. Assume that the angle of incidence of the ray is small.

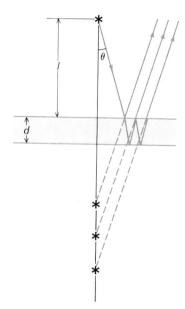

Fig. 37.34 Ray of light reflected and refracted by a plate of glass.

*11. The bottom half of a beaker of depth 20 cm is filled with water ($n = 1.33$), and the top half is filled with oil ($n = 1.48$). If you look into this beaker from above, how far below the upper surface of the oil does the bottom of the beaker seem to be?

*12. Because of refraction in the plate of glass, an object viewed through an ordinary window will seem somewhat nearer than its actual distance.
  (a) Consider the rays that strike the glass at a small angle of incidence (nearly normal). Show that if an object is at a distance l from a window, the image is at a distance that is shorter by an amount $\Delta l = (1 - 1/n)d$, where n is the index of refraction and d the thickness of the glass.
  (b) What change in distance does this formula give if the windowpane is ordinary glass with $d = 2.0$ mm and $n = 1.5$?
  (c) What change in distance does this formula give if the windowpane is heavy plate glass with $d = 8.0$ mm and $n = 1.5$?

*13. The highly reflective paint used on highway signs contains small glass beads which reverse the direction of propagation of a light ray, throwing it back toward the source of light (see Figure 37.35a). The reversal of direction will occur only for those light rays incident on the sphere of glass at a selected distance from the axis, a distance that depends on the index of refraction of the glass. Show that if the reversal of direction is to occur at all, the index of refraction of the glass must be in the range $1 < n \leq \sqrt{2}$. (Hint: Consider the light ray shown in Figure 37.35b; what index of refraction does the reversal of this light ray require?)

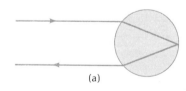

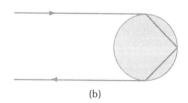

Fig. 37.35 (a) Path of a light ray that reverses direction in a transparent sphere. (b) Path of a light ray that penetrates at maximum distance from the axis of the sphere.

14. An optical fiber is made of a thin strand of glass of index of refraction 1.5. If a ray of light is to remain trapped within this fiber, what is the largest angle it may make with the surface of the fiber?

15. (a) A transparent medium of index of refraction n_1 adjoins a transparent medium of index of refraction n_2. Assuming $n_1 > n_2$, show that the

critical angle for total internal reflection of a ray attempting to leave the first medium is

$$\theta = \sin^{-1}\left(\frac{n_2}{n_1}\right)$$

(b) A layer of kerosene ($n_2 = 1.2$) floats on a water surface. In this case, what is the critical angle for total internal reflection within the water?

*16. A signal rocket explodes at a height of 200 m above a ship on the surface of a smooth lake. The explosion sends out sound waves in all directions. Since the speed of sound in water (1500 m/s) is larger than the speed of sound in air (340 m/s), a sound wave can suffer total reflection at a water surface if it strikes at a sufficiently large angle of incidence. At what minimum distance from the ship will a sound wave from the explosion suffer total reflection?

*17. A layer of oil, of index of refraction n', floats on a surface of water. A ray of light coming from below attempts to pass from the water to the oil and from there to the air above. What is the maximum angle of incidence of the ray on the water–oil surface that will permit the ultimate escape of the ray into the air? Does you answer depend on n'? (Hint: Use the formula derived in Problem 15.)

*18. A seagull sits on the (smooth) surface of the sea. A shark swims toward the seagull at a constant depth of 5 m. How close (measured horizontally) can the shark approach before the seagull can see it clearly?

*19. To discover the percentage of sucrose (cane sugar) in an aqueous solution, a chemist determines the index of refraction of the solution very precisely and then finds the percentage in a table giving the dependence of index of refraction on sucrose concentration. The chemist determines the index of refraction by immersing a glass prism in the sucrose solution and measuring the critical angle for total internal reflection of a light ray inside the glass prism.
(a) Suppose that with a prism of index of refraction 1.6640 the critical angle is 57.295°. Use the result of Problem 15 to find the index of refraction of the sucrose solution.
(b) Use the following table, interpolating if necessary, to find the concentration of sucrose to four significant figures.

Concentration	n
40.00%	1.3997
40.10	1.3999
40.20	1.4001
40.30	1.4003

**20. A prism of glass of index of refraction 1.50 has angles of 45°, 45°, and 90°. A ray of light is incident on one of the short faces at an angle θ (see Figure 37.36). What is the maximum value of θ for which the ray of light will suffer total internal reflection at the long face?

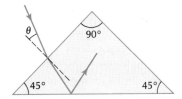

Fig. 37.36

**21. Rainbows are produced by the refraction of sunlight by drops of water. Figure 37.37 shows a ray of light entering a spherical drop of water. The ray

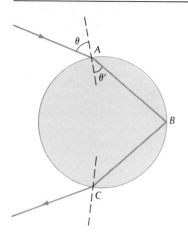

Fig. 37.37 Path of light ray in a drop of water.

is refracted at A, reflected at B, and refracted at C. The angle of incidence at A (between the ray and the normal to the surface) is θ and the angle of refraction is θ'.

(a) By geometry, show that the angles of incidence and reflection at B coincide with θ' and that the angles of incidence and refraction at C coincide with θ' and θ, respectively.

(b) Show that the angular deflection of the ray from its path is $\theta - \theta'$ at A, $\pi - 2\theta'$ at B, and $\theta - \theta'$ at C. These angles are all in radians, and they are measured clockwise from the incident path at each point.

(c) The total angular deflection of the ray by the raindrop is $\Delta = 2(\theta - \theta') + \pi - 2\theta'$. A rainbow will form when all the rays within an infinitesimal range $d\theta$ of angles of incidence suffer the same angular deflection, i.e., when the derivative $d\Delta/d\theta = 2 - 4d\theta'/d\theta$ is zero. If this condition is satisfied, the rays sent back by the raindrops are concentrated, producing a bright zone in the sky. Show that the critical angle θ_c, at which $d\Delta/d\theta = 0$, is given by

$$\cos^2 \theta_{\text{crit}} = \tfrac{1}{3}(n^2 - 1)$$

where n is the index of the refraction of water.

(d) The index of refraction for red light in water is 1.330. Find θ_{crit} and find Δ in degrees and minutes of arc. Draw a diagram showing a red ray coming from the Sun, hitting the drop, and reaching the eye of a rainbow watcher.

(e) The index of refraction for violet light in water is 1.342. Find θ_{crit} and find Δ. On top of the preceding diagram, draw a violet ray coming from the Sun, hitting the drop (at a different point), and reaching the eye of the rainbow watcher. Will the watcher see the red color above or below the violet?

Section 37.4

22. The preferential directions of two adjacent sheets of Polaroid make an angle of 45°. A beam of polarized light, whose direction of polarization coincides with the preferential direction of the *second* sheet, is incident on the *first* sheet. By what factor is the intensity of the transmitted beam emerging from the second sheet reduced compared to the intensity of the incident beam? Assume that the sheets act as ideal polarizing filters.

23. Two sheets of Polaroid are placed on top of one another. Unpolarized light is perpendicularly incident on the sheets. By what factor is the intensity of the emerging light reduced (relative to the incident light) if the polarizing directions of the sheets differ by an angle of 30°? 45°? 60°?

24. What is the Brewster angle for light incident on a water surface from above? For light incident on the surface from below?

25. Show that whenever a ray of light is incident on a glass–air interface from within the glass, the Brewster angle is smaller than the critical angle for total internal reflection.

*26. Prove that if a ray of light is incident on a surface at the Brewster angle, the reflected ray and the refracted ray are perpendicular to each other (see Figure 37.38).

*27. Prove that if a ray of light is incident on the upper (external) surface of a plate of glass at the Brewster angle, the refracted ray reaches the lower (internal) surface of the plate at the Brewster angle for this surface. (Hint: Use the result stated in Problem 26.)

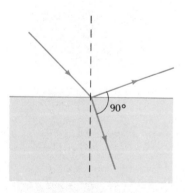

Fig. 37.38

*28. Consider an unpolarized light beam incident on the stack of three plates shown in Figure 37.28. Assume that the reflected ray at each glass–air surface (or air–glass surface) removes 15% of the light polarized at right angles to the plane of incidence. What fraction of the light of this polarization emerges on the far right, in the residual beam? What fraction of the *total* amount of residual light does this amount to? How many plates would you need if you wanted to reduce this fraction of the total to 5% or less?

*29. Figure 37.39 shows a prism cut from a calcite crystal. This kind of prism is used as a polarizer; it is called a **Nicol prism.** The angles of the prism are 68°, 90°, and 22°. A layer of a transparent material (Canada balsam) of index of refraction 1.55 has been applied to the far side of the prism. A ray of light incident from the left from a direction parallel to the base of the prism separates into two rays of indices of refraction 1.658 and 1.486 inside the prism. Show that the former ray (polarized at right angles to the plane of incidence) will suffer total internal reflection at the far surface of the prism, but that the latter (polarized in the plane of incidence) will not.

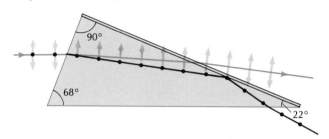

Fig. 37.39 A Nicol prism.

*30. Two sheets of Polaroid are arranged as polarizer and analyzer. Suppose that the preferential direction of the second sheet is rotated by an angle ϕ about the direction of incidence and then rotated by an angle α about the vertical direction (see Figure 37.40). If unpolarized light of intensity I_0 is incident from the left, what is the intensity of the light emerging on the right?

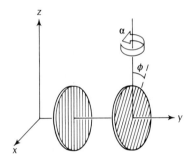

Fig. 37.40 Two sheets of Polaroid.

*31. If the preferential directions of two adjacent sheets of Polaroid are at right angles, they will completely block a light beam. However, if you insert a third sheet of Polaroid between the other two, then some light will pass through. Derive a formula for the dependence of the intensity of the transmitted light as a function of the angle that the preferential direction of the inserted sheet makes with that of the first sheet. Assume that the incident light is unpolarized and that the sheets act as ideal polarizing filters. For what orientation of the inserted sheet is the transmitted intensity maximum?

***32. Consider an infinite number of adjacent sheets of Polaroid, each tilted by an infinitesimal angle with respect to the preceding sheet (Figure 37.41). Show that such a stack rotates the plane of polarization of the light wave through a finite angle without loss of intensity. Assume that each sheet of Polaroid acts as an ideal polarizing filter.

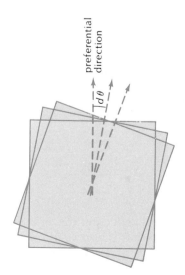

Fig. 37.41 A stack of sheets of Polaroid.

CHAPTER 38

Mirrors, Lenses, and Optical Instruments

Geometrical optics **Geometrical optics** deals with the behavior of light under the assumption that, in any uniform transparent material, light propagates along straight lines. This is a good approximation if the wavelength of the light is very small compared with the sizes of the bodies in and around which light propagates. Under these conditions, we can adequately describe the propagation of light by rays that are rectilinear, except when reflected at a mirror or when refracted at the interface between two different materials. Thus, in geometrical optics, the wave properties of light do not show up explicitly (although these wave properties enter into the derivations of the laws of reflection and refraction). The restriction to a small wavelength is needed to avoid extra changes of the direction of propagation of light by diffraction, that is, the bending and spreading of waves around obstacles. We already briefly mentioned diffraction in Chapter 17, and we will discuss it in detail in a later chapter.

In the context of geometrical optics, the effects of mirrors, lenses, prisms, or other optical devices on light can be discovered by calculating the trajectories of rays through the device. Such calculations are called ray tracing. In essence, this involves nothing but the application of the laws of reflection and refraction.

38.1 Spherical Mirrors

We already stated the law of reflection for light incident on a flat mirror. The same law of reflection is also valid for a curved mirror, provided the angles of incidence and reflection are reckoned relative to the normal erected at the point where the ray arrives at the mirror.

A mirror curved like the surface of a sphere can focus a beam of light to a point. Figure 38.1 shows a **concave** spherical mirror and wave fronts of light incident and reflected on this mirror; the reflected waves converge to a point, the **focal point** of the mirror. We can describe the direction of propagation of the waves by rays; Figure 38.2 shows the incident and the reflected rays. The reflected rays converge at the focal point.

The focal point of the spherical mirror is halfway between the mirror and the center of the spherical surface. For a proof, we use Figure 38.3, which shows the path of a single ray of light. The focal point is

Concave mirror

Focal point

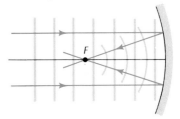

Fig. 38.1 A concave spherical mirror focuses an incident plane wave on a point.

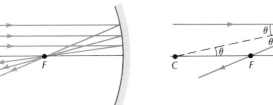

Fig. 38.2 Reflection of parallel rays by a concave spherical mirror.

Fig. 38.3 Reflection of a single ray.

the intersection of this ray with the axial line CA. To find the distance FA, called the **focal length,** we begin with the observation that in the isosceles triangle CFQ the length CF equals FQ. Under the assumption that the angle θ is small (equivalently, that the incident ray is near the axial line), the length FQ is approximately equal to FA. Hence

$$CF = FA \qquad (1)$$

and thus the point F is halfway between the mirror (A) and the center (C) of the spherical surface. The focal length FA is therefore one-half of the radius of the spherical surface. Designating the former by f and the latter by R, we can write

$$\boxed{f = \tfrac{1}{2}R} \qquad (2)$$

Focal length of spherical mirror

Note that this focusing of a beam of light rests on the approximation that the beam is narrow, so all the rays are near the axial line. If a ray strikes the mirror at some appreciable distance from the axial line, then the reflected ray will miss the focal point (Figure 38.4). This defect in the focusing properties of a mirror is called **spherical aberra-**

Spherical aberration

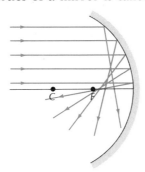

Fig. 38.4 Reflection of rays far from the axis of the mirror. Such rays miss the focal point.

Convex mirror

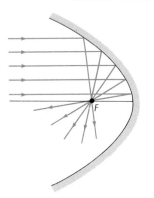

Fig. 38.5 Reflection of parallel rays by a paraboloidal mirror. All rays converge at the focal point.

tion. For better focusing, it is necessary to replace the spherical mirror by a paraboloidal mirror (Figure 38.5).

Figure 38.6 shows a **convex** spherical mirror. Parallel rays incident on this mirror diverge upon reflection. If we extrapolate the divergent rays to the far side of the mirror, they all seem to come from a single point, the focal point of the convex mirror. An argument similar to that given above establishes that the focal length is again one-half of the radius of the spherical surface,

$$f = -\tfrac{1}{2}R \qquad (3)$$

A negative sign has been inserted in Eq. (3) to indicate that the focal point is on the far side of the mirror.

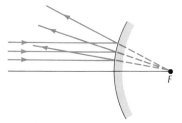

Fig. 38.6 Reflection of parallel rays by a convex spherical mirror.

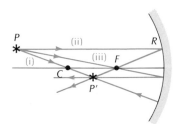

Fig. 38.7 A point source of light P in front of a concave mirror. The rays (i), (ii), and (iii) intersect at the image P'

Both concave and convex mirrors will form images of objects placed in front of them. Figure 38.7 shows a point source of light in front of a concave mirror. To find the position of the image, we must trace some of the rays of light. The three rays that are easiest to trace are (i) the ray PC through the center, (ii) the ray PR parallel to the axis, and (iii) the ray PF through the focal point. The first of these rays strikes the mirror perpendicularly and is therefore reflected on itself; the second ray passes through the focus after being reflected; and the third ray emerges parallel to the axis after being reflected. These three rays are called **principal rays**. All these rays, and any other rays originating at P, come together at P'. This point is the image of the point source. Note that to locate the image, two out of the three rays mentioned above are already sufficient — the third is redundant but serves as a useful check.

Principal rays

If the source of light is an extended object, then we must find the image of each of its points. For instance, a luminous object in the shape of an arrow has an image as shown in Figure 38.8. We can easily verify this by drawing the principal rays that emerge from diverse points, say, the tip of the arrow, the midpoint of the arrow, the tail of the arrow, and so on. The image of an object placed in front of a convex mirror can be found by a similar construction with principal rays, but they must be extrapolated beyond the mirror (Figure 38.9).

The above ray-tracing technique based on the principal rays is a graphical method for finding the image of the source. A large, carefully drawn diagram can provide a reasonably accurate graphical solution; but even a small sketch is helpful in obtaining a rough preview of what to expect from an exact algebraic calculation.

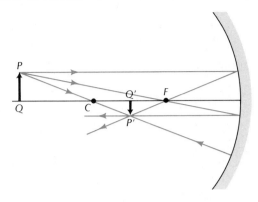

Fig. 38.8 An object PQ in the shape of an arrow and its image $P'Q'$ formed by a concave mirror.

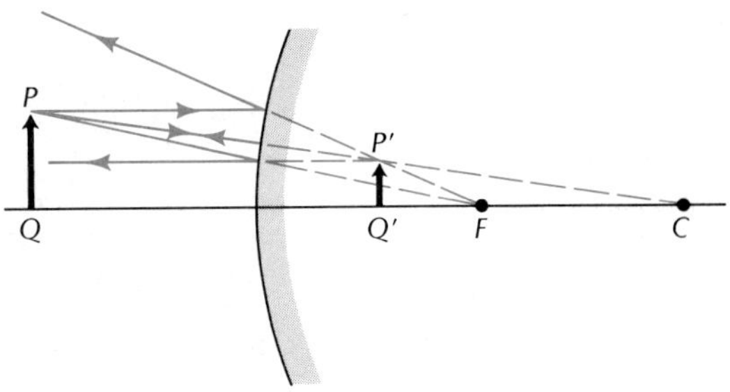

Fig. 38.9 An object PQ and its image $P'Q'$ formed by a convex mirror.

The position of the image can be calculated algebraically by means of the **mirror equation**

$$\frac{1}{s} + \frac{1}{s'} = \frac{1}{f} \qquad (4) \qquad \textit{Mirror equation}$$

Here s is the distance from the object to the mirror and s' is the distance from the image to the mirror. The distance s or s' is positive if the object or image is in front of the mirror; the distance s or s' is negative if the object or image is behind the mirror.[1] As we already indicated above, f is positive for a concave mirror, negative for a convex mirror.

For a derivation of Eq. (4), we make use of Figure 38.10, which shows an object PQ, its image $P'Q'$, and two rays. The ray PCP' passes

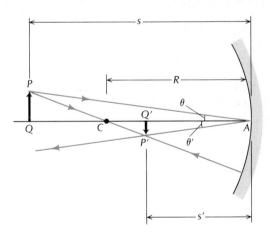

Fig. 38.10 The angles θ and θ' are equal; hence the right triangles PQA and $P'Q'A$ are similar.

through the center of the spherical surface and is reflected on itself; the ray PAP' strikes the center of the mirror and is reflected symmetrically with respect to the axial line, so the angles θ and θ' are equal. The triangles PQA and $P'Q'A$ are similar; hence

---

[1] The object can be behind the mirror if what serves as object for the mirror is actually an image produced by another mirror or by a lens.

$$\frac{PQ}{P'Q'} = \frac{s}{s'} \tag{5}$$

The triangles PQC and $P'Q'C$ are also similar; hence

$$\frac{PQ}{P'Q'} = \frac{QC}{Q'C} \tag{6}$$

or, since $QC = s - R$ and $Q'C = R - s'$,

$$\frac{PQ}{P'Q'} = \frac{s - R}{R - s'} \tag{7}$$

Combining Eqs. (5) and (7), we obtain

$$\frac{s}{s'} = \frac{s - R}{R - s'} \tag{8}$$

By a bit of algebraic manipulation, we can rearrange this equation to read

$$\frac{1}{s} + \frac{1}{s'} = \frac{2}{R} \tag{9}$$

Obviously, Eq. (9) is equivalent to Eq. (4).

EXAMPLE 1. A candle is placed 41 cm in front of a concave spherical mirror of radius 60 cm. Where is the image?

SOLUTION: With $s = 41$ cm and $f = 30$ cm, Eq. (4) gives

$$\frac{1}{41 \text{ cm}} + \frac{1}{s'} = \frac{1}{30 \text{ cm}} \tag{10}$$

or $s' = 112$ cm. The positive value of s' indicates that the image is on the same side of the mirror as the object.

Real image

The image in the preceding example is a **real image.** The light rays not only seem to come from this image, but they actually do. As Figure 38.11 shows, the light rays pass through the image and diverge from it, just as they diverge from the object. The real image is in front of the mirror. Visually, it gives the impression of a ghostly replica of the object floating in midair.

38.2 Thin Lenses

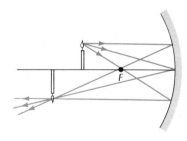

Fig. 38.11 The real image lies in the space in front of the mirror.

A lens made of a refracting material with two spherical surfaces will focus a parallel beam of light to a point (Figure 38.12). For a thin lens, the focal length is given by the **lens-maker's formula**

Lens-maker's formula

$$\boxed{\frac{1}{f} = (n - 1)\left(\frac{1}{R_1} + \frac{1}{R_2}\right)} \tag{11}$$

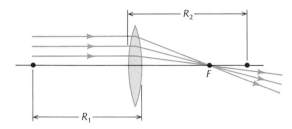

Fig. 38.12 Refraction of rays by a convex lens. The focal length is positive.

where n is the index of refraction of the material of the lens, and R_1, R_2 are the radii of the two spherical surfaces of the lens. This equation is based on the assumption that the lens is thin (its thickness is small compared with R_1, R_2), and that the incident rays are near the axial line. A ray that enters the lens at some appreciable distance from the axis will miss the focus; this is a form of aberration analogous to spherical aberration of a mirror.

Equation (11) can be derived by tracing rays through the lens, taking into account their refraction at the two curved surfaces. For the sake of simplicity, let us begin by considering a lens with one spherical surface and one flat surface (a plano-convex lens; see Figure 38.13). After deriving the focal length of this kind of lens, we can obtain the focal lens of a lens with two spherical surfaces by regarding such a lens as two adjacent plano-convex lenses, placed back to back. Figure 38.13 shows an incident ray parallel to the axial line of the lens. They ray crosses the flat surface on the left of the lens without refraction and reaches the curved surface on the right at an angle of incidence θ'. From the figure we see that $\tan \theta' = y/R_1$; since the angle θ' is small, the tangent approximately equals the angle, and

$$\theta' = y/R_1 \tag{12}$$

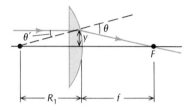

Fig. 38.13 Refraction of a ray by a plano-convex lens. The spherical surface of the lens has a radius R_1. The transverse distance y is assumed to be small compared with R_1, and the angles θ and θ' are also small.

According to the law of refraction, $n \sin \theta' = \sin \theta$, or approximately

$$n \, \theta' = \theta \tag{13}$$

From these equations we find that the downward angle of deflection of the ray is

$$\theta - \theta' = (n-1)\,\theta' = (n-1)\frac{y}{R_1} \tag{14}$$

The focal distance, or the distance at which the ray intersects the axial line, is then approximately

$$f = \frac{y}{\theta - \theta'} = \frac{R_1}{n-1} \tag{15}$$

To compare this with the lens-maker's formula (11), we must take into account that $R_2 = \infty$ for a plano-convex lens; in this special case, Eq. (11) then coincides with Eq. (15).

As was already mentioned, we can regard a doubly convex lens as two adjacent plano-convex lenses. By tracing rays through such a system of two lenses, we can show that the net focal length of the system is given by

$$\frac{1}{f} = \frac{1}{f_1} + \frac{1}{f_2} \qquad (16)$$

(This result is valid for thin lenses; we will leave the proof of this result to Problem 17.) If the focal length of one lens is $f_1 = R_1/(n-1)$ and the focal length of the other lens is $f_2 = R_2/(n-1)$, then the net focal length given by Eq. (16) is

$$\frac{1}{f} = (n-1)\left(\frac{1}{R_1} + \frac{1}{R_2}\right) \qquad (17)$$

which is the general lens-maker's formula.

Note that the focusing of parallel rays by a lens depends on the fact that the rays at some distance from the axial line strike the surface of the lens with larger angles of incidence than rays near the axial line. Thus, the far rays are refracted through a larger angle, that is, they are bent more sharply toward the axial line. This is, of course, exactly what is required to make these distant rays cross the axial line at the same point (the focus) as the near rays.

Equation (11) may also be applied to a concave lens (Figure 38.14). In this case, the radii R_1, R_2 must be reckoned as *negative,* and the focal distance f is then also negative. The meaning of a negative value of f is the same as in the case of mirrors: parallel rays incident on the lens diverge when they emerge from the lens, and the focal point is the point at which the extrapolated rays appear to intersect. Furthermore, Eq. (11) can be applied to a concave–convex lens (Figure 38.15). Whether such a lens produces net convergence or divergence depends on whether the positive radius (convex) or the negative radius (concave) is smaller; for instance, the lens of Figure 38.15 produces convergence.

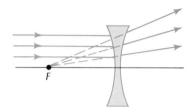

Fig. 38.14 Refraction of rays by a concave lens. The focal length is negative.

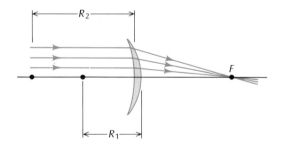

Fig. 38.15 Refraction of rays by a concave–convex lens. The convex surface (right surface) has a smaller radius than the concave surface (left surface). In Eq. (11), the radius of the former surface is reckoned positive ($R_1 > 0$), and the radius of the latter surface is negative ($R_2 < 0$). The sum $1/R_1 + 1/R_2$ is positive.

Note that a lens has *two* focal points at equal distances right and left of the lens. The point on the right of a converging lens is the focus for a parallel beam coming from the left and, conversely, the point on the left is the focus for a parallel beam coming from the right.

To find the image of an object placed near the lens, we can use a ray-tracing technique similar to that used for mirrors. Figure 38.16 shows three rays that are easy to trace: (i) the ray PQP' that starts parallel to the axis and ultimately passes through the focus F, (ii) the ray PCP' that passes undeflected through the center of the lens, and (iii) the ray $PQ'P'$ that passes through the focus F' and emerges parallel to the axis. These are the **principal rays** of the lens. All these rays intersect at the point P', the image point. As in the case of mirrors, two of the above three rays are already sufficient to locate the image. And, of

Principal rays

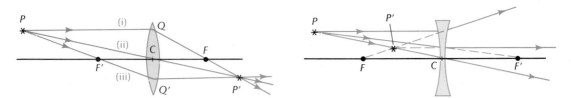

Fig. 38.16 (left) A point source of light P in front of a convex lens. The rays (i), (ii), and (iii) intersect at the image P'.

Fig. 38.17 (right) A point source of light P in front of a concave lens, and the image P'.

course, the same ray-tracing technique can be applied to concave lenses (see Figure 38.17).

The equation to be used for the algebraic calculation of the image distances is the same as Eq. (4),

$$\boxed{\frac{1}{s} + \frac{1}{s'} = \frac{1}{f}} \tag{18}$$

Lens equation

but the sign conventions are slightly different. The object distance s is positive if the object is on the near side of the lens and negative if it is on the far side; the image distance s' is positive if the image is on the far side of the lens and negative if it is on the near side. In this context, the *near* side is the side from which the light rays are incident on the lens, and the *far* side is the other side. The sign of f is positive for a convex lens; negative, for a concave lens.

Although the derivation of the lens equation (18) can be based on a geometric argument similar to that used for the mirror formula (4), we can bypass this labor by a trick. We begin by noting that a concave mirror is equivalent to one-half of a convex lens placed directly in front of a flat mirror. If the concave mirror and the (entire) convex lens have the same value of f, then the two arrangements shown in Figures 38.18a and b have exactly the same optical properties — in both cases the image distances are the same. If we now remove the flat mir-

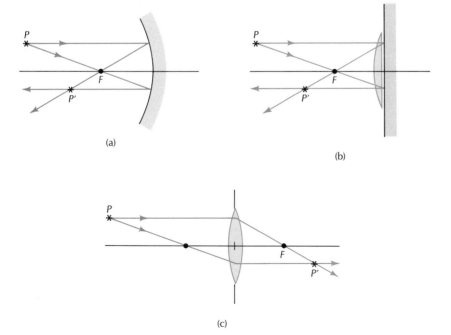

Fig. 38.18 (a) A point source of light and its image formed by a concave mirror. (b) A similar image is formed by one-half of a convex lens placed in front of a flat mirror. Note that each ray has to pass through the one-half lens twice: once before reflection by the mirror, once after. The deflection suffered by a ray in two passages through one-half of a lens is the same as that in a single passage through the entire lens. (c) A similar image is also formed by the entire lens, but the image is now on the other side of the lens.

ror in Figure 38.18a and replace the one-half lens by the entire lens, the image distance will remain the same, but the image will form on the opposite side of the lens, that is, the sign of the image distance will be reversed. Consequently, the same equation (4) must apply to the concave mirror and the convex lens; the only difference is that the sign of the image distance is reversed. This reversal of sign has already been taken into account in our description of the sign conventions associated with Eqs. (4) and (18) — for the mirror, s' is taken as *positive* if the image is on the near side of the mirror, whereas for the lens, s' is taken as *negative* if it is on the near side of the lens. There is, of course, a similar correspondence between a convex mirror and a concave lens.

EXAMPLE 2. A convex lens of focal length 25 cm is placed at a distance of 10 cm from a printed page. What is the image distance? How much larger is the image of the page than the page?

SOLUTION: With $s = 10$ cm and $f = 25$ cm, Eq. (18) gives

$$\frac{1}{10 \text{ cm}} + \frac{1}{s'} = \frac{1}{25 \text{ cm}}$$

which yields $s' = -16.7$ cm. The negative sign indicates that the image is on the near side of the lens (Figure 38.19). The image is virtual.

Since the triangles $P'Q'C$ and PQC are similar, the sizes of image and object are in the ratio

$$\frac{P'Q'}{PQ} = \frac{Q'C}{QC} = \frac{-s'}{s} \tag{19}$$

$$= \frac{16.7 \text{ cm}}{10 \text{ cm}} = 1.67$$

that is, the image is larger than the object by a factor of 1.67. This is the principle involved in the magnifying glass.

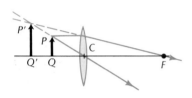

Fig. 38.19 An object PQ and its image $P'Q'$.

Equation (19) is a general result for the magnification produced by a lens or by a mirror: The size of the image differs from the size of the object by a factor

Magnification

$$\boxed{M = -\frac{s'}{s}} \tag{20}$$

The sign of M tells us something about the character of the image. If M is positive, then the image is virtual and upright, as in Example 2. If M is negative, then the image is real and inverted.

EXAMPLE 3. A convex lens of focal length 20 cm and a concave lens of focal length 50 cm are separated by a distance of 10 cm. A light bulb is placed 40 cm to the left of the convex lens (see Figure 38.20a). Where does this system of two lenses form an image of the light bulb?

SOLUTION: We begin by inquiring where the convex lens by itself would form an image of the light bulb. According to the lens equation, with $f = 20$ cm and $s = 40$ cm,

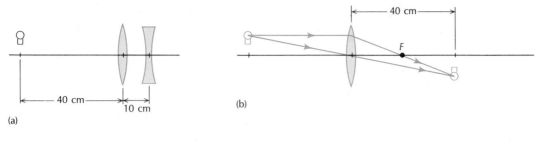

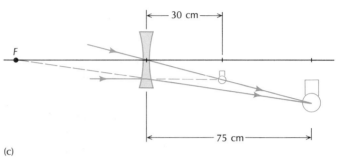

Fig. 38.20 (a) A light bulb placed in front of a system consisting of two lenses, one convex and one concave. (b) The convex lens by itself would produce a first image at the position shown. (c) If we pretend that this first image acts as "object" for the second lens, it will form a final image on the far right.

$$\frac{1}{40 \text{ cm}} + \frac{1}{s'} = \frac{1}{20 \text{ cm}} \tag{21}$$

From this, we find $s' = 40$ cm (see Figure 38.20b).

As a next step, we pretend that this first image formed by the convex lens acts as "object" for the concave lens. This procedure would be quite obvious if the first image were to the left of the concave lens; it is less obvious when, as in the present case, the image is to the right, and the rays arriving at the concave lens are not coming *from* the image, but going *toward* it. The justification is that, as Figure 38.20c shows, the construction of the final image from the "object" can be performed entirely with rays entering the concave lens from the *left*.

The "object" is at a distance $s = -30$ cm from the concave lens (this distance must be reckoned as negative since it is on the far side of the lens). According to the lens equation applied to the concave lens,

$$\frac{1}{-30 \text{ cm}} + \frac{1}{s'} = -\frac{1}{50 \text{ cm}} \tag{22}$$

from which we obtain $s' = 75$ cm. Thus, the final image of the light bulb is 75 cm to the right of the concave lens (see Figure 38.20b). The image is real and inverted.

COMMENTS AND SUGGESTIONS: The step-by-step procedure for tracing rays through the two lenses in this example can be generalized to any number of lenses (and mirrors) arranged in tandem. To find where such a system of lenses forms the final image of an object, begin by calculating where the first lens by itself forms an image. Then use this first image as "object" for the second lens, and find the corresponding second image. Then use this second image as "object" for the next lens, and so on.

In this calculation, as in all calculations with lenses and mirrors, it is essential to keep careful track of the signs of the object and image distances and the signs of the focal lengths. The sign conventions that go with our lens and mirror equations can be summarized as follows:

Sign conventions

Mirrors:
  f is positive for concave mirror, negative for convex mirror
  s or s' is positive if object or image is in front of mirror, negative if behind

Lenses:
  f is positive for convex lens, negative for concave lens
  s is positive if on near side of lens, negative if on far side
  s' is positive if on far side of lens, negative if on near side

(The *near* side of the lens is the side from which the light is incident on the lens; the *far* side is the other side.)

Remember that the distance between one lens and the next must be taken into account whenever the image produced by one lens is used as "object" for the next. To prevent confusion in the manipulation of all these distances and their signs, draw all the intermediate images on a diagram, such as Figure 38.20b. It is also often helpful to include the principal rays in the diagram, and thereby construct a graphical solution of the problem. This serves as a useful check on the algebraic solution and makes it easier to spot mistakes. However, when there are several lenses in the system, the graphical solution tends to get messy.

If the magnification produced by a system of lenses is needed, it can be calculated by taking the product of the magnifications produced by all the individual lenses.

38.3 The Photographic Camera and the Eye

The simple lenses described in the preceding section suffer from diverse kinds of aberrations. Parallel rays that strike the lens at an appreciable distance from the axial line will miss the focal point; this is **spherical** aberration, similar to the spherical aberration of a mirror. Rays of different colors will be refracted differently by the glass of the lens, and they will therefore be focused differently; this is **chromatic aberration.** Furthermore, the images of objects of large size will display complicated three-dimensional distortions. All these troublesome aberrations can be partially eliminated by combining several simple lenses of different shapes made of different kinds of glass into a compound lens. For instance, Figure 38.21 shows the Zeiss "Tessar" lens, one of the best-known photographic lenses. In such a compound lens, the individual lenses are carefully designed so that their aberrations tend to cancel mutually.

Spherical and chromatic aberrations

Fig. 38.21 Zeiss "Tessar" lens. The outer lenses (color) are made of crown glass, and the inner lenses (gray) of flint glass.

Most optical instruments — cameras, magnifiers, microscopes, and telescopes — employ compound lenses. However, in the following discussion of optical instruments, we will schematically represent the compound lenses by single thin lenses of appropriate focal lengths.

Camera

The lens of the **camera** forms a real image of the object on the photographic film and thereby imprints this image on the film (Figure 38.22). The distance between the lens and the film is adjustable, so that the image can always be made to fall on the film, regardless of the object distance. A shutter controls the exposure time during which light is admitted to the camera. For a given exposure time, the amount of light entering the camera is proportional to the area of the lens. Thus, a large lens permits photography with dim light. The size of a camera lens is commonly labeled by the **f number,** which is defined as the ratio of the focal length of the lens to its diameter. For instance, a lens of focal length 55 mm and diameter 32 mm has an f number of 55:32, or 1.7. A lens of small f number is said to be "fast" because it collects sufficient light for a photograph within a short exposure time. Good cameras have an adjustable iris diaphragm that can be used to block part of the area of the lens and thereby alter the effective f number. If the diaphragm is closed down, so only a small central portion of the lens remains unblocked, the camera will require a long exposure time, but it will have a large depth of field — it simultaneously forms sharp images for objects spanning a large range of object distances. This is so because the rays emitted by any object, near or far, then enter the camera within a narrow cone, and they therefore intersect at the image within a narrow cone; thus, these rays will be close together on the photographic film even if the position of the image does not fall exactly on the film.

A **pinhole camera** can be regarded as an ordinary camera with the iris diaphragm closed down to a point. The lens can then be discarded since the rays pass through its center, where they suffer no deflection (Figure 38.23). To obtain a sharp image, we must use an extremely small pinhole; hence, such a camera requires very bright light or a very long exposure time. Figure 38.24 is a picture taken with a pinhole camera.

f number

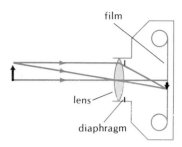

Fig. 38.22 A photographic camera.

Pinhole camera

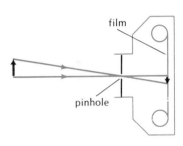

Fig. 38.23 A pinhole camera.

Fig. 38.24 Photograph taken with a pinhole camera. (Courtesy C. C. Jones, Union College.)

In principle, the **eye** is similar to a camera. The lens of the eye forms a real image on the retina, a delicate membrane packed with light-sensitive cells which send nerve impulses to the brain. Figure 38.25 shows a section through the human eye. The diameter of the eyeball is typically 2.3 cm. The cornea and the aqueous humor act as a lens; they provide most of the refraction for rays entering the eye. The crystalline lens merely provides the fine adjustment of focal length required to make the image of an object fall on the retina, regardless of the ob-

Eye

ject distance. The crystalline lens is flexible; its focal length is adjusted by the ciliary muscles. If the eye is viewing a distant object, the muscles are relaxed and the lens is fairly flat, with a long focal length. If the eye is viewing a nearby object, the muscles are contracted and the lens is more rounded, with a shorter focal length. The shortest attainable focal length determines the shortest distance at which an object can be placed from the eye and still be seen sharply. This shortest distance is called the **near point** of the eye. For a normal young adult, the near point is typically 25 cm. With advancing age, the lens loses its flexibility and the near point recedes; for instance, at an age of 60 years, the near point is typically around 200 cm.

The three most common optical defects of the eye are nearsightedness, farsightedness, and astigmatism. In a **nearsighted** (myopic) eye, the focal length is excessively short, even when the ciliary muscles are completely relaxed. Thus, parallel rays from a distant object come to a focus in front of the retina and fail to form a sharp image on the retina — vision of distant objects is blurred. This condition can be corrected by eyeglasses with divergent lenses. In a **farsighted** (hyperopic) eye, the focal length is excessively long, even when the ciliary muscles are fully contracted (in other words, the near point of the eye is farther away than normal). Hence rays from a nearby object converge toward an image beyond the retina and fail to form a sharp image on the retina. This condition can be corrected by eyeglasses with convergent lenses. In old age, both of these conditions often occur simultaneously, through the loss of flexibility of the crystalline lens and the weakening of the ciliary muscles. The correction then requires bifocal lenses, with a lower convergent portion for near vision, and an upper divergent portion for far vision. **Astigmatism** is an inability to focus simultaneously light rays arriving in different planes, for instance, light rays arriving in the vertical plane and light rays arriving in the horizontal plane. This is caused by a slight horizontal or vertical flattening of the cornea — instead of being curved spherically, the cornea is slightly out of round, with more curvature in one direction than in the other. Although astigmatism is quite common, it is often so slight as to be unnoticeable. Correction of this condition requires a lens with a cylindrical surface, which focuses rays of light in the, say, vertical plane, but does not deflect rays of light in the horizontal plane.

Nearsightedness, farsightedness, and astigmatism

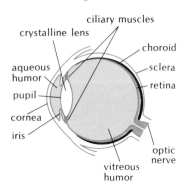

Fig. 38.25 The human eye (natural size). The space between the cornea and the crystalline lens is filled with a transparent jelly (the aqueous humor). The main body of the eye is also filled with a transparent jelly (the vitreous humor). The indices of refraction of the humors are 1.34, nearly the same as for water. The index of refraction of the crystalline lens is 1.44. The sclera is the thick white outer casing of the eye. The choroid is a pigmented black membrane that absorbs stray light, like the black paint in cameras.

38.4 The Magnifier and the Microscope

In order to see fine detail with the naked eye, we must bring the object very close to the eye, so that the angular size of the object is large and, correspondingly, the image on the retina is large (Figure 38.26). This means we want to bring the object to the near point, at a typical distance of 25 cm for the eye of a young adult. To see finer detail, we need a **magnifier.** This consists of a strongly convergent lens placed adjacent to the eye (Figure 38.27).[2] Such a lens permits us to bring the object closer to the eye and thereby increase the size of the image on the retina.

Magnifier

---

[2] Note that such a magnifier is *not* the same thing as a magnifying glass. In common use, the magnifying glass is placed at an appreciable distance from the eye, near the object to be magnified, because this maximizes the magnification. A magnifying glass can be regarded as a magnifier (in the technical sense of this word) only if it is placed next to the eye.

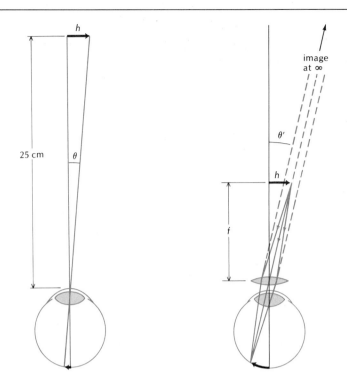

Fig. 38.26 (left) The angular size of the object determines the size of the image on the retina. Here the object has been placed at a distance of 25 cm from the eye.

Fig. 38.27 (right) The magnifier is adjacent to the eye. The object has been placed at a distance slightly shorter than the focal distance, so the eye sees the image at infinity. Note that, compared with that of Figure 38.26, the image on the retina is enlarged.

The angular magnification of the magnifier is defined as the ratio of the angular size of the image at infinity produced by the magnifier (as in Figure 38.27) to the angular size of the object seen by the naked eye at the standard distance of 25 cm (as in Figure 38.26). Since the angles in Figures 38.26 and 38.27 are small, the angles θ and θ' are approximately equal to the tangents:

$$\theta \cong \tan \theta \cong h/25 \text{ cm} \tag{23}$$

and

$$\theta' \cong \tan \theta' = h/f \tag{24}$$

where h is the size of the object. Taking the ratio of these angles, we find

$$\boxed{[\text{angular magnification}] = \theta'/\theta = \frac{25 \text{ cm}}{f}} \tag{25}$$

Angular magnification of magnifier

This tells us the magnification relative to the (typical) naked eye; that is, it tells us how much better the magnifier is than the naked eye. For example, a magnifier with $f = 5$ cm has an angular magnification of 25 cm/5 cm = 5. Note that this result is valid only under the assumptions that the magnifier is placed adjacent to the eye and that the object is placed near the focus of the magnifier, so the image is at infinity. The second of these assumptions is not crucial — if the object is placed somewhat closer than the focus, the magnification will be changed only slightly. But the first assumption is crucial — if the magnifier is placed at some appreciable distance from the eye, then the magnification will be quite different!

Microscope

Angular magnification of microscope

The **microscope** consists of two lenses: the objective and the ocular, or eyepiece. Both of these lenses have very short focal lengths. The objective is placed near the object, and it forms a real, magnified image of the object. This image serves as object for the ocular, which acts as a magnifier and forms a virtual image at infinity (Figure 38.28). Thus, both the objective and the ocular contribute to the magnification of the microscope. The net angular magnification of the microscope is the angular magnification of the ocular multiplied by the magnification of the objective. The angular magnification of the ocular is given by Eq. (25); and the magnification of the objective is given by Eq. (20), where s and s' are, respectively, the object and image distances for the objective (these distances are shown in Figure 38.28). Hence the net angular magnification of the microscope is

$$[\text{angular magnification}] = \frac{25 \text{ cm}}{f_{oc}} \times \frac{s'}{s} \qquad (26)$$

(Here we have discarded a negative sign.) Note that, as in the case of the magnifier, this tells us the magnification relative to the (typical) naked eye. Good microscopes operate at magnifications of up to 1400. Although higher magnifications can be achieved, this serves little purpose because the diffraction of the light waves at the objective limits the detail that can be resolved. To overcome this limitation, we need to use waves of shorter wavelength than light, such as the electron waves used in electron microscopes.

38.5 The Telescope

A simple astronomical **telescope** consists of an objective of very long focal length and an ocular of short focal length. These two lenses are separated by a distance (nearly) equal to the sum of their individual focal lengths, so their focal points coincide. The objective forms a real image of a distant object. This image serves as object for the ocular, which forms a magnified virtual image at infinity (Figure 38.29).

To find the angular magnification produced by this telescope, we begin by noting that the lens equation with $s = \infty$ applied to the objective gives

$$\frac{1}{\infty} + \frac{1}{s'} = \frac{1}{f_{ob}}$$

that is,

$$s' = f_{ob} \qquad (27)$$

This means that the image is at the focal distance, and verifies that the location of the image (FP') is correctly shown in Figure 38.29. We can therefore use the geometric relationships contained in this figure. The angular magnification is the ratio of the angles θ' and θ that represent, respectively, the angular sizes of the object and the final image viewed by the eye. Since both angles are small,

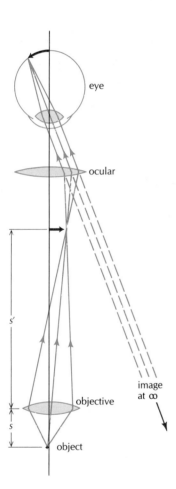

Fig. 38.28 Arrangement of lenses in a microscope. The object to be magnified is below the objective and the eye is just above the ocular. The ocular forms a virtual image at infinity, and the lens of the eye focuses on the retina the parallel rays emerging from the ocular.

$$[\text{angular magnification}] = \frac{\theta'}{\theta}$$

$$\cong \frac{\tan \theta'}{\tan \theta} = \frac{FP'/FB}{FP'/FA} = \frac{FA}{FB} \quad (28)$$

and thus the angular magnification is the ratio of the focal length of the objective to that of the ocular,

$$\boxed{[\text{angular magnification}] = \frac{f_{\text{ob}}}{f_{\text{oc}}}} \quad (29)$$

Angular magnification of telescope

For example, an astronomical telescope with $f_{\text{ob}} = 120$ cm and $f_{\text{oc}} = 2.5$ cm has an angular magnification of 120 cm/2.5 cm = 48.

Many astronomical telescopes are reflecting telescopes in which a concave mirror plays the role of the objective. The mirror forms a real image that serves as object for the ocular. Of course, the ocular must be placed in front of the mirror (and therefore blocks out some of the light). Figure 38.30 shows this arrangement. Note that Figure 38.30 is essentially the same as Figure 38.29 folded over at the position of the

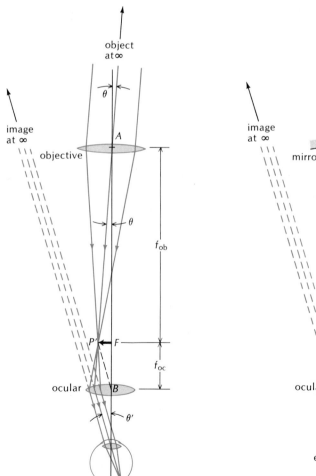
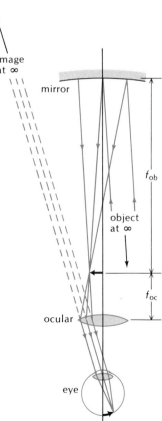

Fig. 38.29 (left) An astronomical telescope. The object is at a large distance above. The observer's eye is below the ocular. The ocular forms a virtual image at infinity, and the lens of the eye focuses on the retina the parallel rays emerging from the ocular.

Fig. 38.30 (right) A reflecting telescope. The object has been placed below, to facilitate comparison with Figure 38.29.

Fig. 38.31 The 200-in. telescope on Palomar Mountain. The mirror can be seen under the observation cage.

objective lens; hence, the geometry of the light rays in reflecting and refracting astronomical telescopes is essentially the same.

Lenses or mirrors of large diameters are required for astronomical telescopes, because the ability of the telescope to detect distant objects in the sky depends on the amount of light it collects. Lenses tend to absorb some of the light passing through them, and lenses of large diameter are harder to manufacture than mirrors of good quality, free of aberrations. Hence, the largest astronomical telescopes all use mirrors. For instance, the telescope on Palomar Mountain, California, uses a mirror of diameter 5.1 m (200 in.) and of a focal length 16.8 m; the mirror is parabolic for the sake of better imaging (Figure 38.31).

The mirror of this telescope has a mass of 30 tons and was manufactured by the laborious grinding of an even larger slab of glass. Modern improvements in the design and manufacture of large mirrors and their mechanical supports have resulted in much lighter mirrors. The mirrors are now manufactured by a spin-casting technique. Molten glass is placed in a shallow bucket, and this is rotated around a vertical axis at constant speed; under these conditions, the surface of the liquid spontaneously adopts a parabolic equilibrium configuration. If the glass is allowed to cool while it spins, it solidifies into a perfect parabolic mirror. Some mirrors of excellent quality of a diameter as large as 1.8 m have already been cast by this technique, and larger mirrors are in the making.

A trend in modern telescope design is to use several mirrors in parallel, all aimed at the same spot. For instance, Figure 38.32 shows the Multiple-Mirror Telescope (MMT) on Mount Hopkins, Arizona. The light is collected by six 1.8-m mirrors mounted in a single framework, and it is brought together to a common focus by a system of small aux-

Fig. 38.32 The Multiple-Mirror Telescope on Mount Hopkins.

iliary mirrors. Several larger telescopes with multiple mirrors are under construction, among them, the Very Large Telescope (VLT) at the European Southern Observatory in Chile and the Keck Telescope on Mauna Kea, Hawaii. The first of these will use four 8-m mirrors in separate mountings in a linear array, and the second will use a 10-m mirror consisting of 36 fitting hexagonal pieces, each of which is individually steerable.

SUMMARY

Focal length of spherical mirror: $f = \pm\frac{1}{2}R$ (f is positive for concave mirror, negative for convex)

Mirror equation: $\dfrac{1}{s} + \dfrac{1}{s'} = \dfrac{1}{f}$ (s or s' is positive if object or image is in front of mirror, negative if behind)

Lens-maker's formula: $\dfrac{1}{f} = (n-1)\left(\dfrac{1}{R_1} + \dfrac{1}{R_2}\right)$ (f is positive for convex lens, negative for concave)

Lens equation: $\dfrac{1}{s} + \dfrac{1}{s'} = \dfrac{1}{f}$ (s is positive if on near side of lens, negative if on far side; s' is positive if on far side, negative if on near side)

Magnification of lens or mirror: $M = -s'/s$

Angular magnification of magnifier: $25 \text{ cm}/f$

Angular magnification of microscope: $(25 \text{ cm}/f_{oc}) \times (s'/s)$

Angular magnification of telescope: f_{ob}/f_{oc}

QUESTIONS

1. What is the minimum size of a mirror hanging on a wall such that you can see your entire body when standing in front of it?

2. Artists are notorious for making mistakes when drawing or painting mirror images. What is wrong with the position and orientation of the mirror images shown in the cartoon in Figure 38.33?

Fig. 38.33

3. Two parallel mirrors are face to face. Describe what you see if you stand between these mirrors.

4. Figure 38.34 shows spots of sunlight on a wall in the shade of a tree. The spots were made by sunlight that has passed through very small gaps between the leaves of the tree. Explain why all the spots are round and of nearly the same size, even though the gaps are of irregular shape and size. (Hint: The Sun is round.)

5. At amusement parks you find mirrors that make you look very short and fat or very tall and thin. What kinds of mirrors achieve these effects?

6. Store owners often install convex mirrors at strategic locations in their stores to supervise the customers. What is the advantage of a convex mirror over a flat mirror?

7. You look toward a lens or a mirror, and you see the image of an object. How can you tell whether this image is real or virtual?

8. Hand mirrors are sometimes concave, but never convex. Why?

9. If you place a book in front of a concave mirror, will it show you mirror writing? Does the answer depend on the distance of the book from the mirror?

10. You place an object in front of a concave mirror, at a distance smaller than the focal length. Is the image real or virtual? Erect or inverted? Magnified or reduced? What if the distance is greater than the focal length? What if the mirror is convex?

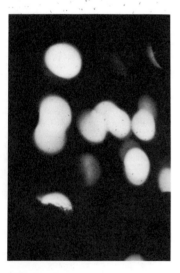

Fig. 38.34

11. How could you make a lens that focuses sound waves?

12. A convex lens is made of glass of index of refraction 1.2. If you immerse this lens in water, will it produce convergence or divergence of incident parallel rays?

13. If you place a small light bulb at the focus of a convex lens and look toward the lens from the other side, what will you see?

14. Consider Figure 38.16. How do we know that the ray $PQ'P'$ emerges parallel to the axis of the lens?

15. Are the distances to the two focal points of a thick lens necessarily the same?

16. Suppose you place a magnifying glass against a flat mirror and look into the glass. What do you see if your face is very near the glass? If it is not very near?

17. Consider a beam of parallel rays incident on a convex lens at a small angle with the axis of the lens (Figure 38.35). Show that these parallel rays will be focused at a point in a plane that is perpendicular to the axis and passes through the focal point. This plane is called the **focal plane**.

Focal plane

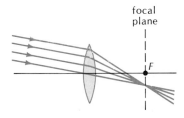

Fig. 38.35 The dashed line indicates the focal plane, seen edge on.

18. You place an object in front of a convex lens at a distance smaller than the focal length. Is the image real or virtual? Erect or inverted? Magnified or reduced? What if the distance is greater than the focal length? What if the lens is concave?

19. Lenses, like mirrors, suffer from aberration. Consider rays incident on a convex lens, parallel to the axis. If the point of incidence is far from the axis, would you expect the ray to pass in front of the focus or behind?

20. Figure 38.36 shows a large **Fresnel lens** used in the lantern of a lighthouse. The lens consists of annular segments, each with a curved surface similar to the curved surface of an ordinary lens. Why is this arrangement better than a single curved surface?

21. The telescopes built by Galileo Galilei consisted of a convex objective lens and a concave ocular lens. The lenses are arranged so that the focal point of the objective coincides with the focal point of the ocular (Figure 38.37). Explain how this telescope produces an angular magnification. Show that the eye sees an erect image. (Hint: The concave lens placed *before* the point F in Figure 38.37 has the same effect as the convex lens placed *beyond* the point F in Figure 38.29 — the lens produces an image at infinity. Is this image erect or inverted?)

Fig. 38.36 Fresnel lens of the lighthouse at Point Reyes, California.

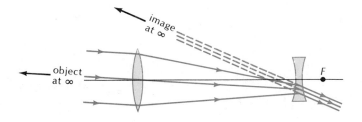

Fig. 38.37 Galilean telescope.

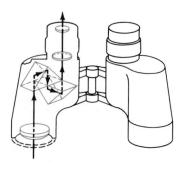

Fig. 38.38 Arrangement of prisms in a binocular.

22. Binoculars use prisms to reflect the light back and forth (Figure 38.38). What is the purpose of these prisms, and what is their advantage over mirrors?

PROBLEMS

Section 38.1

1. In the head lamp of an automobile, a light bulb placed in front of a concave spherical mirror generates a parallel beam of light. If the distance between the light bulb and the mirror is 4 cm, what must the radius of the mirror be?

2. At what distance from a concave mirror of radius R must you place an object if the image is to be at the same position as the object?

3. A concave mirror has a radius of curvature R. If you want to form a real image, within what range of distances from the mirror must you place the object? If you want to form a virtual image, within what range of distances must you place the object?

4. A woman's hand mirror is to show a (virtual) image of her face magnified 1.5 times when held at a distance of 20 cm from the face. What must be the radius of curvature of a spherical mirror that will serve the purpose? Must it be concave or convex?

5. The surface of a highly polished doorknob of brass has a radius of curvature of 4.5 cm. If you hold this doorknob 15 cm away from your face, where is the image that you see? By what factor does the size of the image differ from the size of your face?

6. A candle is placed at a distance of 15 cm in front of a concave mirror of radius 40 cm. Where is the image of the candle? Draw a diagram showing the candle, the mirror, and the image.

*7. Mariners in distress can use a mirror reflecting sunlight to attract the attention of ships or aircraft. This works over long distances, provided the mirror is aimed accurately. One simple aiming method relies on a mirror silvered on both sides, with a small hole drilled through its center (see Figure 38.39). A handbook gives the following instructions for aiming the reflected sunlight at a target:

> Hold the mirror in one hand so that it roughly faces a point midway between the Sun and the target, and sight the target through the hole. Place your other hand below the mirror so that the beam of sunlight passing through the hole falls on your palm and makes a bright spot. In the back side of the mirror, watch the bright spot of sunlight on your palm. While keeping the target in sight, tilt the mirror so that the bright spot disappears in the hole.

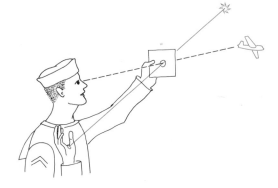

Fig. 38.39

Show that the sunlight reflected by the front side of the mirror will then strike the target. (Hint: Draw a diagram showing the rays of sunlight to your eye and to the target when the mirror is aimed exactly midway between the Sun and the target; draw another diagram showing these rays when the mirror is slightly misaimed.)

*8. A concave mirror of radius 30 cm faces a second concave mirror, of radius 24 cm. The distance between the mirrors is 80 cm, and their axes coincide. A light bulb is suspended between the mirrors, at a distance of 20 cm from the first mirror.
  (a) Where does the first mirror form an image of the light bulb?
  (b) Where does the second mirror form an image of this image?

*9. A concave mirror of radius 60 cm faces a convex mirror of the same radius. The distance between the mirrors is 50 cm, and their axes coincide. A candle is held between the mirrors, at a distance of 10 cm from the convex mirror. Consider rays of light that reflect first off the concave mirror and then off the convex mirror. Where do these rays form an image?

Section 38.2

10. The crystalline lens of a human eye has two convex surfaces with radii of curvature of 10 mm and 6.0 mm. The index of refraction of its material is 1.45. Treating it as a thin lens, calculate its focal length when removed from the eye and placed in air.

11. A thin lens of flint glass with $n = 1.58$ has one concave surface of radius 15 cm and one flat surface.
  (a) What is the focal length of this lens?
  (b) If you place this lens at a distance of 40 cm from a candle, where will you find the image of the candle?

12. A thin, symmetric, convex lens of crown glass with index of refraction $n = 1.52$ is to have a focal length of 20 cm. What are the correct radii of the spherical surfaces of the lens?

13. A slide projector has a lens of focal length 13 cm. The slide is at a distance of 2.0 m from the screen. What must be the distance from the slide to the lens if a sharp image of the slide is to be seen on the screen?

14. The convex lens of a magnifying glass has a focal length of 20 cm. At what distance from a postage stamp must you hold this lens if the image of the stamp is to be twice as large as the stamp?

15. If you place a convex lens of focal length 18 cm at a distance of 30 cm from a small light bulb, where will you find the image of the light bulb? Is this a real or virtual image? Is it upright or inverted?

16. Show that if a thin lens of index of refraction n is immersed in a fluid (for example, water) of index of refraction n_0, then the lens-maker's formula (11) must be modified as follows:

$$\frac{1}{f} = \left(\frac{n}{n_0} - 1\right)\left(\frac{1}{R_1} + \frac{1}{R_2}\right)$$

(Hint: In the law of refraction, only the ratio of the indices of refraction is relevant.)

*17. Show that if two thin lenses of focal lengths f_1 and f_2 are placed next to one another (in contact), the net focal length f is given by

$$\frac{1}{f} = \frac{1}{f_1} + \frac{1}{f_2}$$

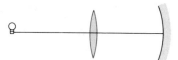
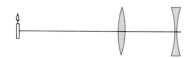

Fig. 38.40 Light bulb, convex lens, and concave mirror.

Fig. 38.41 Candle, convex lens, and concave lens.

*18. A convex lens of focal length 25 cm is at a distance of 60 cm from a concave mirror of focal length 20 cm. A light bulb is 80 cm from the lens (Figure 38.40).
 (a) Where does the lens form an image of the light bulb?
 (b) Where does the mirror form an image of this image?

*19. Two lenses, one concave and one convex, have equal focal lengths of 30 cm. The lenses are separated by a distance of 10 cm. A candle is 20 cm from the convex lens (Figure 38.41).
 (a) Where does the convex lens form an image?
 (b) Where does the concave lens form an image of this image?

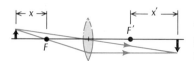

Fig. 38.42

*20 Prove that the lens equation can be put in the form $xx' = f^2$, where x and x' are the object and image distances measured from the focal points (see Figure 38.42). This is called the **Newtonian form** of the lens equation.

*21. To measure the focal length of a thin concave lens, a physicist places this lens adjacent to a thin convex lens of focal length 12.0 cm. She finds that when the joined lenses face the Sun, the rays of sunlight are focused on a spot 22.0 cm behind the lenses. What focal length for the concave lens can she deduce from this?

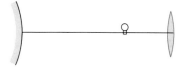

Fig. 38.43 Light bulb, convex mirror, and convex lens.

*22. A light bulb is 15 cm in front of a convex mirror of radius 10 cm. A convex lens of focal length 25 cm is 5 cm beyond the light bulb (Figure 38.43). Where do you see the light bulb if you look through the convex lens at the mirror?

*23. Two convex lenses of focal lengths 40 cm and 60 cm are separated by a distance of 10 cm. Where is the focal point of this system for a parallel beam of light incident from the left? For a beam incident from the right?

*24. Three equal convex lenses of focal length 60 cm each are separated by distances of 12 cm from one to the next. If a small light bulb is placed 20 cm in front of the left lens, where will the final image be? What is the magnification of the final image?

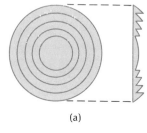

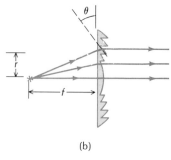

Fig. 38.44 (a) Fresnel lens, frontal view and cross section. (b) A ray passing through one of the annular segments.

**25. Figure 38.44a shows a Fresnel lens consisting of a large number of annular segments each of which is an annular portion of an ordinary lens. Fresnel lenses of diameters more than 1 m are commonly used in the lamps of lighthouses where ordinary lenses without segments would be much too thick and too heavy. Consider one of these annular segments (Figure 38.44b); for the sake of simplicity, assume that the width of the segment is infinitesimal. Derive a formula for the angle of inclination θ of the surface of this segment in terms of the index of refraction n, the distance r of the segment from the axis of the lens, and the focal length of the lens.

**26. A thick slab of glass has a spherical surface of radius R (Figure 38.45). Refraction by such a surface will result in the formation of an image within the

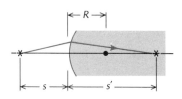

Fig. 38.45

glass if a source is placed outside the glass. Show that, with the object and image distances as in Figure 38.45, the equation relating the object and image distances is

$$\frac{1}{s} + \frac{n}{s'} = \frac{n-1}{R}$$

**27. A large, flattened drop of water sits on a plastic credit card. The drop is 3 mm thick at its center, and its upper surface has a radius of 7 mm. When viewed through this drop, by what factor is the small print on the credit card magnified? (Hint: Use the result stated in Problem 26. For refraction by a spherical curved surface, the magnification is the ratio of the distances of image and object reckoned from the center of curvature, not the ratio of the distances s and s' shown in Figure 38.45. Why?)

**28. A *thick* lens of glass of index of refraction $n = 1.5$ has radii of curvature $R_1 = 30$ cm and $R_2 = 20$ cm on the right and the left, respectively (Figure 38.46). The thickness of the lens (at its center) is 10 cm. Where is the focal point of this lens for a parallel beam of light incident from the left? For a beam incident from the right? (Hint: Use the result stated in Problem 26.)

**29. The "crystal" ball of a fortune-teller has a diameter of 15 cm. It is made of glass of index of refraction 1.60. At what distance from the center will this sphere bring an incident beam of sunlight to a focus?

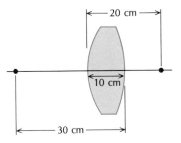

Fig. 38.46

**30. When a spherical paperweight of glass of diameter 6.0 cm sits on the page of a book, it magnifies the print under it by a factor of 1.8. What is the index of refraction of the glass?

**31. A semicylindrical rod of glass of index of refraction 1.55 has a radius of 6.0 cm. A penny coin is placed in contact with the flat face of the rod, near the center. If you look into the rod from the opposite side, you see an elliptically distorted image. What is the ratio of the major and minor axes of the ellipse?

**32. Consider, again, the arrangement of the semicylindrical rod of glass and the penny coin described in the preceding problem. The position at which you perceive the coin when you look into the rod with your binocular vision depends on how the rod is oriented relative to your eyes. At what position do you perceive the coin if the rod is oriented horizontally across your field of view, that is, parallel to an imaginary line from one of your eyes to the other? At what position do you perceive the coin if the rod is oriented vertically in your field of view?

***33. A thick spherical shell of glass of index of refraction 1.6 has an outer radius of 6.0 cm and an inner radius of 3.0 cm. Where does this shell focus a beam of sunlight incident along a diameter?

Section 38.3

34. The lens of a 35-mm camera has a focal length of 55 mm. The distance of the lens from the film is adjustable over a range from 55 mm to 62 mm. Over what range of object distances (measured from the lens) is this camera capable of producing sharp pictures?

35. A miniature Minox camera has a lens of focal length 15 mm. This camera can be focused on an object as close as 20 cm, or as far away as infinity. What must be the distance from the lens to the film if the camera is set for 20 cm? What if the camera is set for infinity?

36. The light meter of a 35-mm camera with a lens of f number 1.7 indicates that the correct exposure time for a photograph is $\frac{1}{250}$ s. If the iris diaphragm is closed down so that the f number becomes 4, what will be the correct exposure time?

*37. Pretend that the cornea and the crystalline lens of the human eye act together as a single thin lens placed at a distance of 2.2 cm from the retina (Fig-

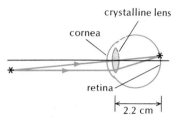

Fig. 38.47 Lens of human eye forms image on retina.

ure 38.47). This lens is deformable; it can change its focal length by changing its shape.
  (a) What must be the focal length if the eye is viewing an object at a very large distance?
  (b) What must be the focal length if the eye is viewing an object at a distance of 25 cm?

*38. In a nearsighted eye the (relaxed) lens has an abnormally short focal length, and consequently the eye fails to form an image of a distant object on the retina. This defect can be corrected with a contact lens. Pretend that both the lens of the eye and the contact lens are thin lenses so that the formula given in Problem 17 applies.
  (a) Suppose that the focal length of the eye is 2.0 cm. What must be the focal length of the contact lens if it is to increase the net focal length to 2.2 cm? Should it be convergent or divergent?
  (b) The radius of curvature of the side of the contact lens next to the eye should be −0.80 cm so as to fit tightly on the cornea. What must be the radius of curvature of the other side? The contact lens is made of plastic with an index of refraction 1.33.

Section 38.4

39. Equation (25) gives the angular magnification for a magnifier if the object is so placed that the image is at infinity. Show that if the object is so placed that the image is at a distance of 25 cm, then the angular magnification is $1 + 25$ cm$/f$.

40. A microscope has an objective of focal length 4.0 mm. This lens forms an image at a distance of 224 mm from the lens. If we want to attain a net angular magnification of 550, what choice must we make for the angular magnification of the ocular?

41. A microscope has an objective of focal length 1.9 mm and an ocular of focal length 25 mm. The distance between these lenses is 180 mm.
  (a) At what distance must the object be placed from the objective so that the ocular forms an image at infinity, as shown in Figure 38.27?
  (b) What is the net angular magnification of this microscope?

Section 38.5

42. A telescope has an objective of focal length 160 cm and an ocular of focal length 2.5 cm. If you look into the *objective* (that is, into the wrong end) of this telescope, you will see distant objects *reduced* in size. By what factor will the angular size of objects be reduced?

43. An amateur astronomer uses a telescope with an objective of focal length 90.0 cm and an ocular of focal length 1.25 cm. What is the angular magnification of this telescope?

44. The large reflecting telescope on Palomar Mountain has a mirror of focal length 1680 cm. If this telescope is operated with an ocular of focal length 1.25 cm, what is the angular magnification?

45. It has been proposed that the six 1.8-m mirrors in the Multiple-Mirror Telescope be replaced by a single 6.5-m mirror, which would fit into the existing mounting. By what factor would this replacement enhance the amount of light collected by the telescope?

***46. Show that if a liquid, such as molten glass, is placed in a rotating bucket, the surface of the liquid will adopt the shape of a parabola. As described in Section 38.5, this result is of practical importance in the manufacture of large glass mirrors. (Hint: Consider a small fluid element, at the surface. The forces on this element are gravity and the pressure of the fluid, which is in the direction normal to the surface. Use the equation for uniform circular motion of the fluid element to obtain an expression for the slope of the fluid surface as a function of the radius of the circle.)

CHAPTER 39

Interference

Geometrical optics relies on the assumption that the wavelength of light is much smaller than the sizes of the mirrors or lenses and the separations between them. The propagation of light can then be described by rays that are rectilinear, except when reflected or refracted by the surfaces of materials.

Wave optics, or physical optics, deals with the propagation of light in the general case, without any restrictive assumptions regarding the wavelengths or the sizes of the bodies through or around which the light is propagating. When the wavelength of the light is comparable to the size of an obstacle or aperture in its path, the light displays its wave properties explicitly through the phenomena of interference and diffraction. As we saw in Chapter 17, interference is the constructive or destructive combination of two or more waves meeting at one place; diffraction is the bending and spreading of waves around obstacles. Phase relationships among the waves play a decisive role in these phenomena, and rays therefore do not provide, by themselves, an adequate description of the behavior of the waves. Even in the simplest problems of wave optics, the information contained in the rays of a wave must be supplemented by information about the phases along these rays.

In this chapter we will examine the interference of light waves and of other electromagnetic waves. Like all electric and magnetic fields, the fields of electromagnetic waves obey the principle of superposition: if two waves meet at some point, the resultant electric or magnetic field is simply the vector sum of the individual fields. If two waves of equal amplitude meet crest to crest, they combine and produce a wave of doubled amplitude; if they meet crest to trough, they cancel and give a wave of zero amplitude. The former case is called **constructive interference** and the latter **destructive interference.** We will encounter both cases of interference in the following sections. The first

Thomas Young, *1773–1829, English physicist, physician, and Egyptologist. Young worked on a wide variety of scientific problems, ranging from the structure of the eye and the mechanism of vision to the decipherment of the Rosetta stone. He revived the wave theory of light and recognized that interference phenomena provide proof of the wave properties of light.*

Constructive and destructive interference

section involves the interference between two waves propagating in opposite directions; the other sections involve waves propagating in the same or nearly the same direction.

39.1* The Standing Electromagnetic Wave

Suppose that an electromagnetic wave strikes a mirror perpendicularly and is totally reflected. The space in front of the mirror will then contain two overlapping waves: the incident wave and the reflected wave. Figure 39.1 shows the mirror in the y–z plane. We will assume that the incident wave is a plane harmonic wave, polarized in the y direction. The electric fields of the incident and the reflected waves are then

$$E_{\rm in} = E_0 \cos(\omega t - \omega x/c) \qquad (1)$$

$$E_{\rm ref} = -E_0 \cos(\omega t + \omega x/c) \qquad (2)$$

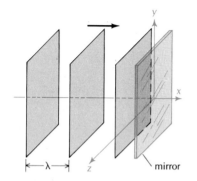

Fig. 39.1 Wave fronts of a plane electromagnetic wave striking a mirror.

Both these electric fields are parallel to the y axis. The first equation represents a wave of frequency ω traveling toward the right, and the second a wave traveling toward the left. A negative sign has been inserted in front of the right side of Eq. (2) so as to satisfy the boundary condition at the mirror; since the mirror at $x = 0$ is a conducting surface, the net electric field must vanish at this point at all times:

$$E_{\rm tot} = E_{\rm in} + E_{\rm ref} = E_0 \cos \omega t - E_0 \cos \omega t = 0 \qquad (3)$$

The negative sign in Eq. (2) means that the reflecting surface not only reverses the direction of propagation of the incident wave but also reverses the electric field. It can be shown that this reversal of the electric field is a general feature of reflection whenever the index of refraction of the medium in which the wave is propagating is smaller than the index of refraction of the medium off which the wave reflects.[1]

In the region $x < 0$, the superposition of the incident and reflected waves gives

$$E_{\rm tot} = E_{\rm in} + E_{\rm ref}$$

$$= E_0[\cos(\omega t - \omega x/c) - \cos(\omega t + \omega x/c)] \qquad (4)$$

With the trigonometric identity $\cos(\alpha - \beta) - \cos(\alpha + \beta) = 2 \sin \alpha \sin \beta$, this simplifies to

$$E_{\rm tot} = 2E_0 \sin \omega t \sin \omega x/c \qquad (5)$$

Standing wave

This is a **standing wave,** that is, a wave whose peaks do not travel right or left but remain fixed in space while the entire wave increases and decreases in unison — the wave pulsates (compare Section 16.6). The maxima of the wave are at the points where

---

* This section is optional.

[1] As was pointed out in Section 27.3, a conducting medium (such as the metal of a mirror) can be regarded as having an infinite value of κ and hence, according to Eq. (37.5), an infinite value of n.

$$\frac{\omega x}{c} = \frac{\pi}{2}, \frac{3\pi}{2}, \frac{5\pi}{2}, \ldots \tag{6}$$

Since the wavelength of the wave is $\lambda = 2\pi c/\omega$, we can also write this as

$$x = \tfrac{1}{4}\lambda, \tfrac{3}{4}\lambda, \tfrac{5}{4}\lambda, \ldots \tag{7}$$

At these points, the incident and reflected waves interfere constructively during one part of the cycle and destructively during another part of the cycle, so the standing wave oscillates with an amplitude $2E_0$. The minima of the wave are at the points where

$$x = 0, \tfrac{1}{2}\lambda, \lambda, \tfrac{3}{2}\lambda, \ldots \tag{8}$$

Here the waves interfere destructively at all times, so the standing wave has zero amplitude.

The magnetic fields associated with the waves described by Eqs. (1) and (2) are in the z direction:

$$B_{\text{in}} = \frac{E_0}{c} \cos\left(\omega t - \frac{\omega x}{c}\right) \tag{9}$$

$$B_{\text{ref}} = \frac{E_0}{c} \cos\left(\omega t + \frac{\omega x}{c}\right) \tag{10}$$

Note that there is no negative sign in Eq. (10); this is so because the right-hand rule tells us that if the electric field is in the negative y direction [see Eq. (2)] and the wave is traveling in the negative x direction, the magnetic field must be in the positive z direction.

The superposition of the magnetic fields gives

$$B_{\text{tot}} = B_{\text{in}} + B_{\text{ref}} = 2\frac{E_0}{c} \cos \omega t \cos \omega x/c \tag{11}$$

This, of course, is also a standing wave. Its maxima are at the points where

$$\frac{\omega x}{c} = 0, \pi, 2\pi, 3\pi, \ldots \tag{12}$$

$$x = 0, \tfrac{1}{2}\lambda, \lambda, \tfrac{3}{2}\lambda, \ldots \tag{13}$$

and its minima at

$$x = \tfrac{1}{4}\lambda, \tfrac{3}{4}\lambda, \tfrac{5}{4}\lambda, \ldots \tag{14}$$

Thus the maxima of the magnetic field are displaced a distance of $\tfrac{1}{4}$ wavelength from the maxima of the electric field. Furthermore, note that the time dependence of the magnetic field ($\cos \omega t$) is a quarter of a cycle out of phase with the time dependence of the electric field

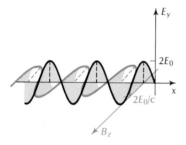

Fig. 39.2 Electric (black) and magnetic (color) fields of the standing wave plotted as a function of position (at a fixed time). The electric field is in the y direction and the magnetic field in the z direction.

(sin ωt). Figure 39.2 shows a plot of the electric and magnetic fields of the standing wave.

The interference between the incident and the reflected waves can be readily observed in a standing microwave. Such a wave can be set up by aiming a microwave generator at a metallic plate, which acts as a mirror. The positions of the maxima and minima of the electric and magnetic fields can be investigated by probing the fields with small antennas connected to a microwave receiver.

39.2 Thin Films

The interference between a light wave incident on a mirror and the light wave reflected by the mirror is very difficult to observe. As we found in the preceding section, the two waves traveling in opposite directions form a standing wave with destructive interference at points one-half wavelength apart [see Eq. (8)]. Since the wavelength of visible light is quite small, our eyes cannot perceive these individual minima (or maxima) of the standing light wave — we see only the average intensity without noticeable interference effects.

However, very spectacular interference effects may become visible when a light wave is reflected by a thin film, such as a thin film of oil floating on water (Figure 39.3). When the wave strikes the upper surface of the film, it will set up a multitude of reflected waves due to reflection at the upper surface, reflection at the lower surface, and multiple zigzags between the surfaces. These reflected waves all travel in the same direction, and they can interfere destructively or constructively over a large region of space. Note that we are now interested only in the interference between the several reflected waves — the incident wave plays no direct role in this.

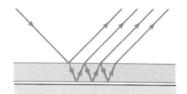

Fig. 39.3 Incident ray and multiple reflected and refracted rays produced by a thin film of oil on water.

To find the conditions for constructive and destructive interference between the waves reflected by a thin film, let us make the simplifying assumption that the direction of propagation of the wave is nearly perpendicular to the surface of the film. The two most intense waves are those that suffer only one reflection: the wave that reflects only at the upper surface and the wave that reflects only at the lower surface. Figure 39.4 shows the rays corresponding to these two waves. Under what conditions will these waves interfere constructively in the space above the film? Obviously, the wave that is reflected at the lower surface has to travel an extra distance to emerge from the film. If the thickness of the film is d, and if the direction of propagation is nearly perpendicular to the film, then the extra distance the wave has to travel is $2d$. Provided that this extra distance is equal to one, two, three, etc., wavelengths, the wave reflected at the lower surface will meet crest to crest with the wave reflected from the upper surface. The condition for constructive interference is therefore

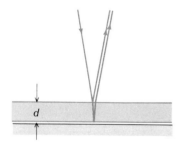

Fig. 39.4 Incident ray and reflected rays for nearly perpendicular incidence and reflection.

Constructive and destructive interference for wave reflected by thin film

$$\boxed{2d = \lambda,\ 2\lambda,\ 3\lambda,\ \ldots} \tag{15}$$

Likewise, the condition for destructive interference is

$$\boxed{2d = \tfrac{1}{2}\lambda,\ \tfrac{3}{2}\lambda,\ \tfrac{5}{2}\lambda,\ \ldots} \tag{16}$$

Thus, depending on the thickness of the film and on the wavelength, the reflected wave can be either very strong or very weak. Note that the wavelength λ in Eqs. (15) and (16) is the wavelength of the light within the film; the wavelength outside the film will differ from this by a factor depending on the index of refraction.

EXAMPLE 1. A film of kerosene 4500 Å thick floats on water. White light, a mixture of all visible colors, is vertically incident on this film. Which of the wavelengths contained in the white light will give maximum intensity upon reflection? Which will give minimum intensity? The index of refraction of kerosene is 1.2.

SOLUTION: For maximum intensity we need

$$\lambda = 2d, \tfrac{2}{2}d, \tfrac{2}{3}d, \ldots \tag{17}$$

$$= 9000 \text{ Å}, 4500 \text{ Å}, 3000 \text{ Å}, \ldots$$

These are the wavelengths in kerosene; to obtain the wavelengths in air, we must multiply by the index of refraction of kerosene, $n = 1.2$. Of the resulting wavelengths, the only one in the visible region is 1.2×4500 Å $= 5400$ Å.

For minimum intensity we need

$$\lambda = 4d, \tfrac{4}{3}d, \tfrac{4}{5}d, \ldots \tag{18}$$

$$= 18{,}000 \text{ Å}, 6000 \text{ Å}, 3600 \text{ Å}, \ldots$$

Upon multiplication by 1.2, we find that the only wavelength in air in the visible region is 1.2×3600 Å $= 4320$ Å.

A wavelength of 5400 Å corresponds to a yellow-green color. The film will therefore be seen to have this color in reflected light.

COMMENTS AND SUGGESTIONS: As this example illustrates, in all calculations of interference in thin films, it is necessary to take into account that the wavelengths of light inside and outside the film differ by a factor of n. The conditions (15) and (16) for constructive and destructive interference involve the wavelength *inside* the film.

The shimmering color displays seen on oil slicks and on soap bubbles arise from such interference effects. Different portions of an oil or soap film usually have different thicknesses, and they therefore give constructive interference for different wavelengths. This results in a pattern of bright colored bands, or colored **fringes**.

Fringes

Note that in the derivation of Eqs. (15) and (16) we have not taken into account the reversal, or change of phase, of the electric field upon reflection. As was stated in the preceding section, this change of phase will occur if (and only if) the index of refraction of the medium in which the wave is propagating is smaller than the index of refraction of the medium off which the wave reflects. In the case of a kerosene film resting on water, both the wave reflected at the upper surface and the wave reflected at the lower surface suffer this change of phase; consequently, the *relative* phase between the waves is unaffected. However, in the case of an oil film or soap film suspended in air, the wave reflected at the lower surface does not suffer a change of phase; this introduces an extra phase difference of 180° between the two waves. Such a phase difference has the same effect as an extra path difference of one-half wavelength. Consequently, Eqs. (15) and (16) must then be interchanged: the former equation now applies to destructive interference and the latter to constructive interference.

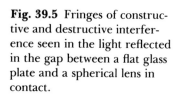

Fig. 39.5 Fringes of constructive and destructive interference seen in the light reflected in the gap between a flat glass plate and a spherical lens in contact.

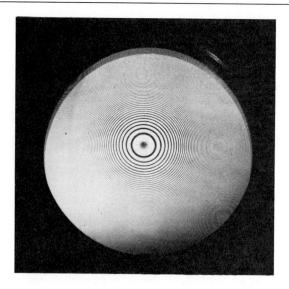

Similar interference effects can also arise in a narrow gap between two adjacent glass surfaces; such a gap can be regarded as a thin film of air. For instance, Figure 39.5 shows a photograph of the interference fringes produced by the thin film of air between a flat glass plate and a lens of large radius of curvature. The convex surface of the lens is in contact with the plate at the center, but leaves a gradually widening gap for increasing distances from the center. The photograph was taken with monochromatic light. At the bright fringes, the width of the gap is such as to give constructive interference of the reflected light. At the dark fringes, the width is such as to give destructive interference. The rings in Figure 39.5 are called **Newton's rings.**

Newton's rings

Thin films are of great practical importance in the manufacture of optical instruments. The lenses of high-quality instruments are often coated with a thin film of a transparent dielectric so that undesirable reflection of light is eliminated. Of course, with one thin film we can achieve destructive interference at only one wavelength; but with several layers of thin films, we can achieve destructive interference at several wavelengths.

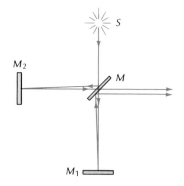

Fig. 39.6 Paths of rays in a Michelson interferometer. For the sake of clarity, rays are shown reaching the mirrors M_1 and M_2 with a small angle between them; they are actually parallel, and they overlap.

39.3 The Michelson Interferometer

The Michelson interferometer takes advantage of the interference between two light waves to accomplish an extremely precise comparison of two lengths. Figure 39.6 is a schematic diagram of the essential parts of such an interferometer. The apparatus consists of two arms at the ends of which are mounted mirrors M_1 and M_2. Light waves from a monochromatic source S fall on a semitransparent mirror M (a half-silvered mirror). This mirror splits the light wave into two parts: one part continues straight ahead and reaches mirror M_1; the other part is reflected and reaches mirror M_2. These mirrors reflect the waves back toward the central mirror M, and upon reflection or transmission by this mirror, the waves emerge from the interferometer. When they emerge, they interfere constructively or destructively. Suppose that the lengths MM_1 and MM_2 differ by d. Then one of the waves must travel an extra distance 2d, and the condition for the constructive interference is

$$2d = 0, \lambda, 2\lambda, \ldots \quad (19)$$

and for destructive interference it is

$$2d = \tfrac{1}{2}\lambda, \tfrac{3}{2}\lambda, \tfrac{5}{2}\lambda, \ldots \quad (20)$$

Constructive and destructive interference in Michelson interferometer

To achieve this interference, the mirrors M_1 and M_2 must be very precisely aligned, so the image of M_1 seen in M is exactly parallel to M_2. The alignment can be achieved by means of adjusting screws on the backs of the mirrors.

The mirror M_1 is usually mounted on a carriage that can be driven along a track by means of a carefully machined screw. If the mirror is slowly moved inward or outward, the interference of the emerging waves will change back and forth between constructive and destructive whenever the mirror is displaced by one-quarter wavelength — and the intensity of the emerging light will change back and forth between maxima and minima, which are seen as bright and dark fringes. Thus, the displacement of the mirror can be measured very precisely by counting fringes and fractions of fringes. This measurement expresses the displacement in terms of the wavelength of the light. Modern interferometers used in metrology, such as that illustrated in Figure 1.10, are designed to count the fringes automatically with a photoelectric device. Such an interferometer is capable of counting 19,000 fringes per second, and its mirror is capable of a total displacement of up to 1 m.

Interferometers have played an important role in the test of the dependence of the speed of light on the motion of the Earth. This test is the famous **Michelson–Morley** experiment, which was first performed in 1881 and came to be of deep significance for the theory of Special Relativity. The principle behind this experiment is as follows: If light were to propagate through the interplanetary "medium" in a manner analogous to the propagation of sound through air, then we would expect that the motion of the Earth toward or away from a light wave would affect the speed of light relative to the Earth, just as the motion of a train toward or away from a source of sound affects the speed of sound relative to the train. Such an alteration of the speed of light could be detected with an interferometer by orienting one of the arms parallel to the direction of motion of the Earth and the other arm perpendicular. A difference in the speeds of light c along the arms would entail a difference in the corresponding wavelengths ($\lambda = 2\pi c/\omega$) and alter the conditions (19) and (20) for bright and dark fringes. The easiest way to detect the speed difference is by rotating the interferometer so that the arm that has been parallel to the motion becomes perpendicular and vice versa. If the speeds were different in the two directions, this rotation could shift the fringes from bright to dark or vice versa.

Michelson and Morley found that there was no observable fringe shift to within the accuracy of their experiment. Taking into account possible experimental errors, they established that the effect of the

Albert Abraham Michelson, *1852–1931, American experimental physicist, professor at the Case School of Applied Science and at the University of Chicago. He made precise measurements of the speed of light and used interferometric methods to determine the wavelength of spectral lines in terms of the standard meter. He first performed the "Michelson-Morley" experiment on his own in 1881 and then repeated it several times in collaboration with E. W. Morley, with increasing accuracy. Michelson received the Nobel Prize in 1907.*

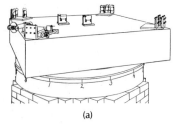

(a)

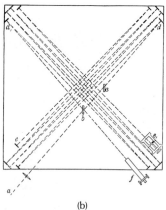

(b)

Fig. 39.7 (a) The interferometer used in the experiment of Michelson and Morley. (b) The many mirrors reflect the light beams back and forth several times, increasing the path length of the light.

motion of the Earth on the speed of light was at most ±5 km/s. This value is substantially less than the speed of the Earth around the Sun (~30 km/s), and the experiment proved beyond all reasonable doubt that the propagation of light through space is *not* analogous to the propagation of sound through air. Figure 39.7 shows Michelson and Morley's interferometer.

39.4 Interference from Two Slits

A very clear experimental demonstration of interference effects in light can be performed with two small light sources placed near each other. The light waves spread out from the sources, run into one another, and interfere constructively or destructively, giving rise to a pattern of bright and dark zones. These interference effects were discovered by Thomas Young around 1800, and in conjunction with diffraction effects (see next chapter), they convinced physicists of the wave nature of light.

If the interference pattern is to remain stationary in space, the two light sources must be **coherent,** that is, they must emit waves of the same frequency and the same phase (a *constant* phase difference is also acceptable; the crucial thing is that the relative phase must not fluctuate in time). Such coherent sources are easily manufactured by aiming a monochromatic wave with plane or spherical wave fronts at an opaque plate with two small slits or holes; the waves diverging from the two slits are then coherent because they arise from a single original wave (Figure 39.8). An extended object, such as an ordinary light bulb, can be used to illuminate the slits, provided it is placed at a *very large* distance. The light bulb will then effectively act as a point source, illuminating the slits with light waves that consist of a succession of nearly plane wave fronts. If the light bulb is placed too close to the slits, then light waves arrive at the slit from several directions at once, and this tends to wash out the interference pattern, because waves of different directions of incidence will produce overlapping zones of brightness and darkness. For convenience, the light bulb is sometimes placed fairly near the slits; but it must then be covered with a shield perforated with a single pinhole so that the emerging light comes from just one point on the surface of the light bulb; this again makes the light source into a point source. In modern practice, a laser is often used to illuminate the slits, because it provides a very intense plane wave, which makes even very faint interference and diffraction effects visible (lasers and the light emitted by them will be discussed in Interlude X).

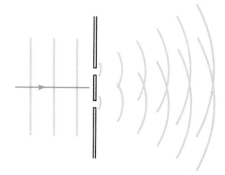

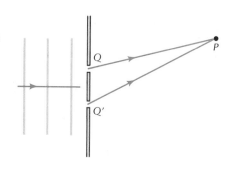

Fig. 39.8 (left) A plane light wave strikes a plate with two very narrow slits. The slits act as two coherent light sources. Light waves diverge from the two slits.

Fig. 39.9 (right) The waves reaching the point P have different path lengths QP and $Q'P$.

If the slits in the plate are very narrow, less than one wavelength in width, they will act as pointlike sources. As expected for a pointlike source, the wave diverges radially and spreads out over a wide angle beyond each slit (see Figure 39.8). This spreading of the wave is a diffraction effect; we will study this in detail in the next chapter.

To find the interference maxima and minima in the space beyond the slits, we need to examine the path difference between the rays from each of the slits. Figure 39.9 shows the light wave incident on the plate from the left and the rays QP and $Q'P$ leading from the slits to a point P on the right. We will assume that the light source generating the incident wave is either a laser or else some other light source placed very far from the plate, so the incident wave is a plane wave (this assumption is implicit in all the calculations of this chapter). The waves emerging from the slits and reaching P interfere constructively if the difference between the lengths QP and $Q'P$ is zero, or one wavelength, or two wavelengths, etc.; and they interfere destructively if this difference is one-half wavelength, or three-halves wavelength, etc.

We can obtain a simple formula for the angular position of the maxima and minima if we make the additional assumption that the point P is at a very large distance from the plate (to be precise, QP is very large compared with QQ'). If so, then the rays QP and $Q'P$ are nearly parallel, and, as Figure 39.10 shows, the difference between the lengths of these rays is approximately $d \sin \theta$, where d is the separation between the slits and θ the angle between the line toward P and the perpendicular midline of the plate. Our condition for maximum intensity is then

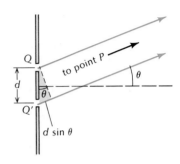

Fig. 39.10 If P is far away, QP and $Q'P$ are nearly parallel. From the small colored triangle we find that the lengths QP and $Q'P$ then differ by $d \sin \theta$.

$$\boxed{d \sin \theta = 0, \lambda, 2\lambda, \ldots} \qquad (21)$$

Maxima and minima for two-slit interference pattern

Likewise, the condition for minimum intensity is

$$\boxed{d \sin \theta = \tfrac{1}{2}\lambda, \tfrac{3}{2}\lambda, \tfrac{5}{2}\lambda, \ldots} \qquad (22)$$

These equations give the angular positions of the interference maxima and minima.

The zones of high intensity have the shape of beams fanning out in the space beyond the slits. The beams of high intensity are separated by lines of zero intensity; these are called nodal lines, analogous to the nodal points produced by interference in a standing wave on a string (see Chapter 16). Figure 39.11 is a photograph of such a pattern of fanlike beams produced by the interference of water waves spreading

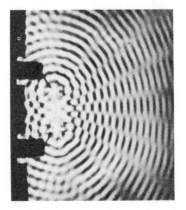

Fig. 39.11 Interference between water waves spreading out from two coherent pointlike sources in a ripple tank.

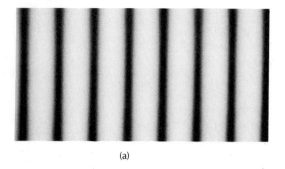
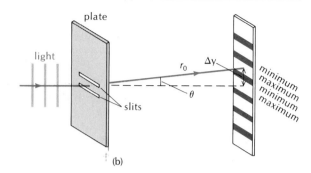

Fig. 39.12 (a) A photographic film placed beyond two illuminated narrow slits records a regular pattern of bright and dark fringes. (Courtesy C. C. Jones, Union College.) (b) Placement of the photographic film beyond the slits.

Fraunhofer approximation

out from two pointlike sources [the formulas (21) and (22) apply to water waves and to any other kinds of waves]. With light waves, it is difficult to photograph the entire pattern of beams at once; instead, we must be content with Figure 39.12, which shows the pattern of bright and dark fringes recorded on a photographic film that intercepts the beams at some fixed distance beyond the slits.

In the above calculation we made the approximation that the distance to the point at which we observe the intensity is very large compared with the separation between the slits. This approximation, which led us to the expression $d \sin \theta$ for the difference between the lengths of the rays, is called the **Fraunhofer approximation.** Sometimes a lens is placed beyond the slits so as to focus their light on the point P. The approximation $d \sin \theta$ for the difference between the lengths of the rays is then justified even if the point P is not at a very large distance, because the lens brings together rays that are parallel when leaving the slits. For the sake of simplicity, we will assume throughout the following discussion that P is at a very large distance and that no lens is required.

EXAMPLE 2. Two narrow slits separated by a distance of 0.12 mm are illuminated with light of wavelength 5890 Å from a sodium lamp. What is the angular position of the first lateral maximum in the interference pattern? If the light is intercepted by a photographic film placed 2.00 m beyond the slits, what is the distance along the film between this maximum and the central maximum?

SOLUTION: With $d = 1.2 \times 10^{-4}$ m and $\lambda = 5.89 \times 10^{-7}$ m, Eq. (21) gives

$$\sin \theta = \frac{\lambda}{d} = \frac{5.89 \times 10^{-7} \text{ m}}{1.2 \times 10^{-4} \text{ m}} = 4.9 \times 10^{-3} \qquad (23)$$

The corresponding angle is $\theta = 4.9 \times 10^{-3}$ radian.

The distance between the points with $\theta = 0$ and $\theta = 4.9 \times 10^{-3}$ radian on the photographic film is (see Figure 39.12b)

$$\Delta y \cong r_0 \theta = 2.00 \text{ m} \times 4.9 \times 10^{-3} = 9.8 \times 10^{-3} \text{ m}$$

Thus, the maxima are separated by nearly 1 cm.

COMMENTS AND SUGGESTIONS: In this example, and in many other examples of interference and diffraction, the angle specifying the position of a maximum or a minimum is small, and the sine of the angle is therefore almost equal to the angle (expressed in radians). We will often find it convenient to rely on this approximate equality when using Eqs. (21) and (22) for calculations of angular positions.

Let us now calculate the intensity distribution as a function of angle. The electric fields of the spherical waves spreading out from points in the two slits[2] and reaching the point P are [compare Eq. (36.1)]

$$E_1 = \frac{A}{r_1} \cos\left(\omega t - \frac{\omega r_1}{c}\right) \quad (24)$$

and

$$E_2 = \frac{A}{r_2} \cos\left(\omega t - \frac{\omega r_2}{c}\right) \quad (25)$$

In these equations, A is a constant that depends on the strength of the wave incident on the slits, r_1 is the distance QP, and r_2 is the distance $Q'P$. With the same large-distance approximation as above, we have

$$r_1 = r_0 - \frac{d}{2} \sin \theta \quad (26)$$

$$r_2 = r_0 + \frac{d}{2} \sin \theta \quad (27)$$

where r_0 is the distance from the midpoint of the two slits to P (Figure 39.13). We can substitute these approximations into the cosine functions in the numerators of Eqs. (24) and (25). We should also substitute these approximations into the denominators of these equations, but these denominators are not all that sensitive to small changes in the distances and the crude approximation $r_1 \cong r_2 \cong r_0$ suffices there:

$$E_1 = \frac{A}{r_0} \cos\left(\omega t - \frac{\omega r_0}{c} + \frac{\omega d}{2c} \sin \theta\right) \quad (28)$$

$$E_2 = \frac{A}{r_0} \cos\left(\omega t - \frac{\omega r_0}{c} - \frac{\omega d}{2c} \sin \theta\right) \quad (29)$$

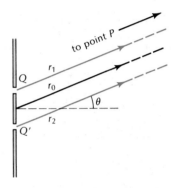

Fig. 39.13 Rays from the slits to the point P.

The superposition of the two electric fields E_1 and E_2 then gives $E = E_1 + E_2$

$$= \frac{A}{r_0} \left[\cos\left(\omega t - \frac{\omega r_0}{c} + \frac{\omega d}{2c} \sin \theta\right) + \cos\left(\omega t - \frac{\omega r_0}{c} - \frac{\omega d}{2c} \sin \theta\right)\right] \quad (30)$$

With the trigonometric identity $\cos(\alpha + \beta) + \cos(\alpha - \beta) = 2 \cos \alpha \cos \beta$ this becomes

$$E = \frac{2A}{r_0} \cos\left(\omega t - \frac{\omega r_0}{c}\right) \cos\left(\frac{\omega d}{2c} \sin \theta\right) \quad (31)$$

The factor $\cos(\omega t - \omega r_0/c)$ is the usual oscillating function of space and time, characteristic of a wave propagating outward. The factor

---

[2] In the following calculation, we will consider only the waves emanating from one point in each slit (the points in the plane of Figure 39.13); other points in the slits contribute similar waves.

$\cos[(\omega d/2c)\sin\theta]$ indicates how the amplitude of this wave depends on the position angle θ.

The intensity of the wave is proportional to E^2. Since we are interested only in the time-average intensity, we can replace the factor $\cos^2(\omega t - \omega r_0/c)$ appearing in E^2 by its time-average value $\tfrac{1}{2}$. Hence

$$[\text{intensity}] \propto \frac{1}{r_0^2}\cos^2\!\left(\frac{\omega d}{2c}\sin\theta\right) \qquad (32)$$

With $\lambda = 2\pi c/\omega$, we can also write this as

$$\boxed{[\text{intensity}] \propto \frac{1}{r_0^2}\cos^2\!\left(\frac{\pi d}{\lambda}\sin\theta\right)} \qquad (33)$$

Intensity for two-slit interference pattern

Fig. 39.14 (below, left) Intensity as a function of the angle θ for the light arriving at a screen placed beyond two narrow slits. The distance between the slits is $d = 8\lambda$.

Fig. 39.15 (below, right) Zones of maximum intensity and minimum intensity on a curved screen. All points of this screen are at the same distance r_0.

This formula describes the intensity pattern produced by the double slits. The factor $\cos^2[(\pi d/\lambda)\sin\theta]$ gives the distribution of intensity as a function of angle. Figure 39.14 is a plot of this factor. This plot represents the intensity of light that reaches different angles at some constant distance r_0 from the midpoint of the slits. If we want to measure this amount of light by intercepting it with a screen or a photographic film, we have to bend the screen along a circular arc of radius r_0 (Figure 39.15); however, the observations are usually restricted to small angles θ, and then a flat screen will serve almost equally well.

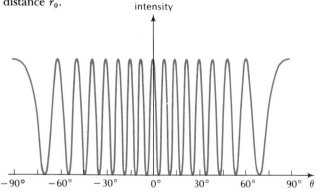

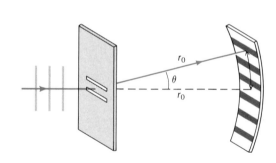

39.5 Interference from Multiple Slits

It is easy to see that Eq. (21) also applies to the case of a wave incident on multiple slits. For instance, Figure 39.16 shows an opaque plate with three evenly spaced slits. If the waves emerging from *adjacent* slits interfere constructively, then all the waves emerging from *all* three slits interfere constructively. Hence the condition for maximum intensity is the same as Eq. (21).

However, the condition for minimum intensity is *not* Eq. (22). In the case of three slits, destructive interference of waves from adjacent slits will lead to cancellation of the waves originating from one pair of slits, but the wave from the remaining slit will not suffer cancellation. The condition for destructive interference between three waves (of equal amplitude) is that the phases differ by 120° from one wave to the next;

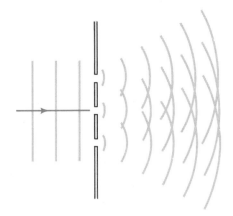

Fig. 39.16 A plane wave strikes a plate with three very narrow slits.

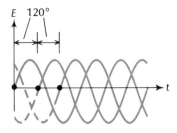

Fig. 39.17 Electric field as a function of time at the point P. The three waves from the three slits differ in phase by 120°; their sum is zero.

the sum of two of the waves then always cancels against the third wave (Figure 39.17). Hence a triple slit will yield a minimum if $d \sin \theta = \lambda/3$, where d is the separation between adjacent slits. The triple slit will also yield a minimum if $d \sin \theta = 2\lambda/3$; this increased value of the path difference merely amounts to an extra shift of each wave crest by one-third of a period and does not alter the relative distribution of these crests (compare Figures 39.17 and 39.18). Pursuing this argument, we find further minima at $d \sin \theta = 4\lambda/3$, $5\lambda/3$, and so on. Figure 39.19 is a plot of the intensity as a function of θ for the case of three slits (for small values of θ). Between any two adjacent **principal maxima** with positions specified by Eq. (21), there are two minima. Between the two minima there is a weak maximum, a **secondary** maximum. Figure 39.20 is a photograph of this interference pattern.

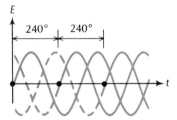

Fig. 39.18 The three waves differ in phase by 240°; their sum is zero.

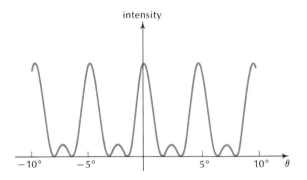

Fig. 39.19 Intensity as a function of the angle θ for light arriving at a screen placed beyond three narrow slits. The distance between one slit and the next is $d = 12\lambda$.

Fig. 39.20 A photographic film placed beyond the three slits shows the strong principal maxima and the weaker secondary maxima. (Courtesy C. C. Jones, Union College.)

For the case of N slits, we can summarize the important features of the interference pattern as follows. The positions of the principal maxima are given by

Maxima and minima for multiple-slit interference pattern

$$\boxed{d \sin \theta = n\lambda \qquad n = 0, 1, 2, \ldots}\qquad(34)$$

The positions of the minima are given by

$$\boxed{d \sin \theta = m\lambda/N} \qquad \begin{array}{l} m = 1, 2, 3, \ldots \\ \text{(with } m = N, 2N, \ldots, \text{excluded)} \end{array} \qquad(35)$$

The secondary maxima are located (approximately) halfway between the minima.

Arrangements of multiple slits, with a large number of slits, are called **gratings**. They are commonly used in spectroscopy laboratories to analyze light into its colors. If a light beam containing several wavelengths passes through such multiple slits, the maxima for these different wavelengths will form beams at different angles — the beams of long-wavelength light will be found at larger angles than the beams of short-wavelength light [compare Eq. (34)]. Thus, the grating separates the light according to color and produces a spectrum in much the same way that a prism does. There is, however, one important difference between the spectra formed by a prism and by a grating: in the prism, the long-wavelength light (red) suffers the least deflection; in the grating, the long-wavelength light suffers the most deflection.

Grating

The grating will produce one complete spectrum for each value of n in Eq. (34), that is, each principal maximum, except the central maximum, gets spread out by color. These spectra are called the first-order spectrum ($n = 1$), the second-order spectrum ($n = 2$), and so on. Sometimes these spectra overlap, for example, the red end of the second-order spectrum may show up at the same angle as the blue end of the third-order spectrum (see Figure 39.21).

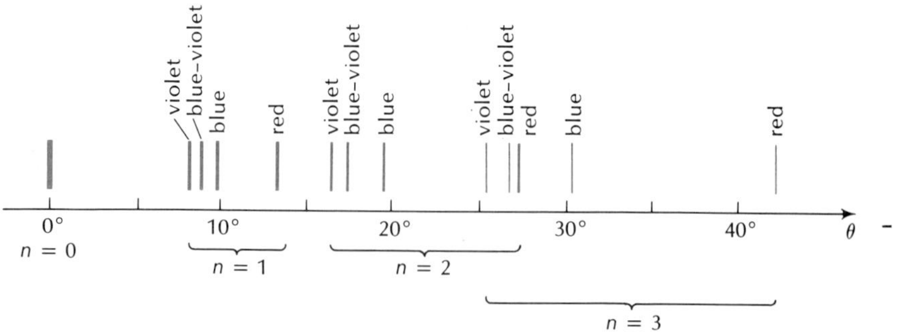

Fig. 39.21 First-, second-, and third-order spectra of hydrogen light produced by a grating of N slits. The lines correspond to the principal maxima; the secondary maxima are weak and can be ignored. Each spectrum consists of a violet, blue-violet, blue, and red spectral line (compare Figure 37.23b). The pattern of spectral lines for negative values of θ (negative n) is similar.

To achieve a very sharp separation of colors within each spectrum, we must use a very large number of slits. This makes the principal maxima very narrow and therefore reduces the overlap between maxima formed by two colors differing but little in wavelength. To understand how this helps, note that the distance from the center of a principal maximum to the next minimum is, according to Eq. (35),

Width of principal maximum

$$\boxed{\Delta\theta = \frac{1}{N}\frac{\lambda}{d}} \qquad(36)$$

where we assume that θ is small, so $\sin\theta \cong \theta$. For a change $\Delta\lambda$ in wavelength, the change in the angular position of the principal maximum is, according to Eq. (34),

$$\Delta\theta = n\frac{\Delta\lambda}{d} \tag{37}$$

If the angular shift (37) is equal to the angular width (36), then the changed wavelength will produce a maximum at the minimum of the original wavelength. Under these conditions, the two maxima are obviously well separated and can be told apart very clearly (Figure 39.22). Setting Eqs. (37) and (36) equal, we therefore find that the wavelength difference that can be clearly resolved by our arrangements of slits is given by

$$\frac{1}{N}\frac{\lambda}{d} = n\frac{\Delta\lambda}{d} \tag{38}$$

or

$$\boxed{\frac{\lambda}{\Delta\lambda} = Nn} \tag{39}$$

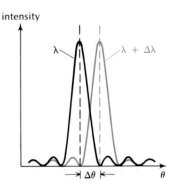

Fig. 39.22 Two distinct principal maxima produced by two different spectral lines of wavelengths λ and $\lambda + \Delta\lambda$. The maximum produced by one of these wavelengths coincides with the minimum produced by the other.

Resolving power

The ratio $\lambda/\Delta\lambda$ is called the **resolving power** of the grating. The formula (39) shows that if we want to detect a small wavelength difference, we need a large value of N.

EXAMPLE 3. Among the wavelengths emitted by atoms of iron are $\lambda = 5005.72$ Å and $\lambda = 5006.13$ Å. If we want to separate these two wavelengths clearly in the first-order spectrum of a grating, what is the minimum number of slits required?

SOLUTION: The wavelength difference is $\Delta\lambda = 0.41$ Å and $n = 1$. Hence Eq. (39) gives

$$N = \frac{\lambda}{\Delta\lambda} = \frac{5006\text{ Å}}{0.41\text{ Å}} = 1.2 \times 10^4$$

This means that we need at least 12,000 slits.

Since it is difficult to cut a large number of slits in an opaque plate, gratings are usually manufactured by cutting fine parallel grooves in a glass or metal surface with a diamond stylus guided by a special ruling machine. When illuminated by a plane light wave, the edges of the grooves act as coherent light sources in much the same way as slits, but the light emerges on the same side as the incident light. Such a grating is called a **reflection grating.** High-quality gratings used in spectroscopy have 100,000 or more grooves with distances of about 10^{-6} m between them.

Radiotelescopes, consisting of a regular array of evenly spaced antennas, act as "gratings" for radio waves; the same formulas for the directions of maximum intensity apply to these as to ordinary gratings used with light. Of course, there is a difference between the operation of a radiotelescope and that of an ordinary grating: in the former the radio waves from a distant point *enter* the antennas and interfere con-

Radiotelescope

structively or destructively within the radio receiver, whereas in the latter the light waves *emerge* from the slits and travel to a distant point where they interfere. Nevertheless, the condition for, say, maximum intensity can still be expressed in terms of the direction of the wave by Eq. (34), because this condition hinges only on the phase relationships among the parts of the wave, and these relationships are independent of the direction of propagation (for instance, in Figure 39.16 it makes no difference whether the waves are traveling toward or away from the slits).

EXAMPLE 4. One branch of the Very Large Array (VLA) radiotelescope at Socorro, New Mexico, has nine antennas arranged on a straight line with a distance of 2.0 km between one antenna and the next (Figure 39.23). These antennas are all connected to a single radio receiver by waveguides of equal length; the receiver then registers a maximum intensity if the radio waves incident on all antennas are in phase. Radio waves of wavelength 21 cm from a pointlike source in the sky strike this radiotelescope. When a source is at the zenith, the intensity in the radio receiver is maximum. What must be the angular displacement of a source from the zenith to make the intensity minimum?

SOLUTION: According to Eq. (36), the angular distance from the center of a principal maximum to the adjacent minimum is

$$\Delta\theta = \frac{1}{N}\frac{\lambda}{d} = \frac{1}{9}\frac{0.21 \text{ m}}{2.0 \times 10^3 \text{ m}} = 1.2 \times 10^{-5} \text{ radian}$$

This angle is about 3 seconds of arc.

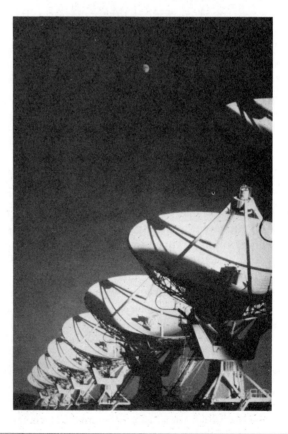

Fig. 39.23 The VLA radiotelescope. The branch in the center consists of nine antennas. The distance between the antennas can be adjusted by rolling them along the tracks. Here the antennas are shown close together.

SUMMARY

Constructive interference: Waves meet crest to crest.

Destructive interference: Waves meet crest to trough.

Change of phase by reflection: 180° if medium off which wave reflects has higher index of refraction

Two-slit interference pattern:
$$\text{maxima:} \quad d \sin \theta = 0, \lambda, 2\lambda, \ldots$$
$$\text{minima:} \quad d \sin \theta = \tfrac{1}{2}\lambda, \tfrac{3}{2}\lambda, \tfrac{5}{2}\lambda, \ldots$$

Multiple-slit interference pattern:
$$\text{principal maxima:} \quad d \sin \theta = 0, \lambda, 2\lambda, \ldots$$
$$\text{minima:} \quad d \sin \theta = \lambda/N, 2\lambda/N, \ldots$$

Width of principal maximum of grating: $\Delta\theta = \dfrac{1}{N}\dfrac{\lambda}{d}$

Resolving power of grating: $\dfrac{\lambda}{\Delta\lambda} = Nn$

QUESTIONS

1. According to Eq. (11), the magnetic field of the standing wave is maximum at the surface of the mirror. The magnetic field is zero inside the mirror. What is the direction of the current that must flow along the surface of the mirror to produce this sudden drop of the magnetic field?

2. The light wave inside the tube of a laser is a standing wave, reflected at both ends of the tube. If the distance between the ends is L, what condition must the wavelength of the standing wave satisfy?

3. Would you expect sound waves reflected by a wall to produce a standing wave? How could you test this experimentally?

4. The radar wave emitted by a stationary police radar unit is reflected by an approaching automobile. Does this set up a standing wave?

5. In a traveling electromagnetic wave, the maxima of the electric and magnetic fields occur at the same place. A standing wave is a superposition of two traveling waves. How, then, can the maxima of the electric and magnetic fields in this wave occur at different places?

6. The maxima and the minima of the electric and magnetic fields in a standing wave produced by a microwave generator can be detected with an electric dipole antenna or a magnetic dipole antenna connected to a radio receiver (Figure 39.24). How must you orient each of these antennas relative to the directions of the electric and magnetic fields for maximum sensitivity?

7. When two waves interfere destructively at one place, what happens to their energy?

8. When two coherent waves of equal intensity interfere constructively at one place, the energy density at that place becomes four times as large as the energy density of each individual wave. Is this a violation of the law of conservation of energy?

9. Suppose that N waves of equal intensity meet at one place. If the waves are coherent, the net intensity is N^2 times that of each individual wave. If the

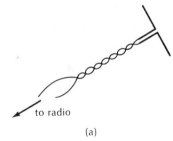
(a)

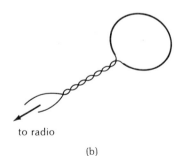
(b)

Fig. 39.24 (a) Electric dipole antenna. (b) Magnetic dipole antenna.

waves are incoherent, the (average) intensity is N times that of each individual wave. Explain, and give an example of each case.

10. Why do we not see colored fringes in *thick* oil slicks?

11. When light strikes a window pane, some rays will be reflected back and forth between the two glass surfaces. Why do these reflected rays not produce visible colored interference fringes?

12. Suppose that a lens is covered with an antireflective coating that eliminates the reflection of perpendicularly incident light of some given color. Will this coating also eliminate the reflection of this light incident at an angle?

13. Is it possible to cover the surface of an aircraft with an antireflective coating so that it does not reflect radar waves of wavelength, say, 5 cm?

14. Consider the light *transmitted* by a thin film. For a light wave with a direction of propagation perpendicular to the surface of the film, what is the condition for constructive interference between the direct wave and the wave that is reflected once by the lower surface and once by the upper surface? Would you expect that this condition for constructive interference in transmission coincides with the condition for destructive interference in reflection?

15. Why is the central spot in Newton's rings (Figure 39.5) dark?

16. Explain how Newton's rings may be used as a sensitive test of the rotational symmetry of a lens.

17. Two flat plates of glass are in contact at one edge and separated by a thin spacer at the other edge (Figure 39.25a). Explain why we see parallel interference fringes in the reflected light if we illuminate these plates from above (Figure 39.25b).

18. In the experiment that sought to test the dependence of the speed of light on the motion of the Earth, Michelson and Morley used an interferometer with very long arms (about 11 m, obtained by multiple reflections back and forth between sets of mirrors). Why does this make the instrument more sensitive?

19. If you stand next to your TV receiver, your body will sometimes affect the reception. Why?

20. Consider (1) sunlight, (2) sunlight passed through a monochromatic filter selecting one wavelength, (3) light from a neon tube, (4) light from a laser, (5) starlight passed through a monochromatic filter, and (6) radio waves emitted by a radio station. Which of these kinds of light or electromagnetic radiation is sufficiently coherent to give rise to an interference pattern when used to illuminate two slits such as shown in Figure 39.8?

21. When a pair of stereo loudspeakers are installed, the terminals of the loudspeakers should be connected to the amplifier in the same way, so that the loudspeakers are in phase. What would happen to sound waves of long wavelength if the loudspeakers were out of phase?

22. What happens to the interference pattern plotted in Figure 39.14 if $d < \lambda$?

23. Suppose we cover the left side of one of the slits in Figure 39.10 with a glass plate. Since the wavelength in glass is shorter than that in air, the path in the glass includes more wavelengths than the path in air, and the phases of the two waves emerging from the slits will be different. If the phase difference between these waves is exactly 180°, how does this change the location of maxima and minima?

24. Consider the interference pattern shown in Figure 39.14. How would this pattern change if the entire interference apparatus were immersed in water?

25. Suppose that a plane wave is incident at an angle on a plate with two nar-

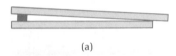
(a)

(b)

Fig. 39.25 (a) Two flat plates of glass, separated by a thin wedge of air. (b) Interference fringes.

row slits (Figure 39.26). In what direction will we then find the central maximum?

26. Several radio antennas are arranged at regular intervals along a straight line; the antennas are connected to the same radio transmitter, so they radiate coherently. How can this array be used to concentrate the radio emission in a selected direction?

27. When a crystal is illuminated with X rays, each atom acts as a pointlike source of scattered X rays. The typical separation between adjacent atoms in a crystal is 1 Å. Roughly what must be the wavelength of the X rays if they are to exhibit distinct interference effects?

Fig. 39.26 The plane wave is incident on the plate at a slant.

PROBLEMS

Section 39.1

1. A light wave of wavelength 6943 Å from a laser strikes a mirror at normal incidence. This will set up a standing wave in front of the mirror.
    (a) At what distance from the mirror will the electric field have nodes? Antinodes?
    (b) At what distance from the mirror will the magnetic field have nodes? Antinodes?

2. In 1887, Hertz gave the first direct experimental demonstration of the existence of the electromagnetic waves that had been predicted by Maxwell's theory. For this experiment, Hertz placed a high-frequency spark generator at some distance in front of a large vertical zinc plate. The wave emitted by the spark reflected off the zinc plate, setting up a standing wave. Hertz found that when the generator was operating at a frequency of about 4×10^7 cycles per second, the distance between a node and an antinode in the standing-wave pattern was about 2 m. What value of the speed of propagation of the electromagnetic disturbances could he deduce from this?

3. With a microwave generator, you can set up a standing electromagnetic wave between two large, flat, parallel plates of metal. The standing wave must have zero electric field at the surfaces of the plates at all times. If your microwave generator produces waves of frequency 2×10^9 Hz, what minimum separation between the plates do you need for a standing wave? Where does this standing wave have a maximum amplitude of the electric field? Of the magnetic field?

*4. The tube of a laser has a mirror at each end (Figure 39.27). A standing electromagnetic wave fills the space between these mirrors. This wave must satisfy the boundary condition that its electric field is zero at all times at the position of the mirrors.
    (a) Show that the distance between the mirrors and the wavelength of the wave must be related by

$$d = \left(\frac{n+1}{2}\right)\lambda \qquad n = 0, 1, 2, \ldots$$

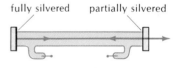

Fig. 39.27 Tube of a laser with mirrors at its ends.

   (b) A He–Ne laser with $d = 30.00$ cm operates at $\lambda = 6328$ Å. What is the value of n? How many minima and maxima does the standing wave have between the two mirrors?
   (c) At what distance from each of the mirrors is the nearest maximum of the electric field? The nearest maximum of the magnetic field?

Section 39.2

5. The wall of a soap bubble floating in air has a thickness of 4000 Å. If sunlight strikes the wall perpendicularly, what colors in the reflected light will be

strongly enhanced as seen in air? The index of refraction of the soap film is 1.35.

6. A lens made of flint glass with an index of refraction of 1.61 is to be coated with a thin layer of magnesium fluoride with an index of refraction of 1.38.
   (a) How thick should the layer be so as to give destructive interference for the perpendicular reflection of light of wavelength 5500 Å seen in air?
   (b) Does your choice of thickness permit constructive interference for the reflection of light of some other wavelength in the visible spectrum? (If it does, you ought to make a better choice.)

*7. A thin oil slick, of index of refraction 1.3, floats on water. When a beam of white light strikes this film vertically, the only colors enhanced in the reflected beam seen in air are orange-red (about 6500 Å) and violet (about 4300 Å). From this information, deduce the thickness of the oil slick.

Fig. 39.28

*8. Two flat parallel plates of glass are separated by thin spacers so as to leave a gap of width d (Figure 39.28). If light of wavelength λ is normally incident on these plates, what is the condition for constructive interference between the rays reflected by the lower surface of the top plate and the upper surface of the bottom plate?

*9. A layer of oil of thickness 2000 Å floats on top of a layer of water of thickness 4000 Å resting on a flat metallic mirror. The index of refraction of the oil is 1.24 and that of the water 1.33. A beam of light is normally incident on these layers. What must be the wavelength of the beam if the light reflected by the top surface of the oil is to interfere destructively with the light reflected by the mirror?

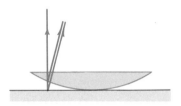

Fig. 39.29 Lens resting on flat glass plate.

**10. A lens with one flat surface and one convex surface rests on a flat plate of glass (Figure 39.29). A light ray, normally incident on the lens, will be partially reflected by the curved surface of the lens and partially reflected by the flat plate of glass. The interference between these reflected rays will be constructive or destructive, depending on the height of the air gap between the lens and the plate. The interference gives rise to the pattern of Newton's rings, shown in Figure 39.5.
   (a) Why is the center of the pattern dark?
   (b) Show that the radius of the nth bright ring is

$$r = \sqrt{n\lambda R - n^2\lambda^2/4}$$

where λ is the wavelength of the light and R is the radius of the convex surface of the lens.
   (c) What is the radius of the first dark ring if $\lambda = 5000$ Å, $R = 3.0$ m?

Section 39.3

11. The **Fabry–Perot interferometer** consists of two parallel half-silvered mirrors. A ray of light entering the space between the mirrors may pass straight through, or be reflected once or several times by each mirror (Figure 39.30). Show that the condition for constructive interference of the emerging light (at large distance from the mirrors) is

$$2d \cos \theta = 0, \lambda, 2\lambda, \ldots$$

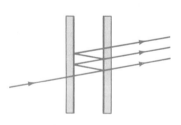

Fig. 39.30 Fabry–Perot interferometer.

where d is the distance between the mirrors, λ the wavelength of light, and θ the angle of incidence of the light.

12. The interferometer of the Bureau International de Poids et Mesures can count 19,000 bright fringes (maxima) per second. To achieve this count rate, what must be the speed of motion of the moving mirror (mirror M_1 in Figure 39.6)? Assume that the interferometer operates with krypton light of wavelength $\lambda = 6058$ Å.

13. An **etalon** consists of two mirrors held a fixed distance apart by means of a rigid support (Figure 39.31a). The distance between the two mirrors can serve as a standard of length. To measure this distance in terms of wavelengths of light, the etalon is installed in a Michelson interferometer, replacing the fixed mirror M_2 (Figure 39.31b). For a start, the distances MM_1 and MM_3 are made exactly equal; then the mirror M_1 is slowly moved outward, producing a sequence of interference maxima and minima until the distances MM_1 and MM_4 are exactly equal. [The exact equality of distances can be checked by replacing the usual monochromatic light source of the interferometer with a white light source. White light will give an interference maximum if and only if the distance d in Eq. (19) is exactly zero, because only then will Eq. (19) be satisfied simultaneously for *all* wavelengths.] Suppose that you operate a Michelson interferometer with krypton light of wavelength 6057.802 Å, and you count 36,484.8 interference maxima while moving the mirror M_1 from its initial to its final position. To within six significant figures, what is the length of the etalon?

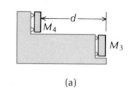

(a)

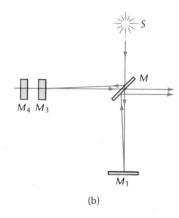

(b)

Fig. 39.31 (a) An etalon. (b) Etalon (M_3, M_4) installed in Michelson interferometer.

*14. The index of refraction of air for monochromatic light of some given wavelength can be determined very precisely with a Michelson interferometer one of whose arms (MM_1 in Fig. 39.6) is in air while the other (MM_2) is in a vacuum tank. With the vacuum tank filled with air, the arms are first adjusted so they are exactly equal.[3] Then the vacuum tank is slowly evacuated, while at the same time the length of the arm is gradually increased so as to maintain constructive interference (this increase of length maintains a fixed number of wavelengths in the arm). Show that the index of refraction of air can be expressed in terms of the fractional change of length:

$$n - 1 = \frac{\Delta L}{L}$$

Suppose that for light of wavelength 5800 Å, an interferometer with arms of length 40.000 cm requires a readjustment of 0.01106 cm to maintain constructive interference while the vacuum tank is being evacuated. What is the index of refraction of air at this wavelength?

Section 39.4

15. Microwaves of wavelength 2.0 cm from a radio transmitter are aimed at two narrow parallel slits in an aluminum plate. The slits are separated by a distance of 5.0 cm. At what angles at some distance beyond the plate will we find maxima in the interference pattern?

16. Suppose that the two slits, the film, and the light source described in Example 2 are immersed in water. What will now be the distance between the central maximum and the first lateral maximum?

17. A piece of aluminum foil with two narrow slits is being illuminated with red light of wavelength 6943 Å from a laser. This yields a row of evenly spaced bright bands on a screen placed 3.00 m beyond the slits. The interval between the bright bands is 1.4 cm. What is the distance between the two slits?

18. Consider the water waves shown in Figure 39.11. With a ruler, measure the wavelength of the waves and the distance between the sources; with a protractor, measure the angular positions of the nodal lines (minima). Do these measured quantities satisfy Eq. (22)?

---

[3] Exact equality of the lengths of the arms can be verified by checking for constructive interference with *white* light. For white light, containing many wavelengths, constructive interference is possible only if the lengths are equal, that is, $d = 0$ in Eq. (39.16).

19. When a beam of monochromatic light is incident on two narrow slits separated by a distance 0.15 mm, the angle between the central beam and the third lateral maximum in the interference pattern is 0.52°. What is the wavelength of the light?

20. An interferometric radiotelescope consists of two antennas separated by a distance of 1.0 km. The two antennas feed their signals into a common receiver tuned to a frequency of 2300 MHz. The receiver will detect a maximum (constructive interference) if the wave sent out by a radio source in the sky arrives at the two antennas with the same phase. What possible angular positions of the radio source will result in such a maximum? Reckon the angular position from the vertical line erected at the midpoint of the antennas. Treat the antennas as pointlike.

*21. Two radio beacons are located on an east–west line and separated by a distance of 6.0 km. The radio beacons emit synchronous (in phase) sinusoidal waves of a frequency of 1.0×10^5 Hz. The navigator of a ship wants to determine his position relative to the radio beacons. His radio receiver indicates zero signal strength at the position of the ship. What is the angular bearing of the ship relative to the radio beacons? Assume that the distance between the ship and the radio beacons is much larger than 6 km. Note: This problem has *several* answers.

*22. Light of wavelength λ is obliquely incident on a pair of narrow slits separated by a distance d. The angle of incidence of the light on the slits is ϕ (Figure 39.32).

(a) Show that the diffracted light emerging at an angle θ interferes constructively if

$$|d \sin \theta - d \sin \phi| = 0, \lambda, 2\lambda, \ldots$$

and destructively if

$$|d \sin \theta - d \sin \phi| = \tfrac{1}{2}\lambda, \tfrac{3}{2}\lambda, \tfrac{5}{2}\lambda, \ldots$$

(b) Show that if θ is small, the angular separation between the interference maxima and minima is independent of the angle ϕ.

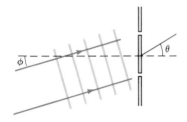

Fig. 39.32

*23. The radio wave from a transmitter to a receiver may follow either a direct path or else an indirect path involving a reflection on the ground (Figure 39.33). This can lead to destructive interference of the two waves and a consequent fading of the radio signal at certain locations. Suppose that a transmitter and a receiver operating at 98 MHz are both at a height of 60 m on tall buildings with bare ground between. What is the maximum (finite) distance between the buildings that will lead to destructive interference? Be careful to take into account the phase change upon reflection on the ground.

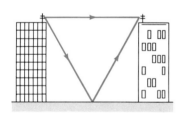

Fig. 39.33 Antennas on two buildings.

*24. Two radio beacons emit waves of frequency 2.0×10^5 Hz. The beacons are on a north–south line, separated by a distance of 3.0 km. The southern beacon emits waves $\tfrac{1}{4}$ of a cycle later than the northern beacon. Find the angular directions for constructive interference. Reckon the angles relative to the east–west line, and assume that the distance between the beacons and the point of observation is large.

*25. A device called **Lloyd's mirror** produces interference between a ray reaching a vertical screen directly and a ray reaching the screen after reflection by a horizontal mirror. Show that in terms of the distances z, z_0, and l defined in Figure 39.34, the condition for constructive interference is

$$\sqrt{l^2 + (z+z_0)^2} - \sqrt{l^2 + (z-z_0)^2} = \tfrac{1}{2}\lambda, \tfrac{3}{2}\lambda, \tfrac{5}{2}\lambda, \ldots$$

and that for destructive interference is

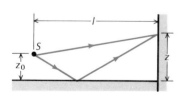

Fig. 39.34 Lloyd's mirror.

$$\sqrt{l^2 + (z + z_0)^2} - \sqrt{l^2 + (z - z_0)^2} = 0, \lambda, 2\lambda, \ldots.$$

Note that in this calculation you must take into account the reversal, or change of phase, of the wave during reflection.

26. Light of wavelength 6943 Å from a ruby laser is incident on two narrow parallel slits cut in a thin sheet of metal. The slits are separated by a distance of 0.11 mm. A screen is placed 1.5 m beyond the slits. Find the intensity, relative to the central maximum, at a point on the screen 1.2 cm to one side of the central maximum.

**27. Two radio beacons transmit waves of the same phases and frequencies. The transmitters are on the x axis, at $\pm x_0$. Show that the interference is constructive at those points of the x–y plane satisfying the condition

$$\sqrt{(x + x_0)^2 + y^2} - \sqrt{(x - x_0)^2 + y^2} = n\lambda$$

where $n = 0, 1, 2, \ldots$. Show that for a given nonzero value of n, this is the equation of a hyperbola. (To show this, use either graphical methods to plot the curve or else your knowledge of analytic geometry.)

**28. In the first application of interferometric methods in radio astronomy, Australian astronomers observed the interference between a radio wave arriving at their antenna on a direct path from the Sun and on a path involving one reflection on the surface of the sea (Figure 39.35). Assuming that the radio waves have a frequency of 6.0×10^7 Hz and that the radio receiver is at a height of 25 m above the level of the sea, what is the least angle of the source above the horizon that will give destructive interference of the waves at the receiver?

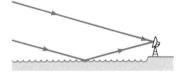

Fig. 39.35

Section 39.5

29. A grating has 5000 lines per centimeter. What are the angular positions of the principal maxima produced by this grating when illuminated with light of wavelength 6500 Å?

30. Sodium light with wavelengths 5889.9 Å and 5895.9 Å is incident on a grating with 5500 lines per centimeter. A screen is placed 3.0 m beyond the grating. What is the distance between the two spectral lines in the first-order spectrum on the screen? In the second-order spectrum?

31. The red line in the spectrum of hydrogen has a wavelength of 6563 Å; the blue line in this spectrum has a wavelength of 4340 Å. If hydrogen light falls on a grating of 6000 slits per centimeter, what will be the angular separation (in degrees) of these two spectral lines as seen in the first-order spectrum?

32. One of the gratings made by H. Rowland — a celebrated maker of gratings, some of which remain unsurpassed to this day for their uniformity — had 14,438 lines per inch and was 6.0 in. wide. What was the resolving power of this grating in the first order?

33. A grating has 40,000 lines uniformly spaced over a width of 8.0 cm. What is the width of the principal maxima formed by this grating with light of wavelength 5500 Å?

*34. A thin curtain of fine batiste consists of vertical and horizontal threads of cotton forming a net that has a regular array of square holes. While looking through this curtain at the red (6700 Å) taillight of an automobile, a physicist notices that the taillight appears as a multiple vertical array of images (an array of principal maxima). The angular separation between adjacent images is 2×10^{-3} radian. From this information deduce the vertical spacing between the threads in the batiste curtain.

*35. A good grating cut in speculum metal has 5900 lines per centimeter. If this grating is illuminated with white light ranging over wavelengths from 4000 Å to 7000 Å, it will produce a spectrum ranging over some interval of angles. From what angle to what angle does the first-order spectrum extend? The second-order spectrum? The third-order spectrum? Do these angular intervals overlap? Is the third-order spectrum complete?

*36. Interference effects can be detected with sound waves passing through a picket fence. Suppose that the fence consists of vertical boards or rods separated by a distance of 20 cm. Suppose you stand on one side of the fence, and a friend stands on the other side of the fence, at some reasonably large distance at an angle of 30° away from the direction perpendicular to the fence. If you clap your hands, producing white noise, what wavelengths and frequencies will your friend hear strongly?

*37. A reflection grating consists of a metal plate in which a large number of closely spaced parallel grooves have been cut.
  (a) Show that if light of wavelength λ is obliquely incident on this grating at an angle ϕ (see Figure 39.36), the light returned by the grating at an angle θ interferes constructively if

  $$|d \sin \theta - d \sin \phi| = 0, \lambda, 2\lambda, \ldots$$

  (b) For what angle of incidence ϕ will the direction of the emerging light overlap with the direction of the incident light?
  (c) For what angle θ does the reflection grating give constructive interference with white light, consisting of a mixture of many wavelengths?

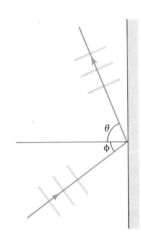

Fig. 39.36

*38. Consider the Rowland grating described in Problem 32. In the interference pattern generated by this grating, how many secondary maxima are there between two adjacent principal maxima?

*39. A carbon arc emits, among others, two spectral lines of wavelengths 4267.02 Å and 4267.27 Å. You wish to resolve these spectral lines in the second-order spectrum of a grating of total width 5.0 cm. How many slits per centimeter do you need in your grating?

CHAPTER 40

Diffraction

We saw in Chapter 17 that waves can bend and spread around obstacles. For instance, a water wave will spread out beyond a gap in a breakwater (see Figure 17.29), and it will spread around an island (see Figure 17.32). Such deviations from rectilinear propagation constitute **diffraction.** These deviations become quite large if the size of the gap, island, or other obstruction is of the same order of magnitude as the wavelength.

Since light is a wave, it displays diffraction effects. But the wavelength of light is very short, and hence these effects are difficult to detect (Figure 40.1). The diffraction effects become pronounced only when light passes through extremely narrow slits or when it strikes extremely small opaque obstacles. Diffraction effects with light had been noticed in the seventeenth century. Nevertheless, Newton held to the belief that light is of a corpuscular nature and consists of a stream of particles. The importance of diffraction effects in establishing the wave nature of light was not fully appreciated until 1818, when Augustin Fresnel mathematically formulated the wave theory of diffraction. The brilliant success of this theory finally convinced physicists that light is a wave. In this chapter we will calculate the intensity pattern produced by a light wave diffracted by a narrow slit.

Diffraction

Fig. 40.1 Diffraction of light around a razor blade. The diffracted light generates a complex pattern of fine fringes at the edges of the shadow. This photograph was prepared by illuminating the razor blade with a distant light source and throwing its shadow on a distant screen.

40.1 Diffraction by a Single Slit

Figure 40.2 shows a plane light wave approaching a slit in an opaque plate. To find the distribution of light in the space beyond the slit, we will use the following prescription, called the **Huygens–Fresnel Principle:**

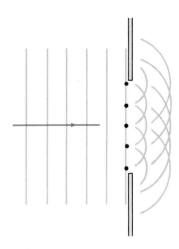

Fig. 40.2 A plane wave approaches a slit in an opaque plate. Each point on the wave front at the slit gives rise to a spherical wave that spreads out beyond the slit.

Huygens-Fresnel Principle

Augustin Fresnel (frenel), 1788–1827, French physicist and engineer. His brilliant experimental and mathematical investigations firmly established the wave theory of light. Fresnel was commissioner of lighthouses, for which he designed large compound lenses (Fresnel lenses) as a replacement for systems of mirrors.

Pretend that each point of the wave front reaching the slit can be regarded as a point source of light emitting a spherical wave; the net wave in the region beyond the slit is simply the superposition of all these waves.

This prescription has some obvious similarities to Huygens' Construction. However, the latter is merely a geometric construction for finding the successive positions of the wave fronts and does not yield any information about the distribution of intensity in different directions, whereas the aim of our new prescription is precisely the calculation of this distribution of intensity. Although the new prescription has a strong intuitive appeal, it turns out that it is not all that easy to justify it rigorously. The trouble is that light waves are not really sources of light waves — only accelerated charges are sources of light waves. The intensity pattern on the far side of the slit arises from the superposition of the original wave with all the waves radiated by the electric charges sitting in the opaque plate when accelerated by the original wave. Thus, our prescription lacks a simple physical basis. Nevertheless, it gives the right answer, or almost the right answer. This can be shown by a somewhat sophisticated mathematical argument, to be presented in Section 40.3.

We begin our calculation by finding the positions of the maxima and minima in the diffraction pattern of the slit. Figure 40.3 shows some rays associated with the waves that spread out from the points of a wave front at the slit. All the rays lead to a point P in the space beyond the slit. As in our calculation of the two-slit pattern, we will assume that P is very far away, so all the rays are nearly parallel. These rays can be divided into two equal groups: those that originate from the upper half of the slit and those that originate from the lower. Rays from the first group have a shorter distance to travel to the point P than rays from the second group. Consider the uppermost ray from the first group and the uppermost ray from the second. The path difference between these is $(a/2) \sin \theta$, where a is the width of the slit. If

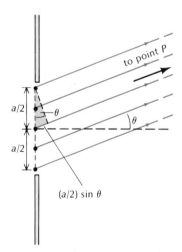

Fig. 40.3 Rays from points in the slit to the point P. From the small colored triangle we find that the path difference between the uppermost ray and the middle ray is $(a/2)\sin \theta$.

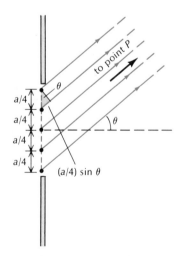

Fig. 40.4 The path difference between the uppermost ray and the ray starting at a distance $a/4$ below the uppermost ray is $(a/4)\sin \theta$.

this path difference is $\tfrac{1}{2}\lambda$, the two waves will interfere destructively. Furthermore, pairs of rays that originate at an equal distance below each of these two uppermost rays in the two groups will also have a path difference of $\tfrac{1}{2}\lambda$, and the waves will, again, interfere destructively. This establishes that all the waves cancel in pairs when

$$\frac{a}{2}\sin\theta = \tfrac{1}{2}\lambda \qquad (1)$$

or

$$a\sin\theta = \lambda \qquad (2)$$

This is the condition for the first minimum.

To find the next minimum, we divide the rays into four equal groups and consider the uppermost ray from the first and the second groups (see Figure 40.4). These rays, and other pairs of rays, will interfere destructively if their path difference is $\tfrac{1}{2}\lambda$:

$$\frac{a}{4}\sin\theta = \tfrac{1}{2}\lambda \qquad (3)$$

or

$$a\sin\theta = 2\lambda \qquad (4)$$

By continuing this argument, we find a general condition for minima:

$$\boxed{a\sin\theta = \lambda,\ 2\lambda,\ 3\lambda,\ \ldots} \qquad (5)$$

Minima for a single slit

As regards the maxima, there is, of course, a strong central maximum ($\theta = 0$). The secondary maxima cannot be found by any simple argument; roughly, their position is halfway between the successive minima. On a photographic film placed at some distance, the maxima and minima show up as a pattern of bright and dark fringes (Figure 40.5).

An essential ingredient in this calculation is the Fraunhofer approximation, that is, the distance to the point at which we observe the intensity must be very large compared with the width of the slit, so the simple formulas $(a/2)\sin\theta$ or $(a/4)\sin\theta$ for the path difference between the rays from different parts of the slit are applicable. If this approximation is not satisfied, the calculation of the intensity pattern becomes much more complicated. As an alternative to the Fraunhofer approximation, it is sometimes possible to use the **Fresnel approxima-**

Fresnel approximation

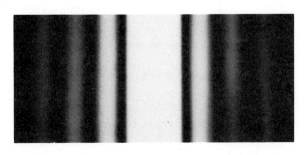

Fig. 40.5 A photographic film placed beyond an illuminated slit records a strong central maximum and successively weaker secondary maxima. (Courtesy C. C. Jones, Union College.)

tion, according to which the distance to the point of observation must be larger than the width of the slit, but need not be *very* large. (The precise distinction between the Fraunhofer and the Fresnel approximations is that in the former the rays from the aperture to the point of observation are parallel, whereas in the latter the rays may make small angles with each other.) We will not deal with this alternative approximation here. In the immediate vicinity of the slit, neither the Fraunhofer nor the Fresnel approximation is satisfied, and the only recourse is an exact calculation based on Maxwell's equations. This leads to an extremely difficult, almost intractable mathematical problem, which has been solved only in a few rare cases involving slits or apertures of exceptionally simple shape.

Note that in the limiting case of very short wavelength, the angles of the minima and maxima in the diffraction pattern shrink toward zero, that is, $\theta \to 0$ as $\lambda \to 0$ in Eq. (5). Under these conditions, the light propagates rectilinearly through the slit, without any significant diffraction. Thus, in the limiting case of very short wavelength, wave optics merges into geometrical optics, and light propagates along rectilinear rays.

EXAMPLE 1. Equation (5) applies not only to light waves but also to radio waves and other waves. Suppose that radio waves from a TV transmitter with a wavelength of 0.80 m strike the wall of a large building. In this wall there is a very wide window with a height of 1.4 m (a horizontal slit). The wall is opaque to radio waves and the window is transparent. What is the angular width of the central maximum of the diffraction pattern formed by the waves inside the building?

SOLUTION: According to Eq. (5), the first minimum is at

$$\theta = \sin^{-1}(\lambda/a) = \sin^{-1}(0.80 \text{ m}/1.4 \text{ m}) = 34.8°$$

The central maximum extends from $-34.8°$ to $+34.8°$, i.e., it has a width of $69.6°$.

To calculate the complete intensity distribution as a function of angle, we must sum, or rather integrate, all the spherical waves originating from all the points of the wave front at the slit.[1] A short segment dy of the slit (Figure 40.6) contributes an amplitude

$$dE = \frac{A}{r} \sin\left(\omega t - \frac{\omega r}{c}\right) dy \qquad (6)$$

at the point *P*. This expression is a spherical wave starting at the point y [compare Eq. (39.24)]. The constant A is not determined by our prescription, but this is of no importance if we concentrate our attention on the intensity *pattern*, that is, the distribution of relative intensity.

With the usual assumption that the distance r is large, the rays are nearly parallel, and we can approximate r as in Eq. (39.26),

$$r = r_0 - y \sin \theta \qquad (7)$$

where r_0 is the distance measured from the midpoint of the slit. The

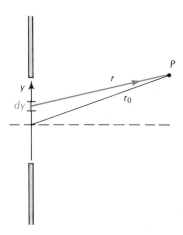

Fig. 40.6 A short segment dy of the slit.

[1] As in the calculation of Section 39.4, we consider only the waves emanating from points in the slit that are in the plane of Figure 40.6; other points in the slit contribute similar waves, but do not affect our final results.

distance r appears in both the denominator and the numerator of Eq. (6). The denominator is not very sensitive to the value of r, and the crude approximation $r \cong r_0$ suffices there. The numerator is more sensitive, and we must use the better approximation (7). Thus

$$dE = \frac{A}{r_0} \cos\left(\omega t - \frac{\omega r_0}{c} + \frac{\omega y}{c} \sin \theta\right) dy \tag{8}$$

We can easily integrate this over y; the limits of integration are $y = \pm a/2$:

$$E = \int_{-a/2}^{a/2} \frac{A}{r_0} \cos\left(\omega t - \frac{\omega r_0}{c} + \frac{\omega y}{c} \sin \theta\right) dy \tag{9}$$

$$= \frac{A}{r_0} \frac{1}{(\omega/c)\sin \theta} \left[\sin\left(\omega t - \frac{\omega r_0}{c} + \frac{\omega a}{2c} \sin \theta\right) \right.$$

$$\left. - \sin\left(\omega t - \frac{\omega r_0}{c} - \frac{\omega a}{2c} \sin \theta\right) \right] \tag{10}$$

We can combine the two terms in the bracket by means of the trigonometric identity

$$\sin \alpha - \sin \beta = 2 \cos \tfrac{1}{2}(\alpha + \beta) \sin \tfrac{1}{2}(\alpha - \beta) \tag{11}$$

This yields

$$E = \frac{2A}{r_0(\omega/c)\sin \theta} \cos\left(\omega t - \frac{\omega r_0}{c}\right) \sin\left(\frac{\omega a}{2c} \sin \theta\right) \tag{12}$$

To find the intensity of light, we must evaluate E^2. A calculation similar to what we did for the double slit [see Eqs. (39.31)–(39.33)] gives us

$$\boxed{[\text{intensity}] \propto \frac{1}{r_0^2} \frac{\sin^2 [(\pi a/\lambda) \sin \theta]}{\sin^2 \theta}} \tag{13}$$

Intensity for single-slit diffraction pattern

This is the formula for the intensity pattern produced by a single slit. Figure 40.7 is a plot of the intensity as a function of θ for the special case $a = 3\lambda$. As expected, there is a maximum at $\theta = 0$ and there are minima at the positions given by Eq. (5).

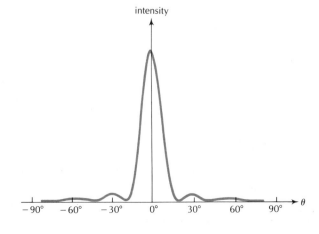

Fig. 40.7 Intensity as a function of θ for light arriving at a screen placed behind a narrow slit of width $a = 3\lambda$.

Note that the width of the central peak *increases* as the width of the slit *decreases*. The central peak extends from the first minimum on the left to the first minimum on the right (see Figure 40.7). The angular half-width of the peak is therefore given by the formula for the first minimum,

$$\sin\theta = \frac{\lambda}{a} \tag{14}$$

This angle θ increases as a decreases. If $a < \lambda$, then there is no minimum, that is, the central peak is so wide that it covers all angles.

Our two-slit calculation of the preceding chapter was based on the implicit assumption that each slit is very narrow ($a \ll \lambda$), so the diffraction pattern is very wide. Each slit by itself would then give essentially uniform illumination over all angles. If so, the light distribution is completely determined by the interference of the waves that spread out from the two slits.

However, if the slits are not very narrow, then the light distribution is determined by an interplay of diffraction at each slit and interference between slits. It turns out that the resulting pattern is the product of an interference curve [Eq. (39.33)] and a diffraction curve [Eq. (13)]. For example, Figure 40.8 shows the intensity pattern produced by two slits of width $a = 8\lambda$ separated by a center-to-center distance $d = 26\lambda$; Figure 40.9 is a photograph of this intensity pattern.

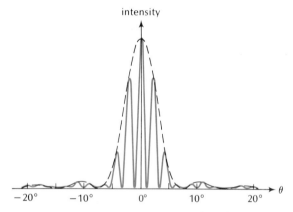

Fig. 40.8 Intensity as a function of θ for light arriving at a screen placed behind two slits of width $a = 8\lambda$ separated by a center-to-center distance of $d = 26\lambda$. The dashed curve (black) is the diffraction curve for a single slit.

Fig. 40.9 A photographic film placed beyond the two slits records a complex pattern of bands. (Courtesy C. C. Jones, Union College.)

40.2 Diffraction by a Circular Aperture; Rayleigh's Criterion

In principle, the diffraction of light by a circular aperture is similar to the diffraction by a slit (a very long rectangular aperture). To calculate the light distribution, we must sum the waves originating at all points of a wave front at the circular aperture. The evaluation of the resulting integral is a bit messy, and we will not attempt it. Figure 40.10 shows a plot of the intensity distribution as a function of the usual angle θ; the distribution is, of course, axially symmetric. The overall

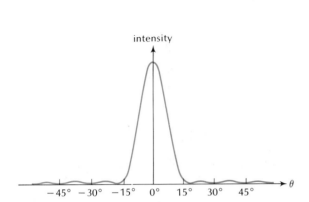

Fig. 40.10 Intensity as a function of θ for light arriving at a screen placed behind a circular aperture of diameter $a = 4\lambda$. Note that this is somewhat similar to Figure 40.7.

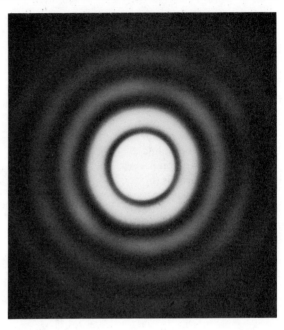

Fig. 40.11 A photographic film placed beyond a circular aperture records a strong central maximum and annular secondary maxima. (Courtesy C. C. Jones, Union College.)

shape of the curve plotted in Figure 40.10 is quite similar to that of Figure 40.7; there is a strong central maximum and a sequence of minima and secondary maxima. The angular position of the first minimum is given by

$$a \sin \theta = 1.22\lambda \tag{15}$$

First minimum for circular aperture

where a is the diameter of the circular aperture.

Figure 40.11 is a photograph of the diffraction pattern produced by a circular aperture.

Many optical instruments — telescopes, microscopes, cameras, and so on — have circular apertures, and these will diffract light. For instance, the objective lens of an astronomical telescope will act like a circular opening in a plate; the parallel wave fronts arriving from some distant star will suffer diffraction effects at this aperture, and produce an intensity distribution such as that plotted in Figure 40.10. The image of the star as seen through this telescope will then be not a bright point, but a disk surrounded by concentric rings as in Figure 40.11. For example, seen through a telescope with an objective lens 6 cm in diameter, stars look like small disks, about 2×10^{-5} radian (or 4 seconds of arc) across.

This spreading out of the image puts a limit on the detail that can be perceived through the telescope. If two stars are very close together, their images tend to merge, and it may be impossible to tell them apart. Figures 40.12a–d show the images produced by a pair of pointlike light sources upon diffraction by a circular aperture. In the first of these figures, the angular separation of the sources is so small that the two images look like one image. In the second figure, the angular separation is large enough to give a clear indication of the existence of two separate images.

John William Strutt, 3rd Baron Rayleigh, *1842–1919, English physicist, professor at Cambridge and at the Royal Institution. He is best known for his extensive mathematical investigations of sound and of light. He also investigated the behavior of gases at high densities, and he discovered argon; for this he was awarded the Nobel Prize in 1904.*

Fig. 40.12 When the light waves from two pointlike sources arrive at a circular aperture simultaneously, each set of light waves will produce its own diffraction pattern. If the angular separation between the two sources is small, the diffraction patterns overlap. In photograph (a), the angular separation is very small. In photographs (b), (c), and (d), the angular separation is progressively larger. In the first photograph, the two light sources are not resolved (not distinct). In the second photograph, they are just barely resolved, according to Rayleigh's criterion. (Courtesy C. C. Jones, Union College.)

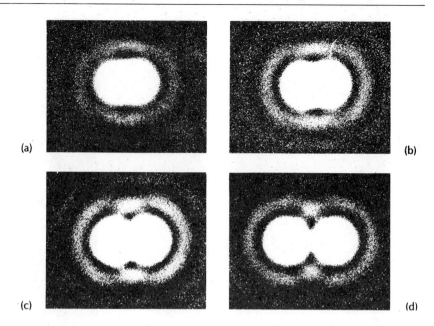

The angular separation of the sources in Figure 40.12b is such that the central maximum of the diffraction pattern of one source coincides with the minimum of the diffraction pattern of the other source. Since we are now dealing with small angles, Eq. (15) tells us that this angular separation must be

Rayleigh's criterion

$$\Delta\theta = 1.22 \frac{\lambda}{a} \qquad (16)$$

We will regard this as the critical angle that decides whether the two sources are clearly distinguishable: the telescope (or other optical device) can resolve the sources if their angular separation is larger than $1.22\,\lambda/a$, and it cannot resolve them if the separation is smaller. This is **Rayleigh's criterion.**

EXAMPLE 2. The star ζ Orionis is a binary star, that is, it consists of two stars very close together. The angular separation of the stars is 2.8 seconds of arc. Can the stars be resolved with a telescope having an objective lens 6 cm in diameter? Assume that the wavelength of the starlight is 5500 Å.

SOLUTION: According to Rayleigh's criterion, a telescope of this aperture can resolve stars as close as

$$\Delta\theta = 1.22 \frac{\lambda}{a} = 1.22 \times \frac{5.5 \times 10^{-7} \text{ m}}{0.06 \text{ m}}$$

$$= 1.1 \times 10^{-5} \text{ radian} \qquad (17)$$

This is equal to 2.3 seconds of arc. Hence the telescope can resolve the double stars.

COMMENTS AND SUGGESTIONS: Keep in mind that the formula (16) for Rayleigh's criterion always gives the angle $\Delta\theta$ in radians. For the conversion of this angle into seconds of arc, use the equivalence 1 second of arc = $\frac{1}{3600}° = 4.85 \times 10^{-6}$ radian.

The ability of a telescope to resolve stars or other objects of small angular separation improves with the size of the telescope — a telescope of 30-cm diameter can resolve angular separations as small as 0.5 seconds of arc. However, beyond 30 cm, the resolution of an Earth-bound telescope does not improve any further with size. The trouble is that fluctuations in the density of the atmosphere produce slight perturbations in the paths of light rays coming down from the sky. This causes a jitter in the apparent position of the star. Even under favorable atmospheric conditions — designated as **good seeing** by astronomers — the jitter is usually at least 0.5 seconds of arc. Thus, the very large aperture of the 5.1-m (200-in.) telescope on Palomar Mountain does not improve the resolution, but it does collect a very large amount of light, and therefore permits the observation of very faint objects in the sky.

The Space Telescope, to be launched sometime soon, will be placed in orbit above the atmosphere of the Earth, and it will therefore not be affected by atmospheric seeing conditions. This telescope (Figure 40.13) has an aperture of 2.4 m, and it will achieve an angular resolution of about 0.1 second of arc, close to the limit set by Rayleigh's criterion. Thus, the Space Telescope will achieve higher resolution than any Earth-bound telescope operating with visible light. Furthermore, it will be able to observe at ultraviolet and infrared wavelengths, which are blocked by the atmosphere.

For Earth-bound radiotelescopes, the atmosphere poses no problem — radio waves do not suffer from the effects of atmospheric density fluctuations.[2] Hence, with increasing size, the resolution of a radiotelescope improves indefinitely. Figure 40.14 shows the large radiotelescope at Arecibo, Puerto Rico. The concave "mirror" of this telescope has an aperture of 300 m and a radius of curvature which is also 300 m. The shortest wavelength at which this telescope has been

Fig. 40.13 The Space Telescope.

Fig. 40.14 Arecibo radiotelescope.

---

[2] The density fluctuations occur over such small distances that radio waves do not notice them.

operated is 4 cm. For this wavelength, Rayleigh's criterion gives a limiting angular resolution of

$$\theta = 1.22 \frac{\lambda}{a} = 1.22 \times \frac{0.04 \text{ m}}{300 \text{ m}}$$

$$= 1.6 \times 10^{-4} \text{ radian}$$

which is about 30 seconds of arc.

A substantial improvement in resolution is attainable by employing two radiotelescopes separated by a large distance. If the signals detected by these two telescopes are brought into superposition, then the system will act essentially like the two-slit system of Section 39.4. The limiting angular resolution is roughly the angle between a maximum and the adjacent minimum,

$$\Delta\theta \cong \frac{\lambda}{d} \tag{18}$$

where d is the distance between the telescopes. In current applications of this method, called very-long-baseline interferometry (VLBI), astronomers employ radiotelescopes separated by up to 10,000 km. When the radiotelescopes are separated by such intercontinental distances, it is not practical to connect both antennas to the same, single radio receiver. Instead, each antenna is connected to its own radio receiver, recordings of the received radio signal are made on magnetic tapes, and the signals are later superposed electronically. Since it is essential that the synchronism of the signals be preserved, extremely accurate atomic clocks are needed at each radiotelescope to tag the times at which the recordings are made. Angular resolutions of better than 0.001 seconds of arc have been achieved by this method (see Figure 40.15). An array of ten matched radiotelescopes (the Very-Long-Baseline Array, or VLBA), to be employed exclusively for interferometry, is under construction. This array will stretch for 8000 km across the United States, from Hawaii to Puerto Rico.

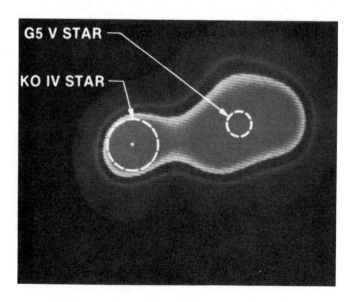

Fig. 40.15 This picture of the radio emission from the binary star UX Arietis was prepared by superposition of the signals from six radiotelescopes. The images of the two stars are well resolved, even though the separation between them is only 0.002 seconds of arc.

40.3* Babinet's Principle

In this last section we will present the theoretical justification for the Huygens–Fresnel Principle. According to this principle, the amplitude of the light in the region beyond an opaque plate with an aperture of a given shape (single slit, double slit, circular hole, etc.) can be calculated by pretending that the points of the aperture act as point sources of light. To gain some insight into the actual mechanism that produces the amplitude of the light, consider first an opaque plate whose aperture has been closed with a plug that exactly fits the opening (Figure 40.16). Obviously, when we place such a plugged plate in the path of a light wave, the amplitude of the light beyond the plate will be zero. It is instructive to inquire, How does the plate stop the light wave? The absence of light beyond the plate is a cancellation effect. When the wave strikes the plate, it shakes the electrons within the atoms and causes them to radiate light of the same wavelength as that of the incident light. The original wave passes through the plate and reaches the region beyond the plate. The radiated waves from all the electrons also reach this region. The net amplitude of the light in this region is then the sum of the contribution of the original wave plus the contribution from all the electrons — in order that the plugged plate be truly opaque, these two contributions must cancel exactly.

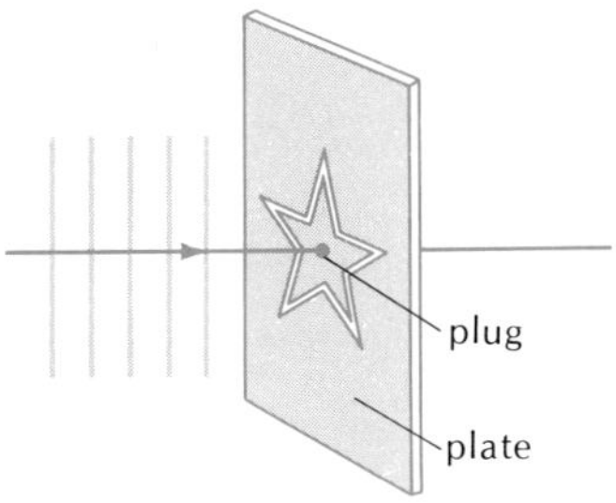

Mathematically, we can express this as follows: At some point beyond the plate, the electric field contributed by the radiating electrons has an amplitude $\mathbf{E}_{rad}$, and the electric field contributed by the original wave has an amplitude $\mathbf{E}_0$. These must be opposite,

$$\mathbf{E}_{rad} = -\mathbf{E}_0 \qquad (19)$$

The field contributed by the electrons can be split into two parts: that due to the electrons in the perforated plate (without the plug) and that in the plug. Thus, Eq. (19) becomes[3]

$$\boxed{\mathbf{E}_{rad,\,plate} + \mathbf{E}_{rad,\,plug} = -\mathbf{E}_0} \qquad (20)$$

This relationship between the fields radiated from the perforated plate and the plug is called **Babinet's Principle**.[4] The plug can be regarded as a plate of a shape **complementary** to that of the perforated plate. With this terminology, Babinet's Principle simply states that the sum of the fields radiated from a plate and from its complement equals the negative of the field of the original wave.

Fig. 40.16 An opaque plate with an aperture. The aperture has been closed with a plug.

Jacques Babinet, *1794–1872, French physicist, professor at the Collège Louis-le-Grand in Paris. He worked on the theory of diffraction, the optics of meteorological phenomena, and the design of optical instruments.*

---

* This section is optional.

[3] This equation is not exact. An unperforated plate, made of a single piece of material, is not completely equivalent to a perforated plate plus a separate plug. The motion of the electrons on the perforated plate and on the plug is subject to the constraint that the electrons cannot cross the boundary separating the perforated plate and the plug; the motion on the unperforated plate is not subject to this constraint. This difference leads to small deviations from Eq. (20); but these deviations are important only in regions fairly near the plate–plug boundary. As in the preceding sections, we will deal only with points at a large distance from the plate, and there Eq. (20) is a good approximation.

[4] For another version of Babinet's Principle, see below.

With these preliminaries, we are now ready to examine the amplitude beyond the plate when the plug is *absent*. This amplitude has a contribution from the radiating electrons in the perforated plate and a contribution from the original wave:

$$\mathbf{E}_{\text{beyond plate}} = \mathbf{E}_{\text{rad, plate}} + \mathbf{E}_0 \qquad (21)$$

But, in view of the identity (20), the right side can be replaced by $-\mathbf{E}_{\text{rad,plug}}$. We then obtain

$$\mathbf{E}_{\text{beyond plate}} = -\mathbf{E}_{\text{rad, plug}} \qquad (22)$$

Thus, apart from a minus sign, the amplitude beyond the plate happens to be just what we would calculate if we were to pretend that the points in the aperture radiate as point sources (or electrons). Of course, this is exactly what the Huygens–Fresnel Principle tells us to do — we have therefore obtained its justification. The minus sign on the right side of Eq. (22) is left out of account by the Huygens–Fresnel Principle, but this is of no importance, since ultimately we are interested only in the *intensity*, and this depends only on the square of the electric field.

Finally, let us rephrase the Babinet Principle in terms of the amplitude of the diffracted light. Suppose we remove the perforated plate from the light beam and instead place the plug there, that is, we replace the plate by the complementary plate. Then an argument similar to that which led to Eq. (22) yields

$$\mathbf{E}_{\text{beyond plug}} = -\mathbf{E}_{\text{rad, plate}} \qquad (23)$$

By substituting Eqs. (22) and (23) into Eq. (20) we obtain

Babinet's Principle

$$\boxed{\mathbf{E}_{\text{beyond plate}} + \mathbf{E}_{\text{beyond plug}} = \mathbf{E}_0} \qquad (24)$$

This identity is another form of Babinet's Principle. It states that the sum of the amplitudes of light diffracted by a plate and by its complement equals the amplitude of the original wave.

Equation (24) has an interesting application. Suppose that we want to calculate the intensity of light in some region outside of the path of the original light wave (see Figure 40.17). In this region $\mathbf{E}_0 = 0$. Hence Eq. (24) gives

$$\mathbf{E}_{\text{beyond plate}} = -\mathbf{E}_{\text{beyond plug}} \qquad (25)$$

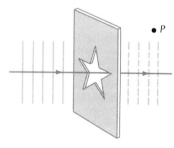

Fig. 40.17 The point P is not in the path of the incident light wave.

Since the intensity of light depends on the square of the electric field, Eq. (25) shows that a plate and its complement produce the same diffraction pattern! For instance, an arrangement of star-shaped holes in a plate and the complementary arrangement of opaque stars suspended in midair produce the same diffraction pattern. Figure 40.18 shows these complementary plates, and Figure 40.19 shows photographs of the diffraction patterns. Of course, the diffraction patterns differ in the central region which is in the path of the original wave; but elsewhere they are identical.

Fig. 40.18 (a) An opaque plate with star-shaped holes. (b) The complementary plate consists of opaque stars. (Courtesy C. C. Jones, Union College.)

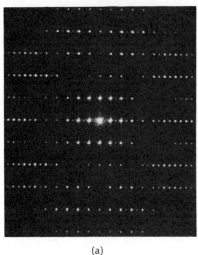
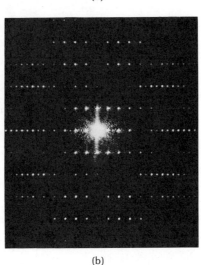

Fig. 40.19 (a) Photograph of the diffraction pattern of the plate shown in Figure 40.18a. (b) Photograph of the diffraction pattern of the complementary plate shown in Figure 40.18b. The two diffraction patterns are identical, except in the central region, where the undiffracted light beam passes through the complementary plate with high intensity. (Courtesy C. C. Jones, Union College.)

SUMMARY

Minima for single slit: $a \sin \theta = \lambda, 2\lambda, 3\lambda, \ldots$

First minimum for circular aperture: $a \sin \theta = 1.22\lambda$

Rayleigh's criterion: $\Delta\theta = 1.22\lambda/a$

Babinet's Principle: $\mathbf{E}_{\text{beyond plate}} + \mathbf{E}_{\text{beyond plug}} = \mathbf{E}_0$

QUESTIONS

1. Gratings used for analyzing light are often called diffraction gratings, but it would be more accurate to call them interference gratings. Why?

2. Consider the diffraction pattern shown in Figure 40.7. How would this pattern change if the entire apparatus were immersed in water?

3. Figure 17.28 shows the diffraction of water waves at the entrance of a harbor. Could this diffraction be eliminated by making the entrance smaller?

4. What would you expect the diffraction pattern of a rectangular aperture to look like?

5. In the center of the shadow of a disk or a sphere there is a small bright spot, called the **Poisson spot** (Figure 40.20). This spot is very faint near the disk but becomes more noticeable at large distances. Qualitatively, explain why the diffraction of the light waves around the edges of the disk gives rise to this spot.

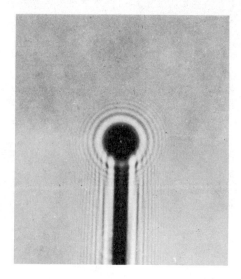

Fig. 40.20 The Poisson spot within the shadow of a pin.

Fig. 40.21 Atoms in a barium titanate crystal.

Fig. 40.22 A portion of the compound eye of a dragonfly. This eye consists of about 28,000 ommatidia. (Courtesy R. Olberg, Union College.)

6. Besides good angular resolution, what other advantage does a telescope of large aperture have over a telescope of small aperture?

7. Spy satellites use cameras with lenses of very large diameter, 30 cm or more. Why are such large diameters necessary?

8. What might be the advantages of building a radio telescope in space, in orbit around the Earth?

9. In order to beam a sound wave sharply in one direction with a loud-hailer, should the horn of the loud-hailer have a small aperture or a large aperture?

10. The manufacturers of the Questar telescope claim that this telescope can distinguish two stars even if their separation is somewhat smaller than that specified by Rayleigh's criterion. Why is this possible?

11. The maximum useful magnification of an optical microscope is determined by diffraction effects at the objective lens. Explain.

12. Other things being equal, how much resolution can you gain by operating an optical microscope with blue light instead of red light? Why can you not operate an ordinary microscope with ultraviolet light?

13. The picture of atoms in Figure 40.21 was taken with an electron microscope using electron waves of extremely short wavelength. Roughly, how short must the wavelength be to make individual atoms visible?

14. The compound eye of insects consists of a large number of small eyes, or ommatidia (Figure 40.22). Each ommatidium is typically 0.03 mm across; it does not form an image, but merely acts as a sensor of the intensity of light arriving from a narrow cone of directions. What are some of the advantages and disadvantages of such a compound eye as compared to the camera eye of vertebrates?

15. The corona formed by diffraction of sunlight by small droplets of water in a cloud consists of a bright ring surrounding the Sun. What is the color of the outer edge of the corona? The inner edge?

16. Antennas used for the transmisson of microwaves in communication links consist of metal dishes. What factors determine the size of these dishes?

17. Very-long-baseline interferometry (VLBI) with two or more radiotelescopes placed on different parts of the Earth has been used for precise determinations of the rate of continental drift and also the rate of rotation of the Earth (sidereal day). Explain how data collected by radiotelescopes monitoring a fixed radio source in the sky can be used for such purposes.

18. What would the diffraction pattern of a narrow opaque strip look like? Of a small opaque disk?

19. How would the diffraction patterns of Figures 40.19a and b change if all the stars in Figure 40.18 were made smaller?

20. What feature of the plate shown in Figure 40.18a determines the spacing between the dots in Figure 40.19a? What feature determines the gradual modulation of the intensity of these dots?

PROBLEMS

Section 40.1

1. Light of wavelength 6328 Å from a He–Ne laser illuminates a single slit of width 0.10 mm. What is the width of the central maximum formed on a screen placed 2.0 m beyond the slit?

2. Consider the water waves shown in Figure 40.23. With a ruler, measure the wavelength of the waves and the length of the gap; with a protractor, measure the angular positions of the two nodal lines (minima). Check whether these quantities satisfy Eq. (5).

3. A sound wave of frequency 820 Hz passes through a doorway of width 1.0 m. What are the angular directions of the minima of the diffraction pattern?

4. Water waves of wavelength 20 m approach a harbor entrance 50 m across at right angles to their path. What is the angular width of the central beam of diffracted waves beyond the entrance?

Fig. 40.23

5. You want to prepare a photograph, such as Figure 40.5, showing diffraction by a single slit. Suppose you use the light of wavelength 5770 Å from a mercury lamp, and you place your photographic film 2.0 m behind the slit. If the width of the central maximum is to be at least 0.5 cm on your film, what width of slit do you require?

*6. Suppose you want to detect diffraction fringes in the intensity pattern produced by *sunlight* passing through a fine slit cut in the blinds covering a window. You have available a filter that blocks all wavelengths except 5500 Å, and you place this filter over the slit.
  (a) If the width of the slit is 0.1 mm, will you be able to detect diffraction fringes? Take into account that the Sun's angular diameter is about $\frac{1}{2}°$.
  (b) If the width of the slit is 0.01 mm?
  (c) If you remove the filter?

*7. Light of wavelength λ is obliquely incident on a slit of width a. The angle of incidence of the light on the slit is ϕ (Figure 40.24).
  (a) Show that the diffracted light emerging at an angle θ interferes destructively if

$$a \sin\theta - a \sin\phi = \lambda, 2\lambda, 3\lambda, \ldots$$

  (b) Show that, for small values of θ, the angular separation between the directions of destructive interference is independent of ϕ.

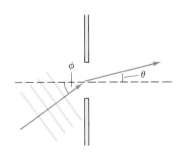

Fig. 40.24

*8. A slit of width 0.11 mm cut in a sheet of metal is illuminated with light of wavelength 5770 Å from a mercury lamp. A screen is placed 4.0 m beyond the slit.
   (a) Find the width of the central maximum in the diffraction pattern on the screen, that is, find the distance between the first minimum on the left and on the right.
   (b) Find the width of the secondary maximum, that is, find the distance between the first minimum and the second minimum on the same side.

*9. Plot intensity vs. angle for the light diffracted by a pair of slits of width $a = 4\lambda$ separated by a center-to-center distance $d = 8\lambda$.

*10. Two narrow slits of width a are separated by a center-to-center distance d. Suppose that the ratio of d to a is an integer, $d/a = n$.
   (a) Show that in the diffraction pattern produced by this arrangement of slits, the nth interference maximum (corresponding to $d \sin \theta = n\lambda$) is suppressed because of coincidence with a diffraction minimum. Show that this is also true for the $2n$th, $3n$th, etc., interference maxima.
   (b) How many interference maxima are there between one diffraction minimum on one side and the next on the same side?

*11. Figure 39.19 shows the intensity curve for the interference pattern produced by light emerging from three very thin slits (three pointlike sources) separated by distances of 12λ. Suppose the three very thin slits are replaced by three new slits of width 6λ. This will change the intensity curve.
   (a) On top of Figure 39.19 (or on a copy of Figure 39.19), plot the intensity curve for the diffraction pattern produced by *one* of these new slits.
   (b) By suitably combining these two curves, obtain the complete intensity curve for the system of the three new slits.

*12. Figure 40.25 is a plot of the intensity of the light on a screen placed beyond an opaque plate with two parallel narrow slits. The distance between the slits is 0.12 mm. Deduce the width of each of the slits. Deduce the wavelength of the light.

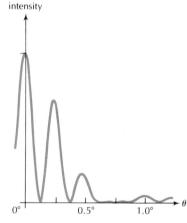

Fig. 40.25 Intensity as a function of θ in the diffraction pattern of two slits.

Section 40.2

13. According to a recent proposal (see Section V.4), solar energy is to be collected by a large power station on an artificial satellite orbiting the Earth at an altitude of 35,000 km. The power is to be beamed down to the surface of the Earth in the form of microwaves. If the microwaves have a wavelength of 10 cm and if the antenna emitting the microwaves is 1.5 km in diameter, what is the angular width of the central beam emerging from this antenna? What will be the transverse dimension of the beam when it reaches the surface of the Earth?

14. When the eye looks at a star (a point of light), diffraction at the pupil spreads the image of the star on the retina into a small disk.
   (a) When opened to maximum size, the diameter of the pupil of a human eye is 7.0 mm. Assuming the starlight has a wavelength of 5500 Å, what is the angular size of the image on the retina?
   (b) The distance from pupil to retina is 23 mm. What is the linear size of the image of the star?
   (c) At the midpoint on the retina (fovea), there are 150,000 light-sensitive cells (rods) per mm^2. How many of these cells are illuminated when the eye looks at a star?

15. A sailor uses a speaking trumpet to concentrate his voice into a beam. The opening at the front end of the speaking trumpet has a diameter of 25 cm. If the sailor emits a sound of wavelength 15 cm (this is very roughly the wavelength a man emits when yelling "eeeee..."), what is the angular width of the central maximum of the beam of sound?

16. The antenna of a small radar transmitter operating at 1.5×10^{10} Hz consists of a circular dish of diameter 1.0 m. What is the angular width of the cen-

tral maximum of the radar beam? What is the linear width at a distance of 5.0 km from the transmitter?

17. The Space Telescope that will be placed into an orbit above the atmosphere of the Earth has an aperature of 2.4 m. According to Rayleigh's criterion, what angular resolution can this telescope achieve with visible light of wavelength 5500 Å? With ultraviolet light of wavelength 1200 Å? How much better is this than the angular resolution of 0.5 seconds of arc achieved by telescopes on the surface of the Earth?

18. The radiotelescope at Jodrell Bank (England) is a dish with a circular aperture of diameter 76 m. What angular resolution can this radiotelescope achieve when operating at a wavelength of 21 cm?

*19. According to newspaper reports, a photographic camera on a "Blackbird" reconnaissance jet flying at an altitude of 27 km can distinguish detail on the ground as small as the size of a man.
   (a) Roughly, what angular resolution does this require?
   (b) According to Rayleigh's criterion, what minimum diameter must the lens of the camera have?

*20. Rumor has it that a photographic camera on a spy satellite can read the license plate of an automobile on the ground.
   (a) If the altitude of the satellite is 160 km, roughly what angular resolution does the camera need to read a license plate? Assume that the reading requires a linear resolution of about 5 cm.
   (b) To attain this angular resolution, what must be the diameter of the aperture of the camera?

*21. A microwave antenna used to relay communication signals has the shape of a circular dish of diameter 1.5 m. The antenna emits waves with $\lambda = 4.0$ cm.
   (a) What is the width of the central maximum of the beam of this antenna at a distance of 30 km?
   (b) The power emitted by the antenna is 1.5×10^3 W. What is the energy flux directly in front of the antenna? What is the energy flux at a distance of 30 km? Assume that the power is evenly distributed over the width of the central beam.

*22. For an optically perfect lens, the size of the focal spot is limited only by diffraction effects. Suppose that a lens of diameter 10 cm and focal length 18 cm is illuminated with parallel light of wavelength 5500 Å. What is the angular width of the central maximum in the diffraction pattern? What is the corresponding linear width at the focal distance?

*23. Kalahari bushmen are said to be able to see the four moons of Jupiter with the naked eye. According to Rayleigh's criterion, what must be the minimum separation between two small light sources placed at the distance of Jupiter if they are to be resolved by the human eye? Compare this with the separations between the moons of Jupiter. Does the limit on the resolving power of the human eye prevent *you* from seeing the moons? The (average) distance to Jupiter is 630×10^6 km; the separations between the moons are given in Chapter 9. Assume that the diameter of the pupil of the eye is 7 mm and that the wavelength of the light is 5500 Å.

*24. Mars has a radius of 3400 km; when Mars is at its closest to the Earth, its distance is 628×10^6 km. Calculate the angular size of Mars as seen from Earth. Estimate the diameter of the objective lens of the telescope of smallest size that will permit you to tell that Mars has a disk, i.e., that the image of Mars is wider than the image of a star.

*25. (a) According to Rayleigh's criterion, what is the angular resolution that the human eye can achieve for light of wavelength 5500 Å? The fully distended pupil of the human eye has a diameter of 7.0 mm.
   (b) Even during steady fixation, the eye has a spontaneous tremor that swings it through angles of 20 or 30 seconds of arc. Compare this an-

gular tremor with the angular resolution that you found in part (a). Would the elimination of the tremor greatly improve the acuity of the eye?

*26. At night, on a long stretch of straight road in Nevada, a truck driver sees the distant headlights of another truck. How close must he be to the other truck in order for his eyes to resolve two headlights? Assume that the pupils of the truck driver have a diameter of 5.0 mm, that the headlights are separated by 1.8 m, and that the light has a wavelength of 5500 Å.

*27. Some spy satellites carry cameras with lenses 30 cm in diameter and with a focal length of 2.4 m.
  (a) What is the angular resolution of the camera according to Rayleigh's criterion? Assume that the wavelength of light is 5500 Å.
  (b) If such a satellite looks down on the Earth from a height of 150 km, what is the distance between two points on the ground that the camera can barely resolve?
  (c) The lens projects images of the two points on a film at the focal plane of the lens. What is the distance between the two images projected on the film?

*28. 7×50 binoculars magnify angles by a factor of 7, and their objective lenses have an aperture of 50-mm diameter.
  (a) According to Rayleigh's criterion, what is the intrinsic angular resolution of these binoculars? Assume that the light has a wavelength of 5500 Å.
  (b) At best, the pupil of your eye has an aperture of 7.0-mm diameter. Compare the angular resolution of your eye divided by a factor of 7 with the intrinsic angular resolution of the binoculars. Which of the two numbers determines the actual angular resolution you can achieve while looking through the binoculars?

*29. When exposed to strong light, the pupil of the eye of a cat narrows to a fine slit, about 0.3 mm across. Suppose that the cat is looking at two white mice 20 m away and separated by a distance of 5 cm. Can the cat distinguish one mouse from the other?

**30. Show that the optimum diameter for the pinhole of a pinhole camera (see Section 38.3) is approximately $\sqrt{2.44\lambda L}$, where λ is the wavelength of the light, and L is the distance from the pinhole to the film. (Hint: Consider light arriving from a distant point source. If the pinhole is excessively small, diffraction at the pinhole spreads the light over a spot diameter $2.44\lambda L/a$ on the film. If the pinhole is excessively large, then the light from a distant point source illuminates a spot diameter a. What choice of a makes the smallest spot on the film?)

Section 40.3

31. A narrow opaque strip, 0.060 mm across, is placed in the light beam of wavelength 4579 Å from a laser. The diffraction pattern formed by this strip on a screen 4.8 m beyond shows bright and dark fringes. What is the distance between the second and the third dark fringe on one side of the central beam?

32. A circular opaque disk of diameter 0.12 mm is placed in the beam of a He–Ne laser emitting a wavelength of 6328 Å. The disk makes a diffraction pattern of bright and dark rings on a screen placed 5.0 m beyond. What is the diameter of the first dark circular ring in this diffraction pattern?

*33. A thin strip of opaque material 1.5 mm wide is illuminated by light of wavelength 6940 Å. At what distance beyond the strip is the width of the central maximum of the diffracted light as large as the width of the geometric shadow (1.5 mm), so that the shadow disappears?

CHAPTER 41

The Theory of Special Relativity

As we saw in Chapter 5, Newton's laws of motion are equally valid in every inertial reference frame. Consequently, no mechanical experiment can detect any intrinsic difference between two inertial reference frames. This is the Newtonian principle of relativity. The example already mentioned in Section 5.6 illustrates this principle very concretely: the behavior of billiard balls on a pool table on a ship steaming away from shore at constant velocity is not different from the behavior of similar billiard balls on the shore. Experiments with such billiard balls aboard the ship will not reveal the uniform motion of the ship. To detect this motion, the crew of the ship must take sightings of points on the shore or use some other navigational technique that fixes the position and velocity of the ship in relation to the shore. Hence, in regard to mechanical experiments, uniform translational motion of our inertial reference frame is always *relative* motion — it can be detected as motion of our reference frame only with respect to another reference frame. There is no such thing as *absolute* motion through space.

The question naturally arises whether the relativity of mechanical experiments also applies to electric, magnetic, optical, and other experiments. Do any of these experiments permit us to detect an absolute motion of our reference frame through space? Albert Einstein answered this question in the negative. He laid down a principle of relativity for *all* laws of physics. Before we deal with the details of Einstein's theory of relativity, we will briefly describe why nonmechanical experiments — and, in particular, experiments with light — might be expected to detect absolute motion which mechanical experiments cannot detect.

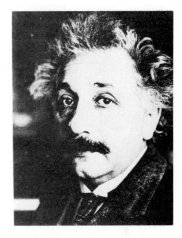

Albert Einstein, *1879–1955, German (and Swiss, and American) theoretical physicist, professor at Zurich, at Berlin, and at the Institute for Advanced Study at Princeton. Einstein was the most celebrated physicist of this century. He formulated the theory of Special Relativity in 1905 and the theory of General Relativity in 1916. In the judgment of Bertrand Russell, "relativity is probably the greatest synthetic achievement of the human intellect up to the present time." Einstein also made incisive contributions to modern quantum theory, for which he received the Nobel Prize in 1921. Einstein spent the last years of his life in an unsuccessful quest for a unified theory of forces that was supposed to incorporate gravity and electricity in a single set of equations.*

41.1 The Speed of Light; the Ether

Since the laws of mechanics are the same in all inertial reference frames, it seems quite natural to assume that the laws of electricity, magnetism, and optics are also the same in all inertial reference frames. But this assumption immediately leads to a paradox concerning the speed of light. As we saw in Chapter 36, light is an oscillating electric and magnetic disturbance propagating through space. We found that the laws of electricity and magnetism permit us to deduce that the speed of propagation of this disturbance must always be 3.00×10^8 m/s, in all reference frames. The trouble with this deduction is that, according to the Galilean addition law for velocity [Eq. (4.58)], the speed of light ought *not* to be the same in all reference frames. For instance, imagine that an alien spaceship approaching the Earth with a speed of, say, 1.00×10^8 m/s flashes a light signal toward the Earth; if this light signal has a speed of 3.00×10^8 m/s in the reference frame of the spaceship, then the Galilean addition law tells us that it ought to have a speed of 4.00×10^8 m/s in the reference frame of the Earth.

To resolve this paradox, we must either give up the notion that the laws of electricity and magnetism (and the values of the speed of light) are the same in all inertial reference frames or else we must give up the Galilean addition law for velocities. Both alternatives are unpleasant: the former means that we must abandon all hope for a principle of relativity embracing electricity and magnetism, and the latter means that together with the Galilean velocity transformation [Eq. (4.58)], we must throw out the Galilean coordinate transformations [Eqs. (4.49)–(4.56)] as well as the intuitively "obvious" notions of absolute time and absolute length from which these transformation equations were derived.

Since the failure of a relativity principle embracing electricity and magnetism seems to be the lesser of two evils, let us first explore this alternative. Let us assume that there exists a preferred inertial reference frame in which the laws of electricity and magnetism take their simplest form, that is, the form expressed in Maxwell's equations. In this reference frame the speed of light has its standard value $c = 3.00 \times 10^8$ m/s (which we will designate by c), whereas in any other reference frame it is larger or smaller according to the Galilean addition law. The propagation of light is then analogous to the propagation of sound. There exists a preferred reference frame in which the equations for the propagation of sound waves in, say, air take their simplest form: the reference frame in which the air is at rest. In this reference frame, sound has its standard speed of 331 m/s. In any other reference frame, the equations for the propagation of sound waves are more complicated, but the velocity of propagation can always be obtained very directly from the Galilean addition law. For instance, if a wind of 40 m/s (a hurricane) is blowing over the surface of the Earth, then sound waves have a speed of 331 m/s relative to the air, but their speed relative to the ground depends on direction — downwind the speed is 371 m/s, whereas upwind it is 291 m/s.

This analogy between the propagation of sound and of light suggests that there exists some pervasive medium that serves as the propagator of light. Presumably this ghostly medium fills all of space, even the interplanetary and interstellar space which is normally regarded as a vacuum. The physicists of the nineteenth century called this hypothetical medium the **ether**, and they attempted to describe light waves as oscillations of the ether. The preferred reference frame in which light has its standard speed is then the reference frame in which the ether is at rest. The existence of such a preferred reference frame would imply that velocity is absolute — the ether frame sets an absolute standard of rest, and the velocity of any body could always be referred to this frame. For instance, instead of describing the velocity of the Earth relative to some material body, such as the Sun, we could always describe its velocity relative to the ether.

Presumably, the Earth has some nonzero velocity relative to the ether. Even if the Earth were at rest in the ether at one instant, this condition could not last, since the Earth continually changes its motion as it orbits around the Sun. The motion of the ether past the Earth was called the *ether wind* by the nineteenth-century physicists. If the Sun is at rest in the ether, then the ether wind would have a velocity opposite to that of the Earth around the Sun — about 30 km/s; if the Sun is in motion, then the ether wind would vary with the seasons — smaller than 30 km/s during one half of the year and greater than 30 km/s during the other half.

Experimenters attempted to detect this ether wind by its effect on the propagation of light. A light wave in a laboratory on the Earth would have a greater speed when moving downwind and a reduced speed when moving upwind or across the wind. If the speed of the ether wind "blowing" through the laboratory is V_O, then the speed of light in this laboratory is $c + V_O$ for a light signal with downwind motion, $c - V_O$ for upwind motion, and $\sqrt{c^2 - V_O^2}$ for motion perpendicular across the wind (Figure 41.1) If V_O has a value of about 30 km/s, then the increase or decrease of the speed of light amounts to only about 1 part in 10^4 — a very sensitive apparatus is required for the detection of this small change.[1]

In a famous experiment first performed in 1881 and often repeated thereafter, A. A. Michelson and E. W. Morley attempted to detect small changes in the speed of light by means of an interferometer (see Section 39.3 for a description of the experiment). Their results were negative. The sensitivity of their original experiment was such that a wind of 5 km/s would have produced a noticeable effect. Since the expected wind is about 30 km/s, the experimental result contradicts the ether theory of the propagation of light. Later, more refined versions of the experiment established that if there is an ether wind, its speed certainly has to be less than 1.5 km/s. What is more, some new experiments with gamma rays[2] have shown that if there is any ether wind, its speed has to be less than about 3 m/s. The experimental evidence therefore establishes conclusively that the motion of the Earth has no effect on the propagation of light. The propagation of light is *not* analogous to the propagation of sound — there is no preferred reference frame. As the Earth moves around the Sun, its velocity changes continuously, and the Earth shifts from one inertial reference frame to another. But all these inertial reference frames appear to be completely equivalent in regard to the propagation of light.

Ether

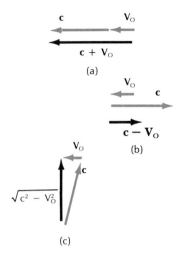

Fig. 41.1 The velocity of light is **c** relative to the ether, and the velocity of the ether is $\mathbf{V}_O$ relative to the laboratory. The velocity of light relative to the laboratory is then the vector sum $\mathbf{c} + \mathbf{V}_O$. (a) If **c** and $\mathbf{V}_O$ are parallel, the magnitude of the vector sum is $c + V_O$. (b) If **c** and $\mathbf{V}_O$ are antiparallel, the magnitude of the vector sum is $c - V_O$. (c) If **c** and $\mathbf{V}_O$ are perpendicular, the magnitude of the vector sum is $\sqrt{c^2 - V_O^2}$.

Michelson–Morley experiment

---

[1] In practice, the experiments make a comparison between the average speeds for *round trips*, upwind and downwind vs. across the wind. The average speeds for these round trips differ by only 1 part in 10^8 — this imposes *extreme* demands on the sensitivity of the apparatus.

[2] Gamma rays are a form of very energetic light (see Interlude IV).

41.2 Einstein's Principle of Relativity

Since both the laws of mechanics and the laws for the propagation of light fail to reveal any intrinsic distinction between different inertial reference frames, in 1905 Einstein boldly proposed a general hypothesis concerning *all* the laws of physics. This hypothesis is the **Principle of Relativity:**

Principle of Relativity

All the laws of physics are the same in all inertial reference frames.

This deceptively simple principle forms the foundation of the theory of Special Relativity. Since Maxwell's equations and the laws for the propagation of light are included in the laws of physics, one immediate corollary of the Principle of Relativity is

The speed of light (in vacuum) is the same in all inertial reference frames; it always has the value $c = 3.0 \times 10^8$ m/s.

As we pointed out in the preceding section, this invariance of the speed of light conflicts with the Galilean addition law for velocities. We will therefore have to throw out this law, and we will also have to throw out the Galilean coordinate transformation on which it is based. In the next section we will see what new coordinate transformation replaces the Galilean transformation.

The invariance of the speed of light also requires that we give up some of our intuitive, everyday notions of space and time. Obviously, the fact that a light signal should always have a speed of 3.0×10^8 m/s, regardless of how hard we try to move toward it or away from it in a fast aircraft or spaceship, does violence to our intuition. This strange behavior of light is possible only because of a strange behavior of length and time in relativistic physics. As we will see later in this section, neither length nor time is absolute — they both depend on the reference frame in which they are measured, and they suffer contraction or dilation when the reference frame changes.

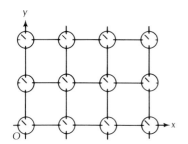

Fig. 41.2 A reference frame consisting of a coordinate grid and synchronized clocks.

Before we can inquire into the consequences of the Principle of Relativity, we must carefully describe the construction of reference frames and the synchronization of clocks. A reference frame is a coordinate grid and a set of clocks (Figure 41.2) which can be used to determine the space and time coordinates of any event, that is, any point in space and time. The grid intersections correspond directly to the space coordinates of the event (with respect to some fiducial point chosen as origin). The time registered by the clock at the event is the time coordinate. Of course, the clocks must all be synchronized with the master clock sitting at the origin of coordinates. This can be accomplished by sending out a flash of light from a point exactly midway between the clock at the origin and the other clock (see Figure 41.3). The two clocks are synchronized if both show exactly the same time when the light from the midpoint reaches them. Note that this synchronization procedure depends on the invariance of the speed of light. If the speed of light were not a universal constant, but were dependent on the reference frame and on the direction of propagation (say, faster toward the right in Figure 41.3 and slower toward the left), then we could not achieve synchronization by the simple procedure with a flash of light from the midpoint.

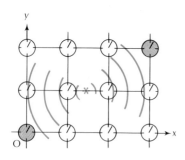

Fig. 41.3 Synchronization procedure for two clocks. A flash of light is sent from the midpoint toward the clock at the origin (lower left) and the other clock (upper right).

One immediate consequence of our synchronization procedure is that *simultaneity is relative,* that is, the simultaneity of two events depends on the reference frame. The following is a concrete example: Suppose that two lightning bolts strike in Boston and in New York, respectively, at exactly 6:00 P.M. Eastern Standard Time. The emissions of brief flashes of light from the lightning bolts are then simultaneous in the reference frame of the Earth. However, in the reference frame of a fast spaceship passing by the Earth in the direction from Boston to New York, these two events are *not simultaneous* — as judged by the clocks on board the spaceship, the lightning in Boston occurs slightly later than the lightning in New York.

To see how this difference between the two reference frames comes about, let us examine the procedure for testing simultaneity. In the reference frame of the Earth, an observer can test for simultaneity by placing herself exactly at the midpoint between Boston and New York (see Figure 41.4); she will then receive flashes of light from the lightning in the two cities at the same instant. Thus, this observer will confirm that, in the reference frame of the Earth, the lightning was simultaneous.

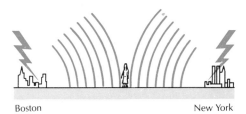

Fig. 41.4 Observer on the ground at the midpoint between Boston and New York.

In the reference frame of the spaceship, another observer can likewise test for simultaneity by placing himself exactly at the midpoint between the places of origin of the two flashes of light. For him, the places of origin of the flashes of light are two grid points in his reference frame. To help us visualize the phenomenon as the observer in the spaceship perceives it, imagine that the front and the rear ends of the (large) spaceship are actually passing through New York and through Boston when the lightning strikes, and that the lightning bolts pass through the ends of the ship, perforating its skin (see Figure 41.5). The observer in the spaceship then regards the perforations as the places of origin of the flashes of light, and, to test for simultaneity, he places himself at the midpoint between these perforations and waits for the arrival of the light at his eyes. Will he receive the flashes of light from the front and the rear of the spaceship at the same instant?[3] The answer is obvious if we adopt the point of view of the observer on

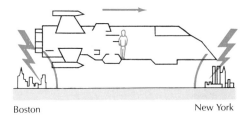

Fig. 41.5 Observer in a fast spaceship, instantaneously at the midpoint between Boston and New York. Light flashes have just been emitted.

---

[3] Note that it makes no difference whether we consider the light as propagating inside or outside the (empty) spaceship. However, as a crutch to our imagination, we find it convenient to suppose that the relevant portions of the light waves in the reference frame of the spaceship actually travel inside the spaceship.

Fig. 41.6 The observer on the spaceship encounters the light flash emanating from New York.

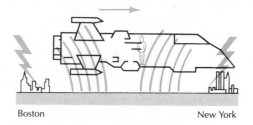

the ground: she sees the spaceship observer at the midpoint between Boston and New York when the lightning strikes, and she then sees him traveling toward New York, that is, she sees him traveling away from the flash of light trying to catch up with him from Boston and toward the flash coming to meet him from New York (see Figure 41.6). Thus, from her point of view, it is quite clear that the flash of light from New York will encounter the spaceship observer before the flash of light from Boston can catch up. But if the spaceship observer receives the flash of light from New York before the flash from Boston, he will conclude that the flashes were not emitted simultaneously — according to his reckoning, the lightning in Boston occurred late!

Although this qualitative argument shows that simultaneity depends on the reference frame, it does not tell us by how much. A quantitative calculation (see next section) shows that for a spaceship traveling at, say, 90% of the speed of light, the lightning in Boston occurs late by about 10^{-3} s.

If simultaneity is relative, then the synchronization of clocks is also relative. In the reference frame of the Earth, clocks in Boston and New York are synchronized, that is, the hands of these clocks reach the 6:00 P.M. position simultaneously. But in the reference frame of the spaceship, the clock in Boston is judged to be late — in the same way as the lightning occurs later, the hands of the clock reach the 6:00 P.M. position later than in New York. Figure 41.7 shows the clocks belonging to the reference frame of the Earth as observed at one instant from the spaceship.

Fig. 41.7 (left) Clocks of the reference frame of the Earth as observed at one instant of spaceship time (see Fig. 41.13 for a more accurate picture).

Fig. 41.8 (right) Clocks of the reference frame of the spaceship as observed at one instant of Earth time (see Fig. 41.12 for a more accurate picture).

The effect is symmetric. In the reference frame of the spaceship, all the clocks on board are synchronized. But, as observed in the reference frame of the Earth, the clocks on the front part of the spaceship are late. Figure 41.8 shows the clocks belonging to the reference frame of the spaceship as observed at one instant from the Earth. Note that in Figure 41.7 we are viewing the reference frame of the Earth moving past the spaceship, and in Figure 41.8 we are viewing the reference frame of the spaceship moving past the Earth. In either case, the clocks on the *leading edge* of the moving reference frame are *late*.

The relativity of synchronization is a direct consequence of the invariance of the speed of light, since our procedures for testing simul-

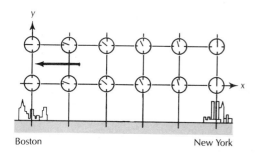

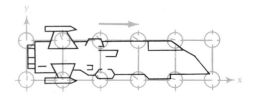

taneity depend crucially on the speed of light. The breakdown of absolute simultaneity implies that time is relative. There exists no absolute time coordinate; instead, there exist only relative time coordinates associated with particular reference frames.

The relativity of time shows up not only in the synchronization of clocks but also in the rate of clocks. A clock on board the spaceship suffers **time dilation**: the rate of the moving clock is slow compared with the rate of identically manufactured clocks at rest on the Earth. To see how this comes about, imagine that the experimenters in the spaceship set up a 300-m racetrack perpendicular to the direction of motion of the spaceship (Figure 41.9).[4] If the experimenters use one of their clocks to measure the time of flight of a light signal that goes from one end of the track to the other and returns, they will find that the light signal takes a time of 600 m/(3.00 × 10^8 m/s) = 2.0 × 10^{-6} s to complete the round trip. But experimenters on the Earth see that the light signal has concurrent vertical and horizontal motions (Figure 41.10). For the experimenters on the Earth, the light signal travels a total distance *larger* than 600 m. Since the speed must still be the standard speed of light, 3.00 × 10^8 m/s, they will find that according to their clocks the light signal now takes a time *longer* than 2.0 × 10^{-6} s to complete the trip. Thus, a given time interval registered by a clock on the spaceship is registered as a longer time interval by the clocks on the Earth. The clock on the spaceship runs slow when judged by the clocks on Earth. Note that the experiment involves *one* clock of the spaceship (the clock at the starting point of the track), but several (synchronized) clocks on the Earth, because the light signal does not return to the point at which it started on Earth.

Time dilation

Fig. 41.9 (left) Spaceship with a "racetrack" for a light pulse.

Fig. 41.10 (right) The trajectory of the light pulse as observed from the Earth.

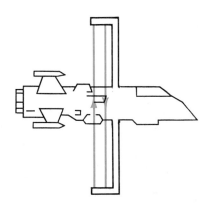
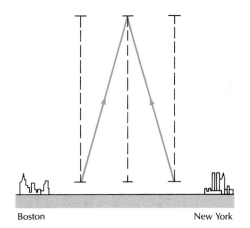

Boston   New York

It turns out that because synchronization is relative, length is also relative. A measuring rod, or any other body, on board the spaceship suffers **length contraction** along the direction of motion — the length of the moving measuring rod will be short when compared with the length of identically manufactured measuring rods at rest on the Earth. The reason for this is that the length measurement of a moving body depends on simultaneity, and since simultaneity is relative, so is length. Suppose that the spaceship carries a measuring rod that has a

Length contraction

---

[4] The track does not have to be perpendicular, but the analysis is simplest if it is.

Fig. 41.11 Spaceship with a measuring rod.

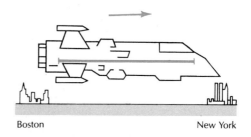

length of, say, 300 km in the reference frame of the spaceship (Figure 41.11). To measure the length of this rod in the reference frame of the Earth, we station observers in the vicinity of New York and Boston with instructions to ascertain the positions of the front and rear ends of the measuring rod at one instant of time, say, 6:00 P.M. Eastern Standard Time. But when the observers on the Earth do this, the observers on the spaceship will claim that the position measurements were not done simultaneously and that the observers in Boston measured the position of the rear end at a later time. In the extra time, the rear end moves an extra distance to the right, and hence the distance between the positions measured for the rear and front ends will be reduced. From the point of view of the observers on the spaceship, it is therefore immediately obvious that the length measured by the observers on the Earth will be *short*. Figures 41.12 and 41.13 show the reference frame of the spaceship moving past the Earth and the reference frame of the Earth moving past the spaceship, respectively. In these figures the length contraction has been included (it was left out in Figures 41.7 and 41.8).

Fig. 41.12 (left) Reference frame of the spaceship as observed at one instant of Earth time (including length contraction).

Fig. 41.13 (right) Reference frame of the Earth as observed at one instant of spaceship time (including length contraction).

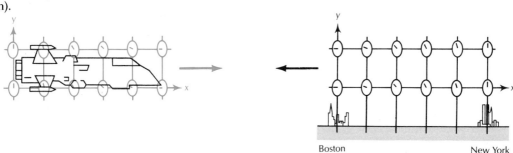

Fig. 41.14 Two identical pieces of pipe. The piece on the right is at rest in the reference frame of the Earth; the piece on the left is at rest in the reference frame of the spaceship.

Incidentally: The contraction effect applies only to lengths along the direction of motion of the spaceship. Lengths perpendicular to the direction of motion are not affected. The proof of this is by contradiction. Imagine that we have two identically manufactured pieces of pipe, one at rest on Earth and one at rest on the spaceship (Figure 41.14). If the motion of the spaceship relative to the Earth were to bring about a transverse contraction of the spaceship pipe, then, by symmetry, the motion of the Earth relative to the spaceship would have to bring about a contraction of the Earth pipe. These contraction effects are contradictory, since in one case the spaceship pipe would fit inside the Earth pipe, and in the other case it would fit outside.

Just like the relativity of time, the relativity of length is a direct consequence of the invariance of the speed of light. The dependence of simultaneity, time intervals, and lengths on the reference frame is a drastic departure from the physics of Newton. It was Einstein's great discovery that Newton's universal, absolute time and absolute length do not exist. This destroys much of the conceptual basis of Newtonian physics and compels us to adopt new definitions of momentum, energy, angular momentum, and so on. In Section 41.7 we will introduce some of these new, relativistic formulas. As we shall see, the differences between the old, nonrelativistic formulas and the new, relativistic formulas are very small if the speeds are small compared with the speed of light. This means that for the motion of airplanes, ships, automobiles, and so on, Newtonian physics, although not exact, is sufficiently accurate for all practical purposes.

41.3 The Lorentz Transformations

In Newtonian physics, the space and time coordinates in one inertial reference frame are related to those in another by the Galilean transformations (see Section 4.5), which rely on absolute time and length. In relativistic physics, we must construct a new set of transformation equations which take into account the relativity of time and length. The new transformation equations are called the **Lorentz transformations.** These transformation equations express the fundamental characteristics of relativistic space and time — they express the geometry of **spacetime.** In the following discussion we will emphasize these geometric properties of the Lorentz transformations by drawing a great many diagrams.

For the sake of simplicity, we will first deal with one-dimensional motion, so that the position x as a function of t gives a complete description of the motion. We recall that in Chapter 2 we introduced the worldline of a particle as the plot of position versus time. In relativistic physics it is customary to arrange the time axis vertically and the space axis horizontally. For example, Figure 41.15 shows the x and t axes associated with the reference frame of the Earth,[5] and it shows the worldline of an automobile traveling in the positive x direction. Before time zero, the automobile was at rest at $x = 0$ and its worldline coincided with the t axis; it then accelerated as indicated by the changing slope of the worldline; finally, it reached a uniform velocity as indicated by the constant slope of the worldline. The worldline consists of individual events, or spacetime points; for instance, the departure of the automobile at $x = 0$, $t = 0$ is an event. In relativistic physics, the complete diagram is called a **spacetime diagram,** or a **Minkowski diagram.**

Now suppose that we wish to describe the motion of the automobile from the point of view of a spaceship zipping past the Earth in the positive x direction at high speed. This spaceship carries its own coordinate grid and clocks, that is, it carries its own inertial reference frame. The motion of the automobile can then be measured with respect to the new x' and t' grid associated with this new reference frame. It is then easy to make a new plot of the worldline of the automobile with respect to the new grid. But instead of replotting all the points of the worldline, let us keep the worldline of Figure 41.15 and rearrange the coordinate grid. The change of the coordinate grid will then graphically represent the change of reference frame.

Spacetime

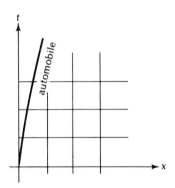

Fig. 41.15 Worldline of an automobile.

Spacetime diagram

---

[5] We will pretend that this is an inertial reference frame.

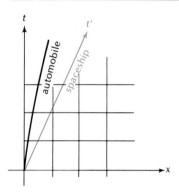

Fig. 41.16 Worldline of a spaceship. This worldline coincides with the t' axis.

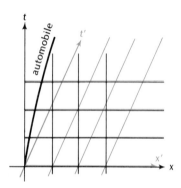

Fig. 41.17 Coordinate grids for the reference frame of the Earth (black) and for the reference frame of the spaceship (colored).

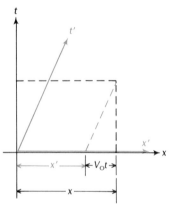

Fig. 41.18 Graphical method of deriving the Galilean equations.

Light-second

How can we draw a new x' and t' grid on top of Figure 41.15 so that the same worldline represents the motion of the automobile with respect to the spaceship? For the sake of practice, we will do this first under the pretense that Newton's notion of absolute space and absolute time is correct.

Figure 41.16 shows the worldline of the midpoint of the spaceship moving with uniform velocity V_O past the Earth. We will regard this midpoint as the origin of the x' coordinates; thus, the slanted line in Figure 41.16 is the worldline of the origin of the spaceship coordinates. Along this line, x' remains permanently zero — only t' increases. Thus, the slanted line is the t' axis of the reference frame of the spaceship. Where is the x' axis? This axis is the set of all those events (or spacetime points) that occur at $t' = 0$. But since in Newtonian physics time is absolute, the clocks of the spaceship and those of the Earth can be synchronized so that $t' = 0$ is the same as $t = 0$. Consequently, the x' axis coincides with the x axis. Figure 41.17 shows the grid of the new x' and t' coordinates; this figure also shows the grid of the old x and t coordinates. The new coordinate grid is slanted with respect to the old one.

Figure 41.17 is a graphical representation of the transformation of coordinates from x, t to x', t'. The old and the new coordinates of any event can be read off by drawing appropriate lines intercepting the axes. From Figure 41.18 we readily see that the mathematical relationship between the values of the old and the new coordinates of an event is

$$t' = t \quad (1)$$

$$x' = x - V_O t \quad (2)$$

where V_O is the velocity of the spaceship relative to the Earth. These equations are the Galilean equations [Eqs. (4.49) and (4.54)]. Although these equations are not new to us, we have succeeded here in deriving them by a graphical method relying on spacetime diagrams. We will now use a similar method to derive the relativistic transformation equations.

The crucial difference between the Galilean and the relativistic transformation equations is that the latter are supposed to leave the speed of a light signal invariant. We begin by plotting the worldline of a light signal. Since the speed of light is inconveniently large, it will be better to introduce new units so that the speed of light has the numerical value $c = 1$. We can do this by taking as our unit of length the distance that light travels in one second. We will call this unit the **light-second:**[6]

$$1 \text{ light-second} = 3.00 \times 10^8 \text{ m/s} \times 1 \text{ s}$$

$$= 3.0 \times 10^8 \text{ m}$$

In terms of this new unit of length, the speed of light is 1 light-second per second:

$$c = 1 \text{ light-second/s} \quad (3)$$

---

[6] This unit is analogous to the light-year, the distance that light travels in 1 year.

Note that when we express any speed in terms of these units, we are actually expressing it as a multiple of the speed of light. For example, an electron moving with a speed of 5.0×10^7 m/s has

$$v = 5.0 \times 10^7 \text{ m/s} \times \frac{1 \text{ light-second}}{3.00 \times 10^8 \text{ m}}$$

$$= 0.166 \frac{\text{light-second}}{\text{s}}$$

or

$$v = 0.166c \tag{4}$$

Since the numerical value of the speed of light in our new units is $c = 1$, we will *omit factors of c or c^2* in all of the equations of this section and Sections 41.4–41.6. For instance, we will write Eq. (4) as

$$v = 0.166$$

Note that this means that our equations are applicable only when the distances and times appearing in them are evaluated in light-seconds and seconds, respectively; we will have to remember this whenever we carry out numerical calculations with these equations (as in Example 1 below).

If the units along the x and t axes of the spacetime diagram are light-seconds and seconds, then the worldline of a light signal has a slope of 1. Figure 41.19 shows the worldlines of several light signals which start at different points along the x axis.

How can we draw new x' and t' axes on top of Figure 41.19 so that the worldline of a light signal has the same speed with respect to both the old and new axes? Figure 41.20 shows the worldline of a light signal that starts at the origin. The worldline lies exactly halfway between the x and t axes. This makes the x and t coordinates of any event of this worldline equal, just as required by the condition that the speed be 1 length unit per 1 time unit. Let us examine the behavior of this light signal from the point of view of a spaceship moving at high speed with respect to the Earth. Figure 41.20 shows the worldline of the midpoint of a spaceship, which coincides with the t' axis. The t' axis is a slanted axis just as in Figure 41.16. What about the x' axis? It is here that the invariance of the speed of light enters our calculations. The worldline of the light signal must lie halfway between the x' and t' axes, so that the x' and t' coordinates of any event on this worldline will be equal, as required by the condition that the speed of light be 1 length unit per 1 time unit. Hence these axes must be symmetrically placed with respect to the worldline light signal — both the x' and t' axes are slanted (Figure 41.21).

The symmetry of the arrangement of the x' and t' axes in Figure 41.21 suggests that the equations relating the old and the new coordinates are

$$x' = x - V_0 t \tag{5}$$

$$t' = t - V_0 x \tag{6}$$

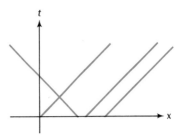

Fig. 41.19 Worldlines of several light signals. Three of these signals travel in the positive x direction; one travels in the negative x direction.

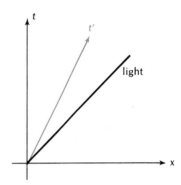

Fig. 41.20 The worldline of the spaceship coincides with the t' axis. The worldline of a light signal that starts at the origin lies halfway between the x and the t axes.

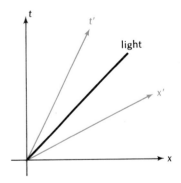

Fig. 41.21 The t' and x' axes are symmetrically placed with respect to the worldline of the light signal.

The first of these equations is the same as Eq. (2); it expresses the slant of the x' axis. The second equation expresses the slant of the t' axis. Note the symmetry between these two equations — the space and time coordinates are simply interchanged between Eq. (5) and (6).

However, the above equations are not yet quite correct. Our choice of units ($c = 1$) assures that the scales along the x and t axes are the same. But this does not mean that the scales along the new axes are the same as those along the old axes. In the case of the Galilean transformations, Eqs. (1) and (2), this problem of scale did not arise; there, distance intervals and time intervals were absolute, and the scale of the new axes could be directly obtained from the old. In the case of the relativistic transformations, we will have to figure out the proper scale by other means.

Since the scale of the new coordinates differs by some unknown factor from that of the old, we will include an extra factor γ in our transformation equations:

$$x' = \gamma(x - V_0 t) \tag{7}$$

$$t' = \gamma(t - V_0 x) \tag{8}$$

Note that the *same* scale factor γ appears in each equation; this is by our choice of equal units for the x' and t' coordinates.

To find an explicit expression for the scale factor γ, we can proceed as follows. First solve Eqs. (7) and (8) for x and t. Regarding x and t as "unknowns" in these equations, we find

$$x = \frac{1}{\gamma(1 - V_0^2)} (x' + V_0 t') \tag{9}$$

$$t = \frac{1}{\gamma(1 - V_0^2)} (t' + V_0 x') \tag{10}$$

These equations are the inverses of Eqs. (7) and (8); the former express the old coordinates in terms of the new, and the latter express the new in terms of the old. But there is an alternative way to express the old coordinates in terms of the new: instead of beginning with rectangular axes for the Earth's reference frame and then constructing slanted axes for the spaceship's reference frame, we could have begun with rectangular axes for the spaceship and then constructed slanted axes for the Earth. This would have led us to a pair of equations similar to Eqs. (7) and (8), with the new and old coordinates exchanged and with one further alteration: the velocity V_0 must be replaced by $-V_0$. This change of sign of the velocity merely reflects the fact that the velocity of the Earth relative to the spaceship is opposite to the velocity of the spaceship relative to the Earth. Thus the equations would have been

$$x = \gamma(x' + V_0 t') \tag{11}$$

$$t = \gamma(t' + V_0 x') \tag{12}$$

By comparing these two equations with Eqs. (9) and (10), we see that they coincide, provided that

$$\gamma = \frac{1}{\gamma} \frac{1}{1 - V_O^2} \qquad (13)$$

or

$$\gamma = \frac{1}{\sqrt{1 - V_O^2}} \qquad (14)$$

Consequently, our transformation equations become

$$x' = \frac{1}{\sqrt{1 - V_O^2}} (x - V_O t) \qquad (15)$$

$$t' = \frac{1}{\sqrt{1 - V_O^2}} (t - V_O x) \qquad (16)$$

Lorentz transformations

and the inverse transformation equations become

$$x = \frac{1}{\sqrt{1 - V_O^2}} (x' + V_O t') \qquad (17)$$

$$t = \frac{1}{\sqrt{1 - V_O^2}} (t' + V_O x') \qquad (18)$$

These equations are the **Lorentz transformations**.[7] They replace the Galilean transformations and accomplish the remarkable feat of making the speed of light the same in all reference frames. Although in the above discussion we labeled the two inertial reference frames as "Earth" and "spaceship," Eqs. (15)–(18) are quite general and give the coordinate transformation between any two inertial reference frames.

We can now put the correct scale on the diagram of Figure 41.21. According to Eq. (15), the line $x' = 1$ intersects the x axis (that is, $t = 0$) at the point

$$x = \sqrt{1 - V_O^2}$$

For instance, if $V_O = \frac{1}{2}$, then the intersection point is

$$x = \sqrt{1 - \tfrac{1}{4}} = 0.866$$

This determines the location of the unit point on the x' axis, and immediately permits us to construct the complete coordinate grid (Figure

Hendrik Antoon Lorentz, *1853–1928, Dutch theoretical physicist, professor at Leiden. He investigated the relationship between electricity, magnetism, and mechanics. In order to explain the observed effect of magnetic fields on emitters of light (Zeeman effect), he postulated the existence of electric charges in the atom, for which he was awarded the Nobel Prize in 1902. He derived the Lorentz transformation equations by some tangled mathematical arguments, but he was not aware that these equations hinge on a new concept of space and time.*

---

[7] In conventional units (x in meters and t in seconds) the Lorentz transformation equations are

$$x' = \frac{1}{\sqrt{1 - V_O^2/c^2}} (x - V_O t)$$

$$t' = \frac{1}{\sqrt{1 - V_O^2/c^2}} \left(t - \frac{V_O x}{c^2}\right)$$

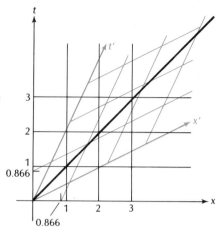

Fig 41.22 Coordinate grids for the reference frame of the Earth (black) and for the reference frame of the spaceship (color).

41.22). The new coordinates of any event can then be read off this grid. Alternatively, the coordinates can be calculated from Eqs. (15) and (16).

The Lorentz transformations treat space and time symmetrically. A change of reference frame intricately mixes the space and time coordinates of an event. What is a pure space interval or a pure time interval in one reference frame becomes a mixture of both space and time intervals in another reference frame. As H. Minkowski wrote, "henceforth space by itself, and time by itself, are doomed to fade away into mere shadows, and only a kind of union of the two will preserve an independent reality."

One interesting conclusion we can draw from the Lorentz transformations is that no signal can propagate with a speed greater than the speed of light. The proof is by contradiction: Suppose that in the reference frame of the Earth we send out a signal of speed greater than that of light. The worldline of this signal is shown in Figure 41.23; the signal is emitted at the point P and is received at the point P' at a later time. But in the reference frame of the spaceship, the point P' has an earlier time coordinate than the point P — hence in this reference frame, the signal is received *before* it is emitted. This is absurd, and rules out the existence of such a signal.

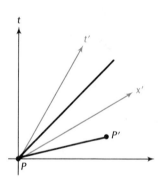

Fig. 41.23 Worldline of a hypothetical signal of speed greater than that of light.

If the relative speed between the two reference frames is small compared with the speed of light, then the Lorentz transformations approximately coincide with the Galilean transformations. In our units, small velocity simply means $V_0 \ll 1$. Hence

$$\sqrt{1 - V_0^2} \cong 1$$

and Eqs. (15) and (16) reduce to

$$x' \cong x - V_0 t \tag{19}$$

$$t' \cong t - V_0 x \tag{20}$$

Furthermore, the term $V_0 x$ in the second equation can be neglected if V_0 and x are velocities and distances of the magnitudes typically encountered on Earth. For instance, suppose that the "spaceship" reference frame is actually that of a supersonic airplane moving at 600 m/s. Expressed as a multiple of the velocity of light, this speed is

$V_O = 2.0 \times 10^{-6}$, a very small number. Hence, even if x has a large value, say, $x = 10^7$ m $= 3.3 \times 10^{-2}$ light-seconds, the term $V_O x$ has a magnitude of only $2.0 \times 10^{-6} \times 3.3 \times 10^{-2}$ s $= 6.6 \times 10^{-8}$ s — this is such a small time interval that it usually can be ignored. We are then entitled to throw out the term $V_O x$ in Eq. (20) and obtain

$$x' \cong x - V_O t \qquad (21)$$

$$t' \cong t \qquad (22)$$

These, of course, are the Galilean transformations. The differences between relativistic physics and Newtonian physics are therefore quite insignificant at low velocities.

Finally, for the sake of completeness, we must write down the equations for the transformation of the transverse coordinates. Since we have assumed that the relative motion of the reference frames is along the direction of the x axis, the transverse directions are those of the y and z axes. In Section 41.2 we found that transverse lengths do not suffer any contraction. Hence the coordinate transformations for y and z must simply be

$$\boxed{y' = y} \qquad (23)$$

$$\boxed{z' = z} \qquad (24)$$

Lorentz transformations for y and z

EXAMPLE 1. Imagine that, as in the example discussed qualitatively in Section 41.2, a spaceship travels in the Boston–New York direction at a speed of 90% of the speed of light, that is, $V_O = 0.90$. At one instant of spaceship time, the clock in New York shows 6:00 P.M. Eastern Standard Time. What time does the clock in Boston show at this instant? The distance between New York and Boston is 290 km.

SOLUTION: For convenience we take the origin of the Earth's reference frame at New York ($x = 0$) and the origin of Earth time at 6:00 P.M. ($t = 0$). According to Eq. (16), the spaceship time corresponding to the event $x = 0$, $t = 0$ in New York is

$$t' = \frac{0 - V_O \cdot 0}{\sqrt{1 - V_O^2}} = 0 \qquad (25)$$

With the x axis in the Boston–New York direction (see Figure 41.24), the x coordinate of Boston is $x = -290$ km $= -2.9 \times 10^5$ m $= -9.7 \times 10^{-4}$ light-seconds. The event in Boston has the same t' coordinate as that in New York, that is, $t' = 0$. Hence Eq. (16) gives

$$0 = \frac{t - V_O x}{\sqrt{1 - V_O^2}} = \frac{t + 0.90 \times 9.7 \times 10^{-4} \text{ s}}{\sqrt{1 - V_O^2}}$$

This leads to

$$t = -0.90 \times 9.7 \times 10^{-4} \text{ s} = -8.7 \times 10^{-4} \text{ s}$$

Thus, the clock in Boston is late; it shows a time 8.7×10^{-4} s *before* 6:00 P.M.

COMMENTS AND SUGGESTIONS: In this calculation, we separately computed the spacetime coordinates of the two events. We can make this calculation more

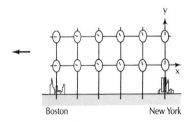

Fig. 41.24 The origin of the reference frame is at New York.

concise by dealing directly with the *differences* between the coordinates. Since Eq. (16) is valid for the coordinates of all spacetime points, it must also be valid for the differences between the coordinates,

$$\Delta t' = \frac{\Delta t - V_O \Delta x}{\sqrt{1 - V_O^2}}$$

The events in Boston and New York occur at one instant of spaceship time; hence, $\Delta t' = 0$ and

$$0 = \frac{\Delta t - V_O \Delta x}{\sqrt{1 - V_O^2}}$$

From this, we can evaluate the time difference Δt in the Earth reference frame, with the same final result as above.

41.4 The Time Dilation

We already discovered in Section 41.2 that a clock in motion relative to some inertial reference frame will run slow as compared with clocks at rest in that reference frame. From the Lorentz transformation equations, we can derive a quantitative formula for this effect. Suppose the clock is at rest in a spaceship that has velocity V_O relative to the Earth. Consider two consecutive ticks of the clock; these ticks are two events between which there are certain coordinate differences. According to Eq. (18), the coordinate differences obey the relation

$$\Delta t = \frac{1}{\sqrt{1 - V_O^2}} (\Delta t' + V_O \Delta x') \qquad (26)$$

Since the clock is at rest in the spaceship, its x' coordinate does not change, that is, $\Delta x' = 0$. Hence

Time dilation (clock in spaceship)

$$\boxed{\Delta t = \frac{1}{\sqrt{1 - V_O^2}} \Delta t'} \qquad \text{clock at rest in spaceship} \qquad (27)$$

This is the **time-dilation** formula. It shows that the time between consecutive ticks measured in the reference frame of the Earth is greater than the time between consecutive ticks measured in the reference frame of the spaceship. As measured by the clocks on the Earth, the clock on the spaceship runs slow by a factor of $1/\sqrt{1 - V_O^2}$.

The time-dilation effect is symmetric: as measured by the clocks on the spaceship, a clock on the Earth runs slow by the same factor,

Time dilation (clock on Earth)

$$\boxed{\Delta t' = \frac{1}{\sqrt{1 - V_O^2}} \Delta t} \qquad \text{clock at rest on Earth} \qquad (28)$$

The derivation of Eq. (28) can be based on Eq. (16) with $\Delta x = 0$.

Figure 41.25 is a plot of the time-dilation factor $1/\sqrt{1-V_O^2}$ as a function of V_O. The time-dilation factor tends to infinity when the speed approaches the speed of light ($V_O \to 1$).

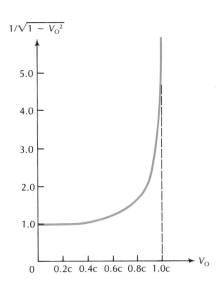

Fig. 41.25 Time-dilation factor as a function of V_O.

The slowing down of the rate of lapse of time applies to all physical processes — atomic, nuclear, biological, and so on. At low speeds, the time-dilation effect is insignificant, but at speeds comparable with the speed of light, it becomes quite large. Very drastic time-dilation effects have been observed in the decay of short-lived elementary particles. For instance, muon particles (see Chapter 46) have an average lifetime of about 2.2×10^{-6} s; but if the muons are moving at high speed through the laboratory, then the internal processes that produce the decay will slow down, and, as judged by the clocks in our laboratory, the muons live a longer time. In accurate experiments performed at the accelerator laboratory of the Organisation Européenne pour la Recherche Nucléaire (CERN), near Geneva, muons with a speed of 99.94% of the speed of light were found to have an average lifetime 29 times as large as the lifetime of muons at rest — in excellent agreement with the value expected from Eq. (27).

EXAMPLE 2. According to Eq. (27), what is the time dilation for such muons? How far do they travel before they decay?

SOLUTION: If the speed is 99.94% of the speed of light, then $V_O = 0.9994$ in our units, and Eq. (27) yields

$$\Delta t = \frac{\Delta t'}{\sqrt{1-(0.9994)^2}} = \frac{\Delta t'}{0.0346} = 28.9 \ \Delta t'$$

which is in agreement with the experimental result.

Since the lifetime of the muons in their own reference frame is $\Delta t' = 2.2 \times 10^{-6}$ s, the lifetime in the laboratory is $\Delta t = 28.9 \times 2.2 \times 10^{-6}$ s, and the distance traveled is

$$\Delta x = V_O \Delta t = 0.9994 \times 28.9 \times 2.2 \times 10^{-6} \text{ s} = 6.3 \times 10^{-5} \text{ light-second}$$

In conventional units this is $c \times 6.3 \times 10^{-5}$ s $= 1.9 \times 10^4$ m.

At everyday speeds the time-dilation effect is extremely small. For example, consider a clock aboard an airplane traveling at 300 m/s over the ground. In our units of light-seconds and seconds this corresponds to $V_0 = 1.0 \times 10^{-6}$. To evaluate the time-dilation factor, it is convenient to take advantage of the Taylor-series expansion

$$\frac{1}{\sqrt{1-V_0^2}} = 1 + \frac{V_0^2}{2} + \cdots \qquad (29)$$

which gives

$$\frac{1}{\sqrt{1-V_0^2}} \cong 1 + 5 \times 10^{-13}$$

that is, the clock in the airplane will slow down by only 5 parts in 10^{13}. However, the detection of such a small change is not beyond the reach of modern atomic clocks. In a decisive experiment, scientists from the National Bureau of Standards placed portable atomic clocks on board a commercial airliner and kept them flying for several days, making a complete trip around the world. Before and after the trip, the clocks were compared with an identical clock that was kept on the ground. The flying clocks were found to have lost time — in one instance, the total time lost because of the motion of the clock was about 10^{-7} s.

The time-dilation effect leads to the famous **twin "paradox,"** which we can state as follows: A pair of identical twins, Terra and Stella, celebrate their, say, twentieth birthday on Earth. Then Stella boards a spaceship that carries her at a speed of $V_0 = 0.99$ to Proxima Centauri, 4 light-years away; the spaceship immediately turns around and brings Stella back to the Earth. According to the clocks on the Earth, this trip takes about 8 years, so that Terra's age will be 28 years when the twins meet again. But Stella has benefited from time dilation — relative to the reference frame of the Earth, the spaceship clocks run slow by a factor

$$\sqrt{1-V_0^2} = \sqrt{1-(0.99)^2} = 0.14$$

Hence, the 8 years of travel registered by the Earth clocks amount to only $8 \times 0.14 = 1.1$ years according to the spaceship clocks, and Stella's biological age on return will be only 21.1 years. Stella will be younger than Terra. The paradox arises when we examine the elapsed times from the point of view of the reference frame of the spaceship. In this reference frame, the Earth is moving; hence the Earth clocks run slow — and Terra should be younger than Stella.

The resolution of this paradox hinges on the fact that our time-dilation formula is valid only if the time of a moving clock is measured from the point of view of an *inertial* reference frame. The reference frame of the Earth is (approximately) inertial, and therefore our calculation of the time dilation of the spaceship clocks is valid. But the reference frame of the spaceship is not inertial — the spaceship must decelerate when it reaches Proxima Centauri, stop, and then accelerate toward the Earth. Therefore, we cannot use the simple time-dilation formula to find the time dilation of the Earth clocks from the point of view of the spaceship reference frame. The "paradox" results from the misuse of this formula.

A detailed analysis of the behavior of the Earth clocks from the point of view of the spaceship reference frame establishes that the Earth clocks run slow as long as the spaceship is moving with uniform velocity, but that the Earth clocks run fast when the spaceship is undergoing its acceleration at Proxima Centauri. The time that the Earth clocks gain during the accelerated portion of the trip more than compensates for the time they lose during the other portions of the trip. This confirms that Stella will be younger than Terra, even from the point of view of the spaceship reference frame.

41.5 The Length Contraction

Suppose that a rigid body, such as a meter stick, is at rest in a spaceship moving relative to the Earth. The length of this body along the direction of motion is $\Delta x'$ in the reference frame of the spaceship. As we pointed out in Section 41.2, to find the length in the reference frame of the Earth, we must measure the position of the forward end and the rear end of the meter stick at the same time t. The measurements at the two ends can be regarded as two events separated by some coordinate difference. According to Eq. (15), coordinate differences obey the transformation law

$$\Delta x' = \frac{1}{\sqrt{1 - V_0^2}} (\Delta x - V_0 \, \Delta t) \tag{30}$$

and since $\Delta t = 0$,

$$\Delta x' = \frac{1}{\sqrt{1 - V_0^2}} \Delta x \tag{31}$$

or

$$\boxed{\Delta x = \sqrt{1 - V_0^2} \, \Delta x'} \qquad \text{body at rest in spaceship} \tag{32}$$ *Length contraction*

This is the formula for **length contraction.** According to this formula, the length of the body measured in the reference frame of the Earth is shorter than the length measured in the reference frame of the spaceship by a factor of $\sqrt{1 - V_0^2}$. This effect is again symmetric: a body at rest on the Earth will suffer from contraction when measured by instruments on board the spaceship.

The length contraction has not been tested directly by experiment. There is no practical method for a high-precision measurement of the length of a fast-moving body. Our best bet might be high-speed photography, but this is nowhere near accurate enough, since the contraction is extremely small even at the highest speeds that we can impart to a macroscopic body.

Incidentally: If we could take a sharp photograph of a macroscopic body zipping by at a speed of, say, $V_0 = 0.8$, the photographic image would show not only the contraction but also a strong distortion of the shape of the body. A photograph never shows the surface of the body as it is, but rather as it was at the instant the light from it began to travel to the camera. The light that makes the photographic image has to reach the camera at the instant its shutter is open. If light from near and far parts of the body is to arrive at the camera at the same instant, the light from the far parts must start out earlier than that from the near parts. Thus the image shows different parts of the body at different times — far parts at an early time and near parts at a late time. The image is therefore a distorted picture of the body. Figure 41.26 is a computer simulation of the photographic image that a camera would record when aimed at some rectangular boxes traveling past the camera at $V_O = 0.8$. The length contraction is clearly noticeable along the midline of the figure, but everywhere else the image is severely distorted. Note that the boxes seem to turn their rear ends toward the camera.

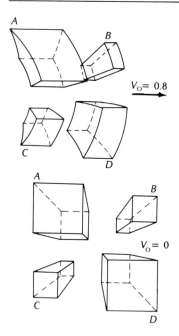

Fig. 41.26 Computer simulation of a photograph of boxes moving a very high speed. [Based on *Am. J. Phys.*, **33**, 534 (1965).]

41.6 The Combination of Velocities

Our search for a new theory of space and time was in part motivated by the conflict between the Galilean addition law for velocities and the invariance of the speed of light. Since the Lorentz transformations were specifically designed to incorporate the invariance of the speed of light, we certainly expect that the relativistic "addition" law, or combination law, for velocities will avoid the above conflict.

We can derive the relativistic combination law for the x component of velocity by simply taking differentials of the Lorentz transformation equations:

$$dx' = \gamma \, (dx - V_O \, dt) \tag{33}$$

$$dt' = \gamma \, (dt - V_O \, dx) \tag{34}$$

Hence

$$\frac{dx'}{dt'} = \frac{dx - V_O \, dt}{dt - V_O \, dx} = \frac{dx/dt - V_O}{1 - V_O \, dx/dt} \tag{35}$$

But dx/dt is the velocity v_x of the particle, light signal, or whatever, with respect to the reference frame of the Earth; dx'/dt' is likewise the velocity v'_x with respect to the reference frame of the spaceship. Hence Eq. (35) may be written

Combination of velocities

$$\boxed{v'_x = \frac{v_x - V_O}{1 - v_x V_O}} \tag{36}$$

This is the relativistic combination law for the x component of the velocity. This law is to be compared with the Galilean equation

$$v'_x = v_x - V_O \tag{37}$$

It is the denominator in Eq. (36) that makes all the difference. For instance, suppose that v_x is the velocity of a light signal propagating along the x axis in the reference frame of the Earth. Then $v_x = 1$, and Eq. (36) gives

$$v'_x = \frac{1 - V_O}{1 - V_O} = 1 \qquad (38)$$

Thus, as required by the Principle of Relativity, the velocity of the light signal has exactly the same magnitude in the reference frame of the spaceship.

The inverse combination law that expresses the velocity v_x in terms of v'_x is

$$v_x = \frac{v'_x + V_O}{1 + v'_x V_O} \qquad (39)$$

The relativistic combination law for light velocities has been explicitly tested in an experiment at CERN involving a beam of very fast pions. These particles decay spontaneously by a reaction that emits a flash of very intense, very energetic light (gamma rays, discussed in Interlude IV). Hence, such a beam of pions can be regarded as a high-speed light source. In the experiment, the velocity of the pions relative to the laboratory was $V_O = 0.99975$. The Galilean combination law would then predict laboratory velocities of $v_x = 1.99975$ for light emitted in the forward direction and of $v_x = 0.00025$ for light emitted in the backward direction. But the experiment confirmed the relativistic combination law — the laboratory velocity of the light had the same magnitude in all directions.

EXAMPLE 3. A spaceship approaching the Earth at a velocity of $V_O = 0.40$ fires a rocket at the Earth. If the velocity of the rocket is $v'_x = 0.80$ in the reference frame of the spaceship, what is its velocity in the reference frame of the Earth?

SOLUTION: By Eq. (39)

$$v_x = \frac{v'_x + V_O}{1 + v'_x V_O} = \frac{0.80 + 0.40}{1 + 0.80 \times 0.40} = 0.91$$

In conventional units this is $0.91c$, or $0.91 \times 3.0 \times 10^8$ m/s $= 2.7 \times 10^8$ m/s.

Equation (36) gives the transformation law only for the component of velocity parallel to the motion of the reference frame (x component). The components of the velocity perpendicular to the motion of the reference frame (y and z components) also have a transformation law different from the Galilean transformation law. Without giving the derivations, we state the results:

$$\boxed{\begin{aligned} v'_y &= \frac{v_y \sqrt{1 - V_O^2}}{(1 - V_O v_x)} \\ v'_z &= \frac{v_z \sqrt{1 - V_O^2}}{(1 - V_O v_x)} \end{aligned}} \qquad (40)$$

Combination of velocities for y and z components

41.7 Relativistic Momentum and Energy

The drastic revision that the theory of Special Relativity imposed on the Newtonian concepts of space and time implies a corresponding revision of the concepts of momentum and energy. The formulas for momentum and energy and the equations expressing their conservation are intimately tied to the transformation equations of the space and time coordinates. To see that this is so, we briefly examine the Newtonian (nonrelativistic) case.

The Newtonian momentum of a particle of mass m and velocity $\mathbf{v}$ is

$$\mathbf{p} = m\mathbf{v} \tag{41}$$

Since the Galilean transformation for velocity is

$$\mathbf{v}' = \mathbf{v} - \mathbf{V}_O \tag{42}$$

the transformation equation for the momentum is

$$\mathbf{p}' = m\mathbf{v}' = m\mathbf{v} - m\mathbf{V}_O = \mathbf{p} - m\mathbf{V}_O \tag{43}$$

Accordingly, the momentum $\mathbf{p}'$ in the new reference frame differs from the momentum $\mathbf{p}$ in the old reference frame by only a constant quantity (a quantity independent of the velocity $\mathbf{v}$ of the particle). Hence, if the total momentum of a system of colliding particles is conserved in one reference frame, it will also be conserved in another frame — and the law of conservation of momentum obeys the Principle of Relativity. This shows that the nonrelativistic formula for momentum and the nonrelativistic transformation law for velocities match in just the right way.

What happens when we replace the nonrelativistic transformation law for velocities [Eq. (42)] by the relativistic law [Eqs. (36) and (40)]? If the law of conservation of momentum is to obey the Principle of Relativity, we must then design a new relativistic formula for momentum that matches the new relativistic transformation law for velocities. It turns out that the correct relativistic formula for momentum is

Relativistic momentum

$$\boxed{\mathbf{p} = \frac{m\mathbf{v}}{\sqrt{1 - v^2/c^2}}} \tag{44}$$

Note that in this formula we have included the proper factors of c, instead of omitting them as in the preceding sections. Thus, the formula is equally valid in units of light-seconds and seconds and in units of meters and seconds; this will make it easier to compare the relativistic and nonrelativistic formulas for momentum. We will forgo the derivation of Eq. (44), but we will verify in Example 6 that this equation does indeed agree with the Principle of Relativity.

If the speed of the particle is small compared with the speed of light, then

$$\sqrt{1 - v^2/c^2} \cong 1$$

and Eq. (44) becomes approximately

$$\mathbf{p} \cong m\mathbf{v} \qquad (45)$$

This shows that, for low speeds, the relativistic and nonrelativistic formulas for momentum coincide. At high speeds the formulas differ drastically — the relativistic momentum becomes infinite as the speed of the particle approaches the speed of light. Figure 41.27 is a plot of the magnitude of **p** as a function of v.

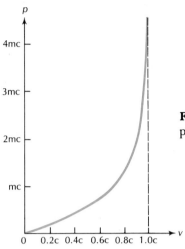

Fig. 41.27 Momentum of a particle as a function of speed.

EXAMPLE 4. An electron in the beam of a TV tube has a speed of 1.0×10^8 m/s. What is the momentum of this electron?

SOLUTION: For this electron, $v/c = (1.0 \times 10^8 \text{ m/s})/(3.0 \times 10^8 \text{ m/s}) = 0.33$. According to Eq. (44), the momentum is then

$$p = \frac{mv}{\sqrt{1 - v^2/c^2}}$$

$$= \frac{9.1 \times 10^{-31} \text{ kg} \times 1.0 \times 10^8 \text{ m/s}}{\sqrt{1 - (0.33)^2}}$$

$$= 9.7 \times 10^{-23} \text{ kg} \cdot \text{m/s}$$

COMMENTS AND SUGGESTIONS: If we had calculated the momentum according to the nonrelativistic equation $p = mv$, we would have obtained 9.1×10^{-23} kg·m/s, and we would have been in error by about 6%.

It turns out that we also need a new formula for kinetic energy. Again, without derivation, we state that the relativistic formula for kinetic energy is

$$\boxed{K = \frac{mc^2}{\sqrt{1 - v^2/c^2}} - mc^2} \qquad (46)$$

Relativistic kinetic energy

To compare this with our old, nonrelativistic formula, we can make use of the Taylor-series expansion:

$$\frac{1}{\sqrt{1-x^2}} = 1 + x^2/2 + \cdots \qquad (47)$$

Thus, if the speed of the particle is small compared with the speed of light, Eq. (46) becomes

$$K \cong mc^2\left(1 + \frac{v^2}{2c^2}\right) - mc^2 = \tfrac{1}{2}mv^2 \tag{48}$$

Again, for low speeds the relativistic and nonrelativistic formulas agree.

The relativistic kinetic energy becomes infinite as the speed of the particle approaches the speed of light. This indicates that, for any particle, the speed of light is unattainable, since it is impossible to supply a particle with an infinite amount of energy. Figure 41.28 is a plot of the kinetic energy as a function of v.

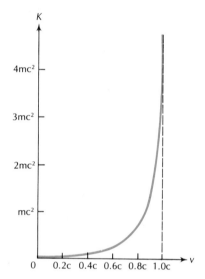

Fig. 41.28 Kinetic energy of a particle as a function of speed.

EXAMPLE 5. The maximum speed that electrons achieve in the Stanford Linear Accelerator is $0.999999999948c$. What is the kinetic energy of an electron moving with this speed?

SOLUTION: For v/c near 1, it is convenient to write

$$\sqrt{1 - v^2/c^2} = \sqrt{1 + v/c}\sqrt{1 - v/c} \cong \sqrt{2}\sqrt{1 - v/c} \tag{49}$$

In our case, the value of $1 - v/c$ is 5.2×10^{-11}. Hence

$$K = mc^2\left(\frac{1}{\sqrt{1 - v^2/c^2}} - 1\right)$$

$$= 9.1 \times 10^{-31} \text{ kg} \times (3.0 \times 10^8 \text{ m/s})^2 \times \left(\frac{1}{\sqrt{2}\sqrt{5.2 \times 10^{-11}}} - 1\right)$$

$$= 8.0 \times 10^{-9} \text{ J}$$

COMMENTS AND SUGGESTIONS: The approximation for $\sqrt{1 - v^2/c^2}$ stated in Eq. (49) is often useful when dealing with speeds close to the speed of light. Most electronic pocket calculators cannot tell the difference between 1 and 0.999999999948, and they would therefore have given a value of 0 for $\sqrt{1 - v^2/c^2}$ in this example.

As we already mentioned in Section 8.6, mass is a form of energy. According to Einstein's mass–energy relation, a particle of mass m at rest has an energy mc^2 [see Eq. (8.25)]; this relation can be derived from the theory of Special Relativity, but, as with other fundamental equations in this section, we will not give a proof. The total energy of a free particle in motion is the sum of its rest-mass energy mc^2 and its kinetic energy K:

$$E = mc^2 + K = mc^2 + \frac{mc^2}{\sqrt{1 - v^2/c^2}} - mc^2 \tag{50}$$

This leads to a simple formula for the total relativistic energy of a free particle:

$$\boxed{E = \frac{mc^2}{\sqrt{1 - v^2/c^2}}} \tag{51}$$

Relativistic total energy

It is easy to verify that this relativistic energy can be expressed as follows in terms of the relativistic momentum:

$$E = \sqrt{c^2 p^2 + m^2 c^4} \tag{52}$$

For a particle moving at a speed close to the speed of light, the first term within the square root in Eq. (52) is much larger than the second term. Hence, for such an ultrarelativistic particle we can neglect $m^2 c^4$ and obtain

$$E \cong cp \tag{53}$$

Thus, the momentum and the energy of an ultrarelativistic particle are directly proportional.

We also mentioned in Section 8.6 that Einstein's mass–energy relation means not only that mass is a form of energy but also that energy has mass. For instance, the mass associated with the energy E of Eq. (51) is

$$m' = \frac{E}{c^2} = \frac{m}{\sqrt{1 - v^2/c^2}} \tag{54}$$

The effective mass of a body in motion is larger by a factor of $1/\sqrt{1 - v^2/c^2}$ than the mass of the body when at rest — a body in motion offers more resistance to acceleration and has more weight than the body at rest.

EXAMPLE 6. Show that, with the relativistic formulas for momentum and energy, the laws of conservation of momentum and energy obey the Principle of Relativity, that is, show that if the total momentum and energy of a system of colliding particles are conserved in one reference frame, then they will also be conserved in any other reference frame.

SOLUTION: For the sake of simplicity, we will deal only with one-dimensional motion, along the x axis. The momentum and energy of a particle are then

$$p_x = \frac{mv_x}{\sqrt{1-v_x^2/c^2}} \quad \text{and} \quad E = \frac{mc^2}{\sqrt{1-v_x^2/c^2}}$$

The relativistic combination law for velocities tells us that the velocity of the particle in a new reference frame moving with velocity V_O along the x axis of the old reference frame is $v_x' = (v_x - V_O)/(1 - v_x V_O/c^2)$. Hence the momentum of the particle in the new reference frame is

$$p_x' = \frac{mv_x'}{\sqrt{1-v_x'^2/c^2}} = \frac{m(v_x - V_O)}{1 - v_x V_O/c^2} \frac{1}{\sqrt{1 - (1/c^2)[(v_x - V_O)/(1 - v_x V_O/c^2)]^2}}$$

$$= \frac{m(v_x - V_O)}{\sqrt{(1 - v_x V_O/c^2)^2 - (v_x - V_O)^2/c^2}}$$

Since $(1 - v_x V_O/c^2)^2 - (v_x - V_O)^2/c^2 = (1 - V_O^2/c^2)(1 - v_x^2/c^2)$, this becomes

$$p_x' = \frac{1}{\sqrt{1 - V_O^2/c^2}} \frac{mv_x}{\sqrt{1 - v_x^2/c^2}} - \frac{V_O}{\sqrt{1 - V_O^2/c^2}} \frac{m}{\sqrt{1 - v_x^2/c^2}}$$

In terms of p_x and E, we can write this as

$$p_x' = \frac{1}{\sqrt{1 - V_O^2/c^2}} p_x - \frac{V_O/c^2}{\sqrt{1 - V_O^2/c^2}} E \tag{55}$$

This equation shows that the momentum p_x' in the new reference frame is a linear superposition of the momentum and energy in the old reference frame, with constant coefficients (the coefficients are independent of the velocity v_x of the particle). The same will therefore hold true for the total momentum of the system of all the colliding particles. Consequently, if the momentum and energy of this system are conserved in the old reference frame, the momentum will necessarily be conserved in the new reference frame.

Note that the conservation of momentum in the new reference frame depends on conservation of *both* momentum and energy in the old reference frame. However, this does not mean that the momentum in the new reference frame is conserved only if the collisions are elastic. The total relativistic energy is conserved even if the collisions are inelastic; such collisions merely convert kinetic energy into rest-mass energy (internal energy of the particles) without changing the total energy.

Equation (55) is a Lorentz transformation equation for momentum. There is also a Lorentz transformation equation for energy:

$$E' = \frac{1}{\sqrt{1 - V_O^2/c^2}} E - \frac{V_O}{\sqrt{1 - V_O^2/c^2}} p_x \tag{56}$$

(we will leave the proof of this to Problem 47). The transformation equations of Eqs. (55) and (56) for momentum and energy are completely analogous to the transformation equations of Eqs. (15) and (16) for position and time. From Eq. (56) we deduce that if the momentum and energy of a system of particles are conserved in the old reference frame, then the energy will necessarily be conserved in the new reference frame.

41.8* Relativity and the Magnetic Field

In Chapter 30, we introduced the law for the magnetic force between moving particles as a separate fundamental law of nature, on a par

---

* This section is optional.

with Coulomb's Law for the electric force or Newton's law of gravitation. However, by means of the theory of Special Relativity, it can be established that the magnetic force is closely related to the electric force — a change of reference frame can transform an electric force partially or completely into a magnetic force, and vice versa. Because of such transformations between electric and magnetic forces, the law for the magnetic force can be regarded as a consequence of Coulomb's Law for the electric force. In this section we will examine a special, simple case of a transformation of a magnetic force into an electric force by a change of reference frame.

Suppose that a positive charge q moves in the magnetic field generated by a current on a straight wire. We will assume that the (instantaneous) velocity **v** of the charge q is parallel to the wire. Figure 41.29 shows the situation in a reference frame in which the wire is at rest. The charge has a velocity **v** toward the right, and the wire carries a current I toward the left. In this reference frame the current generates a magnetic field [see Eq. (30.32)]

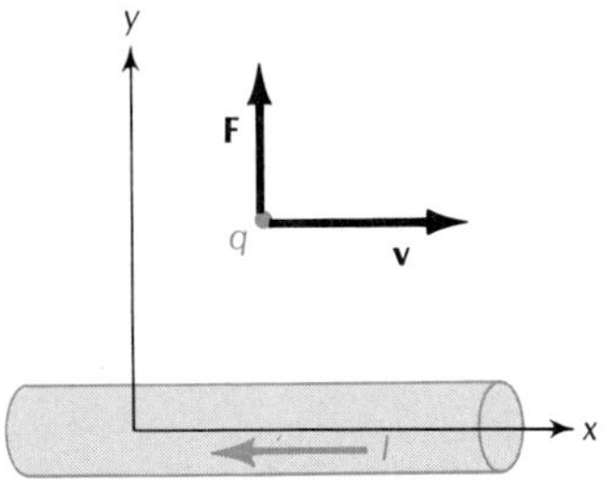

$$B = \frac{\mu_0}{2\pi} \frac{I}{y} \qquad (57)$$

and this magnetic field exerts a magnetic force

$$F = qvB = \frac{\mu_0}{2\pi} \frac{qvI}{y} \qquad (58)$$

on the charge q. The force points radially away from the wire (Figure 41.29).

Next, let us examine the situation in a new reference frame in which the charge q is (instantaneously) at rest. Relative to the old reference frame, this new reference frame moves toward the right with velocity **v**. Figure 41.30 shows the situation in this reference frame. The charge q is at rest, and the wire moves toward the left with velocity −**v**. In this new reference frame, there can be no magnetic force on the charge q, since the velocity of the charge is zero. We are now faced with a paradox: in the old reference frame the charge experiences a magnetic force and hence an acceleration; in the new reference frame the charge experiences no magnetic force, hence no acceleration — and yet accelerations are supposed to be independent of the reference frame!

The resolution of this paradox hinges on a subtle relativistic effect. It turns out that in the new reference frame, the wire generates an *electric field*, and the corresponding electric force on the charge gives it the required acceleration.

To understand where this electric field in the new reference frame comes from, let us begin by asking why there is no electric field in the old reference frame. Obviously, the answer is that in the reference frame of the wire the positive and negative charge densities on the wire are of equal magnitude; hence their electric fields cancel exactly.[8] Figure 41.31 shows these charge distributions in the reference frame of the wire. The negative charge density is due to the free electrons carrying the current; the positive charge density is due to the positive

Fig. 41.29 In the reference frame of the wire, the charge q moves to the right with velocity **v**.

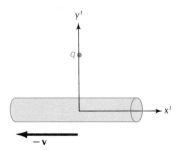

Fig. 41.30 In the reference frame of the charge q, the wire moves to the left with velocity −**v**.

---

[8] For the present purposes, we ignore the small electric field needed to push the current along the wire.

Fig. 41.31 In the reference frame of the wire, the positive and the negative charge densities on the wire are equal.

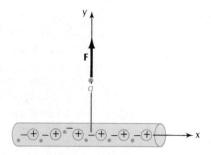

ions fixed in the lattice of the wire. The charge density of the electrons is $-\lambda$ coulomb/m, and that of the ions is $+\lambda$ coulomb/m.

In the reference frame of the wire, the negative charges are in motion and the positive charges are at rest. Since the current is flowing toward the left, the free electrons carrying this current must move toward the right. For the sake of simplicity, let us assume that the velocity of the free electrons coincides with the velocity of the charge q. This is a very special case — and not very likely to happen in reality. It is, of course, possible to deal with different velocities, but it greatly simplifies the calculations if we have to deal with only a single velocity rather than two different velocities.

Now consider the electric charge densities in the new reference frame. The crucial point is that in this reference frame the negative and positive densities will *not be equal*. The inequality arises from the length contraction effect of Special Relativity. We recall from Section 41.5 that if a body has a certain length in its own reference frame, then in any other reference frame the length, along the direction of motion, is shorter by a factor

$$\sqrt{1 - v^2/c^2}$$

where c is the speed of light. Thus, if a given number of positive charges sitting on the wire occupy a length of 1 m in the old reference frame (rest frame of the wire), they will occupy a shorter length of

$$1 \text{ m} \times \sqrt{1 - v^2/c^2}$$

in the new reference frame. Correspondingly, the density of these positive charges will be larger: if the density is λ coulomb/m in the old reference frame, it will be

$$\lambda/\sqrt{1 - v^2/c^2} \text{ coulomb/m}$$

in the new reference frame. For the negative charge distribution, the length contraction has the opposite effect, and the density of negative charges will be smaller: if the density of charge is $-\lambda$ coulomb/m in the old reference frame, it will be

$$-\lambda\sqrt{1 - v^2/c^2} \text{ coulomb/m}$$

in the new reference frame. This is so because in the old reference frame (rest frame of the wire) the charge distribution of electrons is in motion; it therefore is a *contracted* charge distribution. In the new reference frame, the charge distribution is at rest, and it is not contracted. The transformation from the old to the new reference frame therefore is a transformation from a reference frame in which the length of the negative charge distribution is already contracted to a reference frame in which it is not contracted. Correspondingly, the density of the negative charges will be smaller, as was indicated above.

In the new reference frame, the net charge per unit length of the wire is then the sum of *unequal* positive and negative contributions,

$$\lambda_{\text{new}} = \frac{\lambda}{\sqrt{1 - v^2/c^2}} - \lambda\sqrt{1 - v^2/c^2} \tag{59}$$

If the speeds are small compared with the speed of light, we can use the approximations

$$1/\sqrt{1 - v^2/c^2} \cong 1 + \tfrac{1}{2} v^2/c^2 \quad \text{and} \quad \sqrt{1 - v^2/c^2} \cong 1 - \tfrac{1}{2} v^2/c^2$$

so that Eq. (59) becomes

$$\lambda_{\text{new}} = \frac{\lambda}{\sqrt{1 - v^2/c^2}} - \lambda\sqrt{1 - v^2/c^2} \cong \lambda v^2/c^2 \tag{60}$$

Such a charge density along the wire will generate a radial electric field [see Eq. (24.10)]

$$E = \frac{1}{2\pi\varepsilon_0} \frac{\lambda_{\text{new}}}{y} = \frac{1}{2\pi\varepsilon_0 c^2} \frac{\lambda v^2}{y} \tag{61}$$

This electric field exerts an electric force

$$F_{\text{new}} = \frac{1}{2\pi\varepsilon_0 c^2} \frac{q\lambda v^2}{y} \tag{62}$$

on the charge q. This force points away from the wire (Figure 41.32).

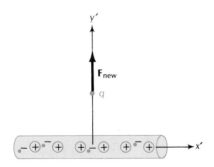

Fig. 41.32 In the reference frame of the charge q, the positive charge density on the wire exceeds the negative charge density, and the charge q experiences a repulsive electric force.

In order to compare this electric force in the new reference frame with the magnetic force in the old reference frame, we note that the product of the velocity v of the electrons and their charge density λ is the current I on the wire. Hence Eq. (62) can be written

$$F_{\text{new}} = \frac{1}{2\pi\varepsilon_0 c^2} \frac{qvI}{y} \tag{63}$$

But, according to Eq. (36.11), $\varepsilon_0 c^2 = 1/\mu_0$, and hence

$$F_{\text{new}} = \frac{\mu_0}{2\pi} \frac{qvI}{y} \tag{64}$$

Thus, the forces in the old and the new reference frames are exactly equal. Note that this result hinges on the fact that the values of ε_0 and μ_0 are exactly right to cancel the factor of c^2 in Eq. (64); again, we see that there is a deep connection between electricity, magnetism, and the speed of light.

What we conclude from the above calculation is then the following: the force that the charges on the wire exert on the positive point charge q is the same in both the old and the new reference frames. The transformation of reference frame does not change the magnitude or direction of the force (and of the acceleration) — it only changes the character of the force from purely magnetic to purely electric. Incidentally: In a reference frame moving toward the right with a speed of, say, $\frac{1}{2}v$, the force would still have the same magnitude and direction, but it would be partially magnetic and partially electric.

Although we obtained these results for only a very special and simple case, the main features are of general validity. If a particle experiences a magnetic force in a given reference frame, a transformation to the rest frame of the particle will make this magnetic force disappear. But in the latter reference frame, charge distributions will appear at the locations of the currents, and the electric force of these charge distributions will replace the original magnetic force. The net force is the same in both reference frames.[9]

Electric and magnetic forces and fields transform into one another if we change the frame of reference. In the rest frame of a charged particle, only electric forces act on the particle. Hence, we can regard the magnetic force that acts on the particle in any other reference frame as resulting from a transformation of the electric force in the rest frame. In this sense, magnetic forces can be regarded as a consequence of electric forces and of relativity.

SUMMARY

Principle of Relativity: All the laws of physics are the same in all inertial reference frames.

Lorentz transformations:[10] $x' = \dfrac{x - V_0 t}{\sqrt{1 - V_0^2}}$

$t' = \dfrac{t - V_0 x}{\sqrt{1 - V_0^2}}$

Time dilation:[10] $\Delta t = \dfrac{1}{\sqrt{1 - V_0^2}} \Delta t'$ (clock at rest in spaceship)

Length contraction:[10] $\Delta x = \sqrt{1 - V_0^2} \, \Delta x'$ (body at rest in spaceship)

---

[9] This invariance of the force is true only if the relative speed of the reference frames is low ($v \ll c$). If the speed is high, then we must take into account the relativistic transformation law for force. It turns out that the transformation of electric and magnetic forces is consistent with that law.

[10] The units of distance and time are light-seconds and seconds, respectively.

Combination of velocities:[10] $v'_x = \dfrac{v_x - V_O}{1 - v_x V_O}$

Relativistic momentum: $\mathbf{p} = \dfrac{m\mathbf{v}}{\sqrt{1 - v^2/c^2}}$

Relativistic kinetic energy and total energy:

$$K = \dfrac{mc^2}{\sqrt{1 - v^2/c^2}} - mc^2$$

$$E = \dfrac{mc^2}{\sqrt{1 - v^2/c^2}}$$

QUESTIONS

1. An astronaut is inside a closed space capsule coasting through interstellar space. Is there any way that the astronaut can measure the speed of the capsule without looking outside?

2. Why did Michelson and Morley use *two* light beams, rather than a single light beam, in their experiment?

3. According to a reliable source,[11] when Einstein was a boy he wondered about the following question: A runner holds a mirror at arm's length in front of his face. Can he see himself in the mirror if he runs at (almost) the speed of light? Answer this question both according to the ether theory and according to the theory of Special Relativity.

4. Consider the piece of paper on which one page of this book is printed. Which of the following properties of the piece of paper are absolute, that is, which are independent of whether the paper is at rest or in motion relative to you? (a) The thickness of the paper, (b) the mass of the paper, (c) the volume of the paper, (d) the number of atoms in the paper, (e) the chemical composition of the paper, (f) the speed of the light reflected by the paper, and (g) the color of the colored print on the paper.

5. Two streetlamps, one in Boston and the other in New York City, are turned on at exactly 6:00 P.M. Eastern Standard Time. Find a reference frame in which the streetlamp in New York was turned on late.

6. According to the theory of Special Relativity, the time order of events can be reversed under certain conditions. Does this mean that a sparrow might fall from the sky before it leaves the nest?

7. Because of the rotational motion of the Earth about its axis, a point on the equator moves with a speed of 460 m/s relative to a point on the North Pole. Does this mean that a clock placed on the equator runs more slowly than a similar clock placed on the pole?

8. According to Jacob Bronowski, author of *The Ascent of Man,* the explanation of time dilation is as follows: If you are moving away from a clock tower at a speed nearly equal to the speed of light, you keep pace with the light that the face of the clock sent out at, say, 11 o'clock. Hence, if you look toward the clock tower, you always see its hands at the 11 o'clock position. Is this explanation correct? What is wrong with it?

9. Suppose you wanted to travel into the future and see what the twenty-fifth century is like. In principle, how could you do this? Could you ever return to the twentieth century?

---

[11] E. F. Taylor and J. A. Wheeler, *Spacetime Physics.*

10. According to the qualitative arguments of Section 41.2, a light signal traveling along a track placed perpendicular to the direction of motion of the spaceship (see Figure 41.9) takes a longer time to complete a round trip when measured by the clocks on the Earth than when measured by the clocks on the spaceship. Would the same be true for a light signal traveling along a track placed parallel to the direction of motion? Explain qualitatively.

11. A cannonball is perfectly round in its own reference frame. Describe the shape of this cannonball in a reference frame relative to which it has a speed of $0.95c$. Is the volume of the cannonball the same in both reference frames?

12. Could we use the argument based on the two identical pieces of pipe (see Figure 41.14) to prove that lengths *along* the direction of motion are not affected? Why not?

13. A rod at rest on the ground makes an angle of $30°$ with the x axis in the reference frame of the Earth. Will the angle be larger or smaller in the reference frame of a spaceship moving along the x axis?

14. In the charming tale "City Speed Limit" by George Gamow,[12] the protagonist, Mr. Tompkins, finds himself riding a bicycle in a city where the speed of light is very low, roughly 30 km/h. What weird effects must Mr. Tompkins have noticed under these circumstances?

15. A long spaceship is accelerating away from the Earth. In the reference frame of the Earth, are the instantaneous speeds of the nose and of the tail of the spaceship the same?

16. Suppose that a *very* fast runner holding a long horizontal pole runs through a barn open at both ends. The length of the pole (in its rest frame) is 6 m, and the length of the barn (in *its* rest frame) is 5 m. In the reference frame of the barn, the pole will suffer length contraction and, at one instant of time, all of the pole will be inside the barn. However, in the reference frame of the runner, the barn will suffer length contraction and all of the pole will never be inside the barn at one instant of time. Is this a contradiction?

17. If the beam from a revolving searchlight is intercepted by a distant cloud, the bright spot will move across the surface of the cloud very quickly, with a speed that can easily exceed the speed of light. Does this conflict with our conclusion of Section 41.3, that nothing can move faster than the speed of light?

18. Figure 41.22 shows the $x'-t'$ coordinate grid for a spaceship moving in the positive x direction. Draw a corresponding diagram showing the coordinate grid for a spaceship moving in the negative x direction.

19. Why can a spaceship not travel as fast as or faster than the speed of light?

20. The arguments of Section 41.8 show that the Eq. (63) for the magnetic force exerted on a charged particle by the current of a straight wire is a direct consequence of the relativistic length contraction. Can we therefore regard the experimental verification of this force law as an experimental verification of the length contraction?

PROBLEMS

Section 41.3

1. (a) In a spacetime diagram, plot the worldline of an electron moving in the positive x direction at a speed of 2.0×10^8 m/s. In your diagram use units of seconds for t and light-seconds for x. What is the slope of this worldline?

---

[12] George Gamow, *Mr. Tompkins in Wonderland*.

(b) In the same diagram plot the worldline of a supersonic aircraft moving in the positive x direction at 500 m/s. What is the slope of this worldline?

2. Suppose that a spaceship moves relative to the Earth at a constant velocity of $V_O = 0.33c$.
   (a) Carefully draw a diagram such as that shown in Figure 41.24. Draw the coordinate grids for x, t and x', t' in different colors, and make sure that you have the correct angles between the axes and the correct scales along the axes.
   (b) On your diagram draw the worldline of an electron of velocity $v_x = 0.66c$. What is the slope of this worldline relative to the x, t axes? Measure the slope of the worldline relative to the x', t' axes and thereby find v'_x. Does your result agree with Eq. (36)?

3. A spaceship moves relative to the Earth at a speed $V_O = 0.5c$. Assume that the spaceship moves along the *negative* x axis.
   (a) Draw a diagram analogous to Figure 14.22. Draw the coordinate grids for x, t and x', t' in different colors, and make sure that you have the correct angles between the axes and the correct scales along the axes.
   (b) Plot the spacetime point $x = 2$, $t = 3$ in your diagram. Read the values of the x' and t' coordinates of this event directly off your diagram. Check that your graphical determination of the values of x' and t' agrees with the computation of these values from Eqs. (15) and (16).

4. The captain of a spaceship traveling away from Earth in the x direction at $V_0 = 0.80c$ observes that a nova explosion occurs at a point with spacetime coordinates $t' = -6.0 \times 10^8$ s, $x' = 6.2 \times 10^8$ light-seconds, $y' = 4.0 \times 10^8$ light-seconds, $z' = 0$ as measured in the reference frame of the spaceship. He reports this event to Earth via radio.
   (a) What are the spacetime coordinates of the explosion in the reference frame of the Earth? Assume that the master clock of the spaceship coincides with the master clock of the Earth at the instant the spaceship passes by the Earth, and that the origin of the spaceship x', y', and z' coordinates is at the midpoint of the spaceship.
   (b) Will the Earth receive the report of the captain before or after astronomers on the Earth see the nova explosion in their telescopes? No calculation is required for this question.

*5. Consider the situation described in Problem 4. Since light takes some time to travel from the nova to the spaceship, the spacetime coordinates that the captain reports are not directly measured but, rather, deduced from the time of arrival and the direction of the nova light reaching the spaceship.
   (a) At what time (t' time) did the nova light reach the spaceship?
   (b) If the captain sends a report to Earth via radio as soon as he sees the supernova, at what time (t time) does the Earth receive the report?
   (c) At what time do Earth astronomers see the nova?

*6. At $11^h0^m0^s$ A.M. a boiler explodes in the basement of the Museum of Modern Art in New York City. At $11^h0^m0.0003^s$ A.M. a similar boiler explodes in the basement of a soup factory in Camden, New Jersey, at a distance of 150 km from the first explosion. Show that, in the reference frame of a spaceship moving at a speed greater than $V_O = 0.60c$ from New York toward Camden, the first explosion occurs *after* the second.

*7. Show that the Lorentz transformations [Eqs. (15) and (16)] can be written in the form

$$x' = x \cosh \theta - t \sinh \theta$$
$$t' = -x \sinh \theta + t \cosh \theta$$

where $\tanh \theta = V_O$. In this form the Lorentz transformation is reminiscent of a rotation (see Section 3.5).

Section 41.4

8. Neutrons have an average lifetime of 15 minutes when at rest in the laboratory. What is the average lifetime of neutrons of a speed of 25% of the speed of light? 50%? 90%?

9. Consider an unstable particle, such as a pion, which has a lifetime of only 2.6×10^{-8} s when at rest in the laboratory. What speed must you give such a particle to make its lifetime twice as long as when at rest in the laboratory?

10. In 1961, the cosmonaut G. S. Titov circled the Earth for 25 h at a speed of 7.8 km/s. According to Eq. (27), what was the time-dilation factor of his body clock relative to the clocks on Earth? By how many seconds did his body clock fall behind during the entire trip?

*11. Muons are unstable particles which — if at rest in a laboratory — decay after a time of only 2.0×10^{-6} s. Suppose that a muon is created in a collision between a cosmic ray and an oxygen nucleus at the top of the Earth's atmosphere, at an altitude of 20 km above sea level.
   (a) If the muon has a downward speed of 2.97×10^8 m/s relative to the Earth, at what altitude will it decay? Ignore gravity in this calculation.
   (b) Without time dilation, at what altitude would the muon have decayed?

*12. In a test of the relativistic time-dilation effect, physicists compared the rates of vibration of nuclei of iron moving at different speeds. One sample of iron nuclei was placed on the rim of a high-speed rotor; another sample of similar nuclei was placed at the center. The radius of the rotor was 10 cm and it rotated at 35,000 rev/min. Under these conditions, what was the speed of the rim of the rotor relative to the center? According to Eq. (27), what was the time-dilation factor of the sample at the rim compared with the sample at the center?

*13. Suppose that a special breed of cat (*Felis einsteinii*) lives for exactly 7 years according to its own body clock. When such a cat is born, we put it aboard a spaceship and send it off at $V_0 = 0.8c$ toward the star Alpha Centauri. How far from the Earth (reckoned in the reference frame of the Earth) will the cat be when it dies? How long after the departure of the spaceship will a radio signal announcing the death of the cat reach us? The radio signal is sent out from the spaceship at the instant the cat dies.

*14. If cosmonauts from the Earth wanted to travel to the Andromeda galaxy in a time of no more than 10 years as reckoned by clocks aboard their spaceship, at what (constant) speed would they have to travel? How much time would have elapsed on Earth after 10 years of time on the spaceship? The distance to the Andromeda galaxy is 2.2×10^6 light-years.

*15. Because of the rotation of the Earth, a point on the equator has a speed of 460 m/s relative to a point at the North Pole. According to the time-dilation effect described in Section 41.4, by what factor do the rates of two clocks differ if one is located on the equator and the other at the North Pole? After 1 year has elapsed, by how many seconds will the clocks differ? Which clock will be ahead?[13]

**16. The star Alpha Centauri is 4.4 light-years away from us. Suppose that we send a spaceship on an expedition to this star. Relative to the Earth, the spaceship accelerates at a constant rate of 0.1 G until it reaches the midpoint, 2.2 light-years from Earth. The spaceship then decelerates at a constant rate of 0.1 G until it reaches Alpha Centauri. The spaceship performs the return trip in the same manner.
   (a) What is the time required for the complete trip according to the clocks

---

[13] The results of this problem cannot be compared directly with experiments, because the clocks also suffer a gravitational time-dilation effect which must be calculated from the theory of General Relativity.

on the Earth? Ignore the time that the spaceship spends at its destination.

(b) What is the time required for the complete trip according to the clocks on the spaceship? Assume that the *instantaneous* time-dilation factor is still $\sqrt{1 - V_O^2}$, even though the speed V_O is a function of time.

Section 41.5

17. According to the manufacturer's specifications, a spaceship has a length of 200 m. At what speed (relative to the Earth) will this spaceship have a length of 100 m in the reference frame of the Earth?

18. What is the percent length contraction of an automobile traveling at 96 km/h?

19. Suppose that a proton speeds by the Earth at $V_O = 0.8c$ along a line parallel to the axis of the Earth.
    (a) In the reference frame of the proton, what is the polar diameter of the Earth? The equatorial diameter?
    (b) In the reference frame of the proton, how long does the proton take to travel from the point of closest approach to the North Pole to the point of closest approach to the South Pole? In the reference frame of the Earth, how long does this take?

*20. Suppose that a meter stick at rest in the reference frame of the Earth lies in the x–y plane and makes an angle of 30° with the x axis. Suppose that one end of the meter stick is at the origin. At a fixed time t, what are the x and y components of the displacement from this end of the meter stick to the other? At a fixed time t', what are the x' and y' components of the displacement from one end of the meter stick to the other in a new reference frame moving with velocity $V_O = 0.7$ in the x direction? What is the angle the meter stick makes with the x' axis of this new reference frame?

*21. A flexible drive belt runs over two flywheels whose axles are mounted on a rigid base (see Figure 41.33). In the reference frame of the base, the horizontal portions of the belt have a speed v and therefore are subject to a length contraction, which tightens the belt around the flywheels. However, in a reference frame moving to the right with the upper portion of the belt, the base is subject to length contraction, which ought to loosen the belt around the flywheels. Resolve this paradox by a qualitative argument. (Hint: Consider the lower portion of the belt as seen in the reference frame of the upper portion.)

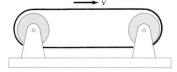

Fig. 41.33

*22. A spaceship has a length of 200 m in its own reference frame. It is traveling at 0.95c relative to the Earth. Suppose that the tail of the spaceship emits a flash of light.
    (a) In the reference frame of the spaceship, how long does the light take to reach the nose?
    (b) In the reference frame of the Earth, how long does this take? Calculate the time directly from the motions of the spaceship and the flash of light; then compare it with the result calculated by applying the Lorentz transformations to the result obtained in (a).

**23. Two spaceships are launched from the Earth at the same instant of time, but from launching pads located at different heights. The difference in the heights of the launching pads is h. In the reference frame of the Earth, the spaceships travel with the same constant acceleration a in the same direction, and their separation therefore always remains exactly h.
    (a) Find the separation between the spaceships as reckoned in the reference of the leading spaceship at the instant when this spaceship attains a speed V_O relative to the Earth. (Hint: Place the origin of the spacetime coordinates at the spaceship at this instant.)
    (b) Show that if the initial height difference is small, then the separation in the reference frame of the spaceship is approximately $h\sqrt{1 - V_O^2}$.

Section 41.6

24. A collision between two gamma rays creates an electron and an antielectron that travel away from the point of creation in opposite directions, each with a speed of $0.95c$ in the laboratory. What is the speed of the antielectron in the rest frame of the electron, and vice versa?

25. A radioactive atom in a beam produced by an accelerator has a speed $0.80c$ relative to the laboratory. The atom decays and ejects an electron of speed $0.50c$ relative to itself. What is the speed of this electron relative to the laboratory, if ejected in the forward direction? If ejected in the backward direction?

*26. Three particles are moving along the positive x axis, in the positive direction. The first particle has a speed of $0.6c$ relative to the second, the second has a speed of $0.8c$ relative to the third, and the third has a speed of $0.5c$ relative to the laboratory. What is the speed of the first particle relative to the laboratory?

*27. Find the inverses of Eq. (40), that is, express v_y and v_z in terms of v'_x, v'_y, and v'_z.

*28. In the reference frame of the Earth, a projectile is fired in the x–y plane with a speed of $0.8c$ at an angle of $30°$ with the x axis. What will be the speed and the angle with the x' axis in the reference frame of a spaceship passing by the Earth in the x direction at a speed of $0.5c$? [Hint: Use Eqs. (36) and (40).]

*29. Derive Eq. (40) for v'_y and v'_z.

Section 41.7

30. A particle has a kinetic energy equal to its rest-mass energy. What is the speed of this particle?

31. The yearly energy expenditure of the United States is about 8×10^{19} J. Suppose that all of this energy could be converted into kinetic energy of an automobile of mass 1000 kg. What would be the speed of this automobile?

32. Derive Eq. (52).

33. What is the kinetic energy of a spaceship of rest mass 50 metric tons moving at a speed of $0.5c$? How many metric tons of a matter–antimatter mixture would have to be consumed to make this much energy available?

34. What is the momentum and what is the kinetic energy of an electron moving at a speed of one-half the speed of light?

*35. What is the percent difference between the Newtonian and the relativistic values for the momentum of a meteoroid reaching the Earth at a speed of 72 km/s?

*36. The speed of an electron in a hydrogen atom is 2.2×10^6 m/s. What is the percent deviation between the Newtonian and the relativistic values of the kinetic energy of an electron at this speed? [Hint: Take the Taylor-series expansion of Eq. (47) one term further.]

*37. Show that the velocity of a relativistic particle can be expressed as follows:

$$\mathbf{v} = \frac{c\mathbf{p}}{\sqrt{m^2c^2 + p^2}}$$

*38. At the Fermilab accelerator, protons are given kinetic energies of 1.6×10^{-7} J. By how many meters per second does the speed of such a proton differ from the speed of light? What is the momentum of such a proton?

*39. Consider the electrons of a speed $0.999999999948c$ produced by the Stanford Linear Accelerator. According to Eq. (54), what is the mass of one of these electrons? By what factor is this mass greater than the mass of an electron at rest?

*40. The most energetic cosmic rays have energies of about 50 J. Assume that such a cosmic ray consists of a proton. By how much does the speed of such a proton differ from the speed of light? Express your answer in meters per second.

*41. Table 46.2 lists several large accelerators that produce high-energy protons or electrons. In each case, calculate the difference $c - v$ between the speed of light and the speed of the particle. [Hint: $1 - v^2/c^2 = (1 + v/c) \times (1 - v/c) \cong 2(1 - v/c)$ if $v \cong c$.]

*42. At the Fermilab accelerator, protons of kinetic energy 1.6×10^{-7} J are made to move along a circle of diameter 2.0 km.
   (a) What is the centripetal acceleration of these protons?
   (b) According to Eq. (54), what is their mass?
   (c) What is the centripetal force on the protons?

*43. Suppose that a spaceship traveling at $0.8c$ through our Solar System suffers an inelastic collision with a small meteoroid of mass 2.0 kg.
   (a) What is the kinetic energy of the meteoroid in the reference frame of the spaceship?
   (b) In the collision all of this kinetic energy suddenly becomes available for inelastic processes that damage the spaceship. The effect on the spaceship is similar to an explosion. How many tons of TNT will release the same explosive energy? One ton of TNT releases 4.2×10^9 J.

*44. A K^0 particle at rest decays spontaneously into a π^+ particle and a π^- particle. What will be the speed of each of the latter? The mass of the K^0 is 8.87×10^{-28} kg, and the masses of the π^+ and π^- particles are 2.49×10^{-28} kg each.

*45. Free neutrons decay spontaneously into a proton, an electron, and an antineutrino:

$$n \rightarrow p + e + \bar{\nu}$$

The neutron has a rest mass 1.6749×10^{-27} kg; the proton, 1.6726×10^{-27} kg; the electron, 9.11×10^{-31} kg; and the antineutrino zero (or nearly zero). Assume that the neutron is at rest.
   (a) What is the energy released in this decay?
   (b) Suppose that in one such decay, this energy is shared between the proton and the antineutrino, so these particles emerge from the point of decay with equal and opposite momenta while the electron remains at rest. What is the momentum of the proton and of the antineutrino?
   (c) What is the kinetic energy of the proton? What is the energy of the antineutrino? [Hint: The antineutrino is always ultrarelativistic; use Eq. (53).]

*46. Derive Eq. (56).

**47. A K^0 particle moving at a speed of $0.60c$ through the laboratory decays into a muon and an antimuon.
   (a) In the rest frame of the K^0, what is the speed of each muon? The mass of the K^0 is 8.87×10^{-28} kg and the masses of the muon and the antimuon are 1.88×10^{-28} kg each.
   (b) Assume that the muon moves in a direction parallel to the original direction of motion of the K^0 and that the antimuon moves in the opposite direction. What are the speeds of the muon and the antimuon with respect to the laboratory?

**48. At the Brookhaven AGS accelerator, protons of kinetic energy 5.3×10^{-9} J are made to collide with protons at rest.
  (a) What is the speed of a moving proton in the laboratory reference frame?
  (b) What is the speed of a reference frame in which the two colliding protons have the same speed (and are moving in opposite directions)?
  (c) What is the energy of each proton in the latter reference frame?

**49. In a given reference frame, a particle of mass m has a momentum with components p_x, p_y, p_z and an energy E. By using the relativistic combination law for velocities, show that the Lorentz transformation equations for the momentum and the energy are

$$p'_x = \frac{p_x - V_0 E/c^2}{\sqrt{1 - V_0^2/c^2}}$$

$$p'_y = p_y$$

$$p'_z = p_z$$

$$E' = \frac{E - V_0 p_x}{\sqrt{1 - V_0^2/c^2}}$$

where, as in Example 6, the velocity of the new reference frame is along the x axis of the old.

Section 41.8

50. A copper wire is carrying a current of 45 A. The wire has 2.7×10^{23} free electrons per meter of length (and an equal number of positive ions).
  (a) What is the average velocity of the free electrons?
  (b) What are the negative and the positive charges per unit length on the wire in the reference frame in which the wire is at rest?
  (c) What are the negative and the positive charges per unit length in the reference frame that has a velocity equal to the average velocity of the free electrons? What is the net charge per unit length in this reference frame?

51. (a) In the reference frame of the laboratory, a proton moves with velocity 1.0×10^7 m/s parallel to a long straight wire at a distance of 0.10 m. The wire carries a current of 15 A in a direction opposite to that of the motion of the proton. What is the magnetic force on the proton.?
  (b) In the reference frame of the proton, what is the magnetic force? What is the electric force? In this reference frame, what must be the charge density on the wire to give the correct value for the electric force?

**52. Repeat the calculation of Section 41.8 of the electric force F_{new} that acts on the charge q in its own rest frame, but do *not* assume that the speed of the charge q coincides with the speed of the electrons on the wire. (Hint: The length contraction factors for the electrons and the ions on the wire are then not equal.)

INTERLUDE IX

GRAVITY AND GEOMETRY*

Few theories can compare in the accuracy of their predictions with Newton's theory of universal gravitation. But in spite of its spectacular successes, Newton's theory is not perfect. As we mentioned in Chapter 9, the planet Mercury wanders slightly ahead of its predicted position. This deviation, called the perihelion precession of Mercury, amounts to 43 seconds of arc per century. The failure of all attempts to explain this small deviation led astronomers to question the validity of Newton's theory. Besides, from the point of view of relativity, there is a serious defect in this theory. According to Newton's law of universal gravitation, any gravitational disturbance propagates instantaneously from one mass to another. For instance, if the Sun suddenly were to grow a large bulge, the gravitational effect of this bulge should be felt at the Earth instantaneously. But this would amount to propagation of a signal with infinite speed, which is impossible according to the theory of Special Relativity.

After the discovery of Special Relativity, Einstein set to work on a new theory of gravitation, and in 1915 he formulated his famous theory of General Relativity. In the preceding chapter we saw that for reference frames in motion, space and time are not absolute; in this chapter we will see that in the vicinity of gravitating bodies, space and time are distorted. In essence, the content of Einstein's new theory of gravitation is that the orbital motion of planets does not arise from a gravitational force, but from the distortion of spacetime. Gravity is due to curved spacetime.

IX.1 THE PRINCIPLE OF EQUIVALENCE

The most important fact about gravity is that, at any given place, all bodies fall at the same rate. This fact was firmly established by the experiments of Galileo Galilei late in the sixteenth century. Galileo not only checked that bodies of unequal weight fall at the same rate, but he also checked that bodies made of different substances fall at the same rate. In his most sensitive experiments, he compared the periods of oscillation of pendulums of equal lengths, but with bobs made of different substances; any inequality of the free-fall accelerations of the bobs would then show up as an inequality of their periods (see Section 15.4). Galileo compared pendulum bobs of cork, lead, and other substances and found them equal. Such pendulum experiments were repeated by Newton with some refinements to compensate for the friction between the pendulum bobs and air.

Experiments testing the universality of free-fall rates for different substances have come to be called **Eötvös experiments,** after Lorand von Eötvös, a Hungarian physicist who, around 1920, did very extensive tests of very high precision (Figure IX.1). Modified versions of these experiments have been used to test the free-fall rates toward the Earth and toward the Sun. The basic method underlying these experiments can most easily be described in the case of free fall toward the Sun: We attach the two masses that are to be compared to the beam of a Cavendish balance. The masses, the Cavendish balance, our laboratory, and the entire Earth are in free fall toward the Sun. If one of the masses is falling faster than the other, then the beam of the balance will tend to turn toward the Sun, that is, the beam will experience a torque. The Eötvös experiment

Fig. IX.1 One of the Cavendish torsional balances used by Eötvös. The balance beam is inside the horizontal box. The suspension fiber is inside the upper vertical tube.

* This chapter is optional.

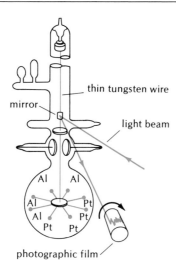

Fig. IX.2 The apparatus of Braginsky.

Fig. IX.3 Astronaut floating freely in Skylab.

attempts to measure this torque. The torque is exactly zero if, and only if, the free-fall rates of the two masses are exactly equal.

In the hope of discovering some small deviation between free-fall rates, Eötvös and his successors tested many common metals and also a wide variety of other substances — snakewood, tallow, asbestos, radioactive strontium, etc. — but they found no evidence for any deviation. The latest and most precise experiments by R. H. Dicke (at Princeton University) and V. B. Braginsky (at Moscow University) were done on gold, platinum, and aluminum. The atoms of gold and aluminum are about as different as can be — one is very heavy and the other very light. Nevertheless, the experiments show that these atoms fall at exactly the same rate to within the instrumental error of the apparatus, about 1 part in 10^{12}.

Figure IX.2 shows the apparatus used by Braginsky. The balance beam is star shaped, with eight prongs carrying small masses of aluminum and platinum. The reason for this curious configuration of the beam is that local gravitational disturbances, such as trucks or streetcars driving on nearby streets, do not disturb it as easily as they would an ordinary single beam. Of course, the beam is placed in a good vacuum (10^{-8} mmHg), and the entire apparatus is carefully kept in thermal isolation. Because there is no air rubbing on the beam, the balance has next to no friction; if set in motion, it swings for several years.

The delicate experiments of Dicke and Braginsky confirm with extreme precision what Galileo already recognized four centuries ago: under the influence of gravity all bodies fall at exactly the same rate. As we mentioned in earlier chapters, one consequence of this universality of the rate of free fall is that free fall simulates the condition of weightlessness, or absence of gravity. For example, Figure IX.3 shows a photograph of the interior of the Skylab satellite in orbit around the Earth. Skylab and the astronauts within it are in free fall. The astronauts feel weightless; they are in a zero-g environment. The reason for this apparent absence of gravitational forces is that any object released by one of the astronauts falls at exactly the same rate as the astronaut, that is, relative to the astronaut it seems not to fall at all.

The freely falling reference frame of Skylab simulates an inertial reference frame: a body at rest remains

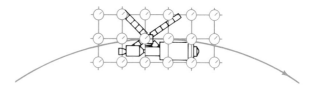

Fig. IX.4 The reference frame of Skylab is in free fall.

at rest and a body in motion continues to move with constant velocity along a straight line (Figure IX.4). A reference frame in free fall under the influence of gravity is called a **local inertial reference frame.** This kind of reference frame must be regarded as having a small size — it can simulate a true inertial reference frame over only a small range of distances. For instance, if we attempt to extend the reference frame of Skylab over an appreciable part of the orbit, then the decrease of the gravitational acceleration with distance from the Earth would cause a free body in the upper part of the reference frame to accelerate less than a body in the lower part — both of these bodies would then have an acceleration relative to the midpoint of the reference frame (Figure IX.5). Although these deviations between the motions of bodies in a local inertial reference frame are most noticeable if the distances between the bodies are large, a very sensitive instrument can detect the deviations even when the distances are small. For instance, a pair of modern, sensitive accelerometers placed at opposite ends of the Skylab satellite could easily have detected the dif-

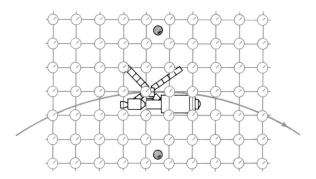

Fig. IX.5 A large reference frame in free fall near the Earth. Bodies (balls) that are above or below the midpoint of this reference frame accelerate relative to the reference frame.

ference in the gravitational accelerations at the two ends. Hence the elimination of gravity by free fall is not quite perfect. But the residual effects are so small that we will ignore them in the following.

The extremely precise experiments on the equality of the rates of free fall establish that, as far as mechanical experiments are concerned, there is no intrinsic distinction between a local inertial reference frame and a true inertial reference frame. This indicates that there is not only a relativity of uniform translational motion but, to some extent, also a relativity of accelerated motion — the astronauts in a freely falling spacecraft cannot detect their acceleration by any mechanical experiments performed within their spacecraft. In 1911, some years before he formulated his new theory of gravitation, Einstein conjectured that this general relativity of motion applies not only to mechanical experiments, but also to any other experiments. He formulated this conjecture as his famous **Principle of Equivalence,** a hypothesis about the character of gravitation:

> In a local inertial reference frame, the effects of gravity are absent; in such a reference frame all the laws of physics are the same as in a true inertial reference frame in interstellar space, far from any gravitating masses. Conversely, in an accelerating reference frame in interstellar space there is artificial gravity; in such a reference frame the laws of physics are just as though the reference frame were at rest on or near a gravitating body.

This principle lies at the basis of the theory of General Relativity. In the following sections we will explore a few consequences of this principle.

IX.2 THE DEFLECTION OF LIGHT AND THE CURVATURE OF SPACE

One of the most obvious conclusions we can draw from the Principle of Equivalence is that the speed of light ought to be the same everywhere in a gravitational field. Imagine that everywhere around the Sun there are small spaceships (Skylabs), instantaneously at rest but in free fall (that is, these spaceships have just been released and they are about to fall toward the Sun, but they have not yet acquired any appreciable velocity). In each of them gravity is effectively absent; and when experimenters aboard them measure the speed of light, the Principle of Equivalence demands that they all obtain the same result, the same as in interstellar space, 3.00×10^8 m/s.

Another conclusion we can draw from the Principle of Equivalence is that light will be deflected by the Sun's gravity. And this leads to a paradox: if light is deflected by the Sun, then light must be slowed down by the Sun — and that contradicts the previous conclusion that the speed of light is constant. Before we explore the consequences of this paradox, let us spell out why we expect that light will be deflected by the Sun and also slowed down by the Sun.

Consider one of the small spaceships in free fall toward the Sun. Since in this spaceship gravity is effectively absent, a light ray will propagate across this spaceship in a straight line, just as in interstellar space. But if the light ray follows a straight path relative to the spaceship, then it must follow a curved path relative to the Sun. Figure IX.6 shows a light ray propagating from

Fig. IX.6 A spaceship in free fall. A light ray propagates from the left side of the spaceship to the right.

one side of the spaceship to the other; the light ray starts at the midpoint of the left side and ends at the midpoint of the right side; since the spaceship accelerates downward, the light ray must also accelerate downward at the same rate if it is to follow a straight (horizontal) path relative to the spaceship. Thus, the Principle of Equivalence leads to the conclusion that in the vicinity of the Sun a light ray must follow a curved path — the Sun "bends" rays of light (Figure IX.7). If the light originates from a star, then an astronomer on

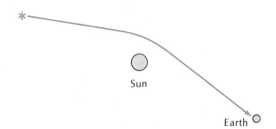

Fig. IX.7 Curved trajectory of light ray passing near the Sun. The bending of this trajectory has been wildly exaggerated in this diagram.

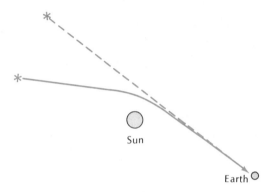

Fig. IX.8 An astronomer on the Earth sees the star at a virtual position different from its real position.

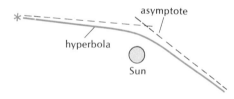

Fig. IX.9 According to Laplace, light should follow a hyperbolic orbit.

the Earth will see this star displaced from its usual position whenever the Sun happens to be near the line of sight (Figure IX.8).

The deflection of light by the gravity of the Sun need not surprise us. Light has energy, hence light has mass. We therefore expect that the Sun will attract light, and we expect that a ray of light passing near the Sun has an hyperbolic orbit (Figure IX.9) just like the orbit of a fast comet. The possibility of this effect was already recognized by the French mathematician and astronomer Laplace more than a century before Einstein. But the effect is very small: a ray passing very close to the Sun is deflected only by a few seconds of arc. It was not until 1919 that astronomers made an attempt to measure this tiny deflection. At that time Einstein had calculated the exact value of the expected deflection from his theory of General Relativity, and this stimulated astronomers to make the attempt.

It is a very difficult measurement to make. The as-

Fig. IX.10 Eclipse instruments at Sobral, Brazil, for the observation of the eclipse of 1919.

tronomer needs to look at a ray of light that comes from a star, passes near the Sun, and reaches the Earth. That means that the astronomer must observe the star when it is near the Sun in the sky; it is of course only possible to do this during a total eclipse of the Sun. Unfortunately, eclipses usually happen at inconvenient places; such as the jungles of Brazil or the deserts of the Sudan, and the primitive conditions at the observation site are not conducive to extremely high precision. All the telescopes, mirrors, cameras, and so on, must be carried to these locations and carefully protected from disturbances while the eclipse photographs are being taken (Figure IX.10). Since 1919, astronomers have gone to great pains with these eclipse observations, and they have sent out a dozen expeditions to remote places, but they have not succeeded in reducing the experimental uncertainties to a satisfactory level. Although all the measurements found a deflection somewhere between 1.5 and 2.0 seconds of arc for a light ray that just grazes the Sun, there are substantial and unexplained differences between the individual measurements.

Rather more reliable measurements of the deflection have recently been made with radio waves. Many objects in the sky are sources of radio waves. Among these, the best suited for the deflection measurement are the *quasars*; these are radio sources of very small angular size, which act as pinpoints of radio waves just as ordinary stars act as pinpoints of light. To measure the deflection of the radio waves, one must wait for some day of the year when the Sun and a quasar are nearly aligned. For the quasar 3C279 this happens every 8th of October, and radio astronomers have taken advantage of this alignment to measure the deflection on several occasions. The best of these measurements indicate that a radio wave just grazing the Sun is deflected by about 1.7 or 1.8 seconds of arc.

The deflection of light or radio waves implies that the speed of these waves is reduced when they go near the Sun. To understand this, look at Figure IX.11,

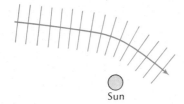

Fig. IX.11 Wave fronts of a light wave passing by the Sun.

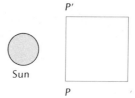

Fig. IX.13 A large square laid out with tightly stretched strings near the Sun. The side PP' nearest the Sun is longer than the other sides, although in this diagram drawn on a flat two-dimensional surface all sides look to be the same length.

which shows wave fronts of a light wave marching along. How do these wave fronts deflect? If every portion of the wave front had the same speed as every other portion, there would be no deflection. If there is to be a deflection, the portion of wave front nearest the Sun must slow down so that this portion begins to lag behind — the wave front gradually turns like a row of marching soldiers swinging to a new direction of advance.

The slowing down of radio waves by the Sun has been confirmed quite directly by means of radar signals sent past the Sun. Radio astronomers have used their giant antennas as transmitters to send radar signals to Mercury and to Venus; the signals then bounce back to Earth. If the target planet happens to be near occultation (that is, nearly eclipsed by the Sun), then these radio waves must pass close to the Sun both on the outbound and inbound portions of the trip (Figure IX.12). The experiments show that the radar echo from Venus is delayed by about 200 microseconds as compared to the echo obtained when the radar pulse does not pass near the Sun. This small time delay can be readily detected by accurate atomic clocks. The delay gives direct evidence that the radio signal is slowed down by gravity. A similar slowing down has also been detected in radio signals sent to the Mariner 6 and 7 spacecraft and retransmitted back to Earth.

So, all our experiments show that light slows down near the Sun — yet according to the Principle of Equivalence, it should not. What is the answer to this paradox? The answer must be that there is more dis-

tance from here to Venus than we suppose there is. Light does not move slower, it merely goes farther and therefore takes longer. In the immediate vicinity of the Sun, distances are longer than expected on the basis of Euclidean geometry. For example, consider a "square" formed by laying out sides at right angles; the segment nearest the Sun is longer than the segment farthest from the Sun (Figure IX.13). Ordinary Euclidean geometry fails — we have a space that is not flat, but a space that is curved and distorted.

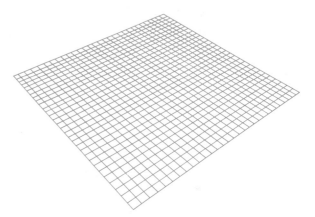

Fig. IX.14 A flat two-dimensional surface.

To shape a mental image or, more precisely, an analogy of this distortion, it is helpful to think of a hypothetical two-dimensional space rather than our real three-dimensional space. Figure IX.14 shows a flat two-dimensional surface; this is the analog of our real three-dimensional space with the Sun absent. Figure IX.15 shows what happens when a "sun" is placed in this two-dimensional space: the space acquires a dip, just as the surface of a waterbed acquires a dip when you sit in the center. The distance between the points P and P' measured along the curved surface is longer than the distance measured along a straight line, and light traveling from P to P' along the curved surface will take longer than what we would expect if we ignored the dip. This resolves the paradox: the speed of light is the same everywhere, but light that passes near the Sun must go an extra distance and hence is de-

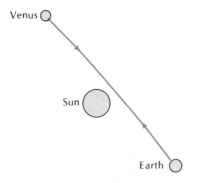

Fig. IX.12 A radar pulse emitted from the Earth reaches Venus and returns to the Earth.

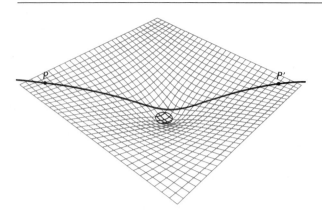

Fig. IX.15 If a "sun" is placed on this surface, the surface becomes curved. The shortest line between two points P and P' is longer than on the flat surface.

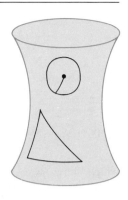

Fig. IX.16 (left) The surface of a sphere is a curved two-dimensional surface.

Fig. IX.17 (right) The surface of a hyperboloid is another curved two-dimensional surface.

layed. How much extra distance? For a light ray grazing the Sun, the extra distance from Earth to Venus is about 30 km.

This compelling argument shows how the Principle of Equivalence leads us to the conclusion that the space near the Sun is curved. Curved space is the starting point of Einstein's theory of General Relativity.

IX.3 THE THEORY OF GENERAL RELATIVITY

The fundamental tenet of Einstein's theory of General Relativity is that gravity is caused by curvature. The Sun, or any other massive body, produces curvature in the space surrounding it, and this curvature affects the motion of light and of planets, making them move in the way they do. The Sun does not exert any force on the light or the planets — it distorts space and also distorts time, and the light or planet merely moves on the "straightest" possible line in this curved spacetime. Thus, Einstein's theory of General Relativity is a geometrical theory of gravitation. John Wheeler has coined the name *geometrodynamics* for Einstein's theory. This name draws attention to the essential feature of the theory: the dynamical, curved spacetime geometry which acts on and reacts to matter.

It is beyond our powers of imagination to visualize a curved three-dimensional space. To visualize a curved three-dimensional space we would have to visualize a fourth or even a fifth dimension into which the three-dimensional space can curve — and we just cannot visualize any four-dimensional object.

Fortunately, we can detect the curvature of a space without stepping into a higher dimension and looking at it from the outside. Consider some little bugs living on a curved two-dimensional surface, for example, some mites living on the surface of a grapefruit. In principle, they can detect the curvature of their living space by experiments within this space. They can do measurements on geometrical figures on their surface and check whether the geometry is Euclidean or something different. If the living space of the bugs is the surface of a sphere, then they will find that the circles and triangles that they draw on their surface do not obey the usual theorems of Euclidean geometry. Figure IX.16 shows the surface of a sphere and shows a circle and its radius (or what the bugs would regard as its radius); obviously the circumference of the circle is shorter than 2π times the radius. Figure IX.16 also shows a triangle on the surface of the sphere; obviously the sum of the interior angles of this triangle is larger than 180°. Incidentally: The sides of this triangle are drawn so that they are the shortest lines connecting the vertex points; that is, the sides are drawn by stretching a string between the vertex points.

Circles and triangles can also be used to detect the curvature of other kinds of curved surfaces. Figure IX.17 shows the surface of a hyperboloid. On this surface, the circumference of a circle is longer than 2π times the radius, and the sum of the interior angles of a triangle is smaller than 180°.

According to General Relativity, the space inside the Sun or any gravitating body has characteristics analogous to those of a spherical surface: circumferences are too small and interior angles of triangles too large. The space outside and near the Sun has characteristics analogous to those of a hyperboloidal surface: circumferences are too large and interior angles of triangles too small. These deviations from Euclidean geometry are implicitly contained in Einstein's equations that relate the curvature of the space to the mass of the body. Einstein's equations are rather complicated, but for a spherical gravitating body of uniform density the deviations from Euclidean geometry demanded by Einstein's equations can be expressed quite simply as follows. A circumference that is too small is equivalent to a radius that is too large. Let us define the radius excess ΔR as the amount by which the radius of the body is too large when compared with its circumference C,

$$\Delta R = R - C/2\pi \qquad (1)$$

Then the Einstein equations state that the radius excess is directly proportional to the mass M of the body:

$$\Delta R \cong \frac{4GM}{3c^2} \qquad (2)$$

[Alternatively, we can define the radius excess as the amount by which the radius is too large when compared with the surface area of the body:

$$\Delta R = R - \sqrt{A/4\pi} \qquad (3)$$

It turns out that this radius excess has the same value as in Eq. (2).] The Einstein equations also yield a formula for the excess angle of a triangle immersed in the body (such as shown in Figure IX.18):

$$\Delta \theta \cong \frac{\sqrt{2}GM}{Rc^2} \qquad (4)$$

When we apply these formulas to the Earth, pretending that the density is constant, we find the following: the circumference of the Earth is smaller than $2\pi R$ by 5.9 mm, the area of the surface of the Earth is

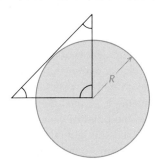

Fig. IX.18 A triangle laid out with tightly stretched strings immersed in a spherical body.

smaller than $4\pi R^2$ by 0.95 km², and the sum of interior angles of a triangle immersed in the Earth is larger than 180° by 2.0×10^{-4} seconds of arc. The corresponding numbers for the Sun cannot be obtained directly from our formulas, because the density of the Sun is far from constant. A calculation taking into account the variation of density leads to the following: the circumference of the Sun is smaller than $2\pi R$ by 23 km, the area of the surface of the Sun is smaller than $4\pi R^2$ by 5.4×10^7 km², and the sum of interior angles of a triangle differs from 180° by 0.62 seconds of arc.

For the Sun and the Earth, these deviations from the expected Euclidean values are so small that it is hopeless to attempt to measure them directly. The curvature of space shows up only indirectly in its effects on the propagation of light mentioned above. It also

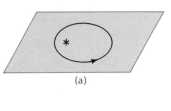

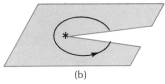

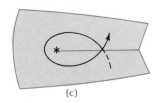

Fig. IX.19 (a) Elliptical orbit of a planet on a flat plane. If we deform the flat plane into a cone (b), the orbit acquires a kink (c).

shows up in an effect on planetary motion: the **perihelion precession**. In flat space Newton's theory tells us that planets move in ellipses around the Sun. But when such an elliptical orbit is made to fit into the curved, bowl-like space surrounding the Sun, the orbit precesses, that is, it changes its orientation a little bit on each revolution. This precession effect can be understood in terms of the curvature of space. The portion of bowl along which a planet moves can be approximated by a conical surface, tangent to the bowl at about the radius of the planetary motion. Figure IX.19 shows an ellipse drawn on a flat plane; to make this flat plane into a cone that approximates the bowl-like curved space, we must cut a wedge out of the plane and join the edges together. But this introduces a kink in the orbit (Figure IX.19c). When the planet reaches this kink, it will overshoot the old elliptical orbit and go into a new elliptical orbit whose major axis has a slightly different orientation. Since this happens each time the planet completes a revolution, the net effect is a gradual change of orientation of the elliptical orbit (Figure IX.20); this is the perihelion precession. For Mercury the precession calculated from General Relativity is 43 seconds of arc per century, in excellent agreement with the observed value.

However, this is not the whole story. The motion of planets, and the motion of light, is not only affected by the curvature of space, but also by the curvature of time. In fact, what we have is not separate space and time but, rather, a single four-dimensional **curved spacetime.** What is meant by curvature of spacetime is even harder to visualize than curvature of space. In

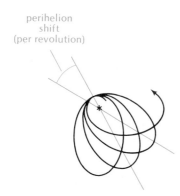

Fig. IX.20 Perihelion precession of the orbit of a planet. The amount of precession has been wildly exaggerated.

the case of space, we can always have recourse to the analogy of a two-dimensional curved space; but in the case of spacetime, there is no such analogy. To get at least some feeling for curved spacetime, let us go back to Figure IX.13 showing a "square." The side near the Sun is longer than the side far from the Sun, that is, curvature of space means that distances near the Sun are longer. Likewise, curvature of spacetime means that time near the Sun is longer. Expressed another way, clocks near the Sun run slower than clocks far from the Sun. And in general, a clock in a strong gravitational field near a large mass runs slower than a clock far from a mass. This effect is called **gravitational time dilation,** or **gravitational red shift.** As we will see in the next section, this effect is a direct consequence of the Principle of Equivalence.

IX.4 THE GRAVITATIONAL TIME DILATION

We know from the theory of Special Relativity that when a clock is in motion relative to another clock, it suffers time dilation. We will now see that the theory of General Relativity predicts an extra kind of time dilation suffered by clocks at rest in a gravitational field.

Consider a stationary laboratory near a gravitating body, say, a laboratory on the surface of the Earth. Suppose that a pair of identical clocks are installed on the floor and on the ceiling of this laboratory (Figure IX.21); the lower clock is then closer to the Earth. We

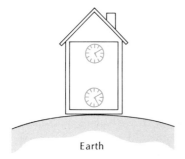

Fig. IX.21 Stationary laboratory with clocks on the Earth.

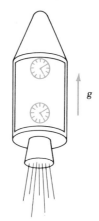

Fig. IX.22 Accelerated laboratory in interstellar space.

want to compare the frequencies of ticking of these clocks. Since the clocks are separated by some distance, we need to send a signal from one clock to the other. Let us couple the lower clock to a radio transmitter that emits a wave of the same frequency as that of the clock; the radio wave travels to the second clock, and we can compare the frequency of the arriving radio wave with the frequency of that clock. This permits us to check the rates of the clocks against each other.

To figure out how these intervals will compare, we appeal to the Principle of Equivalence and replace the stationary laboratory near the Earth by an accelerated laboratory in interstellar space (Figure IX.22). If the latter laboratory has an acceleration $g = 9.8$ m/s², then the behavior of clocks within this accelerated laboratory is the same as within a stationary laboratory on Earth. It is convenient to analyze the motion of the radio waves from the point of view of an inertial reference frame in which the laboratory is initially at rest when the clock on the floor sends out its first radio waves. If the distance between the clocks is h, then the waves take a time h/c to reach the clock on the ceiling. But in this time interval the acceleration g gives the latter clock a velocity gh/c relative to our inertial reference frame. Thus, when the radio waves arrive, the receiver of these waves has a recessional velocity gh/c, and therefore the frequency detected by the receiver is subject to a Doppler shift. This Doppler shift reduces the frequency from a value v at the emitter to a smaller value v' at the receiver. Thus, as compared to the clock on the ceiling, the clock on the floor runs slow.

The time dilation produced by the gravity of the Earth is very small: a clock sitting on the surface of the Earth runs slow by only 1 part in 10^9 compared to a clock far above the surface. However, even these small effects are within reach of extremely precise atomic clocks. In 1972, scientists from the U.S. Naval Observatory demonstrated the gravitational time dila-

Fig. IX.23 Hewlett-Packard portable atomic clocks used to test the gravitational time dilation.

tion by carrying portable atomic clocks to an altitude of about 10,000 m and comparing them with identical clocks kept on the ground. The clocks were carried to this altitude by commercial jets — the experimenters simply bought tickets for a round-the-world trip on Pan Am, TWA, and American Airlines. They took one trip around the Earth in an eastward direction, and one westward, each trip taking about three days, of which about two days were spent airborne. Four portable Hewlett-Packard cesium beam clocks were carried on these trips (Figure IX.23); continual comparisons of these traveling clocks served as protection against occasional sudden, small jumps in rate to which the cesium clocks are prone. When the traveling clocks were brought back down and compared with the clock of the U.S. Naval Observatory in Washington, D.C., it was found that they had gained about 1.5×10^{-7} s.[1]

Although this experiment with traveling clocks is the most direct test of gravitational time dilation, it is not the first test, or the latest, or the most accurate. The time dilation has also been tested by red-shift measurements. Atoms sitting on the surface of the Sun or on the surface of some other star can be regarded as clocks. These atoms tick slower than atoms far from the Sun's surface, and consequently the frequency of the light they emit will be low compared to the frequency of faraway atoms. Hence light emitted by, say, potassium atoms on the Sun will exhibit a red shift when it is received on Earth and is compared to light emitted by potassium atoms on Earth. According to the theory of General Relativity, the frequency should be low by 2 parts in 10^6. The measured values agree with this theoretical value.

Because of the large uncertainties inherent in these measurements, the red-shift observations on sunlight and starlight are not a very precise test of General Relativity. A somewhat better red-shift observation was made on the Earth using a nucleus, rather than an atom, as a clock. A vibrating nucleus emits gamma rays (just as a vibrating atom emits light). The gamma ray emitted by, say, an iron nucleus at the surface of the Earth will exhibit a red shift when it ascends to some height above the surface. Although the red shift is only 2 parts in 10^{15} for a height increase of 20 m, it can be detected by allowing the gamma ray to hit another iron nucleus, which will try to absorb it. The absorption depends very critically on frequency — even a tiny reduction of frequency drastically diminishes the chances of absorption. Physicists at Harvard University performed an experiment of this kind in a 20-m-high tower (Figure IX.24). They placed some radioactive iron (which emits gamma rays) at the bottom of the tower and some extra iron (to serve as detector) at the top. With this they managed to test the red shift to 1%.

Finally, let us mention the most precise red-shift experiment done to date. To obtain a large red shift, it is desirable to send a clock to high altitude above the Earth. So, in 1976, physicists from the Smithsonian Astrophysical Observatory sent a Scout rocket with an atomic clock to a height of 10,000 km above the At-

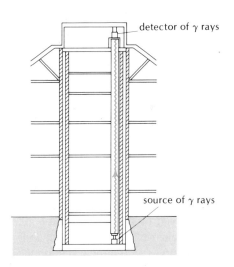

Fig. IX.24 Emitter (bottom) and absorber (top) of gamma rays in the Jefferson Physical Laboratory at Harvard University.

lantic Ocean (Figure IX.25). At this height, the clock is expected to run faster than a clock on the ground by 4.5 parts in 10^{10}. Of course, the rocket clock shatters when the rocket crashes, and the comparison of clock rates has to be made in flight, during the two hours or so that the payload spends in free fall. The rocket clock was linked to a radio transmitter that transmitted the "ticks" of the clock down to the ground. There the ticks were compared with ticks of another, identical

---

[1] To obtain this result, the experimenters had to take into account the special-relativistic time dilation produced by the motion of the clocks (see Chapter 41).

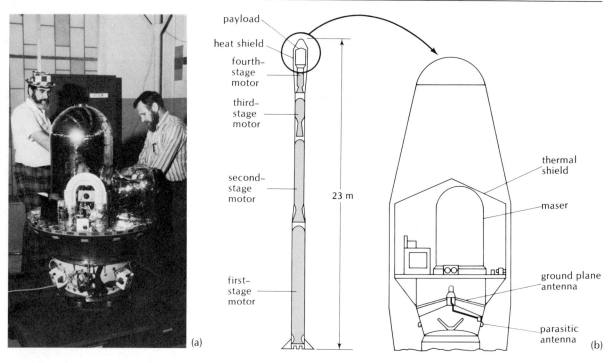

Fig. IX.25 (a) Maser clock (within dome) before installation in the rocket. (b) Position of the maser clock in the bulbous nose cone of the Scout rocket. (Courtesy R. F. C. Vessot and M. L. Levine, Smithsonian Astrophysical Observatory.)

clock kept on the ground. Apart from the tricky technical problems of protecting the rocket clock from the traumatic shock of the takeoff acceleration, the extreme temperature variations encountered in spaceflight, and the changing strength of the Earth's magnetic field, this experiment had to resolve another problem: since the rocket is in continuous motion with a fairly large velocity relative to the ground, the radio signals received on the ground have a Doppler shift. Thus, it is necessary to sort out this Doppler shift from the gravitational red shift (or, rather, blue shift, since the rocket clock will be fast). The Doppler shift was detemined by much the same method as used by the police in their Doppler radar systems. When a beam of radar waves is sent to an oncoming automobile, the waves are received by the surface of the automobile with some Doppler shift, since the automobile is moving toward the waves. When the waves bounce off the automobile and travel back to the policeman's radar receptor, they acquire an extra Doppler shift because the automobile acts as a moving source while reflecting the waves. In the rocket experiment, a radio signal from the ground was aimed at the rocket and retransmitted by the rocket back to the ground by means of a retransmitter, or transponder. This radio signal suffers a Doppler shift both when going out to the rocket and when returning to the ground, that is, it returns with a net Doppler shift of twice the one-way Doppler shift. The signal also suffers a gravitational frequency shift — but this is a red shift on the outgoing part of the trip and a blue shift on the return part; hence the gravitational time dilation does not affect the net frequency of a round-trip signal. The round-trip signal can therefore be used to make the necessary Doppler correction to the clock signal. The result of this elaborate experiment confirmed the theoretical value of the gravitational time dilation to within 0.01%.

IX.5 BLACK HOLES

Within the solar system the effects of Einstein's theory of General Relativity are minuscule — we can only detect the deviations between Newton's theory and Einstein's theory by means of delicate instruments of very high precision. To find drastic deviations, we must seek out places in the sky where the gravitational fields are much stronger than anywhere in the Solar System. The most spectacular relativistic effects are displayed by **black holes.** These are the disembodied remains of stars that have collapsed under their own weight, crushing their material to a singular state of infinite density. After the collapse nothing is left of the star except an extremely intense gravitational field. There is good circumstantial evidence that several such collapsed stars exist in the sky in orbits around ordinary stars belonging to our Galaxy.

A black hole is black because neither light nor anything else can escape the overwhelming grip of its gravitational field. Particles can enter the black hole, but no particle can ever emerge. Electrons, protons, atoms, spaceships, astronauts, or whatever that enter a black hole must abandon all hope of ever escaping again. Once they enter the interior of the black hole, they are trapped; they are inexorably pulled deeper and deeper toward the center. Gravity in a black hole is so strong that even light, which moves at higher speed than anything else, is pulled down.

In terms of escape velocity, we can get some rough understanding of how gravity traps particles in a black hole. Recall that at the surface of the Earth, the escape velocity is 11 km/s (see Section 9.5) — a projectile launched with this velocity, or with more than this velocity, shrugs off the gravitational pull of the Earth and keeps on rising forever; a projectile launched with less than this velocity is gradually brought to a stop by the gravitational pull and then falls back to the Earth. For a body that is more compact, or more massive, than the Earth, the associated escape velocity is more than 11 km/s. For a starlike spherical body of mass M and radius R, the escape velocity will be [see Eq. (9.29)]

$$v = \sqrt{2GM/R} \qquad (5)$$

If the mass of the body is so large (or the radius so small) that this velocity exceeds the speed of light, then light will not be able to escape — the body will appear black. This was already deduced by Laplace some 200 years ago:

> A luminous star, of the same density as the Earth, and whose diameter should be two hundred and fifty times larger than that of the Sun, would not, in consequence of its attraction, allow any of its rays to arrive at us; it is therefore possible that the largest luminous bodies in the universe may, through this cause, be invisible.

What Laplace did not know is something that we learned from the theory of relativity: if light cannot escape, nothing else can either, because nothing can move faster than light. Thus Laplace discovered that the strong gravitational pull might make some "stars" black, but he did not discover that such "stars" had to be holes.

If we take $v = c$ in Eq. (5), we find that the radius within which the mass M must be contained for escape to be impossible is

$$R = 2\frac{GM}{c^2} \qquad (6)$$

This critical radius is called the **Schwarzschild radius.** For a body of mass equal to that of the Sun, the value of the critical radius is

$$R = \frac{2 \times 6.67 \times 10^{-11} \text{ N/m}^2 \cdot \text{kg}^2 \times 2.0 \times 10^{30} \text{ kg}}{(3.0 \times 10^8 \text{ m/s})^2}$$

$$= 3.0 \text{ km}$$

If the radius of the body is smaller than that, any particle approaching closer than 3.0 km will be trapped.

The boundary of the region of no escape that surrounds the black hole on all sides like a ghostly iron curtain, permitting entrance but preventing exit, is called the **one-way membrane,** or the **horizon.** This membrane demarcating the edge of the black hole is not made of anything; it is not a substance, but merely a place, the place of no return. According to Eq. (6), a black hole of mass equal to that of the Sun has a one-way membrane of radius 3.0 km (Figure IX.26); a black hole of a mass equal to that of the Earth would have a one-way membrane of a radius of just 9 mm (Figure IX.27).

Although to talk of escape velocity is helpful, it is also somewhat misleading. According to Newton's theory, a projectile launched in the upward direction from a point inside the black hole first moves up for some distance, then stops, and then falls back. What actually happens in a black hole is quite different. The projectile never even begins to move upward; in a black hole all motion is toward the center. Only Einstein's theory of General Relativity can give an adequate picture of how projectiles move in the interior of a black hole. It turns out that, by a happy coincidence, a careful calculation based on the latter theory yields exactly the same formula (6) for the no-escape radius as our naïve calculation based on Newton's theory.

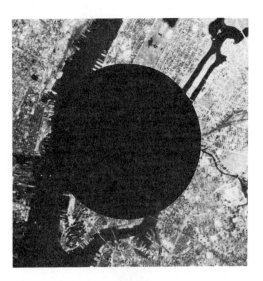

Fig. IX.26 A black hole of radius 3 km on top of New York City.

Fig. IX.27 A black hole of mass equal to that of the Earth would have the size of this dot.

According to General Relativity, the space and time near any gravitating body are curved; near and inside a black hole, the curvature is extreme — space and time are greatly distorted. In terms of the two-dimensional analog described in Section IX.3, a gravitating body or "sun" produces a dip in the surrounding space (Figure IX.28). If we replace the "sun" by a black hole, the dip becomes much deeper (Figure IX.29). Within the dip there is a place of no return (or one-way circle) from beyond which escape is impossible; and at the very center of this circle, the surface has a kink, that is, an infinite distortion. This kink indicates a **singularity,** a place where the curvature and the gravitational force are infinite.

Although Figures IX.28 and IX.29 are suggestive, they do not tell the whole tale. For inside a black hole, space and time are — in a sense — interchanged. The distance from the one-way membrane to the central singularity is not a distance in space but, rather, an interval of time, that is, the singularity is not a point in space, but a point in time. Correspondingly, inside a black hole the gravitational fields are not static; rather, they are time dependent and they evolve, becoming stronger and stronger, until finally they become infinite everywhere in the interior of the black hole. This violent evolution is very quick; in a black hole of one solar mass, it only takes 10^{-5} s. Paradoxically, seen from the outside, the black hole does not appear to evolve at all; it just sits there for ever and ever. The resolution of this paradox hinges on one of the most spectacular features of a black hole: its gravitational time dilation.

As we saw in Section IX.4, a clock placed near any gravitating body will run slow. This gravitational time dilation is one aspect of the distortion of space and time in the vicinity of a gravitating body. Near the Earth the distortion is small, and the time dilation is small. But near a black hole the distortion and the time dilation are very large. At the one-way membrane, the time dilation becomes infinite — a clock placed next to the membrane will be slowed down so much that, compared to a clock far away from the black hole, it will appear to have stopped. If we imagine an astronaut in a space capsule parked right at the edge of the black hole, all his life functions (pulse rate, alpha rhythm, voluntary and involuntary muscular contractions, etc.) will almost be at a standstill — to the scientists at "mission control" on a mother spaceship stationed far away, outside of the strong gravitational field, he will appear to have frozen. Yet the astronaut would not perceive himself as having slowed down; he would merely perceive the surrounding world as having speeded up.

This time dilation resolves the paradox of how the inside of the black hole can change while the outside does not. As judged from outside, any changes at or inside the one-way membrane take an infinite amount of time. Hence, from the outside point of view, these changes never happen, and the black hole seems completely static.

There are several objects in the sky that astrophysicists suspect contain black holes. The best known of these is Cygnus X-1, a strong source of X rays in the constellation Cygnus. Figure IX.30 shows a picture of this object taken with an X-ray telescope installed on the Einstein satellite launched in 1978. Figure IX.31 is a picture of the same object taken with an optical telescope. It seems that Cygnus X-1 is a binary system consisting of two components: one is an ordinary star of about 30 solar masses; this star is the source of the ordinary light visible in Figure IX.31. The other is a very compact body of about 10 solar masses; this is the source of the X rays shown in Figure IX.30. These two components orbit about each other with a period of 5.6 days; the distance between the two components is quite small.

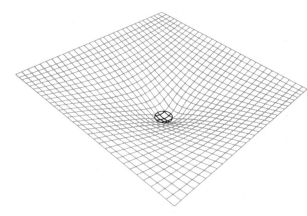

Fig. IX.28 Two-dimensional surface with a "sun" at the center.

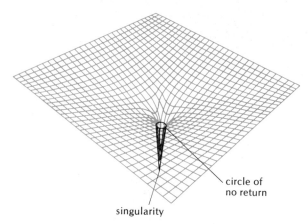

Fig. IX.29 Two-dimensional surface with a black hole at the center. The circle marks the place of no return. At the kink at the bottom, the curvature of the surface is infinite.

The compact body is probably a black hole. It produces the X rays by an **accretion** mechanism. The black hole is so near the surface of the star that it

Fig. IX.30 X-ray picture of Cygnus X-1.

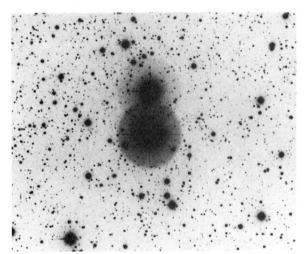

Fig. IX.31 Optical photograph of Cygnus X-1. The star of interest is in the center of this photograph. (Hale Observatories photograph courtesy J. Kristian.)

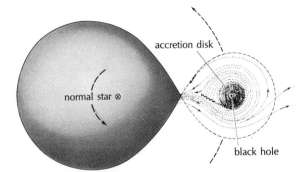

Fig. IX.32 The normal star and the black hole orbit about their common center of mass. The black hole raises a large tidal bulge on the star. Material streaming from the star toward the black hole forms an accretion disk.

raises a large tidal bulge and pulls a steady stream of material off that surface, material that spirals around the black hole, forming a disk (called the accretion disk; Figure IX.32). As the material spirals and gradually falls toward the black hole, it is accelerated by the strong gravitational fields and heated by compression. The material reaches temperatures of up to 100 million degrees Celsius. At such temperatures the violent thermal collisions between the particles of the material release a copious flux of X rays. The key to this mechanism for the production of X rays is the small size or, equivalently, the high density of the black hole. When a mass of 10 Suns is contained in a small region, the gravitational fields in the immediate neighborhood will be very intense; the enormous heating of the material becomes possible only by the action of these intense gravitational fields.

From the characteristics of the observed X-ray flux, astrophysicists have deduced that the size of the compact object is less than 300 km across. To the best of our knowledge, there is no way in which a mass of 10 Suns can be confined in a region 300 km across and remain in equilibrium — at such densities gravity would overwhelm all other forces and would immediately bring about the total collapse of the mass. The compact object in Cygnus X-1 probably collapsed a long time ago and is now a black hole.

There are several other objects similar to Cygnus X-1 in the sky. All these objects are powerful sources of X rays, and astrophysicists believe that they consist of a more or less normal star and a black hole orbiting about each other. The black hole steadily pulls materials off the star, swallows it, and emits the X rays. Note that the X rays do not come from the interior of the black hole; they come from the accretion disk just outside of the black hole. After the in-falling material enters the one-way membrane, any X rays that it emits will remain trapped. The black hole is black, but the material outside it glows with X-ray light.

Some astronomers speculate that there may be a very large number of black holes in our Galaxy and in other galaxies. It may be that most of the mass of some galaxies is in the form of black holes. But the only black holes we can hope to detect with our Earth-bound instruments are those that are swallowing gas or dust from their environment and are producing X rays or other radiation. An isolated black hole is very effectively camouflaged against the black background of outer space. Isolated black holes are the hidden reefs of outer space. They are invisible. They can be detected only by their gravitational pull.

Further Reading

Albert Einstein: Creator and Rebel by B. Hoffmann and H. Dukas (Viking, New York, 1972) is a charming biography that traces how Einstein conceived his theory of General Relativity.

The Riddle of Gravitation by P. G. Bergmann, one of Einstein's co-workers (Scribner, New York, 1968); *Space, Time, and Gravity* by R. M. Wald (University of Chicago Press, Chicago, 1977); and *Cosmology* by E. R. Harrison (Cambridge University Press, Cambridge, 1981) are clear and simple introductions to the theory of General Relativity, curved space, black holes, and the expanding universe.

Space, Time and Gravitation by A. Eddington (Cambridge University Press, Cambridge, 1968) is an old classic that explains the basic concepts of the theory.

Chapters 3–5 of *Principles of Gravitation and Cosmology* by M. Berry (Cambridge University Press, Cambridge, 1976) present some of the mathematics of curved space and General Relativity, concentrating on those aspects that can be handled by simple calculus.

General Relativity by R. Geroch (University of Chicago Press, Chicago, 1978) and *Talking About Relativity* by J. L. Synge (North-Holland, Amsterdam, 1970) are nice introductions to the geometry of four-dimensional spacetime; the former is filled with abundant examples of spacetime diagrams.

Flatland by E. A. Abott (Dover, New York, 1952) is an enchanting tale of the inhabitants of an imagined two-dimensional world and of their difficulties in visualizing the third dimension. *Geometry, Relativity and the Fourth Dimension* by R. v. b. Rucker (Dover, New York, 1977) is a worthy sequel to *Flatland* that provides a gentle introduction to curved space and to four-dimensional spacetime.

The bizarre properties of black holes and the observational evidence for their existence are described in *Black Holes, Quasars, and the Universe* by H. L. Shipman (Houghton Mifflin, Boston, 1980), *Black Holes* by W. Sullivan (Doubleday, Garden City, 1979), *Black Holes in Space* by P. Moore and I. Nicolson (Norton, New York, 1974), and *The Cosmic Frontiers of General Relativity* by W. J. Kaufmann (Little, Brown, Boston, 1977); the last of these books includes a chapter on their weird spacetime geometry and the singularities of rotating black holes. More on the singularities in black holes and in the Big Bang will be found in two books by P. C. W. Davies, *Space and Time in the Modern Universe* (Cambridge University Press, Cambridge, 1977) and *The Edge of the Universe* (Cambridge University Press, Cambridge, 1981).

The Search for Gravity Waves by P. C. W. Davies (Cambridge University Press, Cambridge, 1980) is a concise summary of the properties of gravitational waves and of the recent experiments and observations.

The following articles deal with black holes and with observational evidence for them:

"Introducing the Black Hole," R. Ruffini and J. A. Wheeler, *Physics Today*, January 1971
"Black Holes," R. Penrose, *Scientific American*, May 1972
"The Black Hole in Astrophysics: The Origin of the Concept and Its Role," S. Chandrasekhar, *Contemporary Physics*, Vol. 14, 1974
"The Search for Black Holes," K. S. Thorne, *Scientific American*, December 1974
"X-Ray Emitting Double Stars," H. Gursky and E. P. J. van den Heuvel, *Scientific American*, March 1975

The *Resource Letters GR-1 and GI-1* in the *American Journal of Physics*, February 1968 and June 1982, lists many books and articles on General Relativity at an introductory level and at more advanced levels.

Questions

1. If you were to discover a body that did not accelerate in a gravitational field at the same rate as all other bodies, how would this affect Einstein's theory of General Relativity?

2. The Earth is in free fall in the gravitational field of the Sun, the Moon, and the planets. Does this eliminate all external gravitational effects in the reference frame of the Earth?

3. According to both Newton's and Einstein's theories of gravitation, a ray of light will be deflected by the Sun; but they each assign a different cause to this deflection. Explain.

4. Why do astronomers find it more convenient to measure the deflection of rays by the Sun with radio waves rather than with light waves?

5. Suppose that you measure the speed of light in a laboratory on Earth and that you repeat this measurement in a laboratory in interstellar space, far from any gravitational influences. Will you get the same result in both cases?

6. One of Euclid's postulates states that given a straight line and a separate point, there exists one and only one straight line that passes through the point and does not intersect the given straight line (parallel postulate). What happens to this postulate in the two-dimensional curved spaces shown in Figures IX.16 and IX.17?

7. Is the space in this room flat or curved? If you had available measurement instruments of extreme precision, how could you check your answer?

8. Suppose that you stretch a long massless string so that it passes by the Sun. Would you expect this string to coincide with the path of a light ray?

9. Why did Einstein name his theory of gravitation *General Relativity*?

10. There are two possible circumstances under which an astronaut exploring interstellar space will suffer time dilation relative to his comrades on Earth. What are they?

11. Because of the rotational motion of Earth, a point on the equator has a speed of 460 m/s relative to a point on the pole. Because of the equatorial bulge of Earth, a point on the equator is 21 km farther from the center of Earth than a point on the pole. According to the special relativistic time-dilation effect, will a clock on the equator tick faster or slower than an identical clock on the pole? According to the gravitational time-dilation effect? (It turns out that these two timedilation effects cancel each other if both clocks are at sea level.)

12. Suppose that you dangle a long rope into a black hole from a safe distance. If you attempt to pull the rope out, what will happen?

13. A suicidal scientist jumps off a spaceship into a black hole. What effects does he perceive as he crosses the one-way membrane? What effects do his colleagues watching from the spaceship perceive?

14. According to recent astronomical observations, the X-ray source LMC X-1 is a binary star system one of whose members has a mass about nine times the mass of the Sun and a radius of only a few hundred kilometers. What can you conclude about this star?

CHAPTER 42

Quanta of Light

In the two preceding chapters we explored the wave properties of light. We found that light exhibits interference and diffraction, in agreement with Maxwell's theory, according to which light is a wave consisting of electric and magnetic fields with a smooth distribution of energy. In this chapter we will discover that light has particle properties. We will examine experimental evidence which establishes that a light beam consists of a stream of particle-like energy packets. These energy packets are called **quanta of light,** or **photons.**

The fact that light has the dual attributes of wave and of particle indicates that neither the wave concept of classical physics nor the particle concept of classical physics gives an adequate description of light. We have to think of light as a wave–particle object, sometimes called a **wavicle.** This object sometimes behaves pretty much like a classical wave, sometimes pretty much like a classical particle, and sometimes like a bit of both. Furthermore, we will see in the next chapter that electrons, protons, neutrons, and all the other known "particles" also exhibit such dual attributes of wave and of particle — we must regard all of them as wavicles.

42.1 Blackbody Radiation

The first hint of a failure of classical physics emerged around 1900 from the study of thermal radiation. When we heat a body to high temperature, it gives off a glow; for instance, when we heat a bar of iron to 1300 K, it glows in a deep red color. This glow is thermal radiation ("radiant heat"). The spectrum of thermal radiation is continuous —- if we analyze the light emitted by a glowing body with a prism, we find that the energy is smoothly distributed over all wavelengths.

Spectral emittance

A quantitative description of the distribution of energy over the different wavelengths is provided by the **spectral emittance** S_λ. This is the energy flux (or power per unit area) emitted by the surface of the glowing body per unit wavelength interval; thus, $S_\lambda\, d\lambda$ is the flux emitted in a small wavelength interval $d\lambda$. The spectral emittance is a function of wavelength. Measurements of the thermal radiation emitted by a glowing body show that the energy flux at very long and at very short wavelengths is quite small, and that the energy flux has a maximum at some intermediate wavelength. The position of this maximum depends on the temperature. For example, an iron bar at 1300 K has a maximum emittance at 22,000 Å (infrared), whereas the Sun, with a surface temperature of 5800 K, has a maximum emittance at 5000 Å (blue-green). Figure 42.1 is a plot of the emittance of the Sun as a function of the wavelength; this figure may be regarded as a plot of the intensity distribution in the spectrum that an (ideal) prism produces when exposed to sunlight.

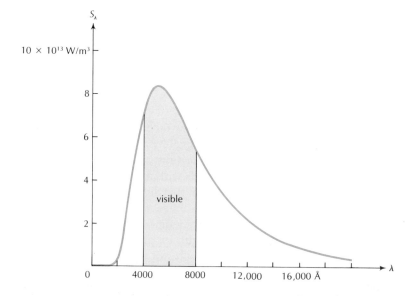

Fig. 42.1 Spectral emittance of the Sun. In this plot, the discrete dark spectral lines (Fraunhofer lines), which result from the blocking out of some of the thermal radiation by gas in the solar atmosphere, have been neglected.

The thermal radiation emerging from the surface of a glowing body is generated within the volume of the body by the random thermal motions of atoms and electrons. Before the radiation reaches the surface and escapes, it is absorbed and reemitted many times, and it attains thermal equilibrium with the atoms and electrons. This equilibration process shapes the continuous spectrum of the radiation, completely washing out all of the original spectral features of the radiation.

The flux of thermal radiation emerging from the surface of a glowing body depends to some extent on the characteristics of the surface. The surface usually permits the escape of only a fraction of the flux reaching it from the inside of the body. Correspondingly, if the body is irradiated with an equal flux of thermal radiation from the outside, the surface permits the ingress of only an equal fraction of this flux, reflecting the rest. (This equality of the emissive and absorptive characteristics of the surface can be deduced by an argument based on the Second Law of Thermodynamics.) Thus we are led to a general rule: *A good absorber is a good emitter; and a poor absorber is a poor emitter.* With this rule we can understand how thermos bottles or dewars provide such excellent thermal insulation. These bottles are constructed with a double glass wall, and the space between these walls is evacuated (Figure 42.2). Heat cannot flow across the evacuated space by conduction or convection; it can flow only by radiation. To inhibit radiation, the glass walls are silvered and thereby made into mirrorlike reflecting surfaces; these highly reflective surfaces are very poor absorbers and emitters of radiation, and they keep the heat transfer between the walls very small.

A body with a perfectly absorbing (and emitting) surface is called a **blackbody**. Such a body would look black under illumination from the outside. But when a blackbody is hot, its surface emits more thermal radiation than that of any other hot body at the same temperature. In practice, the characteristics of an ideal blackbody are most easily achieved by a trick: take a body with a cavity, such as a hollow cube, and drill a small hole in one side of the cube (Figure 42.3). The hole then acts as a blackbody — any radiation incident on the hole from outside will be completely absorbed. Because of this equivalence between a blackbody and a hole in a cavity, the terms *blackbody radiation* and *cavity radiation* are used interchangeably.

The blackbody plays a special role in the study of thermal radiation because its spectral emittance does not depend on the material of which it is made or on any other characteristics of the body — the spectral emittance depends only on the temperature of the body. We can prove this by appealing to the Second Law of Thermodynamics, according to which heat cannot flow spontaneously from a colder body to a warmer body. Consider two cavities at equal temperatures with holes of equal sizes (Figure 42.4). The cavity on the left radiates into the cavity on the right and vice versa. If the flux emitted by the cavity on the left were larger than that emitted by the cavity on the right, the radiative heat transfer would lead to an increase of temperature on the right and a decrease on the left. Heat would then be flowing from a colder to a warmer body, in contradiction to the Second Law. This argument establishes that the fluxes emitted by the two cavities are the same. Furthermore, a slight refinement of this argument establishes that the fluxes in any given small wavelength interval $d\lambda$ are also the same. We need make only a slight alteration in the arrangement shown in Figure 42.4 by inserting a filter for light between the two cavities, a filter that permits the passage of radiation of wavelengths only in the interval λ to $\lambda + d\lambda$. Our thermodynamic argument then leads to the conclusion that the fluxes in this chosen wavelength interval must be the same.

Thus, for a blackbody, the spectral emittance S_λ is a universal function of the wavelength λ and of the temperature T, and of nothing else. In the last years of the nineteenth century, physicists engaged in an intensive experimental and theoretical effort to find the function S_λ.

Blackbody

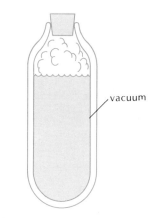

Fig. 42.2 A thermos bottle.

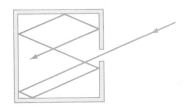

Fig. 42.3 A cavity with a small hole. Any radiation entering the hole is trapped; it will suffer multiple reflections and will ultimately be absorbed.

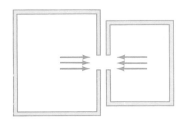

Fig. 42.4 Two cavities with holes of equal sizes.

42.2 Energy Quanta

One of the first attempts at a theoretical derivation of the spectral emittance of the blackbody was made by Lord Rayleigh. Since the energy flux of the radiation emerging from the hole in a cavity is directly proportional to the energy density of the radiation inside the cavity,[1] Rayleigh decided to calculate the latter quantity. He began by noting that the radiation in a cavity consists of a large number of standing waves; Figure 42.5 shows some of these standing waves. Each of these standing waves can be regarded as a mode of vibration of the cavity. Rayleigh then appealed to the equipartition theorem, according to which, at thermal equilibrium, each mode of vibration has an average thermal energy of kT (in Section 19.4 we stated a special case of the equipartition theorem for free translational or rotational motion). Thus, each of the standing waves of Figure 42.5 ought to have an energy kT, and from this he could calculate the spectral emissivity. Although this calculation gave reasonable results at the long-wavelength end of the blackbody spectrum, it gave disastrous results at the short-wavelength end: the number of possible standing-wave modes of very short wavelength is infinitely large, and if each of these modes had an energy kT, the total energy in the cavity would be infinite! This disastrous failure of classical theory has been called the **ultraviolet catastrophe.**

Ultraviolet catastrophe

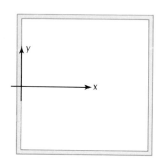

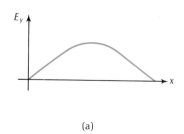
(a)

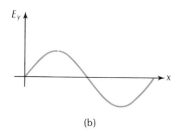
(b)

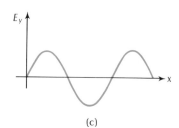
(c)

Fig. 42.5 Some of the possible standing electromagnetic waves in a closed cavity. For the sake of simplicity, only waves with a horizontal direction of propagation are shown. The plots give the electric field as a function of x at one instant of time.

The correct formula for the spectral emittance of a blackbody was finally obtained by Max Planck in 1900. By some inspired guesswork, based on thermodynamics, Planck hit upon the following formula:

Planck's Law

$$S_\lambda = \frac{2\pi c^2 h}{\lambda^5} \frac{1}{e^{hc/kT\lambda} - 1} \tag{1}$$

---

[1] According to Eqs. (36.30) and (36.32), for a plane wave the constant of proportionality between the energy flux (Poynting vector) and the energy density is c, the speed of light. For thermal radiation, with many plane waves traveling in random directions, the constant of proportionality is $c/4$.

where k is Boltzmann's constant and h is **Planck's constant**,[2] a new fundamental constant with the value

$$h = 6.63 \times 10^{-34} \, \text{J} \cdot \text{s} \quad (2)$$

Planck's constant, h

Experimental investigation quickly established that the values of the spectral emittance calculated from Eq. (1) agreed quite precisely with the measured values. Figure 42.6 shows the spectral emittance for several values of the temperature.

At first, Planck had proposed his formula merely as an empirical law that happens to give a good fit to the experimentally measured points. But then he searched for a theoretical explanation of this law. The search led Planck to a revolutionary discovery: the quantization of energy. This discovery was to bring about the overthrow of classical

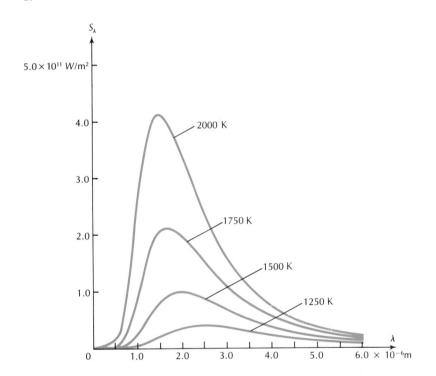

Fig. 42.6 Spectral emittance of a blackbody according to Planck's Law. Each curve is labeled with the appropriate temperature.

Max Planck, *1858–1947, German physicist, professor at Berlin and president of the Kaiser Wilhelm Society for the Advancement of Science (now the Max Planck Institute). Planck made significant contributions to thermodynamics before he became involved with the problem of blackbody radiation. He was a very conservative scientist, and he tried very hard to circumvent the quantization postulate, but without success. Planck received the Nobel Prize in 1918.*

Newtonian physics and the rise of quantum physics. For Planck, the postulate of the quantization of energy was "an act of desperation" which he committed because "a theoretical explanation had to be found at any cost, whatever the price."

Planck's derivation of the blackbody radiation law involves some sophisticated statistical mechanics, and we cannot reproduce the details of his derivation here. We will merely give a rough outline of this derivation. Planck began by making a theoretical model of the walls of the cavity; he regarded the atoms in the walls as small harmonic oscillators with electric charges. Although this is a rather crude model of an atom, it was adequate for his purposes since, by the thermodynamic

---

[2] See Appendix 8 for a more precise value of Planck's constant.

argument of the preceding section, the radiation in a cavity is completely independent of the characteristics of the wall. The random thermal motions of the oscillators result in the emission of electromagnetic radiation. This radiation fills the cavity and acts back on the oscillators. When thermal equilibrium is attained, the average rate of emission of radiation energy by the oscillators matches the rate of absorption of radiation energy. Thus, the oscillators share their energy with the radiation in the cavity, and Planck was able to show that, under equilibrium conditions, the average radiation energy at some frequency v (or at a wavelength $\lambda = c/v$) is directly proportional to the average energy of an oscillator of frequency v. These steps of Planck's calculation involved nothing but classical mechanics. But in the next step of the calculation, Planck departed radically from classical physics. He postulated that the energy of the oscillators is quantized according to the following rule:

In an oscillator of frequency v, the only permitted values of the energy are

$$E = 0, \; hv, \; 2hv, \; 3hv, \ldots \tag{3}$$

All other values of the energy are forbidden. The constant h in Eq. (3) is Planck's constant, the same as appears in Eqs. (1) and (2). The energy hv is called an **energy quantum**. According to the quantization rule, the energy of an oscillator is always some multiple of the basic energy quantum hv,

Energy quantum

Energy quantization of oscillator

$$\boxed{E = nhv, \qquad n = 0, 1, 2, 3, \ldots} \tag{4}$$

Quantum number

The integer n is called the **quantum number** of the oscillator.

With this quantization condition, Planck calculated the average energy of the oscillators; and from that he derived his radiation law. Although we cannot go into the details of this derivation, we can achieve a rough understanding of how Planck's calculation avoids the ultraviolet catastrophe. The thermal energy of the walls of the cavity is shared at random among all the oscillators in these walls. Some of these oscillators have high frequencies, some have low frequencies. For an oscillator of very high frequency, the energy quantum hv is very large. If this oscillator is initially quiescent ($n = 0$), it cannot begin to move unless it acquires one energy quantum; but since this energy quantum hv is very large, the random thermal disturbances will be insufficient to provide it — the oscillator will remain quiescent. Thus, the quantization of energy tends to inhibit the thermal excitation of the high-frequency oscillators. If the high-frequency oscillators remain quiescent, then they will not supply energy to the corresponding high-frequency standing waves in the cavity, and there will be no ultraviolet catastrophe.

Note that for an oscillator with a frequency of $v \simeq 10^{15}/\mathrm{s}$, which is typical for atomic vibrations, the energy quantum is $hv = 6.6 \times 10^{-34}\,\mathrm{J \cdot s} \times 10^{15}/\mathrm{s} = 6.6 \times 10^{-19}\,\mathrm{J}$. Since this is a very small amount of energy, quantization does not make itself felt at a macroscopic level. But quantization plays a pervasive role at the atomic level.

Unfortunately, Planck could not offer any basic justification for his postulate of quantization of energy. His postulate took care of the blackbody radiation law, but raised serious questions about classical physics. Obviously, quantization of energy makes no sense in classical physics — there is nothing in the laws of Newton that would prevent an oscillator from acquiring energy in any amount whatsoever. Planck could justify his postulate only by its consequences; but, in theoretical physics, the end does not justify the means. A deeper explanation of the quantization of energy emerged only much later, with the development of quantum mechanics (see Chapter 43).

From Planck's Law, it can be demonstrated that the spectral emittance of a blackbody has a maximum at a wavelength given by

$$\lambda_{max} = \frac{hc}{4.9651k}\frac{1}{T} = (2.90 \times 10^{-3} \text{ m} \cdot \text{K}) \times \frac{1}{T} \tag{5}$$ *Wien's Law*

This inverse proportionality of the wavelength of maximum emittance and the temperature is called **Wien's Law.**

By integrating Planck's Law we obtain the total flux, or power per unit area, emitted by the blackbody at all wavelengths:

$$S = \int_0^\infty S_\lambda\, d\lambda = \int_0^\infty \frac{2\pi c^2 h}{\lambda^5}\frac{1}{e^{hc/kT\lambda} - 1}\, d\lambda \tag{6}$$

It can be shown that this total flux is proportional to the fourth power of the temperature,

$$\boxed{S = \sigma T^4} \tag{7}$$ *Stefan–Boltzmann Law*

where $\sigma = 2\pi^5 k^4/(15 c^2 h^3) = 5.67 \times 10^{-8} \text{ W}/(\text{m}^2 \cdot \text{K}^4)$.[3] This is called the **Stefan–Boltzmann Law.** The laws of Wien and of Stefan and Boltzmann had both been discovered empirically many years before Planck deduced them from his law and supplied the theoretical expressions for the constants of proportionality.

EXAMPLE 1. On a clear night, the surface of the Earth loses heat by radiation. Suppose that the temperature of the ground is 10°C and that the ground radiates like a blackbody. What is the rate of loss of heat per square meter?

SOLUTION: The absolute temperature of the ground is 283 K. Hence, the Stefan–Boltzmann Law tells us that the radiated flux, or power per unit area, is

$$S = \sigma T^4 = 5.67 \times 10^{-8} \text{ W}/(\text{m}^2 \cdot \text{K}^4) \times (283 \text{ K})^4$$

$$= 364 \text{ W/m}^2$$

This amounts to 87 cal/m² per second.

COMMENTS AND SUGGESTIONS: The sharp temperature drop of the ground during a clear night is due to this heat loss by radiation. If the sky is covered by clouds, much of the radiation from the ground is reflected back by the clouds, and the heat loss is reduced.

---

[3] See Appendix 8 for a more precise value of the Stefan–Boltzmann constant.

EXAMPLE 2. Astronomers sometimes determine the size of a star by a method that relies on the Stefan–Boltzmann Law. Determine the radius of the star Capella from the following data: the flux of the starlight reaching the Earth is 1.2×10^{-8} W/m², the distance of the star is 4.3×10^{17} m, and its surface temperature is 5200 K. Assume the star radiates like a blackbody.

SOLUTION: According to the Stefan–Boltzmann Law, the flux, or power per unit area, emerging from the surface of the star is σT^4. The surface area of the star is $4\pi R^2$, and hence the total emitted power is

$$4\pi R^2 \times \sigma T^4 \tag{8}$$

This must match the total power crossing a sphere of radius 4.3×10^{17} m centered on the star. Since at each point of this sphere the power per unit area is 1.2×10^{-8} W/m², the total power is

$$4\pi (4.3 \times 10^{17} \text{ m})^2 \times 1.2 \times 10^{-8} \text{ W/m}^2 = 2.8 \times 10^{28} \text{ W} \tag{9}$$

Comparing Eqs. (8) and (9), we obtain

$$4\pi R^2 \sigma T^4 = 2.8 \times 10^{28} \text{ W} \tag{10}$$

or

$$R = \left(\frac{2.8 \times 10^{28} \text{ W}}{4\pi \times 5.67 \times 10^{-8} \text{ W/(m}^2 \cdot \text{K}^4) \times (5200 \text{ K})^4} \right)^{1/2}$$

$$= 7.3 \times 10^9 \text{ m}$$

This is about 10 times the radius of the Sun. Capella is a giant star.

42.3 Photons and the Photoelectric Effect

In 1905, Einstein showed that Planck's formula could be understood much more simply in terms of a direct quantization of the energy of the radiation. Planck had postulated that the oscillators in the wall of the cavity have discrete quantized energies, but he had treated the electromagnetic radiation as a smooth, continuous distribution of energy, exactly as it is supposed to be according to classical electromagnetic theory. In contrast, Einstein proposed that electromagnetic radiation consists of particle-like packets of energy. He regarded a wave of some given frequency v as a stream of more or less localized energy packets, each with one quantum of energy hv. The wave then has an energy hv if it contains only one such quantum, $2hv$ if it contains two, and so on. The particle-like energy packets in light are called **photons**. The thermal radiation in a cavity, with waves traveling randomly in all directions, can then be regarded as a gas of photons. Einstein applied statistical mechanics to calculate the energy spectrum of this gas, and he thereby obtained the energy spectrum of the cavity radiation.

Photon

EXAMPLE 3. The energy flux of sunlight reaching the surface of the Earth is 1.0×10^3 W/m². How many photons reach the surface of the Earth per square meter per second? For the purposes of this calculation assume that all the photons in sunlight have an average wavelength 5000 Å.

SOLUTION: The energy of a photon of wavelength 5000 Å is

$$E = h\nu = \frac{hc}{\lambda}$$

$$= \frac{6.63 \times 10^{-34} \text{ J} \cdot \text{s} \times 3.00 \times 10^8 \text{ m/s}}{5.0 \times 10^{-7} \text{ m}} = 4.0 \times 10^{-19} \text{ J}$$

The energy incident per square meter per second is 1.0×10^3 J. To obtain the number of photons, we must divide this by the energy per photon, 4.0×10^{-19} J. This gives 1.0×10^3 J$/4.0 \times 10^{-19}$ J $= 2.5 \times 10^{21}$ photons per square meter per second.

COMMENTS AND SUGGESTIONS: Because the number of photons in sunlight and in light from other ordinary sources is so large, we do not perceive the grainy character of the energy distribution in light. At the macroscopic level, light from ordinary sources *seems* to have a smooth, continuous distribution of energy.

With this concept of light as a stream of photons, Einstein was also able to offer an explanation of the **photoelectric effect.** In the early experiments on the production of radio waves by sparks, Hertz had noticed that light shining on an electrode tended to promote the formation of sparks. Subsequent careful experimental investigations demonstrated that the impact of light on an electrode can eject electrons. The electrons emerge with a kinetic energy which increases directly with the frequency of the light.

Figure 42.7 is a schematic diagram of the apparatus used in the investigation of the photoelectric effect. Light from a lamp illuminates an electrode of metal (*C*) enclosed in an evacuated tube. Electrons ejected from this electrode travel to the collecting electrode (*A*) and then flow around the external circuit. A galvanometer (*G*) detects this flow of electrons. The kinetic energy of the ejected photoelectrons can be determined by applying a potential difference between the emitting and the collecting electrodes by means of an adjustable source of emf (*V*). With the polarity shown in the figure, the collector has a negative potential relative to the emitter, that is, the collector exerts a repulsive force on the photoelectrons. If the potential energy matches or exceeds the initial kinetic energy of the photoelectrons, then the flow of these electrons will stop. The corresponding potential is called the **stopping potential**; the measured value of this stopping potential gives us the kinetic energy of the electrons:

$$K = eV_{\text{stop}} \quad (11)$$

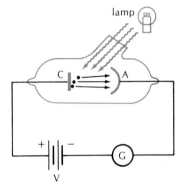

Fig. 42.7 Schematic diagram of the apparatus for the investigation of the photoelectric effect. The light from the lamp ejects electrons from the electrode *C* (cathode), and they travel to the collecting electrode *A* (anode).

Stopping potential

Experimentally, it is found that the kinetic energy determined in this way increases linearly with the frequency of the incident light. For example, Figure 42.8 is a plot of the kinetic energy vs. the frequency of the light for photoelectrons ejected from sodium. Note that if the fre-

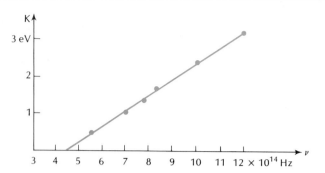

Fig. 42.8 Kinetic energies (in electron-volts) of photoelectrons ejected from sodium by light of different frequencies.

quency is below 4.4×10^{14} Hz, then the light is incapable of ejecting electrons. This frequency is called the **threshold frequency.**

Threshold frequency

Einstein's quantum theory of light accounts for these experimental observations as follows. The electrons in the illuminated electrode absorb photons from the light, one at a time. When an electron absorbs a photon, it acquires an energy $h\nu$. But before this electron can emerge from the electrode, it must overcome the restraining forces that bind it to the metal of the electrode. The energy required for this is called the **work function** of the metal, designated by ϕ. The energy left with the electron is then $h\nu - \phi$, and this must be the kinetic energy of the emerging electron,

Work function

Einstein's photoelectric equation

$$K = h\nu - \phi \qquad (12)$$

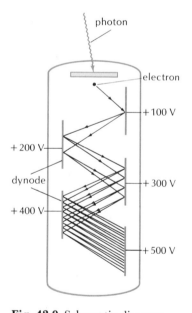

Fig. 42.9 Schematic diagram of a photomultiplier tube. The secondary electrodes are called dynodes. For the purpose of this diagram it has been assumed that each electron impact on a dynode releases two electrons. The arrows show an avalanche of electrons.

Some electrons suffer extra energy losses in collisions within the metal before they emerge; thus, this expression actually gives the *maximum* possible kinetic energy with which electrons can emerge.

Equation (12) is Einstein's photoelectric equation. It shows that the kinetic energy is indeed a linearly increasing function of the frequency, in agreement with the data of Figure 42.8. According to Eq. (12), the slope of the straight line in Figure 42.8 should equal Planck's constant.

Einstein's photoelectric equation was verified in detail by a long series of meticulous experiments by R. A. Millikan (the data in Figure 42.8 are due to him). In order to obtain reliable results, Millikan found it necessary to take extreme precautions to avoid contamination of the surface of the photosensitive electrode. Since the surfaces of metals exposed to air quickly accumulate a layer of oxide, he developed a technique for shaving the surfaces of his metals in a vacuum, by means of a magnetically operated knife.

The results of these experiments gave strong support to the quantum theory of light. This success of Einstein's theory was all the more striking in view of the failure of the classical wave theory of light to account for the features of the photoelectric effect. According to the wave theory, the crucial parameter that determines the ejection of a photoelectron should be the intensity of light. If an intense electromagnetic wave strikes an electron, it should be able to jolt it loose from the metal, regardless of the frequency of the wave. Furthermore, the kinetic energy of the ejected electron should be a function of the intensity of the wave. The observational evidence contradicts these predictions of the wave theory: a wave with a frequency below the threshold frequency never ejects an electron, regardless of its intensity; and, furthermore, the kinetic energy depends on the frequency, not on the intensity. High-intensity light ejects more photoelectrons, but does not give the individual electrons more kinetic energy.

Today, the photoelectric effect finds many practical applications in sensitive electronic devices for the detection of light. For instance, in a photomultiplier tube, an incident photon ejects an electron from an electrode; this electron is accelerated toward a second electrode (called a dynode; see Figure 42.9), where its impact ejects several secondary electrons; these, in turn, are accelerated toward a third electrode, where their impact ejects tertiary electrons, and so on. Thus, one electron from the first electrode generates an avalanche of electrons. In a high-gain photomultiplier tube, a pulse of 10^9 electrons emerges from the last electrode, delivering a measurable pulse of current to an external circuit. In this way, the photomultiplier tube can detect the arrival of individual photons. Some sensitive television cameras, such as the image orthicon, rely on the same multiplier principle to convert the arrival of a photon at a photosensitive faceplate into a measurable pulse of current.

EXAMPLE 4. The work function for platinum is 9.9×10^{-19} J. What is the threshold frequency for the ejection of photoelectrons from platinum?

SOLUTION: If an electron absorbs a photon at the threshold frequency, it will just barely have enough energy to overcome the binding forces holding it in the metal, and it will emerge with zero kinetic energy. By Eq. (12) this corresponds to

$$0 = h\nu - \phi$$

or

$$\nu = \frac{\phi}{h} = \frac{9.9 \times 10^{-19} \text{ J}}{6.63 \times 10^{-34} \text{ J} \cdot \text{s}} = 1.5 \times 10^{15} \text{ Hz}$$

42.4 The Compton Effect

Very clear experimental evidence for the particle-like behavior of photons was uncovered by A. H. Compton in 1922. Compton had been investigating the scattering of X rays by a target of graphite. When he bombarded the graphite with monochromatic X rays, he found that the scattered (deflected) X rays had a wavelength somewhat larger than that of the original X rays. Compton soon recognized that this effect could be understood in terms of collisions of photons with electrons, collisions in which the photons behave like particles.

In this collision the electron of a carbon atom can be regarded as free because the force binding the electron to the atom is insignificant compared with the force exerted by the incident photon. When the photon bounces off the electron, the electron recoils and thereby picks up some of the photon's energy — the deflected photon is left with reduced energy. Since the energy of the photon is $E = h\nu = hc/\lambda$, a reduction of energy implies an increase of wavelength. Qualitatively, we expect that those photons deflected through the largest angles should lose the most energy and therefore emerge with the longest wavelength. And this is just what Compton found in his experiments.

Arthur Holly Compton, *1892–1962, American experimental physicist, professor at the University of Chicago. For his discovery of the Compton effect, he received the Nobel Prize in 1927.*

For a quantitative discussion of the photon–electron collision, we need an expression for the momentum of a photon. From Eq. (36.42) we know that the momentum and the energy of light are related by $p = E/c$. A photon of energy $E = h\nu$ therefore has a momentum

Momentum of photon

$$\boxed{p = \frac{h\nu}{c}} \tag{13}$$

Alternatively, we can derive this equation from the relativistic formula for the momentum of a particle. Since the speed of the photon is the speed of light, it must be regarded as an ultrarelativistic particle. For such a particle, Eq. (41.53) tells us that $p = E/c$, which leads, again, to $p = h\nu/c$. [Note that according to Eq. (41.51), the energy of a particle with a speed equal to the speed of light can be finite only if the mass is zero; thus, photons must be regarded as particles of zero mass.]

Before the collision, the electron is at rest and the photon has a frequency ν and momentum $h\nu/c$. After the collision, the electron has a recoil momentum $m_e v$ and the photon has a reduced frequency $\nu - \Delta\nu$ and momentum $h(\nu - \Delta\nu)/c$. Figure 42.10a shows the initial momentum vector of the photon and the final momentum vectors of the electron and the photon. Conservation of momentum demands that the sum of the two final momentum vectors equal the initial momentum vector; the three momentum vectors therefore form a triangle (Figure 42.10b). Applying the law of cosines to this triangle, we find

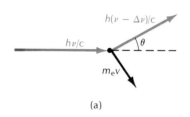

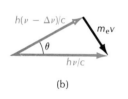

Fig. 42.10 (a) Momentum vector of photon before the collision and momentum vectors of photon and electron after the collision. Before the collision, the photon has a momentum $h\nu/c$ and the electron is at rest; after the collision, the photon has a momentum $h(\nu - \Delta\nu)/c$ and the electron has a momentum $m_e v$. (b) The triangle of the momentum vectors.

$$(m_e v)^2 = \left(\frac{h\nu}{c}\right)^2 + \left(\frac{h(\nu - \Delta\nu)}{c}\right)^2 - \frac{2h\nu}{c}\frac{h(\nu - \Delta\nu)}{c}\cos\theta \tag{14}$$

or

$$(m_e v)^2 = \left(\frac{h\nu}{c}\right)^2 \left[1 + \left(1 - \frac{\Delta\nu}{\nu}\right)^2 - 2\left(1 - \frac{\Delta\nu}{\nu}\right)\cos\theta\right] \tag{15}$$

$$= \left(\frac{h\nu}{c}\right)^2 \left[2 - 2\frac{\Delta\nu}{\nu} + \left(\frac{\Delta\nu}{\nu}\right)^2 - 2\left(1 - \frac{\Delta\nu}{\nu}\right)\cos\theta\right] \tag{16}$$

If we neglect the term $(\Delta\nu/\nu)^2$, we can rewrite this as

$$\tfrac{1}{2} m_e v^2 = \frac{1}{m_e}\left(\frac{h\nu}{c}\right)^2 \left(1 - \frac{\Delta\nu}{\nu}\right)(1 - \cos\theta) \tag{17}$$

Conservation of energy demands that the kinetic energy of the electron equals the energy lost by the photon,

$$\tfrac{1}{2} m_e v_e^2 = h\,\Delta\nu \tag{18}$$

Comparing Eqs. (17) and (18), we see that

$$\Delta\nu = \frac{h\nu^2}{m_e c^2}\left(1 - \frac{\Delta\nu}{\nu}\right)(1 - \cos\theta) \tag{19}$$

or

$$\frac{c\,\Delta\nu}{\nu(\nu-\Delta\nu)} = \frac{h}{m_e c}(1-\cos\theta) \qquad (20)$$

The complicated expression on the left side of this equation is simply the change of wavelength:

$$\frac{c\,\Delta\nu}{\nu(\nu-\Delta\nu)} = \frac{c}{\nu-\Delta\nu} - \frac{c}{\nu} = (\lambda + \Delta\lambda) - \lambda = \Delta\lambda \qquad (21)$$

where we have taken into account that a decreased frequency $\nu - \Delta\nu$ implies an increased wavelength $\lambda + \Delta\lambda$. Our equation for the change of wavelength of the photon as a function of the angle of deflection is then

$$\boxed{\Delta\lambda = \frac{h}{m_e c}(1-\cos\theta)} \qquad (22)$$

Wavelength shift for photon in Compton effect

This result, first obtained by Compton, remains valid even if the speed of recoil of the electron is relativistic [in fact, the relativistic calculation shows that Eq. (22) is *exact*].

The constant of proportionality in Eq. (22) is called the **Compton wavelength**; its value is

$$\frac{h}{m_e c} = 2.43 \times 10^{-12}\text{ m} = 0.0243\text{ Å}$$

However, in spite of its name, this is not the wavelength of anything.

Figure 42.11 is a plot of $\Delta\lambda$ as a function of θ. For $\theta = 180°$ (a head-on collision), the wavelength shift is maximum, $\Delta\lambda = 2h/(m_e c) = 0.0485$ Å; but even this maximum is quite small. Such a small wavelength shift is difficult to detect unless the wavelength of the X rays is itself quite small, so $\Delta\lambda$ is an appreciable fraction of λ. In his experiments, Compton used X rays of a wavelength $\lambda = 0.71$ Å; thus, the maximum wavelength shift amounted to about 7% of the wavelength.

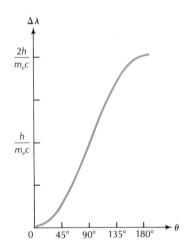

Fig. 42.11 Wavelength shift of the photon as a function of the deflection angle.

42.5 Wave vs. Particle

From the photoelectric effect and the Compton effect we learned that photons have particle properties. On the other hand, we know that light displays interference and diffraction phenomena, which prove that the photons also have wave properties. Thus photons are neither classical particles nor classical waves. They are some new kind of object, unknown to classical physics, with a subtle combination of both wave and particle properties. Arthur Eddington has coined the name **wavicle** for this new kind of object. It is difficult to achieve a clear understanding of the character of a wavicle, because these objects are very remote from our everyday experience. We all have an intuitive grasp of the concept of a classical particle or of a classical wave from

Wavicle

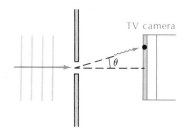

Fig. 42.12 A diffraction experiment. A light beam is incident from the left on a plate with a narrow slit. A sensitive TV camera detects photons on the right.

our experience with, say, billiard balls and water waves, but we have no such experience with wavicles.

We can gain some insight into the interplay between the particle behavior and the wave behavior of a photon by a new, detailed examination of a simple diffraction experiment. Figure 42.12 shows a light beam striking a plate with a narrow slit. In the usual diffraction experiments described in Chapter 40, we installed a screen or a photographic film at the far right, and with this we recorded the intensity of the light in the diffraction pattern. Instead, in our new arrangement we will install the faceplate of a very sensitive TV camera in place of the customary screen. With this, we can detect the individual photons of the light in the diffraction pattern.

If the incident light beam has a very low intensity, so that there is only one photon passing through the slit at a time, we can watch the photons arriving one by one at the faceplate of our TV camera. Figure 42.13a shows a typical pattern of impacts of 30 photons. The pattern seems quite random. If the photons behaved like classical particles, they would travel along a straight line and would reach only those points on the faceplate that are within the geometric image of the slit. The widely scattered impacts prove that the photons are certainly not traveling along such straight lines. Figure 42.13b shows the pattern of accumulated impacts for 300 photons; and Figure 42.13c shows it for 3000 photons. In these figures we can recognize a tendency of the photons to cluster in bandlike zones. These zones correspond to the maxima of the diffraction pattern predicted by the wave theory of light. Finally, Figure 42.13d shows the pattern of accumulated impacts for a very large number of photons; this is simply the familiar intensity pattern of light diffracted by a slit (see also Figure 40.5).

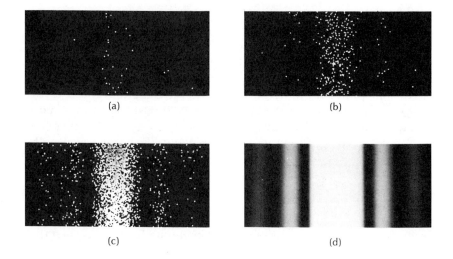

Fig. 42.13 (a) Pattern of impacts of 30 photons (simulation). (b) Pattern of impacts of 300 photons (simulation). (c) Pattern of impacts of 3000 photons (simulation). (d) Intensity pattern of light recorded on a photographic plate by a very large number of photons. The intensity of light at each point is proportional to the number of photons arriving at that point.

From this diffraction experiment we learn that the behavior of the photons is governed by a probabilistic law. The point of impact of an individual photon on the faceplate is unpredictable. Only the average distribution of impacts of a large number of photons is predictable: the distribution of photons matches the intensity distribution calculated from the wave theory of light. Thus, the probability that a photon arrives at a given point on the faceplate is proportional to the intensity of the wave at that point. Since the intensity of an electromagnetic wave is proportional to the square of its electric field, we can write the proportionality of probability and intensity as

$$\boxed{[\text{probability for photon at point } P] \propto E^2(P)} \qquad (23)$$

Probability interpretation of wave

Here we have a connection between the wave and the particle aspects of the photon: *the intensity of the photon wave at some point determines the probability that there is a photon particle at that point.* This probability interpretation of the intensity of the wave was discovered by Max Born.

Our single-slit experiment can also teach us something about the limitations that quantum theory imposes on the ultimate precision of measurement of the position of a wavicle. Suppose we have a light wave consisting of one photon and we want to measure the position of this photon. Of course, the position has x, y, and z components; we will concentrate on the y component, perpendicular to the direction of propagation. Figure 42.14 shows the wave propagating in the horizontal direction; the y direction is vertical. To determine this vertical position of the photon, we use a narrow slit placed in the path of the wave. If the photon succeeds in passing through this slit, then we will have achieved a determination of the vertical position to within an uncertainty

$$\Delta y = a \qquad (24)$$

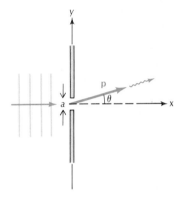

Fig. 42.14 Photon passes through a narrow slit and emerges at an angle θ.

where a is the width of the slit. If the photon fails to pass through this slit, then our measurement is inconclusive and will have to be repeated.[4]

By making the slit very narrow, we can make the uncertainty of our determination of the y coordinate very small. But this has a surprising consequence for the y component of the momentum of the photon: if we make the uncertainty in the y coordinate small, we will make the uncertainty in the y component of the momentum large. To see how this comes about, let us recall that, according to our preceding discussion of the single-slit experiment, the photon suffers diffraction by the slit and emerges at some angle θ (Figure 42.14). This angle θ is unpredictable; all we can say about the photon after it emerges from the slit is that it will be heading toward some point within the diffraction pattern. Thus, the direction of motion of the photon is uncertain. As a rough measure of the magnitude of this uncertainty in direction, we can take the angular width of the central diffraction maximum (most of the intensity of the photon wave is gathered within the region of this central maximum, and hence the photon is most likely to be found in this region). According to Eq. (40.5), this angular width, measured from the central maximum to the first minimum, is $\Delta\theta = \sin^{-1}(\lambda/a)$. If we assume that the angle is small, we can approximate this width by

$$\Delta\theta \cong \lambda/a \qquad (25)$$

The y component of the momentum is $p_y = p \sin \theta$; for a small angle, we can approximate this by $p_y = p\theta$. The uncertainty in p_y is then

Max Born, *1882–1970, German, and later British, theoretical physicist, professor at Göttingen and at Edinburgh. He was awarded the Nobel Prize somewhat tardily in 1954 for his discovery of the probabilistic interpretation of quantum waves in 1926.*

---

[4] An alternative method for measuring the position of a photon is to observe its impact on the faceplate of a sensitive TV camera. But this measurement is destructive (the photon is absorbed by the photoelectron and disappears). Besides, the uncertainties in this measurement are not easily analyzed.

$$\Delta p_y = p\,\Delta\theta \cong p\lambda/a \tag{26}$$

But, from Eq. (13),

$$p = \frac{h\nu}{c} = \frac{h}{\lambda} \tag{27}$$

which leads to

$$\Delta p_y \cong h/a \tag{28}$$

Thus, if the slit is very narrow, the uncertainty in the y component of the momentum will be very large! Comparing Eqs. (28) and (24), we find that

$$\Delta y\,\Delta p_y \cong h \tag{29}$$

Werner Heisenberg, *1901–1976, German theoretical physicist, professor at Leipzig, and later director of the Max Planck Institute for Physics in Munich. He was one of the founders of the new quantum mechanics, and received the Nobel Prize in 1932.*

This equation states that Δy and Δp_y cannot both be small; if one is small then the other must be large, so that their product equals (or exceeds) Planck's constant.

Although we obtained Eq. (29) by examining the special case of a position measurement by means of a slit, it turns out that other methods of measurement are not much better than a slit — any such measurements of position result in uncertainties of roughly the same magnitude as those we found for a slit. In general, any kind of position measurement is subject to the inequality

Heisenberg's uncertainty relation for y and p_y

$$\boxed{\Delta y\,\Delta p_y \geq h/4\pi} \tag{30}$$

This is one of **Heisenberg's uncertainty relations.** There are corresponding relations for the other components of position and momentum. The Heisenberg uncertainty relations tell us that there exist ultimate, insuperable limitations in the precision of our measurements. At the macroscopic level, the quantum uncertainties in our measurements can be neglected. But at the atomic level, these quantum uncertainties are often so large that it is completely meaningless to speak of the position or momentum of a wavicle.

EXAMPLE 5. A slit of width 0.01 mm is placed across the path of a plane light wave. What are the uncertainties in the transverse position and momentum of a photon that passes through the slit?

SOLUTION: The uncertainty in the transverse position is 0.01 mm, or 1.0×10^{-5} m. The uncertainty in the transverse momentum is, according to Eq. (28),

$$\Delta p_y \cong \frac{h}{a} = \frac{6.63 \times 10^{-34}\,\text{J}\cdot\text{s}}{1 \times 10^{-5}\,\text{m}} = 6.6 \times 10^{-29}\,\text{kg}\cdot\text{m/s}$$

COMMENTS AND SUGGESTIONS: These are the values of the uncertainties for a photon just beyond the slit. When this photon moves farther away from the slit, its uncertainty in transverse position becomes progressively larger, because the wave spreads out. Thus, $\Delta y\,\Delta p_y \cong h$ at the slit, but $\Delta y\,\Delta p_y > h$ in the region beyond the slit.

SUMMARY

Planck's Law:
$$S_\lambda = \frac{2\pi c^2 h}{\lambda^5} \frac{1}{e^{hc/kT\lambda} - 1}$$

Planck's constant: $h = 6.63 \times 10^{-34}\,\text{J}\cdot\text{s}$

Energy quantization of an oscillator: $E = nh\nu$

Wien's Law: $\lambda_{\max} \propto \dfrac{1}{T}$

Stefan–Boltzmann Law: $S \propto T^4$

Energy and momentum of a photon: $E = h\nu,\ p = h\nu/c = h/\lambda$

Kinetic energy of photoelectron: $K = h\nu - \phi$

Wavelength shift of photon (Compton effect):
$$\Delta \lambda = \frac{h}{m_e c}(1 - \cos\theta)$$

Probability interpretation of wave:

[probability for presence of photon] ∝ [intensity of wave]

Heisenberg's uncertainty relation for y and p_y:
$$\Delta y\,\Delta p_y \geq h/4\pi$$

QUESTIONS

1. Is the light emitted by a neon tube thermal radiation? The light emitted by an ordinary incandescent light bulb?

2. Does your body emit thermal radiation?

3. In Example 1 we calculated the radiative heat loss from the ground during a clear night. Qualitatively, how would a cover of snow on the ground affect the result?

4. The insulation used in the walls of homes consists of a thick blanket of fiber glass covered on one side by a thin aluminum foil. What is the purpose of these two layers?

5. Black velvet looks much blacker than black paint. Why?

6. For protection against the heat of sunlight, parts of the Lunar Lander (and some other spacecraft) were wrapped in shiny aluminum foil. Why is shiny foil useful for this purpose?

7. If you look into a kiln containing pottery heated to a temperature equal to that of the walls of the kiln, you can scarcely see the pottery. Explain.

8. According to Figure 42.6, at what wavelength is the spectral emittance maximum for a body at 2000 K? At 1750 K? At 1500 K? At 1250 K? Do these wavelengths satisfy Wien's Law?

9. The quantization of electric charge is consistent with classical physics, but the quantization of energy is not. Does this make sense?

10. Suppose that Planck's constant were much larger than it is, say, 10^{34} times larger. What strange behavior would you notice in a simple harmonic oscillator consisting of a mass hanging on a spring?

11. Consider a seconds pendulum, that is, a pendulum that has a period of 2 seconds. What is the magnitude of one energy quantum for such a pendulum? Would you expect that quantum effects are noticeable in such a pendulum?

12. Suppose that two stars have the same size, but the temperature of one is twice that of the other. By what factor will the thermal power radiated by the hotter star be larger than that radiated by the cooler star?

13. Why do we not notice the discrete quanta of light when we look at a light bulb?

14. Day-glo paints achieve their exceptionally bright orange or red color by converting short-wavelength photons into long-wavelength (red) photons. Why can we not make such a paint in a blue or violet color?

15. Figure 42.15 shows a plot of current vs. applied potential for the photoelectric current emitted by the surface of a metal illuminated with light of a given wavelength. Qualitatively, explain why the current is zero if $V < -V_{stop}$, explain why the current levels off for a large positive V, and explain why the curves differ for different intensities of the light.

16. When light of a given wavelength ejects photoelectrons from the surface of a metal, why is it that not all of these photoelectrons emerge with the same kinetic energy?

17. Can a particle of mass zero ever be at rest?

18. According to Eq. (22), a photon suffers a maximum change of wavelength in a collision with an electron if it emerges at an angle $\theta = 180°$, and a minimum change of wavelength (no change) if it emerges at an angle $\theta = 0°$. Is this reasonable?

19. Suppose that a photon and an electron have the same momentum. Which has the larger energy, taking into account both the rest-mass energy and the kinetic energy?

20. Can the Compton effect occur with visible light? Would it be observable?

21. Photons of short wavelength are more particle-like than photons of long wavelength. Why?

22. Give an example of an experiment in which photons behave like waves. Give an example of an experiment in which they behave like particles.

23. What happens to the y momentum that a photon gains or loses in the diffraction experiment described in Figure 42.14?

24. If photons were classical particles, what pattern of impact points would we find in the diffraction experiment that led to the results described in Figures 42.13a–d?

25. According to Eq. (23), we can predict only the *probability* that a photon will be found at some given point. Does this mean that quantum physics is not deterministic?

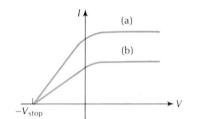

Fig. 42.15 I vs. V for two different values of the intensity of light: (a) high intensity and (b) low intensity.

PROBLEMS

Sections 42.1 and 42.2

1. Incandescent light bulbs have a tungsten filament whose temperature is typically 3200 K. At what wavelength does such a filament radiate a maximum flux? Assume that the filament acts as a blackbody.

2. Interplanetary and interstellar space is filled with thermal radiation of a temperature of 3 K left over from the Big Bang.

(a) At what wavelength is the flux of this radiation maximum?

(b) What is the power incident on the surface of the Earth due to this radiation?

3. The tungsten filament of a light bulb is a wire of diameter 0.080 mm and length 5.0 cm. The filament is at a temperature of 3200 K. Calculate the power radiated by the filament. Assume the filament acts as a blackbody.

4. The star Procyon B is at a distance of 11 light-years from Earth. The flux of its starlight reaching us is 1.7×10^{-12} W/m², and the surface temperature of the star is 6600 K. Calculate the size of the star.

*5. Prove that the spectral emissivity given by Eq. (1) does have a maximum at $\lambda_{max} = hc/(4.9651kT)$. (Hint: Substitute this wavelength into the mathematical condition for a maximum.)

*6. The spectral emittance of the Sun is maximum at 5000 Å. By what factor is the emittance smaller at 7000 Å? At 4000 Å?

*7. Show that the flux radiated by a blackbody in a frequency interval dv is

$$S_v \, dv = \frac{2\pi h}{c^2} \frac{v^3}{e^{hv/kT} - 1} \, dv$$

Does the maximum of S_v coincide with the maximum of S_λ?

*8. Derive the Stefan–Boltzmann Law

$$S \propto T^4$$

from Planck's Law. [Hint: Consider the integral given in Eq. (6); change the variable of integration to $x = hc/\lambda kT$ and show that the result has the form $S = [\text{constant}] \times T^4$; you do not have to evaluate the constant.]

*9. At the Earth, the flux of sunlight per unit area facing the Sun is 1.34×10^3 W/m². The Earth absorbs heat from the sunlight and reradiates heat as thermal infrared radiation. For equilibrium, the power arriving from the Sun must equal the average power radiated by the surface of the Earth. What average surface temperature for the Earth can you deduce from this? Assume that the Earth radiates like a blackbody.

*10. Deduce the average surface temperature of Pluto by the method described in Problem 9. You will find data on Pluto in the table printed on the endpapers.

*11. If you stand naked in a room, your skin and the walls of the room will exchange heat by radiation. Suppose the temperature of your skin is 33°C; the total area of your skin is 1.5 m². The temperature of the walls is 15°C. Assume your skin and the walls behave like blackbodies.

(a) What is the rate at which your skin radiates heat?

(b) What is the rate at which your skin absorbs heat? What is your net rate of loss of heat?

12. In molecules, the atoms can vibrate about their equilibrium positions. For instance, in the H_2 molecule, the hydrogen atoms vibrate about their equilibrium positions with a frequency of 1.13×10^{14} Hz. What is the magnitude of the energy quantum for this oscillating system? Note that although there are two masses in this system, they must be regarded as a single oscillator because the vibrational motions of the two masses are always equal.

13. If we take Planck's model of the walls of a blackbody cavity seriously, we will have to assume that the oscillating masses are electrons. Imagine an electron oscillating with a frequency of 2×10^{15} Hz under the influence of a springlike force. What is the amplitude of oscillation of this electron if its energy of oscillation is one energy quantum? Two energy quanta?

*14. Consider a mass of 1.0 g attached to a spring of spring constant 3.0×10^{-2} N/m. What is the frequency of oscillation of this system? What is the magnitude of an energy quantum? What is the amplitude of oscillation if the energy of the system equals one quantum?

*15. We know from Chapter 15 that at small amplitudes a pendulum behaves like an oscillator. Suppose that a pendulum consists of a mass of 0.10 kg attached to a (massless) string of length 1.0 m.
   (a) Taking into account the quantization of energy, what is the least (nonzero) amount of energy that this pendulum can have?
   (b) What is the amplitude of oscillation of the pendulum with this least amount of energy?

**16. In Problem 24.21, you will find a description of the Thomson model of the hydrogen atom.
   (a) What is the frequency of oscillation of the electron in this atom?
   (b) The electron can be regarded as an oscillator. Show that if the energy of the electron is one quantum, the amplitude of oscillation exceeds the radius (0.5 Å) of the atom.

Section 42.3

17. For each of the following kinds of electromagnetic waves, find the energy of a photon: FM radio wave of wavelength 3 m, infrared light of 1×10^{-5} m, visible light of 5×10^{-7} m, ultraviolet light of 1×10^{-7} m, X rays of 1×10^{-10} m.

18. You are lying on a beach, tanning in the sun. Roughly how many photons strike your skin in one hour? Assume that the energy flux of sunlight is as described in Example 3.

19. Photons of green light have a wavelength of 5500 Å. What is the energy and what is the momentum of one of these photons?

20. Show that if we express the energy of a photon in keV and the wavelength in angstroms, then

$$E = 12.4/\lambda$$

21. A radio transmitter radiates 10 kW at a frequency of 8.0×10^5 Hz. How many photons does the transmitter radiate per second?

22. The energy flux in the starlight reaching us from the bright star Capella is 1.2×10^{-8} W/m². If you are looking at this star, how many photons per second enter your eyes? The diameter of your pupil is 0.70 cm. Assume that the average wavelength of the light is 5000 Å.

23. An incandescent light bulb radiates 40 W of thermal radiation from a filament of temperature 3200 K. Estimate the number of photons radiated per second; assume that the photons have an average wavelength equal to the λ_{max} given by Wien's Law.

24. The energy density of starlight in intergalactic space is 10^{-15} J/m³. What is the corresponding density of photons? Assume the average wavelength of the photons is 5000 Å.

*25. If you want to make a very faint light beam that delivers only 1 photon per square meter per second, what must be the amplitude of the electric field in this light beam? The wavelength of the light is 5000 Å.

*26. Show that the flux of photons, or the number of photons per unit area and unit time, emitted by a blackbody in the frequency interval $d\nu$ is

$$\frac{2\pi}{c^2} \frac{\nu^2}{e^{h\nu/kT} - 1} d\nu$$

*27. A photon has an energy of 5 eV in the reference frame of the laboratory. What is the energy of this photon in the reference frame of a proton moving through the laboratory at a speed of $\frac{1}{2}c$ in the same direction as the photon?

28. According to Figure 42.8, what is the work function of sodium? Express your answer in electron-volts.

29. The work function of potassium is 2.26 eV. What is the threshold frequency for the photoelectric effect in potassium?

30. The work functions of K, Cr, Zn, and W are 2.26, 4.37, 4.24, and 4.49 eV, respectively. Which of these metals will emit photoelectrons when illuminated with red light ($\lambda = 7000$ Å)? Blue light ($\lambda = 4000$ Å)? Ultraviolet light ($\lambda = 2800$ Å)?

31. By inspection of Figure 42.8, find the slope of the line in eV/Hz. Convert these units into J·s, and verify that the slope is the same as Planck's constant.

*32. The binding energy of an electron in a hydrogen atom is 13.6 eV. Suppose that a photon of wavelength 400 Å strikes the atom and gives up all of its energy to the electron. With what kinetic energy will the electron be ejected from the atom?

*33. In an X-ray tube, fast-moving electrons strike a metal target, and the violent decelerations suffered by the electrons result in the emission of X rays.
  (a) Show that if the kinetic energy of the incident electrons is K, then the shortest possible wavelength, or the **cutoff wavelength,** of the X rays is

$$\lambda = hc/K$$

  (Hint: The maximum energy that an electron can give to an X-ray photon is all of its kinetic energy.)
  (b) What is the cutoff wavelength for electrons of 20,000 eV?

Section 42.4

34. X rays emitted by molybdenum have a wavelength of 0.72 Å. What are the energy and the momentum of one of the photons in these X rays?

35. For an experiment on the Compton effect, you want the X rays emerging at 90° from the incident direction to suffer an increase of wavelength by a factor of two. What wavelength do you need for your incident X rays?

36. In a collision with an initially stationary electron, a photon suffers a wavelength increase of 0.022 Å. What must have been the deflection angle of this photon?

37. X rays of wavelength 0.30 Å are incident on a graphite target. Calculate the wavelength of the X rays that emerge from this collision with a deflection of 60°. Calculate the wavelength of the X rays that emerge from this collision with a deflection of 120°.

*38. What is the maximum energy that a free electron (initially stationary) can acquire in a collision with a photon of energy 4.0×10^3 eV?

*39. In a collision with a free electron, a photon of energy 2.0×10^3 eV is deflected by 90°. What energy does the electron acquire in this collision?

*40. A photon of energy 1.6×10^8 eV collides with a *proton* initially at rest. The photon is deflected by 45°. What is its new energy?

*41. A photon of initial wavelength 0.400 Å suffers two successive collisions with two electrons. The deflection in the first collision is 90° and in the second collision it is 60°. What is the final wavelength of the photon?

**42. A photon of energy 5.0×10^3 eV collides head on with a free electron of energy 2.0×10^3 eV. After the collision the photon moves in a direction opposite to its initial direction. Find the energy of the photon. Find the energy of the electron.

Section 42.5

43. A photon passes through a horizontal slit of width 5×10^{-6} m. What uncertainty in the vertical position will this photon have as it emerges from the slit? What uncertainty in vertical momentum?

*44. Consider a radio wave in the form of a pulse lasting 0.001 s. This pulse then has a length of $0.001 \text{ s} \times c = 3 \times 10^5$ m. Since an individual photon of this radio wave can be anywhere within this pulse, the uncertainty in the position of the photon is $\Delta x = 3 \times 10^5$ m along the direction of propagation.
   (a) According to Heisenberg's relation, what is the corresponding uncertainty in the momentum of the photon?
   (b) What is the uncertainty in the frequency of the photon?

CHAPTER 43

Atomic Structure and Spectral Lines

The photographs reproduced in Figure 43.1 (see also the Prelude) give convincing visual evidence that solids, liquids, and gases are made of atoms, small grains of matter with a diameter of about 10^{-10} m. These photographs were prepared with powerful microscopes of a special design. Unfortunately, none of these microscopes is sufficiently powerful to reveal the inside of the atom. Hence, for the exploration of the internal structure of the atom, we still have to rely on the technique developed by Ernest Rutherford and his associates around 1910: bombard the atom with a beam of particles, and use this beam as a probe to "feel" the interior of the atom.

By 1910, most physicists had come to believe that atoms are made of some combination of positive and negative electric charges, and that the attractions and repulsions between these electric charges are the basis for all the chemical and physical phenomena observed in solids,

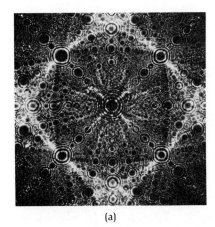

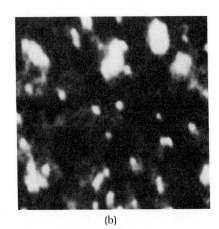

Fig. 43.1 (a) Platinum atoms in the tip of a fine needle as seen with an ion microscope (magnification $\sim 5 \times 10^6 \times$). (b) Uranium atoms (small white dots) seen with an electron microscope (magnification $\sim 5 \times 10^5 \times$).

(a)   (b)

liquids, and gases. Since electrons were known to be present in all of these forms of matter, it seemed reasonable to suppose that each atom consists of a combination of electrons and positive charge. The vibrational motions of the electrons within such an atom would then result in the radiation of electromagnetic waves; this was supposed to account for the emission of light by the atom. However, both the arrangement of the electric charges within the atom and the mechanism that accounts for the characteristic colors of the emitted light remained mysteries until Rutherford's discovery of the nucleus and Niels Bohr's discovery of the quantization of atomic states. In this chapter we will look at these two crucial discoveries.

The exploration of the internal structure of the atom led to the inescapable conclusion that in the atomic realm Newton's laws of motion are not valid. Electrons and other subatomic particles obey new equations of motion which are drastically different from the old equations of motion obeyed by planets, billiard balls, or shotgun pellets. The new theory of motion is called **quantum mechanics;** it rules the realm of the atom. In contrast, the old theory of Newton is called **classical mechanics.** The discovery of an entirely new set of laws of motion was the greatest scientific revolution of this century.

Quantum mechanics

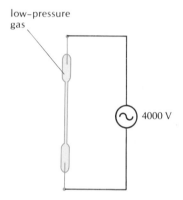

Fig. 43.2 An electric discharge tube. The tube contains gas at a very low pressure. When the terminals are connected to a high-voltage generator, an electric current flows through the gas and makes it glow.

43.1 Spectral Lines

The earliest attempts at a theory of atomic structure ended in failure — they were not able to explain the characteristic colors of the light emitted by atoms. These colors show up very distinctly when a small sample of gas is made to emit light by the application of heat or of an electric current. For instance, if we put a few grains of ordinary salt into a flame, the sodium vapor released by the salt will glow with a characteristic yellow color. If we put neon gas into an evacuated glass tube and connect the ends of the tube to a high-voltage generator (Figure 43.2), the gas will glow with the familiar orange-red color of neon signs.

The light emitted by an atom can be precisely analyzed with a prism; this breaks the light up into its component colors. In the arrangement shown in Figure 43.3, each separate color generates a bright line. These are the **spectral lines.** Each kind of atom has its own discrete spectral lines. The color plate between pages 898 and 899 shows the spectral lines of hydrogen; the numbers next to the spectral

Fig. 43.3 Analysis of light by means of a prism. In this arrangement, each separate color of the light emerging from the slit gives rise to a separate spectral line on the screen.

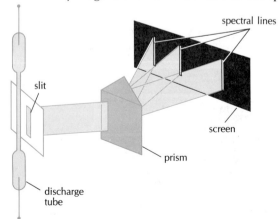

lines give the wavelengths in angstroms. Hydrogen has four spectral lines in the visible region (already mentioned in Chapter 37) and many ultraviolet and infrared lines not visible to the human eye. The color plate also shows the spectral lines of sodium, helium, and mercury. Obviously, the set of spectral lines, or **spectrum,** belonging to hydrogen is unmistakably different from those of sodium, helium, and mercury — the spectrum of an atom can serve as a fingerprint for its identification.

In spectroscopy laboratories, scientists often perform the quantitative analysis of a sample of atoms with absorption lines rather than emission lines. It so happens that an atom capable of emitting light of a given wavelength is also capable of absorbing light of that wavelength. When we illuminate a sample of atoms with white light (a uniform mixture containing all colors or wavelengths), the atoms will absorb light of their characteristic wavelengths, and, upon analyzing the remaining light with a prism, we find dark absorption lines in the continuous background generated by the white light. The last picture in the color plate shows such an absorption spectrum for sodium vapor. The dark lines of this absorption spectrum coincide with the bright lines in the emission spectrum (see the second picture in the color plate).[1]

One advantage of spectroscopy over chemistry is that the analysis can be performed even with minuscule amounts of material. What is more, atoms can be identified at a distance. For example, we can identify the atoms on the surface of the Sun by careful analysis of the distribution of colors in sunlight — we do not need to pluck a sample of atoms from the Sun. The power of this technique is best illustrated by the story of the discovery of helium (the "Sun element"). In 1868, this gas was yet unknown to chemists when an astronomer discovered it on the Sun by means of its light; 30 years later, chemists finally found traces of helium in minerals on the Earth. By spectroscopic techniques, astronomers can identify atoms in remote stars, clouds of interstellar gas, galaxies, and quasars. For instance, Figure 43.4 shows the spectrum of the star Caph in the constellation Cassiopeia; the spectral lines indicate the presence of hydrogen, calcium, iron, manganese, chromium, and so on.

Note that only a small part of the light from a star is in the form of discrete spectral lines from individual atoms on the stellar surface. Most of the starlight is continuously and more or less uniformly distributed over a broad range of wavelengths. This kind of light is thermal radiation produced in the stellar interior. As we know from the preceding chapter, this kind of light does not retain the fingerprint of the atoms that produced it. Light originating in the stellar interior cannot escape directly, but is first tossed back and forth (scattered) many times by the restless atoms of the hot stellar gas. The random motion of these atoms communicates random changes of wavelength to the light, and what finally emerges from the stellar interior is a continuous mixture with a wide range of wavelengths.

Figure 43.5 displays a portion of the spectrum of the light from our Sun. The bright background of white light in this spectrum has a continuous and uniform distribution over the entire range of wavelengths.

Spectrum

Joseph von Fraunhofer, *1787–1826, German optician and physicist. Starting as an apprentice to a glazier, he became a famous maker of optical instruments and a member of the Bavarian Academy of Sciences. He rediscovered the dark lines in the solar spectrum, observed previously by W. H. Wollaston, and introduced many improvements in the design of the spectroscope. He was one of the first makers of optical gratings.*

---

[1] The latter spectrum sometimes has extra lines that are absent in the former spectrum. This is so because the absorption process tends to suppress some lines; they become so faint as to be unnoticeable.

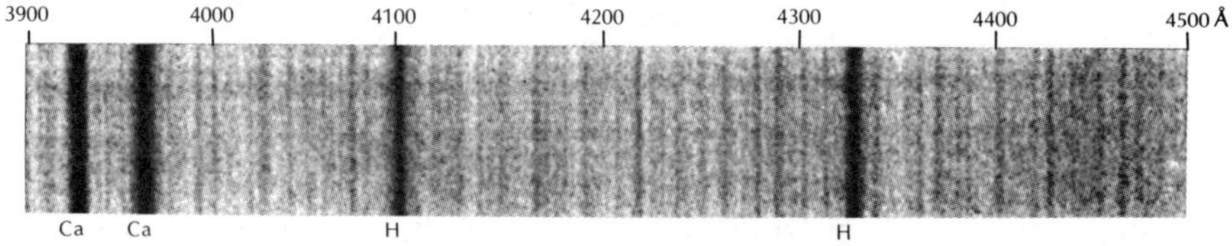

Fig. 43.4 A portion of the spectrum of the star Caph (β Cassiopeiae), from 3900 Å to 4500 Å. The two strong absorption lines on the left are due to ionized calcium. The other two strong lines (middle and right) are due to hydrogen.

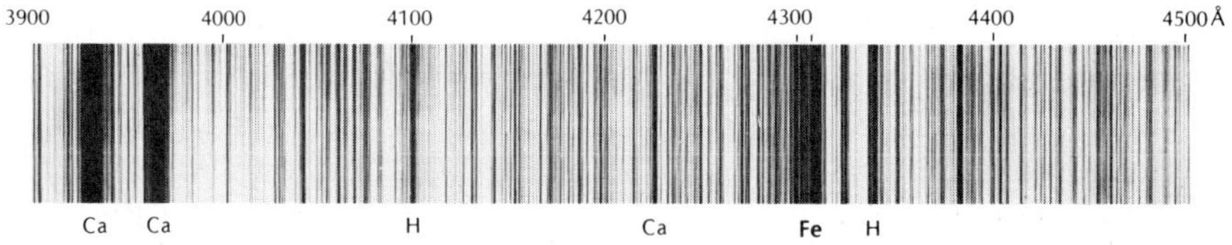

Fig. 43.5 A portion of the spectrum of the Sun, from 3900 Å to 4500 Å. Besides the strong absorption lines of calcium and of hydrogen, similar to those in Figure 43.4, there are also strong lines of iron (close pair, right of center).

Fraunhofer lines

However, there are many dark lines in this spectrum, caused by absorption in the gas surrounding the Sun. These are called the **Fraunhofer lines.**

43.2 The Balmer Series and Other Spectral Series

Careful examination of the sets of spectral lines produced by an element reveals certain systematic regularities in the spacings of the lines. These regularities sometimes become especially striking if, instead of examining only the visible region of the spectrum, we examine both the visible and the adjacent ultraviolet and infrared regions. For instance, Figure 43.6 shows the spectrum of hydrogen in the visible and near-ultraviolet regions, as recorded on a photographic film. We notice immediately that the spacings of the lines and their intensities decrease systematically as we look at shorter and shorter wavelengths. The hydrogen lines in Figure 43.6 are said to form a **spectral series.** In the spectra of other elements, we find similar series; however, the spectra usually contain several overlapping series, and this makes it a bit harder to perceive the regularities in the spacings.

Spectral series

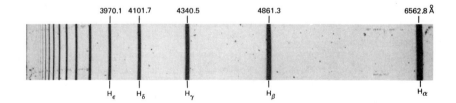

Fig. 43.6 Spectrum of hydrogen in the visible and the near ultraviolet regions. (Courtesy G. Herzberg.)

The systematic pattern in the spacing of the spectral lines of hydrogen suggests that the wavelengths of these lines should be described by some simple mathematical formula. Table 43.1 lists the wavelengths of the first few of these spectral lines; there actually is an infinite number of spectral lines, the spacing between them becoming smaller and smaller at shorter wavelengths. In 1885, Johann Balmer scrutinized the numbers in such a table and discovered that the wavelengths accurately fit the formula

$$\lambda = 911.76 \text{ Å} \times \frac{4n^2}{n^2 - 4} \tag{1}$$

with $n = 3, 4, 5, 6$, etc. This infinite series of spectral lines is called the **Balmer series.** Note that for $n \to \infty$, the wavelength approaches the asymptotic value $\lambda = 3647.0$ Å; this is called the **series limit.**

Balmer's formula was purely descriptive, or phenomenological; it did not explain the atomic mechanism responsible for the production of the spectral lines. Nevertheless, it proved very fruitful because it led to more general formulas describing other series of spectral lines. It is best to rewrite the formula in terms of the frequency,

$$\nu = \frac{c}{\lambda} = \frac{c}{911.76 \text{ Å}} \left(\frac{1}{4} - \frac{1}{n^2} \right) \qquad (2)$$

or

$$\boxed{\nu = cR_\text{H} \left(\frac{1}{2^2} - \frac{1}{n^2} \right)} \qquad (3)$$

where R_H is the **Rydberg constant,**

$$R_\text{H} = \frac{1}{911.76 \text{ Å}} = 109{,}678 \text{ cm}^{-1}$$

Balmer proposed that there might be other series in the hydrogen spectrum, with the 2 in Eq. (3) replaced by 1, or 3, or 4, or 5, etc. This yields the frequencies

$$\nu = cR_\text{H} \left(\frac{1}{1^2} - \frac{1}{n^2} \right) \qquad n = 2, 3, 4, \ldots \qquad (4)$$

$$\nu = cR_\text{H} \left(\frac{1}{3^2} - \frac{1}{n^2} \right) \qquad n = 4, 5, 6, \ldots \qquad (5)$$

$$\nu = cR_\text{H} \left(\frac{1}{4^2} - \frac{1}{n^2} \right) \qquad n = 5, 6, 7, \ldots \qquad (6)$$

$$\nu = cR_\text{H} \left(\frac{1}{5^2} - \frac{1}{n^2} \right) \qquad n = 6, 7, 8, \ldots \qquad (7)$$

These four series of spectral lines were actually discovered many years after Balmer proposed them; they are called, respectively, the **Lyman,** the **Paschen,** the **Brackett,** and the **Pfund series** (Figure 43.7).

Table 43.1 THE BALMER SERIES IN THE HYDROGEN SPECTRUM

Wavelength λ[a]
6564.7 Å
4862.7
4341.7
4102.9
3971.2
3890.2
3836.5
3799.0
etc.

[a] Wavelengths are measured in vacuum.

Balmer series of hydrogen

Rydberg constant

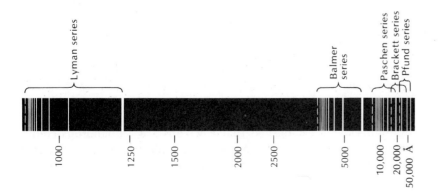

Fig. 43.7 The series of lines in the spectrum of hydrogen.

We can combine all these formulas for the spectral series into the single general formula

Spectral series of hydrogen

$$\nu = cR_H\left(\frac{1}{n_2^2} - \frac{1}{n_1^2}\right) \tag{8}$$

where n_1 and n_2 are positive integers and $n_1 > n_2$.

EXAMPLE 1. According to Eq. (8), what is the shortest wavelength that a hydrogen atom will emit or absorb?

SOLUTION: To find the shortest wavelength, we must choose n_1 and n_2 in Eq. (8), so as to obtain the highest frequency. Obviously, this demands $n_1 = \infty$ and $n_2 = 1$, which gives

$$\nu = cR_H\left(\frac{1}{1} - \frac{1}{\infty}\right) = cR_H \tag{9}$$

or

$$\lambda = c/\nu = 1/R_H = 911.76 \text{ Å}$$

Note that according to Eq. (8) the frequencies of hydrogen are written as differences between two terms, cR_H/n_2^2 and cR_H/n_1^2. Therein lies a crucial clue to the atomic mechanism responsible for the production of the spectral lines — as we will see in Section 43.4, these term differences correspond to energy differences. Examination of the frequencies of the spectral lines of other atoms shows that in all cases we can write the frequencies as differences between two terms, although the terms for these other atoms do not have as simple a mathematical form as those for the hydrogen atom. Furthermore, whenever we take any difference between two terms, there actually exists a spectral line that corresponds to this difference; this rule is called the **Rydberg–Ritz combination principle.**

43.3 The Nuclear Atom

The regularity in the series of spectral lines of the atom must be due to an underlying regularity in the structure of the atom. We may think of an atom as analogous to a musical instrument, such as a flute. The atom can emit only a discrete set of spectral lines, just as the flute can emit only a discrete set of tones which make up a musical scale. The regularity in the spacing of tones in this musical scale is due to an underlying regularity in the structure of the flute — the tube of the instrument has regularly spaced blowholes that determine what kind of standing waves can build up within the tube and what kind of waves will be radiated.

J. J. Thomson, the discoverer of the electron, made one of the first attempts at explaining the emission of light in terms of the structure of the atom. Having established that electrons are a ubiquitous component of matter, Thomson proposed the following picture: An atom consists of a number of electrons, say, Z electrons, embedded in a cloud of positive charge. The cloud is heavy, carrying almost all of the mass of the atom. The positive charge in the cloud is $+Ze$, so it exactly neutralizes the negative charge $-Ze$ of the electrons. In an undisturbed atom the electrons will sit at their equilibrium positions, where the attraction of the cloud on the electrons balances their mutual repulsion (Figure 43.8). But if the electrons are disturbed by, say, a collision, then they will vibrate around their equilibrium positions and emit light. This model of the atom, called the "plum-pudding model," does yield frequencies of vibration of the same order of magnitude as the frequency of light, but it does not yield the observed spectral series; for instance, on the basis of this model, hydrogen should have only one single spectral line, in the far ultraviolet. And in 1910, experiments by Rutherford and his collaborators established conclusively that most of the mass of the atom is not spread out over a cloud — instead, the mass is concentrated in a small kernel, or **nucleus,** at the center of the atom.

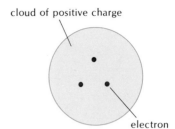

Fig. 43.8 The lithium atom according to the "plum-pudding" model. The three electrons sit at their equilibrium positions.

Rutherford had been studying the emission of alpha particles from radioactive substances. These alpha particles carry a positive charge $2e$, and they have a mass of 6.64×10^{-27} kg, about four times the mass of a proton (alpha particles have the same structure as nuclei of helium atoms; see Section IV.1). Some radioactive substances, such as radioactive polonium and radioactive bismuth, spontaneously emit alpha particles with energies of several million electron-volts. These energetic alpha particles readily pass through thin foils of metal, or thin sheets of glass, or other materials. Rutherford was much impressed by the penetrating power of these alpha particles, and it occurred to him that a beam of these particles can serve as a probe to "feel" the interior of the atom. When a beam of alpha particles strikes a foil of metal, the alpha particles penetrate the atoms and are deflected by collisions with the subatomic structures; the magnitude of these deflections gives a clue about the subatomic structures. For example, if the interior of the atom had the "plum-pudding" structure proposed by J. J. Thomson, then the alpha particles would suffer only very small deflections, since neither the electrons, with their small masses, nor the diffuse cloud of positive charge would be able to disturb the motion of a massive and energetic alpha particle.

The crucial experiments were performed by H. Geiger and E. Marsden working under Rutherford's direction. They used thin foils of gold and of silver as targets and bombarded these with a beam of alpha particles from a radioactive source. After the alpha particles passed through the foil, they were detected on a zinc sulfide screen which registers the impact of each particle by a faint scintillation (Figure 43.9).

Sir Ernest Rutherford, *1871– 1937, British experimental physicist, professor at McGill and at Manchester, and director of the Cavendish Laboratory at Cambridge, where he succeeded J. J. Thomson. Rutherford identified alpha and beta rays. He founded nuclear physics with his discoveries of the nucleus and of transmutation of elements by radioactive decay; he also produced the first artificial nuclear reaction. He was awarded the Nobel Prize in chemistry in 1908.*

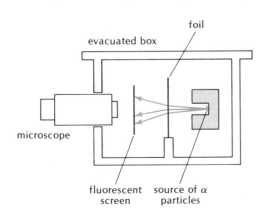

Fig. 43.9 Rutherford's apparatus.

To Rutherford's amazement, some of the alpha particles were deflected by such a large angle that they came out backward. In Rutherford's own words: "It was quite the most incredible event that has ever happened to me in my life. It was almost as incredible as if you fired a 15-inch shell at a piece of tissue paper and it came back and hit you." Rutherford immediately recognized that the large deflection must be produced by a close encounter between the alpha particle and a very small but very massive kernel inside the atom. He therefore proposed the following picture: An atom consists of a small nucleus of charge $+Ze$ containing almost all of the mass of the atom; this nucleus is surrounded by a swarm of Z electrons. Thus, the atom is like a solar system — the nucleus plays the role of sun, and the electrons play the role of planets.

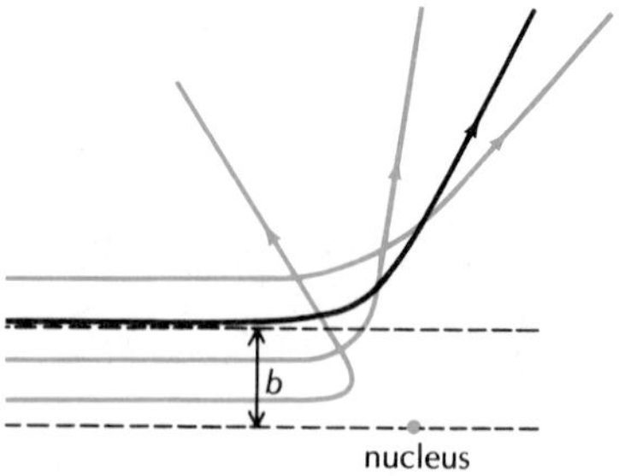

On the basis of this nuclear model of the atom, Rutherford calculated what fraction of the beam of alpha particles should be deflected through what angle. If an alpha particle passes close to the nucleus, it will experience a large electric repulsion and will be deflected by a large angle; if it passes far from the nucleus, it will be deflected by only a small angle. Figure 43.10 shows the trajectories of several alpha particles approaching a nucleus; these trajectories are hyperbolas. The perpendicular distance between the nucleus and the original (undeflected) line of motion is called the **impact parameter**. In order to suffer a large deflection, the alpha particle must hit an atom with a very small impact parameter, 10^{-13} m or less; since the alpha particles in the beam strike the foil of metal at random, only very few of them will score such a close hit.

Fig. 43.10 Hyperbolic trajectories of several alpha particles with different impact parameters passing by a nucleus.

EXAMPLE 2. An alpha particle of energy E approaches a nucleus of charge Ze. The impact parameter of the alpha particle is b. What is the distance of the closest approach between the alpha particle and the nucleus?

SOLUTION: Figure 43.11 shows the hyperbolic orbit of the alpha particle and the point of closest approach. When the alpha particle is far away, its velocity is v and the perpendicular distance between the origin and the line of motion is b; hence the initial angular momentum is mvb. When the alpha particle is at the point of closest approach, its velocity is v' and its position vector is $\mathbf{r}'$; since these vectors are perpendicular, the angular momentum is $mv'r'$. Conservation of angular momentum tells us

$$mv'r' = mvb \qquad (10)$$

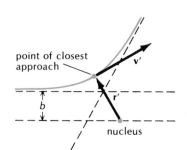

Fig. 43.11 At the point of closest approach, the alpha particle has a velocity $\mathbf{v}'$ perpendicular to its position vector $\mathbf{r}'$.

When the alpha particle is far away its energy is purely kinetic, $E = \tfrac{1}{2}mv^2$. When the alpha particle is at the point of closest approach its energy is a sum of kinetic and potential energies,

$$E = \tfrac{1}{2}mv'^2 + \frac{(2e)(Ze)}{4\pi\varepsilon_0}\frac{1}{r'}$$

Conservation of energy then tells us

$$\tfrac{1}{2}mv'^2 + \frac{2Ze^2}{4\pi\varepsilon_0}\frac{1}{r'} = \tfrac{1}{2}mv^2 \qquad (11)$$

Using Eq. (10), we can eliminate v' from Eq. (11),

$$\tfrac{1}{2}mv^2\frac{b^2}{r'^2} + \frac{2Ze^2}{4\pi\varepsilon_0}\frac{1}{r'} = \tfrac{1}{2}mv^2$$

We can regard this as a quadratic equation for the unknown $1/r'$. The standard formula for the solution of such a quadratic equation then gives

$$\frac{1}{r'} = \frac{-2Ze^2/4\pi\varepsilon_0 \pm \sqrt{(2Ze^2/4\pi\varepsilon_0)^2 + m^2v^4b^2}}{mv^2b^2}$$

Only the positive square-root sign makes sense (the negative sign gives a negative value of $1/r'$). In terms of the energy $E = \frac{1}{2}mv^2$ of the alpha particle, we can express the distance of the closest approach as

$$r' = \frac{Eb^2}{-(Ze^2/4\pi\varepsilon_0) + \sqrt{(Ze^2/4\pi\varepsilon_0)^2 + E^2b^2}} \quad (12)$$

For example, if an alpha particle of energy $E = 1.2 \times 10^{-12}$ J approaches a gold nucleus ($Z = 79$) with an impact parameter of 5.0×10^{-14} m, then

$$\frac{Ze^2}{4\pi\varepsilon_0} = \frac{79 \times (1.6 \times 10^{-19} \text{ C})^2}{4\pi\varepsilon_0} = 1.8 \times 10^{-26} \text{ J} \cdot \text{m}$$

and the distance of closest approach will be

$$r' = \frac{1.2 \times 10^{-12} \text{ J} \times (5.0 \times 10^{-14} \text{ m})^2}{-1.8 \times 10^{-26} \text{ J} \cdot \text{m} + \sqrt{(1.8 \times 10^{-26} \text{ J} \cdot \text{m})^2 + (1.2 \times 10^{-12} \text{ J} \times 5.0 \times 10^{-14} \text{ m})^2}}$$

$$= 6.7 \times 10^{-14} \text{ m}$$

COMMENTS AND SUGGESTIONS: The expression (12) is not applicable in a head-on collision, where $b = 0$. In this exceptional case, $v' = 0$, and it is then easy to check from Eq. (11) that the distance of closest approach is $r' = 2Ze^2/(4\pi\varepsilon_0 E)$.

43.4 Bohr's Theory

Rutherford's experiments did reveal the gross arrangement of the electrons in the atom, but not the details of their motion. Since the electrons make up the outer layers of an atom, their arrangement and motion should determine the chemical properties of the atom and the emission of light. But when physicists tried to calculate the electron motion according to the laws of classical mechanics and electromagnetism, they immediately ran into trouble.

To gain some insight into the source of this trouble, let us examine the case of the hydrogen atom. Suppose that the single electron of this atom is moving, according to the laws of classical mechanics, in a circular orbit of radius $\sim 10^{-10}$ m. The electron would then have a centripetal acceleration which is very large, about 10^{23} m/s^2. Because of this acceleration, the electron would emit high-frequency electromagnetic radiation, that is, it would emit light. The energy carried away by the light must be supplied by the electron. Hence, the emission process has the same effect on the electron as a friction force — it removes energy from the electron. This kind of friction would cause the electron to leave its circular orbit and gradually spiral down toward the nucleus, just as the residual atmospheric friction on an artificial satellite in a low-altitude orbit around the Earth causes it to spiral down toward the ground. A calculation using the laws of classical mechanics and electricity shows that the rate of emission of light by the orbiting electron

Niels Bohr, *1885–1962, Danish theoretical physicist. He worked under J. J. Thomson and Rutherford in England and then became professor at Copenhagen and director of the Institute of Theoretical Physics, for the foundation of which he was largely responsible. After formulating the quantum theory of the atom, he played a leading role in the further development of the new quantum mechanics. He received the Nobel Prize in 1922.*

Bohr's postulates

Quantization of angular momentum

in a hydrogen atom would be quite large. Correspondingly, the rate of energy loss of the electron would be large — the electron would spiral inward and collide with the nucleus within a time as short as 10^{-10} s!

Thus, our classical calculation leads us to the incongruous conclusion that hydrogen atoms, and other atoms, ought to be unstable — all the electrons ought to collapse into the nucleus almost instantaneously. Furthermore, the light that the electron emits during the spiraling motion ought to be a wave of continually increasing amplitude and increasing frequency (in musical terminology, crescendo and glissando); this is so because the closer the electron approaches the nucleus, the larger its acceleration and the higher its frequency of orbital motion. Hydrogen atoms do not behave as this calculation predicts. Hydrogen atoms are stable, and when they do emit light, they emit discrete frequencies (spectral lines) instead of a continuum of frequencies.

These irreconcilable differences between the observed properties of atoms and the calculated properties gave evidence of a serious breakdown of the classical mechanics of Newton and the classical theory of electromagnetism. Although these theories had proved very successful on a macroscopic scale, they were in need of some drastic modification on an atomic scale.

In 1913, Niels Bohr took a drastic step toward resolving these difficulties. He made the radical proposal that, at the atomic level, the laws of classical mechanics and of classical electromagnetism must be replaced or supplemented by other laws. Bohr expressed these new laws of atomic mechanics in the form of several postulates:

1. The orbits and the energies of the electrons in an atom are quantized, that is, only certain discrete orbits and energies are permitted. When an electron is in one of the quantized orbits, it does not emit any electromagnetic radiation; thus, the electron is said to be in a **stationary state.** The electron can make a discontinuous transition, or **quantum jump,** from one stationary state to another. During this transition it does emit radiation.

2. The laws of classical mechanics apply to the orbital motion of the electrons in a stationary state, but these laws do not apply during the transition from one state to another.

3. When an electron makes a transition from one stationary state to another, the excess energy ΔE is released as a single photon of frequency $v = \Delta E/h$.

4. The permitted orbits are characterized by quantized values of the orbital angular momentum L. This angular momentum is always an integral multiple of $h/2\pi$:

$$L = nh/2\pi \qquad n = 1, 2, 3, \ldots \qquad (13)$$

The number n is called the **angular-momentum quantum number.** Let us now see how to calculate the stationary states and the spectrum of the hydrogen atom on the basis of these postulates. For the sake of simplicity, we will assume that the electron moves in a circular orbit around the proton, which remains at rest (Figure 43.12). Since the electric force of attraction between the electron and the proton is $e^2/4\pi\varepsilon_0 r^2$, the equation of motion for the electron in a circular orbit is

$$\frac{m_e v^2}{r} = \frac{1}{4\pi\varepsilon_0} \frac{e^2}{r^2} \tag{14}$$

The orbital angular momentum is $L = m_e vr$. According to Bohr's postulate, this orbital angular momentum must be $h/2\pi$ multiplied by an integer,

$$L = m_e vr = nh/2\pi \qquad n = 1, 2, 3, \ldots \tag{15}$$

or

$$m_e vr = n\hbar \tag{16}$$

where $\hbar$ (pronounced "h bar") is Planck's constant divided by 2π:

$$\hbar = h/2\pi = 1.06 \times 10^{-34} \text{ J} \cdot \text{s}$$

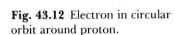

Fig. 43.12 Electron in circular orbit around proton.

From Eq. (16),

$$v^2 = \frac{n^2 \hbar^2}{m_e^2 r^2} \tag{17}$$

which, when substituted into Eq. (14), yields an expression for the radius of the orbit,

$$r = \frac{4\pi\varepsilon_0 n^2 \hbar^2}{m_e e^2}$$

Thus, the radius of the smallest permitted orbit ($n = 1$) is

$$\boxed{a_0 = \frac{4\pi\varepsilon_0 \hbar^2}{m_e e^2} = 0.529 \times 10^{-10} \text{ m} = 0.529 \text{ Å}} \tag{18}$$

Bohr radius

This is called the **Bohr radius**.[2] The radii of the other permitted orbits are multiples of the Bohr radius:

$$\boxed{r = n^2 a_0} \tag{19}$$

Figure 43.13 shows the permitted circular orbits, drawn to scale.

The energy of the electron in one of these orbits is a sum of kinetic and potential energies:

$$E = \tfrac{1}{2} m_e v^2 - \frac{e^2}{4\pi\varepsilon_0 r}$$

$$= \tfrac{1}{2} m_e \left(\frac{n^2 \hbar^2}{m_e^2}\right) \left(\frac{m_e e^2}{4\pi\varepsilon_0 n^2 \hbar^2}\right)^2 - \frac{e^2}{4\pi\varepsilon_0} \left(\frac{m_e e^2}{4\pi\varepsilon_0 n^2 \hbar^2}\right)$$

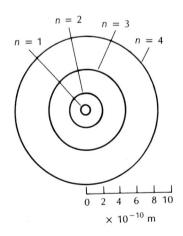

Fig. 43.13 The possible Bohr orbits of an electron in the hydrogen atom.

---

[2] See Appendix 8 for a more precise value of the Bohr radius.

or

$$E = -\frac{m_e e^4}{2(4\pi\varepsilon_0)^2 \hbar^2} \frac{1}{n^2} \tag{20}$$

We will label this energy with the subscript n,

$$E_n = -\frac{m_e e^4}{2(4\pi\varepsilon_0)^2 \hbar^2} \frac{1}{n^2} \tag{21}$$

Thus, the energy of the stationary state of least energy ($n = 1$) is

$$E_1 = -\frac{m_e e^4}{2(4\pi\varepsilon_0)^2 \hbar^2}$$

$$= -2.18 \times 10^{-18} \text{ J} = -13.6 \text{ eV} \tag{22}$$

The energies of the other stationary states are fractions of this energy:

Energies of stationary states of hydrogen

$$\boxed{E_n = -\frac{13.6 \text{ eV}}{n^2}} \tag{23}$$

Energy-level diagram

Figure 43.13 displays these quantized energies in an **energy-level diagram.** Each horizontal line represents one of the energies given by Eq. (23). According to Bohr's assumptions, the electron radiates when it makes a quantum jump from one stationary state to a lower stationary state. Such quantum jumps have been indicated by arrows in Figure 43.14. The stationary state of lowest energy is called the **ground state;** the next one is called the **first excited state;** and so on. Ordinarily, the electron of the hydrogen atom is in the ground state, that is, the circular orbit of radius a_0. This is the configuration of least energy, and it is the configuration into which the atom tends to settle when it is left undisturbed. As long as the atom remains in the ground state, it does not emit light. To bring about the emission of light, we must first kick the electron into one of the excited states, that is, a circular orbit of larger radius. We can do this by heating a sample of atoms or by passing an electric current through the sample. Collisions between the atoms will then disturb the electronic motions and occasionally kick an electron into a larger orbit. From there, the electron will spontaneously jump into a smaller orbit, giving off a quantum of light. Note that the quantum jumps shown in Figure 43.14 form several series: one series consists of all those jumps that end in the ground state, another series consists of all those jumps that end in the first excited state, and so on. These series of jumps give rise to the series of spectral lines: the Lyman series, the Balmer series, and so on.

Ground state and excited states

We can now calculate the frequency of the light emitted in a quantum jump from some initial state i to a final state f. In this jump, the electron releases an energy

$$\Delta E = E_i - E_f$$

$$= \frac{m_e e^4}{2(4\pi\varepsilon_0)^2 \hbar^2} \left(\frac{1}{n_f^2} - \frac{1}{n_i^2} \right) \tag{24}$$

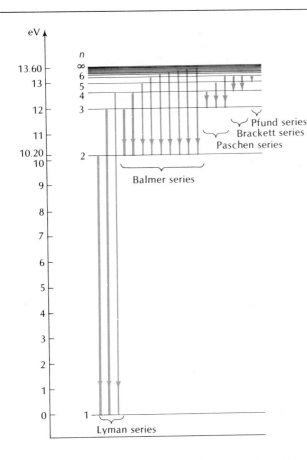

Fig. 43.14 Energy-level diagram for the hydrogen atom. The arrows show the possible quantum jumps for the electron. Note that in this diagram the values of the energies are given relative to the ground state, which is assigned an energy of zero.

According to Bohr's postulate, this energy is radiated as a single photon of frequency $v = \Delta E/h$, that is,

$$v = \frac{E_i - E_f}{h} = \frac{m_e e^4}{4\pi(4\pi\varepsilon_0)^2\hbar^3}\left(\frac{1}{n_f^2} - \frac{1}{n_i^2}\right) \quad (25)$$

Frequency of photon emitted in transition

This equation looks just like the general formula (8) for the frequencies of the spectral series. Comparison of Eqs. (25) and (8) yields the following theoretical formula for the Rydberg constant:

$$R_H = \frac{m_e e^4}{4\pi(4\pi\varepsilon_0)^2\hbar^3 c} \quad (26)$$

Upon insertion of the accurate values of the fundamental constants given in Appendix 8, we find

$$R_H = \frac{9.10953 \times 10^{-31} \text{ kg} \times (1.602189 \times 10^{-19} \text{ C})^4}{4\pi(4\pi \times 8.854178 \times 10^{-12} \text{ F/m})^2 \times (1.054589 \times 10^{-34} \text{ J}\cdot\text{s})^3 \times 2.997925 \times 10^8 \text{ m/s}}$$

$$= 109{,}737 \text{ cm}^{-1}$$

This theoretical value of R_H agrees quite well with the experimental value quoted in Section 43.2.[3]

---

[3] The small disagreement between the theoretical value of R_H given in Eq. (26) and the experimental value given in Section 43.2 is due to the motion of the nucleus of the

EXAMPLE 3. Suppose that the atoms in a sample of hydrogen gas are initially in the ground state. If we illuminate these atoms with white light (from some kind of lamp), what frequencies will the atoms absorb?

SOLUTION: Absorption of light is the reverse of emission. When an electron in an atom absorbs a photon (supplied by the lamp), it jumps from the initial state to a state of higher energy. The energy of the photon must match the energy difference between the states. Thus, the frequencies of the photons that the electrons can absorb when jumping upward from the ground state are exactly those frequencies that they emit when jumping downward into the ground state, that is, the frequencies of the Lyman series.

With slight modifications, the formulas of this section can also be applied to other hydrogen-like systems. For instance, a singly ionized helium atom (a helium atom with one missing electron) has a structure similar to that of a hydrogen atom — it consists of one electron orbiting around a nucleus of charge $2e$. The energies of the stationary states are therefore given by a formula similar to Eq. (21), but with the factor e^4 replaced by $e^2(2e)^2$. Likewise, the energies of the stationary states of a doubly ionized lithium atom (a lithium atom with two missing electrons) are given by Eq. (21) with e^4 replaced by $e^2(3e)^2$, and so on.

More generally, such modified formulas are approximately applicable to the stationary states of the innermost electrons in an atom with several electrons. In such an atom — for instance, an atom of iron — the electrons are distributed over diverse orbits around the nucleus. These orbits, and their energies, cannot be calculated from the simple Bohr theory discussed above. The electrons in this atom interact not only with the nucleus but also with each other. In consequence of these mutual interactions among the electrons, the orbits are much more complicated than in the hydrogen atom, and Bohr's theory fails. However, the innermost orbits retain the simplicity of the hydrogen atom, because the electrons in these orbits are very near the nucleus, and the attractive electric force of the nucleus overshadows the mutual repulsive electric forces among the electrons; thus, the motions of these electrons are approximately similar to the motion in the hydrogen atom. In an atom of atomic number Z, the energies of the stationary states of the innermost electrons are approximately given by a formula similar to Eq. (21) with e^4 replaced by $e^2(Ze)^2$. Transitions between the innermost orbits in an atom of moderate or large value of Z give rise to photons of short wavelengths, that is, X rays. Such X rays emitted by transitions between the innermost orbits of atoms are called **characteristic X rays**. The spectrum of these X rays consists of discrete spectral lines, like the spectral lines emitted by the atom in the visible region. This line spectrum of X rays stands in sharp contrast to the spectrum of X rays produced by Bremsstrahlung, which is continuous, with a smooth distribution of energy over a broad range of wavelengths.

With his theory Bohr attained the goal of explaining the regularities in the spectrum of hydrogen in terms of the regularities of the structure of the atom. By showing that this structure is based on a simple numerical sequence, he fulfilled the ancient dream of Pythagoras of a

Characteristic X rays

atom, which we have neglected in our calculation. A careful calculation that takes into account the motions of electron and proton about their common center of mass eliminates the disagreement.

universe based on simple numerical ratios, a dream that arose from an analogy with musical instruments. Bohr's theory tells us how the atom plays its tune.

43.5 The Correspondence Principle

We saw in Chapter 42 that in an ordinary light wave we usually do not notice the individual quanta of energy — the number of quanta is so large that the energy distribution seems continuous. Likewise, when the electron in a hydrogen atom is in an orbit of a large angular momentum, an increase or decrease of this angular momentum by one quantum $\hbar$ represents such a small fraction of the whole that the change in angular momentum seems continuous. For instance, if an electron in a Bohr orbit with quantum number $n = 5000$ jumps down, step by step, to the next orbit ($n = 4999$), and to the next orbit ($n = 4998$), and so on, the changes of angular momentum, energy, and radius will seem quite continuous because each step is small compared with the remaining angular momentum, energy, and radius. Under these conditions, the electron will behave pretty much like a classical particle — the quantized character of the angular momentum and energy and the discrete character of the orbits will not be very noticeable.

This is an instance of Bohr's **Correspondence Principle.** This principle states that *in the limiting case of large quantum numbers, the results obtained from quantum theory must agree with those obtained from classical theory.*

Correspondence Principle

We can check that the behavior of an electron in a hydrogen atom does obey this principle both qualitatively and quantitatively, by calculating the frequency of the emitted light; we can check that in the limiting case of large n, the results of quantum theory and of classical theory agree. According to Eq. (25), the frequency emitted in a transition from the state n to the state $n - 1$ is

$$\nu = \frac{m_e e^4}{4\pi(4\pi\varepsilon_0)^2 \hbar^3} \left[\frac{1}{(n-1)^2} - \frac{1}{n^2} \right] \tag{27}$$

$$= \frac{m_e e^4}{4\pi(4\pi\varepsilon_0)^2 \hbar^3} \frac{2n-1}{n^2(n-1)^2} \tag{28}$$

If n is very large, $(2n-1)/[n^2(n-1)^2] \cong 2n/n^4 = 2/n^3$, so

$$\nu \cong \frac{m_e e^4}{4\pi(4\pi\varepsilon_0)^2 \hbar^3} \frac{2}{n^3} \tag{29}$$

This is the frequency according to quantum theory. To find the frequency according to classical theory, we note that for a charge in accelerated motion, classical electromagnetism predicts that the frequency of the emitted light coincides with the frequency of the motion. For our electron in a circular orbit, the frequency of the motion is $v/2\pi r$. From Eqs. (16) and (18) we then find that the frequency of the emitted light is

$$v_{\text{class}} = \frac{v}{2\pi r} = \frac{n\hbar/m_e r}{2\pi r} = \frac{n\hbar}{2\pi m_e} \frac{1}{r^2}$$

$$= \frac{n\hbar}{2\pi m_e} \left(\frac{m_e e^2}{4\pi\varepsilon_0 n^2 \hbar^2} \right)^2 \tag{30}$$

If we simplify the right side of this equation, we recognize that it coincides with the right side of Eq. (29), that is, the result of the classical calculation agrees with the result of the quantum-mechanical calculation. Note, however, that this agreement holds only for large values of n; if n is *not* large, then the classical frequency [Eq. (30)] is smaller than the quantum-mechanical frequency [Eq. (28)].

43.6 Quantum Mechanics

Bohr's theory is a hybrid. It relies on some basic classical features (orbits) and grafts onto these some quantum features (quantum jumps, quanta of light). In the 1920s, the cooperative efforts of several brilliant physicists — L. de Broglie, E. Schrödinger, W. Heisenberg, M. Born, P. Jordan, P. A. M. Dirac — established that the remaining classical features had to be eradicated from the theory of the atom. Bohr's semiclassical theory had to be replaced by a new quantum mechanics with an entirely different equation of motion.

Louis Victor, prince de Broglie (de broy), *1892–1987, French theoretical physicist, professor at the University of Paris. He discovered Eq. (43.31) by reasoning that if waves have particle properties, then maybe particles have wave properties. For his discovery of the wave properties of matter he was awarded the Nobel Prize in 1929, after the existence of these wave properties was confirmed experimentally.*

The basis for the new quantum mechanics was laid by de Broglie's discovery that electrons — as well as protons, neutrons, and all the other "particles" found in nature — have not only particle properties but also wave properties. When a beam of electrons is made to pass through an extremely narrow slit, the electrons exhibit diffraction. This means that electrons are neither classical particles nor classical waves. Electrons, just like photons, are a new kind of object with a subtle combination of particle and wave properties. Electrons are wavicles. The wavelength associated with an electron or some other wavicle is inversely proportional to its momentum:

De Broglie wavelength

$$\boxed{\lambda = h/p} \tag{31}$$

This is called the **de Broglie wavelength.** This wavelength is quite small, even for electrons of the lowest energies attainable in experiments with beams of electrons. For instance, if the electron energy is 1 eV, the de Broglie wavelength is 12 Å.

The wave properties of electrons were confirmed experimentally by C. J. Davisson and L. Germer, who observed interference effects with beams of electrons scattered by crystals. In these experiments, the crystal acts as a grating for electron waves. When exposed to the incident electron beam, the atoms in the crystal scatter electrons and thereby serve as pointlike sources of electron waves. Depending on the phase relationships, the waves emanating from these sources interfere constructively or destructively, just like the light waves emanating from an ordinary optical grating. Davisson and Germer found that constructive interference produced strong maxima in selected directions, and they were able to verify that these directions agreed with calculations based on Eq. (31).

In the new quantum mechanics, or wave mechanics, the motion of an electron is described by a wave equation, the **Schrödinger equation**. This Schrödinger equation plays the same role for electrons as the Maxwell equations play for photons. The intensity of the electron wave at some point determines the probability that there is an electron particle at that point [compare Eq. (42.23)]. Furthermore, in consequence of their wave properties, electrons obey the Heisenberg uncertainty relations for position and momentum [see Eq. (42.30)]. These quantum uncertainties are of crucial importance for the behavior of an electron inside an atom. For such an electron, the uncertainty in the position is very large — about as large as the size of the atom. This implies that the electron follows no definite orbit. It is therefore not surprising that the Bohr theory should have failed in all attempts at calculating the electron motion in the helium atom and in other atoms with several electrons; what is surprising is that this theory should have succeeded as well as it did in the case of the hydrogen atom.

Schrödinger equation

EXAMPLE 4. Consider an electron in the ground state of hydrogen. Show that a well-defined orbit is inconsistent with the Heisenberg uncertainty relations.

SOLUTION: If the electron is to follow a well-defined orbit, the uncertainty in its momentum (in any direction) must be much smaller than the magnitude of the momentum. According to Eq. (16), the speed of the electron in the smallest circular orbit ($n = 1$) is $v = \hbar/m_e r$, and the magnitude of the momentum is $p = m_e v = \hbar/r$. For a well-defined orbit we therefore require

$$\Delta p_y \ll \hbar/r \tag{32}$$

Furthermore, we require that the uncertainty in the position must be much smaller than the size of the orbit,

$$\Delta y \ll r \tag{33}$$

Taking the product of Eqs. (32) and (33), we find

$$\Delta y \, \Delta p_y \ll \hbar \tag{34}$$

This is inconsistent with the Heisenberg uncertainty relation [Eq. (42.30)].

In wave mechanics, the quantization of the energy in the hydrogen atom and other atoms is an automatic consequence of the wave properties of the electron. The attractive electric force of the nucleus confines the electron wave to some region near the nucleus and causes the wave to reflect back and forth across the region, forming a standing wave. The different stationary states of the atom correspond to different standing-wave modes. Like the standing waves on a string, the standing electron waves in the atom have a discrete set of wavelengths and eigenfrequencies and, therefore, a discrete set of energies. The ground state of the atom is analogous to the fundamental mode of the string, the first excited state is analogous to the first overtone, and so on. However, whereas the determination of the eigenfrequencies of standing waves on a string is a quite trivial mathematical exercise, the determination of the eigenfrequencies of the electron waves in an atom is a formidable mathematical problem, which requires an investigation of the solutions of the Schrödinger equation.

Schrödinger equation

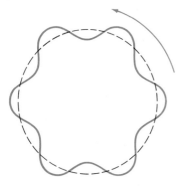

Fig. 43.15 In this example, six wavelengths of the electron wave fit around the circumference of the orbit.

Erwin Schrödinger, *1887–1961, Austrian theoretical physicist, professor at Berlin and at Vienna. Another of the founders of the new quantum mechanics, he received the Nobel Prize in 1933.*

Although here we cannot deal with the mathematical complexities of the Schrödinger equation, we can gain some insight into how electron waves determine the discrete energies in the hydrogen atom by means of the following simple calculation. Let us assume that the electron travels around the nucleus along an orbit of radius r, but instead of thinking of the electron as a particle, as in the Bohr theory, let us think of it as a wave. Figure 43.15 shows a "snapshot" of such a wave at one instant of time. If the wave is to have a well-defined amplitude at all points, it must repeat whenever we go once around the circumference — if it did not, then the wave amplitude would have two different values at a single point, which makes no sense. Hence, the wave is subject to the condition that some integer number of wavelengths must fit around the circumference:

$$2\pi r = n\lambda \tag{35}$$

Although this equation looks like the condition for a standing wave on a string of length $2\pi r$, we are here dealing with a traveling wave, for which the entire wave pattern in Figure 43.15 rotates rigidly around the center. With the de Broglie relation $\lambda = h/p$, Eq. (35) becomes

$$2\pi r = nh/p$$

or

$$rp = \frac{nh}{2\pi} \tag{36}$$

Since rp is the orbital angular momentum, this equation coincides with Bohr's quantization condition for the angular momentum. Thus, the wave picture of the electron implies the quantization of the angular momentum and, therefore, the quantization of the energy. But we must not take this calculation too seriously — its legitimacy is questionable, since it relies in part on the wave picture and in part on the particle picture. Furthermore, analysis of the Schrödinger equation shows that it is not enough to consider the behavior of the wave around the circumference; we must also consider the behavior of the wave outward along each radius. Thus, this simple calculation provides no more than a crude qualitative sketch of the role of the wave properties of electrons in the atom.

Within the atom, electrons always behave very much like waves. But outside the atom, they will sometimes behave pretty much like classical particles. Roughly, we can say that classical mechanics will be a good approximation whenever the quantum uncertainties are small compared with the relevant magnitudes of positions and momenta. For instance, for the electrons in the beam of a TV tube, the quantum uncertainties in the momenta are negligible compared with the magnitude of the momentum. Under these conditions, classical mechanics gives an adequate description of the motion of the electrons.

What we have said about the quantum mechanics of electrons also applies to all other "particles" found in nature. Protons, neutrons, muons, pions, etc., all have wave properties — they all are wavicles. They obey the laws of wave mechanics, and they have quantum uncertainties in their positions and momenta. Composite bodies, such as atoms or molecules, also have wave properties; the translational motions of such bodies are governed by the same laws of wave mechanics as the motion of a wavicle. Strictly speaking, even large macroscopic bodies have wave properties. For example, the translational motion of an automobile is that of a wavicle, and the automobile has some quantum uncertainty in its position. However, it turns out that the quantum uncertainties are very small whenever the mass of the body is large compared with atomic masses — the quantum uncertainty in the position of an automobile is typically no more than about 10^{-18} m, a number that can be ignored for all practical purposes. Hence, for automobiles and other macroscopic bodies, quantum effects are completely insignificant, and classical mechanics gives an excellent description of the motion of these bodies.

SUMMARY

Spectral series of hydrogen:
$$v = cR_H \left(\frac{1}{n_2^2} - \frac{1}{n_1^2} \right); \quad R_H = 109{,}678 \text{ cm}^{-1}$$

Quantization of angular momentum: $L = n\hbar$

Bohr radius: $a_0 = \dfrac{4\pi\varepsilon_0 \hbar^2}{m_e e^2} = 0.529 \text{ Å}$

Energy of stationary states of hydrogen:
$$E_n = -\frac{m_e e^4}{2(4\pi\varepsilon_0)^2 \hbar^2} \frac{1}{n^2}$$
$$= -\frac{13.6 \text{ eV}}{n^2}$$

Frequency of photon emitted in transition: $v = \dfrac{E_i - E_f}{h}$

Correspondence Principle: In the limiting case of large quantum numbers, the results obtained from quantum theory must agree with those obtained from classical theory.

De Broglie wavelength of wavicle: $\lambda = h/p$

QUESTIONS

1. Do the spectral lines seen in a stellar spectrum (for example, Figure 43.4) tell us anything about the chemical composition of the stellar interior?

2. The spectrum of hydrogen shown in the color plate between pages 898 and 899 displays all of the spectral lines simultaneously. Since a hydrogen atom emits only one spectral line at a time, how can all the lines be visible simultaneously?

3. What is the longest wavelength that a hydrogen atom will emit or absorb?

4. The target used in Rutherford's scattering experiment was a very thin foil of metal. What is the advantage of a thin foil over a thick foil in this experiment?

5. How can Rutherford's experiment tell us something about the size of the nucleus?

6. Why is Bohr's postulate of stationary states in direct contradiction with classical mechanics and electromagnetism?

7. If an electron in a hydrogen atom makes a transition from some state to a lower state, does its kinetic energy increase or decrease? Its potential energy? Its orbital angular momentum?

8. At low temperatures, the absorption spectrum of hydrogen displays only the spectral lines of the Lyman series. At higher temperatures, it also displays other series. Explain.

9. The planets move around the Sun in circular orbits. Is their orbital angular momentum quantized?

10. In a muonic atom, a muon orbits around the nucleus. The mass of the muon is 207 times the mass of the electron. What is the Bohr radius for a muonic atom with a hydrogen nucleus?

11. Given that the orbital angular momentum of an atom is quantized, can we conclude that the orbital magnetic moment is also quantized?

Fine-structure constant

12. The quantity $\hbar/(m_e c)$ is the Compton wavelength divided by 2π. The quantity $e^2/(4\pi\varepsilon_0 m_e c^2)$ is called the "classical electron radius." Show that the Bohr radius, the Compton wavelength divided by 2π, and the classical electron radius are in the ratio $1:\alpha:\alpha^2$, where $\alpha = e^2/(4\pi\varepsilon_0 \hbar c)$. The quantity α is called the **fine-structure constant**. What is the numerical value of this constant?

13. Bohr's theory of the hydrogen atom can be adapted to the singly ionized helium atom, that is, the helium atom with one missing electron. To what other ionized atoms can Bohr's theory be adapted?

Complementarity Principle

14. According to the **Complementarity Principle,** formulated by Bohr, a wavicle has both wave properties and particle properties, but these properties are never exhibited simultaneously: if the wavicle exhibits wave properties in an experiment, then it will not exhibit particle properties, and conversely. Give some examples of experiments in which wave or particle properties (but not both simultaneously) are exhibited.

15. Show that photons obey the de Broglie relation.

16. If the de Broglie wavelengths of two electrons differ by a factor of 2, by what factor must their energies differ?

17. According to the de Broglie relation, the wavelength of an electron of very small momentum is very large. Could we take advantage of this to design an experiment that makes the wave properties of the electron obvious?

18. An electron and a proton have the same energy. Which has the longer de Broglie wavelength?

19. Describe the interference pattern expected for an electron wave incident on a plate with two very narrow parallel slits separated by a small distance.

20. Electron microscopes achieve high resolution because they use electron waves of very short wavelength, usually less than 0.1 Å. Why can we not build a microscope that uses *photons* of equally short wavelength?

PROBLEMS

Section 43.2

1. Use Eq. (4) to calculate the wavelengths of the first four lines of the Lyman series.

2. Which of the spectral lines of the Brackett series is closest in wavelength to the first spectral line ($n = 6$) of the Pfund series? By how much do the wavelengths differ?

*3. Show that the spectral lines of the Balmer series all have a higher frequency than the spectral lines of the Paschen series. Do the spectral lines of the Paschen series all have a higher frequency than those of the Brackett series?

*4. When astronomers examine the light of a distant galaxy, they find that all the wavelengths of the spectral lines of the atoms are longer than those of the atoms here on Earth by a common multiplicative factor. This is the *red shift* of light; it is a Doppler shift caused by the motion of recession of the galaxy, away from the Earth. In the light of a galaxy beyond the constellation Virgo, astronomers find spectral lines of wavelengths 4117 Å and 4357 Å.
  (a) Assume that these are two spectral lines of hydrogen, with the wavelengths multiplied by some factor. Identify these lines. What is the factor by which these wavelengths are longer than the normal wavelengths of the two spectral lines?
  (b) What is the speed of recession of the galaxy?

*5. One of the spectral series of the lithium atom is the **principal series,** with the following wavelengths, measured in vacuum: 6710 Å, 3234 Å, 2742 Å, 2563 Å, 2476 Å. These wavelengths approximately fit the formula

$$\frac{1}{\lambda} = R\left[\frac{1}{(1+s)^2} - \frac{1}{(n+p)^2}\right] \qquad n = 2, 3, 4, \ldots$$

where $R = 109{,}729 \text{ cm}^{-1}$ is the Rydberg constant for lithium, and s and p are constants characteristic of the series. Find the values of these constants.

*6. Another of the spectral series of the lithium atom is the **diffuse series,** with the following wavelengths, measured in vacuum: 6105 Å, 4604 Å, 4134 Å, 3916 Å, 3796 Å. These wavelengths approximately fit the formula

$$\frac{1}{\lambda} = R\left[\frac{1}{(2+p)^2} - \frac{1}{(n+d)^2}\right] \qquad n = 3, 4, 5, \ldots$$

where, as in the preceding problem, $R = 109{,}729 \text{ cm}^{-1}$ and p and d are constants.
  (a) Find the values of these constants.
  (b) The principal series (see Problem 5) and the diffuse series of lithium are analogous to two spectral series of hydrogen. Which two series?

Section 43.3

7. What is the distance of closest approach for a 5.5-MeV alpha particle in a head-on collision with a gold nucleus? With an aluminum nucleus?

8. The nucleus of platinum has a radius 6.96×10^{-15} m and an electric charge of $78e$. What must be the minimum energy of an alpha particle in a head-on collision if it is just barely to reach the nuclear surface? Assume the alpha particle is pointlike.

9. An alpha particle of energy 5.5 MeV is incident on a silver nucleus with an impact parameter 8.0×10^{-15} m. Find the distance of closest approach of the particle. Find the speed at the point of closest approach.

*10. Prove that the distance of closest approach given by Eq. (12) can be expressed as

$$r' = \frac{b^2}{-r^*/2 + \sqrt{(r^*/2)^2 + b^2}}$$

where r^* is the distance of closest approach for a head-on collision (with $b = 0$).

**11. A foil of gold, 2.1×10^{-5} cm thick, is being bombarded by alpha particles of energy 7.7 MeV. The particles strike at random over an area of 1 cm² of the foil of gold.
  (a) How many atoms are there within the volume 1 cm² $\times 2.1 \times 10^{-5}$ cm under bombardment? The density of gold is 19.3 g/cm², and the mass of one atom is 3.27×10^{-25} kg.
  (b) It can be shown that to suffer a deflection of more than 30°, an alpha particle must strike within 5.5×10^{-14} m of the center of a gold nucleus. What is the probability for this to happen?
  (c) If 10^{10} alpha particles impact on the foil, how many will suffer deflections of more than 30°?

Section 43.4

12. What is the speed of an electron in the smallest ($n = 1$) Bohr orbit? Express your answer as a fraction of the speed of light.

13. What is the frequency of the orbital motion for an electron in the smallest ($n = 1$) Bohr orbit? In the next ($n = 2$) Bohr orbit? Do either of these frequencies coincide with the frequency of the light emitted during the transition $n = 2$ to $n = 1$?

14. If a hydrogen atom is in the ground state, what is the *longest* wavelength it will absorb?

15. What is the ionization energy of hydrogen (that is, what energy must you supply to remove the electron from the atom when it is in the ground state)? Express the answer in electron-volts.

16. A hydrogen atom emits a photon of wavelength 1026 Å. From what stationary state to what lower stationary state did the electron jump?

17. Suppose that the electron in a hydrogen atom is initially in the second excited state ($n = 3$). What wavelength will the atom emit if the electron jumps directly to the ground state? What two wavelengths will the atom emit if the electron jumps to the first excited state and then to the ground state?

18. Find the orbital radius, the speed, the angular momentum, and the centripetal acceleration for an electron in the $n = 2$ orbit of hydrogen.

19. If you bombard hydrogen atoms in their ground state with a beam of electrons, the collisions will (sometimes) kick atoms into one of their excited states. What must be the minimum kinetic energy of the bombarding electrons if they are to achieve such an excitation?

20. Hydrogen atoms in highly excited states with a quantum number as large as $n = 732$ have been detected in interstellar space by radio astronomers. What is the orbital radius of the electron in such an atom? What is the energy of the electron?

*21. Suppose that a sample of hydrogen atoms, initially in their ground states, is under bombardment by a beam of electrons of kinetic energy 12.2 eV. In an inelastic collision between a hydrogen atom and one of the incident electrons, the hydrogen atom will occasionally absorb all, or almost all, the kinetic energy of the electron and make a transition from the ground state to an excited state. If so, what excited state will the hydrogen atom attain? What are the possible spectral lines that the atom can emit subsequently?

*22. A hydrogen atom is initially in the ground state. In a collision with an argon atom, the electron of the hydrogen atom absorbs an energy of 15.0 eV. With what speed will the electron be ejected from the hydrogen atom?

*23. The singly ionized helium atom (usually designated HeII) has one electron in orbit around a nucleus of charge 2e.
   (a) Apply Bohr's theory to this atom and find the energies of the stationary states. What is the value of the ionization energy, that is, the energy that you must supply to remove the electron from the atom when it is in the ground state? Express the answer in electron-volts.
   (b) Show that for every spectral line of the hydrogen atom, the ionized helium atom has a spectral line of identical wavelength.

*24. Doubly ionized lithium (usually designated LiIII) has one electron in orbit around a nucleus of charge $Z = 3e$. What is the radius of the smallest Bohr orbit in doubly ionized lithium? What is the energy of this orbit?

*25. The muon (or mu meson) is a particle somewhat similar to an electron; it has a charge $-e$ and a mass 206.8 times as large as the mass of the electron. When such a muon orbits around a proton, they form a **muonic hydrogen atom,** similar to an ordinary hydrogen atom, but with the muon playing the role of the electron. Calculate the Bohr radius of this muonic atom and calculate the energies of the stationary states. What is the energy of the photon emitted when the muon makes a transition from $n = 2$ to $n = 1$?

*26. Consider a helium atom in interstellar space in a circular orbit around a meteoroid of mass 4 kg and radius 10 cm under the influence of the gravitational force. We can apply Bohr's theory to this system; the meteoroid plays the role of the nucleus, the atom that of the electron, and the gravitational force that of the electric force. Find formulas analogous to Eqs. (41.19) and (41.21) for the orbital radii and the energies of the permitted circular orbits. Because of the finite size of the meteoroid, the smallest feasible orbit has a radius of 10 cm. What is the quantum number and what is the energy (in eV) of a helium atom in this orbit?

*27. The conventional range of wavelengths for X rays extends from 100 Å to 0.1 Å. Suppose you want to generate characteristic X rays within this range of wavelengths by means of the transition of an internal electron in an atom, from the first excited state to the ground state. What atomic numbers are suitable? Which atoms do these correspond to?

*28. Calculate the wavelengths of the characteristic X rays emitted by the inner electrons in molybdenum atoms in transitions from the first and the second excited states to the ground state. Compare your results with the observed wavelengths, 0.72 Å and 0.61 Å.

*29. When a block of metal serving as target in an X-ray tube is bombarded with a beam of fast electrons, it emits not only Bremsstrahlung but also characteristic X rays of wavelengths 1.67 Å, 1.41 Å, and 1.33 Å. Assuming that these X rays arise in transitions from excited states into the ground state, identify the metal.

*30. Assume that, as proposed by J. J. Thomson, the hydrogen atom consists of a cloud of positive charge e, uniformly distributed over a sphere of radius R. However, instead of placing the electron in static equilibrium at the center of the sphere, assume that the electron orbits around the center with uniform circular motion under the influence of the electric centripetal force $(e^2/4\pi\varepsilon_0)(r/R^3)$. If the angular momentum of this orbiting electron is quantized according to Bohr's theory (so that $L = n\hbar$), what are the radii and the energies of the quantized orbits? What are the frequencies of the photons emitted in transitions from one quantized orbit to another? What must be the value of R if at least two orbits are to fit inside this atom?

**31. In our calculation of the energies of the stationary states of hydrogen we pretended that the proton remains at rest. Actually, both the electron and the proton orbit about their common center of mass. Show that the energies of the stationary states, taking into account this motion of the proton, are given by

$$E_n = -\frac{\mu e^4}{2(4\pi\varepsilon_0)^2 \hbar^2}\frac{1}{n^2}$$

where

$$\mu = \frac{m_e m_p}{m_e + m_p}$$

(Hint: The electron and the proton move in circles of radii

$$r_e = r\frac{m_p}{m_p + m_e} \quad \text{and} \quad r_p = r\frac{m_e}{m_p + m_e}$$

where r is the distance between the electron and the proton. According to Bohr's theory, the net angular momentum of this system of two particles is quantized, $L = n\hbar$.)

32. The atom of **positronium consists of an electron and a positron (or antielectron) orbiting about their common center of mass. According to Bohr's theory, the net angular momentum of this system is quantized, $L = n\hbar$. What is the radius of the smallest possible circular orbit of this system? What is the wavelength of the photon released in the transition $n = 2$ to $n = 1$?

Section 43.5

*33. In principle, Bohr's theory also applies to the motion of the Earth around the Sun. The Earth plays the role of the electron; the Sun, that of the nucleus; and the gravitational force, that of the electric force.
  (a) Find a formula analogous to Eq. (18) for the radii of the permitted circular orbits of the Earth around the Sun.
  (b) The actual radius of the Earth's orbit is 1.50×10^{11} m. What value of the quantum number n does this correspond to?
  (c) What is the radial distance between the Earth's actual orbit and the next larger orbit?

*34. According to classical electrodynamics, an electron in an elliptical orbit emits not only radiation at the orbital frequency but also radiation at multiples of the orbital frequency; thus, if the orbital frequency is ν, the radiation contains the fundamental frequency ν and also the harmonic frequencies 2ν, 3ν, 4ν, etc. Does this contradict the Correspondence Principle? (Hint: Calculate what frequencies the electron will emit in the transition n to $n - 2$, n to $n - 3$, n to $n - 4$, etc.)

Section 43.6

35. What must be the energy of an electron if its wavelength is to equal the wavelength of visible light, about 5500 Å?

36. Find the de Broglie wavelength for each of the following electrons with the specified kinetic energy: electron of 20 keV in a TV tube, conduction electron of 5.4 eV in a metal, orbiting electron of 13.6 eV in a hydrogen atom, orbiting electron of 91 keV in a lead atom.

37. Show that the de Broglie wavelength of an electron can be calculated from the formula

$$\lambda = 12.3/\sqrt{K}$$

where the wavelength λ is expressed in Å and the kinetic energy K is expressed in eV.

38. A photon and an electron each have an energy of 6.0×10^3 eV. What are their wavelengths?

39. An electron microscope operates with electrons of energy 40,000 eV. What is the wavelength of such electrons? By what factor is this wavelength smaller than that of visible light?

40. What is the de Broglie wavelength of a tennis ball of mass 0.060 kg moving at a speed of 1.0 m/s?

41. What is the de Broglie wavelength, in angstroms, of an electron in the ground state of hydrogen? In the first excited state?

*42. What is the de Broglie wavelength of a N_2 molecule in air at room temperature (20°C)? Assume that the molecule is moving with the rms speed of molecules at this temperature.

*43. Interferometric methods permit us to measure the position of a macroscopic body to within $\pm 10^{-12}$ m (see Section 1.3). Suppose we perform a position measurement of such a precision on a body of mass 0.050 kg. What uncertainty in momentum is implied by the Heisenberg relation? What uncertainty in velocity?

*44. Suppose that the velocity of an electron has been measured to within an uncertainty of ± 1 cm/s. What minimum uncertainty in the position of the electron does this imply?

*45. If the position of a parked automobile of mass 2×10^3 kg is uncertain by $\pm 10^{-18}$ m, what is the corresponding uncertainty in its velocity?

*46. The nucleus of aluminum has a diameter of 7.2×10^{-15} m. Consider one of the protons in this nucleus. The uncertainty in the position of this proton is necessarily less than 7.2×10^{-15} m. What is the minimum uncertainty in its momentum and velocity?

*47. Consider an electron in a circular orbit of quantum number n in a hydrogen atom. The orbit is well defined provided that $\Delta p_y \ll p$ and $\Delta y \ll r$. Show that if $n \gg 1$, these requirements are *not* in conflict with the Heisenberg uncertainty relations. Thus, *large* orbits in the hydrogen atom are well defined (this is in accord with the Correspondence Principle).

*48. An electron is confined to an interval from $x = 0$ to $x = L$ along the x axis by two rigid barriers.
  (a) What are the de Broglie wavelengths of the standing electron waves in this interval? Assume that, as for a standing wave on a string, the wavefunction is zero at $x = 0$ and $x = L$.
  (b) What are the energies of the stationary states?
  (c) Evaluate the energies of the first three excited states numerically for $L = 1.0$ Å.

INTERLUDE X

LASER LIGHT*

Man is a visual animal. Man's intake of visual information is about nine times larger than his combined intake of all other kinds of information. It is therefore no wonder that, until a few years ago, we regarded light as illumination — our main direct use of light was to make our environment visible. The most powerful light sources were searchlights and lighthouses with an output of up to 1 kW.

The invention of the laser in 1958 changed all that. We now have available light beams of fearsome intensity — up to 10^{10} kW in short pulses. Light beams from lasers are now used in industry to cut and weld metal, cloth, and plastic; in medicine to cut and cauterize human tissues; and in plasma research to vaporize and ignite the fuel for fusion reactors. Light has become a tool and, in military applications, it has become a weapon.

X.1 STIMULATED EMISSION

Light is emitted by the electrons in the atoms of the light source. As we saw in Chapter 43, the electrons in an atom can move only in certain selected, quantized orbits. As long as the electron stays in one of these orbits, it does not emit light. But when the electron jumps from one orbit to a smaller orbit, it emits a packet of light, or photon, with an energy equal to the energy difference between the two orbits (Figure X.1).

In an ordinary light source, such as a light bulb, the atoms emit their light independently. The atom suffers concussions from thermal collisions, and this excites one or more electrons from lower orbits to higher orbits. At some later, unpredictable time the electron spontaneously jumps back down, emitting a pulse of light. Because the atoms emit at random, the light emerging from an ordinary light bulb is a confused combination of many different waves with no special directions of propagation and no special phase relationships (Figure X.2a). On the average, the total intensity of the light is simply the sum of the intensities contributed by the individual atoms; therefore, the total intensity is simply proportional to the number of atoms.

In a laser, the atoms emit their light in unison. The electrons in different atoms either jump down at the same time or else they jump with a time difference of one or several periods of oscillation of the light wave. Furthermore, the electrons emit their light waves in the same direction. The result is that the light emerging from a laser is a **coherent** combination of waves (Figure X.2b). All the light waves from different atoms

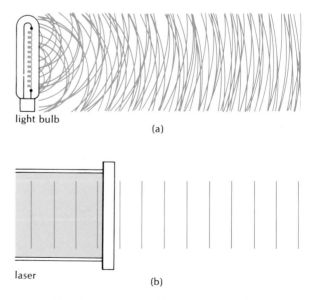

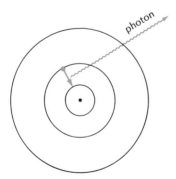

Fig. X.1 The electron in an atom emits a photon during a quantum jump.

Fig. X.2 (a) Light waves emitted by an ordinary light bulb. (b) Light wave emitted by a laser.

* This chapter is optional.

are in phase — the light wave contributed by each atom combines crest to crest with the light waves contributed by the other atoms. The total amplitude is then proportional to the number of atoms, and therefore the total intensity is proportional to the *square* of the number of atoms. Since the number of atoms in even a fairly small light source is more than 10^{16}, the coherent emission from these atoms can be enormously stronger than the incoherent emission.

What keeps the electrons in different atoms in a laser in step is the phenomenon of **stimulated emission.** Imagine that an excited electron in one of the atoms jumps to its lower orbit, releasing a wave of light. As this wave passes by some other atom with an excited electron, it will jiggle the electron; this causes the electron to resonate and to jump down in unison with the wave instead of waiting to jump spontaneously. The passage of the light wave triggers, or stimulates, the emission of an additional coherent light wave.

The mechanism behind this process involves quantum theory and we cannot present the details here, but it turns out that all the additional waves emitted by the atoms in the laser will have the same direction of propagation and the same phase as the original light wave. The waves therefore combine constructively, and they proceed to stimulate the emission of more and more waves from other excited atoms. The process has some features of a chain reaction: whenever an excited atom joins in the emission, it strengthens the wave and thereby increases the probability that the remaining excited atoms will also join in the emission.

The word *laser* is an acronym for *l*ight *a*mplification by *s*timulated *e*mission of *r*adiation. From the above discussion it is obvious that the key to the operation of a laser is the initial presence of a large number of atoms in an excited configuration. This is called a **population inversion** because under normal conditions atoms tend to settle into an unexcited configuration. Indeed, there must be more atoms in the excited configuration than in the unexcited configuration — otherwise the light wave will lose energy by stimulated absorption rather than gain energy by stimulated emission. One method for achieving the required population inversion uses an intense flash of (ordinary) light to lift the electrons into excited orbits; this is called **optical pumping.** Unfortunately, the flash of light that pumps the electrons upward into an excited orbit can also pump them downward; hence direct pumping into an excited orbit does not produce a population inversion. This problem can be circumvented by making the electron jump upward by an indirect route: it first jumps to some high excited orbit, and then spontaneously jumps down into a somewhat lower excited orbit, where it remains for some time awaiting stimu-

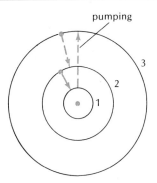

Fig. X.3 Upward and downward jumps of an electron in an atom subjected to optical pumping. A flash of light lifts the electron from the lowest orbit (1) to a high excited orbit (3). From there the electron spontaneously jumps to a somewhat lower excited orbit (2). The lasing action occurs between this orbit and the lowest orbit.

lated emission (Figure X.3). The lasing action begins when a sufficiently large number of electrons have accumulated in these lower excited orbits.

The fundamental theoretical principles involved in the operation of a laser were first published by C. H. Townes and A. L. Schawlow.[1] Their speculations on the generation of coherent light were motivated by an earlier discovery of the generation of coherent microwaves in a maser, a device similar to a laser but operating with radiowaves rather than light waves.

X.2 LASERS

The first laser designed according to the above principles was a **ruby laser** built by T. H. Maiman in 1960 (Figure X.4). This kind of laser, still in common use today, generates light in a long cylindrical crystal of synthetic ruby. The crystal contains corundum, an oxide of aluminum, with a few chromium impurities; the red color of ruby is due to these impurities. In the ruby laser, only the chromium atoms lase. To pump the electrons in these atoms into their excited orbits, the crystal is surrounded by a flash lamp (Figure X.4); the high-intensity light from the flash lamp supplies the energy for the initial upward jump of the electrons. The first electron that engages in the spontaneous emission of light will trigger the stimulated emission of light by other electrons.

In order to ensure that all, or almost all, of the excited atoms participate in stimulated emission, it is best to make the light wave traverse the ruby rod several times, so that any excited atoms not triggered on the first pass are triggered on a later pass. For this pur-

---

[1] **Charles H. Townes,** 1915–, and **Arthur L. Schawlow,** 1921–, American physicists. Townes shared the Nobel Prize in 1964 with the Soviet physicists Nicolai G. Basov and Alexander M. Prochorov for the discovery of the maser. Schawlow was awarded the 1981 Nobel Prize for his work on lasers.

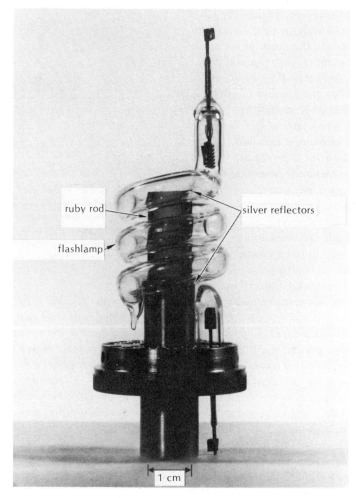

Fig. X.4 The ruby laser of Maiman.

pose, the ends of the rod are polished flat and silvered, making them into mirrors. One end is only partially silvered so that the light can ultimately escape. The back-and-forth reflections drastically enhance the lasing action for those light waves that are emitted exactly parallel to the axis of the rod and that have a wavelength fitting into the length of the rod exactly an integral number of times. For such light waves, the rod acts as a resonant cavity oscillating in one of its normal modes. In this cavity, the light wave is a standing wave and the stimulated emission by the excited atoms feeds more and more energy into this standing wave. Some of the wave leaks out of the partially silvered end of the rod, forming the useful external beam of the laser. The chromium atoms in ruby will lase at one or another of several wavelengths between 6930 Å and 7000 Å, in the red part of the spectrum. The emerging beam is essentially unidirectional, since it originates from light waves traveling exactly parallel to the axis of the rod. Typically, the beam of a laser has an angular spread of a minute of arc or less.

The beam from a ruby laser emerges as a pulse lasting as long as excited atoms remain in the rod, about 10^{-3} s. When the atoms become exhausted, they must be pumped again with a new flash from the flash lamp. For some investigations of the effect of intense light on matter, the concentration of power into a succession of short pulses of laser light is very useful. Neodymium glass lasers designed for thermonuclear fusion experiments generate pulses as short as 10^{-12} s with a peak power of 10^9 kW.

However, for many other applications it is desirable to build lasers with a continuous output of light. Obviously, this requires that, on the average, the electrons in the atoms be pumped up into the excited orbits at the same rate as they jump down by stimulated emission. The most commonly employed laser with a continuous output of light is the **helium–neon laser.** It consists of a glass tube filled with a mixture of helium and neon at low pressure. The tube has a silvered and a partially silvered mirror at the two ends, and also two electrodes (Figure X.5). The lasing action is due to the neon; the helium merely serves to pump the neon. When the electrodes are connected to a high-voltage power supply, a current of electrons flows through the tube. These electrons collide with the he-

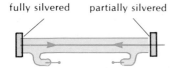

Fig. X.5 Schematic diagram of a helium–neon laser. The terminals are connected to a high-voltage supply.

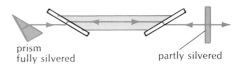

Fig. X.6 Laser with a prism for the selection of one color.

lium atoms and kick their atomic electrons into excited orbits. The excited helium atoms in turn collide with the neon atoms. In such a collision it is very likely that the helium atom will transfer its excess internal energy to the neon atom. This pumps the electron in the neon into an excited orbit suitable for stimulated emission. The neon atoms will lase at a wavelength of 6328 Å (red light) and also at several wavelengths in the infrared region of the spectrum (there exist several techniques to restrict the lasing action to just a single wavelength; for instance, see below).

A laser capable of very high power output is the **carbon dioxide–nitrogen laser.** This laser operates in much the same way as the helium–neon laser — the carbon dioxide molecules lase and the nitrogen molecules transfer energy to them. Lasers of this kind can easily deliver 10 kW of continuous power. But they can also be operated in a pulsed fashion with a much higher power concentrated in very short pulses. The carbon dioxide laser emits infrared light at several wavelengths, but no visible light. The high power and relatively high efficiency ($\sim 30\%$) attained by this laser is largely due to the long wavelength of the emitted infrared light. The downward energy jump during emission by a molecule is small and, correspondingly, the upward energy jump during pumping is also small. The pumping mechanism can supply these small energy quanta much more easily than the larger energy quanta required for lasers operating with shorter-wavelength visible light.

The resonant cavity of a ruby, helium–neon, or carbon dioxide laser has very many different normal modes and, unless special precautions are taken, the atoms will simultaneously lase in all the normal modes whose frequencies coincide with the frequencies of light generated by jumps from the available excited states. This means that the light emerging from the laser is a mixture of several colors. To obtain truly monochromatic light, a special selective device must be attached to the laser. Figure X.6 shows one of the devices that can be used for this purpose. The rear end of the laser tube is left transparent and a prism with one silvered face is placed beyond it. The prism refracts rays of different colors at different angles. When the multicolored light from the laser penetrates the prism, only one of the rays, of a selected color, strikes the silvered face at right angles and is reflected back into the laser. Hence standing waves of only one selected color are possible, and lasing will occur only at this one color. The emerging laser beam will then be perfectly monochromatic. Note that by turning the prism through some angle, we can select a different color for lasing. This is the principle of tunable lasers that produce light of adjustable color.

If laser light illuminates a slightly irregular surface — a wall or a sheet of paper — the patch of light on the surface will look grainy to the eye, appearing to have bright specks and dark specks. This is called the **speckle effect** (Figure X.7). The apparent irregularity of the illumination of the wall is an illusion. The laser beam illuminates the entire patch of surface uniformly, but small irregular points of the surface reflect light with somewhat different phases; thus, the incident light is a uniform plane wave, but the reflected light is a superposition of many spherical waves with somewhat different phases. When these waves reach the retina of the eye or a photographic plate, they interfere constructively or destructively, making bright spots and dark spots.

Incidentally: Looking directly into a laser beam can damage the eye. This danger is of course obvious in the case of powerful lasers that are capable of burning holes in metals, but the danger subsists even in the case of the small milliwatt laser commonly used for optical experiments. The lens of the eye can focus the laser beam on a single spot of the retina, and it will then not take very much power to cause a retinal burn.

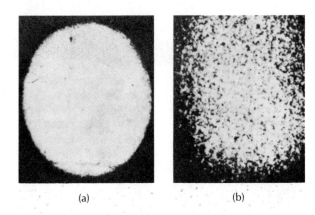

Fig. X.7 Bright spot made by a beam of light on a cement surface: (a) ordinary light from a mercury arc and (b) laser light from a helium–neon laser.

X.3 SOME APPLICATIONS

Laser light is useful because of its directionality, intensity, pure color, and coherence. The applications of laser light exploit one or several of these characteristics.

Surveys and Distance Measurements

Because of its sharp directionality, the beam from a laser is a very convenient tool for laying out straight lines over large distances. The traditional and tedious method for doing this involves setting up a row of markers along the line of sight of a survey telescope. If we replace the telescope by a laser, we can dispense with the cumbersome markers, since the laser beam itself can serve as a marker. For example, to dig a straight horizontal trench, we can aim a laser beam parallel to the ground and proceed to dig in its direction; we can check the depth of the trench by intercepting the beam with a vertical meter stick resting against the bottom of the trench. Laser beams can similarly be used in the construction of large aircraft and ships to check the alignment of ribs and frames.

Laser pulses have found application in rangefinders, especially for military purposes. The rangefinder consists of a laser and a light detector, both linked to a timing device. The laser sends a short pulse of light out to the target and the target reflects part of this pulse back. The travel time for this round trip indicates the distance.

As was mentioned in Section 37.2, this method has also been used with spectacular success in an accurate determination of the Earth–Moon distance. Both the Apollo 11 and the Apollo 14 astronauts installed corner reflectors on the surface of the Moon during their missions (Figure X.8). By means of a telescope at MacDonald Observatory, Texas, a pulse from a powerful laser was aimed at one of these reflectors. By the time the pulse reached the Moon, it had spread out to a diameter of about 3 km, but enough light was reflected back to the Earth to be picked up by a sensitive phototube. The precise measurement of the elapsed time for the round trip gave the distance to the Moon to within about 15 cm, that is, about nine significant figures!

Light-Wave Communications

As we saw in Section 37.3, light pulses can be piped through long thin glass fibers. If the sound of the human voice is encoded as a series of light pulses, then these optical fibers can be used to transmit telephone conversations. In commercial telephone installations of this kind, the light pulses are generated either by small lasers or by light-emitting diodes. The laser light is much more monochromatic than the diode light, and this is an advantage because it helps to preserve the shape of the light pulses. If a light pulse containing a mixture of colors is sent along an optical fiber, the dispersion of the medium will cause the short-wavelength colors to fall behind and the long-wavelength colors to get ahead; this tends to spread the pulse out, making it indistinct. For laser light, the spreading effect is about 10 times smaller than for diode light.

The lasers that generate the pulses for such telephone installations are solid-state devices made of a sandwich of semiconductor, roughly the size of a grain of salt (Figure X.9); their power is about 0.5 mW. In an optical telephone line, each laser feeds 5×10^7 pulses per second into its attached optical fiber; this is enough to encode 672 one-way telephone speeches.

Fig. X.8 Corner reflector placed on the surface of the Moon by the Apollo 14 astronauts.

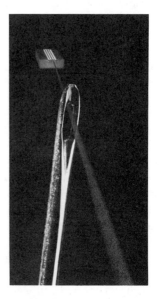

Fig. X.9 Solid-state laser.

Fig. X.10 Robot laser welding machine at an automobile assembly line.

Laser as Torch

Many applications take advantage of the high concentration of energy in laser light to melt, weld, cut, or vaporize materials. For industrial applications, a carbon dioxide laser with a power of several kilowatts makes an excellent welding torch; it has a high welding speed, requires little or no filler metal, and produces joints of excellent quality. Figure X.10 shows a laser welding system in service at an automobile factory. High-power lasers are also used to cut holes in very hard materials, e.g., steel (Figure X.11). In the clothing industry, laser beams are now used to cut out patterns in stacks of several hundreds of layers of cloth all at once.

Fig. X.11 Laser beam cutting a thick steel plate.

In medicine, lasers are used for surgery of the eye. A laser beam aimed through the (transparent) lens of the eye and focused on the retina can "weld" a detached patch of retina into its proper position. In some other surgical operations, a laser beam can serve as a scalpel; this is particularly suitable for operations on blood-rich tissues, such as the liver, where the immediate cauterization of the tissues by the laser beam prevents excessive bleeding.

Laser Interferometry

Lasers make excellent light sources for interference experiments. As we saw in Section 39.4, when an ordinary light source is used to illuminate the slits or holes, a shield with a pinhole must be placed around the light source so as to select waves of a single direction from the confused combination of waves produced by the light source; this, of course, entails a drastic reduction of the light intensity. The light from a laser already has a sharply defined direction, and no pinhole is required.

In an interferometer, such as that of Michelson, the great advantage of laser light is its large-scale coherence. An ordinary light beam, even when selected in one direction, is coherent only over a length of a few meters. This is so because the individual wave trains of light emitted by individual atoms are only about that long. Hence portions of the light beam separated by more than a few meters consist of waves produced by different atoms; these waves are not coherent — they have no regular phase relationships and they will not produce interference patterns (Figure X.12a). When the light beam is split into two by a half-silvered mirror, it will display interference patterns only if each portion travels roughly the same distance to the place of recombination, because only then will the wave pulse of a given atom be able to interfere with itself. On the other hand, in the light beam of a laser all the waves are coherent, and the entire beam is a single coherent wave (Figure X.12b). Therefore any segment

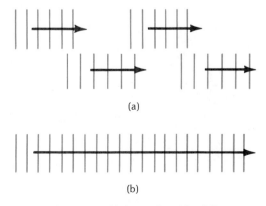

Fig. X.12 (a) The waves of light produced by different atoms are not coherent. (b) The waves produced by a laser are coherent.

of the wave train will display interference with any other segment when they are brought together.

The light from ordinary lasers is nearly monochromatic, but not perfectly monochromatic because the thermal motion of the emitting atoms gives the light a fluctuating Doppler shift. To eliminate or reduce these Doppler shifts, scientists have developed **stabilized lasers** in which a feedback device monitors the wavelength. A laser of this kind can serve as an excellent standard of length and also of frequency.

As mentioned in Chapter 1, stabilized lasers are used to implement the new definition of the meter. For this purpose, the frequency of the laser light must be measured in terms of the frequency of the cesium atomic clock that serves as our standard of time. The wavelength of the laser light can then be calculated by multiplying the inverse of the frequency by the standard value of the speed of light specified in the definition of the meter, $c = 2.99792458 \times 10^8$ m/s. However, it is difficult to measure the frequency of a laser with precision: even for a long-wavelength laser (infrared laser) the frequency is several orders of magnitude higher than that of the cesium clock, and direct comparison is therefore impossible. Scientists at the National Bureau of Standards overcame this obstacle by using several pairs of lasers of different frequencies and cleverly exploiting beat frequencies and harmonic frequencies generated when the light waves struck suitable detectors of light (diodes). With a chain of four pairs of lasers (Figure X.13), they were able to span the range of frequencies from the cesium atomic clock to a methane-stabilized laser, and thereby determine the frequency of the laser light to nine significant figures.

Fig. X.13 Several of the lasers used at the National Bureau of Standards for a precise comparison of frequencies. (Courtesy K. M. Evenson, National Bureau of Standards.)

Laser Spectroscopy

The development of tunable lasers has led to spectacular advances in spectroscopy. **Tunable lasers** operate with molecules of organic dyes, which emit a large number of closely spaced spectral lines; each of the spectral lines contains a spread of frequencies[2] and therefore overlaps to some extent with the adjacent spectral lines. Thus the frequencies available for lasing in an organic dye span a continuous range, and the selective device attached to the laser — such as the prism described in Section X.2 — can be used to tune the laser to any chosen frequency in this range. When the beam from the laser strikes a sample of atoms, it will trigger resonant quantum jumps of the electrons, provided that the frequency of the laser light coincides with the frequency of a spectral line of these atoms. Thus, the laser light will be strongly absorbed if, and only if, its frequency coincides with that of an atomic spectral line; this provides a sensitive and convenient method for the measurement of the frequency of the spectral line. One of the great advantages of this method is that, with a clever arrangement of two laser beams passing through the sample in opposite directions, it is possible to selectively trigger absorption in only those atoms of the sample that have zero velocity. This eliminates the Doppler shift usually present in spectral lines, and therefore permits an extremely precise measurement of their frequencies. Recent measurements of the frequencies of the spectral lines of hydrogen atoms by this technique have led to the best available determination of the Rydberg constant.

Laser Fusion

The most powerful lasers have been developed in an attempt to extract energy from the thermonuclear fusion of heavy hydrogen (deuterium and tritium). At a temperature of 10^8 K, the violent thermal collisions between the nuclei of heavy hydrogen merge them together into helium. This reaction releases a large amount of heat (see Interlude VII for details of this reaction).

In the laser-fusion reactors now under development, a small pellet of deuterium–tritium mixture is dropped into a combustion chamber where the intense beam of a very powerful laser is suddenly focused on it. The sudden influx of energy not only heats the pellet to 10^8 K, but also compresses its inner core, because the sudden vaporization of the outer layers generates high pressure, which causes these layers to explode outward and implode inward. This increases the density of the core by a factor of about

---

[2] All spectral lines contain some small spread of frequencies; in organic dyes, this spread is exceptionally large.

Fig. X.14 Neodymium lasers at Lawrence Livermore Laboratory.

Fig. X.16 Deuterium–tritium pellet placed on the head of a pin.

Fig. X.15 Target chamber at Lawrence Livermore Laboratory. Laser beams, within the tubes, converge toward the center of this chamber.

1000. The high temperature and high density ignite thermonuclear reactions within the pellet. These reactions go on for a short time; then the pellet flies apart. Thus, the fusion energy is released in a miniature explosion — the pellet acts as a miniature H-bomb. The essential difference between the H-bomb and the pellet is the method of ignition; fusion in the H-bomb is ignited by an A-bomb, whereas fusion in the pellet is ignited by the laser beam.

An experimental laser fusion device, called **Shiva**, at the Lawrence Livermore Laboratory employs 20 neodymium lasers focused into a target chamber (Figures X.14 and X.15). The lasers deliver a total power of more than 2×10^{10} kW in a short pulse. The radiant energy strikes a deuterium–tritium pellet about 0.1 mm in diameter. Figure X.16 shows one of these pellets; it is a small glass balloon filled with deuterium and tritium under pressure, and holds an energy equivalent to that of a barrel of oil. Figure X.17 shows the compression of such a pellet. The Shiva device is able to trigger miniature thermonuclear explosions, but the energy output is below the energy input required by the lasers. Livermore has built a new device, called **Nova**, with lasers 10 times more powerful. This device is expected to reach the break-even point — the energy output of the explosions is expected to equal the energy input of the lasers. Of course, it will be necessary to go beyond this point and it will also be neces-

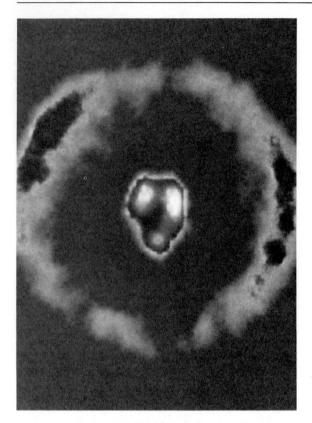

Fig. X.17 X-ray picture of a deuterium–tritium fuel pellet that has been compressed by an intense pulse of light from the Shiva laser at Livermore. The large fuzzy ring shows the initial size of the pellet (compare Figure X.16). The pellet has reached a temperature of 10^7 K and it shines in its own X-ray light. Fusion reactions are occurring within the pellet.

sary to build accessories that convert the heat from the fusion into electric energy.

X.4 HOLOGRAPHY

The large-scale coherence and high intensity of laser light has made it possible to take three-dimensional photographic pictures of objects. The photographic plate, or **hologram,** is prepared by illuminating the object with laser light. Figure X.18 shows the arrangement of light source, object, and photographic plate. In this arrangement, we first split the laser beam into a reference beam and an object beam; after the object beam has been reflected or scattered by the object, we let it interfere with the reference beam at the photographic plate. The plate records the pattern of interference fringes (Figure X.19). After developing the plate, we shine laser light on it. If we then view the plate from the far side, we will see an exact replica of the original object. The image is three dimensional; by moving our eye up, down, or sideways we can bring the top, bottom, or sides of the object into view (Figure X.20). But, of course, the image is virtual — if we try to touch the ghostly thing behind the hologram with a finger, we find that nothing is there.

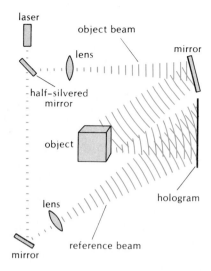

Fig. X.18 Arrangement for making a hologram. All points of the illuminated object act as sources of spherical waves. For the sake of simplicity, only one spherical wave originating at one corner of the object has been shown.

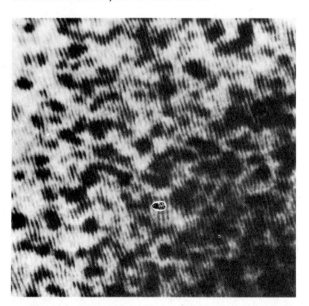

Fig. X.19 A hologram. The dark and the light fringes on this photographic plate record the constructive and the destructive interference of reference and object beams.

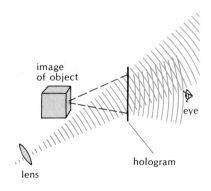

Fig. X.20 Arrangement for viewing a hologram.

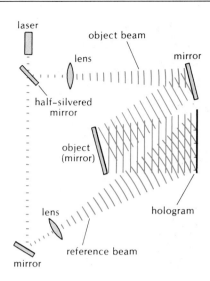

Fig. X.21 Arrangement for making a hologram. The object is a mirror. The marks on the hologram show the points of constructive interference.

The hologram is produced by the interference between the complicated wave fronts emerging from the illuminated object and the plane wave fronts of the reference beam. To understand how this hologram generates a three-dimensional image, suppose that instead of a complicated object we have a very simple object: a plane mirror (Figure X.21). The wave fronts emerging from the illuminated mirror are then simply plane waves. At the photographic plate, these plane waves interfere with the plane waves of the reference beam. Obviously, this interference will give a fringe pattern consisting of parallel bright and dark bands. If the angle of incidence of the object beam is α, then the distance d between one dark fringe and the next is related to the wavelength λ by (Figure X.22)

$$d \sin \alpha = \lambda \qquad (1)$$

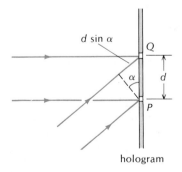

Fig. X.22 The object beam (horizontal) and the reference beam (at an angle α) reaching the photographic plate. If there is constructive interference at P, then there is also constructive interference at Q, provided the extra distance $d \sin \alpha$ equals one wavelength.

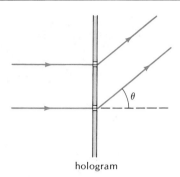

Fig. X.23 Diffracted beam (at an angle θ) emerging from the illuminated hologram.

The developed photographic plate will then look like a grating with a distance d between the slits. If we illuminate this grating with laser light, constructive interference from these slits will yield several diffracted beams of maximum intensity (Figure X.23). The angle of the first-order beam will satisfy the condition [see Eq. (39.21)]

$$d \sin \theta = \lambda \qquad (2)$$

Comparing Eqs. (1) and (2), we recognize that the angle of emergence of the diffracted beam (produced by the illuminated hologram) coincides with the angle of incidence of the original beam (produced by the illuminated mirror). Thus, the hologram *reconstructs* the wave fronts — it generates a light wave that has exactly the same characteristics as the light wave emitted by the object that was photographed. If we place our eyes beyond the hologram (Figure X.20), we will see exactly what we would see if instead of the illuminated hologram we had the illuminated object in front of our eyes.

It can be demonstrated that the same reconstruction of wave fronts obtains if the illuminated object has some complicated shape, so the emitted wave fronts also have some complicated shape. The illuminated hologram will then generate a wave with exactly the same complicated shape, and our eyes cannot tell the difference between this reconstructed wave and the wave emitted by the object itself — when we view the hologram, we believe we see the object with all its three-dimensional features (Figure X.24).

A remarkable aspect of this imaging process is that each piece of a hologram contains enough information for the reconstruction. We can cut a hologram into two or more pieces, and each piece will retain the capability of giving an entire view of the three-dimensional image. However, small pieces provide a view from only a narrow range of directions, like a view through a small window.

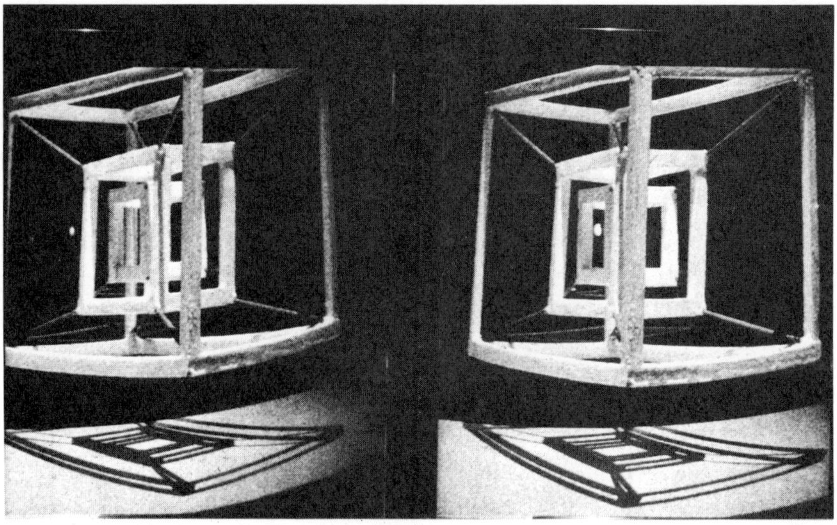

Fig. X.24 These two photos show the holographic image of a piece of modern sculpture photographed from two different directions, giving different points of view. (If you can de-couple your eyes, and view the left image with the left eye and the right image with the right eye, you will be able to perceive these photos as three dimensional.)

Note that, in principle, we could make a hologram with light from an ordinary monochromatic light source (for example, a mercury lamp) instead of light from a laser. As in the investigation of interference and diffraction by thin slits, such a light source must be placed either at a very large distance from the object or else covered with a shield with a single pinhole (see Section 39.4). But this would reduce the intensity of the light and would require very long and very impractical exposure times. Even with light from a laser, the exposure times for holography are much longer than for photography. Because of this, it is usually necessary to take special precautions to hold the object stationary during exposure.

Holographic images make spectacular displays for demonstrations, but they also have many practical applications. They can be used for sensitive measurements of small deformations and displacements. For example, Figure X.25 is a holographic image of the vibrating body of a viola. The interference fringes indicate contours of equal height and therefore reveal the pattern of vibration of the viola. The exposure time in Figure X.25 was much longer than the period of vibration of the viola; since a vibrating body spends most of its time at the turning points of the motion (where the velocity is zero), a time exposure is then essentially equivalent to a double exposure — one exposure at each turning point. When we illuminate such a doubly exposed hologram, it will reconstruct simultaneously the wave fronts emitted by the body at the two separate instants of exposure. These two wave fronts will interfere constructively or destructively, depending on how far the surface of the body has moved between one exposure and the next; thus, in the holographic image, the displacement of the surface becomes visible by interference fringes. Scientists hope that the in-

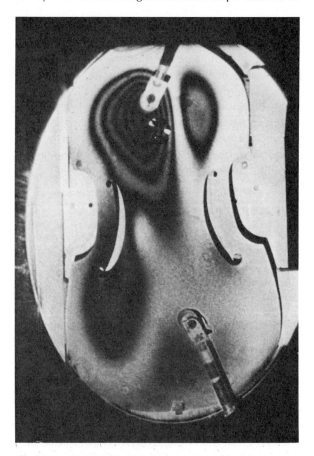

Fig. X.25 Image of the vibrating top plate of a viola, generated by a time-exposure hologram.

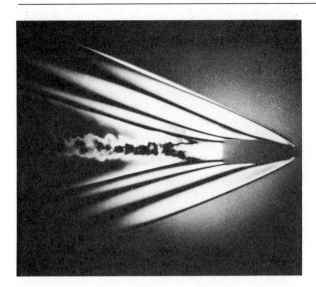

Fig. X.26 Image of shock waves around a bullet in flight, generated by a double-exposure hologram.

vestigation of the pattern of vibration of the viola will lead to improvements in its loudness, dynamic range, and playing ease.

A double-exposure hologram can also make visible small alterations of the density of air. Figure X.26 shows the shock waves surrounding a bullet speeding through air. The hologram was first exposed before the bullet entered the field of view, and then again when the bullet arrived. The compression and rarefaction of the air near the shock fronts alters the index of refraction and hence changes the phase of light between one exposure and the next. When the reconstructed wave fronts from this hologram interfere, the variations of index of refraction show up as bright and dark zones. The exposures were made with high-intensity laser pulses of very short duration so as to "stop" the action at one instant.

Another very promising application of holography is information storage. Obviously, we can store the information of, say, a printed page by taking a holographic photograph of the page. The hologram can store the information at high density, because it can be made as small or smaller than an ordinary microfilm photograph; and it has the further advantage that specks of dust or small defects in the film will not block out the words on the page (remember that each part of the hologram can reconstruct the entire image). In principle, a three-dimensional hologram would permit the most efficient storage of information. In a three-dimensional medium, such as a thick photographic emulsion or a light-sensitive crystal, many different holograms can be stored successively by means of multiple exposures with light waves incident at different angles. To recover any one of the holographic images, it is necessary only to illuminate the medium with a light wave incident at the same angle as the original wave. About 1000 different holograms could be stored in a layer about 1 mm thick. With the right technology, it should be possible to store an entire 3D movie in a cube of transparent material as small as a cube of sugar.

Holography was originally invented by D. Gabor[3] as a means for increasing the power of microscopes. Gabor intended to photograph holographically the image produced by a microscope with light of short wavelength and afterward illuminate the hologram with light of long wavelength. The reconstructed image would then be larger than the original image in the ratio of the wavelengths. This technique of holographic microscopy was employed to make the photograph of the neon atom shown in Figure 22.2. The atom was first illuminated with electron waves of very short wavelength,[4] which generated a hologram on a photographic plate. The hologram was then illuminated with visible light, of a wavelength much larger than that of the electron waves. This yielded the enormous magnification in Figure 22.2.

Further Reading

Lasers and Holography by W. E. Kock (Doubleday, Garden City, 1969) is an elementary introduction to coherent light, its generation by lasers, and its application in holography. *The Amazing Lazer* by B. Bova (Westminster, Philadelphia, 1971) is another elementary, and very readable, introduction.

Engineering Applications of Lasers and Holography by W. E. Kock (Plenum, New York, 1975) and *Laser* by J. Hecht and D. Teresi (Ticknor and Fields, New Haven, 1982) give broad surveys of many practical applications of lasers.

The May 1985 issue of *Physics Today* is devoted to lasers and their applications.

The following magazine articles give more information:

"Laser Light," A. L. Schawlow, *Scientific American*, September 1968

"Applications of Laser Light," D. R. Herriott, *Scientific American*, September 1968

"The Lunar Laser Reflector," J. E. Faller and E. J. Wampler, *Scientific American*, March 1970

"Progress in Holography," E. N. Leith and J. Upatnieks, *Physics Today*, March 1972

"Metal-Vapor Lasers," W. T. Silfvast, *Scientific American*, February 1973

"Ultrafast Phenomena in Liquids and Solids," R. R. Alfano and S. L. Shapiro, *Scientific American*, June 1973

---

[3] **Dennis Gabor,** 1900–1979, British physicist. He received the Nobel Prize in 1971 for the discovery of the principles of holography.

[4] As was pointed out in Chapter 43, on the atomic scale electrons behave as waves; their wavelengths are usually much shorter than that of light. The operation of electron microscopes takes advantage of these electron waves.

"Laser-induced Thermonuclear Fusion," J. Nuckolls, J. Emmet, and L. Wood, *Physics Today*, August 1973

"Fusion Power by Laser Implosion," J. L. Emmet, J. Nuckolls, and L. Wood, *Scientific American*, June 1974

"X-Ray Lasers," G. Chapline and L. Wood, *Physics Today*, June 1975

"Processing Materials with Lasers," E. M. Breinan, B. H. Kear, and C. M. Banas, *Physics Today*, November 1976

"Laser Separation of Isotopes," R. N. Zare, *Scientific American*, February 1977

"Light-Wave Communications," W. S. Boyle, *Scientific American*, August 1977

"Cosmic Masers," D. F. Dickinson, *Scientific American*, June 1978

"Laser Annealing of Silicon," J. M. Poate and W. L. Brown, *Physics Today*, June 1978

"Laser Fusion," C. M. Stickley, *Physics Today*, May 1978

"Guided-Wave Optics," A. Yariv, *Scientific American*, January 1979

"Laser Chemistry," A. M. Ronn, *Scientific American*, May 1979

"Counting the Atoms," G. S. Hurst, M. G. Pane, S. D. Kramer, and C. H. Chen, *Physics Today*, September 1980

"White-Light Holograms," E. N. Leigh, *Scientific American*, October 1980

"Laser Weapons," K. Tsipis, *Scientific American*, December 1981

"Laser Applications in Manufacturing," A. V. La Rocca, *Scientific American*, March 1982

"The Feasibility of Inertial-Confinement Fusion," J. N. Nuckolls, *Physics Today*, September 1982.

"Lasers and Physics: A Pretty Good Hint," A. L. Schawlow, *Physics Today*, December 1982

"Optical Phase Conjugation," V. V. Shkunov and B. Ya. Zel'dovich, *Scientific American*, December 1985.

"Applications of Optical Phase Conjugation," D. M. Pepper, *Scientific American*, January 1986.

"Progress in Laser Fusion," R. S. Craxton, R. L. McCrory, and J. M. Soures, *Scientific American*, August 1986.

The Resource Letter L-1 in the *American Journal of Physics*, October 1981, gives a comprehensive list of references on lasers.

Questions

1. Consider two overlapping waves with equal amplitudes and with a random phase difference. Explain why the superposition of these waves has an average intensity equal to twice the average intensity of each wave.

2. Consider two overlapping waves with equal amplitudes and with equal phases. Explain why the superposition of these waves has an intensity four times as large as the intensity of each individual wave. Does this violate energy conservation?

3. Suppose that a light source consists of 10^{16} atoms. If these atoms radiate coherently, by what factor is the average intensity larger than if they radiate incoherently?

4. Can you guess what the acronym *maser* stands for?

5. Why can the flash of light that pumps electrons upward into excited orbits also pump them downward?

6. It is easiest to attain a population inversion by optical pumping if the lower excited orbit (see orbit 2, Figure X.3) is a **metastable** orbit, that is, an orbit in which the electron tends to remain for a long time without performing a spontaneous downward jump. Why does this help?

7. Explain how a grating can be used as a color-selective device in a laser, instead of the prism shown in Figure X.6.

8. The beam of a laser can be made visible by blowing chalk dust or smoke in its path. Can this also be done with a beam of ordinary, incoherent light?

9. For which of the applications of laser light described in Section X.3 is the coherence essential? For which is the pure color essential?

10. Because of the pull of gravity, a tightly stretched string sags downward slightly, whereas a laser beam "sags" upward (however, on the Earth the amount of this "sag" is too small to be measurable). How does this difference arise?

11. The U.S. Department of Defense is trying to develop a laser gun. What might be the advantages and disadvantages of such a gun over a conventional gun? Explain how a cloud would stop the beam of a laser gun, even though it cannot stop the bullet from a conventional gun.

12. One obstacle to shooting intense laser beams through the atmosphere at a distant target is that the beam heats the air, changing its index of refraction. This causes the beam to spread out laterally, an effect called **thermal blooming.** Explain.

13. Could an intense laser beam be used to propel a rocket? For this purpose would a laser beam be any better than a light beam?

14. If the individual wave trains of light emitted by individual atoms have a length of a few meters in space, how long are they in time?

15. The wavelength of the light emitted by a laser can be determined to nine significant figures. If you use the light from this laser in conjunction with an interferometer to measure a displacement of roughly 1 m, what precision can you expect? Assume that the interferometer introduces no additional uncertainties.

16. The frequency of a methane-stabilized laser is 8.84×10^{13} Hz. By what factor is this larger than the frequency of the cesium atomic clock?

17. Suppose that a hologram has been prepared with green light. If you illuminate this hologram simultaneously with green light and with red light, what do you see?

18. Why is the image produced by a very small piece of a hologram somewhat fuzzy? (Hint: A hologram can be regarded as a grating. According to Section 39.5, the angular resolution of a grating increases in direct proportion with the number of slits, that is, with the size of the grating.)

19. The automatic checkout system found in large supermar-

kets sweeps many beams of laser light over the bottom and sides of a package, so that at least one of the beams is reflected off the product-code stripes with sufficient intensity to register in the light detector. To generate these many beams of light, the system requires a complicated array of mirrors; but instead of actual mirrors, the system uses a hologram. Explain how you could make a hologram that simulates a complicated array of mirrors.

CHAPTER 44

Quantum Structure of Atoms, Molecules, and Solids

The physical and chemical properties of atoms and of molecules depend on the quantum behavior of their electrons. These electrons pervade most of the volume of the atom, and their arrangement in different orbits determines the size of the atom, the ionization energy, the spectrum of light emitted and absorbed, the chemical bonds the atom forms with other atoms, and so on. Likewise, the physical properties of a solid — such as diamond, silicon, silver, copper — depend on the quantum behavior of the electrons in the solid. The spacing of the crystal lattice of the solid, the electric and thermal conductivity, the magnetic properties, and the mechanical properties all hinge on the arrangement of the electrons. For instance, the arrangement of the electrons determines in a very direct way the ability of the solid to conduct an electric current and to conduct heat. As we will see in Section 44.5, the differences among conductors, semiconductors, and insulators hinge on what stationary states are available in the solid and which of these are occupied by electrons.

The arrangement of the electrons in the stationary states of an atom or a solid is subject to an important restriction: no more than two electrons can occupy the same orbital state. This is called the Exclusion Principle. In Section 44.1, we will become acquainted with the spin, or intrinsic angular momentum, of the electron, and in the next section we will see that the Exclusion Principle is intimately linked to the spin. In later sections we will examine the implications of the Exclusion Principle for the arrangement of the electrons in atoms and in solids.

44.1 Principal, Orbital, and Magnetic Quantum Numbers; Spin

In the preceding chapter, we saw that Bohr's theory characterizes the stationary states of the hydrogen atom and their energies by a single quantum number n. However, this simple version of Bohr's theory deals only with circular orbits. We know from the study of planetary orbits (Chapter 9) that the general orbit of a particle moving under the influence of an inverse-square force is an ellipse with focus at the center of attraction. A couple of years after Bohr formulated the quantization condition for circular orbits, Arnold Sommerfeld proposed quantization conditions for elliptical orbits. Sommerfeld's extension of Bohr's theory takes into account that, compared with motion in a circular orbit, motion in an elliptical orbit has an extra variable quantity. For an electron moving in a circular orbit, the radial distance from the nucleus is fixed, and only the angular position of the electron varies; but for an electron moving in an elliptical orbit, both the radial distance and the angular position vary simultaneously. Sommerfeld therefore proposed that besides the Bohr quantization condition, which is associated with the angular motion, there should be an additional quantization condition associated with the radial motion. He established that for an elliptical orbit these two quantization conditions lead not only to a quantization of the size of the orbit, as in Bohr's theory, but also to a quantization of the shape of the orbit.

Figure 44.1 shows the permitted elliptical orbits. These orbits are characterized by two quantum numbers, n and l. The first of these is called the **principal quantum number** and the second the **orbital quantum number**. The principal quantum number n characterizes the size of the orbit (more precisely, the length of the semimajor axis of the ellipse) and determines the energy of the orbit, as specified by Eq. (43.23). The orbital quantum number l characterizes the elongation of the ellipse, or its eccentricity. For a given value of the principal quantum number n, the orbital quantum number l can assume any integer value from 1 to n. In Figure 44.1, the orbits have been labeled with their appropriate values of n and l. Note that the circular orbit always has $l = n$, that is, the two quantum numbers coincide. For $n = 1$, there is only a circular orbit, and no elliptical orbit; for $n = 2$, there is a circular orbit and one elliptical orbit; for $n = 3$, there is a circular orbit and two different elliptical orbits; and so on.

Furthermore, Sommerfeld recognized that the possible orbits of an electron in an atom need not all be in the same plane — the orbits can

Arnold Sommerfeld, *1868–1951, German theoretical physicist, professor at Munich. Sommerfeld was a gifted teacher, and he attracted many brilliant students, among them Pauli and Heisenberg.*

Principal and orbital quantum numbers, n and l

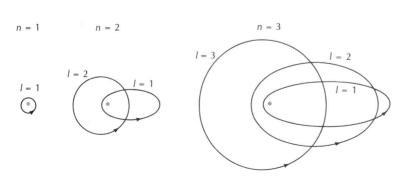

Fig. 44.1 Elliptical orbits in the hydrogen atom for $n = 1$, 2, and 3. The semimajor axes of the two orbits with $n = 2$ are the same, and the semimajor axes of the three orbits with $n = 3$ are the same.

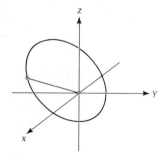

Fig. 44.2 Elliptical orbit tilted relative to the z axis. As the electron moves around this orbit, the radial line from the nucleus to the electron swings up and down, and the angle between this line and the z axis varies periodically.

Magnetic quantum number, m

Spin quantum number, m_s

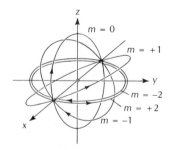

Fig. 44.3 Possible orientations of a circular orbit with $l = 2$, according to Sommerfeld's theory.

Spin

be tilted at different angles. For instance, Figure 44.2 shows an elliptical orbit tilted relative to the z axis (drawn vertically in the figure). For an electron moving in this orbit, the radial distance, the angle measured in the plane of the orbit, and the angle measured relative to the z axis all vary simultaneously. Accordingly, Sommerfeld proposed one more quantization condition associated with the angular up-and-down motion. He established that this additional quantization condition leads to a quantization of the orientation of the orbit in space: the plane of the orbit can assume only one of several discrete orientations. This is called **space quantization.** For a circular or elliptical orbit with orbital quantum number l, there are $2l + 1$ possible orientations, which can be characterized by an additional quantum number m with integer values from $-l$ to l. For instance, if $l = 2$, then the orbit has $2 \times 2 + 1 = 5$ possible orientations characterized by $m = -2, -1, 0, +1$, and $+2$. Figure 44.3 illustrates this case. The quantum number m is called the **magnetic quantum number.** (The reason for this name is that the quantum number m acquires a special significance when the atom is placed in a magnetic field, say, a magnetic field in the z direction; the energy of the orbit then depends not only on the size of the orbit and on the principal quantum number n but also on the orientation of the orbit and, therefore, on the "magnetic" quantum number m.)

Although the three quantum numbers n, l, and m were originally introduced on the basis of semiclassical considerations, the later, more rigorous analysis based on wave mechanics and the Schrödinger equation confirmed that the stationary states are, indeed, characterized by these three quantum numbers. However, wave mechanics demands a slight modification in the scheme of quantum numbers described above: for a given principal quantum number n, the orbital quantum number l ranges from 0 to $n - 1$ (rather than from 1 to n). This implies that the orbital angular momentum of the ground state of hydrogen is zero (rather than $1\hbar$).

In addition to these quantum numbers n, l, and m, one more quantum number is needed for the complete characterization of the stationary states of an electron in the hydrogen atom. This is the **spin quantum number** m_s that characterizes the spin, or intrinsic angular momentum, of the electron. The quantum number m_s was originally proposed by Wolfgang Pauli in an attempt to describe the "hyperfine" structure of the spectral lines. When the spectral lines of hydrogen and of other atoms are examined with a spectroscope of high resolving power, they are often found to consist of pairs, or doublets, of very closely spaced lines. This splitting of the spectral lines implies a corresponding splitting of the atomic energy levels: one or both of the energy levels involved in the transition must be a closely spaced doublet of energy levels. Consequently, besides the quantum numbers n, l, and m, there must be another quantum number that distinguishes between the two energy levels in the doublet. Pauli assigned the values $m_s = +\frac{1}{2}$ and $m_s = -\frac{1}{2}$ to the two energy levels in the doublet, but he offered no explanation of the physical significance of this new quantum number.

Shortly thereafter, Samuel Goudsmit and George Uhlenbeck suggested that the electron has an intrinsic spin angular momentum and that the two values of m_s correspond to different orientations of the axis of spin. They imagined the electron as a small ball of charge spinning about its axis, like the Earth spinning about its axis, with an angular momentum of magnitude $\frac{1}{2}\hbar$, that is, an angular-momentum quantum number $\frac{1}{2}$. If we use the same rule for the possible orientations of the axis of spin as for the possible orientations of the orbit, we find that there are $2 \times \frac{1}{2} + 1 = 2$ possible orientations for the spin. One of the orientations is characterized by $m_s = +\frac{1}{2}$, and the other by $m_s = -\frac{1}{2}$; these two possible spin orientations are called spin up and spin down, respectively.

Goudsmit and Uhlenbeck's simple picture of the spin as due to a rotation of the electron about its axis proved untenable. Instead, modern wave mechanics tell us that the spin is an angular momentum generated by a circulating energy flow in the electron wave (in contrast, the orbital angular momentum is due to the translational motion of the electron wave). Nevertheless, the simple picture of the electron as a rotating ball of charge can serve as a convenient crutch for our imagination. According to this simple picture, we expect that the electron has a magnetic moment, since a piece of charge rotating about an axis amounts to a current loop (see Figure 44.4). And indeed, experiments show that the electron has an intrinsic magnetic moment of magnitude

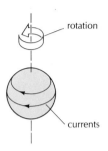

Fig. 44.4 Simple picture of the electron as a rotating ball of charge. The motion of the negative charge in, say, counterclockwise circles is equivalent to a clockwise current of positive charge.

$$\mu_{\text{spin}} = \frac{e\hbar}{2m_e} = 9.27 \times 10^{-24} \text{ A} \cdot \text{m}^2 \qquad (1)$$

The direction of this magnetic moment is opposite to the direction of the spin, as expected for a rotating negative charge distribution. The energy difference between the two levels in an energy doublet in an atom arises from this intrinsic **magnetic moment,** which interacts with the magnetic field generated by the motion of the nuclear charge seen in the reference frame of the electron.

Magnetic moment

The intrinsic magnetic moment of the electron plays a role in all phenomena involving the behavior of atoms placed in external magnetic fields. In Chapter 33, we already saw how the intrinsic magnetic moment of the electron accounts for the magnetic properties of ferromagnetic materials. Furthermore, the intrinsic magnetic moment of the electron makes an important contribution to the splitting of the spectral lines that occurs when an atom is immersed in an external magnetic field. This splitting is called the **Zeeman effect.** The intrinsic magnetic moment and the orbital magnetic moment (see Section 33.1) both interact with the external magnetic field, and thereby change the energies of the stationary states of the atom by an amount depending on the orientation of the spin and of the orbital angular momentum relative to the magnetic field. This results in a splitting of the energy levels into closely spaced multiplets, and in a corresponding splitting of the spectral lines.

Zeeman effect

EXAMPLE 1. Calculate the splitting of the ground-state energy level of a hydrogen atom placed in an external magnetic field of 0.20 T. Estimate the Zeeman splitting of the first spectral line of the Lyman series ($\lambda = 1216$ Å).

SOLUTION: Since, according to wave mechanics, the orbital angular momentum of the ground state of hydrogen is zero, the only angular momentum and the only magnetic moment are those contributed by the spin of the electron. This spin is either parallel or antiparallel ("up" or "down") relative to the direction of the external magnetic field. The potential energy of the magnetic moment in this external magnetic field is then

$$U = \pm \mu_{\text{spin}} B \qquad (2)$$

$$= \pm 9.27 \times 10^{-24} \text{ A m}^2 \times 0.20 \text{ T} = \pm 1.85 \times 10^{-24} \text{ J}$$

where the plus sign is for spin antiparallel to the magnetic field, and the minus sign for spin parallel. The energy difference, or the energy splitting, between these two configurations is then $\Delta E = 2 \times 1.85 \times 10^{-24}$ J $= 3.7 \times 10^{-24}$ J.

The Zeeman effect in the spectral line arises from the energy splittings of both the upper and the lower levels in the transition. Both of these contribute comparable amounts to the Zeeman effect. For a simple estimate, let us ignore the energy splitting of the upper level. Since $E = h\nu = hc/\lambda$, we have approximately

$$\Delta E = -\frac{hc}{\lambda^2} \Delta \lambda \tag{3}$$

or

$$\Delta \lambda = -\frac{\lambda^2}{hc} \Delta E$$

$$= \frac{\pm (1216 \times 10^{-10} \text{ m})^2}{6.63 \times 10^{-34} \text{ J} \cdot \text{s} \times 3.0 \times 10^8 \text{ m/s}} \times 3.7 \times 10^{-24} \text{ J} \tag{4}$$

$$\cong \pm 3 \times 10^{-13} \text{ m} \cong \pm 0.003 \text{ Å}$$

Thus, according to this estimate, the spectral line splits into two spectral lines, differing in wavelength by about 0.003 Å.

COMMENTS AND SUGGESTIONS: A detailed calculation, taking into account the energy splitting of the upper level, shows that the spectral line actually splits into ten spectral lines, but the Zeeman splittings among these are in rough agreement with our simple estimate.

Besides the electron, many other fundamental particles have spins. For example, the proton and the neutron have spins of magnitude $\frac{1}{2}\hbar$, the photon has spin $\hbar$, and so on (a complete tabulation of all the fundamental particles and their spins will be found in Chapter 46).

The quantum numbers n, l, m, and m_s provide a complete characterization of the stationary states of the hydrogen atom. The energy of a stationary state depends mainly on the principal quantum number n; as given by Eq. (43.23), the energy is

$$E = -\frac{13.6 \text{ eV}}{n^2} \tag{5}$$

However, the energy also depends slightly on the orbital quantum number l and on the orientation of the spin relative to the plane of the orbit. This means that Eq. (5) for the energies of the stationary states of the hydrogen atom is not quite accurate. However, the deviations from Eq. (5) are very small, and we will continue to ignore them.

Table 44.1 lists the quantum numbers for the stationary states of the hydrogen atom.[1] These quantum numbers can also be used to characterize the stationary states of atoms other than hydrogen, but the energies of other atoms are not given by the simple formula (5).

---

[1] Strictly, these quantum numbers are appropriate for a hydrogen atom placed in a magnetic field. For a hydrogen atom by itself, the appropriate quantum numbers are n and some intricate combinations of l, m, and m_s. For the sake of simplicity, we will ignore this complication.

Table 44.1 Quantum Numbers of Electronic States

Quantum number	Symbol	Values
Principal	n	$1, 2, 3, \ldots$
Orbital	l	$0, 1, 2, \ldots, n-1$
Magnetic	m	$-l, -l+1, -l+2, \ldots, l-2, l-1, l$
Spin	m_s	$-\tfrac{1}{2}, +\tfrac{1}{2}$

44.2 The Exclusion Principle and the Structure of Atoms

The paramount question in atomic structure is the determination of the detailed arrangement of the electrons in their orbits around the nucleus. The electron arrangement, or **configuration,** determines all the physical and chemical properties of the atom — if the electron configuration is known, all the properties of the atom can be deduced by theoretical considerations. For instance, the observed similarities of chemical properties among select groups of elements must be due to similarities in their electron configurations. Chemists list similar elements in columns in the Periodic Table of elements (see Appendix 9); thus, the noble gases helium, neon, argon, krypton, etc. are listed in one column; the halogens fluorine, chlorine, bromine, etc. are listed in another; and the alkalis hydrogen, lithium, sodium, potassium, etc., in yet another. This pattern of elements displayed in the Periodic Table emerges from the study of electron configurations.

Periodic Table

For the case of the hydrogen atom, the determination of the electron configuration is trivial: the single electron of this atom is in one or another of the stationary states characterized by the quantum numbers n, l, m, and m_s. If the atom is in the ground state, the values of the quantum numbers of the electron configuration are $n = 1$, $l = 0$, $m = 0$, and $m_s = \pm\tfrac{1}{2}$; thus, everything is fixed, except the direction of the spin, which can be up or down.

But for atoms with several electrons, the determination of the electron configuration is not so trivial. It might be tempting to suppose that the ground state of the atom (the state of least energy) is attained by placing all the electrons in the lowest stationary state, with $n = 1$, $l = 0$, $m = 0$, $m_s = \pm\tfrac{1}{2}$, as for the hydrogen atom. But this would imply that all the atoms ought to have a spectrum similar to that of hydrogen, and it would also imply that atoms of large Z ought to be very small, since the Bohr radius for an atom of nuclear charge Ze is a_0/Z [if the nuclear charge is Ze instead of e, then in the denominator of the formula (43.18) for the Bohr radius of hydrogen, we must replace *one* of the factors of e by Ze]. These conclusions are in stark conflict with the observed properties of atoms: the spectra of most atoms are quite different, and the sizes of atoms of large Z — such as lead or bismuth — are considerably larger than hydrogen, which is one of the smallest of all atoms.

Wolfgang Pauli, *1900–1958, Austrian and later Swiss theoretical physicist, professor at Zurich. For his discovery of the Exclusion Principle, he was awarded the Nobel Prize in 1945. Pauli made an important contribution to the theory of beta decay by proposing that the emission of the beta particle is always accompanied by the emission of a neutrino (see Chapter 45).*

The rule that governs the configuration of the electrons in an atom is the **Exclusion Principle,** which was discovered by Pauli:

> *Each stationary state of quantum numbers n, l, m, m_s can be occupied by no more than one electron.*

Exclusion Principle

Since for each orbital state of quantum numbers n, l, m, there are two possible spin states ($m_s = \pm\frac{1}{2}$), we can also rephrase the Exclusion Principle as follows: Each orbital state of quantum numbers n, l, m can be occupied by no more than two electrons.

Pauli originally proposed this principle as an empirical rule, based on the observed features of atomic spectra. It was later established that the Exclusion Principle is intimately linked to the value of the spin of the electron; the Exclusion Principle can be shown to be a necessary consequence of the quantum theory of particles of half-integer spin. Thus, protons and neutrons also obey the Exclusion Principle, a fact of great importance for the configuration of these particles in the interior of the nucleus (see next chapter). In contrast, particles of integer spin, such as photons, do not obey the Exclusion Principle. There is no limit to the number of such particles that can be packed into a given stationary state, for instance, one of the standing-wave states in a cavity filled with blackbody radiation.

For our investigation of the electron configuration of atoms, we will find it convenient to prepare a list of the quantum numbers of all the available states, in order of increasing energy. The states of lowest energy have $n = 1$; there are two such states.

Quantum numbers of atomic states

States with $n = 1$ (K shell):

$$n = 1 \quad l = 0 \quad m = 0 \quad m_s = -\tfrac{1}{2}$$
$$n = 1 \quad l = 0 \quad m = 0 \quad m_s = +\tfrac{1}{2}$$

Next, consider $n = 2$; there are eight available states.

States with $n = 2$ (L shell):

$$n = 2 \quad l = 0 \quad m = 0 \quad m_s = -\tfrac{1}{2}$$
$$n = 2 \quad l = 0 \quad m = 0 \quad m_s = +\tfrac{1}{2}$$
$$n = 2 \quad l = 1 \quad m = -1 \quad m_s = -\tfrac{1}{2}$$
$$n = 2 \quad l = 1 \quad m = -1 \quad m_s = +\tfrac{1}{2}$$
$$n = 2 \quad l = 1 \quad m = 0 \quad m_s = -\tfrac{1}{2}$$
$$n = 2 \quad l = 1 \quad m = 0 \quad m_s = +\tfrac{1}{2}$$
$$n = 2 \quad l = 1 \quad m = +1 \quad m_s = -\tfrac{1}{2}$$
$$n = 2 \quad l = 1 \quad m = +1 \quad m_s = +\tfrac{1}{2}$$

Shells

Likewise, for $n = 3$, there are eighteen available states, and so on. The groups of states of a given value of n are called **shells,** and they are conventionally labeled with the letters K, L, M, etc. Thus, the two states with $n = 0$ form the K shell; the eight states with $n = 1$ form the L shell; the eighteen states with $n = 3$ form the M shell, etc.

According to the Exclusion Principle, each of the states listed above can accommodate one, and only one, electron. Thus, if an atom with Z electrons is in its ground state, the electrons will occupy the first Z of the states in the above list. We can therefore build up the configurations for all the atoms in the Periodic Table of elements by beginning with hydrogen and adding electrons one by one, sequentially filling the states in our list.

The second element in the Periodic Table is helium, which has two electrons. To obtain its electron configuration, we must add one electron to the hydrogen configuration; since the single electron of hydrogen occupies one of the states of the K shell, we can place the second electron in the other available state in the K shell. Helium therefore has a full K shell.

The third element is lithium, with three electrons. When we add one electron to the helium configuration, we must place this third electron in the L shell, in the state with quantum numbers $n = 2$, $l = 0$, $m = 0$, $m_s = -\frac{1}{2}$.

The next element is beryllium, with four electrons. Thus, we must add one more electron in the L, shell, in the state with quantum numbers $n = 2$, $l = 0$, $m = 0$, $m_s = +\frac{1}{2}$.

We can continue in this way, filling up the states in our list one by one. With the tenth element, neon, we will have filled the L shell. And with the eleventh element, sodium, we must place one electron in the M shell, and so on.

This simple procedure for building up the electron configurations of the atoms provides us with an immediate explanation of the similarities of the elements in columns of the Periodic Table. For instance, the similarity in the chemical behavior and the similarity in the spectra of helium and neon can be traced to a similarity of their electron configurations: both these atoms have full shells of electrons. Likewise, the chemical and spectroscopic similarities of hydrogen, lithium, and sodium can be traced to a similarity of their electron configurations: they all have a single electron outside of a full shell of electrons. This single, outer electron tends to come off the atom fairly easily, and in chemical reactions these atoms all tend to lose an electron. In contrast, fluorine and chlorine are one electron short of a full shell, and in chemical reactions they tend to capture an electron to complete their shell.

Thus, the Exclusion Principle in conjunction with a simple counting procedure for the available stationary states is sufficient to account for the broad, qualitative features of the Periodic Table of elements. Detailed calculations, based on wave mechanics, provide quantitative theoretical results for ionization energies, spectral lines, atomic sizes, and so on, in agreement with the observed atomic properties.

44.3 Energy Levels in Molecules

The chemical bonds that bind two or more atoms together in a molecule, such as O_2 or HCl, arise from a rearrangement of outer electrons, or valence electrons, of the atoms. In some molecules — such as O_2 — the outer electrons are shared between the atoms, which produces an attractive force (covalent bond). In some other molecules — such as HCl — one atom loses an electron to the other atom, and the atom with the missing electron is then electrically attracted by the atom with the extra electron (ionic bond). These two mechanisms account for almost all cases of bonding in diatomic molecules. Other mechanisms produce weaker bonds (hydrogen bond, van der Waals bond, metallic bond), which are of importance in some of the more complicated molecules.

Interatomic bonds

The chemical bonds are elastic — they behave rather like springs tying the atoms together. The spring holds the atoms at an average equilibrium distance, but permits the atoms to oscillate back and forth about this average distance. This means that the energy of the molecule is the sum of the electronic energy of the atoms and the vibrational energy of the motion of the atoms in relation to each other. Let

Fig. 44.5 An oscillating diatomic molecule can be represented as two pointlike masses on the ends of a massless spring.

us now focus on the vibrational energy of a molecule and examine its quantization.

The mass of the atom is concentrated in its center, in the nucleus, which is much smaller than the interatomic distances in a molecule. We can therefore schematically represent a molecule — for instance, a diatomic molecule — as a system of pointlike masses connected by massless springs (see Figure 44.5). The atoms oscillate in unison relative to the center of mass, which we can regard as fixed. Thus, the system is an oscillator, and the energy of this oscillator is subject to Planck's quantization condition, Eq. (42.4). If the frequency of oscillation is ν, the energy must be

Vibrational energy of molecule

$$E = nh\nu \qquad n = 0, 1, 2, \ldots \tag{6}$$

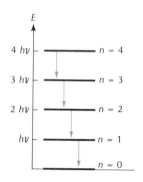

Fig. 44.6 Energy-level diagram for the oscillating molecule. The arrows indicate the possible transitions.

The corresponding energy-level diagram is shown in Figure 44.6. The molecule will emit a photon if it makes a transition from an upper level to a lower level. The vibrational transitions are restricted by a **selection rule**: the transition must proceed from one level to the next, that is, transitions spanning two or more levels in one jump are forbidden. This selection rule can be deduced from wave mechanics. The arrows in Figure 44.6 indicate the permitted transitions. The frequencies of the radiation emitted during all these transitions are therefore the same,

$$\nu_{\text{light}} = \frac{\Delta E}{h} = \frac{h\nu}{h} = \nu \tag{7}$$

which is also the same as the frequency of vibration of the molecule. Typically, the frequencies of vibration of molecules are of the order of 10^{13} Hz, and the wavelengths of the emitted radiation lie in the infrared.

Besides the vibrational motion, the molecule can also perform a rotational motion. For the purposes of this rotational motion, we can regard the molecule as two pointlike masses linked by a massless rigid rod, that is, a dumbbell (see Figure 44.7). If the moment of inertia of the dumbbell about a perpendicular axis through the center of mass is I, then the kinetic energy of rotation is

$$E = \tfrac{1}{2} I \omega^2 \tag{8}$$

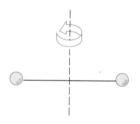

Fig. 44.7 A rotating molecule can be regarded as two pointlike masses linked by a massless rigid rod.

where ω is the angular frequency of the rotation. Let us express this in terms of the angular momentum. Since the angular momentum is $L = I\omega$, the angular frequency is $\omega = L/I$ and Eq. (8) becomes

$$E = \frac{L^2}{2I} \tag{9}$$

The angular momentum is quantized in the usual way, $L = n\hbar$; this immediately implies that the energy of the rotational motion is quantized,

Rotational energy of molecule

$$E = \frac{n^2 \hbar^2}{2I} \qquad n = 0, 1, 2, \ldots \tag{10}$$

Figure 44.8 displays the energy-level diagram for the rotational states of the molecule. The rotational transitions are, again, subject to the selection rule that they must proceed from one level to the next. Such transitions are indicated by the arrows in Figure 44.8.

The typical wavelength emitted in a purely rotational transition lies in the far infrared. However, rotational molecular transitions are often observed in conjunction with a simultaneous electron transition in one of the atoms of the molecule. This increases the energy of the transition and reduces the wavelength of the radiation. As is obvious from Figure 44.8, successive rotational transitions have slightly different energies and wavelengths; thus, they give rise to a group, or sequence, of adjacent spectral lines. This is called a **band spectrum**. Figure 44.9 is a photograph of several spectral bands in the spectrum of the NO molecule.

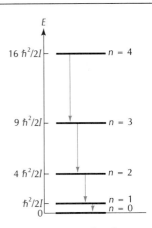

Fig. 44.8 Energy-level diagram for a rotating molecule. The arrows indicate the possible transitions.

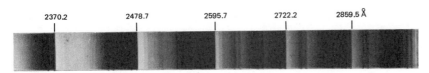

Fig. 44.9 Bands of spectral lines emitted by the NO molecule.

Band spectrum

EXAMPLE 2. The moment of inertia of the HCl molecule about its center of mass is 2.7×10^{-47} kg·m². What is the energy of the first excited rotational state?

SOLUTION: From Eq. (10),

$$E = \frac{\hbar^2}{2I} = \frac{(1.05 \times 10^{-34} \text{ J} \cdot \text{s})^2}{2 \times 2.7 \times 10^{-47} \text{ kg} \cdot \text{m}^2}$$

$$= 2.0 \times 10^{-22} \text{ J} = 1.3 \times 10^{-3} \text{ eV}$$

44.4 Energy Bands in Solids

As we mentioned in Section 22.4, in a metal the outermost, or valence, electrons of the atoms are detached from their atoms, and they are free to wander all over the volume of the metal. However, whenever such a "free" electron passes by an atom, it experiences an attractive force. For an electron moving along a row of atoms in the crystal lattice of a metal, the force will act repetitively, each time the electron passes by an atom (see Figure 44.10). Under special conditions, such a repetitive action of a force can lead to a large cumulative effect. According to wave mechanics, we have to think of the electron as a wave, and an encounter with an atom scatters the wave: some fraction of the wave proceeds in its original direction of motion, and some fraction is reflected. The repetitive scatterings at the atoms in the row will build up a large reflected wave by constructive interference if the phase increment of the wave from one scattering to the next is 360° or some multiple of 360°. This will happen if the extra distance for a round

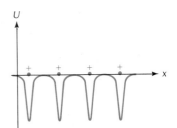

Fig. 44.10 The plus signs mark the positions of the atoms along a row in a crystal. Whenever an electron comes near one of these atoms, it experiences an attractive force, which is here plotted as a negative potential.

trip from one atom to the next and back is equal to one de Broglie wavelength or a multiple of one wavelength. Designating the distance between the atoms by a, we can express the condition for constructive interference of all the reflected waves as

$$2a = \lambda, 2\lambda, 3\lambda, \ldots \tag{11}$$

If this condition is satisfied, the reflected wave will gain more and more strength at each reflection at each atom, and finally match the strength of the incident wave. The result is a standing wave, which travels neither right nor left. This means that the electron cannot move through the lattice!

Since the de Broglie wavelength is related to the momentum of the electron by $\lambda = h/p$ [see Eq. (43.31)], we can express the condition for total reflection in terms of the momentum of the electron:

$$2a = \frac{h}{p}, \frac{2h}{p}, \frac{3h}{p}, \ldots \tag{12}$$

or

$$p = \frac{h}{2a}, \frac{2h}{2a}, \frac{3h}{2a}, \ldots \tag{13}$$

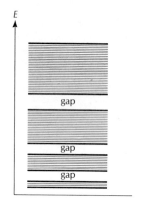

Fig. 44.11 Energy-level diagram for an electron moving in a crystal. The permitted intervals of energies occur in bands, which are separated by forbidden intervals, or gaps.

Energy bands

These values of the momentum are forbidden values, that is, they are values at which the electron cannot move. Corresponding to the forbidden values of the momentum, there are forbidden values of the energy. A more thorough analysis of the wave-mechanical motion of an electron through a crystal lattice reveals that the forbidden energies are actually forbidden energy gaps, that is, forbidden energy intervals. Figure 44.11 shows such forbidden energy gaps on an energy-level diagram. The permitted ranges of the energy, shown shaded in the diagram, are called **energy bands.** The precise widths of the forbidden energy gaps and of the permitted energy bands depend on the details of the crystal lattice, but all crystals with "free" electrons have some kind of band pattern in their energy-level diagram.

As in the case of the electron configuration of atoms, we can deduce the electron configuration of crystals by means of the Exclusion Principle. In a crystal in its ground state, the electrons occupy the available states of lowest energy. To discover the electron configuration, we proceed as before: we take all the "free" electrons and pack them, one by one, into the available energy bands. The lowest energy bands will then be completely filled, but the upper energy band will be either filled or partially filled, depending on the number of "free" electrons and the number of available states. The differences among the electric properties of conductors, semiconductors, and insulators arise from the partial or complete filling of the upper energy band. In a conductor, such as copper or silver, the upper band is only partially filled with electrons (see Figure 44.12). When the electrons in this partially filled band are subjected to an electric field, they absorb energy from the field and make transitions to some of the slightly higher, empty states of the band. Thus, the electrons respond to the electric field, and they begin to carry an electric current. In an insulator, such as diamond, the upper band is completely filled with electrons (see Figure 44.13). When the electrons in this full band are subjected to an electric field,

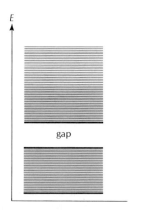

Fig. 44.12 In a conductor, the upper energy band is only partially filled with electrons. The portion of the band filled with electrons is shown in color.

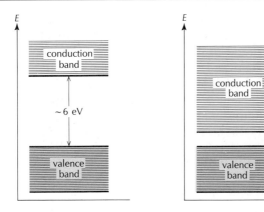

Fig. 44.13 (left) In an insulator, the upper (valence) energy band is completely filled with electrons.

Fig. 44.14 (right) In a semiconductor, the upper (valence) energy band is completely filled with electrons, but the gap between this band and the next is small.

they cannot make transitions into other states in the band, because all of these states are already full, and the Exclusion Principle forbids transitions into already full states. The only way the electrons in the full band can respond to the electric field is by making transitions to the *next*, empty energy band; but this is too difficult, since it requires the absorption of a large amount of energy from the electric field. Consequently, the electrons in the full band will not respond to the electric field. They will not accelerate in the direction of the electric force, and will not begin to carry a current.

In a semiconductor, such as silicon or germanium, the upper band is completely filled with electrons, as in an insulator. However, the energy gap between this band and the next band is much smaller than in an insulator; typically, the width of the energy gap separating the full band from the next empty band is of the order of only 1 eV, whereas in an insulator the gap is 6 eV or more (see Figure 44.14). It is therefore not too difficult for an electron at the top of the full band to make a transition to the next empty band. This means the electrons can respond to the electric field, and they can carry a current. The full band in an insulator or a semiconductor is called the **valence band**, and the empty band above it is called the **conduction band**.

Valence and conduction bands

As we already mentioned in Section 28.4, semiconductors fall into two categories: *n* type and *p* type. In an *n*-type semiconductor, the carriers of current are free electrons that have reached the conduction band. Thus, the mechanism for conduction is the same as in a metallic conductor. However, the resistance of a semiconductor is higher than that of a metal because the semiconductor has fewer free electrons in its conduction band than a metal in its partially filled uppermost band. Also, the semiconductor differs from a metal in that the resistivity *decreases* as the temperature increases. This curious behavior is due to an increase in the number of free electrons — as the temperature increases, more electrons are excited into the conduction band by thermal disturbances, and these extra free electrons more than compensate for the extra resistance experienced by each at the higher temperature.

n-type and p-type semiconductors

In a *p*-type semiconductor, the carriers of current are **holes** of positive charge. This type of semiconductor has a valence band that is almost, but not quite, filled with electrons. Thus, there are holes in the electron distribution, and if these holes move, they will transport charge. To see how such a transport of charge comes about, consider Figure 44.15 showing an array of electrons and positive ions. In Fig-

Holes

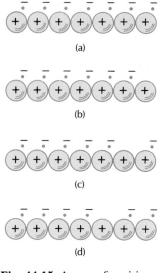

Fig. 44.15 A row of positive ions (balls marked with +) and electrons (color dots marked with —).

Donor and acceptor impurities

ure 44.15a, these electrons and ions form neutral atoms. Suppose that the right end of this array is connected to the positive pole of a battery (not shown) and the left end to the negative pole. If the battery pulls an electron out of the right end, it will leave the array with a hole or missing electron at the position of the first atom (Figure 44.15b). The electrons will then play a game of musical chairs: the electron from the next atom will jump into this hole, leaving a hole at the position of the second atom (Figure 44.15c); and then the electron from the next atom will jump; and so on. The collective motion of the electrons from left to right can be conveniently described as the motion of a hole from right to left. The hole virtually carries positive charge from the right to the left. Apart from some wave-mechanical refinements, this simple picture of moving holes describes the essential features of the mechanism for conduction in a *p*-type semiconductor. In this type of semiconductor, a flow of current is a flow of free holes, and the direction of the current is the same as the direction of motion of the holes. Experimentally, the direction of motion of the charge carriers and the sign of their charge can be verified by the Hall effect, discussed in Section 31.4.

A pure semiconductor contains equal numbers of free electrons (in the conduction band) and free holes (in the valence band), and both of these contribute to the flow of current. However, the density of free electrons and of free holes can be altered drastically by deliberately contaminating the semiconductor with impurities. **Donor** impurities consist of atoms that release their valence electrons when placed in the semiconductor, and they thereby increase the number of free electrons. **Acceptor** impurities consist of atoms that trap electrons when placed in the semiconductor, and they thereby generate holes. Hence, a semiconductor with donor impurities will be *n* type, and one with acceptor impurities will be *p* type. For instance, silicon with arsenic impurities is an *n*-type semiconductor, and silicon with boron impurities is a *p*-type semiconductor.

The behavior of arsenic and of boron atoms when inserted as substitutional impurities in silicon crystals can be understood in terms of the electron structure of these atoms. By inspection of the Periodic Table, we see that arsenic is in column V*A* and silicon in column IV*A*. This tells us that arsenic has one more outer, or valence, electron than silicon; but, except for this extra electron, the structure of arsenic is similar to that of silicon. When an arsenic atom is substituted for one of the silicon atoms in a silicon crystal, it will readily fit into the available space, provided it loses its extra electron. This extra electron becomes a free electron, in the conduction band. Thus, for every arsenic atom introduced into the silicon crystal, we gain a free electron. Likewise, by inspection of the Periodic Table, we see that boron is in column III*A*, which tells us that it has one less outer electron than silicon. When a boron atom is substituted for one of the silicon atoms, it will fit, provided it traps an electron from the neighboring silicon atoms. This leaves a hole in the electron distribution. For every boron atom introduced into the silicon crystal, we gain a free hole.

Since pure silicon has very few free electrons or free holes to begin with, the addition of just a minuscule amount of donor or acceptor impurities will decrease the resistivity by a large factor. For example, the addition of one part per million of arsenic decreases the resistivity by a factor of 10^5.

44.5 Semiconductor Devices

The manipulation of the resistivity of semiconductor materials by intentional contamination with carefully selected impurities plays a crucial role in the manufacture of semiconductor devices, such as diodes and transistors. The silicon crystals used in these devices are usually contaminated, or "doped," with small amounts of substitutional impurities. The impurities are often inserted into the crystal in layers, with different kinds and amounts in each layer. By suitably layering and masking different parts of a silicon crystal, it is possible to construct different electronic devices.

RECTIFIER (DIODE) A rectifier consists of a piece of n-type semiconductor joined to a piece of p-type. The n-type semiconductor has free electrons, and the p-type semiconductor has free holes; when they are joined, some of the free electrons will wander from the n region into the p region, and some of the holes will wander from the p region into the n region. Wherever the electrons and the holes meet, they annihilate each other — the electron falls into the hole and fills it, which means that both the electron and the hole disappear. The annihilation of some electrons and holes leaves residual positive and negative ions near the interface of the two regions, and the electric charges of these ions generate an electric field across the interface (see Figure 44.16). This electric field opposes any further wandering of holes or electrons from one region into the other.

When such a p–n junction is connected to a battery or some other source of emf, it will permit the flow of current from the p region into the n region, but not in the opposite direction. Figure 44.17 shows the p–n junction connected to the source of emf so that the p region is at high potential and the n region at low potential, a conformation called "forward bias." The source of emf pumps a steady flow of electrons into the n region, and it removes electrons from the p region, which is equivalent to pumping holes *into* the p region. The electrons and the holes meet at the interface, and they annihilate. This process can continue indefinitely, and therefore the source of emf can continue to pump current around the circuit indefinitely. Figure 44.18 is a plot of the current vs. the voltage applied to the p–n junction. The current increases steeply with the voltage, because the electric field associated with the applied voltage tends to cancel the internal electric field at the interface, and this makes it easier for the electrons and holes to meet

Rectifier

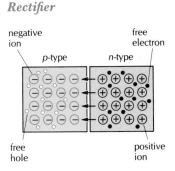

Fig. 44.16 Pieces of p-type and n-type semiconductor in contact. The plus and minus signs represent the ions of the lattice. The black dots represent electrons and the white dots holes. The arrows indicate the electric field generated by the ions at the interface.

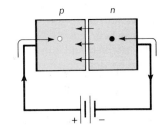

Fig. 44.17 A p–n junction connected to a source of emf. The p region is at high potential and the n region at low potential (forward bias).

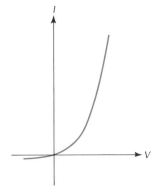

Fig. 44.18 Plot of current vs. voltage for the p–n junction.

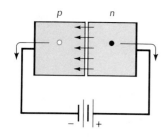

Fig. 44.19 A p–n junction connected to a source of emf. The p region is at low potential and the n region at high potential (reverse bias).

there. Obviously, the current is not simply proportional to the voltage, and the p–n junction does not obey Ohm's Law.

Now, consider what happens if the p–n junction is connected to the source of emf so the p region is at low potential and the n region at high potential, as is shown in Figure 44.19. This conformation is called "reverse bias." The free electrons in the n region then flow away through the wire on the right, and the free holes in the p region flow away through the wire on the left. Consequently, each region is depleted of its charge carriers, and the flow of current stops almost immediately — the p–n junction blocks the current.

The p–n junction is called a rectifier because it can be used to convert an alternating current into a direct current. If the junction is connected to a source of alternating emf, it will pass current only during the "forward" part of the cycle. The alternating positive and negative emf then yields a periodic sequence of positive current pulses (see Figure 44.20). Such solid-state rectifiers find many practical applications; for instance, they are used in the "alternators" that generate DC power in the electrical systems of automobiles.

Transistor

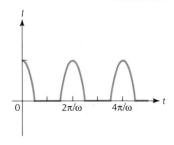

Fig. 44.20 Current passed by the rectifier vs. time. The negative portions of the alternating current are blocked by the rectifier, and only the positive portions remain.

TRANSISTOR (TRIODE) A transistor consists of a thin piece of semiconductor of one type sandwiched between two pieces of semiconductor of the other type. Figure 44.21 illustrates an n–p–n junction transistor. The thin piece in the middle is called the base; and the pieces at the ends are called the emitter and the collector, respectively. The transistor has three terminals which are connected to two sources of emf, V_B and V_C, so the emitter–base junction has a forward bias and the base–collector junction has a reverse bias. In this configuration, the emitter–base junction acts as diode with forward bias, and it permits the flow of electrons from the emitter into the base. However, the electrons that enter the p region fail to annihilate with holes, because the p region is quite thin and contains only a low density of holes, and the electrons pass through it before they have a chance to meet with a hole. The electrons wander to the base–collector junction, and the electric field across this junction (indicated by the arrows in Figure 44.21) pulls the electrons into the collector. They leave the collector via the terminal connected to its end, and they continue around the external circuit, forming the external collector current I_C. Of the electrons that enter the base, a small fraction wanders to the terminal connected to the base, and they leave via the wire connected there, forming the external base current I_B.

The use of the transistor as an amplifier for currents and voltages hinges on the relationship between the collector and the base currents: a small change in the base current I_B or the base voltage V_B leads to a quite large change in the collector current I_C. As in the case of the diode with forward bias, if we increase the base potential V_B, we cause a drastic increase in the flow of electrons entering the base from the emitter. But most of these electrons flow straight through to the collector, and only a small fraction flows to the terminal connected to the base. Thus, the result of an increase of base potential is a large increase of collector current, but only a small increase of base current. For a typical transistor, the ratio of the collector-current increment and the base-current increment is of the order of 100 or 200, and this ratio has a fixed value, over a wide range of currents. The ratio of these current increments is called the gain factor,

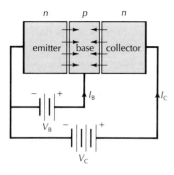

Fig. 44.21 An n–p–n junction transistor. Two sources of emf V_B and V_C are connected to the base and the collector, respectively.

$$[\text{gain factor}] = \frac{\Delta I_C}{\Delta I_B} \qquad (14)$$

This means that whenever we change the base current I_B by some amount (by adjusting the voltage V_B), we will change the collector current by an amount a hundred or so times larger. The transistor amplifies the current — a small input current at the base results in a much larger output current at the collector. For instance, a weak current picked up by a radio antenna can be amplified by sending it through a transistor. Further amplification can be achieved by connecting several transistors in tandem, so the output of each serves as input for the next.

In an integrated circuit, such as the one shown in Figure 44.22, many transistors and other circuit elements of extremely small size are built up on a single crystal of silicon. The small transistors are manufactured not by sticking together separate pieces of n- and p-type material, but by diffusing suitable concentrations of acceptor and of donor impurities into different layers of the silicon crystal.

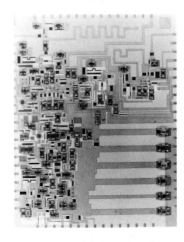

Fig. 44.22 This integrated circuit contains 146 resistors, 74 capacitors, and 33 transistors. (Approximately one-half natural size.)

Light-emitting diode

LIGHT-EMITTING DIODE (LED) In principle, a light-emitting diode is simply a p–n junction operated with forward bias. At such a junction, electrons arriving from the n region meet holes arriving from the p region, and they annihilate; that is, the electrons jump into the holes. But this jump is a transition of the electron from a state of high energy in the conduction band to a state of lower energy in the valence band, a transition that releases energy. In gallium arsenide and some other semiconducting materials, the released energy takes the form of a photon of visible light (see Figure 44.23). Thus, the p–n junction emits light when an electric current passes through it. Such light-emitting diodes have many practical applications in luminous displays in the dials of measuring instruments, watches, electronic calculators, clocks, automobile speedometers, and so on.

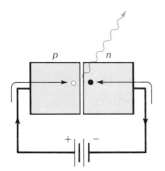

Fig. 44.23 A p–n junction used as a light-emitting diode.

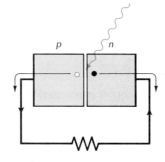

Fig. 44.24 A p–n junction used as a solar cell.

SOLAR CELL A solar cell is simply a light-emitting diode operating in reverse. When sunlight is absorbed at the p–n junction, it excites an electron from the valence band to the conduction band, creating a free electron and a free hole (see Figure 44.24). The electric field at the junction then pulls the electron toward the n region and the hole toward the p region. This means negative charge flows into the n region and from there into the external wire connected on the left; while positive charge flows into the p region and from there into the external wire connected on the right (see Figure 44.24). Thus, sunlight striking the junction generates an electric current in the external circuit.

Solar cell

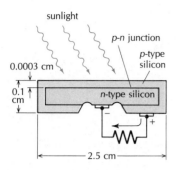

Fig. 44.25 Cross-sectional view of a solar cell.

Solar cells are commonly manufactured out of p-type silicon and n-type silicon. Figure 44.25 shows the structure of such a solar cell. The p-type silicon is placed in a thin, transparent layer on top of the n-type, so sunlight can reach the p–n junction. The emf of such a silicon solar cell is only about 0.6 V, and the current it delivers is fairly small. For a solar cell of the dimensions indicated in Figure 44.25, with an area of about 5 cm², the current delivered in full sunlight is about 0.1 A, and the efficiency for conversion of sunlight into electric energy is less than 10%. However, some solar cells of a new design attain efficiencies of up to 27%. Further increases of efficiency can be obtained by stacking solar cells in layers, so any sunlight not exploited in the top layer is exploited in the next layer.

Many of the artificial satellites used for communications and other purposes obtain their power from arrays of solar cells, or solar panels (see Figure 44.26). The satellites can take good advantage of the somewhat higher intensity of sunlight above the atmosphere. A few power stations for the practical exploitation of sunlight have been built on the surface of the Earth. The largest of these power plants produces about 7200 kW (see Figure V.15).

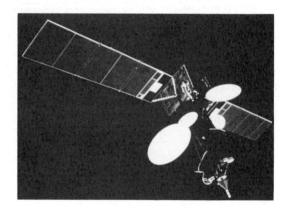

Fig. 44.26 Intelsat V satellite with solar panels extended.

SUMMARY

Electron spin: $\frac{1}{2}\hbar$

Electron magnetic moment: $\dfrac{e\hbar}{2m_e}$

Quantum numbers of atomic states: n, l, m, m_s

Exclusion Principle: Each stationary state of quantum numbers n, l, m, m_s can be occupied by no more than one electron.

Vibrational energies of molecule: $E = nh\nu$

Rotational energies of molecule: $E = \dfrac{n^2 \hbar^2}{2I}$

QUESTIONS

1. Consider the possible Sommerfeld orbits with $n = 1$, $l = 1$ in a hydrogen atom. Draw a diagram showing the orientations of these orbits for all the different values of the quantum number m.

2. According to wave mechanics, what are the possible values of l if $n = 0$? If $n = 1$? If $n = 2$?

3. The ρ meson is a particle of spin $\hbar$. What are the possible values of m_s for the ρ meson?

4. If there were no Exclusion Principle, what would be the electron configuration of lithium?

5. The bond in the NaCl molecule is similar to the bond in the KBr molecule. Explain this similarity.

6. Rotational transitions in a molecule give a band spectrum, but vibrational transitions do not. Explain.

7. Silicon and germanium look like silvery metals, but diamond does not look like a metal. Explain this difference qualitatively.

8. How does n-type silicon differ from p-type?

9. If you dope silicon with phosphorus impurities, will the silicon become n type or p type?

10. If you dope germanium with gallium, will the germanium become n type or p type?

11. What kind of valve in a hydraulic circuit is analogous to a diode rectifier?

12. Figure 44.27 shows a full-wave rectifier consisting of four diodes connected together. This rectifier not only blocks the negative portion of an entering alternating current but also reverses this portion, so the current is positive at all times. Describe the flow of current through the four diodes when the entering alternating current is positive and when it is negative.

13. Do the current I_c and the voltage V_c in a transistor obey Ohm's Law?

14. A spring-loaded butterfly valve in a large water pipe is controlled by a stream of water from a separate pipe (see Figure 44.28). Is this a reasonable hydraulic analog of a transistor? How could you use such a device to amplify a water current?

15. How many solar cells of the type shown in Figure 44.25 do you need to hook up in series to charge a 12-volt automobile battery?

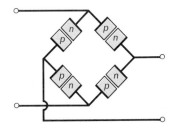

Fig. 44.27 Four diodes connected to form a full-wave rectifier.

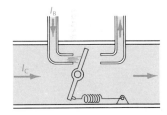

Fig. 44.28 The butterfly valve in the large pipe is hinged at the center. The stream of water from the small pipe strikes the upper portion of the butterfly valve and pushes it open.

PROBLEMS

Section 44.1

1. Pretend that the electron is a small sphere of uniform density of radius 2.8×10^{-15} m.
   (a) What is the moment of inertia of the electron according to this model?
   (b) What angular velocity of rotation is required to give the sphere an angular momentum of $\hbar/2$? What is the corresponding speed of rotation of a point on the equator of the sphere? Does this model make any sense?

2. An electron in an atom is in a state of quantum numbers $l = 2$, $m = 2$, $m_s = +\tfrac{1}{2}$. What is the total angular momentum of the electron? What is the total magnetic moment of the electron?

3. What are the possible values of the orbital angular momentum of the hydrogen atom in its first excited state? Taking into account the spin, what is the maximum possible value of the total angular momentum? What is the corresponding value of the magnetic moment of the atom?

*4. The electron of a hydrogen atom is in a state of orbital angular momentum $3\hbar$. Assume that the spin angular momentum is parallel to the orbital angular momentum, so the total angular momentum has a fixed magnitude of $3\hbar + \tfrac{1}{2}\hbar$. If the atom is placed in a magnetic field of 0.33 T, into how many energy levels will the original energy level split? What will be their energies relative to the original level?

*5. If a sample of sodium atoms is placed in a magnetic field, the spectral lines will exhibit the Zeeman effect. Estimate the splitting of the wavelength of the yellow spectral line of sodium ($\lambda = 5898$ Å) in a magnetic field of 1 T. (Hint: This spectral line is analogous to the first spectral line of the Lyman series.)

*6. In the sunlight coming from sunspot regions of the Sun's surface, astronomers observe Zeeman splittings of the spectral lines. Given that the blue Balmer line ($\lambda = 4342$ Å) displays a splitting of 0.03 Å, roughly estimate the magnetic field at the sunspot.

Section 44.2

7. How many possible electron states (including states of different spins) are there in the M shell ($n = 3$) of a hydrogen atom?

8. List the quantum numbers of all the electrons of a boron atom in its ground state.

9. List the quantum numbers of all the electrons of a carbon atom in its ground state.

10. List the quantum numbers of all the electrons of a magnesium atom in its ground state.

11. List the quantum numbers of all the electrons of a Na^+ ion in its ground state.

*12. Show that the shell of quantum number n in the hydrogen atom comprises $2n^2$ electron states (including states of different spins).

*13. Show that when a shell is completely filled with electrons, the z component of the net angular momentum and the z component of the net magnetic moment are zero. (More generally, it can be shown that the x and y components are also zero; but this requires some further knowledge about how to evaluate these components.)

*14. What are the orbital angular momentum and the spin angular momentum of the sodium atom? What is the magnetic moment of the sodium atom? (Hint: Use the result stated in the preceding problem.)

*15. A fluorine atom can be regarded as a neon atom with a "hole" in its electron distribution. Given that the net angular momentum of any shell completely filled with electrons is zero (see Problem 13), what are the orbital angular momentum and the spin angular momentum of the fluorine atom?

*16. Table 33.1 lists the magnetic moments of some atoms.
   (a) Explain why He and Ne have zero magnetic moment. (Hint: See Problem 13.)
   (b) Explain why H, Li, and Na have a magnetic moment of 9.27×10^{-24} A·m².

*17. In the lithium atom, the excited states are obtained by lifting the outermost electron into a higher orbit, while leaving the two inner electrons undisturbed. Accordingly, what are the quantum numbers n and l for the first excited state of lithium? For the second excited state? (Hint: The energy of the orbit does not depend on its orientation, but it does depend on its size and shape. Ignore the spin.)

Section 44.3

18. (a) The binding energy of the NaCl crystal (salt) is 183 kcal/mole; this is the energy required to dissociate the crystal into separate ions. This crystal is cubic, as is illustrated in Figure I.25. Pretend that each ion forms a bond with only the four nearest ions. What is the energy per Na^+–Cl^- bond? Express the answer in electron-volts.

(b) The distance between each ion and the nearest ion is 2.81 Å. What is the Coulomb energy of a pair of ions separated by this distance? Express the answer in electron-volts. Explain why the answers obtained in (a) and in (b) are of the same order of magnitude, although not exactly equal.

19. The frequency of vibration of the H_2 molecule is 1.3×10^{14} Hz. What are the energies of the vibrational states? What is the frequency of the emitted radiation? The wavelength?

*20. The frequency of vibration of the HCl molecule is 8.6×10^{13} Hz. Regard this molecule as two masses connected by a spring, and assume the Cl atom remains at rest.
   (a) What is the effective spring constant of the spring?
   (b) Suppose we replace the hydrogen in the molecule by deuterium or by tritium. What are the frequencies of vibration of the DCl molecule or the TCl molecule? The masses of D (or ^{2}H) and of T (or ^{3}H) atoms are given in the chart of isotopes in Chapter 45.

*21. In the oxygen molecule, the distance between the two nuclei is 2.0 Å.
   (a) What is the moment of inertia of the molecule for rotation about the perpendicular axis through the center of mass?
   (b) What is the wavelength of the photon emitted in a purely rotational transition from $n = 1$ to $n = 0$? From $n = 2$ to $n = 1$?

*22. What are the ratios of the frequencies emitted in rotational transitions in a molecule from the first excited state to ground state, from the second to the first, from the third to the second, from the nth to the $(n - 1)$th?

*23. According to spectroscopic measurements, the energy difference between the first and the second excited rotational states of the N_2 molecule is 1.98×10^{-22} J. Deduce the moment of inertia of the molecule. Deduce the center-to-center distance between the N atoms.

*24. The distance between the K and the Br nuclei in the KBr molecule is 2.82 Å, and the center of mass is at a distance of 0.93 Å from the Br nucleus. What is the moment of inertia of a KBr molecule rotating about its center of mass? What are the energies of the first, second, and third excited rotational states? Express these energies in electron-volts.

**25. The distance between the two nuclei in the H_2 molecule is 0.74 Å.
   (a) What are the energies of the rotational states? Express the energies in electron-volts.
   (b) Suppose we replace one of the hydrogen atoms by a deuterium atom (D, or ^{2}H). What are the energies of the rotational states of the DH molecule? By what factor do these energies differ from those of the H_2 molecule?

Section 44.4

26. Consider a crystal with a spacing of 1.0 Å between one atom and the next. What are the de Broglie wavelengths and the energies at which the electron wave cannot propagate through this crystal? Express the energies in electron-volts.

27. In silicon, the energy gap between the valence and the conduction bands is 1.1 eV. If we want to excite an electron from the top of the valence band to the conduction band by means of a photon, what is the maximum permitted wavelength for the photon?

*28. A germanium crystal is doped with 7×10^{21} arsenic donors per cubic meter. Suppose that a strip of this crystal, 1.0 cm wide and 0.01 cm thick, is placed face-on to a magnetic field 2.0 T, and a current of 0.50 A is made to flow along the strip. What is the Hall potential difference between opposite sides of this strip? Assume that all the donors are ionized.

*29. The following table gives the resistivity of n-type germanium at different donor concentrations:

Donor concentration, per m³	10^{20}	10^{22}	10^{24}
Resistivity, $\Omega \cdot$ m	0.175	0.0020	0.000067

For each case, calculate the average collision time τ (see Section 28.3); assume that all the donors are ionized. Explain qualitatively why the collision time is longer when the donor concentration is larger.

Section 44.5

30. Figure 44.21 shows a circuit diagram for an n–p–n transistor. Draw the analogous diagram for the p–n–p transistor, and explain how a small current I_B leads to a large current I_C.

*31. When a transistor is connected to a circuit such as that shown in Figure 44.29, it serves as an amplifier of voltage. The voltage-gain factor is defined as the ratio of the output voltage (measured across the resistor R_C) to the input voltage V_B. Evaluate this ratio if $R_B = 3000 \; \Omega$ and $R_C = 6000 \; \Omega$. Assume that the current-gain factor for this transistor is 100 and that the internal resistance of the emitter–base junction (a diode with forward bias) is negligible.

*32. We want to connect two transistors in tandem, so the net voltage amplification of the combination is the product of the individual voltage amplifications of the two transistors. Design a circuit that will accomplish this.

*33. A solar cell delivers 0.1 A at 0.6 V. How many such solar cells do you need, and how must you connect them, to obtain 2A at 6V for charging a battery?

*34. A solar cell of area 5 cm² facing the Sun delivers 0.1 A at 0.6 V. What is the power delivered by this solar cell? Compare with the incident power of sunlight (1.4 kW/m²), and deduce the efficiency for the conversion of energy of light into electric energy.

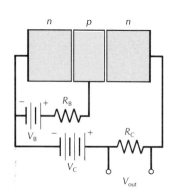

Fig. 44.29 A transistor acting as a voltage amplifier. The source of emf V_B provides the input signal, and the voltage across the two free terminals constitutes the output signal.

CHAPTER 45

Nuclei

Rutherford's first experiments on the bombardment of atoms with a beam of alpha particles established that the nucleus of the atom is very small but contains most of the mass of the atom. The nucleus is therefore very dense, and it must be made of massive particles packed very tightly together. In later experiments, Rutherford proceeded to explore the structure of the nucleus, again using a beam of alpha particles as a probe. He found that if the projectiles were energetic enough to penetrate the nucleus, they would often split it into two pieces, two smaller nuclei. The smallest such piece that could be split off was found to be a proton, and Rutherford therefore conjectured that all nuclei contain protons. The simplest nucleus is that of the atom of hydrogen, which consists of a single proton. However, the nuclei of all other atoms have more mass and less charge than expected if protons were their only constituents — there must be some neutral particles in these nuclei or, alternatively, some electrically neutral combination of particles of opposite charges. The mystery of the neutral constituent in nuclei was not solved until 1932, when J. Chadwick discovered the **neutron,** a particle of about the same mass as the proton but without electric charge. This discovery led to the modern view of the nucleus as a tightly packed conglomerate of protons and neutrons (see Figure 45.1).

Since the average distance between the protons in the nucleus is quite small, the repulsive electric force among the nuclear protons is very large. This force would burst the nucleus apart if there were not an extra, even larger, attractive force holding the protons and the neutrons together. This extra force is the **nuclear force,** or the **"strong" force.** Acting on two adjacent protons in a nucleus, this attractive force is about 100 times as large as the repulsive electric force. Thus, the strong force completely overwhelms the electric force. How-

James Chadwick, *1891–1974, English physicist, professor at Liverpool. For his discovery of the neutron, he received the Nobel Prize in 1935.*

Neutron

Fig. 45.1 A carbon nucleus consisting of six protons (color) and six neutrons (white).

ever, in heavy nuclei — such as uranium — with a large number of protons and a large total electric charge, the electric repulsion becomes important, and it can cause the nucleus to split apart, or to fission. The fission of uranium, as manifested in the explosion of a nuclear bomb, provides a spectacular demonstration of a victory of the electric force over the strong force.

45.1 Isotopes

Nucleons

Nuclei are made of protons and neutrons. Generically, these two kinds of constituents of the nucleus are called **nucleons.** Table 45.1 lists the main properties of protons and neutrons. It is instructive to compare these particles with the electron. Both the proton and the neutron have masses about 1800 times as large as that of the electron.[1] Their spins are $\frac{1}{2}\hbar$, the same as that of the electron. In contrast to the electron, which is a pointlike particle of no discernible size, the proton and the neutron are small balls, of a radius of about 10^{-15} m.

Both the proton and the neutron have magnetic moments, of a magnitude roughly 1000 times smaller than that of the electron. The magnetic moment of the proton is in the same direction as the spin, and the magnetic moment of the neutron is in the opposite direction. (Since the neutron is electrically neutral, its magnetic moment poses a puzzle. The answer to this puzzle is that, although the neutron has a net charge of zero, it contains a core of positive charge and an outer layer with an equal amount of negative charge; the rotation of these charge distributions accounts for the magnetic moment.)

Table 45.1 THE NUCLEONS

Nucleon	Mass	Spin	Radius	Magnetic moment
Proton	$m_p = 1.00728$ u	$\frac{1}{2}\hbar$	1×10^{-15} m	1.41×10^{-26} A · m²
Neutron	$m_n = 1.00866$ u	$\frac{1}{2}\hbar$	1×10^{-15} m	0.97×10^{-26} A · m²

The number of protons in the nucleus of a (neutral) atom of a given element matches the number of its electrons, that is, it matches the atomic number Z of the element. For example, the carbon atom has six electrons, and it has six protons in its nucleus.

All the atoms of a given chemical element, such as carbon, have exactly the same chemical properties, because they all have exactly the same number of electrons and the same electron configuration. However, the atoms of a chemical element can differ in mass, because their nuclei can have different numbers of neutrons. Thus, all carbon atoms have six protons in their nuclei, but some have six neutrons, some have seven, some have eight, and so on. Atoms with the same number of protons in their nuclei but with different numbers of neutrons are called **isotopes.** Carbon has eleven known isotopes, designated ^{8}C, ^{9}C, ^{10}C, ^{11}C, ^{12}C, ^{13}C, ^{14}C, ^{15}C, ^{16}C, ^{17}C, and ^{18}C. The superscript on the

Isotopes

---

[1] Expressed in atomic mass units, the electron mass is 5.49×10^{-4} u.

chemical symbol (for instance, the superscript "12" in "^{12}C") indicates the sum of the number of protons and the number of neutrons; this sum is called the **mass number**. If we designate the mass number by the symbol A, then

Mass number, A

$$A = N + Z \tag{1}$$

where N is the number of neutrons and Z is the number of protons. Since the mass of each proton and each neutron is approximately one atomic mass unit (u), the mass number is approximately equal to the mass of the nucleus in atomic mass units.

Natural samples of atoms of carbon or any other chemical element contain characteristic percentages of different isotopes. Natural carbon, as found in coal, is a mixture of 98.9% of the isotope ^{12}C and 1.1% of the isotope ^{13}C. Carbon dioxide, as found in air, contains not only the isotopes ^{12}C and ^{13}C but also a very small amount (about 2.4×10^{-10} %) of the isotope ^{14}C. The other isotopes of carbon do not occur naturally; they can be produced only artificially by transmutation of elements, or "alchemy," in a nuclear reactor or an accelerator.

All chemical elements have several isotopes (see the excerpt from the chart of isotopes reproduced in Table 45.2). Hydrogen has three isotopes (^{1}H, or ordinary hydrogen; ^{2}H, or deuterium; ^{3}H, or tritium). Helium has five isotopes, lithium has six, and so on. Some of these isotopes occur in nature; others can be produced only by artificial means.

The masses listed on the chart of isotopes are given in atomic mass

Table 45.2 EXCERPT FROM THE CHART OF ISOTOPES[a]

[a] The number Z, increasing vertically along the chart, is the number of protons in the isotope; it coincides with the atomic number. The number N, increasing horizontally, is the number of neutrons. In each box, the number directly below the symbol for the isotope gives the abundance in percent for naturally occurring isotopes, or else the half-life for unstable, artificially produced isotopes (the half-life is the time required for one-half of a sample of unstable isotope to decay). The Greek letters indicate the emissions that accompany the decay: β^{-} rays (electrons), β^{+} rays (antielectrons), or γ rays. The bottom number gives the mass of the neutral atom (nucleus plus Z electrons) in atomic mass units.

units. By definition, the mass of the ^{12}C isotope is exactly 12 atomic mass units,

$$[\text{mass of } ^{12}\text{C atom}] = 12 \text{ atomic mass units} = 12 \text{ u} \tag{2}$$

In kilograms the atomic mass unit is, approximately,

$$1 \text{ u} = 1.66055 \times 10^{-27} \text{ kg}$$

Note that all the *masses listed on the chart are those of the atom;* if we want the mass of the nucleus, we must subtract the masses of the electrons ($m_e = 5.49 \times 10^{-4}$ u).

According to the definition (2), the carbon atom is used as a standard of mass for the measurement of atomic masses. Very precise measurements of the masses of atoms are performed with mass spectrometers, which compare the mass of one atom with that of another by comparing the circular motions of ions of these atoms in a constant magnetic field. Such direct mass measurements are supplemented by indirect measurements which deduce masses from the energy released in nuclear reactions. The precision of atomic mass measurements is remarkable — some isotope masses have been measured, relative to carbon, to within nine significant figures! Since such mass measurements rely on classical physics, their high precision is a final tribute to Newton.

Most of the isotopes listed on the chart are unstable; they decay by a spontaneous nuclear reaction and transmute themselves into another element. The decay is accompanied by the emission of alpha rays, beta rays, or gamma rays. We will examine these decay processes in Section 45.3.

Experiments on the bombardment of nuclei with alpha particles and other particles indicate that the size of the nucleus is proportional to the cube root of the mass number. Specifically, these experiments indicate that the radius of a nucleus of mass number A is

Enrico Fermi, *1901–1954, Italian and later American physicist, professor at Rome and at Chicago. He worked on experimental and theoretical investigations of beta decay, the artificial production of isotopes by neutron bombardment, for which he received the 1938 Nobel Prize, and the fission of uranium. Fermi was one of the leaders of the Manhattan Project, and he provided the first experimental demonstration of a chain reaction.*

Nuclear radius

$$\boxed{R = (1.2 \times 10^{-15} \text{ m}) \times A^{1/3}} \tag{3}$$

Nuclear radii can be conveniently expressed in terms of a new unit of length, the fermi,

Fermi, fm

$$1 \text{ fermi} = 1 \text{ fm} = 10^{-15} \text{ m}$$

For example, the ^{12}C nucleus has a radius of

$$R = 1.2 \text{ fm} \times (12)^{1/3} \text{ m} = 2.7 \times 10^{-15} \text{ m} = 2.7 \text{ fm}$$

whereas the ^{238}U nucleus has a radius

$$R = 1.2 \text{ fm} \times (238)^{1/3} \text{ m} = 7.4 \times 10^{-15} \text{ m} = 7.4 \text{ fm}$$

The proportionality between R and $A^{1/3}$ implies that the number of nucleons per unit volume is the same for all nuclei,

$$\frac{A}{(4\pi/3)R^3} = \frac{A}{(4\pi/3)(1.2 \times 10^{-15} A^{1/3})^3 \text{ m}^3}$$

$$= 1.38 \times 10^{44} \text{ nucleons/m}^3$$

Since the mass of each nucleon is about 1.7×10^{-27} kg, the mass density of the nuclear material is

$$1.7 \times 10^{-27} \text{ kg} \times \frac{1.38 \times 10^{44}}{\text{m}^3} = 2.3 \times 10^{17} \text{ kg/m}^3 \tag{4}$$

This means that one cubic centimeter, or 10^{-6} m³, of nuclear material would have a mass of 230 million tons! Note that the average volume per nucleon is $1/(1.38 \times 10^{44})$ m³. We can think of this volume as a cube enclosing the nucleon; the edge of the cube or, equivalently, the center-to-center distance from one nucleon to its nearest neighbor, is therefore the cube root of the volume, $1/(1.38 \times 10^{44} \text{ m}^3)^{1/3} \cong 2 \times 10^{-15}$ m. By comparing this with the radius of a proton or neutron, about 1×10^{-15} m, we see that inside the nucleus the nucleons are so tightly packed together that they almost touch (see Figure 45.1).

45.2 The Strong Force and the Nuclear Binding Energy

Since the protons within a nucleus are at such short distances from one another, they exert very large repulsive electric forces on one another. Two neighboring protons, separated by a center-to-center distance of $\simeq 2 \times 10^{-15}$ m, experience an electric repulsive force of

$$F = \frac{1}{4\pi\varepsilon_0} \frac{e^2}{r^2} = \frac{1}{4\pi\varepsilon_0} \times \frac{(1.6 \times 10^{-19} \text{ C})^2}{(2 \times 10^{-15} \text{ m})^2} \tag{5}$$

$$= 58 \text{ N}$$

This is, roughly, twice the weight of this book (both volumes); acting on a mass of only 10^{-27} kg, this force is of colossal magnitude.

Obviously, some extra force must be present in the nucleus to prevent it from instantaneously bursting apart under the influence of the mutual electric repulsion of the protons. This extra force is the **strong force**, already mentioned in Section 6.1. This force acts equally between any two nucleons, regardless of whether they are protons or neutrons (the force is "charge independent"). Figure 45.2 is a plot of the potential energy associated with the strong nucleon–nucleon force. The force depends on the orientation of the spins; the plot in Figure 45.2 is for nucleons of parallel spins. The energy unit in the plot is the MeV,

$$1 \text{ MeV} = 10^6 \text{ eV} = 1.602 \times 10^{-13} \text{ J} \tag{6}$$

The MeV is a convenient energy unit in nuclear physics, because typical energy changes in nuclear phenomena are one or several MeV. For comparison, we recall from Chapter 43 that typical energy changes in atomic phenomena are one or several electron-volts. Thus, nuclear energies are typically a million times as large as atomic energies.

Since the force is the negative of the derivative of the potential energy, we see from the plot of the potential energy that the strong force

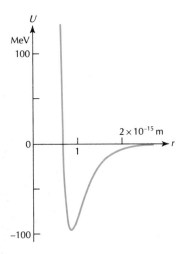

Fig. 45.2 Potential energy for the strong force acting between two nucleons.

is attractive over a range of internucleon distances from $\cong 1 \times 10^{-15}$ m to $\cong 2 \times 10^{-15}$ m. In this range of distances, the strong force is much larger than the electric force, as much as 100 times larger. The strong force is repulsive for internucleon distances less than $\cong 1 \times 10^{-15}$; this means the nucleons have a hard core that resists interpenetration. For distances larger than $\cong 2 \times 10^{-15}$ m, the strong force decreases abruptly and finally vanishes. Thus, in contrast to the electric force, which fades only gradually and reaches out to large distances, the strong force cuts off sharply and has only a short range. In order to feel the strong force, the nucleons must be touching or almost touching; that is, the force acts only between nearest neighbors in the nucleus.

In consequence of the short-range character of the strong force, a nucleon deep inside the nucleus does not experience any net force — the nucleon interacts only with its nearest neighbors, and since these pull it with equal force in almost all directions, the net force on the nucleon is zero or nearly zero (see Figure 45.3). However, a nucleon at the nuclear surface has neighbors only on the side that is toward the interior, and hence these will exert a net force pulling the nucleon inward (see Figure 45.3). Altogether, this means that nucleons are more or less free to wander about the interior of the nucleus, but whenever they approach the nuclear surface, strong forces pull them back and prevent their escape.

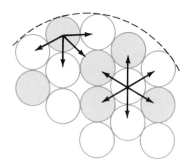

Fig. 45.3 Forces on a nucleon at the nuclear surface, and forces on a nucleon in the nuclear interior.

This suggests that the nucleons in the nucleus behave somewhat like the molecules in a drop of water; such molecules are free to wander throughout the volume of the drop, but when they approach the water surface, intermolecular forces hold them back. This similarity between nuclei and drops of water rests on a similarity of the laws of force. The intermolecular force and potential have general features rather similar to those displayed in Figure 45.2; the force is attractive over a short range and then becomes strongly repulsive when the molecules begin to interpenetrate. The hard, repulsive core of the potential makes the water nearly incompressible, whereas the short-range attraction provides a cohesive force that prevents water droplets from falling apart. The balance of attraction and repulsion encourages water molecules to stay at a particular distance from one another, and this gives water a particular, uniform density.

Because of the similarities between a liquid and the nuclear material, the nucleus can be crudely regarded as a droplet of incompressible "nuclear fluid" of uniform density. The fluid is, of course, made of nucleons, but for some purposes we can ignore the individual nucleons, and we can calculate the properties of nuclei in terms of the gross properties of a liquid. For example, the spherical shape adopted by most nuclei can easily be understood as follows: Any nucleon located on the surface of a globule of nuclear fluid experiences an inward force pulling it back into the volume, and consequently the fluid tends to shrink its exposed surface to the smallest value compatible with its (fixed) volume. Since a sphere has the least surface area for a given volume, the globule of fluid will take the shape of a spherical droplet.

In a stable nucleus, the repulsive electric forces among the protons are held in check by the attractive strong forces. To achieve this balance of forces, the presence of neutrons is an advantage: a nucleus with more neutrons will have a larger size and, therefore, a larger average distance between pairs of protons — the neutrons in the nucleus dilute the repulsive effect of the electric force. Consequently, all stable nuclei, with the exception of hydrogen and one isotope of helium, contain at least as many neutrons as protons; heavy nuclei, such as uranium, contain substantially more neutrons than protons.

Figure 45.4 is a plot of the number of neutrons vs. the number of protons (N vs. Z). On this plot, black dots indicate the stable nuclei and colored dots indicate unstable nuclei, that is, radioactive isotopes. Note that there is no stable nucleus beyond bismuth ($Z = 83$). However, several elements beyond bismuth have some isotopes with very long half-lives; these are therefore almost stable, and they occur naturally.

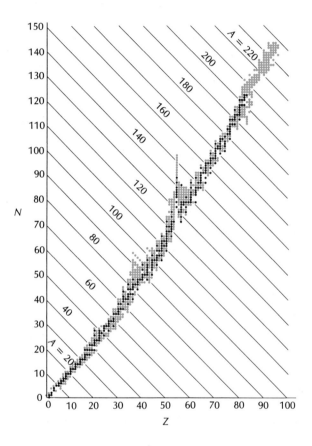

Fig. 45.4 Number of protons (Z) vs. number of neutrons (N) for stable nuclei (black dots) and unstable nuclei (color dots).

The energy stored in a nucleus is a sum of the potential energies contributed by the electric and the strong forces and the kinetic energies of the nucleons. This "stored" energy is negative, since energy is *released* when the nucleus is assembled out of its constituent nucleons. Conversely, energy must be supplied to take the nucleus apart into its constituent nucleons. The negative of the "stored" energy is called the **binding energy** (B.E.); this is the energy released during the assembly of the nucleus out of its constituent nucleons. Figure 45.5 is a plot of B.E./A, the binding energy divided by the number of nucleons, or the average binding energy per nucleon. The curve plotted in Figure 45.5 is called the **curve of binding energy.**

Binding energy

The binding energy of a typical nucleus is a rather large amount of energy. As may be seen from Figure 45.5, the average binding energy per nucleon is in the vicinity of 8 MeV for almost all nuclei; thus, a nucleus with a mass number A typically has a binding energy of about $A \times 8$ MeV. To put this number in perspective, we may compare it

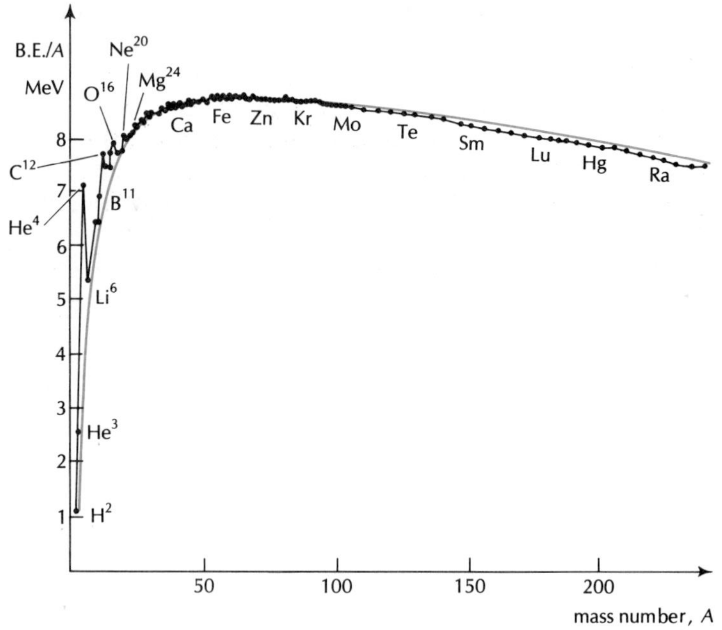

Fig. 45.5 Average binding energy per nucleon vs. mass number.

with the rest-mass energy of the nucleons. Each nucleon has a mass of about one atomic mass unit. The energy corresponding to one atomic mass unit is $1 \text{ u} \times c^2 = 1.6606 \times 10^{-27}$ kg $\times (2.9979 \times 10^8 \text{ m/s})^2 = 1.492 \times 10^{-10}$ J or, in MeV units [see Eq. (6)],[2]

$$1 \text{ u} \times c^2 = 931.5 \text{ MeV} \tag{7}$$

Thus, each nucleon has a rest-mass energy of about 930 MeV, and the A nucleons in the nucleus have a rest-mass energy of about $A \times 930$ MeV. The ratio of binding energy to rest-mass energy is then about $8/(930) \simeq 0.009$, which means the binding energy is about 1% of the rest-mass energy!

The mass associated with the binding energy is B.E./c^2; this mass is carried away by the energy released during the assembly of the nucleus from its constituent protons and neutrons. The mass of a typical nucleus is therefore about 1% less than the sum of the masses of these protons and neutrons. The mass difference is called the **mass defect,**

Mass defect

$$\begin{aligned} \text{B.E.}/c^2 &= [\text{mass defect}] \\ &= [\text{mass of } N \text{ neutrons and } Z \text{ protons}] \\ &\quad - [\text{mass of nucleus}] \end{aligned} \tag{8}$$

Experimental values of the binding energy of a nucleus usually are obtained via the mass defect. A precise measurement of the mass of the nucleus is compared with the sums of the masses of the constituent protons and neutrons; the difference, or mass defect, then gives the binding energy according to Eq. (8). This is how the experimental points in the plot of the curve of binding energy were obtained.

---

[2] We have retained four significant figures in this calculation because we will be needing a precise value of the equivalence between energy and mass later on.

EXAMPLE 1. What is the nuclear binding energy of the isotope ^{238}U? Express the energy in MeV. The mass of an atom of this isotope is 238.0508 u.

SOLUTION: Uranium-238 has 92 protons and 146 neutrons. According to Eq. (8),

$$\text{B.E.}/c^2 = [92m_p + 146m_n] - [\text{mass of nucleus}] \tag{9}$$

The mass of the uranium nucleus is the mass of the uranium atom minus the mass of the 92 electrons of this atom. Hence

$$\text{B.E.}/c^2 = [92m_p + 146m_n] - ([\text{mass of atom of uranium}] - 92m_e) \tag{10}$$

But 92 proton masses plus 92 electrons masses equals 92 hydrogen masses, and we therefore obtain the following convenient formula for the binding energy:

$$\text{B.E.}/c^2 = 92M_H + 146m_n - [\text{mass of atom of uranium}] \tag{11}$$

With $M_H = 1.007825$ u and $m_n = 1.00866$ u, this yields

$$\text{B.E.}/c^2 = 92 \times 1.007825 \text{ u} + 146 \times 1.00866 \text{ u} - 238.0508 \text{ u}$$

$$= 1.933 \text{ u}$$

The binding energy is therefore

$$\text{B.E.} = 1.933 \text{ u} \times c^2$$

Since $u \times c^2 = 931.5$ MeV [see Eq. (7)], this is

$$\text{B.E.} = 1.933 \times 931.5 \text{ MeV} = 1801 \text{ MeV}$$

COMMENTS AND SUGGESTIONS: The formula (11) for the binding energy of the uranium nucleus can be generalized to any other nucleus with Z protons and N neutrons:

$$\text{B.E.}/c^2 = ZM_H + Nm_n - [\text{mass of atom}] \tag{12}$$

This formula is handy for calculation of the binding energy from the atomic masses listed in tables or charts of isotopes.

45.3 Radioactivity

Radioactivity was discovered in 1896 by Henri Becquerel, who noticed that samples of uranium salt emitted invisible rays which could penetrate through sheets of opaque materials and make an imprint on a photographic plate. Subsequent investigations established that uranium, and many other radioactive substances, emit three kinds of rays: **alpha rays, beta rays,** and **gamma rays.** Of these, the alpha rays are the least penetrating; they can be stopped by a thick piece of paper. The beta rays are more penetrating; they can pass through a foil of lead or a plate of aluminum. The gamma rays are the most penetrating; they can pass through a thick wall of concrete. When these three kinds of rays are emitted into a magnetic field, the alpha and beta rays are de-

Alpha, beta, and gamma rays

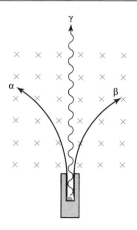

Fig. 45.6 Alpha, beta, and gamma rays emitted by a radioactive source. When these rays enter a magnetic field (perpendicular to the plane of the page), the alpha and beta rays are deflected in opposite directions, and the gamma rays are not deflected.

flected in opposite directions, whereas the gamma rays proceed without deflection (see Figure 45.6). This simple experiment demonstrates that alpha rays are positively charged, beta rays negatively charged, and gamma rays are neutral.

By more detailed experiments, Becquerel identified the beta rays as high-speed electrons. Some years later, Rutherford investigated the nature of the alpha rays and demonstrated that they are identical to nuclei of helium. Thus, electrons and helium nuclei are, somehow, manufactured within the sample of uranium or other radioactive substance, and they are ejected at high speeds. Gamma rays are very energetic photons emitted by the radioactive substance; the energy of these gamma-ray photons is typically a thousand times as large as that of X-ray photons.

The "manufacture" and ejection of these high-speed electrons and helium nuclei occurs in the nuclei of the radioactive substance, in nuclear-decay reactions. In the following, we will discuss the broad features of these nuclear reactions.

ALPHA DECAY   The alpha particle has the same structure as the helium nucleus; that is, it consists of two protons and two neutrons. When a radioactive nucleus ejects an alpha particle with its two protons and neutrons, the radioactive nucleus does not create these protons and neutrons out of nothing — the nucleus merely takes two of its own protons and neutrons and spits them out. Since the nucleus loses two protons and neutrons, both the atomic number and the mass number of the nucleus decrease. The atomic number decreases by two and the mass number by four. Thus, the original isotope becomes a different isotope of a different chemical element. The following examples of alpha decays illustrate such transmutation of chemical elements:

Alpha decay

$$^{238}\text{U} \rightarrow {}^{234}\text{Th} + \alpha \tag{13}$$

$$^{226}\text{Ra} \rightarrow {}^{222}\text{Rn} + \alpha \tag{14}$$

In the first of these alpha-decay reactions, uranium is transmuted into thorium; and in the second, radium into radon. The original nucleus in a decay reaction is called the **parent,** and the resulting nucleus is called the **daughter.**

Parent and daughter

The alpha-decay reaction can be regarded as a fission, or splitting, of the nucleus into two smaller nuclei. For instance, the alpha decay of the uranium nucleus can be regarded as a fission into a thorium nucleus and a helium nucleus. This fission occurs spontaneously, because of an instability in the original nucleus. Heavy nuclei, such as uranium or radium, contain a large number of protons which exert repulsive electric forces on each other. Although the repulsive electric force is balanced by the attractive strong force, the balance of these forces is rather precarious because the electric force easily reaches from one end of the nucleus to the other, whereas the strong force acts only between adjacent nucleons and therefore, in a heavy, big nucleus, cannot reach directly from one end to the other. Thus, any accidental, spontaneous elongation of the nucleus can shift the balance in favor of the electric force — the nucleus then elongates more and more and ultimately burst apart into two fragments. In true fission, the two fragments are of approximately equal size. Alpha decay is an extreme case of fission, with two fragments of very unequal size. The ejection of an alpha particle is strongly favored over ejection of, say, a hydrogen nu-

Antoine Henri Becquerel, *1852–1908, French physicist, professor at the École Polytechnique. He was awarded the 1903 Nobel Prize for his discovery of radioactivity.*

cleus or a lithium nucleus, because the helium nucleus is an exceptionally tightly bound nucleus. In consequence of this large binding energy, the formation of an alpha particle, just before its ejection, makes more energy available for driving the fission reaction.

The kinetic energy of the alpha particles ejected in reactions such as (13) and (14) can be calculated from the masses of the participating nuclei and the mass of the alpha particle.

EXAMPLE 2. Calculate the energy released in the alpha decay of ^{238}U, and calculate the kinetic energy of the alpha particle ejected in this reaction. The atomic masses of the isotopes of uranium, thorium, and helium in this reaction are 238.0508 u, 234.0436 u, and 4.0026 u, respectively.

SOLUTION: The energy released in the reaction (13) is simply the difference between the binding energy after the reaction and the (smaller) binding energy before the reaction. We already know the binding energy before the reaction, the binding energy of the uranium nucleus [see Example 1],

$$[\text{B.E. uranium}] = 1801 \text{ MeV} \qquad (15)$$

The binding energies of the thorium nucleus and of the alpha particle can be calculated from Eq. (12):

$$[\text{B.E. thorium}]/c^2 = 90M_H + 144m_n - [\text{mass of atom of thorium}]$$

$$[\text{B.E. alpha}]/c^2 = 2M_H + 2m_n - [\text{mass of atom of helium}]$$

If we evaluate the binding energies from these formulas, we find, in MeV,

$$[\text{B.E. thorium}] = 1777 \text{ MeV} \qquad (16)$$

$$[\text{B.E. alpha}] = 28.3 \text{ MeV} \qquad (17)$$

The net final binding energy is therefore 1777 MeV + 28 MeV = 1805 MeV. Comparing this with the initial binding energy, 1801 MeV, we see that the energy released in the decay reaction is 4 MeV.

Since the thorium nucleus is much heavier than the alpha particle, it remains at rest, and the alpha particle carries away almost all of the energy. Thus, the kinetic energy of the alpha particle will be 4 MeV.

COMMENTS AND SUGGESTIONS: Uranium actually emits several groups of alpha particles, with slightly different energies. This is possible because the thorium nucleus can retain some of the energy and reach an excited state. The above calculation gives the energy of the alpha particle emitted if the thorium nucleus reaches its ground state.

In many cases of alpha decay, the daughter nucleus is also unstable, and decays some time after its formation, either by alpha decay or by beta decay. The daughter of the daughter then decays, and so on. The sequence of daughters descending from the original parent is called a **radioactive series.** The series ends when it reaches a stable isotope. For instance, the decay of uranium initiates a radioactive series, which ultimately ends with a stable isotope of lead.

Radioactive series

BETA DECAY The simplest beta-decay reaction is the decay of the neutron. The free neutron is unstable, and it decays into a proton, an electron, and an antineutrino:

Beta decay

$$n \to p + e^- + \bar{\nu} \tag{18}$$

Neutrino

In this equation, the symbol $\bar{\nu}$ (pronounced "nu bar") represents the antineutrino. Neutrinos and antineutrinos are particles of mass zero and spin $\frac{1}{2}\hbar$, which travel at the speed of light. Thus, they are somewhat similar to photons. However, in contrast to photons, which interact with electric charges, neutrinos and antineutrinos do not interact directly with electric charges — in fact, they hardly interact with anything at all, and they pass through the entire bulk of the Earth with little hindrance. As their names indicate, neutrinos and antineutrinos are antiparticles of each other — they can annihilate each other, producing a flash of light (gamma rays).

For a free neutron, the decay reaction (18), on the average, takes a time of 15 minutes.[3] For a neutron in a nucleus, the decay may proceed at a faster rate or at a slower rate, depending on whether the nucleus promotes the reaction by supplying extra energy or inhibits the reaction by withdrawing energy. In stable nuclei, the decay of the neutron is completely suppressed, because the nucleus withdraws a prohibitive amount of energy from the reaction. The suppression mechanism hinges on the Exclusion Principle: since the protons of the nucleus fill all the states of low energy, the extra proton released in the reaction (18) must go into an empty state of high energy, and therefore the presence of the nuclear protons effectively brings about a withdrawal of energy available for the reaction. From this we see that the stability of nuclei depends on a delicate symbiotic relationship between neutrons and protons — the presence of protons stabilizes the neutrons against decay, and the presence of neutrons stabilizes the protons against the disruptive electric forces that repel one proton from another.

When the reaction occurs in a nucleus, the net effect is the conversion of a neutron into a proton, which increases the atomic number of the nucleus by one, while leaving the mass number unchanged. Thus, the original isotope is transmuted into an isotope of the next chemical element. A few examples of beta decays are the following:

$$^{60}\text{Co} \to {}^{60}\text{Ni} + e^- \tag{19}$$

$$^{90}\text{Sr} \to {}^{90}\text{Y} + e^- \tag{20}$$

$$^{131}\text{I} \to {}^{131}\text{Xe} + e^- \tag{21}$$

In writing these reactions, we have omitted the antineutrino emitted by the nucleus, because this antineutrino is next to impossible to detect. Its existence, however, can be inferred from the conservation of energy. In the beta decay of, say, radioactive strontium [Eq. (20)], the electron emerges sometimes with one energy and sometimes with another. This is because the energy of the decay is shared between the electron and the neutrino, and sometimes one of these particles carries off most of the energy, sometimes the other. Hence, the energy of the electron is unpredictable — it can be anything from zero up to a maximum amount that corresponds to the electron carrying away *all* the energy of the decay. In fact, it was on the basis of this variability of the

---

[3] The average lifetime of free neutrons is 15 min, but the half-life (see next section) is 10.6 min.

energy of the ejected electron that Pauli first proposed the existence of the neutrino, since this was the only way to preserve the law of conservation of energy. When Pauli made this proposal in 1931, there was no direct evidence for the existence of an extra particle, and Pauli's proposal was an act of faith. Neutrinos and antineutrinos were detected only much later, in experiments with nuclear reactors.

Besides the beta-decay reactions of the type (19)–(21), which involve the ejection of an electron, there also are beta-decay reactions involving the ejection of an antielectron, or positron. The antielectron has the same mass and spin as the electron, but the opposite elecric charge. An example of a decay reaction with ejection of an antielectron is

$$^{22}\text{Na} \rightarrow {}^{22}\text{Ne} + e^+ \tag{22}$$

The antielectron ejected in this decay will sooner or later collide with one of the abundant atomic electrons in the surrounding environment and annihilate with it, producing gamma rays.

All the beta-decay reactions are instigated by a new kind of force, the **"weak" interaction.** This is one of the four fundamental forces found in matter (see Section 6.1). Whereas the electromagnetic and the strong forces play an important role in determining the structure of atoms and of nuclei, the weak force plays no role in this. Its only effect is to engender the beta-decay reactions (19)–(22) and other, similar reactions.

"Weak" interaction

EXAMPLE 3. What is the maximum kinetic energy of the beta rays emitted in neutron decay?

SOLUTION: The energy for the decay is supplied by the mass difference between the neutron and the proton. With the masses given in Table 45.1, the mass difference is

$$1.00866 \text{ u} - 1.00728 \text{ u} = 0.00138 \text{ u}$$

According to Eq. (7), this mass difference corresponds to an energy difference

$$0.00138 \times 931.5 \text{ MeV} = 1.29 \text{ MeV}$$

The proton, being much heavier than the electron, remains at rest, or nearly at rest. The decay energy is therefore shared between the electron and the antineutrino. If the antineutrino carries away none of the energy, then the electron will acquire all the energy. Thus, the maximum possible electron energy is 1.29 MeV. However, this energy includes the rest-mass energy of the electron; i.e., 1.29 MeV is the total (relativistic) energy of the electron. The kinetic energy of the electron is the total energy minus the rest-mass energy. The rest-mass energy of the electron is

$$m_e c^2 = 5.49 \times 10^{-4} \text{u} \times c^2 = 5.49 \times 10^{-4} \times 931.5 \text{ MeV} = 0.511 \text{ MeV}$$

Hence, the kinetic energy of the electron is

$$K = 1.29 \text{ MeV} - 0.511 \text{ MeV} = 0.78 \text{ MeV}$$

GAMMA EMISSION Gamma rays are emitted by nuclei when nucleons make transitions from one stationary nuclear state to another. The emission of a gamma ray by transitions of nucleons is similar to the emission of visible photons or of X rays by transitions of the atomic

Gamma emission

electrons. When a nucleus suffers alpha decay or beta decay, it is often left in an excited state, and it then eliminates the excitation energy in the form of a gamma ray. Thus, gamma emission is usually a two-step process, with an alpha or beta decay preceding the gamma emission. For example, the beta decay of ^{60}Co leads to subsequent gamma emission according to the sequence of reactions

$$^{60}\text{Co} \rightarrow {}^{60}\text{Ni} + e^-$$
$$\hookrightarrow {}^{60}\text{Ni} + \gamma \qquad (23)$$

The isotope ^{60}Co is commonly employed in high-intensity industrial irradiation cells, called cobalt cells. But, as indicated by Eq. (23), the emitter of the gamma rays is nickel, not cobalt.

Geiger counter

Alpha, beta, and gamma rays are invisible, but they can be detected with Geiger counters and with scintillation counters. Figure 45.7 shows a schematic diagram of a Geiger counter. It consists of a conducting cylindrical tube with a fine wire placed along the axis. The tube contains air or argon at low pressure. A high voltage, about a thousand volts, is applied between the tube and the wire; this creates a strong electric field in the vicinity of the wire. The voltage is adjusted so it is just short of what is required to initiate an electric discharge between the wire and the tube. When an alpha, beta, or gamma ray passes through the tube, it disrupts the atoms of the gas and produces ions (see Interlude IV for more discussion of the mechanism of ionization). The presence of these ions triggers a sudden electric discharge in the tube. The pulse of current from this discharge is registered, or "counted," by electronic circuits connected to the tube.

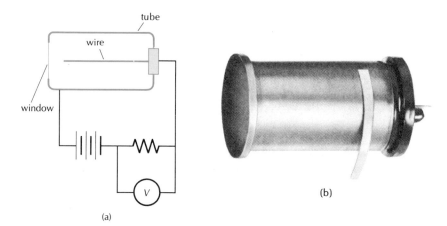

Fig. 45.7 (a) Schematic diagram of a Geiger counter. When an electric discharge occurs in the tube, a pulse of current flows through the resistor. This pulse is detected by the voltmeter. (b) Tube of a Geiger counter.

Scintillation counter

In modern laboratories, scintillation counters are widely preferred over Geiger counters. Scintillators are crystals (for example, sodium iodide and anthracene), liquids, or plastics that emit weak flashes of light when alpha, beta, or gamma rays pass through them. The rays disturb electrons in the scintillator material, raising them to high energy levels, from which they make downward transitions with the emission of photons. These photons, which are usually too faint to be seen by the eye, can be intercepted by a sensitive photomultiplier tube placed adjacent to the volume of scintillator. In such a photomultiplier tube (see Section 42.3), the incident photons produce a detectable pulse of current.

45.4 The Law of Radioactive Decay

In radioactive alpha or beta decay, the original isotope, or parent, becomes transmuted into another isotope, or daughter. If we have a given initial amount of parent material, say, 1 gram of radioactive strontium, some nuclei decay at one time and some at another, and the parent material disappears only gradually. In the case of radioactive strontium, one-half of the original amount disappears in 29 years, one-half of the remainder in the next 29 years, one-half of the new remainder in the next 20 years, and so on. Hence the amount left at times 0, 29, 58, 87 years, etc. is 1, $\frac{1}{2}$, $\frac{1}{4}$, $\frac{1}{8}$ gram, etc. The time interval of 29 years is called the **half-life**, or $t_{1/2}$, of strontium.

Mathematically, we can represent the amount of strontium at different times by the formula

$$n = n_0 \left(\frac{1}{2}\right)^{t/t_{1/2}} \tag{24}$$

Marie Sklodowska Curie, 1867–1934, and **Pierre Curie,** 1859–1906, French physicists and chemists, professors at the Sorbonne. For their discovery of the radioactive elements radium and polonium, they shared the 1903 Nobel Prize with Becquerel. On the death of Pierre, Marie succeeded him in his professorship at the Sorbonne. She received a second Nobel Prize, in chemistry, for further work on radium. Her daughter, Irène Joliot-Curie, shared the 1935 Nobel Prize with Frédéric Joliot-Curie for their work on production of radioactive substances by bombardment with alpha particles.

where n is the number of strontium nuclei at time t, and n_0 is the number at the initial time. As expected, the formula (24) yields $n = n_0$ at $t = 0$, $n = n_0 \times 1/2$ at $t = 29$ years, $n = n_0 \times 1/4$ at $t = 58$ years, etc. Note that here we are measuring the amount of strontium by the number of nuclei, rather than by the mass; the number of nuclei and the mass of strontium are, of course, proportional (1 gram of strontium corresponds to 6.70×10^{21} nuclei).

Since $1/2 = e^{-\ln 2}$, we can rewrite Eq. (24) as

$$n = n_0 e^{-t(\ln 2)/t_{1/2}}$$

or as

$$n = n_0 e^{-t/\tau} \tag{25}$$

Law of radioactive decay

where

$$\tau = \frac{t_{1/2}}{\ln 2} = \frac{t_{1/2}}{0.693} \tag{26}$$

Mean lifetime and half-life

Equation (25) is called the **law of radioactive decay,** and the quantity τ is called the **mean lifetime** (it can be demonstrated that τ is, indeed, the average lifetime of the radioactive nuclei in the sample). Figure 45.8 is a plot of n vs. t for the decay of 1 gram of strontium.

With appropriate values of the lifetime, the law (25) is, of course, applicable to the decay of any radioactive isotope. Half-lives of different radioisotopes vary over a wide range. Some have half-lives of several billions of years; others have half-lives as short as a fraction of a second. Table 45.3 lists the half-lives of some radioisotopes. The chart of isotopes in Table 45.2 lists some more half-lives.

Since each decay produces one alpha or beta ray, the rate of emission of rays by the parent material is equal to the **decay rate,** or the number of parent nuclei that decay per second. The instantaneous

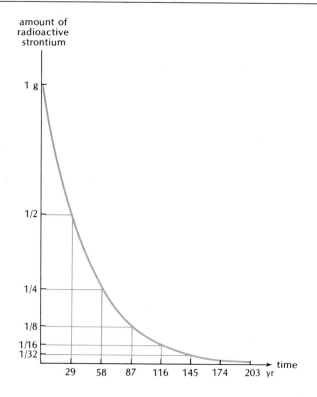

Fig. 45.8 Amount of remaining strontium vs. time. In this plot, the amount of strontium is measured in grams. The initial amount is 1 gram, which corresponds to 6.7×10^{21} nuclei. After 29 years the remaining amount is $\frac{1}{2}$ gram, which corresponds to 3.3×10^{21} nuclei.

decay rate is obtained by differentiating n with respect to time:

$$-\frac{dn}{dt} = -\frac{d}{dt} n_0 e^{-t/\tau} \tag{27}$$

or

Decay rate

$$\boxed{-\frac{dn}{dt} = \frac{n_0}{\tau} e^{-t/\tau} = \frac{0.693}{t_{1/2}} n_0 e^{-t/\tau}} \tag{28}$$

Thus, the rate of decay is directly proportional to the number of parent nuclei and inversely proportional to the half-life.

The SI unit for radioactive decay rate is the **becquerel** (Bq),

Becquerel, Bq

$$1 \text{ becquerel} = 1 \text{ Bq} = 1 \frac{\text{disintegration}}{\text{s}} \tag{29}$$

Table 45.3 SOME RADIOISOTOPES

Radioisotope	Radioactivity	Half-life
^{14}C	β	5730 years
^{22}Na	β,γ	2.6 years
^{60}Co	β,γ	5.27 years
^{90}Sr	β,γ	28.8 years
^{131}I	β,γ	8.04 days
^{226}Ra	α,γ	1600 years
^{238}U	α,γ	4.5×10^9 years

In practice, a larger unit is traditionally preferred; this larger unit is the **curie** (Ci),

Curie, Ci

$$1 \text{ curie} = 1 \text{ Ci} = 3.70 \times 10^{10} \frac{\text{disintegrations}}{\text{s}} \qquad (30)$$

EXAMPLE 4. What is the decay rate of 1 gram of radioactive strontium, ^{90}Sr?

SOLUTION: The atomic mass of this strontium isotope is 89.9 g; thus, 1 gram of strontium is 1/89.9 mole, or $1/89.9 \times 6.02 \times 10^{23}$ atoms $= 6.7 \times 10^{21}$ atoms. According to Eq. (28), the decay rate is

$$-\frac{dn}{dt} = \frac{0.693}{29 \text{ years}} \times n = \frac{0.693}{29 \text{ years}} \times 6.7 \times 10^{21}$$

$$= \frac{0.693}{29 \times 3.16 \times 10^{7} \text{ s}} \times 6.7 \times 10^{21}$$

$$= 5.1 \times 10^{12} \text{ /s} = 5.1 \times 10^{12} \text{ Bq}$$

Thus, 1 gram of strontium emits 5.1×10^{12} beta rays per second.

45.5 Nuclear Reactions

Radioactive decay is a spontaneous nuclear reaction. Other kinds of nuclear reactions can be initiated by bombarding a nucleus with a projectile, such as an alpha particle or another nucleus. The first reaction of this kind was discovered in 1919 by Rutherford, who bombarded nitrogen nuclei with fast-moving alpha particles and found that this resulted in the formation of an isotope of oxygen and the ejection of protons. Since the alpha particle is a helium nucleus and the proton is a hydrogen nucleus, we can write the reaction as

$$^{4}\text{He} + {}^{14}\text{N} \rightarrow {}^{17}\text{O} + {}^{1}\text{H} \qquad (31)$$

This reaction achieved the first artificial transmutation of chemical elements.

In these early experimental investigations of nuclear reactions, Rutherford used alpha particles from natural radioactive sources. Nowadays, experimental physicists use much more energetic projectiles produced by **accelerators**. These machines, popularly called atom smashers, produce intense beams of protons, electrons, alpha particles, deuterons, or ions, and they accelerate these projectiles to high energies by pushing on them with suitable electric forces. The first such accelerators were built in 1932 by R. J. Van de Graaff and by J. D. Cockcroft and E. T. S. Walton (Figure 45.9); both of these accelerators produced beams of energetic protons. By bombarding nuclei with the protons from their machine, Cockcroft and Walton succeeded in inducing the first nuclear transmutation with artificially accelerated particles. Soon thereafter, E. O. Lawrence invented the cyclotron — a very efficient accelerator in which the particles travel round and round

Accelerators

Fig. 45.9 The accelerators of Van de Graaff (left) and of Cockcroft and Walton (right). The large metallic sphere is charged with an electrostatic generator and provides a strong electric field for the acceleration of protons.

a circular (or spiral) path under the influence of magnetic forces, while at the same time they acquire higher and higher energies from the carefully timed push of an electric force (Figure 45.10; for some more details on the operation of the cyclotron, see Section 31.3). The gigantic accelerators used in particle physics today are direct descendants of the cyclotron.

Fig. 45.10 One of Lawrence's cyclotrons.

In the typical nuclear-reaction experiment, the beam of projectiles is aimed at a stationary target consisting of a thin sheet or foil of material with the chosen target nuclei, for example a foil of gold, or aluminum, or copper. The collision of a projectile with one of the target nuclei triggers a nuclear reaction and also kicks the target nucleus out of the foil. Detectors placed around the point of impact of the beam on the target intercept the reaction products ejected in the collision. By means of appropriate combinations of detectors, it is possible to identify the reaction products and to measure their kinetic energies. From such investigations of nuclear reactions, we can extract information about the structure of the nucleus and about the forces acting on the nuclear constituents.

Nuclear reactions, of course, obey the usual conservation laws for energy (including rest-mass energy), momentum, angular momentum (including spin), and electric charge. Besides, nuclear reactions obey a conservation law for mass number, also called the conservation law for baryon number: *In any nuclear reaction, the net mass number after the reaction is the same as before the reaction.* This conservation law means that for every proton that disappears in a nuclear reaction, a neutron must

Conservation of mass number

be created and vice versa. It is easy to verify that all the reactions listed in the preceding sections obey the conservation law for the net mass number. Several other esoteric conservation laws have been discovered in high-energy nuclear physics, but we cannot deal with these here, as they involve some subtle aspects of quantum physics.

Let us examine some details of the conservation of energy in nuclear reactions. For the sake of simplicity, we will focus on reactions with just two final reaction products, such as the reaction (31). Suppose that the mass of the projectile is m_1, the mass of the target nucleus is m_2, and the masses of the two nuclei produced in the reaction are m'_2 and m'_3. As in Section 11.4, we designate the change of rest-mass energy by Q:

$$Q = (m_1 + m_2 - m'_1 - m'_2)c^2 \qquad (32)$$

Q value of reaction

The total energy of each particle or nucleus is the sum of its kinetic energy and its rest-mass energy. Thus, the law of conservation of energy for the reaction is

$$K + Q = K' \qquad (33)$$

Reactions with a positive value of Q convert some of the initial rest-mass energy into kinetic energy; such reactions are called **exoergic**. Reactions with a negative value of Q convert some of the initial kinetic energy into rest-mass energy; these reactions are called **endoergic**.

Exoergic and endoergic reactions

An endoergic reaction is not viable unless the initial kinetic energy is sufficiently large. The initial kinetic energy must provide not only the increase in rest-mass energy but also the energy for the (constant) motion of the center of mass. This means that only a portion of the initial kinetic energy of the projectile is available for conversion into rest-mass energy. According to Eq. (11.29), in a one-dimensional collision, the available portion of the initial kinetic energy (the "internal" kinetic energy) is

$$\frac{1}{2} \frac{m_1 m_2}{m_1 + m_2} (v_1 - v_2)^2$$

or, since the target is at rest,

$$\frac{1}{2} \frac{m_1 m_2}{m_1 + m_2} v_1^2 \qquad (34)$$

Thus, the available kinetic energy is smaller than the actual kinetic energy of the projectile by a factor of $m_2/(m_1 + m_2)$. For the reaction to proceed, this available kinetic energy must be at least as large as $-Q$,

$$\frac{1}{2} \frac{m_1 m_2}{m_1 + m_2} v_1^2 \geq -Q \qquad (35)$$

or

$$\frac{1}{2} m_1 v_1^2 \geq \frac{m_1 + m_2}{m_2} (-Q) \qquad (36)$$

Threshold energy

Although we have obtained this result for the minimum required kinetic energy on the assumption that the motion of all particles is one dimensional, the same result applies when the particles are allowed to move in two dimensions, because any motion in a direction transverse to the direction of incidence of the projectile requires even more kinetic energy and hence does not help the minimum kinetic energy. Note that if the projectile has exactly the minimum kinetic energy for initiating the endoergic reaction, then the reaction products will emerge from the reaction with no relative motion, that is, they will emerge together, at a speed equal to that of the center of mass.

The minimum kinetic energy required to initiate an endoergic reaction is called the **threshold energy.** When the energy of the projectile has this minimum value, the reaction is energetically viable, but the reaction rate is usually extremely low. To attain a significant (and observable) reaction rate, it is usually necessary to increase the projectile somewhat above the threshold energy.

EXAMPLE 5. Consider the following reaction initiated by a deuteron (^{2}H) incident on a carbon nucleus:

$$^2\text{H} + ^{12}\text{C} \rightarrow \,^4\text{He} + ^{10}\text{B}$$

(a) From the masses given in the chart of isotopes, find the Q value of this reaction. (b) Find the threshold energy required for the deuteron.

SOLUTION: (a) The Q value is

$$Q = (m_{^2\text{H}} + m_{^{12}\text{C}} - m_{^4\text{He}} - m_{^{10}\text{B}})c^2$$

The masses in this equation are nuclear masses. However, we can include electron masses and complete each nuclear mass to an atomic mass, since these electron masses cancel:

$$Q = [(m_{^2\text{H}} + m_e) + (m_{^{12}\text{C}} + 6m_e) - (m_{^4\text{He}} + 2m_e) - (m_{^{10}\text{B}} + 5m_e)]c^2 \tag{37}$$

The masses in parentheses are now atomic masses, for which we can substitute the values listed in the chart of isotopes:

$$Q = [2.01410 \text{ u} + 12.0000 \text{ u} - 4.0026 \text{ u} - 10.01294 \text{ u}]c^2 \tag{38}$$

$$= -0.00144 \text{ u} \times c^2 = -0.0144 \times 931.5 \text{ MeV}$$

$$= -1.34 \text{ MeV}$$

(b) According to Eq. (36), the threshold energy is

$$\frac{m_1 + m_2}{m_2} \times (-Q) = \frac{2.01 \text{ u} + 12.00 \text{ u}}{12.00 \text{ u}} \times 1.34 \text{ MeV} = 1.57 \text{ MeV}$$

COMMENTS AND SUGGESTIONS: In any nuclear reaction in which the reactants and the reaction products are nuclei or neutrons, the inclusion of electron masses in the evaluation of the Q value [see Eq. (37)] is always valid, since charge conservation in the nuclear reaction guarantees that equal numbers of electrons will be added and subtracted when we complete each nuclear mass to an atomic mass. Hence the Q value for such a reaction can always be directly evaluated with the atomic masses, as in Eq. (38).

SUMMARY

Radius of nucleus: $R = (1.2 \times 10^{-15} \text{ m}) \times A^{1/3}$

Energy equivalent of atomic mass unit: $1 \text{ u} \times c^2 = 931.5 \text{ MeV}$

Mass defect and binding energy: $\text{B.E.}/c^2 = [\text{mass defect}]$
$= [\text{mass of } N \text{ neutrons and } Z \text{ protons}] - [\text{mass of nucleus}]$

Law of radioactive decay: $n = n_0 e^{-t/\tau}$

$$\tau = \frac{t_{1/2}}{0.693}$$

Decay rate: $-\dfrac{dn}{dt} = \dfrac{n}{\tau} = \dfrac{0.693}{t_{1/2}} n$

Becquerel: $1 \text{ Bq} = 1 \text{ disintegration/s}$

Curie: $1 \text{ Ci} = 3.70 \times 10^{10} \text{ disintegrations/s}$

Q value of reaction: $Q = (m_1 + m_2 - m_1' - m_2')c^2$

Threshold energy: $\frac{1}{2} m_1 v_1^2 \geq \dfrac{m_1 + m_2}{m_2}(-Q)$

QUESTIONS

1. Qualitatively, why would you expect the magnetic moment of the proton to be smaller than that of the electron?

2. How many protons and neutrons are there in the nucleus of the isotope ^{16}O? ^{56}Fe? ^{238}U?

3. What isotope has 82 protons and 122 neutrons in its nucleus?

4. Naturally occurring magnesium has an atomic mass of 24.305. On the basis of this information, can you conclude that natural magnesium contains a mixture of several isotopes? Can you guess which isotopes?

5. What isotope has 17 protons and 18 neutrons in its nucleus? What isotope has 18 protons and 17 neutrons? Such isotopes in which the numbers of protons and neutrons are exchanged are called **mirror nuclei**. Find another example of mirror nuclei on the chart of isotopes.

Mirror nuclei

6. By inspection of Figure 45.4, determine the ratio of neutrons to protons in stable light nuclei ($Z \cong 20$) and in stable heavy nuclei ($Z \cong 80$).

7. According to Figure 45.5, which isotope has the largest binding energy per nucleon? Which has the least binding energy and is therefore capable of releasing the most energy in a nuclear reaction?

8. What isotope is formed by the alpha decay of ^{210}Po? ^{239}Pu?

9. What isotope is formed by the beta decay of ^{85}Kr? ^{63}Ni?

10. What isotope is formed by the positron beta decay, or beta plus decay, of ^{22}Na? ^{64}Cu?

11. According to Figure 45.8, at what time will the remaining amount of radioactive strontium have fallen to $\frac{1}{10}$ of the initial amount?

12. Hydrogen bombs operate with tritium, a radioactive isotope of hydrogen with a half-life of 5.5 years. When a hydrogen bomb is held in storage, its tritium gradually decays and must be replenished with fresh tritium every few

years. According to a recent proposal, arms control could be achieved by halting the production of tritium. Every few years, the superpowers would then have to take some of their bombs out of service, extract the remaining tritium, and use it to replenish their other bombs. Suppose that the United States and the Soviet Union each have 2000 hydrogen bombs now. Without fresh tritium, how many bombs will each superpower have in 11 years? In 22 years?

13. Atmospheric carbon dioxide contains the stable isotope ^{12}C and also the radioactive isotope ^{14}C, with a half-life of 5730 years. When a plant or an animal dies, it ceases to inhale, and the ^{14}C in its body then decreases according to the law of radioactive decay. Thus, fresh wood has the same abundance of ^{14}C relative to ^{12}C as air, but fossilized wood (coal) has no ^{14}C. Suppose that in a piece of wood discovered in an ancient tomb, the relative abundance of ^{14}C has decreased to $\frac{1}{4}$ of the initial relative abundance. What age can you deduce for the piece of wood?

14. Alpha particles are more massive than beta particles, yet for equal kinetic energies, the alpha particles are stopped by a thinner layer of material than the beta particles. Can you explain this? (Hint: Think of the electric forces that act on the particles in a collision with an atom.)

PROBLEMS

Section 45.1

1. According to Eq. (3), what is the nuclear radius of the smallest of the isotopes of carbon? The largest of these isotopes?

2. The largest known nucleus is that of the isotope^{266} of element 109, which has not yet been given a name. What is the radius of this nucleus?

3. Neutron stars are made (almost) entirely of neutrons, and they have approximately the same density as a nucleus. What is the radius of a neutron star of mass 0.5 times the mass of the Sun?

*4. What fraction of the volume of your body is filled with nuclear material? The average density of your body is about 1000 kg/m^3.

*5. A uranium nucleus of radius 7.4×10^{-15} m fissions into two equal spherical pieces. What is the radius of each piece?

*6. Suppose you bombard a target of aluminum with alpha particles. If, in a head-on collision with a stationary aluminum nucleus, an alpha particle is to just barely touch the nuclear surface before being halted by the repulsive electric force, what must be the initial kinetic energy of the alpha particle? Express your answer in MeV.

*7. Suppose that an alpha particle of energy 4.4 MeV collides head-on with a stationary gold nucleus. What is the distance of closest approach? Does the alpha particle make contact with the surface of the nucleus?

8. The deuteron, or nucleus of the isotope ^2H, consists of a proton and a neutron. The spin of the deuteron is 1 $\hbar$. What prediction can you make for the magnetic moment of the deuteron?

*9. When an atom is placed in an external magnetic field, the magnetic moment of the nucleus will interact with the magnetic field. The different quantized orientations of the nuclear spin in this magnetic field therefore will have different energies, and the nucleus can make transitions between these energy levels, with the emission of photons. Suppose that a nucleus has spin $\frac{1}{2} \hbar$ and a magnetic moment 1.41×10^{-26} A · m^2. Suppose this nucleus is immersed in an external magnetic field of 0.2 T.
(a) What are the possible orientations of the spin, and what are the energy levels of these states of different orientations?

(b) If the nucleus makes a transition from the upper state to the lower state, what is the frequency of the emitted photon? What kind of electromagnetic radiation is this?

*10. The interaction of the magnetic moment of the nucleus with the magnetic moment of the electrons (orbital plus spin) splits each of the energy levels of an atom into a multiplet of closely spaced energy levels. Consider the case of a hydrogen atom in the state $n = 0$, $l = 0$. This state splits into two states, one of which has the electron spin parallel to the spin of the nucleus, and the other has the electron spin antiparallel to the spin of the nucleus.
   (a) Which of these two states has the higher energy?
   (b) When hydrogen atoms make transitions from this state of higher energy to the state of lower energy, they emit radio waves of wavelength 21 cm (this is the famous 21-cm line in the radio spectrum of hydrogen, of great importance in radio astronomy). Calculate the energy difference between the two states; express your answer in electron-volts.

Section 45.2

11. The binding energy of the electron in the ground state of the hydrogen atom is 13.6 eV. Calculate the corresponding mass defect of the hydrogen atom; express this in atomic mass units. (Since the mass defects associated with atomic binding energies are small, they are usually ignored in nuclear physics.)

*12. According to Figure 45.5, what is the binding energy of ^{56}Fe? From this binding energy and the masses of the proton and the neutron, calculate the mass of the isotope (in atomic mass units).

*13. When two protons are in contact, their center-to-center distance is 2.0×10^{-15} m. Roughly, evaluate the magnitude of the nuclear attractive force at this distance from the potential energy plotted in Figure 45.2. Compare the magnitude of this force with that of the repulsive electric force.

*14. What are the mass defect (in atomic mass units) and the nuclear binding energy (in MeV) of the isotope ^{4}He? The mass of ^{4}He is given in the chart of isotopes.

*15. What are the mass defect (in atomic mass units) and the nuclear binding energy (in MeV) of the isotope ^{12}C?

*16. How much energy (in MeV) is required to remove one neutron from the nucleus of the isotope ^{14}N? (Hint: Compare the binding energies of ^{14}N and ^{13}N; the masses of these isotopes are given in the chart of isotopes.)

*17. Nuclei with 2, 8, 14, 20, 28, 50, 82, or 126 protons or neutrons are called **magic nuclei** because they have exceptionally large binding energies and are exceptionally stable. Compare the binding energy of the magic nucleus ^{16}O with that of the nonmagic nucleus ^{16}F. The masses of these isotopes are given in the chart of isotopes.

**18. According to the alpha-particle model of the nucleus, some nuclei consist of an arrangement of alpha particles. For instance, ^{12}C consists of three alpha particles arranged at the vertices of an equilateral triangle, and ^{16}O consists of four alpha particles arranged at the vertices of a tetrahedron.
   (a) Calculate the binding energies of ^{12}C and of ^{4}He from their masses, and find the energy per alpha–alpha bond.
   (b) With your value for the energy per alpha–alpha bond, predict the binding energy of ^{16}O and compare your prediction with the binding energy calculated from the mass of this isotope.

Section 45.3

19. The alpha decay of uranium [see Eq. (13)] is the first step in a radioactive series of decays. The next four steps are a beta decay, another beta decay, another alpha decay, and another alpha decay. What are the daughter nuclei produced in these four steps?

*20. Find the maximum kinetic energy of the beta ray emitted in the reaction

$$^{16}\text{N} \rightarrow {}^{16}\text{O} + e^-$$

Use the values of the masses given in the chart of isotopes, and assume that the beta ray carries away all the available energy. Be careful to take into account the electron masses in the isotopes in your calculation.

*21. What isotope is formed in the beta decay of ^{14}C? Calculate the maximum kinetic energy of the beta ray emitted in this decay. Use the values of the masses given in the chart of isotopes, and assume that the beta ray carries away all the available energy. Be careful to take into account the electron masses in the isotopes in your calculation.

*22. The alpha decay of ^{210}Po results in ^{206}Pb. Calculate the energy of the emitted alpha particle. The masses of these two isotopes are 209.9829 u and 205.9745 u, respectively. Assume that the ^{206}Pb nucleus is formed in its ground state.

*23. Consider the motion of the alpha particle emitted in the decay of ^{238}U [see Eq. (13)].
   (a) Suppose that, at the initial instant, the alpha particle is at rest at the surface of the residual nucleus ^{234}Th. What kinetic energy will the alpha particle gain from the electric repulsion as it moves radially outward to a very large distance? Express your answer in MeV.
   (b) The actual measured kinetic energy of alpha particles emitted in this decay is 4.20 MeV. The discrepancy between this measured kinetic energy and the kinetic energy calculated in (a) is due to a wave-mechanical effect. Instead of starting its outward motion at the nuclear surface, at the highest point of the electric potential, the alpha particle "tunnels" under the electric potential and begins its motion at some distance from the nuclear surface. Extrapolate the motion of a 4.20 MeV alpha particle backward, and find at what distance from the ^{234}Th nucleus it is at rest.

Section 45.4

24. The isotope ^{238}U found on the Earth was originally synthesized in nuclear reactions in supernovas that exploded in our Galaxy about 6.8×10^9 years ago and scattered debris through the Galaxy, some of which became trapped in the Earth during its formation, about 4.6×10^9 years ago. What fraction of the original amount of uranium synthesized in the Galaxy is still in existence? What fraction of the amount of uranium trapped in the Earth during its formation is still in existence? The half-life of ^{238}U is 4.5×10^9 years.

25. The half-life of radiophosphorus (^{32}P) is 14.5 days. If you initially have a sample of 1.0 microgram of radiophosphorus, after how many days will the remaining amount be 0.1 microgram? 0.01 microgram? 0.001 microgram?

26. The activity of an industrial ^{60}Co irradiation cell is 1.0×10^{16} Bq when the cell is new. What will be the activity after the cell has been in use for two years? The half-life of ^{60}Co is 5.3 years.

*27. What is the activity of 1 gram of pure radium (^{226}Ra)? The half-life of radium is 1600 years.

*28. The partial eradication of the thyroid in patients suffering from hyperthyroidism can be accomplished by injecting a compound containing radioactive iodine ^{131}I into the body; the iodine then concentrates in the thyroid and kills its cells. If the thyroid is to be subjected to an activity of 0.1 Ci, how many grams of ^{131}I should be injected? The half-life of ^{131}I is 8.04 days.

*29. Table 45.2 lists the half-lives of ^{22}Na, ^{60}Co, and ^{90}Sr. For each of these radioisotopes, calculate what amount (in grams) will give an activity of 1 millicurie.

*30. To determine the half-life of ^{210}Po, a physicist uses a Geiger counter to measure the alpha activity of a sample of this isotope at the initial time and then again 30 days later. The first measurement gives 2.0×10^{-4} Ci, and the second measurement gives 1.7×10^{-4} Ci. What is the half-life? What was the initial amount of ^{210}Po (in grams)?

*31. Both ^{140}Ba and ^{141}Ce are beta emitters, with half-lives of 12.8 days and 32.5 days, respectively. Suppose you have a sample consisting initially of a mixture with 1 microgram of each radioisotope. What is the net activity of this sample at the initial time? After 10 days? 20 days? 30 days?

**32. Prove that if the half-life of some radioisotope is $t_{1/2}$, then the mean lifetime is $t_{1/2}/\ln 2$. $\left(\text{Hint: } \int_0^\infty t\, e^{-t/\tau}\, dt = \tau^2\right)$

Section 45.5

33. Which of the following reactions are forbidden by the conservation laws for electric charge or mass number?

$$^2\text{H} + {}^{12}\text{C} \rightarrow {}^4\text{He} + {}^9\text{B}$$
$$^2\text{H} + {}^{63}\text{Cu} \rightarrow {}^{63}\text{Zn} + n + {}^1\text{H}$$
$$^1\text{H} + {}^{52}\text{Cr} \rightarrow {}^{52}\text{Mn} + n$$
$$^4\text{He} + {}^{10}\text{B} \rightarrow {}^{13}\text{C} + n$$
$$n + {}^{27}\text{Al} \rightarrow {}^{24}\text{Na} + {}^4\text{He}$$
$$n + {}^{238}\text{U} \rightarrow {}^{121}\text{Ag} + {}^{118}\text{Pd}$$

34. If neutrons had a somewhat smaller mass, then the (slow) electrons in atoms could combine with the protons in the nucleus according to reaction

$$e + p \rightarrow n + \nu$$

How much smaller would the mass of the neutron have to be to make this reaction viable? What consequences would this reaction have for the existence of atoms and the existence of life?

35. Bombardment of lithium with protons results in the reaction

$$^1\text{H} + {}^7\text{Li} \rightarrow {}^7\text{Be} + n$$

What is the Q value for this reaction? What is the threshold energy for this reaction?

36. Consider the reaction

$$^1\text{H} + {}^3\text{H} \rightarrow {}^3\text{He} + n$$

(a) What is the Q value for this reaction?

(b) If we initiate this reaction by bombarding a tritium target with protons, what is the threshold energy?

(c) If we initiate this reaction by bombarding a hydrogen target with tritium nuclei, what is the threshold energy?

*37. The following sequence of reactions, called the proton–proton chain, serves as source for the thermal energy of the Sun:

$$^1\text{H} + {}^1\text{H} \rightarrow {}^2\text{H} + e^+ + \nu$$
$$^1\text{H} + {}^2\text{H} \rightarrow {}^3\text{He} + \gamma$$
$$^3\text{He} + {}^3\text{He} \rightarrow {}^4\text{He} + {}^1\text{H} + {}^1\text{H}$$

(a) For each of these reactions, calculate the energy released; express your answer in MeV. Note that the neutrino (ν) and the gamma ray (γ) have no mass, and they can therefore be included in the released energy.

(b) If each of the first two reactions proceeds twice and the last reaction proceeds once, the net result is the consumption of four protons and the formation of one helium nucleus. What is the energy released per consumed proton?

(c) The Sun releases thermal energy at the rate of 3.9×10^{26} W. What is the rate at which the Sun consumes protons? Express your answer in kg of protons per second.

(d) What is the rate at which the Sun releases neutrinos?

*38. Direct experimental measurement of the Q value of a reaction can be used to determine mass differences between nuclei. For instance, consider the reaction

$$^2\text{H} + {}^{14}\text{N} \rightarrow {}^{15}\text{N} + {}^1\text{H}$$

From direct measurement of the energy of the emerging proton, it is known that the Q value of this reaction is 8.61 MeV. From this, determine the mass difference (in atomic mass units) between the ^{15}N and ^{14}N isotopes. Assume that the masses of the ^{1}H and ^{2}H isotopes are known, as listed in the chart of isotopes.

**39. Show that for a one-dimensional collision Q can be expressed in terms of the kinetic energies of the projectile and *one* of the reaction products, as follows:

$$Q = \tfrac{1}{2} m_1' v_1'^2 \left(1 + \frac{m_1'}{m_2'}\right) - \tfrac{1}{2} m_1 v_1^2 \left(1 - \frac{m_1}{m_2'}\right) - \frac{m_1 m_1'}{m_2'} v_1 v_1'$$

**40. Deuterons incident on a tritium target produce the exoergic reaction

$$^2\text{H} + {}^3\text{H} \rightarrow {}^4\text{He} + n$$

(a) What is the Q value of this reaction?

(b) Suppose that the deuteron energy is 6.00 MeV. If the collision is one dimensional, what is the energy of the emerging neutron?

INTERLUDE XI

NUCLEAR FISSION*

In heavy, large nuclei — such as uranium — the balance between the attractive strong force and the repulsive electric force is quite precarious. Although the strong force between two adjacent nucleons in the nucleus can easily exceed the electric force, the strong force has only a short range, and it cannot bridge the distance from one side of a large nucleus to the other. Hence the interaction of the distant parts of a large nucleus is dominated by the repulsive electric force. The dominance of the repulsive electric force becomes more pronounced if the nucleus suffers some extra elongation, either spontaneous or caused by an external disturbance, such as the impact of an external neutron. The electric force then overcomes the strong force, and the nucleus splits, or fissions, into two or more fragments. The fission of uranium, as manifested in the explosion of a nuclear bomb (Figure XI.1), provides a spectacular demonstration of the victory of the electric force over the strong force.

Fig. XI.1 The mushroom cloud of a nuclear explosion of low yield.

* This chapter is optional.

During fission, the large electric energy stored within the nucleus is (partially) converted into kinetic energy of the fission fragments. Nuclear bombs and nuclear reactors derive almost all their power from this conversion of electric energy into kinetic energy. Thus, the "nuclear" energy released in fission devices is primarily electric energy.

XI.1 FISSION

From our discussion of the curve of binding energy (Figure 45.5), we know that for most nuclei the binding energy per nucleon is in the vicinity of 8 MeV, whereas for heavy nuclei (say, $A > 200$) the binding energy per nucleon is somewhat smaller. The reduction of the binding energy is due to the increasing importance of the electric repulsion. This indicates that it is energetically favorable for a heavy nucleus to split up into two fragments, forming two lighter nuclei. For example, consider the splitting of a ^{238}U nucleus into two equal fragments:

$$^{238}\text{U} \rightarrow {}^{119}\text{Pd} + {}^{119}\text{Pd} \qquad (1)$$

This is a **symmetric fission** reaction: a nucleus of mass number 238 and charge $92e$ splits into two nuclei, each of mass number 119 and charge $46e$.

The binding energy of the ^{238}U nucleus is approximately 1800 MeV and that of each ^{119}Pd nucleus is 990 MeV. The energy released in the reaction (1) is therefore

$$-1800 \text{ MeV} + 2 \times 990 \text{ MeV} = 180 \text{ MeV} \qquad (2)$$

Thus, this fission reaction is *permitted* by energy conservation — but what is permitted is not compulsory. The process of fission begins with a slight elongation of the nucleus (Figure XI.2b); the elongation then increases, developing two bulges joined by a waist (Figures XI.2c and d); finally, the waist pinches off and the two fragments separate (Figures XI.2e and f). Obviously, the attractive strong forces resist the elongation of the nucleus, and it is only when the nucleus splits (Figures XI.2e and f) that these strong forces ir-

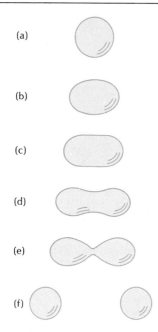

Fig. XI.2 Stages of deformation in the fission of a nucleus.

splitting into two unequal fragments. However, the mass difference between the fragments is usually not all that large (typically 30%), and the energy released is not all that far from the value calculated in Eq. (2).

The energy of 180 MeV released by the reaction (1) represents the net work done by the electric and the strong forces during the fission process. This energy will emerge in the form of kinetic energy of the fission fragments, kinetic energy that is impressed on the fragments by the repulsive electric force. The electric force does positive work (it pushes the fragments apart, increasing their kinetic energy), while the strong force does negative work (it attempts to pull the fragments together, decreasing their kinetic energy). Thus, the initial electric energy of the nucleus must not only supply all the kinetic energy of the fragments, but it must also supply the energy absorbed by the strong force.

The total energy released as a consequence of the fission of a uranium nucleus is actually somewhat larger than 180 MeV. The fission fragments are very unstable — they contain both an unacceptable excess of internal (vibrational) energy and an unacceptable excess of neutrons. To eliminate these excesses, the fission fragments undergo a series of radioactive decays involving the ejection of neutrons, beta rays, gamma rays, and neutrinos. Some of these particles are ejected almost instantaneously, while others suffer a slight delay. These secondary emissions release roughly an extra 20 MeV and bring the total energy yield per fission to about 200 MeV. Table XI.1 shows how the total energy is distributed, on the average, among fission fragments and other ejecta.

revocably lose out to the repulsive electric forces. Figure XI.3 shows a plot of the potential energy of the drop of nuclear fluid as a function of its deformation; zero deformation corresponds to Figure XI.2a; extreme deformation corresponds to Figure XI.2f, in which the drop has split into two well-separated drops. Although the energy of the undeformed drop is higher than the energy of the separated drops, the intermediate stages of deformation have the highest energy; thus, there is a potential barrier which tends to prevent fission.

In the case of ^{238}U, spontaneous fission is a rare occurrence; only 1 nucleus in 2×10^6 will decay by spontaneous fission, the others will decay by alpha emission. Furthermore, when spontaneous fission does occur, it is rarely symmetric [as in Eq. (1)]; more often than not the fission is asymmetric, the nucleus

Table XI.1 Distribution of Energy in Fission

Fission fragments (kinetic energy)	165 ± 5 MeV
Neutrons (kinetic energy)	5 ± 0.5
Gamma rays, instantaneous	7 ± 1
Beta rays, delayed	7 ± 1
Gamma rays, delayed	6 ± 1
Neutrinos, delayed	10
Total energy per fission	200 ± 6 MeV

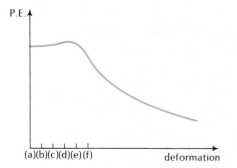

Fig. XI.3 Potential energy as a function of deformation. The marks on the horizontal axis correspond to the stages of deformation shown in Figure XI.2. The potential energy is maximum at stage (d).

Note that, according to this table, the average kinetic energy of the fission fragments is somewhat lower than was calculated above [see Eq. (2)]; this is so because asymmetric fission, which predominates, yields somewhat lower kinetic energies than symmetric fission.

We can gain a better appreciation of the large amount of energy released in fission by looking at the following numbers. The total energy released by the complete fission of 1 kg of uranium (a lump slightly larger than a golf ball) is 200 MeV times the number of atoms in 1 kg of uranium, that is,

$$200 \frac{\text{MeV}}{\text{atom}} \times 6.02 \times 10^{23} \frac{\text{atoms}}{\text{mole}} \times \frac{1000}{238} \text{ moles}$$

$$= 5.1 \times 10^{26} \text{ MeV}$$

$$= 8.1 \times 10^{13} \text{ J}$$

This is equivalent to the energy released in the burning of 2 million liters of gasoline. It is also equivalent to the energy released in the explosion of about 20,000 tons of TNT.[1]

XI.2 CHAIN REACTIONS

The rate of spontaneous fission in uranium is very low. Consequently the rate of release of energy is also very low and quite inadequate for applications such as nuclear reactors or nuclear bombs. However, the rate of fission increases drastically if the uranium is exposed to a flux of neutrons. The absorption of an incident neutron by a uranium nucleus will deform the nucleus and set it in violent vibration which may split it apart (Figure XI.4). This results in neutron-induced fission reactions, such as

$$\text{neutron} + {}^{238}\text{U} \rightarrow {}^{145}\text{Ba} + {}^{94}\text{Kr} \quad (3)$$

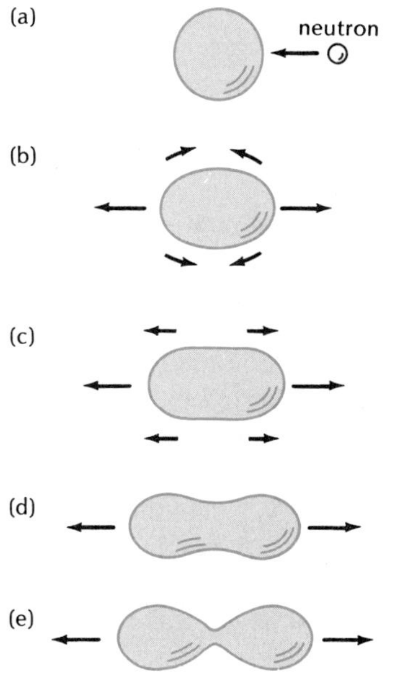

Fig. XI.4 Impact of a neutron on a uranium nucleus causes vibrations that lead to fission.

[1] One ton (2000 lb, or 910 kg) of TNT releases 10^9 cal $= 4.2 \times 10^9$ J.

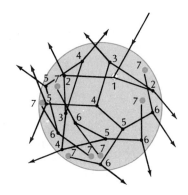

Fig. XI.5 A chain reaction in a supercritical mass of fissionable material. The reaction starts at (1) with the arrival of a neutron from the outside. The diagram shows the first seven steps of the chain reaction. Each fission absorbs one neutron and releases two; occasionally a neutron escapes from the surface of the mass and is lost.

The rate of neutron-induced fission depends on the supply of neutrons and on their kinetic energy. What makes fission reactions such as in Eq. (3) interesting is that the reaction itself supplies neutrons — both fission fragments in the reaction (3) are very neutron rich and they almost instantaneously eject two or three neutrons. Hence the net reaction is

$$\text{neutron} + {}^{238}\text{U} \rightarrow \text{fission fragments}$$
$$+ \text{ 2 or 3 neutrons} \quad (4)$$

The neutrons released in this first fission can then strike other uranium nuclei and induce their fission, and the neutrons released there can induce further fissions, and so on. If no neutrons, or only a few neutrons, are lost from this avalanche, then the result is a self-sustaining **chain reaction** (Figure XI.5). The rate of fission and the rate of release of energy grow drastically with time. For example, if on the average two neutrons released in each fission reaction succeed in generating further fission reactions and further neutrons, then the numbers of fission reactions in successive steps of the chain will be 2, 4, 8, 16, 64, ... If this geometric growth continues unchecked, the rate of release of energy will become explosive.

In the case of ^{238}U, conditions are not favorable for sustaining a chain reaction. Uranium-238 is a fairly stable nucleus; it will fission only if struck a hard blow by an energetic neutron — the kinetic energy of the incident neutron must be at least 1.2 MeV. The neutrons released by the fission of a uranium nucleus are energetic enough (see Table XI.1), and when they collide with other nuclei they will occasionally induce a fission; but by far the most likely outcome of a collision is inelastic scattering, that is, the neutron bounces off the nucleus with some loss of kinetic energy. Successive collisions of this kind gradually reduce the kinetic energy of the neutrons and remove their ability to induce fission — the neutrons are effectively lost from the fission chain.

In the case of ^{235}U, conditions for sustaining a chain reaction are much more favorable. This nucleus is less stable than ^{238}U; it will fission even if the kinetic energy of the incident neutron is very low. In fact, at low kinetic energy the fission reaction is more likely than at high kinetic energy; the reason for this is that the binding energy released when a neutron is captured by a ^{235}U nucleus is by itself quite sufficient to trigger the fission and, the kinetic energy of the neutron not being needed, it is advantageous for the neutron to move at low speed, since this lengthens the time it spends in the vicinity of the nucleus, and therefore enhances the probability that a reaction will take place.

Besides ^{235}U, two other isotopes exist in which chain reactions are practicable. Both resemble ^{235}U in that the spontaneous decay rate is low (so they can be held in storage without serious loss) and in that the rate of induced fission favors chain reactions: one is ^{233}U and the other is ^{239}Pu, an isotope of plutonium. The latter is very fissionable, but it is not found in ores on the Earth; it can only be obtained by artificial means, by nuclear alchemy.

In a given mass of ^{235}U or ^{239}Pu, neutrons produced by spontaneous fission or stray neutrons coming from elsewhere can initiate the first step of the chain reaction. Whether the reaction keeps going depends on the **multiplication factor,** that is, the factor by which the number of neutrons increases between one step and the next along the fission chain. If no neutrons are lost from the fission chain, then the multiplication factor is simply equal to the average number of neutrons released per fission; but if some neutrons are absorbed by impurities or escape without inducing a fission, then the multiplication factor will be smaller. The mass of fissionable material is said to be in a **critical** condition if the multiplication factor is unity; in this case the chain reaction merely proceeds at a constant rate — as in a nuclear reactor. The mass is **supercritical** if the multiplication factor exceeds unity; in that case the chain reaction proceeds at an ever-increasing runaway rate leading to an explosion — as in a nuclear bomb.

Neutrons can be lost from the fission chain by several mechanisms. For example, if the ^{235}U is not pure but contains an admixture of ^{238}U, then neutrons will be lost when they are absorbed by ^{238}U nuclei in transmutation reactions without fission. Even if the ^{235}U is pure, neutrons will still be lost when they escape from the surface of the piece of ^{235}U material. The rate of escape of the neutrons depends on the size and shape of the piece of ^{235}U material. The best shape is a sphere, since in this case the surface is as distant as possible from the bulk of the material. Furthermore, neutrons are less likely to escape from a large sphere than from a small sphere — in the former a neutron released at some average point within the bulk of the material has a longer distance to travel to the surface and is therefore more likely to trigger a fission reaction while on the way. This indicates that a sphere of fissionable material must have a minimum size if it is to maintain fission. For ^{235}U, the sphere of minimum size has a diameter of 18 cm; the corresponding minimum mass — called the **critical mass** — is 53 kg.

We can significantly diminish the critical mass if we surround the sphere of fissionable material by a neutron "reflector" that prevents the loss of neutrons. The reflector must be made of some substance whose nuclei strongly scatter neutrons, but do not absorb them. In a nuclear bomb, a thick shell of beryllium metal makes a good neutron reflector; neutrons that escape from the sphere of fissionable material collide with the beryllium nuclei and bounce back into the fissionable material. In a nuclear reactor, the moderator (see Section XI.5) acts as neutron reflector.

We can further diminish the critical mass by compressing the fissionable material to higher than normal densities. For example, compression of the material to twice its normal density diminishes the critical mass by a factor of 4. In a denser material, a neutron seeking to escape encounters more nuclei along its path to the surface and therefore is more likely to trigger a fission on the way.

XI.3 THE BOMB

The simplest fission bomb, or **A-bomb,** consists of two pieces of ^{235}U such that separately their masses are less than the critical mass, but jointly their masses add up to more than the critical mass. To detonate such a bomb, the two pieces of ^{235}U, initially at a safe distance from one another, are suddenly brought closely together. The assembly of the two subcritical masses into a single supercritical mass must be carried out very quickly; if the two masses are brought together slowly, a partial explosion (predetonation) will push them apart prematurely, before the chain reaction can release its full energy — the explosion fizzles. The simplest device used for the assembly of the two pieces of uranium consists of a gun which propels one piece of uranium toward the other at high speed (Figure XI.6); the propellant is an ordinary chemical high explosive.

A more sophisticated fission bomb consists of a (barely) subcritical mass of ^{239}Pu; if this is suddenly compressed to a higher than normal density, it will become supercritical. The sudden compression is achieved by the preliminary explosion of a chemical high explosive, such as TNT. If this explosive has been

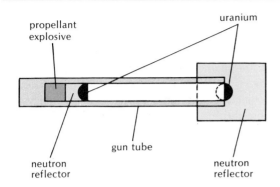

Fig. XI.6 A fission bomb using a gun device.

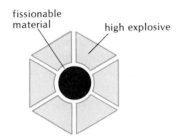

Fig. XI.7 An implosion device.

carefully arranged in a shell around a sphere of ^{239}Pu (Figure XI.7), then its detonation will crush the sphere of ^{239}Pu into itself; this implosion of the plutonium very suddenly brings its density to the supercritical value and triggers the chain reaction. The implosion device is used with ^{239}Pu because this isotope has a strong tendency to predetonate; if one were to use the gun device to bring together two subcritical masses of ^{239}Pu, the chain reaction would start while the masses were still moving toward one another; the consequent premature explosion would push the masses apart and prevent a full development of the chain reaction. The implosion device assembles the supercritical mass much faster, and therefore avoids the problem of a premature explosion.

During World War II, a scientific–military–industrial complex, known as the Manhattan Engineer District, produced three A-bombs: one plutonium bomb exploded at Alamogordo, New Mexico, on July 16, 1945, another plutonium bomb exploded at Nagasaki, Japan, on August 9, 1945, and one uranium bomb exploded at Hiroshima, Japan, on August 6, 1945. All of these bombs had yields of about 20 kilotons, that is, an explosive energy equivalent to that of 20,000 tons of TNT. This is the energy released by the fission of 1 kg of uranium (or plutonium). Hence these devices were quite inefficient — only a small fraction of the total mass of fissionable material actually underwent fission; the rest was merely scattered in all directions, blown apart before it had a chance to react.

The uranium used for bombs has to be highly purified, "weapons-grade" ^{235}U. Uranium ores contain a mixture of 99.3% of the undesirable isotope ^{238}U and only 0.7% of the isotope ^{235}U. Since these isotopes are chemically identical, their separation is very laborious. The separation process depends on the small difference in the masses: in a gaseous compound of uranium, such as UF_6, at a given temperature, the molecules containing ^{235}U have a slightly higher average speed than the molecules containing ^{238}U, and they will diffuse slightly faster through a porous membrane; hence such a membrane acts as a (partial) filter that separates ^{235}U from ^{238}U. This is the basis for the gaseous-diffusion process, which is still the main source of highly enriched ^{235}U.

Plutonium is not found in nature, except in insignificant trace amounts. It is made artificially, by transmutation of uranium in a nuclear reactor.

As can be seen from Figures XI.8 and XI.9, the

Fig. XI.8 The bomb dropped on Hiroshima. This device was 3.0 m long and had a mass of 3200 kg.

Fig. XI.9 The bomb dropped on Nagasaki. This device was 3.3 m long and had a mass of 4500 kg.

bombs used in World War II were rather cumbersome. Advances in the technology of shaping chemical high explosives used for the implosion of plutonium have made it possible to construct bombs in which the lump of plutonium is only about as large as a golf ball; the overall diameter of such a bomb is less than half a meter, and yet it releases an amount of energy in the kiloton range.

Much higher yields can be achieved by taking advantage of nuclear **fusion.** From the curve of binding energy (Figure 45.5) we see that the binding energy of light nuclei is relatively low — when two such light nuclei are made to fuse together to form a heavier nucleus, energy will be released. Fusion is the opposite of fission: in the former two nuclei merge into one, in the latter one nucleus splits into two. Furthermore, the energy released in fusion is strong energy whereas the energy released in fission is electric energy. The strong force favors fusion since, by merging, the nuclei reduce their surface area. The electric repulsion between the two nuclei opposes fusion, but in the case of light nuclei the strong force overcomes this opposition.

The heat given off by the Sun is due to a fusion reaction called *hydrogen burning*: hydrogen nuclei fuse together to make helium nuclei. This reaction involves several intermediate steps, which were discovered by theoretical calculations by H. Bethe.[2] This reaction cannot be duplicated on Earth because it will proceed only at extremely high temperatures and pressures, such as are found near the center of the Sun. However, some fusion reactions involving deuterium and tritium (the isotopes ^{2}H and ^{3}H) can be made to work on Earth:

$$^2H + {}^2H \rightarrow {}^3He + n \quad (5)$$

$$^2H + {}^2H \rightarrow {}^3H + p \quad (6)$$

$$^2H + {}^3H \rightarrow {}^4He + n \quad (7)$$

$$5{}^2H \rightarrow {}^3He + {}^4He + p + 2n + 24.3 \text{ MeV}$$

The net result of the three reactions taken together is the disappearance of five ^{2}H nuclei and the formation of ^{3}He, ^{4}He, one free proton, and two neutrons, with the release of 24.3 MeV. The amount of energy released per nucleon of reactant is 24.3 MeV per 10 nucleons, or 2.43 MeV per nucleon, whereas for the fission of uranium the energy released per nucleon of reactant is 200 MeV per 235 nucleons, or 0.85 MeV per nucleon. Thus the fusion of a given mass of ^{2}H will yield about three times as much energy as the fisson of an equal mass of ^{235}U.

[2] **Hans A. Bethe,** 1906–, German, later American, physicist. He formulated the theory of nuclear fusion in stars in 1937 and received the Nobel Prize for this in 1967; during the war he was director of the theoretical-physics division at Los Alamos.

The reactions (5)–(7) are called **thermonuclear** because they will proceed only at very high temperature and pressure. The requisite temperatures and pressures are attained at the place of explosion of an A-bomb. Hence, the fusion reactions can be initiated by exploding a fission bomb next to a mass of heavy hydrogen; this results in self-sustained explosive "burning" of the hydrogen nuclei. This so-called **H-bomb** is really a fission–fusion device, in which fission triggers fusion. What is more, the fusion reactions release a large number of energetic neutrons [see Eqs. (5) and (7)] which can be exploited to further enhance the violence of the explosion: the trick is to surround the fusion bomb by a blanket of cheap, natural uranium, consisting of mainly ^{238}U; although this isotope will not maintain a chain reaction, it will fission when exposed to the large flux of neutrons from the fusion. This kind of H-bomb is a fisson–fusion–fission device; typically, one-half of the total energy yield is due to fusion, one-half is due to fission. The fission of a large amount of uranium leaves behind a residue of highly radioactive fission products; hence fission–fusion–fission bombs are **dirty,** that is, they generate a large volume of radioactive fallout.

The total energy yield of an H-bomb is of the order of one or several megatons[3] — roughly a thousand times the yield of an A-bomb. Explosions of up to 60 megatons have been tried with great success, and there seems to be no limit to the suicidal madness that nature will let us get away with.

XI.4 THE EFFECTS OF NUCLEAR WEAPONS

The energy released by a nuclear bomb takes several forms as it emerges from the place of explosion: blast and shock in air, thermal radiation (including light), immediate nuclear radiation (neutrons and gamma rays), and delayed nuclear radiation (residual radioactivity, fallout). The exact distribution of the energy among

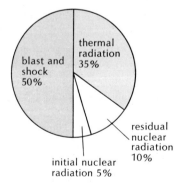

Fig. XI.10 Distribution of energy released in a typical air burst at an altitude below 30,000 m. (From S. Glasstone, ed., *The Effects of Nuclear Weapons.*)

[3] One megaton = 1000 kilotons.

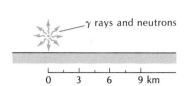

Fig. XI.11 Chronological development of an air burst of 1 megaton; time 5×10^{-6} s.

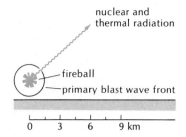

Fig. XI.12 Chronological development of an air burst of 1 megaton; time 1.8 s. (Figures XI.12–XI.16, based on S. Glasstone, ed., *The Effects of Nuclear Weapons*.)

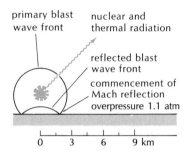

Fig. XI.13 Chronological development of an air burst of 1 megaton; time 4.6 s.

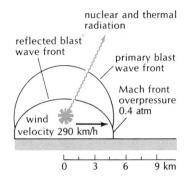

Fig. XI.14 Chronological development of an air burst of 1 megaton; time 11 s.

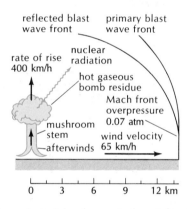

Fig. XI.15 Chronological development of an air burst of 1 megaton; time 37 s.

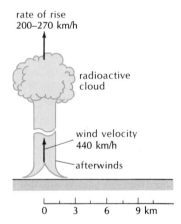

Fig. XI.16 Chronological development of an air burst of 1 megaton; time 110 s.

these several forms depends on the type of bomb and on the altitude at which it explodes. For an air burst — an explosion in air at low or medium altitude — the energy distribution is typically as shown in Figure XI.10.

The chronological sequence of phenomena in the air burst of a 1-megaton bomb is as follows (Figures XI.11–XI.16).

Within a few microseconds after the initiation of the explosion, the nuclear reactions will have released their energy. These reactions also release a large number of neutrons and gamma rays which immediately escape from the scene of the explosion (Figure XI.11). Although these neutrons and gamma rays are gradually absorbed by the air, their energy is high and they penetrate the air for well over a mile with lethal intensity.

The energy released in the explosion raises the fragmented residues of the bomb to an enormous temperature — tens of millions of degrees — and vaporizes them instantly. The incandescent vapor radiates X rays which are quickly absorbed by the surrounding air, heating it. The heated air and the vaporized bomb residues form a spherical, hot, luminous mass of gas — the **fireball**. Figure XI.17 is a high-speed photograph of such a fireball. For a 1-megaton explosion, at 1.8 s

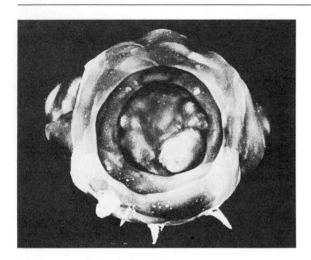

Fig. XI.17 Fireball of a nuclear explosion of low yield. This high-speed photograph was taken from a large distance by means of a telescope. (Courtesy H. E. Edgerton, MIT.)

after time zero, the fireball is 1.9 km across (Figure XI.12); it continues to grow to a maximum diameter of 2.3 km. The fireball attains a temperature of 6000 to 7000 K — hotter than the surface of the Sun. It emits intense thermal radiation, including intense light. Meanwhile a **blast wave,** caused by the sudden increase of pressure, travels outward from the center of explosion.

At 5 s after time zero, the blast wave has spread to a radius of 3 km and begins to reflect from the ground (Figure XI.13). The direct and the reflected wave merge and reinforce each other near the ground — they form a single, strong wave front called the **Mach front.** The overpressure at the wave front is 1.1 atm at this time. The emission of thermal radiation gradually decreases as the fireball cools, but significant amounts continue to be emitted until 10 s after time zero.

At 11 s after time zero, the blast wave has spread out to a radius of 5 km and the overpressure at the wave front has dropped to 0.4 atm (Figure XI.14). Behind the front, surface winds of 290 km blow radially outward for a few seconds. Most of the thermal radiation from the fireball (about 80%) will already have been emitted by this time.

At 37 s after time zero, the blast wave reaches out to 15 km, with an overpressure of about 0.07 atm (Figure XI.15). Meanwhile, the hot gas of the fireball begins to rise like a hot-air balloon, at about 400 km/h. The rising gas generates a vertical updraft as well as horizontal afterwinds that rush radially inward along the surface. These winds push dirt and dust into a column or stem below the fireball. As the fireball expands and cools, the vaporized bomb residues condense and form a large radioactive cloud. This cloud and the stem below take on the characteristic mushroom shape.

The mushroom grows, reaching a height of 11 km 110 s after the explosion (Figure XI.16). At this time, the mushroom cloud is still shooting upward at about 240 km/h while along the surface the afterwinds reach 300 km/h or more. Ultimately, the mushroom cloud attains a height of 22 km. Then it gradually begins to dissipate.

For a 1-megaton bomb bursting at an altitude of, say, 2000 m, the violent effects of thermal radiation, nuclear radiation, and blast cover an area of radius 16 km around ground zero. Figure XI.18 shows a circle of radius 16 km superimposed on an aerial photo of New York City; circles of radii 2.4 and 6.9 km are also shown in this photo. Within a radius of 2.4 km, the flux of the neutrons emitted by the initial nuclear reactions is so intense that any exposed victim absorbs a lethal dose of ionizing radiation. However, the physiological effects of overexposure to ionizing radiation are delayed; and the victims within this distance are more likely to suffer almost instantaneous death from the blast wave, which reaches them in a few seconds. Out to a radius of 6.9 km the blast wave has an overpressure of more than 0.34 atm, enough to level most buildings and crush the victims with flying and falling debris. Beyond 6.9 km, the blast wave damages buildings more or less severely, but usually does not destroy them completely.

The flash of thermal radiation emitted by the fireball poses the most frightful danger in the zone from 6.9 to 16 km from ground zero. The radiant heat is so intense that even at a distance of 16 km, exposed combustible material will be ignited. At 16 km, exposed human skin suffers second-degree burns (blisters); at 10 km human

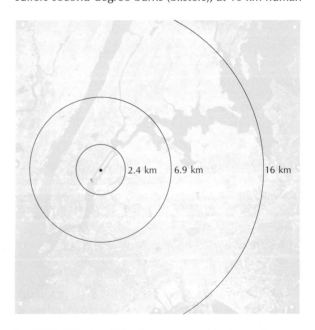

Fig. XI.18 Critical radii for damage caused by 1-megaton explosion over New York City.

skin is charred throughout its entire thickness. A victim caught in the open will be burnt fatally if at a distance of less than 13 km; beyond this distance survival depends on the area of unprotected skin exposed to the flash of heat and light.

Since the lens of the eye focuses light on the retina, any victim looking at the fireball will suffer severe retinal damage. At distances of up to 30 km, the result is total blindness; even at distances of 300 or 400 km, retinal burns and partial loss of vision are likely.

Besides these primary effects of a nuclear explosion, there are secondary effects. A 1-megaton air burst over a city will set off numerous flash fires within a circle of a 16-km radius. Such a large number of fires ignited over a wide area may cause a **firestorm:** hot air from the fires rises and creates both an updraft and strong surface winds (80 to 160 km/h) which fan the flames. Consequently, annihilation of life and destruction of buildings may be total, or nearly total, within a 16-km radius.

Another delayed effect is **fallout,** which consists of the radioactive residues in the mushroom cloud mixed with the dirt and dust of the mushroom stem. Wherever this contaminated dust falls on the surface of the Earth, the ground will become radioactive. Winds may carry the contaminated dust hundreds of miles as it gradually settles; the result is a fallout plume which covers a large area with dangerous radioactive contamination. The details of the distribution of fallout depend on the wind speed and direction; they also depend on the type of bomb and on the altitude of explosion — fallout is more intense for a surface burst (at ground level) than for an air burst. For a 1-megaton surface burst, whose yield is one-half from fission and one-half from fusion, fallout will be lethal over a downwind area about 240 km long and 30 km wide (Figure XI.19). This assumes that the wind is steady at about 24 km/h and that the victims are not sheltered from the ionizing radiation.

The basement of a house offers some shelter from the radiation emitted by the radioactive dust blanketing the ground and the roof of the house — the radiation penetrating an underground basement is $\frac{1}{25}$ to $\frac{1}{100}$ of the outside radiation. In general, a massive layer of any material provides shielding against the radiation. Crude **fallout shelters** can be improvised with a layer of any dense material; 10 cm of concrete, 18 cm of earth, or 36 cm of books all give about the same protection (Figure XI.20). The radioactivity of the fallout decreases with time (as a rough rule, a sevenfold increase in time reduces the intensity of the radiation by a factor of 10); but, depending on details of the fallout pattern, it may not be safe to leave the shelter for several weeks.

In view of the awesome immediate effects of a nu-

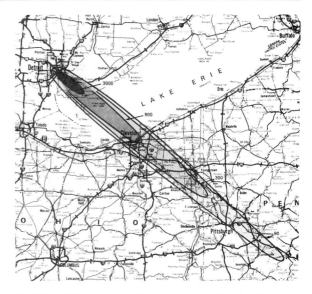

Fig. XI.19 Fallout pattern for 1-megaton surface burst in Detroit, with 24-km/h northwesterly wind. The radiation dose accumulated in a 7-day period is lethal within the second largest gray area. (From *The Effects of Nuclear War,* Office of Technology Assessment, Congress of the United States.)

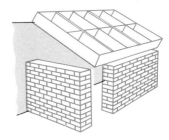

Fig. XI.20 A simple fallout shelter in a basement. The bins are to be filled up with bricks or sand. (From *Radiological Defense Textbook,* Defense Civil Preparedness Agency.)

clear explosion, it is easy to underrate the effects of fallout. It should be kept in mind that even in the case of a limited nuclear attack on the United States — in which bombs are only directed against military installations rather than cities — there might be more than 12 million civilian casualties from the effects of fallout alone (Figure XI.21). Besides these casualties from early, local fallout, there would be some further casualties from long-term, worldwide fallout involving isotopes of strontium (^{90}Sr) and cesium (^{137}Cs). These isotopes are produced in abundance by the fission of uranium, and they have very long half-lives; therefore, they remain a hazard for a long time.

In this discussion of the effects of nuclear weapons we have concentrated on an explosion in air. If, instead, the explosion happens in space, above the Earth's atmosphere, the effects are somewhat different. There is no fireball in space, and no blast wave.

Fig. XI.21 Fallout pattern for a massive nuclear attack on military installations in the United States, assuming typical westerly winds. More than one-half of the inhabitants of the colored areas would die. (Based on *Scientific American*, November 1976.)

The burst of neutrons, gamma rays, and X rays released by the explosion travels outward freely, without absorption. When the burst of gamma rays and X rays strikes the upper layer of the atmosphere, it suddenly ejects a large number of electrons from the air molecules. The violent acceleration of these electrons generates a strong pulse of electromagnetic radiation, called the **electromagnetic pulse** (EMP). This electromagnetic pulse can cause widespread destruction in the circuitry of radios, telephone installations, avionics, and other electronic communication and control equipment. Military planners view this secondary effect of nuclear explosions with grave concern.

XI.5 NUCLEAR REACTORS

In a nuclear reactor, a fission chain reaction takes place under controlled conditions. The condition of the fission material in the reactor is *critical* rather than supercritical, that is, the number neutrons and the number of fissions along each step of the chain reaction remains constant instead of increasing geometrically.

The most common type of reactor operates with "enriched" uranium consisting of a few percent ^{235}U mixed with 90-odd percent ^{238}U. Such a uranium mixture cannot by itself maintain a chain reaction — the ^{238}U soaks up too many of the fission neutrons. However, if the uranium is surrounded by a substance capable of slowing down the (fast) fission neutrons to a low speed, then a chain reaction becomes viable. The substance that slows down the neutrons is called the **moderator**. The role of the moderator in a nuclear reaction is analogous to that of a catalyst in a chemical reaction. The moderator enhances the nuclear reaction because slow neutrons are much more efficient at maintaining a chain reaction than fast neutrons. There are two reasons for this: slow neutrons are less likely to be absorbed by ^{238}U (the reaction rate for the capture of neutrons by ^{238}U is small at low neutron energies), and slow neutrons trigger the fission of ^{235}U more readily than fast neutrons (the reaction rate for induced fission of ^{235}U is large at low neutron energies).

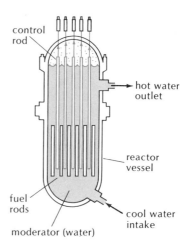

Fig. XI.22 Nuclear reactor (schematic).

Inside the reactor the uranium is usually placed in long fuel rods, and these are immersed in the bulk of the moderator (Figure XI.22). Fast neutrons released by fissions travel from the fuel rods into the moderator; there they lose their kinetic energy by collisions with the moderator's nuclei; and then they wander back into one or another of the fuel rods and induce further fissions. An efficient moderator must absorb the kinetic energy of the neutrons, but not absorb the neutrons themselves. The three best moderators are ordinary water (H_2O), heavy water (D_2O), and graphite (pure carbon). These materials are very efficient at slowing down neutrons because they contain nuclei of low mass, that is, nuclei of mass equal to, or not much larger than, the mass of the neutron. As we saw in Section 11.2, the collision between two particles of equal masses transfers all of the kinetic energy to the initially stationary particle, whereas the collision between particles of unequal masses transfers only a fraction of the kinetic energy. After a few collisions with the moderator nuclei, the neutrons will have lost almost all their kinetic energy; they will retain only a kinetic energy equal to the kinetic energy of the random thermal motion of the moderator nuclei, that is, a kinetic energy of $\frac{3}{2}kT$ [where T is the temperature of the moderator; see Eq. (19.24)]. Such neutrons are called **thermal neutrons**. Although some of these neutrons will escape from the reactor, in a critical reactor enough return to the uranium to keep the chain reaction going.

The configuration of the reactor — the size, number, and location of uranium rods, and the amount and shape of moderator — must be designed so that the reactor is critical or just barely supercritical. The number of neutrons can be finely adjusted to the point of criticality by control rods made of boron or cadmium; these substances readily soak up neutrons, and, by pushing the control rods in or pulling them out of the reactor, more or fewer neutrons can be removed from the fission chain, bringing the reactor from subcritical, to critical, or even to supercritical (brief instants of supercriticality are called **excursions**).

Both heavy water and graphite are such good moderators that in their presence even natural uranium, with its small percentage of ^{235}U, can maintain a chain reaction. The first nuclear reactor was built at the University of Chicago in 1942 under the direction of Enrico Fermi;[4] it contained natural uranium as fuel and graphite as moderator. Several similar reactors were built at Hanford, Washington, shortly afterward, as part of the Manhattan Project. These reactors were used as **converters,** that is, they were used to convert ^{238}U into ^{239}Pu by the following sequence of reactions: Within the rods of natural uranium, some of the neutrons from the fission of ^{235}U are soaked up by ^{238}U, transmuting it into ^{239}U. The isotope ^{239}U then spontaneously undergoes two successive beta decays, which transmute it into ^{239}Pu. The few kilograms of ^{239}Pu for the Alamogordo and Nagasaki A-bombs were obtained by these means.

Nowadays reactors are extensively used to provide intense beams of neutrons for research; to produce radioisotopes for scientific, industrial, and medical applications (see Interlude IV); and to generate mechanical or electric power. The latter reactors are called **power reactors.** In them, the heat energy released by the fission reactions is removed from the reactor core by a circulating coolant; this heat is then transferred to steam, which drives a steam turbine. Thus, the nuclear reactor serves the same purpose as the furnace of a conventional steam engine — and uranium replaces coal or oil.

Most power reactors in operation or under construction in the United States have water-filled cores; the water acts simultaneously as coolant and as moderator. Figure XI.23 shows a schematic diagram of a **pressurized-water reactor** (PWR), the most common type of power reactor. The reactor core with its fuel rods is enclosed in a massive reactor vessel. Water, driven by powerful pumps, circulates through the reactor vessel; this water is kept at a high pressure (150 atm) which prevents it from boiling even though the temperature within the reactor vessel is 320°C. The hot water circulates from the reactor vessel to a steam generator where it passes through a system of pipes immersed in water at a lower pressure. The low-pressure water boils off as steam, which drives a turbine;

[4] **Enrico Fermi,** 1901–1954, Italian, later American, physicist. He was awarded the Nobel Prize in 1938 for his discovery of nuclear reactions initiated by neutron bombardment. The element *fermium* was named in his memory.

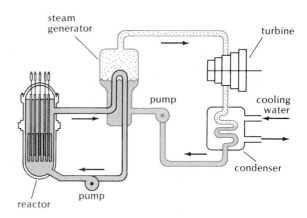

Fig. XI.23 Schematic diagram of a nuclear power plant with a pressurized-water reactor.

after the steam has done its work, it condenses and returns to the steam generator. Meanwhile the high-pressure water circulates back to the reactor vessel in a closed loop.

In a **boiling-water reactor** (BWR) the coolant is allowed to boil directly within the reactor vessel. Steam emerges from the top of the reactor and is fed into a turbine; it then condenses and circulates back to the reactor.

Figure XI.24 shows the reactor vessel for a large power plant capable of generating 1200 MW of electricity; the height of the reactor vessel is 18 m, the

Fig. XI.24 Reactor vessel being installed in a nuclear power plant.

thickness of the steel walls is 25 cm at its thinnest, and the mass of the vessel is 1400 metric tons when empty. Smaller reactors are used for the propulsion of submarines, aircraft carriers, and icebreakers. For a submarine, a reactor power plant has the obvious advantage that it does not require oxygen to "burn" its fuel; besides, the reactor only needs to be refueled at very long intervals.

Both pressurized-water reactors and boiling-water reactors operate with ordinary water; they are **light-water reactors.** By contrast, **heavy-water reactors** operate with heavy water (D_2O) as moderator. Although heavy water is very expensive — the 680 metric tons of heavy water required for a single large reactor cost $80 million — it is such a good moderator that cheap, natural uranium can be used as fuel (see above). A type of heavy-water reactor developed in Canada (CANDU), contains heavy water as moderator and uses ordinary water, in a separate system of pipes, as coolant.

Another type of reactor, first developed in the United Kingdom and in France, is the **gas-cooled reactor** (GR). It contains graphite as moderator and uses gas as coolant. Gases that suit this purpose are carbon dioxide (low-temperature operation, 380°C) and helium (high-temperature operation, 760°C). Since graphite is a good moderator, such reactors can be fueled by natural uranium; however, structural problems arising from the large size of these reactors (20-m diameter) make it awkward to rely on natural uranium, and modern high-temperature gas-cooled reactors are fueled with highly enriched uranium, which permits a more compact design.

Power reactors now in operation obtain their energy from the fission of the isotope ^{235}U. Since the world's supply of this isotope is limited (see Interlude V), it would be desirable to build reactors that consume some other fissionable nucleus. One obvious choice is the very fissionable isotope ^{239}Pu. Although this isotope does not occur naturally, it can be readily manufactured by transmutation of the very abundant isotope ^{238}U. In fact, since the fuel of all power reactors is a mixture of ^{238}U and ^{235}U, the manufacture of ^{239}Pu is an automatic side effect of the operation of power reactors; the ^{239}Pu can subsequently be extracted by chemical reprocessing of the spent uranium fuel. A reactor fueled with ^{239}Pu not only makes good use of a material that would otherwise go to waste, but, if the reactor is supplied with a quantity of ^{238}U, it can also manufacture its own ^{239}Pu. What is more, the number of neutrons released in the fusion of ^{239}Pu is sufficiently large so in an efficiently designed reactor slightly more than one of the neutrons released in an average fission reaction can be diverted from the fission chain to the transmutation of ^{238}U. This implies that the reactor produces *more* ^{239}Pu (from ^{238}U) than it consumes (from its original supply).

A reactor that produces more fissionable material than it consumes is called a **breeder.** Once the fuel cycle of breeder reactors has been started with an initial load of ^{239}Pu, only the abundant and cheap ^{238}U needs to be supplied to keep the fuel cycle going. Essentially, breeders extract energy indirectly from ^{238}U; breeders would therefore be able to generate power for as long as our (abundant) supply of ^{238}U lasts (see Interlude V). Much effort has been expended on the development of breeders and a few experimental reactors have already been built. However, these reactors are afflicted with design and safety problems that have not yet been satisfactorily resolved.[5]

Further Reading

Nuclear Science and Society by B. L. Cohen (Doubleday, New York, 1974) is an excellent introduction to nuclear energy and its applications.

A concise and clear survey of nuclear physics will be found in *Secrets of the Nucleus* by J. S. Levinger (McGraw-Hill, New York, 1967). Very brief discussions of nuclear physics can be found in chapters in the books *Energy, Ecology, and the Environment* by R. Wilson and W. J. Jones (Academic, New York, 1974) and *The Atom and Its Nucleus* by G. Gamow (Prentice-Hall, Englewood Cliffs, 1965). *The Atomic Energy Deskbook,* edited by J. F. Hogerton (Reinhold, New York, 1963), and *Sourcebook on Atomic Energy,* edited by S. Glasstone (Van Nostrand Reinhold, Princeton, 1967), are encyclopedias containing technical information.

The standard reference on nuclear explosions is *The Effects of Nuclear Weapons,* edited by S. Glasstone (United States Atomic Energy Commission, Washington, 1962). This contains a wealth of detail on nuclear explosions, their thermal radiation, blast, and radioactivity, and the injuries and damage wrought on people and buildings. *The Effects of Nuclear War* by the Office of Technology Assessment (Congress of the United States, Washington, D.C., 1979) presents case studies of hypothetical nuclear attacks on Soviet and U.S. industrial and civilian targets. *Radiological Defense* by the Defense Preparedness Agency (Department of Defense, 1974) is a textbook on protective measures against fallout. *Arsenal: Understanding Weapons in the Nuclear Age* by K. Tsipis (Simon and Schuster, New York, 1983) is a lucid survey of the principles and effects of nuclear bombs, the technology of delivery systems, and strategic implications.

The Cold and the Dark: The World After Nuclear War by P. R. Ehrlich, C. Sagan, D. Kennedy, and W. O. Roberts (Norton, New York, 1984) discusses the long-term worldwide consequences of nuclear war, especially the severe climatic changes ("nuclear winter") caused by the reduction of sunlight by the dust and smoke accumulated in the atmosphere, and the catastrophic effects of this on plants and animals. *Last*

---

[5] For a discussion of reactor safety, see Section V.6.

Aid, edited by E. Chivian, S. Chivian, R. J. Lifton, and J. E. Mack (Freeman, San Francisco, 1982), is a chilling examination of the medical implications of nuclear war. *The Fate of the Earth* by J. Schell (Knopf, New York, 1982) gives a frightening description of the global disasters that would follow a nuclear war and makes an impassioned plea for total disarmament. *Nuclear Nightmares* by N. Calder (Viking, New York, 1979) is a journalist's investigation of the threat of nuclear war; it deals with weapons technology, proliferation, and scenarios that might lead to war. *Survival and the Bomb,* edited by E. P. Wigner (Indiana University Press, Bloomington, 1969), presents the arguments for the implementation of a civil-defense program.

Manhattan Project by S. Groueff (Little, Brown, Boston, 1967) is a historical account of the development of the first A-bombs. *Enrico Fermi* by E. Segré (University of Chicago Press, Chicago, 1970) is a biography of the eminent physicist who designed and built the first nuclear reactor and triggered the first chain reaction. *Hans Bethe: Prophet of Energy* by J. Bernstein (Basic Books, New York, 1979) is a brilliant biography of another eminent physicist who discovered the cycles of nuclear reactions that release energy in stars. *The Curve of Binding Energy* by J. McPhee (Ballantine Books, New York, 1973) is a splendidly written account of the design and manufacturing problems involved in the production of small nuclear weapons.

The following is a list of magazine articles dealing with nuclear reactors, nuclear weapons, and related questions:

"Energy from Breeder Reactors," F. L. Culler and W. O. Harms, *Physics Today,* May 1972

"Natural-Uranium Heavy-Water Reactors," H. C. McIntyre, *Scientific American,* October 1975

"Civil Defense in Limited War — A Debate," A. A. Broyles and E. P. Wigner vs. S. D. Drell, *Physics Today,* April 1976

"A Natural Fission Reactor," G. A. Cowan, *Scientific American,* July 1976

"Limited Nuclear War," S. D. Drell and F. von Hippel, *Scientific American,* November 1976

"Superphénix: A Full-Scale Breeder Reactor," G. A. Vendryes, *Scientific American,* March 1977

"Nuclear Power and Nuclear-Weapons Proliferation," E. J. Moniz and T. L. Neff, *Physics Today,* April 1978

"Enhanced-Radiation Weapons," F. M. Kaplan, *Scientific American,* May 1978

"The Prompt and Delayed Effects of Nuclear War," K. L. Lewis, *Scientific American,* July 1979

"Catastrophic Releases of Radioactivity," S. A. Fetter and K. Tsipis, *Scientific American,* April 1981

"Gas-Cooled Nuclear Power Reactors," H. M. Agnew, *Scientific American,* June 1981

"Freeze on Nuclear Weapons Development and Deployment: Pro–Con," H. Feiveson and F. von Hippel vs. H. W. Lewis, *Physics Today,* January 1983

"Arms Limitation Strategies," H. F. York, *Physics Today,* March 1983

"Effects of Nuclear Weapons," L. Sartori, *Physics Today,* March 1983

"The Nuclear Arsenals of the US and the USSR," B. G. Levi, *Physics Today,* March 1983

"The Uncertainties of a Preemptive Nuclear Attack," M. Bunn and K. Tsipis, *Scientific American,* November 1983

"Weapons and Hope," F. J. Dyson, *The New Yorker,* February 6, 13, 20, and 27, 1984

"The Climatic Effects of Nuclear War," R. P. Turco, O. B. Toon, T. P. Ackerman, J. B. Pollack, and C. Sagan, *Scientific American,* August 1984

"Rethinking Nuclear Power," R. L. Lester, *Scientific American,* March 1986

"Third-Generation Nuclear Weapons," T. B. Taylor, *Scientific American,* April 1987

Questions

1. According to Figure 45.5, which isotope has the largest binding energy per nucleon? Which has the least binding energy and is therefore capable of releasing the most energy in a nuclear reaction?

2. By what factor is the amount of energy released per kilogram of reactant in a typical fission reaction larger than in a typical chemical reaction? (Hint: Compare fission of uranium with explosion of TNT.)

3. In 1933, Ernest Rutherford, the discoverer of the nucleus, declared that "the energy produced by the breaking down of the atom is a very poor kind of thing. Anyone who expects a source of power from the transformation of these atoms is talking moonshine." Taking into consideration that fission was unknown at the time, do you think Rutherford was making a fair assessment? What about heat released by natural radioactivity?

4. Free neutrons are unstable; they decay with an average lifetime of about 15 min. What effect does this have on a fission chain reaction?

5. Why is a critical mass needed for fission but not for fusion?

6. George Gamow, author of many delightful books on physics, was fond of saying that natural uranium is just as useless for carrying out a nuclear chain reaction as soaking-wet logs are for building a campfire. Explain this analogy.

7. A widely used modern method for the separation of the isotopes ^{235}U and ^{238}U relies on high-speed centrifuges. How does centrifugation affect a test tube full of a gaseous compound with molecules containing these isotopes?

8. Since the critical mass of ^{235}U is 53 kg, the Hiroshima bomb must have contained about that much uranium. If all of this uranium had undergone fission, what would have been the energy released?

9. Compare the shapes of the bombs in Figures XI.8 and XI.9. Why does the Hiroshima bomb have an elongated shape, and the Nagasaki bomb a rounded shape?

10. Why are high temperatures required to initiate fusion, but not to initiate fission?

11. Per kilogram of reactant, fusion releases about three times as much energy as fission. But the energy yield of a typical H-bomb is about a hundred times larger than that of a typical A-bomb. How can this be?

12. If an H-bomb is not surrounded by a blanket of natural uranium, it is a cleaner bomb, producing a smaller amount of radioactive fallout and a smaller energy yield. However, such a bomb, called a neutron bomb, releases a large number of energetic neutrons. Compare the effects of a neutron bomb with those of an ordinary H-bomb.

13. If a blast wave of overpressure 1 atm strikes you, what is the force on the front of your body?

14. If you are at a distance of 12 km from the place of a nuclear explosion, you have available about 30 s between the flash of light and the arrival of the blast wave. Look around the room you are in. Where could you take shelter in 30 s?

15. Contamination of food with radioactive strontium poses a severe hazard because strontium is chemically similar to calcium. Explain.

16. One of the effects of a large-scale nuclear war would be the accumulation in the atmosphere of dust and smoke from explosions and fires. Such a blanket of dust and smoke, covering the entire Earth for several months, would reduce the amount of sunlight reaching the ground, perhaps by 50% or more, so the climate would become much cooler. Recent calculations suggest that even in summer the temperature would remain below freezing, a phenomenon that has been called the **nuclear winter**. What agricultural and other problems would arise from such a drastic modification of the climate?

17. Helium should make a good moderator, since it does not absorb neutrons and has a low mass. Why do we not use helium as moderator in a nuclear reactor?

18. Why must the fuel rods in a nuclear reactor be thin?

19. The uranium in the fuel rods in a nuclear reactor is enclosed in a metal pipe (cladding). What are the requirements that the material of the pipe must meet?

20. In a nuclear power plant, the water that passes through the reactor core flows through one loop, and the water that passes through the turbine flows in a separate loop (Figure XI.23). Why do we not use a single loop that directly connects the reactor core to the turbine, as in an ordinary coal-burning power plant?

21. If you wanted to build a nuclear reactor of very small size, say, to power an artificial satellite, what isotope would you use?

22. The following question was asked by a reader of the *New York Times* (February 22, 1983): "Why is it less dangerous if the nuclear reactor of a falling satellite (such as the Soviet Cosmos 1402) burns up in space than if it falls to earth in one piece? Aren't radioactive atoms just as present one way as another?" How would you answer?

23. Breeder reactors generate more fuel than they consume. Does this violate the law of conservation of energy?

24. One of the disadvantages of breeder reactors is that their fuel can be diverted to the manufacture of bombs. Why is this not a problem with ordinary reactors?

CHAPTER 46

Elementary Particles

The search for the elementary, indivisible building blocks of matter is the most fundamental problem in physics. Early in this century physicists discovered that the atom is not an elementary, indivisible unit — each atom consists of electrons orbiting around a nucleus. And then they discovered that the nucleus is not an elementary, indivisible unit — each nucleus consists of protons and neutrons packed tightly together. In the 1930s, physicists began to build accelerating machines producing beams of energetic protons or electrons that could serve as projectiles; with these atom smashers physicists could split the nucleus. In the 1950s, they built much larger and more powerful accelerating machines; with these new machines they attempted to split the proton and the neutron. But the result of these attempts was chaos: when bombarded by very energetic projectiles, the proton and the neutron do not split into any simple subprotonic pieces. Instead, the violent collisions of protons and neutrons generate a variety of new, exotic particles by the conversion of kinetic energy into mass. For want of a better name, these new particles were called "elementary particles." However, most of these elementary particles are more massive and more complicated than protons and neutrons — they are obviously not elementary, indivisible units. Only in recent years has some order emerged from this chaos. Physicists have found convincing circumstantial evidence that protons, neutrons, and other elementary particles are made of very small, compact subunits. The subprotonic building blocks are called **quarks**.

46.1 The Tools of High-Energy Physics

Protons and neutrons are much "harder" than atoms or nuclei. A projectile of a bombarding energy of a few eV can shatter an atom, and a projectile of a bombarding energy of a few MeV can shatter a nucleus. But to make a dent in a proton, we need a bombarding energy of a few hundred or a thousand MeV. Elementary-particle physicists like to measure the energy of their projectiles in billion electron-volts (GeV) or in trillion electron-volts (TeV). Expressed in joules, these energy units are

$$1 \text{ GeV} = 10^9 \text{ eV} = 1.6 \times 10^{-10} \text{ J}$$

$$1 \text{ TeV} = 10^{12} \text{ eV} = 1.6 \times 10^{-7} \text{ J}$$

The acceleration of a projectile to such a high energy requires large, sophisticated accelerator machines. High energy — one or several GeV — characterizes the realm of particle physics. Collisions between particles at these high energies are capable of creating a large variety of new particles by the conversion of energy into mass.

The two largest accelerating machines are the proton synchrotrons at the Fermi National Accelerator Laboratory (Fermilab, near Chicago) and at the Organisation Européenne pour la Recherche Nucleaire (CERN, on the Swiss–French border near Geneva). Figure 46.1 shows an overall view of Fermilab; the accelerator is buried underground in a circular tunnel 6 km in circumference. The CERN accelerator is slightly larger, about 7 km in circumference. The Fermilab accelerator produces a beam of protons with an energy of 1000 GeV and a speed of 99.99995% of the speed of light. The protons travel in an evacuated circular beam pipe (Figure 46.2). Large magnets placed along this pipe exert forces on the protons, preventing their escape — the protons move as if on a circular racetrack. At regular intervals the pipe is joined to electromagnetic cavities connected to high-voltage oscillators; in each of these cavities, oscillating electric fields impel the protons to higher energy. After several hundred thousand circuits around the racetrack, the protons reach their final energy of 1000 GeV.

Fig. 46.1 (left) Panoramic view of Fermilab.

Fig. 46.2 (right) View of the underground tunnel housing the Tevatron accelerator at Fermilab. The long row of magnets encases the beam pipe.

Before the protons are allowed to enter the giant circular racetrack, they must pass through several smaller preliminary accelerators. At Fermilab there are three preliminary accelerators. The protons generated by a proton gun are first given an energy of about 1 MeV by an electrostatic generator; then their energy is raised to 200 MeV by a linear accelerator; and then it is raised further to 8 GeV by a "small" circular accelerator. Only after the protons have passed through these stages do they enter the main ring.

The next largest accelerating machine is the Stanford Linear Accelerator (SLAC), 3 km long. This accelerator produces a beam of electrons with an energy of 50 GeV. The speed of electrons of this energy is within 2 cm/s of the speed of light; thus, the Stanford machine holds the speed record for artificially accelerated particles.

Table 46.1 list some other large accelerators, completed or under construction. An even larger accelerator, the Superconducting Super Collider (SSC), is being planned in the United States. This monster, to be built in the 1990s, will have a circumference of 90 km, and it will accelerate protons and antiprotons to 20 TeV.

Once the particles have been given their maximum energy, they are guided out of the accelerator by steering magnets and made to crash against a target consisting of a block of metal or a tankful of liquid. Within the material of the target, the high-energy particles collide violently with the protons or neutrons of the nuclei. The reactions that take place in these collisions create an abundance of new particles by conversion of energy into mass. Unfortunately, not all of the kinetic energy of the incident particles can participate in these reactions. As we saw in Section 11.4, the kinetic energy associated with the motion of the center of mass remains constant during collisions and is therefore not available for inelastic reactions; only the "internal" kinetic energy associated with motion relative to the center of mass is available for inelastic reactions and can be used up in such reactions. According to Eq. (11.27), this available energy for a proton of kinetic energy 1000 GeV colliding with a stationary proton is 500 GeV. However, this calculation ignores relativistic effects; and since the speed of the incident proton is very close to the speed of light, these effects are actually very important. A correct relativistic calculation indicates the

Table 46.1 SOME LARGE ACCELERATORS

Accelerator	Start of operation	Projectiles[a]	Maximum Kinetic energy
Stanford Linear Accelerator (SLAC), California	1961	Electrons	50 GeV
Deutsches Elektronen Synchrotron (DESY), Germany	1973	Electrons	26
CERN Super Proton Synchrotron (SPS), Switzerland–France	1976	Protons	450
Fermilab Tevatron, Illinois	1984	Protons	1000
KEK TRISTAN Accelerator, Japan	1987	Electrons	30
CERN Large Electron–Positron Storage Ring (LEP)	1989	Electrons	60
Serpukhov UNK, U.S.S.R.	1995	Protons	3000
Superconducting Super Collider (SSC), U.S.A.	1996?	Protons	20,000

[a] In all of these accelerators, particles are made to collide head on with antiparticles of equal energy.

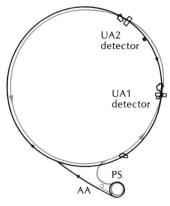

Fig. 46.3 In the SPS accelerator at CERN, high-energy protons and antiprotons travel in opposite directions around a ring of circumference 7 km. The protons and antiprotons are produced by a medium-energy accelerator (PS, lower right). The antiprotons are temporarily stored in an accumulator ring (AA), while the protons are sent directly into the large ring. When enough antiprotons have accumulated, they are also sent into the large ring, and both kinds of particles are accelerated to their final, high energy. The orbits of the protons (black) and the antiprotons (color) are slightly separated, but they intersect at two points, where the detectors UA 1 and UA 2 are located.

available energy is much less — it is only 43 GeV. Thus, collisions between a high-energy proton and a stationary proton are quite inefficient.

The efficiency improves drastically if two high-energy protons are made to collide head on. The available energy is then the sum of the energies of the two protons. At CERN, a beam of protons and a beam of antiprotons of 450 GeV each are made to travel in opposite directions around the accelerator in slightly different orbits that cross at two places (Figures 46.3 and 46.4). Where the two orbits cross, the two beams meet nearly head on. Several other accelerator laboratories have recently installed similar storage rings with which they can bring about head-on collisions between particles. Head-on collisions will also be exploited by the Superconducting Super Collider.

Fig. 46.4 The UA 1 detector at CERN. This detector surrounds the point of intersection of the proton and antiproton orbits.

When a beam of high-energy particles crashes into a stationary target or into an oncoming beam, the violent reactions generate a wide variety of new particles. In order to observe the particles that emerge from the scene of these collisions, physicists use several kinds of particle detectors. Some of these — scintillation counters and Čerenkov counters — signal the passage of each electrically charged particle by giving off brief (and weak) flashes of light; these flashes of light are detected by sensitive photomultiplier tubes. Other detectors — bubble chambers and spark chambers — render the track of an electrically charged particle visible and record these tracks on photographs.

A **bubble chamber** is a tank filled with a superheated liquid, usually liquid hydrogen (a liquid is said to be superheated when its temperature is slightly above the boiling point; see Section 20.5). Such a liquid is unstable — it is about to start boiling but it will usually not start until some disturbance triggers the formation of the first few bubbles. A charged particle zipping through the chamber provides just the kind of disturbance the liquid is waiting for — a fine stream of bubbles forms in the wake of the particle's passage. High-speed cameras can take a picture of these bubble tracks before they disperse and disappear in the turmoil of subsequent widespread bubbling and boiling of the liquid.

Figure 46.5 shows the BEBC bubble chamber at CERN and its ancillary equipment. The bubble chamber is surrounded by a large electromagnet, which aids in the identification of the particles passing through the bubble chamber. The magnetic field generated by this magnet pushes the particles into curved orbits as they pass through the chamber. The direction of the curvature depends on the sign of the electric charge, and the magnitude of the curvature depends on the momentum of the particles; thus, measurement of the curvature of a bubble track tells us the sign of the electric charge of the particle and the momentum of the particle. Figure 46.6 shows the bubble chamber itself during construction. The portholes are for cameras and for flash lamps. Photographs are taken simultaneously with several cameras at different angles so as to obtain a stereoscopic view of the particle tracks. The chamber shown here can go through two complete boiling cycles per second; each time a completely different set of tracks will be recorded on the photograph. Sometimes experimenters must take several hundred thousand photographs before finding one with the special kind of collision they are looking for.

In the photograph in Figure 46.7 we see a high-energy beam of positive pions (π^+) entering the chamber. The tracks of the pions, entering the field of view from the left, are slightly curved upward by the magnetic field. Pions are particles heavier than electrons but lighter than

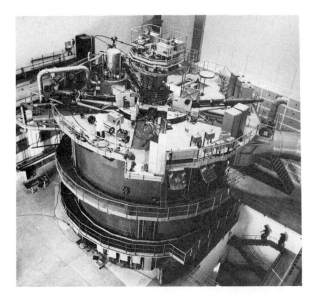

Fig. 46.5 The Big European Bubble Chamber (BEBC) at CERN, surrounded by the large magnet which almost completely hides it from view.

Fig. 46.6 The BEBC before its installation in the magnet.

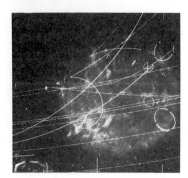

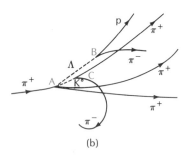

Fig. 46.7 (a) Photograph of tracks of particles in a bubble chamber and (b) tracing of the most interesting of these tracks. Neutral particles do not make visible tracks in the bubble chamber, but their extrapolated tracks have been indicated by dashed lines in the tracing. An incident π^+ collided with one of the protons in a hydrogen atom at the point A. The new particles created in this collision are responsible for the tracks shown in the tracing.

protons; they are generated in great abundance whenever protons strike nuclei. A beam of pions can be produced by simply aiming a beam of protons from the accelerator at a target (a thick chunk of metal); many pions then come flying out of the far side of the target, and they can be assembled in a beam and guided to the bubble chamber by an array of electromagnets.

At point A in Figure 46.7, one of the pions strikes one of the protons sitting at rest at the center of a hydrogen atom. Both the pion and the proton are destroyed by this violent collision — they cease to exist. In their place two new pions, a kaon (K^0), and a lambda particle (Λ) emerge from the scene of the accident. Both the kaon and the lambda particle are electrically neutral, and hence they leave no visible tracks in the bubble chamber. Nevertheless, it is possible to reconstruct their trajectories because, after a short while, both of these particles decay spontaneously into new particles that do leave tracks — at point B the lambda decays into a proton and a negative pion (π^-), and at point C the kaon decays into two pions. We can summarize these reactions as follows:

$$\pi^+ + p \rightarrow \Lambda + K^0 + \pi^+ + \pi^+ \qquad (1)$$
$$\rightarrow \pi^+ + \pi^- \qquad (2)$$
$$\rightarrow p + \pi^- \qquad (3)$$

The net result of these reactions is the conversion of one proton and one pion into one proton and five pions. Thus, the mass after the reactions is much larger than the mass before, that is, the Q value of the reaction is negative [see Eq. (11.27)]. The excess mass comes from the conversion of energy into mass — some of the kinetic energy of the incident pion has been converted into mass.

This conversion of kinetic energy into mass plays a crucial role in the discovery of new particles. Almost all the new particles discovered during the last 20 years are considerably heavier than protons and neutrons. Physicists need powerful accelerators to produce the large kinetic energies that must be supplied for the manufacture of these new heavy particles.

Although bubble chambers yield the best pictures of particle tracks, they are very complex, very large, and very expensive machines. **Spark chambers** yield somewhat cruder pictures of particle tracks, but they are much simpler. A spark chamber consists of several parallel screens or thin plates of metal, each separated from the next by a gap of a few centimeters (Figure 46.8). The plates are connected to a high-voltage supply. Any charged particle passing through the chamber ionizes the air and thereby triggers the formation of electric sparks between the screens. A succession of sparks marks the passage of the particle through the chamber (Figure 46.9). Cameras record these sparks photographically; usually mirrors are placed around the spark chamber so that a single photograph can simultaneously record several views of the track of sparks seen from several angles.

In recent years, spark chambers have been supplanted by **multiwire chambers,** built with grids of wires individually connected to high-voltage supplies, instead of plates. The formation of sparks near the wires is detected by individual electronic sensors on the wires. The signals from the sensors permit the determination of the position of the spark

Fig. 46.8 Spark chamber.

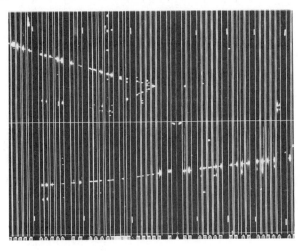

Fig. 46.9 Tracks of particles in a spark chamber.

relative to the wire grid, that is, the determination of the spatial coordinates of the spark. The signals from such a multiwire chamber are sent directly to a computer, which immediately reconstructs the trajectory of the particle in space and displays a picture of the trajectory. The UA 1 detector at CERN (see Figure 46.4) incorporates several large multiwire chambers. (Figure 46.15, on page 1138, is an example of a picture of trajectories reconstructed from signals collected by the chambers in this detector.)

46.2 The Multitude of Particles

As physicists built accelerators of higher and higher energy, they discovered more and more new particles. By now, close to 300 kinds of particles are known and a few extra particles are found every year. Whenever a new, more powerful accelerator makes available more energy for collisions, more particles and heavier particles are produced —there seems to be no end in sight.

The known particles fall into three groups: leptons, baryons, and mesons.

There are six different **leptons** (see Table 46.2). Among these, the electron (e) is the most familiar. The muon (μ) is very similar to an electron; it has the same electric charge as an electron but its mass is about 200 times as large. The tau (τ) is an esoteric particle, discovered only recently in high-energy experiments at several accelerator labora-

Leptons

Table 46.2 THE LEPTONS

Particle	Mass	Electric charge	Spin
e	0.511 MeV	−1	$\frac{1}{2}$
ν_e	0	0	$\frac{1}{2}$
μ	105.7	−1	$\frac{1}{2}$
ν_μ	0	0	$\frac{1}{2}$
τ	1784	−1	$\frac{1}{2}$
ν_τ	0	0	$\frac{1}{2}$

tories. The neutrinos (ν_e, ν_μ, and ν_τ) are particles of zero mass and zero electric charge. Although muons and neutrinos are not constituents of ordinary atoms, they are both quite abundant in nature. Large numbers of muons are generated at the top of the Earth's atmosphere by the impact of cosmic rays; these muons are secondary cosmic rays that penetrate to sea level and hit our bodies, doing some radiation damage (see Interlude IV). Large numbers of neutrinos are released by the nuclear reactions in the core of the Sun; these neutrinos penetrate through the Earth and through our bodies, but they do very little damage — our bodies and even the entire bulk of the Earth are nearly transparent to neutrinos.

In Table 46.2 the masses of the particles are expressed in MeV energy units (the "mass" listed in this table is really mass $\times c^2$), and the electric charge and the spin are expressed, respectively, in multiples of the proton charge e and in multiples of Planck's constant $\hbar$.

Besides the leptons of Table 46.2, there are six antileptons: the antielectron, the antimuon, the antitau, and the three antineutrinos. These antiparticles have an opposite electric charge and some other opposite properties, but they have exactly the same mass and spin as the corresponding particles. (The notation for an antiparticle is a bar over the letter or, alternatively, a superscript indicating the electric charge; thus, the notation for the antielectron is $\bar{e}$ or e^+.)

Baryons

The **baryons** are the most numerous group of particles (see Table 46.3). The proton and the neutron are baryons; in the table they have been listed as a pair under the single entry N(939) because of close similarities (the proton and the neutron are much more similar to each other than to any other particle). For the same reason, other baryons have also been listed as pairs, triplets, or quadruplets under a single entry. The first column of the table gives the baryon name consisting of a letter and a number in parentheses; the number is (approximately) the baryon mass expressed in MeV energy units. The second column gives the alternative electric charges of the several baryons that share a common name, and the third column gives the spin.

For every baryon in Table 46.3 there exists an antibaryon. As in the case of leptons, these antiparticles have an opposite charge but the same mass and spin as the corresponding particles.

Mesons

Finally, the **mesons** are another numerous group of particles. They are listed in Table 46.4. Mesons with close similarities are listed under a single entry; for instance, $\pi(140)$ includes the positive pion π^+, the negative pion π^-, and the neutral pion π^0. Note that all mesons have spin 0, or 1, or 2, etc., whereas all baryons have spin $\frac{1}{2}$, or $\frac{3}{2}$, or $\frac{5}{2}$, etc. This is the distinctive difference between mesons and baryons. For every meson there exists an antimeson; however, these antiparticles have already been included in Table 46.4. For example, the antiparticle to the π^+ is the π^- (and vice versa). The antiparticle to the π^0 is the π^0; this means that when two π^0 mesons meet, they can annihilate each other.

Physicists know much more about these particles than their mass, electric charge, and spin. All the baryons and mesons are endowed with several quantum numbers representing diverse esoteric quantities that obey newly discovered conservation laws (see next section). Furthermore, physicists know the reaction rates and the decay rates of the particles. Most of the particles are unstable; they decay, spontaneously falling apart into several other particles. The only absolutely stable particles are the electron, the proton, and the neutrinos. However, in the terminology of high-energy physics, a particle is called "stable" if it

Table 46.3 THE BARYONS[a]

Particle (and mass)[b]	Electric charge	Spin	Particle (and mass)[b]	Electric charge	Spin
N(939)	0, 1	$\frac{1}{2}$	Λ(1690)	0	$\frac{3}{2}$
N(1440)	0, 1	$\frac{1}{2}$	Λ(1800)	0	$\frac{1}{2}$
N(1520)	0, 1	$\frac{3}{2}$	Λ(1810)	0	$\frac{1}{2}$
N(1535)	0, 1	$\frac{1}{2}$	Λ(1820)	0	$\frac{5}{2}$
N(1650)	0, 1	$\frac{1}{2}$	Λ(1830)	0	$\frac{5}{2}$
N(1675)	0, 1	$\frac{5}{2}$	Λ(1890)	0	$\frac{3}{2}$
N(1680)	0, 1	$\frac{5}{2}$	Λ(2100)	0	$\frac{7}{2}$
N(1700)	0, 1	$\frac{1}{2}$	Λ(2110)	0	$\frac{5}{2}$
N(1710)	0, 1	$\frac{1}{2}$	Λ(2350)	0	$\frac{9}{2}$
N(1720)	0, 1	$\frac{3}{2}$	Σ(1193)	1, 0, −1	$\frac{1}{2}$
N(2190)	0, 1	$\frac{7}{2}$	Σ(1385)	1, 0, −1	$\frac{3}{2}$
N(2220)	0, 1	$\frac{9}{2}$	Σ(1660)	1, 0, −1	$\frac{1}{2}$
N(2250)	0, 1	$\frac{9}{2}$	Σ(1670)	1, 0, −1	$\frac{3}{2}$
N(2600)	0, 1	$\frac{11}{2}$	Σ(1750)	1, 0, −1	$\frac{1}{2}$
Δ(1232)	2, 1, 0, −1	$\frac{3}{2}$	Σ(1775)	1, 0, −1	$\frac{5}{2}$
Δ(1620)	2, 1, 0, −1	$\frac{1}{2}$	Σ(1915)	1, 0, −1	$\frac{5}{2}$
Δ(1700)	2, 1, 0, −1	$\frac{3}{2}$	Σ(1940)	1, 0, −1	$\frac{3}{2}$
Δ(1900)	2, 1, 0, −1	$\frac{1}{2}$	Σ(2030)	1, 0, −1	$\frac{7}{2}$
Δ(1905)	2, 1, 0, −1	$\frac{5}{2}$	Σ(2250)	1, 0, −1	?
Δ(1910)	2, 1, 0, −1	$\frac{1}{2}$	Ξ(1318)	0, −1	$\frac{1}{2}$
Δ(1920)	2, 1, 0, −1	$\frac{3}{2}$	Ξ(1530)	0, −1	$\frac{3}{2}$
Δ(1930)	2, 1, 0, −1	$\frac{5}{2}$	Ξ(1690)	0, −1	?
Δ(1950)	2, 1, 0, −1	$\frac{7}{2}$	Ξ(1820)	0, −1	$\frac{3}{2}$
Δ(2420)	2, 1, 0, −1	$\frac{11}{2}$	Ξ(1950)	0, −1	?
Λ(1116)	0	$\frac{1}{2}$	Ξ(2030)	0, −1	?
Λ(1405)	0	$\frac{1}{2}$	Ω⁻(1672)	−1	$\frac{3}{2}$
Λ(1520)	0	$\frac{3}{2}$	Λ_c^+(2285)	0	$\frac{1}{2}$
Λ(1600)	0	$\frac{1}{2}$	Σ_c(2455)	2, 1, 0	$\frac{1}{2}$
Λ(1670)	0	$\frac{1}{2}$	Ξ_c^+(2460)	1	$\frac{1}{2}$

[a] Data in Tables 46.3 and 46.4 from *Review of Particle Properties*, by The Berkeley Particle Data Group, April 1988.
[b] The number in parentheses is the mass expressed in MeV.

Table 46.4 THE MESONS

Particle (and mass)	Electric charge	Spin	Particle (and mass)	Electric charge	Spin
π(140)	1, 0, −1	0	ψ(3770)	0	1
η(549)	0	0	ψ(4040)	0	1
ρ(770)	1, 0, −1	1	ψ(4160)	0	1
ω(783)	0	1	ψ(4415)	0	1
η′(958)	0	0	Υ(9460)	0	1
f_0(975)	0	0	χ_{b0}(9860)	0	?
a_0(980)	1, 0, −1	0	χ_{b1}(9892)	0	?
φ(1020)	0	1	χ_{b2}(9913)	0	?
h_1(1170)	0	1	Υ(10023)	0	1
b_1(1235)	1, 0, −1	1	χ_{b0}(10235)	0	?
a_1(1260)	1, 0, −1	1	χ_{b1}(10255)	0	?
f_2(1270)	0	2	χ_{b2}(10270)	0	?
η(1280)	0	0	Υ(10355)	0	1
f_1(1285)	0	1	Υ(10580)	0	1
a_2(1320)	1, 0, −1	2	Υ(10860)	0	1
f_0(1400)	0	0	Υ(11020)	0	1
f_1(1420)	0	1	$K_\pm$(494)	1, −1	0
η(1430)	0	0	K^0(498)	0, 0	0
f_0'(1525)	0	2	K^*(892)	1, 0, −1	1
f_1(1530)	0	1	K_1(1270)	1, 0, −1	1
f_0(1590)	0	0	K_1(1400)	1, 0, −1	1
ω_3(1670)	0	3	K_0^*(1430)	1, 0, −1	0
π_2(1670)	1, 0, −1	2	K^*(1415)	1, 0, −1	1
φ(1680)	0	1	K_2^*(1430)	1, 0, −1	2
ρ_3(1690)	1, 0, −1	3	K^*(1715)	1, 0, −1	1
ρ(1700)	1, 0, −1	1	K_2(1770)	1, 0, −1	2
f_2(1720)	0	2	K_3^*(1780)	1, 0, −1	3
f_2(2010)	0	2	K_4^*(2075)	1, 0, −1	1
f_4(2050)	0	4	D^0(1864)	0, 0	0
f_2(2300)	0	2	$D^\pm$(1869)	1, −1	0
f_2(2340)	0	2	$D^{*\pm}$(2010)	1, −1	1
η_c(2980)	0	0	D^{*0}(2010)	0, 0	1
J/ψ(3097)	0	1	$D_s^\pm$(1969)	1, −1	0
χ_0(3415)	0	0	$D_s^{*\pm}$(2113)	1, −1	0
χ_1(3510)	0	1	D_J(2420)	1, −1	?
χ_2(3555)	0	2	$B^\pm$(5278)	1, −1	0
ψ(3686)	0	1	B^0(5279)	0, 0	0

lives long enough for physicists to do experiments on or with the particle. Thus, particles that live only 10^{-10} s or even 10^{-14} s are regarded as "stable," because such particles live long enough to be assembled in a beam and shot at a target. If this is stable, what is unstable? Most of the particles listed in Tables 46.3 and 46.4 have lifetimes of about 10^{-23} s. Such particles are regarded as unstable.

A particle that lives only 10^{-23} s is incapable of making a visible track in a bubble chamber. Thus, such an unstable particle cannot be detected directly, but its existence can be inferred from a careful study of the rates of reactions of stable particles engaged in collisions. The unstable particle participates in these reactions as an intermediate, ephemeral state, which causes a characteristic increase of the reaction rate whenever the energy of the stable particles coincides with the energy required for the production of the unstable particle. Because of their effects on reaction rates, the unstable particles are often called **resonances**.

Resonances

46.3 Interactions and Conservation Laws

The reactions that occur among the particles in a high-energy collision are governed by the four fundamental forces: the "strong" force, the electromagnetic force, the "weak" force, and the gravitational force. However, at the microscopic level, particles are subject to quantum uncertainties in position and velocity, and the force acting on them is not well defined. Hence, physicists prefer to speak of four fundamental kinds of **interactions,** instead of forces. Mathematically, these interactions can be described by formulas that specify the amount of energy for each interaction — energies remain meaningful even at a microscopic level, and therefore energies are more relevant than forces. Nevertheless, for the purpose of the following qualitative discussion, let us continue to speak of forces, even though this intuitive concept is ambiguous and ought to be replaced by a more precise and sophisticated mathematical concept of interaction energies.

Hadrons

The strong force acts on baryons and mesons, but not on leptons. The particles that interact via the strong force are called **hadrons;** thus, baryons and mesons are hadrons, but leptons are not.

The electromagnetic force acts primarily on charged particles, but it also acts on neutral particles — such as the neutron — which contain an internal distribution of electric charge.

The weak force acts on leptons, baryons, and mesons. However, its effect on baryons and mesons is often hidden behind the much larger effects produced by the strong or electromagnetic forces. To see the purest manifestation of the weak force, we have to examine reactions involving leptons. The weak force is deeply involved in many reactions that bring about the decay of unstable particles. For example, the weak force is responsible for the decay of the neutron.

The gravitational force is of no direct interest in particle physics. Although all particles and all forms of energy interact gravitationally, the gravitational effects produced by individual particles are too feeble to be of any significance at the energies available in laboratories on Earth.

Table 46.5 lists the strength of each of the fundamental forces and also the range, or the maximum distance over which this force can reach from one particle to another. This table contains the same information as Table 6.1, but in the new table we have expressed the strengths of the forces relative to that of the strong force to which we arbitrarily have assigned a strength of 1.[1]

Table 46.5 THE FOUR FUNDAMENTAL FORCES

Force	Acts on	Relative strength	Range
Strong	Baryons and mesons (hadrons)	1	$\sim 10^{-15}$ m
Electromagnetic	Charged particles	10^{-2}	Infinite
Weak	Leptons, baryons, and mesons	10^{-6}	$\sim 10^{-18}$ m
Gravitational	All forms of matter (all forms of energy)	10^{-38}	Infinite

---

[1] The strengths of the forces depend on the energies of the particles. The values in the table are appropriate for low energies.

The rate of a reaction among elementary particles depends primarily on the strength of the force that drives the reaction. For instance, the reaction (1), initiated by the collision of a pion with a proton, is an extremely fast reaction driven by the strong force. The time scale of the reaction can be estimated as follows: the incident pion travels at nearly the speed of light, and hence passes by or through the target proton very quickly; since the diameter of the proton is of the order of 10^{-15} m, the time available for the interaction can be calculated by dividing this distance by the speed of light. This gives an extremely short time of about 10^{-23} s.

The strong interaction can bring about not only particle creation but also particle decay. The time scale for such decays is about the same as the time scale for creation, about 10^{-23} s. Hence those particles that decay strongly lead a very ephemeral existence; they are very unstable and burst apart within an instant of their creation. For example, the ρ meson decays into two pions (π) by the strong reaction

$$\rho^- \to \pi^- + \pi^0 \qquad (4)$$

within 10^{-23} s. The ρ meson is a resonance particle. Since it lasts only such a short time, it cannot be detected directly; its existence can only be inferred from the correlations observed in the energies of the decay products.

Reactions caused by the electromagnetic force take longer. Figure 46.10 shows the decay of a pion (π^0) into an electron (e), antielectron ($\bar{\text{e}}$), and gamma ray (γ) by the electromagnetic reaction

$$\pi^0 \to \text{e} + \bar{\text{e}} + \gamma \qquad (5)$$

which takes about 10^{-14} s (this meson can also decay into two gamma rays, which is a somewhat faster reaction). Although 10^{-14} s is a short time by ordinary standards, it is a long time by the standards of high-energy physics, and the pion is regarded as stable.

The reactions involving the weak force take much longer. Figure 46.10 also shows the decay of a lambda baryon (Λ) into a proton (p) and a pion (π^-),

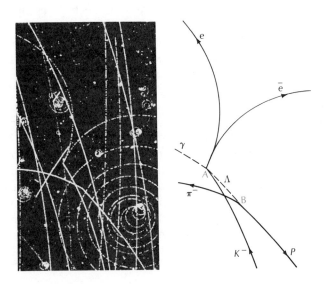

Fig. 46.10 (left) Photograph of tracks of particles in a bubble chamber and (right) tracing based on the photograph. Neutral particles do not make visible tracks in the bubble chamber, but their extrapolated trajectories have been indicated by dashed lines in the tracing. A Λ baryon and a π^0 were created in a collision at A. The π^0 decayed immediately into an e, $\bar{\text{e}}$, and γ. The tracks of the e and $\bar{\text{e}}$ can be seen diverging from the point of decay, but the γ made no track in the photograph. The Λ subsequently decayed at B.

$$\Lambda \to p + \pi^- \tag{6}$$

This reaction is weak and takes about 10^{-10} s. By the standards of high-energy physics, this is a *very* long time. Moving at close to the speed of light, the Λ travels a distance of several centimeters between its creation and its decay.

Note that the above reactions, with time scales of 10^{-23}, 10^{-14}, and 10^{-10} s for strong, electromagnetic, and weak forces, are intended only as more or less typical examples. Rates of other reactions involving these forces can differ by several, or even by many, powers of 10 from the above. Thus, the decay of the neutron into a proton, an electron, and an antineutrino,

$$n \to p + e + \bar{\nu} \tag{7}$$

is a weak reaction which is extremely slow — on the average, it takes about 15 min. This delay in the reaction is to be blamed on the small amount of energy available for the decay process. Although there is some overlap among the rates of the reactions generated by the different kinds of forces, as a rough rule we can say that the strong force generates the fastest reactions, the electromagnetic force slower reactions, and the weak force the slowest reactions[2]. The gravitational force between individual particles is so feeble that it is incapable of generating reactions that can be observed in our laboratories.

All the forces, and all the reactions that they produce, obey the usual conservation laws for energy, momentum, angular momentum, and electric charge. The reactions also obey the conservation law for **baryon number,** which is a generalization of the conservation law for mass number familiar from nuclear physics. Each baryon has a baryon number of $+1$, each antibaryon -1, and all other particles have baryon number 0. The conservation law for baryon number then simply states that the net baryon number remains unchanged in any reaction. For example, consider the reaction (1). The baryon numbers for π^+, p, Δ and K^0 are 0, $+1$, $+1$, and 0, respectively; hence in the reaction (1), the net baryon number is $0 + 1$ before and $1 + 0 + 0 + 0$ after, that is, the net baryon number remains unchanged.

Baryon number

Besides these conservation laws for familiar quantities, experiments with high-energy particles have led to the discovery of new conservation laws involving several esoteric quantities, such as lepton number, isospin, strangeness, and parity.

Lepton number

Lepton number is for leptons what the baryon number is for baryons. Each lepton has a lepton number $+1$, each antilepton -1, and all other particles have lepton number 0. The net lepton number remains unchanged in any reaction.

Isospin

Isospin is somewhat more complicated because it is a quantity with several components; that is, isospin is a vector quantity. The conservation law for isospin states that the net sum of the isospin vectors of all the particles involved in a reaction remains unchanged. Obviously, this is somewhat similar to the conservation law for the momentum vector or for the angular momentum vector. However, the isospin is not a vector in our ordinary three-dimensional space but, rather, a vector in an abstract mathematical space that theoretical physicists invented as a computational tool for the description of the "strong" force.

---

[2] The reaction rates depend on energy. This comparison of the reaction rates is valid for the typical energies attained in laboratories.

Strangeness is similar to the baryon and lepton numbers. Each hadron has a strangeness number: the proton has strangeness 0, the kaon has +1, the lambda has −1, the pion has 0, etc. The conservation law for strangeness states that the net strangeness number remains unchanged in any reaction. For example, in reaction (1) the net strangeness is $0 + 0$ before and $-1 + 1 + 0 + 0$ after; that is, there is no change.

Strangeness

Parity characterizes the behavior of a quantum-mechanical wave under a reversal of the x, y, and z coordinates. Such a reversal is physically equivalent to forming a mirror image of the wave. It can be shown that the mirror image of the quantum-mechanical wave for a stationary state is either equal to the original wave (parity +1) or else equal to the negative of the original wave (parity −1). Conservation of parity means that the net parity (the product of all the individual parities) of all the particles participating in a reaction is unchanged.

Parity

The conservation laws for energy, momentum, angular momentum, electric charge, baryon number, and lepton number are absolute — no violation of any of them has ever been discovered. By contrast, the conservation laws for isospin, strangeness, and parity are *approximate* conservation laws — they are valid for some reactions, but fail in some others. This, of course, raises the question of what possible meaning can be attached to a "law" that works sometimes and fails sometimes. The answer is that whether a conservation law is obeyed or not depends on the kind of interaction, or the kind of force, that drives the reaction. The reactions caused by the strong force obey all the conservation laws, but reactions caused by the electromagnetic or weak forces do not. It is usually easy to tell what force is involved in a reaction: reactions involving the strong force tend to be fast; reactions involving the other forces tend to be (relatively) slow. For example, reaction (4) is a fast reaction brought about by the strong force, whereas reaction (6) is a slow reaction brought about by the weak force.

Table 46.6 lists the conservation laws obeyed by the strong, electromagnetic, and weak forces.

Table 46.6 FORCES AND CONSERVED QUANTITIES

Force	Energy, momentum, and angular momentum	Charge	Baryon number	Lepton number	Strangeness	Parity	Isospin
Strong	✓	✓	✓	✓	✓	✓	✓
Electromagnetic	✓	✓	✓	✓	✓	✓	
Weak	✓	✓	✓	✓			

46.4 Fields and Quanta

According to classical theory, forces are mediated by fields. Distant particles do not act on one another directly; rather, each particle generates a field of force and this field acts on the other particles. As we saw in Section 23.2, the existence of fields is required by conservation of energy and momentum. Fields play the role of storehouses of en-

Fig. 46.11 The two spiraling tracks in this bubble-chamber photograph were made by an electron and an antielectron. These particles were created by a high-energy gamma ray in a collision with the electron of a hydrogen atom in the bubble chamber. The long, slightly curved downward track was made by the recoiling electron.

ergy and momentum; the energy and momentum stored in the fields balance any excess or deficit in the energy and momentum of the interacting particles engaged in (nonuniform) motion. The most spectacular example of the conversion of particle energy into field energy occurs in the annihilation of matter with antimatter: if an electron collides with an antielectron, the two particles annihilate one another, giving off a burst of very energetic light, or gamma rays. In this annihilation the energy of the particles — including their rest-mass energy — is completely converted into field energy. The reverse reaction is also possible: if a gamma ray collides with a charged particle, it can create an electron–antielectron pair. In such a pair creation, the energy of the gamma ray is converted into the energy of the pair of particles (Figure 46.11).

Each of the four fundamental forces is mediated by fields of its own. Hence there are gravitational fields, electromagnetic fields, strong fields, and weak fields. On a macroscopic scale, the gravitational and electric fields of everyday experience are smooth functions of space and time; these functions can be calculated by classical field theory. The electric-field problems that we solved in Chapter 23 are examples of such classical calculations. However, on a microscopic scale, physics is ruled by quantum theory and not by classical theory. According to quantum theory, the energy stored in fields is not smoothly distributed; rather the energy is found in **quanta,** that is, small packets or lumps of energy.

In Chapter 42, we became acquainted with the quanta of the electromagnetic field; these quanta are the photons. Each of the other fundamental fields also has quanta of its own. Table 46.7 lists the quanta of all the four fundamental fields. Like photons, the quanta of the gravitational field, or gravitons, are massless. Our belief in gravitons rests largely on theoretical considerations; these particles have not yet been detected and, given the extremely small strengths of gravitational forces, it is unlikely that they will be directly observed soon.

The quanta of the strong field, or pions, and the quanta of the weak

Table 46.7 Fields and Their Quanta

Field	Quanta	Mass[a]
Gravitational	Gravitons	0 MeV
Weak	W particles	82,000
	Z particles	93,000
Electromagnetic	Photons	0
Strong	Pions and some other mesons	140

[a] The masses are expressed in energy units, as in Tables 46.3 and 46.4.

Richard Phillips Feynman, *1918–1988, American physicist, professor at the California Institute of Technology. His invention of the Feynman diagram revolutionized the computations of relativistic quantum processes. He shared the 1965 Nobel Prize with* **Julian Schwinger,** *1918–, American physicist, and* **Sin-Itiro Tomonaga,** *1906–1979, Japanese physicist, for work on quantum electrodynamics.*

field, or W and Z particles, are endowed with mass. The notion that a "solid" particle such as a pion should play the role of quantum of a field seems a bit bizarre. But quantum theory teaches us that pions, like all other particles, have a dual character: they are both waves and particles (see Section 43.6). In their character of waves, pions are described by a wave field; taking into account that the energy of this field must be quantized, it turns out that the corresponding quanta are nothing but the pions themselves. This identification of particles and quanta holds in general: *all particles are quanta of fields.* Pions are quanta of the **pion field,** electrons are quanta of the **electron field,** protons are quanta of the **proton field,** and so on. Unfortunately, this means that there is one kind of field for each kind of particle — there are very many kinds of fields. Since it is unlikely that the many, many particles we know of are all elementary, it is unlikely that the many, many fields are fundamental. Thus, if the proton is a composite structure made of three quarks (see Section 46.6), then the proton field is also a composite structure made of a combination of three quark fields.

In view of this multitude of new fields, the question arises whether each of the new fields will act as mediator and carry some kind of force between particles. The answer is yes: all fields can play an intermediate role and mediate forces between particles. For example, the electron field can act between two photons, permitting them to exert forces on one another. Diverse meson fields (ρ-meson field, ω-meson field, etc.) can act between two protons, permitting them to exert strong forces in addition to the forces generated via the pion field; in fact, the diversity of these additional corrections is one of the reasons why the theoretical analysis of the strong force is so dreadfully difficult.

At the quantum level, we can picture the field of force generated by a particle as a swarm of quanta buzzing around the particle. For example, we can picture the electric field surrounding an electron, or any other charged particle, as a swarm of photons. The swarm is in a state of everlasting activity — the charged particle continually emits and reabsorbs the photons of the swarm. Emission is creation of a photon; absorption is annihilation of a photon. Hence we can say that the electric field arises from the continual interplay of three fundamental processes: creation, propagation, and annihilation of photons. The action of one charged particle on another involves a sequence of these three fundamental processes: a photon is emitted by one particle, propagates through the intervening distance, and is absorbed by the other particle. This exchange process can be represented graphically by a diagram showing the worldlines of the particles (Figure 46.12); such a

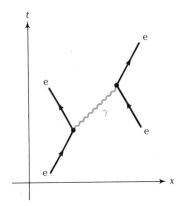

Fig. 46.12 Feynman diagram representing the exchange of a virtual photon between two electrons. The solid black lines indicate the worldlines of the two electrons, with the t axis plotted vertically and the x axis plotted horizontally. The wavy colored line indicates the worldline of the photon. The electron on the left emits this photon and the electron on the right absorbs it.

Feynman diagram
Virtual photon

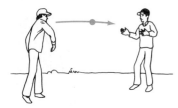

Fig. 46.13 Two boys throw a ball back and forth.

Fig. 46.14 Two boys throw a boomerang back and forth.

Quantum electrodynamics

Paul Adrien Maurice Dirac, *1902–1984, English physicist, professor at Cambridge. He formulated the relativistic quantum theory of the electron and predicted the existence of antiparticles. For his contributions to theoretical physics, he shared the 1933 Nobel Prize with E. Schrödinger.*

diagram is called a **Feynman diagram.** The photon exchanged between the two electrons is called a **virtual photon,** because it lasts only a very short time and, being reabsorbed by an electron, is undetectable by any direct experiment. The steady attractive or repulsive force between two charged particles is generated by continual repetition of this exchange process. This is action-by-contact with a vengeance — at a fundamental level, all forces reduce to local acts of creation and destruction involving particles in direct contact.

In terms of a simple analogy, we can easily understand how the exchange of particles brings about forces. Imagine two boys tossing a ball back and forth between them (Figure 46.13); it is intuitively obvious that this produces a net repulsive force between the boys due to the recoil they suffer when throwing or catching the ball. Our intuition suggests that no such exchange process can ever produce attraction. However, imagine two Australian boys tossing a boomerang back and forth between them (Figure 46.14); it is then obvious that this produces an attractive force between the boys.[3] Whether a photon exchanged between two charges behaves like a ball or like a boomerang depends on the signs of the charges. Quantum calculations, which take into account the wave nature of all the particles involved, show that the net force is attractive for unlike charges and repulsive for like charges, as it should be.

This theory of electric forces is called **quantum electrodynamics** (or QED); it was developed by P. A. M. Dirac, R. P. Feynman, J. Schwinger, S. Tomonaga, and other physicists between 1930 and 1950. Quantum electrodynamics is the most accurate theory in all of physics, and in all of science. Some theoretical calculations in quantum electrodynamics have been carried out to ten significant figures, and delicate experimental measurements have confirmed these calculations exactly.

The gravitational, weak, and strong forces are also generated by the exchange of virtual particles: gravitons, W and Z particles, and pions (and some other mesons). It is a general rule of quantum theory that the range of the force is inversely related to the mass of the particle that serves as the carrier of the force. Thus, the photons and the gravitons that are the carriers of the electromagnetic and the gravitational force have zero mass — the ranges of these forces are infinite. The pions that are the carriers of the strong force have a fairly large mass — the range of this force is short. The W and Z particles that are the carriers of the weak force have a very large mass — the range of this force is very short.

46.5 The Unified Electroweak Force

It has long been one of the aspirations of theoretical physicists to achieve the unification of the forces of nature. Electromagnetism is the

---

[3] One of the flaws in this analogy is that the momentum of the boomerang changes during its flight through the air. But this is not a fatal flaw because we can assume that the initial and the final direction of motion are the same (a 180° turn), so that the boomerang merely borrows momentum from the air, without any *net* change.

best-known example of a **unified field theory,** that is, a theory that treats electric and magnetic forces as two aspects of a single underlying force. This unification of electricity and magnetism is a consequence of the theory of relativity, just as is the unification of space and time. When Einstein formulated the theory of relativity and proved that space and time transform into one another under Lorentz transformations, he also proved that electric and magnetic fields transform into one another. In Section 41.8, we examined a special instance of such a transformation; we saw that an electric field transforms into a combination of electric and magnetic fields when we go from one inertial reference frame to another. Thus, the magnetic force is not fundamentally different from the electric force; these two forces are merely two aspects of a single force called the electromagnetic force, just as space and time are two aspects of a single entity called spacetime.

Unified field theory

Some years ago, S. Weinberg, A. Salam, and S. Glashow[4] achieved the unification of weak and electromagnetic forces. This new theory has shown that, in spite of the seemingly drastic differences between their characters, the weak and electromagnetic forces are basically the same — they are merely two aspects of a single **electroweak force.**

Electroweak force

If we seek to unify the weak and electromagnetic forces, we must regard the carriers of these forces — the quanta whose exchange generates the forces — as closely related. This would seem to conflict with the large mass difference between these particles: the photon is massless, but the W and Z particles are the heaviest particles known. The unified electroweak theory attributes this mass difference to a minor asymmetry (a "broken symmetry") between the photon and the W and Z particles. The theory asserts that perfect symmetry between these particles can be restored by giving them very high energies, in excess of 100 GeV; at such high energies, the photon and the W and Z particles should become essentially identical. Such high energies are difficult to achieve in our laboratories, but they were readily available during the very early stages of the Big Bang, when the universe was younger than 10^{-10} s and had a temperature in excess of 10^{15} K. It is believed that at these early times, there was no difference between the photon and the W and Z particles, and there was no difference between the electromagnetic and weak forces.

One notable consequence of the unified theory of weak and electromagnetic forces is that it provides a natural explanation of the quantization of electric charge. As we saw in Section 22.3, the electric charge on all known particles is some integral multiple of the basic unit of charge, $e = 1.60 \times 10^{-19}$ C. It can be shown that this quantization rule emerges as an immediate consequence of the unified theory.

From the experimental point of view, an early success of the unified theory was its prediction of some hitherto unknown forces between baryons and neutrinos. The weak forces between electrons and neutrinos can be accounted for by exchange of two kinds of W particles: a positively charged particle W^+ and a negatively charged particle W^-. However, the unified electroweak theory demands the existence of one more particle: the electrically neutral Z, also called Z^0. The force generated by the exchange of these neutral Z particles is called the

---

[4] **Stephen Weinberg,** 1933–, American physicist, **Abdus Salam,** 1926–, Pakistani physicist, and **Sheldon Lee Glashow,** 1932–, American physicist, shared the Nobel Prize in 1979 for the unified theory of weak and electromagnetic forces.

neutral-current force. The existence of this new force was soon confirmed by experiments on high-energy baryon–neutrino collisions at CERN.

The most recent and most impressive success of the unified theory was its prediction of the masses of the W and Z particles. The unified theory predicts the masses of the W and Z particles by the following argument: If the weak and electromagnetic forces are essentially the same, then they must also have the same strength. The fact that the experimentally observed strengths seem quite different (see Table 46.5) is attributed to the masses of the W and Z particles — under certain conditions a force of large strength can have the appearance of a force of small strength if the particle that carries the force is very massive. A theoretical calculation shows that at a fundamental level the weak and the electromagnetic forces can be regarded as having the same strength provided that the W and the Z particles have masses of about 80 and 90 times the mass of a proton. These large masses explain both the strength of the weak force and its short range.

The W and Z particles were detected in 1982 in experiments at the proton–antiproton collider at CERN.[5] The experiments involved the observation of about a billion head-on collisions between protons and antiprotons of the same energy, 270 GeV. A few dozen W and Z particles were produced in these collisions (Figure 46.15). The measured masses of the W and the Z particles are, respectively, 82 GeV and 93 GeV. These measured values are within 2 GeV of the theoretically predicted values. This excellent agreement constitutes a brilliant confirmation of the unified theory of weak and electromagnetic interactions.

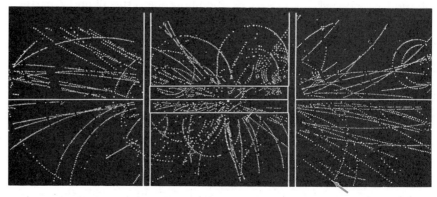

Fig. 46.15 Tracks of particles produced in a very energetic head-on collision between a proton of 270 GeV that entered from the right and an antiproton of 270 GeV that entered from the left. In this collision a W⁻ particle was created. It immediately decayed into an electron and a neutrino; the track of the former (indicated by the arrow) can be seen emerging toward the lower right.

46.6 Quarks

Let us now return to our initial question. What are the ultimate, indivisible building blocks of matter? We know of nearly 300 particles. It is inconceivable that all these hundreds of particles are *elementary* particles. It is likely that most of them, or maybe all of them, are composite particles made of just a few truly elementary building blocks.

---

[5] **Carlo Rubbia,** 1934–, Italian physicist, and **Simon van der Meer,** 1925–, Dutch physicist, shared the Nobel Prize in 1984 for the development of the proton-antiproton collider at CERN and the discovery of the W and Z particles.

To discover these building blocks, physicists have tried to break protons into pieces by bombarding them with projectiles of very high energy. Unfortunately, if the energy of the projectile is large enough to make a dent in a proton, then it is also large enough to create new particles during the collision. This abundant creation of particles confuses the issue — we can never be quite sure which of the pieces that come flying out of the scene of the collision are newly created particles and which are fragments of the original proton. In fact, none of the pieces ever found in such collision experiments seem likely candidates for an elementary building block. Typically, the particles that emerge from a collision between a high-energy projectile and a proton are pions, kaons, lambdas, deltas, etc., all of which seem to be even less elementary than a proton.

Although brute-force collision experiments have failed to fragment protons into elementary building blocks, somewhat more subtle experiments have provided us with some evidence that distinct building blocks do indeed exist inside protons. At the Stanford Linear Accelerator, very high-energy electrons were shot at protons; these electrons served as probes to "feel" the interior of the protons. The experiments showed that occasionally the bombarding electrons were deflected through large angles, bouncing off sharply from the interior of a proton. These deflections indicate the presence of some lumps or hard kernels in the interior of the proton, just as, in Rutherford's experiments, the deflections of alpha particles by atoms indicated the presence of a hard kernel (nucleus) in the interior of the atom. Protons and all the other baryons and mesons seem to be composite bodies made of several distinct pieces. In contrast, electrons and the other leptons seem to be indivisible bodies with no internal structure. In recent experiments the electron has been probed with beams of extremely energetic particles to within 10^{-18} m of its center. Even at these extremely short distances, no substructures of any kind were found. Thus, the electron seems to be a pointlike body with no size at all, a truly elementary particle.

Even before the experimental evidence for lumps inside protons became available, theoretical physicists had noticed that particles could be classified into groups or families of similar particles on the basis of their quantum numbers and their behavior in reactions. To explain these similarities, they had proposed theories in which all the known baryons and mesons are regarded as constructed out of a few fundamental building blocks. According to these theories, the similarity between particles in a given family reflects the similarity of their internal construction, just as the similarities between atoms in a group of the Periodic Table reflect the similarity of their internal construction.

The most successful of these theories is the quark model proposed by M. Gell-Mann[6] and by G. Zweig. In this model, all particles are constructed of three kinds of fundamental building blocks called **quarks**. Gell-Mann took the word *quark* from *Finnegan's Wake*, a book by James Joyce quopping with weird words. The three quarks are usually labeled **up, down,** and **strange,** or simply **u, d,** and **s.** They all have spin $\frac{1}{2}$ and electric charges of $\frac{2}{3}$, $-\frac{1}{3}$, and $-\frac{1}{3}$, respectively (see Table 46.8). Of course, each quark — like any other particle — has an antiparticle, of opposite electric charge.

Quarks

---

[6] **Murray Gell-Mann,** 1929–, American physicist, received the 1969 Nobel Prize for his work on the classification of elementary particles.

Table 46.8 Three Quarks

Quark	Mass[a]	Electric charge	Spin
u	5 MeV	$\frac{2}{3}$	$\frac{1}{2}$
d	9	$-\frac{1}{3}$	$\frac{1}{2}$
s	180	$-\frac{1}{3}$	$\frac{1}{2}$

[a] Based on theoretical estimates.

Fig. 46.16 Structure of the proton: two u quarks and one d quark. The sizes of the quarks have not been drawn to scale. Their sizes are probably much smaller than the size of the proton.

Fig. 46.17 Structure of the neutron: two d quarks and one u quark.

Fig. 46.18 Structure of a positive pion (π^+): one u quark and one d antiquark.

To make the ordinary particles out of the quarks, the latter must be glued together in diverse ways. For example, a proton is made of two u quarks and one d quark (Figure 46.16). A neutron is made of two d quarks and one u quark (see Figure 46.17). A positive pion is made of one u quark and one d antiquark (Figure 46.18), and so on. By exercising some care in what quarks are glued together to make what particle, we can build up all the known baryons and mesons and explain their quantum numbers and their similarities. Besides, we can predict some properties, such as magnetic moments and reaction rates, of the composite particles from the (assumed) properties of the quarks. The experimental confirmation of these predictions is strong circumstantial evidence for the quark model.

There is only one snag: in spite of prodigious efforts, experimental searches for quarks have been unsuccessful. Experimenters have looked for quarks in the debris of the very energetic collisions of particles from accelerators. They have looked for quarks in the debris of the even more energetic collisions caused by the impact of cosmic rays on the atmosphere of the Earth; they have looked for quarks in samples of water, air, dust, meteorites, lava, limestone, iron, graphite, niobium, plankton, seaweed, mineral oil, vegetable oil, moon rocks, etc. In all these searches the distinctive feature that would permit the unambiguous identification of a quark is its *fractional charge* — in contrast to normal particles, which all have an electric charge of 0, ±1, ±2, etc., the quarks have charges of $\pm\frac{1}{3}$ or $\pm\frac{2}{3}$.

The searches have failed to turn up any quarks. Only in a very few instances was anything found that could *possibly* have been a quark, and even in these exceptional cases the evidence was not sufficiently convincing to establish the existence of the quark beyond reasonable doubt.

Physicists now suspect that quarks are permanently confined inside the ordinary particles so that there is no way to break a quark out of, say, a proton. Although the details of the confinement mechanism are not known, it seems that the quarks are held in place by an exceptionally strong force, which prevents their escape. This new force is called the "color" force; we will discuss its properties in the next section.

46.7 Color and Charm

The simple quark model with three quarks (and three antiquarks) described in the preceding section suffers from a few deficiencies, which required some modifications of the model. The first such modification arose from the investigation of the quantum states of quarks within ordinary particles. Several apparently identical quarks are often found in the *same* quantum state. This is an unacceptable violation of the Exclusion Principle. To avoid this violation, physicists postulated that each of the quarks exists in three varieties, and that whenever two apparently identical quarks are found in the same quantum state, they actually are of different varieties. The varieties of quarks are characterized by a new property called **color.** Of course, this "color" has nothing to do with real color; it is merely a (somewhat unimaginative) name for a new property of matter. The different quark colors are **red, green,** and **blue.** Thus, there is a *red* u quark, a *green* u quark, and a *blue* u quark, and so on. The antiquarks have anticolors; the different antiquark colors are anti*red,* anti*green,* and anti*blue.*

Color

Color is a very subtle property of matter; it usually remains hidden inside the ordinary particles. All the normal particles are "colorless" — they consist of several quarks with an equal mixture of all three colors. For instance, one of the three quarks inside the proton is *red,* one is *green,* and one is *blue.* Nevertheless, color plays a crucial role in the theory of the forces that confine the quarks inside the ordinary particles.

The quarks are confined by extremely strong mutual attractive forces. These forces between quarks are color forces — the source of these forces is color just as the source of electric forces is electric charge. Each of the three varieties of color (*red, green,* and *blue*) is analogous to a kind of electric charge. A body is color neutral, or "colorless," if it contains equal amounts of all three colors or if it contains equal amounts of color and anticolor, just as a body is electrically neutral if it contains equal amounts of positive and negative charge.

The color force is a fundamental force that should be included in our table of fundamental forces instead of the strong force (Table 46.5). The color force is closely related to the strong force — the latter is actually a special instance of the former. The relationship between the color force and the strong force is analogous to the relationship between the electric force and the intermolecular force. As we saw in Section 22.1, the force between two electrically neutral atoms or molecules is a residual electric force resulting from an imperfect cancellation among the attractions and repulsions of the charges in the two atoms or molecules. Likewise, the strong force between, say, two "colorless" protons is a residual color force resulting from an imperfect cancellation among the attractions and repulsions of the quarks in the two protons. Thus, the "strong" force between protons is no more than a pale reflection of the much stronger color forces acting within each proton.

The color force has some very remarkable properties. All the forces listed in Table 46.5 *decrease* with distance, but the color force remains constant as the distance between the quarks increases. This persistence of the color force brings about the confinement of quarks. If one of the quarks in, say, a proton is somewhat separated from its companions by the violent impact of a collision, the color attraction pulls it back into its original position as soon as the collision ends. The quarks behave as though linked by rubber strings. During a collision, the rubber strings stretch, but after the collision they again contract. This analogy is perhaps fairly close to the truth; according to one theory, the color force is communicated from one quark to another along tightly bundled field lines concentrated in thin channels, rather like strings.

Although the color force keeps the quarks on a short leash and pulls them back sharply whenever they wander too far apart, the force allows the quarks to move rather freely within certain limits. This behavior is in accord with the string analogy — if the quarks stretch the strings, they experience strong restraining forces, but if they stay within the limits set by the relaxed lengths of the strings, they feel almost no force. This situation has been described by the picturesque phrases "infrared slavery" and "ultraviolet freedom" (in this context *infrared* refers to long distances and *ultraviolet* refers to short distances).

At a fundamental level, the color force between quarks is due to an exchange of virtual particles between the quarks. The particle that acts as the carrier of the color force is the **gluon.** Figure 46.19 shows a Feynman diagram representing the exchange of a gluon between two quarks. Such an exchange of a gluon between two colored quarks is analogous to the exchange of a photon between two charged particles (see Figure 46.12). However, the gluon exchange is a rather more complicated process than photon exchange — the gluons themselves have color, whereas the photons do not have any electric charge. Whenever a quark emits a gluon, this gluon takes color away from the quark and thereby changes the color of the quark. For this reason, the quark entering a vertex in Figure 46.19 has a different color than the quark leaving the vertex. The color of the gluon is the difference between the colors of these quarks; for example, the gluon emerging from the left vertex in Figure 46.19 has colors *green* and anti*blue,* that is, it has two colors. Thus, the exchange of a gluon between two quarks entails an exchange of colors. The force between, say, the three quarks within a proton is due to continual, repetitive exchanges of gluons between the quarks. During this process, the color of each quark changes again and again; for instance, the u quark within the proton is sometimes *red,* sometimes *green,* and sometimes *blue.*

Gluons

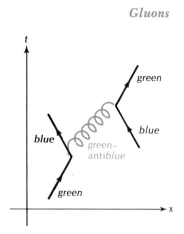

Fig. 46.19 Exchange of a gluon between two quarks. The quark on the left suffers a color change from *green* to *blue;* that on the right a color change from *blue* to *green.*

Quantum chromodynamics

The theory of the color force is called **quantum chromodynamics** (or QCD). The equations of this theory are much more complicated than those of quantum electrodynamics. Because the calculations in chromodynamics are exceedingly difficult and tedious, most of the theoretical predictions that have been obtained so far are qualitative rather than quantitative. Nevertheless, experiments performed at the DESY accelerator in 1978 did reveal some circumstantial evidence confirming the existence of the gluons. In these experiments, physicists attempted to manufacture gluons in electron–antielectron collisions of extremely high energy. Since the gluons only last for a short instant, there is no hope of seeing them directly. But chromodynamics predicts that in some violent electron–antielectron collisions, a quark, an antiquark, and a gluon will be produced. Each of these three particles decays into several pions; when emerging from the scene of the collision, these pions should then be arranged in three distinct jets spurting out at distinct angles. The experiments confirmed the existence of such triple jets. Figure 46.20 shows a picture of such jets produced at the proton–antiproton collider at CERN.

Charm

Another modification of the simple quark model with three quarks arose from the theory of the unification of the electromagnetic and weak forces. In order to make this theory fit some experimental data, physicists had to postulate the existence of a fourth quark, different from the u, d, and s quarks. This new hypothetical quark was labeled **charmed.**

The hypothesis of the charmed quark soon received firm experimental support. In 1974, a team of experimenters under the direction

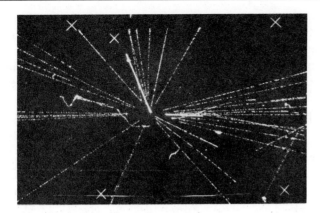

Fig. 46.20 Tracks of particles produced in a very energetic head-on collision between a proton of 270 GeV that entered from the left and an antiproton of 270 GeV that entered from the right. Among the multitude of particles created in this collision, we can distinguish three jets: one jet of three particles in the lower right, one of five particles in the upper center, and one of four particles in the upper left.

of S. Ting at the Brookhaven accelerator and an independent team under the direction of B. Richter[7] at the Stanford accelerator discovered the J/ψ meson and some other related ψ mesons (see Table 46.4), all of which have exceptionally long lifetimes. It seems that these particles contain charmed quarks with an electric charge of $\frac{2}{3}$ and an (estimated) mass of 4800 MeV.

Unfortunately, the proliferation of quarks did not stop with four quarks. In 1977 a team of experimenters at Fermilab discovered the Υ mesons. These are by far the most massive mesons known (see Table 46.4). It seems that each of these contains a new and very massive quark. This fifth kind of quark has been labeled **bottom.** Theoretical considerations suggest that there should also exist a sixth quark, labeled **top.** But the existence of this sixth quark has not yet been confirmed experimentally.

Each of the six quarks (up, down, strange, charmed, top, bottom) comes in three varieties of color (*red, green, blue*); furthermore, for each quark there exists an antiquark, which comes in three varieties of "anticolor" (anti*red*, anti*blue*, and anti*green*). Altogether, this amounts to 36 quarks — and the proliferation of quarks will apparently not end with this. Some new theories postulate the existence of several dozen quarks (plus their color varieties plus their antiquarks). This continuing proliferation raises the question of whether matter really has such a large number of elementary building blocks. Pushing forward our search for the ultimate building blocks, we have uncovered layers of structures within layers of structure — nuclei within atoms, protons and neutrons within nuclei, and quarks within protons and neutrons. Is there another layer within quarks?

Supplement: Further Reading

This chapter introduces only the highlights of particle physics. More detail is available in many good books and articles, such as the following:

Elementary Particles by C. N. Yang (Princeton University Press, Princeton, 1962), *The Discovery of Subatomic Particles* by S. Weinberg (Freeman, San Francisco, 1983), *From X-Rays to Quarks* by E. Segrè

---

[7] **Samuel Chao Chung Ting,** 1936–, and **Burton Richter,** 1931–, American physicists, were awarded the Nobel Prize in 1976 for their discovery of the J/Ψ.

(Freeman, San Francisco, 1980), and *The Particle Explosion* by F. Close, M. Marten, and C. Sutton (Oxford University Press, New York, 1987) give accounts of the history of the discovery of the subatomic particles; of these books, the last is the most complete, and it is lavishly illustrated with a splendid selection of color photographs. The exciting story of the recent discoveries of quarks and other exotic particles is told with journalistic verve in *The Key to the Universe* by N. Calder (Viking, New York, 1977).

The Cosmic Onion by F. Close (American Institute of Physics, New York, 1983) is a concise, up-to-date introduction to particle physics. Other excellent introductions with a minimum of mathematics are *The Particle Hunters* by Y. Ne'eman and Y. Kirsch (Cambridge University Press, Cambridge, 1986), *From Atoms to Quarks* by J. S. Trefil (Scribner's, New York, 1980), *The Great Design* by R. K. Adair (Oxford University Press, New York, 1987), *Longing for the Harmonies* by F. Wilczek and B. Devine (Norton, New York, 1988), and *The Particle Play* by J. C. Polkinghorne (Freeman, San Francisco, 1979). Further, more elementary, books are *Fearful Symmetry* by A. Zee (McMillan, New York, 1986), *The Stuff of Matter* by H. Fritzsch (Basic Books, New York, 1983), and *What Is the World Made Of?* by G. Feinberg (Doubleday, Garden City, 1977). *The Cosmic Code* by H. R. Pagels (Simon and Schuster, New York, 1982) explores the mysteries of particle physics and of quantum physics; it is beautifully written.

The unified theory of electromagnetic and weak interactions is described in some of the above books and also in *Story of the W and Z* by P. Watkins (Cambridge University Press, Cambridge, 1986) and in *The Moment of Creation* by J. S. Trefil (Scribner's, New York, 1983). The former book provides a detailed account of the experiments that led to the discovery of the W and Z particles. A more intimate and more revealing account of these experiments will be found in *Nobel Dreams* by G. Taubes (Random House, New York, 1986); this gives a rare, behind-the-scenes glimpse of physicists at work, and portrays both their successes and their failures.

Many articles dealing with particle physics have been published in *Scientific American* and in *Physics Today:*

"Field Theory," F. J. Dyson, *Scientific American,* April 1953
"Unified Theories of Elementary Particle Interactions," S. Weinberg, *Scientific American,* July 1974
"The Detection of Neutral Weak Currents," A. K. Mann and C. Rubbia, *Scientific American,* December 1974
"Light as a Fundamental Particle," S. Weinberg, *Physics Today,* June 1975
"Quarks with Color and Flavor," S. L. Glashow, *Scientific American,* October 1975
"The Search for New Families of Elementary Particles," D. B. Kline, A. K. Mann, and C. Rubbia, *Scientific American,* January 1976
"The Confinement of Quarks," Y. Nambu, *Scientific American,* November 1976
"Fundamental Particles with Charm," R. F. Schwitters, *Scientific American,* October 1977
Page 1144–1145"The Tevatron," R. R. Wilson, *Physics Today,* October 1977
"Heavy Leptons," M. L. Perl and W. T. Kirk, *Scientific American,* March 1978
"When Is a Particle?" S. D. Drell, *Physics Today,* June 1978
"The Upsilon Particle," L. M. Lederman, *Scientific American,* October 1978

"The Spin of the Proton," A. D. Krisch, *Scientific American,* May 1979

"The Bag Model of Quark Confinement," K. A. Johnson, *Scientific American,* July 1979

"The Next Generation of Particle Accelerators," R. R. Wilson, *Scientific American,* January 1980

"The Inner Structure of the Proton," M. Jacob and P. Landshoff, *Scientific American,* March 1980

"Gauge Theories of the Forces between Elementary Particles," G. 'tHooft, *Scientific American,* June 1980

"Antiproton–Proton Colliders, Intermediate Bosons," D. Cline and C. Rubbia, *Physics Today,* August 1980

"Unified Theory of Elementary-Particle Forces," H. Georgi and S. L. Glashow, *Physics Today,* September 1980

"A Unified Theory of Elementary Particles and Forces," H. Georgi, *Scientific American,* April 1981

"The Decay of the Proton," S. Weinberg, *Scientific American,* June 1981

"The Development of Field Theory in the Last 50 Years," W. F. Weisskopf, *Physics Today,* November 1981

"U.S. Particle Accelerators at Age 50," R. R. Wilson, *Physics Today,* November 1981

"The Search for the Intermediate Vector Bosons," D. B. Cline, C. Rubbia, and S. van der Meer, *Scientific American,* March 1982

"Quarkonium," E. D. Bloom and G. J. Feldmann, *Scientific American,* May 1982

"Glueballs," K. Ishikawa, *Scientific American,* November 1982

"The Lattice Theory of Quark Confinement," C. Rebbi, *Scientific American,* February 1983

"The Structure of Quarks and Leptons," H. Harari, *Scientific American,* April 1983

"Particles with Naked Beauty," N. B. Mistry, R. A. Poling, and E. H. Thorndike, *Scientific American,* July 1983

"The SSC: A Machine for the Nineties," S. L. Glashow and L. M. Lederman, *Physics Today,* March 1985

"A New Level of Structure," O. W. Greenberg, *Physics Today,* September 1985

"Popular and Unpopular Ideas in Particle Physics," M. L. Perl, *Physics Today,* December 1985

"The Superconducting Supercollider," J. D. Jackson, M. Tigner, and S. Wojicki, *Scientific American,* March 1986

The *Resource Letter NP-1* in the *American Journal of Physics,* February 1980, gives a long list of further references at a more advanced level.

SUMMARY

Particles: Leptons

Baryons $\Big\}$ hadrons
Mesons

Interactions:

	Strengths:	
Strong		1
Electromagnetic		10^{-2}
Weak		10^{-6}
Gravitational		10^{-38}

Conserved quantities:

Absolute: Energy
Momentum
Angular momentum
Electric charge
Baryon number
Lepton number

Approximate: Isospin
Strangeness
Parity

Quarks: u, d, s, c, b, t

QUESTIONS

1. Why are high-energy accelerators necessary for the production and discovery of new, massive particles?

2. Physicists are planning to construct the SSC, a 20-TeV accelerator, which will cost some 4 billion dollars. Can such an expenditure be justified?

3. Neutrons do not make tracks in a bubble chamber. Why not?

4. The names *baryon, meson,* and *lepton* come from the Greek *barys* (heavy), *mesos* (middle), and *leptos* (thin, slender). These names were originally intended to indicate the masses of the particles. According to the lists of particles and masses given in this chapter, is it true that the baryons have the largest masses and the leptons the smallest?

5. How does the antiproton differ from the proton? The antineutron from the neutron?

6. Why does a particle that lives only 10^{-23} s not make a track in a bubble chamber? (Hint: Suppose the particle moves at the maximum conceivable speed; how far will it travel in 10^{-23} s?)

7. The strengths of the fundamental forces depend on the energies of the particles. In the case of the gravitational force, the strength increases with the energy. Why would you expect this to be true?

8. The boomerang analogy described in Figure 46.14 is defective in that the boomerang requires the presence of air. What would be the motion of a boomerang in vacuum?

9. In Figure 46.20, a large number of particles emerge in the longitudinal direction (toward the right and the left). Why is this expected, whereas the emergence of particles in the transverse direction (upward and downward) is surprising? (Hint: Consider a head-on collision between two aircraft; which way do you expect most fragments to spurt out?)

10. According to some recent speculations, quarks are made of smaller constituents variously called prequarks, preons, or rishons (from the Hebrew word for first or primary). Can you think of a better name for the constituents of the quarks?

PROBLEMS

Section 46.1

1. The planned SSC accelerator will have a radius of 14 km and will produce protons of momentum 20 TeV/c, or 1.1×10^{-14} kg·m/s. What magnetic

field is required to hold the protons in a circular orbit of this radius? What is the period of the orbital motion? (Hint: The speed of the protons is nearly equal to the speed of light.)

2. What is the Q value for the reaction (1)? Can this reaction be initiated by a pion of very low kinetic energy? The Λ baryon in this reaction is $\Lambda(1116)$.

3. The relativistic formula $\sqrt{2mc^2(2mc^2 + K)}$ gives the energy available for inelastic reactions when a particle of mass m and kinetic energy K is incident on a stationary particle of the same mass. Use this formula to calculate the available energy (in GeV) for an antiproton incident on a stationary proton for the following cases:
(a) The kinetic energy of the incident antiproton is very low ($K << mc^2$).
(b) The kinetic energy of the incident antiproton is 1 TeV (as in the Tevatron).
(c) The kinetic energy of the incident antiproton is 20 TeV (as in the SCC).

Section 46.2

4. Which is the most massive particle listed in the tables of Section 46.2? Express the mass of this particle in atomic mass units and compare the mass with that of the helium atom and that of the lithium atom.

5. Count the number of particles (including antiparticles) in the tables of leptons, baryons, and mesons.

Section 46.3

6. Verify that each of the reactions (2), (3), and (6) conserves baryon number. Do these reactions conserve strangeness?

7. Which of the following reactions are forbidden by an absolute conservation law?

$$\pi^+ + p \rightarrow \Lambda + K^0$$

$$K^- + p \rightarrow K^- + p + \pi^0$$

$$\pi^- + n \rightarrow \pi^- + \pi^0 + \pi^0$$

$$K^- + n \rightarrow \Sigma^- + \pi^0$$

$$e + \nu \rightarrow \pi^- + \pi^0$$

8. Is strangeness conserved in the following reactions?

$$\pi^+ + n \rightarrow K^+ + \Lambda$$

$$\Lambda \rightarrow p + \pi^-$$

$$K^0 \rightarrow \pi^+ + \pi^-$$

9. Show that the emission of a photon by a free electron is impossible because it conflicts with energy conservation. (Hint: Consider the emission process in the reference frame in which the electron is initially at rest.)

10. Show that the annihilation of an electron and an antielectron into one single photon ($e + \bar{e} \rightarrow \gamma$) is impossible, because it conflicts with conservation of momentum. (Hint: Consider the reaction from the reference frame in which the two electrons have opposite velocities of equal magnitudes.)

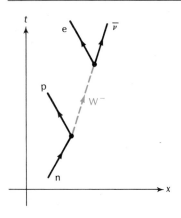

Fig. 46.21

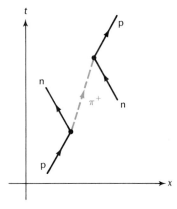

Fig. 46.22

Sections 46.4 and 46.5

11. The W. particle can have either a positive charge (W^+) or a negative charge (W^-). Figure 46.21 shows the Feynman diagram for the decay of the neutron (n) via exchange of a W^-; the end products are a proton (p), an electron (e), and an antineutrino ($\bar{\nu}$). Can you guess the Feynman diagram for the decay of the antineutron?

12. Figure 46.22 shows the Feynman diagram for the exchange of a π^+ between a proton and a neutron; note that the proton changes into a neutron, and vice versa. Draw corresponding diagrams for the exchange of a π^- and of a π^0.

*13. According to a speculative theory of the strong interactions, the proton should be unstable and decay with a lifetime of about 10^{33} years. Consider the protons in a mass of 10^6 kg of water. How many of these protons will decay in one year?

Sections 46.6 and 46.7

14. The negative pion (π^-) is made of two quarks. What kinds of quarks are these?

15. How many quarks are there in a hydrogen atom? In a water molecule? (The oxygen nucleus contains eight protons and eight neutrons.)

16. How many quarks are created in the net reaction (1)–(3)?

17. A particle is made of one d quark and one u antiquark. What is the electric charge of this particle? What is this particle?

18. Given that each gluon has a color and an anticolor, how many different kinds of gluons are there?

19. A glueball consists of two or more gluons bound together in a composite structure (without quarks). The net color of glueballs is neutral; they always contain equal net amounts of a color and its anticolor. Suppose that a glueball consists of two gluons, one of which is *green*-anti*blue*. What must be the colors of the other gluon? Suppose that a glueball consists of three gluons, one *green*-anti*blue*, one *red*-anti*green*. What must be the colors of the third gluon?

APPENDIX 1: INDEX TO TABLES

Note: In the two-volume edition, all pages after 570 are in Vol. 2. Interlude V is the last in Vol. 1.

Accelerators, some large 1123
Accidents, reactor V-13
Acute radiation doses, effects of IV-9
Angular momenta, some 313
Areas A-4
Artificial satellites of the Earth, the first 221
Atmospheric pollutants from fossil fuels V-II
Atoms A-29, P-14
Balmer series in the hydrogen spectrum 1053
Baryons 1129
Charges, electric, of some particles 577
Charges, electric, of electrons, protons, and neutrons 573
Chemical elements and periodic table P-14
Chromatic musical scale 441
Comets, some 227
Conductivities, some thermal 525
Conserved quantities, forces and 1133
Constants, fundamental, best values of A-27
Conversion efficiencies, energy V-2
Conversion factors A-22
Densities of some fluids 467
Derivatives, some A-11
Derived units, names of A-21
Diamagnetic materials, permeabilities of some 817
Dielectric constants of some materials 669
Earth endpaper
Electric charges of some particles 577
Electric charges of protons, electrons, and neutrons 573
Electric fields, some 592
Electric forces (qualitative) 573
Elements, periodic table of P-14
Energies, some 195
Energy and entropy V-2
Energy conversion efficiencies V-2
Energy in fission, distribution of XI-2
Energy storage VIII-10
Entropy, energy and V-2
Expansion, coefficients of 519
Fields and their quanta 1135
Fission, distribution of energy in XI-2
Fluids, densities of some 467

Forces and conserved quantities 1133
Forces, some 105
Forces, the four fundamental 1130
Fossil fuels, atmospheric pollutants from V-II
Fossil fuels, worldwide supply of V-4
Friction coefficients 136
Fuels, supply of nuclear V-5
Fuels, worldwide supply of fossil V-4
Fundamental constants, best values of A-27
Fundamental forces, the four 125
Fusion, heats of 527
g, variation with latitude 38
Gases, specific heats of some 531
Greek alphabet endpaper
Heats of fusion and vaporization 527
Hydrogen spectrum, Balmer series 1053
Indices of refraction, of some materials 904
Inertia, some moments of 309
Insulators, resistivities of 692
Integrals, some A-15
Isotopes, chart of P-17, 1097
Jupiter, main moons of 221
Kinetic energies, some 172
Leptons 1127
Lightning data VI-10
Magnetic fields, some 736
Magnetic moments of some atoms and ions 809
Mathematical symbols and formulas A-2
Mesons 1129
Metals, resistivities of 690
Moments, magnetic 809
Moments of inertia, some 309
Moon endpaper
Moons of Jupiter, main 221
Moons of Saturn, some 239
Musical scale, chromatic 441
Nuclear fuels, supply of V-5
Nucleons 1096
Paramagnetic materials, permeabilities of some 811
Penetrating radiations IV-3
Perimeters, areas, and volumes A-4
Period of simple pendulum, increase of 392

Periodic table of the elements A-29, P-14
Permeabilities of some paramagnetic materials 811
Planets 220
Plasmas, some VII-2
Pollutants, atmospheric, from fossil fuels V-11
Powers, some 200
Prefixes for units A-21
Pressures, some 473
Quanta, fields and their 1135
Quarks, three 1140
Radiation doses, effects of acute IV-9
Radiation doses from diagnostic X rays IV-10
Radiation doses per inhabitant of the United States, average annual IV-10
Radiations, penetrating IV-3
Radioactive sources, strengths of IV-5
Radioisotopes 1110
RBE values IV-8
Reactor accidents V-13
Refraction, indices of, of some materials 904
Resistivities of insulators 692
Resistivities of metals 690
Resistivities of semiconductors 692
Satellites of the Earth, the first artificial 221
Saturn, some moons of 239
Semiconductors, resistivities of 692
Simple pendulum, increase of period of 392
Sound intensities, some 440
Sound, speed of, in some materials 444
Specific heats of some gases 531
Specific heats, some 516
Speed of sound in some materials 444
Storage, energy VIII-10
Sun endpaper
Superconductors, some VIII-2
Temperatures, some 500
Thermal conductivities, some 525
Vaporization, heats of 527
Volumes A-4

APPENDIX 2: MATHEMATICAL SYMBOLS AND FORMULAS

A2.1 Symbols

$a = b$ means a equals b
$a \neq b$ means a is not equal to b
$a > b$ means a is greater than b
$a < b$ means a is less than b
$a \geq b$ means a is not less than b
$a \leq b$ means a is not greater than b
$a \propto b$ means a is proportional to b
$a \cong b$ means a is approximately equal to b
$a \sim b$ means a is of the order of magnitude of b, i.e., a is within a factor of 10 or so of b
$a \gg b$ means a is much larger than b
$a \ll b$ means a is much less than b
$\Sigma_i a_i$ stands for the sum $a_1 + a_2 + a_3 + a_4 + \cdots$
$n!$ (or "n factorial") stands for the product $1 \cdot 2 \cdot 3 \cdots n$
$\pi = 3.14159\ldots$
$e = 2.71828\ldots$

A2.2 The Quadratic Equation

The quadratic equation $ax^2 + bx + c = 0$ has two solutions:

$$x = \frac{-b \pm \sqrt{b^2 - 4ac}}{2a} \tag{1}$$

A2.3 Some Approximations

The following approximations are valid for small values of x, that is, $x \ll 1$:

$$(1 + x)^n \cong 1 + nx \tag{2}$$

$$(1 + x)^{\frac{1}{2}} \cong 1 + \frac{x}{2} \tag{3}$$

$$\frac{1}{(1 + x)} \cong 1 - x \tag{4}$$

$$\frac{1}{(1 + x)^{\frac{1}{2}}} \cong 1 - \frac{x}{2} \tag{5}$$

These approximations can be derived from the binomial expansion

$$(1+x)^n = 1 + nx + \frac{n(n-1)}{2!}x^2 + \frac{n(n-1)(n-2)}{3!}x^3 + \cdots \quad (6)$$

by neglecting all powers of x, except the first power.

A2.4 The Exponential and the Logarithmic Functions

The **exponential function** $\exp(x)$ is defined by the following infinite series:

$$\exp(x) = 1 + x + \frac{x^2}{2!} + \frac{x^3}{3!} + \frac{x^4}{4!} + \cdots \quad (7)$$

This function is equivalent to raising the constant $e = 2.71828 \ldots$ to the power x,

$$\exp(x) = e^x \quad (8)$$

The **natural logarithm** $\text{Ln } x$ is the inverse of the exponential function, so

$$x = e^{\ln x} \quad (9)$$

and

$$x = \ln(e^x) \quad (10)$$

Natural logarithms obey the usual rules for logarithms,

$$\ln(x \cdot y) = \ln x + \ln y \quad (11)$$

$$\ln\left(\frac{x}{y}\right) = \ln x - \ln y \quad (12)$$

$$\ln(x^a) = a \ln x \quad (13)$$

Note that

$$\ln e = 1 \quad (14)$$

and

$$\ln 10 = 2.3026 \ldots \quad (15)$$

If we designate the base-10, or **common logarithm**, by $\log x$, then the relationship between the two kinds of logarithm is as follows:

$$\ln x = \ln(10^{\log x}) = (\log x)(\ln 10) = 2.3026 \log x \quad (16)$$

APPENDIX 3: PERIMETERS, AREAS, AND VOLUMES

[perimeter of a circle of radius r] $= 2\pi r$
[area of a circle of radius r] $= \pi r^2$
[area of a triangle of base b, altitude h] $= hb/2$
[surface of a sphere of radius r] $= 4\pi r^2$
[volume of a sphere of radius r] $= 4\pi r^3/3$
[curved surface of a cylinder of radius r, height h] $= 2\pi rh$
[volume of a cylinder of radius r, height h] $= \pi r^2 h$

APPENDIX 4: BRIEF REVIEW OF TRIGONOMETRY

A4.1 Angles

The angle between two intersecting straight lines is defined as the fraction of a complete circle included between these lines (Figure A4.1). To express the angle in **degrees,** we assign an angular magnitude of 360° to the complete circle; any arbitrary angle is then an appropriate fraction of 360°. To express the angle in **radians,** we assign an angular magnitude of 2π radian to the complete circle; any arbitrary angle is then an appropriate fraction of 2π. For example, the angle shown in Figure A4.1 is $\frac{1}{8}$ of a complete circle, that is, 45°, or $\pi/4$ radian. In view of the definition of angle, the length of arc included between the two intersecting straight lines is proportional to the angle θ between these lines; if the angle is expressed in radians, then the constant of proportionality is simply the radius:

$$s = r\theta \qquad (1)$$

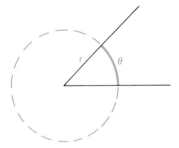

Fig. A4.1 The angle θ in this diagram is $\theta = 45°$.

Since 2π radian $= 360°$, it follows that

$$1 \text{ radian} = \frac{360°}{2\pi} = \frac{360°}{2 \times 3.14159} = 57.2958° \qquad (2)$$

Each degree is divided into 60 minutes of arc (arcminutes), and each of these into 60 seconds of arc (arcseconds). In degrees, minutes of arc, and seconds of arc, the radian is

$$1 \text{ radian} = 57° \; 17' \; 44.8'' \qquad (3)$$

A4.2 The Trigonometric Functions

The trigonometric functions of an angle are defined as ratios of the lengths of the sides of a right triangle erected on this angle. Figure A4.2 shows an acute angle θ and a right triangle, one of whose angles coincides with θ. The adjacent side OQ has a length x, the opposite side

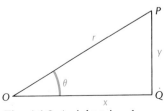

Fig. A4.2 A right triangle.

QP a length y, and the hypotenuse OP a length r. The **sine, cosine, tangent, cotangent, secant,** and **cosecant** of the angle θ are then defined as follows:

$$\sin \theta = y/r \qquad (4)$$

$$\cos \theta = x/r \qquad (5)$$

$$\tan \theta = y/x \qquad (6)$$

$$\cot \theta = x/y \qquad (7)$$

$$\sec \theta = r/x \qquad (8)$$

$$\csc \theta = r/y \qquad (9)$$

EXAMPLE 1. Find the sine, cosine, and tangent for angles of 0°, 90°, and 45°.

SOLUTION: For an angle of 0°, the opposite side is zero ($y = 0$), and the adjacent side coincides with the hypotenuse ($x = r$). Hence

$$\sin 0° = 0 \qquad \cos 0° = 1 \qquad \tan 0° = 0 \qquad (10)$$

For an angle of 90°, the adjacent side is zero ($x = 0$), and the opposite side coincides with the hypotenuse ($y = r$). Hence

$$\sin 90° = 1 \qquad \cos 90° = 0 \qquad \tan 90° = \infty \qquad (11)$$

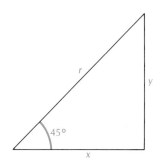

Fig. A4.3 A right triangle with an angle of 45°.

Finally, for an angle of 45° (Figure A4.3), the adjacent and the opposite sides have the same length ($x = y$) and the hypotenuse has a length of $\sqrt{2}$ times the length of either side ($r = \sqrt{2}x = \sqrt{2}y$). Hence

$$\sin 45° = \frac{1}{\sqrt{2}} \qquad \cos 45° = \frac{1}{\sqrt{2}} \qquad \tan 45° = 1 \qquad (12)$$

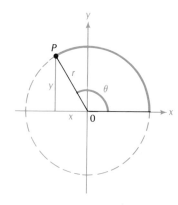

Fig. A4.4 The angle θ in this diagram is larger than 90°.

The definitions (4)–(9) are also valid for angles greater than 90°, such as the angle shown in Figure A4.4. In the general case, the quantities x and y must be interpreted as the rectangular coordinates of the point P. For any angle larger than 90°, one or both of the coordinates x and y are negative. Hence some of the trigonometric functions will also be negative. For instance,

$$\sin 135° = \frac{1}{\sqrt{2}} \qquad \cos 135° = -\frac{1}{\sqrt{2}} \qquad \tan 135° = -1 \qquad (13)$$

Figure A4.5 shows plots of the sine, cosine, and tangent vs. θ for the range 0°–360°.

If the angle θ is small, say, $\theta < 0.2$ radian, or $\theta < 10°$, then the length of the opposite side is approximately equal to the length of the circular arc (Figure A4.6). If we express the angle in radians, this length is [see Eq. (1)]

$$y \cong s = r\theta \qquad (14)$$

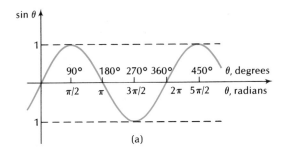

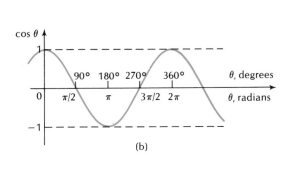

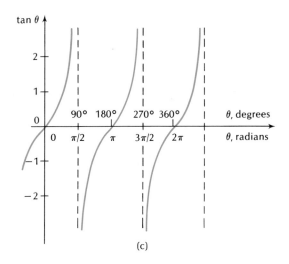

Fig. A4.5

and therefore

$$\sin \theta \cong \theta \tag{15}$$

To obtain a corresponding approximation for $\cos \theta$, we use the Pythagorean theorem to express the adjacent side in terms of the hypotenuse r and the opposite side $y \cong r\theta$,

$$x = \sqrt{r^2 - y^2} \cong \sqrt{r^2 - r^2\theta^2} \tag{16}$$

$$\cong r\sqrt{1 - \theta^2} \tag{17}$$

For small θ, Eq. (A2.3) gives us the approximation

$$\sqrt{1 - \theta^2} \cong 1 - \theta^2/2 \tag{18}$$

so

$$\cos \theta = x/r \cong 1 - \theta^2/2 \tag{19}$$

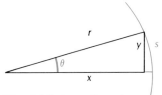

Fig. A4.6 A small angle.

When using the approximate formulas (15) and (19), we must always remember that the angle θ is expressed in radians!

The inverse trigonometric functions **sin⁻¹**, **cos⁻¹**, and **tan⁻¹** give the angle that corresponds to a specified value of the sine, cosine, or tangent, respectively. Thus, if

$$\sin \theta = u$$

then

$$\theta = \sin^{-1} u$$

A4.3 Trigonometric Identities

From the definitions (4)–(9) we immediately find the following identities:

$$\tan \theta = \sin \theta / \cos \theta \tag{20}$$

$$\cot \theta = 1/\tan \theta \tag{21}$$

$$\sec \theta = 1/\cos \theta \tag{22}$$

$$\csc \theta = 1/\sin \theta \tag{23}$$

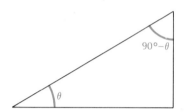

Fig. A4.7 A right triangle with angles θ and $90° - \theta$.

Figure A4.7 shows a right triangle with angles θ and $90° - \theta$. Since the adjacent side for the angle θ is the opposite side for the angle $90° - \theta$ and vice versa, we see that the trigonometric functions also obey the following identities:

$$\sin(90° - \theta) = \cos \theta \tag{24}$$

$$\cos(90° - \theta) = \sin \theta \tag{25}$$

$$\tan(90° - \theta) = \cot \theta = 1/\tan \theta \tag{26}$$

According to the Pythagorean theorem, $x^2 + y^2 = r^2$. With $x = r \cos \theta$ and $y = r \sin \theta$ this becomes $r^2 \cos^2 \theta + r^2 \sin^2 \theta = r^2$, or

$$\cos^2 \theta + \sin^2 \theta = 1 \tag{27}$$

The following are a few other trigonometric identities, which we state without proof:

$$\sec^2 \theta = 1 + \tan^2 \theta$$

$$\csc^2 \theta = 1 + \cot^2 \theta$$

$$\sin 2\theta = 2 \sin \theta \cos \theta$$

$$\cos 2\theta = 2 \cos^2 \theta - 1$$

$$\sin \tfrac{1}{2}\theta = \sqrt{(1 - \cos \theta)/2}$$

$$\cos \tfrac{1}{2}\theta = \sqrt{(1 + \cos \theta)/2}$$

$$\sin(\alpha + \beta) = \sin \alpha \cos \beta + \cos \alpha \sin \beta$$

$$\cos(\alpha + \beta) = \cos \alpha \cos \beta - \sin \alpha \sin \beta$$

$$\tan(\alpha + \beta) = \frac{\tan \alpha + \tan \beta}{1 - \tan \alpha \tan \beta}$$

$$\sin \alpha + \sin \beta = 2 \sin \tfrac{1}{2}(\alpha + \beta) \cos \tfrac{1}{2}(\alpha - \beta)$$

$$\cos \alpha + \cos \beta = 2 \cos \tfrac{1}{2}(\alpha + \beta) \cos \tfrac{1}{2}(\alpha - \beta)$$

A4.4 The Laws of Cosines and of Sines

In an arbitrary triangle the lengths of the sides and the angles obey the laws of cosines and of sines. The **law of cosines** states that if the lengths of two sides are A and B and the angle between them is γ (Figure A4.8), then the length of the third side is given by

$$C^2 = A^2 + B^2 - 2AB \cos \gamma \tag{28}$$

The **law of sines** states that the sines of the angles of the triangle are in the same ratio as the lengths of the opposite sides (Figure A4.8):

$$\frac{\sin \alpha}{A} = \frac{\sin \beta}{B} = \frac{\sin \gamma}{C} \tag{29}$$

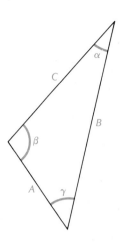

Fig. A4.8 An arbitrary triangle.

Both of these laws are very useful in the calculation of unknown lengths or angles of a triangle.

APPENDIX 5: BRIEF REVIEW OF CALCULUS

A5.1 Derivatives

We saw in Section 2.3 that if the position of a particle is some function of time, say $x = x(t)$, then the instantaneous velocity of the particle is the derivative of x with respect to t,

$$v = \frac{dx}{dt} \tag{1}$$

This derivative is defined by first looking at a small increment Δx that results from a small increment Δt, and then evaluating the ratio $\Delta x/\Delta t$, in the limit when both Δx and Δt tend toward zero. Thus

$$\frac{dx}{dt} = \lim_{\Delta t \to 0} \frac{\Delta x}{\Delta t} \tag{2}$$

Graphically, in a plot of the worldline, the derivative dx/dt is the slope of the straight line tangent to the (curved) worldline at the time t (see Figure 2.6).

In general, if $f = f(u)$ is some given function of a variable u, the **derivative** of f with respect to u is defined by

$$\frac{df}{du} = \lim_{\Delta u \to 0} \frac{\Delta f}{\Delta u} \tag{3}$$

In a plot of f vs. u, this derivative is the slope of the straight line tangent to the curve representing $f(u)$ (see Figure A5.1).

Starting with the definition (3) we can find the derivative of any function (provided the function is sufficiently smooth so the derivative exists!). For example, consider the function $f(u) = u^2$. If we increase u to $u + \Delta u$, the function $f(u)$ increases to

$$f + \Delta f = (u + \Delta u)^2 \tag{4}$$

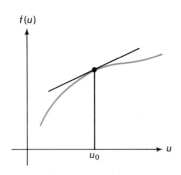

Fig. A5.1 The derivative of $f(u)$ at u_0 is the slope of the straight line tangent to the curve at u_0.

and therefore

$$\Delta f = (u + \Delta u)^2 - f = (u + \Delta u)^2 - u^2$$

$$= 2u\,\Delta u + (\Delta u)^2 \tag{5}$$

The derivative df/du is then

$$\frac{df}{du} = \lim_{\Delta u \to 0} \frac{\Delta f}{\Delta u} = \lim_{\Delta u \to 0} \frac{2u\,\Delta u + (\Delta u)^2}{\Delta u} \qquad (6)$$

$$= \lim_{\Delta u \to 0} (2u) + \lim_{\Delta u \to 0} (\Delta u) \qquad (7)$$

The second term on the right side vanishes in the limit $\Delta u \to 0$; the first term is simply $2u$. Hence

$$\frac{df}{du} = 2u \qquad (8)$$

or

$$\frac{d}{du}(u^2) = 2u \qquad (9)$$

This is one instance of the general rule for the differentiation of u^n:

$$\frac{d}{du}(u^n) = nu^{n-1} \qquad (10)$$

This general rule is valid for any positive or negative number n, including zero. The proof of this rule can be constructed by an argument similar to that above. Table A5.1 lists the derivatives of the most common functions.

Table A5.1 SOME DERIVATIVES

$\dfrac{d}{du} u^n = nu^{n-1}$

$\dfrac{d}{du} \ln u = \dfrac{1}{u}$

$\dfrac{d}{du} e^u = e^u$

$\dfrac{d}{du} \sin u = \cos u$ (where u is in *radians*)

$\dfrac{d}{du} \cos u = -\sin u$ (where u is in *radians*)

$\dfrac{d}{du} \tan u = \sec^2 u$ (where u is in *radians*)

$\dfrac{d}{du} \cot u = -\csc^2 u$ (where u is in *radians*)

$\dfrac{d}{du} \sec u = \tan u \sec u$ (where u is in *radians*)

$\dfrac{d}{du} \csc u = -\cot u \csc u$ (where u is in *radians*)

$\dfrac{d}{du} \sin^{-1} u = 1/\sqrt{1-u^2}$ (where u is in *radians*)

$\dfrac{d}{du} \cos^{-1} u = -1/\sqrt{1-u^2}$ (where u is in *radians*)

$\dfrac{d}{du} \tan^{-1} u = \dfrac{1}{1+u^2}$ (where u is in *radians*)

A5.2 Other Important Rules for Differentiation

i. Derivative of a constant times a function:

$$\frac{d}{du}(cf) = c\frac{df}{du} \qquad (11)$$

For instance,

$$\frac{d}{du}(6u^2) = 6\frac{d}{du}(u^2) = 6 \times 2u = 12u$$

ii. Derivative of the sum of two functions:

$$\frac{d}{du}(f+g) = \frac{df}{du} + \frac{dg}{du} \qquad (12)$$

For instance,

$$\frac{d}{du}(6u^2 + u) = \frac{d}{du}(6u^2) + \frac{d}{du}(u) = 12u + 1$$

iii. Derivative of the product of two functions:

$$\frac{d}{du}(f \times g) = g\frac{df}{du} + f\frac{dg}{du} \qquad (13)$$

For instance,

$$\frac{d}{du}(u^2 \sin u) = \sin u \frac{d}{du}u^2 + u^2 \frac{d}{du}\sin u$$

$$= \sin u \times 2u + u^2 \times \cos u$$

iv. Chain rule for derivatives: If f is a function of g and g is a function of u, then

$$\frac{d}{du}f(g) = \frac{df}{dg}\frac{dg}{du} \qquad (14)$$

For instance, if $g = 2u$ and $f(g) = \sin g$, then

$$\frac{d}{du}\sin(2u) = \frac{d\sin(2u)}{d(2u)}\frac{d(2u)}{du}$$

$$= \cos(2u) \times 2$$

A5.3 Integrals

We have learned that if the position of a particle is known as a function of time, then we can find the instantaneous velocity by differentiation. What about the converse problem: if the instantaneous velocity is known as a function of time, how can we find the position? In Section 2.5 we learned how to deal with this problem in the special case of motion with constant acceleration. The velocity is then a fairly simple function of time [see Eq. (2.17)]

$$v = v_0 + at \tag{15}$$

and the position deduced from this velocity is [see Eq. (2.22)]

$$x = x_0 + v_0 t + \tfrac{1}{2} a t^2 \tag{16}$$

where x_0 and v_0 are the initial position and velocity at the initial time $t_0 = 0$. Now we want to deal with the general case of a velocity that is an arbitrary function of time,

$$v = v(t) \tag{17}$$

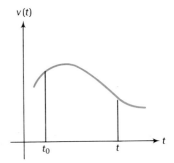

Fig. A5.2 Plot of a function $v(t)$.

Figure A5.2 shows what a plot of v vs. t might look like. At the initial time t_0, the particle has an initial position x_0 (for the sake of generality we now assume that $t_0 \neq 0$). We want to find the position at some later time t. For this purpose, let us divide the time interval $t - t_0$ into a large number of small time intervals, each of the duration Δt. The total number of intervals is N, so $t - t_0 = N \, \Delta t$. The first of these intervals lasts from t_0 to $t_0 + \Delta t$; the second from $t_0 + \Delta t$ to $t_0 + 2 \, \Delta t$; etc.

In Figure A5.3 the beginnings and the ends of these intervals have been marked t_0, t_1, t_2, etc., with $t_1 = t_0 + \Delta t$, $t_2 = t_0 + 2\Delta t$, etc. If Δt is sufficiently small, then during the first time interval the velocity is approximately $v(t_0)$; during the second, $v(t_1)$; etc. This amounts to replacing the smooth function $v(t)$ by a series of steps (see Figure A5.3). Thus, during the first time interval, the displacement of the particle is approximately $v(t_0) \, \Delta t$; during the second interval, $v(t_1) \, \Delta t$; etc. The net

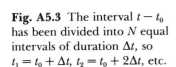

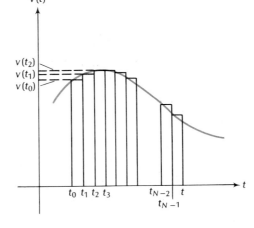

Fig. A5.3 The interval $t - t_0$ has been divided into N equal intervals of duration Δt, so $t_1 = t_0 + \Delta t$, $t_2 = t_0 + 2\Delta t$, etc.

displacement of the particle during the entire interval $t - t_0$ is the sum of all these small displacements,

$$x(t) - x_0 \cong v(t_0) \, \Delta t + v(t_1) \, \Delta t + v(t_2) \, \Delta t + \cdots \tag{18}$$

Using the standard mathematical notation for summation, we can write this as

$$x(t) - x_0 \cong \sum_{i=0}^{N-1} v(t_i) \, \Delta t \tag{19}$$

We can give this sum the following graphical interpretation: since $v(t_i) \Delta t$ is the area of the rectangle of height $v(x_i)$ and width Δt, the sum is the net area of all the rectangles shown in Figure A5.3, i.e., it is approximately the area under the velocity curve. Note that if the velocity is negative, the area must be reckoned as negative!

Of course, Eq. (19) is only an approximation. To find the exact displacement of the particle we must let the step size Δt tend to zero (while the number of steps N tends to infinity). In this limit, the step-like horizontal and vertical line segments in Figure A5.3 approach the smooth curve. Thus,

$$x(t) - x_0 = \lim_{\substack{\Delta t \to 0 \\ N \to \infty}} \sum_{i=0}^{N-1} v(t_i) \, \Delta t \tag{20}$$

In the notation of calculus, the right side of Eq. (20) is usually written in the following fashion:

$$x(t) - x_0 = \int_{t_0}^{t} v(t') \, dt' \tag{21}$$

The right side is called the **integral** of the function $v(t)$. The subscript and the superscript on the integration symbol $\int$ are called, respectively, the lower and the upper limit of integration; and t' is called the variable of integration (the prime on the variable of integration t' merely serves to distinguish that variable from the limit of integration t). Graphically, the integral is the exact area under the velocity curve between the limits t_0 and t in a plot of v vs. t (see Figure A5.4). Areas below the t axis must be reckoned as negative.

In general, if $f(u)$ is a function of u, then the integral of this function is defined by a limiting procedure similar to that described above for special case of the function $v(t)$. The integral over an interval from $u = a$ to $u = b$ is

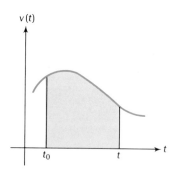

Fig. A5.4 The area under the velocity curve.

$$\int_{a}^{b} f(u) \, du = \lim_{\substack{\Delta u \to 0 \\ N \to \infty}} \sum_{i=0}^{N-1} f(u_i) \, \Delta u \tag{22}$$

where $u_i = a + N \Delta u$. As in the case of the integral of $v(t)$, this integral can again be interpreted as an area: it is the area under the curve between the limits a and b in a plot of f vs. u.

For the explicit evaluation of integrals we can take advantage of the connection between integrals and antiderivatives. An **antiderivative** of a function $f(u)$ is simply a function $F(u)$ such that $dF/du = f$. For example, if $f(u) = u^n$ and $n \neq -1$, then an antiderivative of $f(u)$ is $F(u) = u^{n+1}/(n+1)$. The fundamental theorem of calculus states that the integral of any function $f(u)$ can be expressed in terms of antiderivatives:

$$\int_{a}^{b} f(u) \, du = F(b) - F(a) \tag{23}$$

In essence, this means that integration is the inverse of differentiation. We will not prove this theorem here, but we remark that such an inverse relationship between integration and differentiation should not come as a surprise. We have already run across an obvious instance of such a relationship: we know that velocity is the derivative of the position, and we have seen above that the position is the integral of the velocity.

We will sometimes write Eq. (23) as

$$\int_a^b f(u)\, du = [F(u)]_a^b \qquad (24)$$

where the notation $[\ ]_a^b$ means that the function enclosed in the brackets is to be evaluated at a and at b, and these values are to be subtracted. For example, if $n \neq -1$,

$$\int_a^b u^n\, du = \left[\frac{u^{n+1}}{n+1}\right]_a^b = \frac{b^{n+1}}{n+1} - \frac{a^{n+1}}{n+1} \qquad (25)$$

Table A5.2 lists some frequently used integrals. In this table, the limits of integration belonging with Eq. (24) have been omitted for the sake of brevity.

Table A5.2 SOME INTEGRALS

$\int u^n\, du = \dfrac{u^{n+1}}{n+1}$ for $n \neq -1$

$\int \dfrac{1}{u}\, du = \ln u$ for $u > 0$

$\int e^{ku}\, du = \dfrac{e^{ku}}{k}$

$\int \ln u\, du = u \ln u - u$

$\int \sin(ku)\, du = -\dfrac{1}{k}\cos(ku)$ (where ku is in *radians*)

$\int \cos(ku)\, du = \dfrac{1}{k}\sin(ku)$ (where ku is in *radians*)

$\int \dfrac{du}{1+ku} = \dfrac{1}{k}\ln(1+ku)$

$\int \dfrac{du}{\sqrt{k^2 - u^2}} = \sin^{-1}\left(\dfrac{u}{k}\right)$

$\int \dfrac{du}{\sqrt{u^2 \pm k^2}} = \ln(u + \sqrt{u^2 \pm k^2})$

$\int \sqrt{k^2 - u^2}\, du = \dfrac{1}{2}\left[u\sqrt{k^2-u^2} + k^2 \sin^{-1}\left(\dfrac{u}{k}\right)\right]$

$\int \dfrac{du}{k^2 + u^2} = \dfrac{1}{k}\tan^{-1}\left(\dfrac{u}{k}\right)$

$\int \dfrac{du}{u\sqrt{k^2 \pm u^2}} = -\dfrac{1}{k}\ln\left(\dfrac{k + \sqrt{k^2 \pm u^2}}{u}\right)$

A5.4 The Most Important Rules for Integration

i. Integral of a constant times a function:

$$\int_a^b cf(u)\, du = c \int_a^b f(u)\, du$$

For instance,

$$\int_a^b 5u^2\, du = 5 \int_a^b u^2\, du = 5\left(\frac{b^3}{3} - \frac{a^3}{3}\right)$$

ii. Integral of a sum of two functions:

$$\int_a^b [f(u) + g(u)]\, du = \int_a^b f(u)\, du + \int_a^b g(u)\, du$$

For instance,

$$\int_a^b (5u^2 + u)\, du = \int_a^b 5u^2\, du + \int_a^b u\, du = 5\left(\frac{b^3}{3} - \frac{a^3}{3}\right) + \left(\frac{b^2}{2} - \frac{a^2}{2}\right)$$

iii. Change of limits of integration:

$$\int_a^b f(u)\, du = \int_a^c f(u)\, du + \int_c^b f(u)\, du$$

$$\int_a^b f(u)\, du = -\int_b^a f(u)\, du$$

iv. Change of variable of integration:
If u is a function of v, then

$$\int_a^b f(u)\, du = \int_{v(a)}^{v(b)} f(u) \frac{du}{dv}\, dv$$

For instance, with $u = v^2$,

$$\int_a^b u^3\, du = \int_a^b v^6\, du = \int_{\sqrt{a}}^{\sqrt{b}} v^6 (2v)\, dv$$

Finally, let us apply these general results to some specific examples of integration of the velocity.

EXAMPLE 1. A particle with constant acceleration has the following velocity as a function of time [compare Eq. (15)]:

$$v(t) = v_0 + at$$

where v_0 is the velocity at $t = 0$.
By integration, find the position as a function of time.

SOLUTION: According to Eq. (21), with $t_0 = 0$,

$$x(t) - x_0 = \int_0^t v(t')\, dt' = \int_0^t (v_0 + at')\, dt'$$

Using rule ii and rule i, we find that this equals

$$x(t) - x_0 = \int_0^t v_0\, dt' + \int_0^t at'\, dt' = v_0 \int_0^t dt' + a \int_0^t t'\, dt'$$

The first entry listed in Table A5.2 gives $\int dt' = t'$ (for $n = 0$) and $\int t'\, dt' = t'^2/2$ (for $n = 1$). Thus,

$$x(t) - x_0 = v_0 [t']_0^t + a[t'^2/2]_0^t$$

$$= v_0 t + at^2/2 \qquad (27)$$

This, of course, agrees with Eq. (16).

EXAMPLE 2. The instantaneous velocity of a projectile traveling through air is the following function of time:

$$v(t) = 655.9 - 61.14t + 3.26t^2$$

where $v(t)$ is measured in meters per second and t is measured in seconds. Assuming that $x = 0$ at $t = 0$, what is the position as a function of time? What is the position at $t = 3.0$ s?

SOLUTION: With $x_0 = 0$ and $t_0 = 0$, Eq. (21) becomes

$$x(t) = \int_0^t (655.9 - 61.14t' + 3.26t'^2)\, dt'$$

$$= 655.9 \int_0^t dt' - 61.14 \int_0^t t'\, dt' + 3.26 \int_0^t t'^2\, dt'$$

$$= 655.9\, [t']_0^t - 61.14[t'^2/2]_0^t + 3.26\, [t'^3/3]_0^t$$

$$= 655.9t - 61.14t^2/2 + 3.26t^3/3$$

When evaluated at $t = 3.0$ s, this yields

$$x(1.0) = 655.9 \times 3.0 - 61.14 \times (3.0)^2/2 + 3.26 \times (3.0)^3/3$$

$$= 1722 \text{ m}$$

EXAMPLE 3. The acceleration of a mass pushed back and forth by an elastic spring is

$$a(t) = B \cos \omega t \tag{28}$$

where B and ω are constants. Find the position as a function of time. Assume $v = 0$ and $x = 0$ at $t = 0$.

SOLUTION: The calculation involves two steps: first we must integrate the acceleration to find the velocity, then we must integrate the velocity to find the position. For the first step we use an equation analogous to Eq. (21),

$$v(t) - v_0 = \int_{t_0}^{t} a(t') \, dt' \tag{29}$$

This equation becomes obvious if we remember that the relationship between acceleration and velocity is analogous to that between velocity and position. With $v_0 = 0$ and $t_0 = 0$, we obtain from Eq. (29)

$$v(t) = \int_0^t B \cos \omega t' \, dt' = B \left[\frac{1}{\omega} \sin \omega t' \right]_0^t$$

$$= \frac{B}{\omega} \sin \omega t \tag{30}$$

Next,

$$x(t) = \int_0^t v(t') \, dt' = \int_0^t \frac{B}{\omega} \sin \omega t' \, dt' = \frac{B}{\omega} \left[-\frac{1}{\omega} \cos \omega t' \right]_0^t$$

$$= -\frac{B}{\omega^2} \cos \omega t + \frac{B}{\omega^2} \tag{31}$$

A5.5 The Taylor Series

Suppose that $f(u)$ is a smooth function of u in some neighborhood of a given point $u = a$, so the function has continuous derivatives of all orders. Then the value of the function at an arbitrary point near a can be expressed in terms of the following infinite series, where all the derivatives are evaluated at the point a:

$$f(u) = f(a) + \frac{df}{du}(x-a) + \frac{1}{2!}\frac{d^2f}{du^2}(x-a)^2 + \frac{1}{3!}\frac{d^3f}{du^3}(x-a)^3 + \cdots \tag{32}$$

This is called the **Taylor series** for the function $f(u)$ about the point a. The series converges, and is valid, provided x is sufficiently close to a. How close is "sufficiently close" depends on the function f and on the point a. Some functions, such as $\sin u$, $\cos u$, and e^u, are extremely well behaved, and their Taylor series converge for any choice of x and of a. The Taylor series gives us a convenient method for the approximate evaluation of a function.

EXAMPLE 4. Find the Taylor series for sin u about the point $u = 0$.

SOLUTION: The derivatives of sin u at $u = 0$ are

$$\frac{d}{du} \sin u = \cos u = 1$$

$$\frac{d^2}{du^2} \sin u = \frac{d}{du} \cos u = -\sin u = 0$$

$$\frac{d^3}{du^3} \sin u = \frac{d}{du} (-\sin u) = -\cos u = -1$$

$$\frac{d^4}{du^4} \sin u = \frac{d}{du} (-\cos u) = \sin u = 0, \quad \text{etc.}$$

Hence Eq. (32) gives

$$\sin u = 0 + 1 \times (u - 0) + \frac{1}{2!} \times 0 \times (u - 0)^2 + \frac{1}{3!} \times (-1) \times (u - 0)^3$$

$$+ \frac{1}{4!} \times 0 \times (u - 0)^4 + \cdots$$

$$= u - \tfrac{1}{6} u^3 + \cdots$$

Note that for very small values of u, we can neglect all higher powers of u, so $\sin u \cong u$, which agrees with the approximation given in Appendix 4.

APPENDIX 6: THE INTERNATIONAL SYSTEM OF UNITS (SI)

A6.1 Base Units

The SI system of units is the modern version of the metric system. The SI system recognizes seven fundamental, or base, units for length, mass, time, electric current, thermodynamic temperature, amount of substance, and luminous intensity.[a] The following definitions of the base units were adopted by the Conférence Générale des Poids et Mesures in the years indicated:

Meter (m) "The metre is the length of the path travelled by light in vacuum during a time interval of 1/299 792 458 of a second." (Adopted in 1983.)

Kilogram (kg) "The kilogram is . . . the mass of the international prototype of the kilogram." (Adopted in 1889 and in 1901.)

Second (s) "The second is the duration of 9 192 631 770 periods of the radiation corresponding to the transition between the two hyperfine levels of the ground state of the cesium-133 atom." (Adopted in 1967.)

Ampere (A) "The ampere is that constant current which, if maintained in two straight parallel conductors of infinite length, of negligible circular cross section, and placed one meter apart in vacuum, would produce between these conductors a force equal to 2×10^{-7} newton per meter of length." (Adopted in 1948.)

Kelvin (K) "The kelvin . . . is the fraction 1/273.16 of the thermodynamic temperature of the triple point of water." (Adopted in 1967.)

Mole "The mole is the amount of substance of a system which contains as many elementary entities as there are atoms in 0.012 kilogram of carbon-12." (Adopted in 1967.)

Candela (cd) "The candela is the luminous intensity, in a given direction, of a source that emits monochromatic radiation of frequency 540×10^{12} Hz and that has a radiant intensity in that direction of $\frac{1}{683}$ watt per steradian." (Adopted in 1979.)

Besides these seven base units, the SI system also recognizes two supplementary units of angle and solid angle:

Radian (rad) "The radian is the plane angle between two radii of a circle which cut off on the circumference an arc equal in length to the radius."

Steradian (sr) "The steradian is the solid angle which, having its vertex in the center of a sphere, cuts off an area equal to that of a [flat] square with sides of length equal to the radius of the sphere."

---

[a] At least two of the seven base units of the SI system are redundant. The mole is merely a certain number of atoms or molecules, in the same sense that a dozen is a number; there is no need to designate this number as a unit. The candela is equivalent to $\frac{1}{683}$ watt per steradian; it serves no purpose that is not served equally well by watt per steradian. Two other base units could be made redundant by adopting new definitions of the unit of temperature and of the unit of electric charge. Temperature could be measured in energy units because, according to the equipartition theorem, temperature is proportional to the energy per degree of freedom. Hence the kelvin could be defined as a derived unit, with 1 K $= \frac{1}{2} \times 1.38 \times 10^{-23}$ joule per degree of freedom. Electric charge could also be defined as a derived unit, to be measured with a suitable combination of the units of force and distance, as is done in the cgs system.

Furthermore, the definitions of the supplementary units — radian and steradian — are gratuitous. These definitions properly belong in the province of mathematics and there is no need to include them in a system of physical units.

A6.2 Derived Units

The derived units are formed out of products and ratios of the base units. Table A6.1 lists those derived units that have been glorified with special names. (Other derived units are listed in the tables of conversion factors in Appendix A.7.)

Table A6.1 NAMES OF DERIVED UNITS

Quantity	Derived Unit	Name	Symbol
frequency	1/s	hertz	Hz
force	kg · m/s²	newton	N
pressure	N/m²	pascal	Pa
energy	N · m	joule	J
power	J/s	watt	W
electric charge	A · s	coulomb	C
electric potential	J/C	volt	V
electric capacitance	C/V	farad	F
electric resistance	V/A	ohm	Ω
conductance	A/V	siemens	S
magnetic flux	V · s	weber	Wb
magnetic field	V · s/m²	tesla	T
inductance	V · s/A	henry	H
temperature	K	degree Celsius	°C
luminous flux	cd · sr	lumen	lm
illuminance	cd · sr/m²	lux	lx
radioactivity	1/s	becquerel	Bq
absorbed dose	J/kg	gray	Gy
dose equivalent	J/kg	sievert	Sv

A6.3 Prefixes

Multiples and submultiples of SI units are indicated by prefixes, such as the familiar *kilo, centi,* and *milli* used in *kilometer, centimeter,* and *millimeter,* etc. Table A6.2 lists all the accepted prefixes. Some enjoy more popularity than others; it is best to avoid the use of uncommon prefixes, such as *atto* and *exa,* since hardly anybody will recognize those.

Table A6.2 PREFIXES FOR UNITS

Multiplication factor	Prefix	Symbol
10^{18}	exa	E
10^{15}	peta	P
10^{12}	tera	T
10^{9}	giga	G
10^{6}	mega	M
10^{3}	kilo	k
10^{2}	hecto	h
10	deka	da
10^{-1}	deci	d
10^{-2}	centi	c
10^{-3}	milli	m
10^{-6}	micro	μ
10^{-9}	nano	n
10^{-12}	pico	p
10^{-15}	femto	f
10^{-18}	atto	a

APPENDIX 7: CONVERSION FACTORS

The units for each quantity are listed alphabetically, except that the SI unit is always listed first. The numbers are based on "American National Standard; Metric Practice" published by the Institute of Electrical and Electronics Engineers, 1982.

ANGLE

1 radian = $57.30° = 3.438 \times 10^3{}' = \frac{1}{2\pi}$ rev $= 2.063 \times 10^5{}''$
1 degree (°) $= 1.745 \times 10^{-2}$ **radian** $= 60' = 3600'' = \frac{1}{360}$ rev
1 minute of arc (') $= 2.909 \times 10^{-4}$ **radian**$= \frac{1}{60}° = 4.630 \times 10^{-5}$ rev $= 60''$
1 revolution (rev) $= 2\pi$ **radian** $= 360° = 2.160 \times 10^4{}' = 1.296 \times 10^6{}''$
1 second of arc ('') $= 4.848 \times 10^{-6}$ **radian** $= \frac{1}{3600}° = \frac{1}{60}{}' = 7.716 \times 10^{-7}$ rev

LENGTH

1 meter (m) $= 1 \times 10^{10}$ Å $= 6.685 \times 10^{-12}$ AU $= 100$ cm $= 1 \times 10^{15}$ Fm $= 3.281$ ft $= 39.37$ in. $= 1 \times 10^{-3}$ km $= 1.057 \times 10^{-16}$ light-year $= 1 \times 10^6 \mu$m $= 5.400 \times 10^{-4}$ nmi $= 6.214 \times 10^{-4}$ mi $= 3.241 \times 10^{-17}$ pc $= 1.094$ yd
1 angstrom (Å) $= 1 \times 10^{-10}$ m $= 1 \times 10^{-8}$ cm $= 1 \times 10^{-5}$ fm $= 3.281 \times 10^{-10}$ ft $= 1 \times 10^{-4} \mu$m
1 astronomical unit (AU) $= 1.496 \times 10^{11}$ m $= 1.496 \times 10^{13}$ cm $= 1.496 \times 10^8$ km $= 1.581 \times 10^{-5}$ light-year $= 4.848 \times 10^{-6}$ pc
1 centimeter (cm) $= 0.01$ m $= 1 \times 10^8$ Å $= 1 \times 10^{13}$ fm $= 3.281 \times 10^{-2}$ ft $= 0.3937$ in. $= 1 \times 10^{-5}$ km $= 1.057 \times 10^{-18}$ light-year $= 1 \times 10^4 \mu$m
1 fermi (fm) $= 1 \times 10^{-15}$ m $= 1 \times 10^{-13}$ cm $= 1 \times 10^5$ Å
1 foot (ft) $= 0.3048$ m $= 30.48$ cm $= 12$ in. $= 3.048 \times 10^5 \mu$m $= 1.894 \times 10^{-4}$ mi $= \frac{1}{3}$ yd
1 inch (in.) $= 2.540 \times 10^{-2}$ m $= 2.54$ cm $= \frac{1}{12}$ ft $= 2.54 \times 10^4 \mu$m $= \frac{1}{36}$ yd
1 kilometer (km) $= 1 \times 10^3$ m $= 1 \times 10^5$ cm $= 3.281 \times 10^3$ ft $= 0.5400$ nmi $= 0.6214$ mi $= 1.094 \times 10^3$ yd
1 light-year $= 9.461 \times 10^{15}$ m $= 6.324 \times 10^4$ AU $= 9.461 \times 10^{17}$ cm $= 9.461 \times 10^{12}$ km $= 5.879 \times 10^{12}$ mi $= 0.3066$ pc
1 micron, or micrometer (μm) $= 1 \times 10^{-6}$ m $= 1 \times 10^4$ Å $= 1 \times 10^{-4}$ cm $= 3.281 \times 10^{-6}$ ft $= 3.937 \times 10^{-5}$ in.
1 nautical mile (nmi) $= 1.852 \times 10^3$ m $= 1.852 \times 10^5$ cm $= 6.076 \times 10^3$ ft $= 1.852$ km $= 1.151$ mi
1 statute mile (mi) $= 1.609 \times 10^3$ m $= 1.609 \times 10^5$ cm $= 5280$ ft $= 1.609$ km $= 0.8690$ nmi $= 1760$ yd
1 parsec (pc) $= 3.086 \times 10^{16}$ m $= 2.063 \times 10^5$ AU $= 3.086 \times 10^{18}$ cm $= 3.086 \times 10^{13}$ km $= 3.262$ light-years
1 yard (yd) $= 0.9144$ m $= 91.44$ cm $= 3$ ft $= 36$ in. $= \frac{1}{1760}$ mi

TIME

1 second (s) = 1.157×10^{-5} day = $\frac{1}{3600}$ h = $\frac{1}{60}$ min = 1.161×10^{-5} sidereal day = 3.169×10^{-8} yr

1 day = 8.640×10^4 s = 24 h = 1440 min = 1.003 sidereal days = 2.738×10^{-3} yr

1 hour (h) = 3600 s = $\frac{1}{24}$ day = 60 min = 1.141×10^{-4} yr

1 minute (min) = 60 s = 6.944×10^{-4} day = $\frac{1}{60}$ h = 1.901×10^{-6} year

1 sidereal day = 8.616×10^4 s = 0.9973 day = 23.93 h = 1.436×10^3 min = 2.730×10^{-3} yr

1 year (yr) = 3.156×10^7 s = 365.24 days = 8.766×10^3 h = 5.259×10^5 min = 366.24 sidereal days

MASS

1 kilogram (kg) = 6.024×10^{26} u = 5000 carats = 1.543×10^4 grains = 1000 g = 1×10^{-3} t = 35.27 oz. = 2.205 lb = 1.102×10^{-3} short ton = 6.852×10^{-2} slug

1 atomic mass unit (u) = 1.6605×10^{-27} kg = 1.6605×10^{-24} g

1 carat = 2×10^{-4} kg = 0.2 g = 7.055×10^{-3} oz. = 4.409×10^{-4} lb

1 grain = 6.480×10^{-5} kg = 6.480×10^{-2} g = 2.286×10^{-3} oz. = $\frac{1}{7000}$ lb

1 gram (g) = 1×10^{-3} kg = 6.024×10^{23} u = 5 carats = 15.43 grains = 1×10^{-6} t = 3.527×10^{-2} oz. = 2.205×10^{-3} lb = 1.102×10^{-6} short ton = 6.852×10^{-5} slug

1 metric ton, or tonne (t) = 1×10^3 kg = 1×10^6 g = 2.205×10^3 lb = 1.102 short ton = 68.52 slugs

1 ounce (oz.) = 2.835×10^{-2} kg = 141.7 carats = 437.5 grains = 28.35 g = $\frac{1}{16}$ lb

1 pound (lb)[b] = 0.4536 kg = 453.6 g = 4.536×10^{-4} t = 16 oz. = $\frac{1}{2000}$ short ton = 3.108×10^{-2} slug

1 short ton = 907.2 kg = 9.07×10^5 g = 0.9072 t = 2000 lb

1 slug = 14.59 kg = 1.459×10^4 g = 32.17 lb

AREA

1 square meter (m²) = 1×10^4 cm² = 10.76 ft² = 1.550×10^3 in.² = 1×10^{-6} km² = 3.861×10^{-7} mi² = 1.196 yd²

1 barn = 1×10^{-28} m² = 1×10^{-24} cm²

1 square centimeter (cm²) = 1×10^{-4} m² = 1.076×10^{-3} ft² = 0.1550 in.² = 1×10^{-10} km² = 3.861×10^{-11} mi²

1 square foot (ft²) = 9.290×10^{-2} m² = 929.0 cm² = 144 in.² = 3.587×10^{-8} mi² = $\frac{1}{9}$ yd²

1 square inch (in.²) = 6.452×10^{-4} m² = 6.452 cm² = $\frac{1}{144}$ ft²

1 square kilometer (km²) = 1×10^6 m² = 1×10^{10} cm² = 1.076×10^7 ft² = 0.3861 mi²

1 square statute mile (mi²) = 2.590×10^6 m² = 2.590×10^{10} cm² = 2.788×10^7 ft² = 2.590 km²

1 square yard (yd²) = 0.8361 m² = 8.361×10^3 cm² = 9 ft² = 1296 in.²

VOLUME

1 cubic meter (m³) = 1×10^6 cm³ = 35.31 ft³ = 264.2 gal. = 6.102×10^4 in.³ = 1×10^3 liters = 1.308 yd³

1 cubic centimeter (cm³) = 1×10^{-6} m³ = 3.531×10^{-5} ft³ = 2.642×10^{-4} gal. = 6.102×10^{-2} in.³ = 1×10^{-3} liter

1 cubic foot (ft³) = 2.832×10^{-2} m³ = 2.832×10^4 cm³ = 7.481 gal. = 1728 in.³ = 28.32 liters = $\frac{1}{27}$ yd³

1 gallon (gal.)[c] = 3.785×10^{-3} m³ = 0.1337 ft³

---

[b] This is the "avoirdupois" pound. The "troy" or "apothecary" pound is 0.3732 kg, or 0.8229 lb avoirdupois.

[c] This is the U.S. gallon; the U.K. and the Canadian gallon are 4.546×10^{-3} m³, or 1.201 U.S. gallons.

1 cubic inch (in.^{3}) = 1.639×10^{-5} m^3 = 16.39 cm^3 = 5.787×10^{-4} ft^3
1 liter (l) = 1×10^{-3} m^3 = 1000 cm^3 = 3.531×10^{-2} ft^3
1 cubic yard (yd^3) = 0.7646 m^3 = 7.646×10^5 cm^3 = 27 ft^3 = 202.0 gal.

DENSITY

1 kilogram per cubic meter (kg/m^3) = 1×10^{-3} g/cm^3 = 6.243×10^{-2} lb/ft^3 = 8.345×10^{-3} lb/gal. = 3.613×10^{-5} lb/in.^{3} = 8.428×10^{-4} short ton/yd^3 = 1.940×10^{-3} slug/ft^3
1 gram per cubic centimeter (g/cm^3) = 1×10^3 kg/m^3 = 62.43 lb/ft^3 = 8.345 lb/gal. = 3.613×10^{-2} lb/in.^{3} = 0.8428 short ton/yd^3 = 1.940 slug/ft^3
1 lb per cubic foot (lb/ft^3) = 16.02 kg/m^3 = 1.602×10^{-2} g/cm^3 = 0.1337 lb/gal. = 1.350×10^{-2} short ton/yd^3 = 3.108×10^{-2} slug/ft^3
1 pound-mass per gallon (1 lb/gal.) = 119.8 kg/m^3 = 7.481 lb/ft^3 = 0.2325 slug/ft^3
1 short ton per cubic yard (short ton/yd^3) = 1.187×10^3 kg/m^3 = 74.07 lb/ft^3
1 slug per cubic foot (slug/ft^3) = 515.4 kg/m^3 = 0.5154 g/cm^3 = 32.17 lb/ft^3 = 4.301 lb/gal.

SPEED

1 meter per second (m/s) = 100 cm/s = 3.281 ft/s = 3.600 km/h = 1.944 knot = 2.237 mi/h
1 centimeter per second (cm/s) = 0.01 m/s = 3.281×10^{-2} ft/s = 3.600×10^{-2} km/h = 1.944×10^{-2} knot = 2.237×10^{-2} mi/h
1 foot per second (ft/s) = 0.3048 m/s = 30.48 cm/s = 1.097 km/h = 0.5925 knot = 0.6818 mi/h
1 kilometer per hour (km/h) = 0.2778 m/s = 27.78 cm/s = 0.9113 ft/s = 0.5400 knot = 0.6214 mi/h
1 knot, or nautical mile per hour = 0.5144 m/s = 51.44 cm/s = 1.688 ft/s = 1.852 km/h = 1.151 mi/h
1 mile per hour (mi/h) = 0.4470 m/s = 44.70 cm/s = 1.467 ft/s = 1.609 km/h = 0.8690 knot

ACCELERATION

1 meter per second squared (m/s^2) = 100 cm/s^2 = 3.281 ft/s^2 = 0.1020 G
1 centimeter per second squared, or Gal (cm/s^2) = 0.01 m/s^2 = 3.281×10^{-2} ft/s^2 = 1.020×10^{-3} G
1 foot per second squared (ft/s^2) = 0.3048 m/s^2 = 30.48 cm/s^2 = 3.108×10^{-2} G
1 G = 9.807 m/s^2 = 980.7 cm/s^2 = 32.17 ft/s^2

FORCE

1 newton (N) = 1×10^5 dynes = 0.1020 kp = 0.2248 lbf = 1.124×10^{-4} short ton-force
1 dyne = 1×10^{-5} N = 1.020×10^{-6} kp = 2.248×10^{-6} lbf = 1.124×10^{-9} short ton-force
1 kilopond, or kilogram force (kp) = 9.807 N = 9.807×10^5 dynes = 2.205 lbf = 1.102×10^{-3} short ton-force
1 pound-force (lbf) = 4.448 N = 4.448×10^5 dynes = 0.4536 kp = $\frac{1}{2000}$ short ton-force
1 short ton-force = 8.896×10^3 N = 8.896×10^8 dynes = 907.2 kp = 2000 lbf

ENERGY

1 joule (J) = 9.478×10^{-4} Btu = 0.2388 cal = 1×10^7 ergs = 6.242×10^{18} eV = 0.7376 ft·lbf = 2.778×10^{-7} kW·h

1 **British thermal unit** (Btu)[d] $= 1.055 \times 10^3$ J $= 252.0$ cal $= 1.055 \times 10^{10}$ ergs $= 778.2$ ft·lbf $= 2.931 \times 10^{-4}$ kW·h

1 **calorie** (cal)[e] $= 4.187$ J $= 3.968 \times 10^{-3}$ Btu $= 4.187 \times 10^7$ ergs $= 3.088$ ft·lbf $= 1 \times 10^{-3}$ kcal $= 1.163 \times 10^{-6}$ kW·h

1 **erg** $= 1 \times 10^{-7}$ J $= 9.478 \times 10^{-7}$ Btu $= 2.388 \times 10^{-8}$ cal $= 6.242 \times 10^{11}$ eV $= 7.376 \times 10^{-8}$ ftf·lb $= 2.778 \times 10^{-14}$ kW·h

1 **electron-volt** (eV) $= 1.602 \times 10^{-19}$ J $= 1.602 \times 10^{-12}$ erg $= 1.182 \times 10^{-19}$ ft·lbf

1 **foot-pound-force** (ft·lbf) $= 1.356$ J $= 1.285 \times 10^{-3}$ Btu $= 0.3239$ cal $= 1.356 \times 10^7$ ergs $= 8.464 \times 10^{18}$ eV $= 3.766 \times 10^{-7}$ kW·h

1 **kilocalorie** (kcal), or **large calorie** (Cal) $= 4.187 \times 10^3$ J $= 1 \times 10^3$ cal

1 **kilowatt-hour** (kW·h) $= 3.600 \times 10^6$ J $= 3412$ Btu $= 8.598 \times 10^5$ cal $= 3.6 \times 10^{13}$ ergs $= 2.655 \times 10^6$ ft·lbf

POWER

1 **watt** (W) $= 3.412$ Btu/h $= 0.2388$ cal/s $= 1 \times 10^7$ ergs/s $= 0.7376$ ft·lbf/s $= 1.341 \times 10^{-3}$ hp

1 **British thermal unit per hour** (Btu/h) $= 0.2931$ W $= 7.000 \times 10^{-2}$ cal/s $= 0.2162$ ft·lbf/s $= 3.930 \times 10^{-4}$ hp

1 **calorie per second** (cal/s) $= 4.187$ W $= 14.29$ Btu/h $= 4.187 \times 10^7$ erg/s $= 3.088$ ft·lbf/s $= 5.615 \times 10^{-3}$ hp

1 **erg per second** (erg/s) $= 1 \times 10^{-7}$ W $= 2.388 \times 10^{-8}$ cal/s $= 7.376 \times 10^{-8}$ ft·lbf/s $= 1.341 \times 10^{-10}$ hp

1 **foot-pound per second** (ft·lbf/s) $= 1.356$ W $= 0.3238$ cal/s $= 4.626$ Btu/h $= 1.356 \times 10^7$ ergs/s $= 1.818 \times 10^{-3}$ hp

1 **horsepower** (hp)[f] $= 745.7$ W $= 2.544 \times 10^3$ Btu/h $= 178.1$ cal/s $= 550$ ft·lbf/s

1 **kilowatt** (kW) $= 1 \times 10^3$ W $= 3.412 \times 10^3$ Btu/h $= 238.8$ cal/s $= 737.6$ ft·lbf/s $= 1.341$ hp

PRESSURE

1 **newton per square meter** (N/m²), or **pascal** (Pa) $= 9.869 \times 10^{-6}$ atm $= 1 \times 10^{-5}$ bar $= 7.501 \times 10^{-4}$ cmHg $= 10$ dynes/cm² $= 2.953 \times 10^{-4}$ in. Hg $= 2.089 \times 10^{-2}$ lbf/ft² $= 1.450 \times 10^{-4}$ lbf/in.² $= 7.501 \times 10^{-3}$ torr

1 **atmosphere** (atm) $= 1.013 \times 10^5$ N/m² $= 76.00$ cmHg $= 1.013 \times 10^6$ dynes/cm² $= 29.92$ in. Hg $= 2.116 \times 10^3$ lbf/ft² $= 14.70$ lbf/in.²

1 **bar** $= 1 \times 10^5$ N/m² $= 0.9869$ atm $= 75.01$ cmHg

1 **centimeter of mercury** (cmHg) $= 1.333 \times 10^3$ N/m² $= 1.316 \times 10^{-2}$ atm $= 1.333 \times 10^{-2}$ bar $= 1.333 \times 10^4$ dynes/cm² $= 0.3937$ in. Hg $= 27.85$ lbf/ft² $= 0.1934$ lbf/in.² $= 10$ torr

1 **dyne per square centimeter** (dyne/cm²) $= 0.1$ N/m² $= 9.869 \times 10^{-7}$ atm $= 7.501 \times 10^{-5}$ cmHg $= 2.089 \times 10^{-3}$ lbf/ft² $= 1.450 \times 10^{-5}$ lbf/in.²

1 **inch of mercury** (in. Hg) $= 3.386 \times 10^3$ N/m² $= 3.342 \times 10^{-2}$ atm $= 2.540$ cmHg $= 0.4912$ lbf/in.²

1 **kilopond per square centimeter** (kp/cm²) $= 9.807 \times 10^4$ N/m² $= 0.9678$ atm $= 9.807 \times 10^5$ dynes/cm² $= 14.22$ lbf/in.²

1 **pound per square inch** (lbf/in.², or psi) $= 6.895 \times 10^3$ N/m² $= 6.805 \times 10^{-2}$ atm $= 6.895 \times 10^4$ dynes/cm² $= 2.036$ in. Hg $= 7.031 \times 10^{-2}$ kp/cm²

1 **torr**, or **millimeter of mercury** (mmHg) $= 1.333 \times 10^2$ N/m² $= 0.1$ cmHg

---

[d] This is the "International Table" Btu; there are several other Btus.

[e] This is the "International Table" calorie, which equals exactly 4.1868 J. There are several other calories; for instance, the thermochemical calorie, which equals 4.184 J.

[f] There are several other horsepowers; for instance, the metric horsepower, which equals 735.5 J.

ELECTRIC CHARGE[g]
1 coulomb (C) ↔ 2.998×10^9 statcoulombs, or esu of charge ↔ 0.1 abcoulomb, or emu of charge

ELECTRIC CURRENT
1 ampere (A) ↔ 2.998×10^9 statamperes, or esu of current ↔ 0.1 abampere, or emu of current

ELECTRIC POTENTIAL
1 volt (V) ↔ 3.336×10^{-3} statvolt, or esu of potential ↔ 1×10^8 abvolts, or emu of potential

ELECTRIC FIELD
1 volt per meter (V/m) ↔ 3.336×10^{-5} statvolt/cm ↔ 1×10^6 abvolts/cm

MAGNETIC FIELD
1 tesla (T), or **weber per square meter** (Wb/m²) ↔ 1×10^4 gauss

ELECTRIC RESISTANCE
1 ohm (Ω) ↔ 1.113×10^{-12} statohm, or esu of resistance ↔ 1×10^9 abohms, or emu of resistance

ELECTRIC RESISTIVITY
1 ohm-meter (Ω · m) ↔ 1.113×10^{-10} statohm-cm ↔ 1×10^{11} abohm-cm

CAPACITANCE
1 farad (F) ↔ 8.988×10^{11} statfarads, or esu of capacitance ↔ 1×10^{-9} abfarad, or emu of capacitance

INDUCTANCE
1 henry (H) ↔ 1.113×10^{-12} stathenry, or esu of inductance ↔ 1×10^9 abhenrys, or emu of inductance

---

[g] The dimensions of the electric quantities in SI units, electrostatic units (esu), and electromagnetic units (emu) are different; hence the relationships among these units are correspondences (↔) rather than equalities (=).

APPENDIX 8: BEST VALUES OF FUNDAMENTAL CONSTANTS

The values in the following table were taken from the report of the CODATA Task Group on fundamental constants by E. R. Cohen and B. N. Taylor, *The 1986 Adjustment of the Fundamental Physical Constants*, CODATA Bulletin No. 63, November 1986. The digits in parentheses are the one-standard deviation uncertainty in the last digits of the given value.

Quantity	Symbol	Value	Units	Relative uncertainty (parts per million)
UNIVERSAL CONSTANTS				
speed of light in vacuum	c	299792458	ms^{-1}	(exact)
permeability of vacuum	μ_o	$4\pi \times 10^{-7}$	NA^{-2}	
		$=12.566370614\ldots$	$10^{-7}\,NA^{-2}$	(exact)
permittivity of vacuum	ε_o	$1/\mu_o c^2$		
		$=8.854187817\ldots$	$10^{-12}\,Fm^{-1}$	(exact)
Newtonian constant of gravitation	G	6.67259(85)	$10^{-11}\,m^3\,kg^{-1}\,s^{-2}$	128
Planck constant	h	6.6260755(40)	$10^{-34}\,Js$	0.60
in electron volts		4.1356692(12)	$10^{-15}\,eVs$	0.30
	$\hbar = h/2\pi$	1.05457266(63)	$10^{-34}\,Js$	0.60
in electron volts		6.5821220(20)	$10^{-16}\,eVs$	0.30
ELECTROMAGNETIC CONSTANTS				
elementary charge	e	1.60217733(49)	$10^{-19}\,C$	0.30
magnetic flux quantum, $h/2e$	Φ_o	2.06783461(61)	$10^{-15}\,Wb$	0.30
Josephson frequency–voltage ratio	$2e/h$	4.8359767(14)	$10^{14}\,HzV^{-1}$	0.30
quantized Hall conductance	e^2/h	3.87404614(17)	$10^{-5}\,\Omega^{-1}$	0.045
Bohr magneton, $e\hbar/2m_e$	μ_B	9.2740154(31)	$10^{-24}\,JT^{-1}$	0.34
in electron volts		5.78838263(52)	$10^{-5}\,eVT^{-1}$	0.089
nuclear magneton, $e\hbar/2m_p$	μ_N	5.0507866(17)	$10^{-27}\,JT^{-1}$	0.34
in electron volts		3.15245166(28)	$10^{-8}\,eVT^{-1}$	0.089
ATOMIC CONSTANTS				
fine-structure constant, $e^2/4\pi\varepsilon_o\hbar c$	α	7.29735308(33)	10^{-3}	0.045
inverse fine-structure constant	α^{-1}	137.0359895(61)		0.045
Rydberg constant, $\frac{1}{2}m_e c\alpha^2/h$	$R\infty$	10973731.534(13)	m^{-1}	0.0012
Bohr radius, $4\pi\varepsilon_o\hbar^2/m_e e^2$	a_o	0.529177249(24)	$10^{-10}\,m$	0.045
quantum of circulation	$h/2m_e$	3.63694807(33)	$10^{-4}\,m^2 s^{-1}$	0.089
	h/m_e	7.27389614(65)	$10^{-4}\,m^2 s^{-1}$	0.089
Electron				
electron mass	m_e	9.1093897(54)	$10^{-31}\,kg$	0.59
		5.48579903(13)	$10^{-4}\,u$	0.023
$m_e c^2$ in electron volts		0.51099906(15)	MeV	0.30
electron–proton mass ratio	m_e/m_p	5.44617013(11)	10^{-4}	0.020
electron specific charge	$-e/m_e$	$-1.75881962(53)$	$10^{11}\,Ckg^{-1}$	0.30
Compton wavelength, $h/m_e c$	λ_C	2.42631058(22)	$10^{-12}\,m$	0.089
$\lambda_C/2\pi = \alpha a_o$	$\bar\lambda_C$	3.86159323(35)	$10^{-13}\,m$	0.089
classical electron radius, $e^2/4\pi\varepsilon_o m_e c^2$	r_e	2.81794092(38)	$10^{-15}\,m$	0.13
Thomson cross section, $(8\pi/3)r_e^2$	σ_e	0.66524616(18)	$10^{-28}\,m^2$	0.27
electron magnetic moment	μ_e	928.47701(31)	$10^{-26}\,JT^{-1}$	0.34
in Bohr magnetons	μ_e/μ_B	1.001159652193(10)		1×10^{-5}
in nuclear magnetons	μ_e/μ_N	1838.282000(37)		0.020
electron magnetic moment anomaly, $\mu_e/\mu_B - 1$	a_e	1.159652193(10)	10^{-3}	0.0086
electron g-factor, $2(1 + a_e)$	g_e	2.002319304386(20)		1×10^{-5}

Best Values of Fundamental Constants (APP. 8)

Quantity	Symbol	Value	Units	Relative uncertainty (parts per million)
Muon				
muon mass	m_μ	1.8835327(11)	10^{-28} kg	0.61
		0.113428913(17)	u	0.15
$m_\mu c^2$ in electron volts		105.658389(34)	MeV	0.32
muon–electron mass ratio	m_μ/m_e	206.768262(30)		0.15
muon magnetic moment	μ_μ	4.4904514(15)	10^{-26} JT^{-1}	0.33
in Bohr magnetons,	μ_μ/μ_B	4.84197097(71)	10^{-3}	0.15
muon magnetic moment anomaly				
$[\mu_\mu/(e\hbar/2m_\mu)] - 1$	a_μ	1.1659230(84)	10^{-3}	7.2
muon g-factor, $2(1+a_\mu)$	g_μ	2.002331846(17)		0.0084
Proton				
proton mass	m_p	1.6726231(10)	10^{-27} kg	0.59
		1.007276470(12)	u	0.012
$m_p c^2$ in electron volts		938.27231(28)	MeV	0.30
proton–electron mass ratio	m_p/m_e	1836.152701(37)		0.020
proton specific charge	e/m_p	9.5788309(29)	10^7 Ckg^{-1}	0.30
proton Compton wavelength, $h/m_p c$	$\lambda_{C,p}$	1.32141002(12)	10^{-15} m	0.089
proton magnetic moment	μ_p	1.41060761(47)	10^{-26} JT^{-1}	0.34
in Bohr magnetons	μ_p/μ_B	1.521032202(15)	10^{-3}	0.010
in nuclear magnetons	μ_p/μ_N	2.792847386(63)		0.023
Neutron				
neutron mass	m_n	1.6749286(10)	10^{-27} kg	0.59
		1.008664904(14)	u	0.014
$m_n c^2$ in electron volts		939.56563(28)	MeV	0.30
neutron–electron mass ratio	m_n/m_e	1838.683662(40)		0.022
neutron–proton mass ratio	m_n/m_p	1.001378404(9)		0.009
neutron Compton wavelength, $h/m_n c$	$\lambda_{C,n}$	1.31959110(12)	10^{-15} m	0.089
neutron magnetic moment	μ_n	0.96623707(40)	10^{-26} JT^{-1}	0.41
in Bohr magnetons	μ_n/μ_B	1.04187563(25)	10^{-3}	0.24
in nuclear magnetons	μ_n/μ_N	1.91304275(45)		0.24
Deuteron				
deuteron mass	m_d	3.3435860(20)	10^{-27} kg	0.59
		2.013553214(24)	u	0.012
$m_d c^2$ in electron volts		1875.61339(57)	MeV	0.30
deuteron–electron mass ratio	m_d/m_e	3670.483014(75)		0.020
deuteron–proton mass ratio	m_d/m_p	1.999007496(6)		0.003
deuteron magnetic moment	μ_d	0.43307375(15)	10^{-26} JT^{-1}	0.34
in Bohr magnetons,	μ_d/μ_B	0.4669754479(91)	10^{-3}	0.019
in nuclear magnetons,	μ_d/μ_N	0.857438230(24)		0.028
PHYSICO-CHEMICAL CONSTANTS				
Avogadro constant	N_A	6.0221367(36)	10^{23} mol^{-1}	0.59
atomic mass constant, $m_u = \frac{1}{12}m(^{12}C)$	m_u	1.6605402(10)	10^{-27} kg	0.59
$m_u c^2$ in electron volts		931.49432(28)	MeV	0.30
Faraday constant	F	96485.309(29)	C mol^{-1}	0.30
molar gas constant	R	8.314510(70)	J mol^{-1}K^{-1}	8.4
Boltzmann constant, R/N_A	k	1.380658(12)	10^{-23} JK^{-1}	8.5
in electron volts		8.617385(73)	10^{-3} eVK^{-1}	8.4
molar volume (ideal gas), RT/p				
$T = 273.15K$, $p = 101325Pa$	V_m	22.41410(19)	liter/mol	8.4
Loschmidt constant, N_A/V_m	n_o	2.686763(23)	10^{25} m^{-3}	8.5
Stefan-Boltzmann constant,				
$(\pi^2/60)k^4/\hbar^3 c^2$	σ	5.67051(19)	10^{-8} Wm^{-2}K^{-4}	34
Wien displacement law constant	b	2.897756(24)	10^{-3} mK	8.4

APPENDIX 9: THE CHEMICAL ELEMENTS AND THE PERIODIC TABLE

Table A9.1 The Periodic Table of the Chemical Elements[h]

IA																	0
1 H 1.00794	IIA											IIIA	IVA	VA	VIA	VIIA	2 He 4.00260
3 Li 6.941	4 Be 9.01218											5 B 10.81	6 C 12.011	7 N 14.0067	8 O 15.9994	9 F 18.998403	10 Ne 20.179
11 Na 22.98977	12 Mg 24.305	IIIB	IVB	VB	VIB	VIIB	←	VIII	→	IB	IIB	13 Al 26.98154	14 Si 28.0855	15 P 30.97376	16 S 32.06	17 Cl 35.453	18 Ar 39.948
19 K 39.0983	20 Ca 40.08	21 Sc 44.9559	22 Ti 47.88	23 V 50.9415	24 Cr 51.996	25 Mn 54.9380	26 Fe 55.847	27 Co 58.9332	28 Ni 58.69	29 Cu 63.546	30 Zn 65.38	31 Ga 69.72	32 Ge 72.59	33 As 74.9216	34 Se 78.96	35 Br 79.904	36 Kr 83.80
37 Rb 85.4678	38 Sr 87.62	39 Y 88.9059	40 Zr 91.22	41 Nb 92.9064	42 Mo 95.94	43 Tc (98)	44 Ru 101.07	45 Rh 102.9055	46 Pd 106.42	47 Ag 107.8682	48 Cd 112.41	49 In 114.82	50 Sn 118.69	51 Sb 121.75	52 Te 127.60	53 I 126.9045	54 Xe 131.29
55 Cs 132.9054	56 Ba 137.33	57–71 Rare Earths	72 Hf 178.49	73 Ta 180.9479	74 W 183.85	75 Re 186.207	76 Os 190.2	77 Ir 192.22	78 Pt 195.08	79 Au 196.9665	80 Hg 200.59	81 Tl 204.383	82 Pb 207.2	83 Bi 208.9804	84 Po (209)	85 At (210)	86 Rn (222)
87 Fr (223)	88 Ra 226.0254	89–103 Actinides	104 Rf (261)	105 Ha (260)	106 (263)	107 (262)	108 (265)	109 (266)									

Rare Earths (Lanthanides)	57 La 138.9055	58 Ce 140.12	59 Pr 140.9077	60 Nd 144.24	61 Pm (145)	62 Sm 150.36	63 Eu 151.96	64 Gd 157.25	65 Tb 158.9254	66 Dy 162.50	67 Ho 164.9304	68 Er 167.26	69 Tm 168.9342	70 Yb 173.04	71 Lu 174.967

Actinides	89 Ac 227.0278	90 Th 232.0381	91 Pa 231.0359	92 U 238.0289	93 Np 237.0482	94 Pu (244)	95 Am (243)	96 Cm (247)	97 Bk (247)	98 Cf (251)	99 Es (252)	100 Fm (257)	101 Md (258)	102 No (259)	103 Lr (260)

[h] In each box, the upper number is the *atomic number*. The lower number is the *atomic mass*, i.e., the mass (in grams) of one mole or, alternatively, the mass (in atomic mass units) of one atom. Numbers in parentheses denote the atomic masses of the most stable or best-known isotope of the element; all other numbers represent the average masses of a mixture of several isotopes as found in naturally occurring samples of the element.

Table A9.2 THE CHEMICAL ELEMENTS

Element	Chemical symbol	Atomic number	Atomic mass[i]
Hydrogen	H	1	1.00794 u
Helium	He	2	4.00260
Lithium	Li	3	6.941
Beryllium	Be	4	9.01218
Boron	B	5	10.81
Carbon	C	6	12.011
Nitrogen	N	7	14.0067
Oxygen	O	8	15.9994
Fluorine	F	9	18.998403
Neon	Ne	10	20.179
Sodium	Na	11	22.98977
Magnesium	Mg	12	24.305
Aluminum	Al	13	26.98154
Silicon	Si	14	28.0855
Phosphorus	P	15	30.97376
Sulfur	S	16	32.06
Chlorine	Cl	17	35.453
Argon	Ar	18	39.948
Potassium	K	19	39.0983
Calcium	Ca	20	40.08
Scandium	Sc	21	44.9559
Titanium	Ti	22	47.88
Vanadium	V	23	50.9415
Chromium	Cr	24	51.996
Manganese	Mn	25	54.9380
Iron	Fe	26	55.847
Cobalt	Co	27	58.9332
Nickel	Ni	28	58.69
Copper	Cu	29	63.546
Zinc	Zn	30	65.38
Gallium	Ga	31	69.72
Germanium	Ge	32	72.59
Arsenic	As	33	74.9216
Selenium	Se	34	78.96
Bromine	Br	35	79.904
Krypton	Kr	36	83.80
Rubidium	Rb	37	85.4678
Strontium	Sr	38	87.62
Yttrium	Y	39	88.9059
Zirconium	Zr	40	91.22
Niobium	Nb	41	92.9064
Molybdenum	Mo	42	95.94
Technetium	Tc	43	(98)
Ruthenium	Ru	44	101.07
Rhodium	Rh	45	102.9055
Palladium	Pd	46	106.42
Silver	Ag	47	107.8682
Cadmium	Cd	48	112.41
Indium	In	49	114.82
Tin	Sn	50	118.69
Antimony	Sb	51	121.75
Tellurium	Te	52	127.60
Iodine	I	53	126.9045

[i] Numbers in parentheses denote the atomic masses of the most stable or best-known isotope of the element; all other numbers represent the average masses of a mixture of several isotopes as found in naturally occurring samples of the element.

Table A9.2 THE CHEMICAL ELEMENTS

Element	Chemical symbol	Atomic number	Atomic mass
Xenon	Xe	54	131.29
Cesium	Cs	55	132.9054
Barium	Ba	56	137.33
Lanthanum	La	57	138.9055
Cerium	Ce	58	140.12
Praseodymium	Pr	59	140.9077
Neodymium	Nd	60	144.24
Promethium	Pm	61	(145)
Samarium	Sm	62	150.36
Europium	Eu	63	151.96
Gadolinium	Gd	64	157.25
Terbium	Tb	65	158.9254
Dysprosium	Dy	66	162.50
Holmium	Ho	67	164.9304
Erbium	Er	68	167.26
Thulium	Tm	69	168.9342
Ytterbium	Yb	70	173.04
Lutetium	Lu	71	174.967
Hafnium	Hf	72	178.49
Tantalum	Ta	73	180.9479
Tungsten	W	74	183.85
Rhenium	Re	75	186.207
Osmium	Os	76	190.2
Iridium	Ir	77	192.22
Platinum	Pt	78	195.08
Gold	Au	79	196.9665
Mercury	Hg	80	200.59
Thallium	Tl	81	204.383
Lead	Pb	82	207.2
Bismuth	Bi	83	208.9804
Polonium	Po	84	(209)
Astatine	At	85	(210)
Radon	Rn	86	(222)
Francium	Fr	87	(223)
Radium	Ra	88	226.0254
Actinium	Ac	89	227.0278
Thorium	Th	90	232.0381
Protactinium	Pa	91	231.0359
Uranium	U	92	238.0289
Neptunium	Np	93	237.0482
Plutonium	Pu	94	(244)
Americium	Am	95	(243)
Curium	Cm	96	(247)
Berkelium	Bk	97	(247)
Californium	Cf	98	(251)
Einsteinium	Es	99	(252)
Fermium	Fm	100	(257)
Mendelevium	Md	101	(258)
Nobelium	No	102	(259)
Lawrencium	Lr	103	(260)
Rutherfordium	Rf	104	(261)
Hahnium	Ha	105	(260)
?		106	(263)
?		107	(262)
?		108	(265)
?		109	(266)

APPENDIX 10: FORMULA SHEETS

Chapters 1–21:

$v = dx/dt$

$a = dv/dt = d^2x/dt^2$

$x = x_0 + v_0 t + \tfrac{1}{2}at^2$

$a(x - x_0) = \tfrac{1}{2}(v^2 - v_0^2)$

$A_x = A \cos \theta_x$

$A = \sqrt{A_x^2 + A_y^2 + A_z^2}$

$\mathbf{A} \cdot \mathbf{B} = AB \cos \phi$
$= A_x B_x + A_y B_y + A_z B_z$

$|\mathbf{A} \times \mathbf{B}| = AB \sin \phi$

$a = v^2/r$

$\mathbf{v'} = \mathbf{v} - \mathbf{V_O}$

$\mathbf{ma} = \mathbf{F}$

$\mathbf{p} = m\mathbf{v}$

$w = mg$

$f_k = \mu_k N$

$f_s \leq \mu_s N$

$F = -kx$

$W = \mathbf{F} \cdot \Delta \mathbf{r}$

$W = \int \mathbf{F} \cdot d\mathbf{r}$

$K = \tfrac{1}{2}mv^2$

$U = mgz$

$E = K + U = \text{[constant]}$

$U(P) = -\int_{P_0}^{P} \mathbf{F} \cdot d\mathbf{r} + U(P_0)$

$U = \tfrac{1}{2}kx^2$

$E = mc^2$

$P = dW/dt$

$P = \mathbf{F} \cdot \mathbf{v}$

$F = GMm/r^2$

$v^2 = GM/r$

$g = GM_E/R_E^2$

$U = -GMm/r$

$\mathbf{r}_{CM} = \dfrac{1}{M}\int \mathbf{r}\rho \, dV$

$\mathbf{I} = \int_0^{\Delta t} \mathbf{F}\, dt$

$v_1' = \dfrac{m_1 - m_2}{m_1 + m_2} v_1; \quad v_2' = \dfrac{2m_1}{m_1 + m_2} v_1$

$\omega = d\phi/dt$

$\alpha = d\omega/dt = d^2\phi/dt^2$

$v = R\omega$

$K = \tfrac{1}{2}I\omega^2$

$I = \int \rho R^2 \, dV$

$I_{CM} = MR^2 \text{ (hoop)}; \tfrac{1}{2}MR^2 \text{ (disk)};$
$\tfrac{2}{5}MR^2 \text{ (sphere)}; \tfrac{1}{12}ML^2 \text{ (rod)}$

$I = I_{CM} + Md^2$

$\mathbf{L} = \mathbf{r} \times \mathbf{p}$

$\dfrac{d\mathbf{L}}{dt} = \mathbf{r} \times \mathbf{F}$

$L_z = I\omega$

$I\alpha = \tau_z$

$P = \tau_z \omega$

$x = A\cos(\omega t + \delta)$

$T = 2\pi/\omega; \quad \nu = 1/T = \omega/2\pi$

$m\, d^2x/dt^2 = -kx$

$\omega = \sqrt{k/m}$

$\omega = \sqrt{g/l}; \ T = 2\pi\sqrt{l/g}$

$\omega = \sqrt{mgl/I}$

$\omega = \sqrt{\kappa/I}$

$y = A\cos k(x - vt) = A\cos(kx - \omega t)$

$\lambda = 2\pi/k; \quad v = \nu/\lambda; \quad \omega = 2\pi\nu$

$v = \sqrt{F/\mu}$

$P \propto v\omega^2 A^2$

$\nu_{\text{beat}} = \nu_1 - \nu_2$

$\nu' = \nu(1 \pm V_R/v)$

$\nu' = \nu/(1 \mp V_E/v)$

$\sin \theta = v/V_E$

$p - p_0 = -\rho gz$

$\tfrac{1}{2}\rho v^2 + \rho gz + p = \text{[constant]}$

$pV = NkT$

$T_C = T - 273.15$

$v_{\text{rms}} = \sqrt{3kT/m}$

$\lambda = \dfrac{1}{4\pi R^2 (N/V)}$

$r = \sqrt{n}\, \lambda$

$pV^\gamma = \text{[constant]}; \quad \gamma = C_p/C_V$

$\Delta E = \Delta Q - \Delta W$

$e = 1 - T_2/T_1$

$S(A) = \int_{A_0 \text{ rev.}}^{A} \dfrac{dQ}{T} + S(A_0)$

$S(B) - S(A) \geq \int_{A \text{ irrev.}}^{B} \dfrac{dQ}{T}$

$g = 9.81 \text{ m/s}^2$
$G = 6.67 \times 10^{-11} \text{ N} \cdot \text{m}^2/\text{kg}^2$
$M_E = 5.98 \times 10^{24} \text{ kg}$
$R_E = 6.37 \times 10^6 \text{ m}$

$m_e = 9.11 \times 10^{-31} \text{ kg}$
$m_p = 1.67 \times 10^{-27} \text{ kg}$
$c = 3.00 \times 10^8 \text{ m/s}$

$N_A = 6.02 \times 10^{23}/\text{mole}$
$k = 1.38 \times 10^{-23} \text{ J/K}$
$1 \text{ cal} = 4.19 \text{ J}$

Chapters 22–46:

$F = \dfrac{1}{4\pi\varepsilon_0} \dfrac{qq'}{r^2}$

$E = \dfrac{1}{4\pi\varepsilon_0} \dfrac{q'}{r^2}$

$E = \sigma/2\varepsilon_0$

$p = lQ$

$\tau = \mathbf{p} \times \mathbf{E}$

$U = -\mathbf{p} \cdot \mathbf{E}$

$\oint E_n \, dS = \dfrac{Q}{\varepsilon_0}$

$V(P_2) - V(P_1) = -\displaystyle\int_{P_1}^{P_2} \mathbf{E} \cdot d\mathbf{l}$

$V = \dfrac{1}{4\pi\varepsilon_0} \dfrac{q'}{r}$

$\dfrac{\partial V}{\partial x} = -E_x, \quad \dfrac{\partial V}{\partial y} = -E_y, \quad \dfrac{\partial V}{\partial z} = -E_z$

$U = \tfrac{1}{2} Q_1 V_1 + \tfrac{1}{2} Q_2 V_2 + \tfrac{1}{2} Q_3 V_3 + \cdots$

$u = \tfrac{1}{2} \varepsilon_0 E^2$

$C = Q/\Delta V$

$C = \varepsilon_0 A/d$

$E = E_{\text{free}}/\kappa$

$\oint \kappa E_n \, dS = \dfrac{Q_{\text{free}}}{\varepsilon_0}$

$u = \tfrac{1}{2} \kappa \varepsilon_0 E^2$

$I = \Delta V/R$

$R = \rho l/A$

$P = I\mathscr{E}$

$P = I\, \Delta V$

$\mathbf{F} = \dfrac{\mu_0}{4\pi} \dfrac{qq'}{r^2} \mathbf{v} \times (\mathbf{v}' \times \hat{\mathbf{r}})$

$\mathbf{B} = \dfrac{\mu_0}{4\pi} \dfrac{q'}{r^2} (\mathbf{v}' \times \hat{\mathbf{r}})$

$\mathbf{F} = q\mathbf{v} \times \mathbf{B}$

$d\mathbf{B} = \dfrac{\mu_0}{4\pi} I \dfrac{d\mathbf{l} \times \hat{\mathbf{r}}}{r^2}$

$\oint \mathbf{B} \cdot d\mathbf{l} = \mu_0 I$

$B = \mu_0 I_0 n$

$r = \dfrac{p}{qB}$

$d\mathbf{F} = I \, d\mathbf{l} \times \mathbf{B}$

$\mu = I \cdot [\text{area of loop}]$

$\boldsymbol{\tau} = \boldsymbol{\mu} \times \mathbf{B}$

$U = -\boldsymbol{\mu} \cdot \mathbf{B}$

$\mathscr{E} = vBl$

$\mathscr{E} = -\dfrac{d\Phi_B}{dt}$

$\Phi_B = \displaystyle\int \mathbf{B} \cdot d\mathbf{S}$

$\Phi_B = LI$

$\mathscr{E} = -L \dfrac{dI}{dt}$

$U = \tfrac{1}{2} LI^2$

$u = \dfrac{1}{2\mu_0} B^2$

$\mu = \dfrac{e}{2m_e} L$

$B = \kappa_m B_{\text{free}}$

$\omega_0 = 1/\sqrt{LC}$

$Z = \sqrt{R^2 + \left(\omega L - \dfrac{1}{\omega C}\right)^2}$

$Z = 1 \Big/ \sqrt{\dfrac{1}{R^2} + \left(\dfrac{1}{\omega L} - \omega C\right)^2}$

$\mathscr{E}_2 = \mathscr{E}_1 \dfrac{N_2}{N_1}$

$\oint \mathbf{B} \cdot d\mathbf{l} = \mu_0 I + \mu_0 \varepsilon_0 \dfrac{d\Phi}{dt}$

$E_\theta = \dfrac{1}{4\pi\varepsilon_0} \dfrac{qa \sin\theta}{c^2 r}$

$B = E_\theta/c$

$\mathbf{S} = \dfrac{1}{\mu_0} \mathbf{E} \times \mathbf{B}$

$P_x = U/c$

$v = \sqrt{\dfrac{1 - v/c}{1 + v/c}}\, v_0$

$v = c/n$

$\sin\theta = n \sin\theta'$

$f = \pm \tfrac{1}{2} R$

$\dfrac{1}{s} + \dfrac{1}{s'} = \dfrac{1}{f}$

Interference minima:
$d \sin\theta = \tfrac{1}{2}\lambda, \tfrac{3}{2}\lambda, \tfrac{5}{2}\lambda, \ldots$

Interference maxima:
$d \sin\theta = 0, \lambda, 2\lambda, \ldots$

Diffraction minima:
$a \sin\theta = \lambda, 2\lambda, 3\lambda, \ldots$

$a \sin\theta = 1.22\lambda$

$x' = \dfrac{x - V_O t}{\sqrt{1 - V_O^2}}; \quad t' = \dfrac{t - V_O x}{\sqrt{1 - V_O^2}}$

$\Delta t = \dfrac{\Delta t'}{\sqrt{1 - V_O^2}}$

$\Delta x = \sqrt{1 - V_O^2} \, \Delta x'$

$v_x' = \dfrac{v_x - V_O}{1 - v_x V_O}$

$\mathbf{p} = \dfrac{m\mathbf{v}}{\sqrt{1 - v^2/c^2}}; \quad E = \dfrac{mc^2}{\sqrt{1 - v^2/c^2}}$

$E = h\nu$

$p = h\nu/c$

$\Delta y\, \Delta p_y \geq \hbar/2$

$L = n\hbar$

$E_n = -\dfrac{m_e e^4}{2(4\pi\varepsilon_0)^2 \hbar^2} \dfrac{1}{n^2} = -\dfrac{13.6 \text{ eV}}{n^2}$

$\lambda = h/p$

$\mu_{\text{spin}} = \dfrac{e\hbar}{2m_e}$

$E = \dfrac{n^2 \hbar^2}{2I}$

$R = (1.2 \times 10^{-15} \text{ m}) \times A^{1/3}$

$n = n_0\, e^{-t/\tau}; \quad \tau = t_{1/2}/0.693$

$e = 1.60 \times 10^{-19}$ C
$\varepsilon_0 = 8.85 \times 10^{-12}$ F/m

$\mu_0 = 1.26 \times 10^{-6}$ H/m
$c = 1/\sqrt{\mu_0 \varepsilon_0} = 3.00 \times 10^8$ m/s

$m_e = 9.11 \times 10^{-31}$ kg
$m_p = 1.67 \times 10^{-27}$ kg
$h = 2\pi\hbar = 6.63 \times 10^{-34}$ J·s

APPENDIX 11: ANSWERS TO PROBLEMS

Chapter 1

2. 6×10^{-5} m
4. 8×10^{-7}, 4×10^{-9}, 3×10^{-13}, and 8×10^{-14} in.
6. 6.9×10^{8} m
8. 7.5×10^{5} m
10. 1.00×10^{7} m; 9.01×10^{6} m
12. 6.3×10^{6} m
14. 1.4×10^{17} s
16. 0.25 minute of arc, 0.46 km
18. 0.134%, 99.866%
20. 0.021%, 99.979%
22. (a) 8.4×10^{24}; (b) 4.3×10^{46}; (c) 1.6×10^{3}
24. 9.22×10^{56}
26. 6.7×10^{27}
28. 8.3 light-minutes; 1.3 light-seconds
30. 8.9×10^{3} kg/m^{3}, 5.6×10^{2} lb/ft^{3}, 0.32 lb/in.^{3}
32. 73×10^{-3} m^{3}
34. 5.0×10^{-3} m^{3}/s; 5.0 kg/s
36. 2.3×10^{8} tons/cm^{3}
38. 6.0×10^{7} tons/cm^{3}
40. 5.4×10^{3}, 5.2×10^{3}, 5.5×10^{3}, 3.9×10^{3}, 1.2×10^{3}, 0.63×10^{3}, 1.3×10^{3}, 0.17×10^{3}, and 1.1×10^{3} kg/m^{3}, respectively, for Mercury, Venus, Earth, Mars, Jupiter, Saturn, Uranus, Neptune, and Pluto
42. 171 km
44. 8.9 m; 9300 tons

Chapter 2

2. 23 mi/h
4. 1.9×10^{10} years
6. 13 m/s, 0.77 s
8. (a) 13.8 s, 388 m; (b) 72 m
10. 6.27 m/s; 0 m/s
12. 32.4 m/s
14. 3.0×10^{3} m/s^{2}
16. (a) 4.25 m, 3.0 m/s, -6.0 m/s^{2}; (b) 2.0 m, -6.0 m/s, -6.0 m/s^{2}; (c) -1.5 m/s, -6.0 m/s^{2}
18. (b) 1720 m; (c) approximately 1720 m
20. (b) 1.6 s and also at times earlier or later than this by a multiple of 3.14s, ± 2.0 m/s, 0 m/s^{2}; (c) 0 s and also at times earlier or later than this by a multiple of 3.14s, 0 m/s, ± 2.0 m/s^{2}
22. 2.4 m/s^{2}
24. 4.0 years; 6.2×10^{8} m/s
26. 350 m/s^{2}; yes
28. 7.1 m/s^{2}; 3.7 s
30. (a) 65.2 ft/s^{2}; (b) at constant acceleration the distance would have been 345 yd; (c) 319 mi/h
32. (a) 7.3 m/s; (b) 0.53 m/s^{2}, 15 m/s
34. 76 km/h
36. 0.48 m/s^{2}; 14 s
38. 8.8 s; 86 m/s
40. 6.8×10^{3} m
42. 849 m/s
44. 33 m/s; 2.2×10^{3} m/s^{2}
46. 1.6×10^{4} m/s^{2}
48. 1.88×10^{3} m/s; 255 s
50. 802 m/s; 1.89 s
52. (a) $n\sqrt{2h/g}$; (b) $\tfrac{3}{4} h$; (c) $\tfrac{2}{3} h$

Chapter 3

2. 609 m at 38° west of north
4. 11.2 km at 2.3° north of east
6. 3.4 m at 22° west of north
10. (b) 2.12×10^{11} m
14. $4\hat{x} + 5\hat{y} + 3\hat{z}$; 7.1 m; 9.4 m
16. 5690 km
18. 9.2 km; 7.7 km
20. (b) $C_x = 2$ cm, $C_y = 5$ cm
22. $A_x = 4.2$, $A_y = 0.5$, $A_z = \pm 4.2$
24. 3.74
26. 5.9; 32°, $-60°$, 80°
28. $17\hat{x} + 3\hat{y} - 16\hat{z}$
30. -2.24×10^{6} m^{2}
32. -304 m^{2}
34. 2.4; 1.8
36. $0.600\hat{x} + 0.097\hat{y} + 0.794\hat{z}$
38. 54.7°

40. 47.3 m² down
42. $-12\hat{x} - 14\hat{y} - 9\hat{z}$
44. $0.45\hat{x} - 0.59\hat{y} - 0.67\hat{z}$
48. (a) 2.80 km, 4.15 km; (b) 3.59 km, 3.48 km
52. rotate coordinate system by an angle $\theta = -26.6°$
54. (a) $x'' = x'$, $y'' = y' \cos \phi + z' \sin \phi$, $z'' = -y' \sin \phi + z' \cos \phi$; (b) $x'' = x \cos \theta + y \sin \theta$, $y'' = -x \sin \theta \cos \phi + y \cos \theta \cos \phi + z \sin \phi$, $z'' = x \sin \theta \sin \phi - y \cos \theta \sin \phi + z \cos \phi$

Chapter 4

2. 29.9 km/s; 19.0 km/s
4. (a) $8\hat{x} + 10\hat{y}$, 12.8 m/s; (b) $4\hat{x} + 6\hat{y}$, 7.2 m/s²
6. (a) $2\hat{x} + (5 + 8t)\hat{y} + (-2 - 6t)\hat{z}$; (b) $8\hat{y} - 6\hat{z}$; 10 m/s² at 37° down from y axis
8. 0.97 m
10. 135 km/h
12. 65.8 m/s; 93.4 m/s
14. (a) 7.25°; (b) 13 m
16. (a) 2.5×10^4 m; (b) 5.0×10^4 m; (c) no
18. 0.11 m
20. (a) 22 m/s; (b) 3.5 rev/s
22. 65 m; 14 liters
24. (a) 71 m/s; (b) 4 m; (c) 8 m
26. 43°; 35 ft/s
28. yes, by 0.5 s
30. (a) 270 m; (b) air resistance
32. (a) 6.0 m; (b) 1.5 m; (c) 1.1 s
34. 7.8 m/s; 63°
36. no; 12.2 m
38. $R + gR^2/u^2 + (1 - g^2R^2/u^4)u^2/2g$; 4.4 m
40. 9.95°; 205 m, or 42 minutes of arc
42. 73 m/s; 5.4×10^4 m/s²
44. 8.9 m/s²
46. 2.2×10^3 m/s²
48. 9.45 m/s², 29.7 m/s²; -0.36 m/s², 39.5 m/s²
50. 13 m/s at 40° from vertical
52. 27 m/s at 68° from vertical
54. 329 m/s
56. 27 km/h at 43° east of north; 33°
58. (a) 50 km; (b) 33 km, 67 km
60. 15 km/h at 15° east of north
62. $v = 4v_0^2 t/(4v_0^2 t^2 + h^2)^{1/2}$
64. 528 km/h

Chapter 5

2. 17 m/s²
4. 6.6×10^3 N, 12 times weight
6. 1.2×10^4 N
8. 1.8×10^3 m/s²; 1.3×10^5 N
10. 1.46×10^6 N
12. 2.9×10^3 N, 1.8×10^3 N
14. $-2773 + 296t$ measured in N
16. no, since with 1150 children on each side the tension would be 150,000 N
18. 4.7×10^{20} N at 25° with the Sun–Moon line
20. 770 N toward dock
22. 1.2×10^3 m/s²; 23 m/s
24. 6.9×10^5 N; 2.8×10^3 N
26. 180 N
28. 8.2×10^3 N
30. 165 N at 20° downward from horizontal
32. 150 N
34. 1.6×10^4 kg·m/s; 8.1 km/h
36. 5.7 kg·m/s; 8.7 kg·m/s
38. 7.4×10^2 N
40. $p_x = 8.5 \times 10^{-23}$ kg·m/s, $p_y = 3.4 \times 10^{-23}$ kg·m/s

Chapter 6

2. (a) 0.0010; (b) no, since price is based on mass
4. 29 N, 128 N
6. 9.2×10^2 N; 1.4×10^2 N
8. 4.1 m/s
10. (b) [weight] = 7.4×10^2 N, [normal force] = 6.0×10^2 N, [resultant] = 4.2×10^2 N; (c) 5.6 m/s²
12. 64 m; 5.1 s
14. 9.8 m/s²
16. 7.0°
18. $a_1 = g(m_1 - 2m_2)/(m_1 - 4m_2)$; $-a_1$; $-2a_1$
20. $a_1 = g(4m_2m_3 - m_1m_2 - m_1m_3)/(4m_2m_3 + m_1m_2 + m_1m_3)$, $a_2 = g(3m_1m_3 - 4m_2m_3 - m_1m_2)/(4m_2m_3 + m_1m_2 + m_1m_3)$, $a_3 = g(3m_1m_2 - 4m_2m_3 - m_1m_3)/(4m_2m_3 + m_1m_2 + m_1m_3)$, $T_1 = 2T_2 = 8m_1m_2m_3g/(4m_2m_3 + m_1m_2 + m_1m_3)$
22. 4.3×10^3 N
24. 3.9 m/s²
26. 40 m
28. 53 m
30. 0.17 m/s²
32. 1.9×10^2 N; 7.4×10^3 N
34. 1.2×10^2 m; 6.5 s
36. 27°
38. 4.4 m/s², 1.3 N
40. $T = \mu_k mg/\sqrt{1 + \mu_k^2}$
42. $a_1 = F/m_1 - \mu_1 g$, $a_2 = \mu_1 m_1 g/m_2 - \mu_2 g(m_1 + m_2)/m_2$
44. 6.8×10^2 N
46. 74 N/m
48. 0.15 m
50. 180 N/m, 360 N/m
54. 3.4 s
56. 4.4×10^2 N
58. 7.6×10^2 N; 8.1×10^2 N
60. 76 km/h
62. outer; inner
64. 3.0 m/s
66. 68°
68. $v = \sqrt{gl \tan \theta \sin \theta}$
70. 28°
72. $T = mv^2/(2\pi r)$
74. $\phi = \tan^{-1}[\tan \theta/(1 - v_E^2/gR_E)] - \theta$ or approximately $(\sin \theta \cos \theta)v_E^2/gR_E$, where R_E and v_E are the equatorial radius and speed of the Earth and θ is the latitude angle; 0.099°
76. $v_0^2 = (gR^2 \tan \alpha)/(R + L \sin \alpha)$; 0.95 m/s

Chapter 7

2. 4.93×10^3 J
4. 8.8×10^3 J
6. 2.4×10^5 J; 3.6×10^2 J/s
8. 6 J
10. 2.6×10^3 J
12. 24 J
14. 40 J
16. (a) 1.2×10^4 J; (b) 290 N, 1.2×10^4 J
18. (a) 7.1×10^3 N; (b) 2.2×10^5 J, 8.1×10^3 N
20. curved ramp: $W = mgR[(1 - \sqrt{2}/2) + \mu_k\sqrt{2}/2]$; straight ramp: $W = mgR(1 - \sqrt{2}/2)[1 + \mu_k \cot(45°/2)]$, which is the same
22. 2.66×10^{33} J
24. 1.3×10^3 J, 5.8×10^3 J; 22 times
26. (a) 4.0×10^5 J; (b) 2.5×10^4 J; (c) 1.2×10^6 J
28. 9.79×10^6 J, 5.72×10^6 J; 4.07×10^6 J
30. $h/2$; $3h/4$
32. $W = -\frac{1}{2}a(x_2^2 - x_1^2) + \frac{1}{4}b(x_2^4 - x_1^4)$; $\Delta K = W$
34. (a) 1.2×10^4 N; (b) 39 m; (c) 4.7×10^5 J; (d) 4.7×10^5 J
36. 1.9×10^5 J
38. 2.2×10^9 J, 6.4×10^9 J
40. 8.2×10^6 m^3
42. 5.1 m; yes, since the jumper does some extra work with his arms
44. 99 m/s; 9.8×10^{10} J; 23 tons
46. (a) 2.5×10^4 N; (b) 1.2×10^7 J; (c) 15 m/s
48. 50 J, 17 J
50. 1.6 m/s; 26°
52. 9.9 m/s; 30 m/s
54. 48° from top

Chapter 8

4. $U = K/(3x^3)$
6. 217 J
8. (a) 5 J; (b) −4 J, no
10. (a) 47 J; (b) 0.89 m
12. 0.26 m
14. 1.9 s; 9.6 m; 2.9×10^5 J
16. $t = v/\mu_k g$; $x = v^2/2\mu_k g$; $mv^2/2$; mv^2
18. $\mathbf{F} = (2K/x^3)\hat{\mathbf{x}}$
20. 0.19 Å, 0.80 Å
22. (a) none and 0.2 m for E_1, 3.1 m and 0.3 m for E_2, 1.3 m and 0.5 m for E_3; (b) absolute maximum at 0.9 m, absolute minimum at turning points, local maximum at 2.2 m, local minimum at 1.7 m; (c) unbound for E_1, bound for E_2 and E_3
24. 11 eV
26. 2.2×10^6, 1.1×10^7, 2.7×10^6, 5.8×10^5, 3.6×10^6, and 9.8×10^6 J; bus; snowmobile
28. 1.65 kcal/kg, 2.81 kcal/kg, 2.82 kcal/kg
30. 1.0×10^6 eV
32. 9.4×10^8 eV
34. 540 kcal
36. 18 kW·h
38. 1.1×10^3 W; 0 W
40. 0.61 hp
42. 746 J
44. 9.1 gal./h
46. 4.2×10^5 W
48. 2.5×10^3 km^2
50. (a) 1.7×10^{10} J; (b) 17 min; (c) 17 min, 4 km
52. 1.2×10^{-4} W
54. 32 km/h
56. 52 hp
58. 37%
60. (a) 3.2×10^4 W; (b) 7.8×10^2 W; (c) 3.1×10^4 W
62. (a) $45.36 (61.14 + 6.52t)(655.9 - 61.14t + 3.26t^2)$ in joules; (b) 9.76×10^6 J, 5.71×10^6 J; (c) 1.35×10^6 W
64. 2.0×10^5 W
66. (a) 1.08×10^4 N (weight), 1.05×10^4 N (lift), 2.43×10^3 N (friction); (b) 4.85×10^3 (push), same weight, lift, friction; (c) 300 hp

Chapter 9

2. 4.7×10^{-35} N
4. 8.88 m/s^2, 3.71 m/s^2, 3.71 m/s^2
6. 4.34×10^{20} N, 1.99×10^{20} N; 6.32×10^{20} N toward Sun; 2.35×10^{20} N toward Sun
8. 4.0×10^{-4} radian/s, or 0.23 rev/h
10. 2×10^8 years, 3×10^5 m/s
12. Lincoln, Nebraska, 22.6° west of New York City
14. $m_1/m_2 = 1.6$
16. 3.0×10^{10} m
18. (a) 7.5×10^3 m/s, 8.32×10^3 m/s; (b) 3.94×10^8 J, 4.85×10^8 J
20. 8.21×10^3 m/s
22. 3.5 days
24. 2.66×10^{33} J, -5.31×10^{33} J; -2.66×10^{33} J
26. (a) 6.0×10^{-14} m/s^2, 9.1×10^{-14} m/s^2; (b) 1.1×10^5 m/s, 1.6×10^5 m/s; (c) 3.4×10^{51} J, 5.1×10^{51} J, 7.7×10^{51} J, yes
28. (a) 1.1×10^4 m/s; (b) 1.2×10^{11} J, 29 tons; (c) 1.2×10^5 m/s^2
30. (a) -1.1×10^{11} J, -2.2×10^{11} J, -1.1×10^{11} J; (b) yes, yes
32. (a) 1.1×10^4 m/s; (b) 2.4×10^3 m/s
34. (a) I and III elliptical, II circular
36. 16 km/s
38. (a) yes; (b) yes
40. (a) 4.21×10^4 m/s; (b) 7.19×10^4 m/s, 1.23×10^4 m/s
42. 7.78×10^7 m
46. (a) 2.9×10^3 m/s; (b) 2.6×10^3 m/s; (c) 0.71 year; (d) 44° ahead of Earth; 256° from initial position
48. (c) 23 km/s
50. 9.0×10^{-3} m/s^2; larger
52. 1.0 rev/min

Chapter 10

2. 1.36 m/s
4. 24 km/h; 89 km/h, 41 km/h
6. 150 bullets
8. 5.3×10^2 kg·m/s; 0.18 m/s; no

10. $\sqrt{2}v$, at 135° with respect to the direction of the other fragments
12. (a) 42 km/h at 56° east of north; (b) 3.6×10^5 J
14. 3.5×10^2 N
16. (a) 0.10 kg/s; (b) 2.3 kg·m/s, 2.3 N
18. $v_n = \sum_{k=1}^{n} mu/(M - km)$
20. 0.19 m from center of seesaw
22. 2.7×10^{-4} m
24. 2.28 Å from H
26. (950, 180, 820)
28. on diagonal of cube, at $\sqrt{3}/3 L$ from vertex
30. 0.333 m above bottom stick
32. $L/(64/\pi - 4)$ from center of cube, toward hole
36. 950 m
38. 9.0×10^7 J
40. 2.4×10^3 m/s
42. 950 kg
44. 1.0 m
46. 1.8×10^4 m/s, at 127°
48. (a) 9.2×10^6 J, 0; (b) 9.2×10^6 J, 1.7×10^6 J, extra energy comes from explosive chemical reactions
50. (a) 3.4×10^5 J, 3.6×10^5 J; (b) 3.4×10^5 J, 0
52. 11 tons/s
54. 4.4×10^3 m/s

Chapter 11

2. -7.8×10^8 N; 1.1 m/s^2
4. 1.0×10^3 m/s^2; 6.5×10^4 N
6. 9.3 kg·m/s, 24° up
8. (a) -0.27 m/s, 0.53 m/s; (b) 0.019 J, 0.0021 J, 0; 0.017 J
10. 14 m/s
12. 2.6 km/h, 12.6 km/h
14. (a) 18 m/s; (b) 30 m/s
16. 1.1 m/s
18. 39 m/s
20. 0.57 J
22. 4.0×10^{-13} J
24. (a) $h/9$, $4h/9$; (b) h, 0
26. $v/3$ to left, $2v/9$ to right, $8v/9$ to right
28. 13.5 m/s
30. (a) 3.0 m/s; (b) 2.1×10^2 m/s^2
32. 21 m/s
34. $v/5$ to left, $2\sqrt{3}v/5$ up 30° to right, and $2\sqrt{3}v/5$ down 30° to right
36. 2.6×10^7 m/s
38. 45° each, 4.0×10^{-13} J each
40. 1.9×10^5 eV, 70°
42. 1.25×10^{-13} J
44. 4.3×10^{-11} J

Chapter 12

2. 7.3×10^{-5} radian/s; 460 m/s, 350 m/s
4. 160 km/h horizontal; 0; 113 km/h down 45°
6. 5.50×10^{-3} m/s; 0.509 m/s; 0.221 m/s
8. (b) 5.4×10^{-3} m
10. 12 radian/s^2
12. 9.6×10^{-22} radian/s^2
14. 1.21×10^{-10} m
16. 6.50×10^{-46} kg·m^2
18. 0.44 kg·m^2
20. -1.1×10^{22} kg·m^2
22. $0.38 M_E R_E^2$
24. $\frac{1}{12} ML^2 \sin^2 \theta$
28. $(83/256) MR^2$
30. $\frac{1}{2} MR^2 - (r^2 M/R^2)(\frac{1}{2} r^2 + d^2)$
32. $\frac{1}{2} MR^2 - (2r^2 M/R^2)(r^2 + \frac{1}{2} R^2)$
34. $0.338 M_E R_E^2$
36. (a) 2.2×10^2 kg·m^2; (b) 4.4×10^3 J
38. 3.44×10^5 J, 3.91%
40. 2.5 km
42. 2.74×10^3 J; 13.1 J; 4.74×10^{-3}
44. $\frac{1}{2} MR^2$
46. $\frac{1}{6} Ml^2$
48. $\frac{1}{2} M (R^4 - r^4 - 8hr^3/3\pi - r^2 h^2/2)/(R^2 - r^2)$
52. 5.3×10^{-13} kg·m^2/s
54. 2.1×10^{12} kg·m^2/s in north direction
56. (a) 1.1×10^{14} kg·m^2/s; (b) 4.3×10^{36} kg·m^2/s
58. (a) 0 kg·m^2/s; (b) 2.7×10^{40} kg·m^2/s, 5.4×10^{40} kg·m^2/s, 2.7×10^{40} kg·m^2/s, no
60. $2\hbar/[(1 - 2\sqrt{2}/3) m_e a_0] = 7.6 \times 10^7$ m/s; $2\hbar/[(1 + 2\sqrt{2}/3) m_e a_0] = 2.2 \times 10^6$ m/s
62. 0.052 kg·m^2/s^2
64. 5.6×10^{41} J·s; 3.1×10^{43} J·s; 1.8%
66. 1.8×10^{22} kg·m^2/s^2

Chapter 13

2. 400 N
4. 1.2×10^4 N; 7.2×10^3 N·m; depress
6. 2.7×10^4 N·m
8. 9.7 m/s^2
10. 8.2×10^2 N
12. (a) 110 J·s; (b) 34 kg·m^2/s^2; (c) 34 N·m
14. (a) 76°; (b) 290 N, 250 N
16. 1.6×10^8 J·s east or west; 1.1×10^4 J·s/s; 1.1×10^4 N·m; 9.5×10^3 N on each side
18. (a) $\omega_1 R_1/R_2$; (b) $(T - T')R_1$, $(T - T')R_2$; (c) $(T - T')R_1 \omega_1$, $(T - T')R_2 \omega_2$, yes
20. 1.1 kcal/min
22. 27 N·m
24. 4.6 radian/s
26. (a) 1140 N·m; (b) 2.1 m/s^2
28. (a) 2.4×10^{19} J·s; (b) 3.0×10^{-19} radian/s
30. 7.2°
32. (a) 5.0 m/s; (b) 8.7×10^{-3} radian/s; (c) 19 times
34. [height of end] = $m^2 v^2/[2g(M/3 + m)(M/2 + m)]$
36. $v_{CM} = mv/(m + M)$, $\omega = v_{CM}/[\frac{1}{3} l + \frac{1}{4} l(1 + m/M)^2 + \frac{1}{4} lm/M]$
40. 9.6 m/s^2
42. 100 N in forward direction; 300 N in forward direction
44. $2a/7$
46. $a = g(\sin \theta - \mu_k \cos \theta)$; $\alpha = (2/R) g \mu_k \cos \theta$
48. $a = (g \sin \theta)(m + 4M)/(m + 6M)$
50. 0.37 radian/s
52. 0.85 radian/s

Chapter 14

2. 1.2×10^4 N; 4.6×10^4 N
4. 99 N at 27° right of vertical at the top, and at 27° left of vertical at the bottom
6. 2.3 N
8. 9.0×10^3 N, 8.9×10^3 N; the second
10. $0.408\, Mg$
12. $Mg\sqrt{1 - R^2/(R+l)^2}$, $MgR/(R+l)$
14. $0.11\, Mg$
16. (a) $U = 5.3 \times 10^2$ J $\times \{\sin(\theta + \tan^{-1} 2) +$ [constant]$\}$; (b) 27°; (c) 56 J
18. 1.04 book lengths; infinite
20. $\sqrt{Rgl/2h} = 18$ m/s
22. 7.4 m/s²
24. (a) 8.8 m/s²; (b) 5.4 m/s²; (c) 3.7 m/s²
26. $T_{AB} = T_{DE} = 0.375$, $T_{AF} = T_{EH} = -0.530$, $T_{BF} = T_{CG} = T_{DH} = 0.250$, $T_{FG} = T_{GH} = -0.375$ in units of Mg
28. $0.577\, Mg$ tension in cables; $0.289\, Mg$ compression in top beam, $\tfrac{3}{8} Mg$ and $\tfrac{1}{8} Mg$ tension at top and at bottom of side beams
30. $2\, mg$, $mg/\sqrt{3}$, $mg/\sqrt{3}$
32. $\tan^{-1}(1/2\mu_s)$
36. 1.7×10^3 N; 1.5×10^3 N
38. 1.9×10^3 N
40. 7.4×10^2 N
42. 8.9
44. 393
46. 34.6
48. 2.9×10^4 N/m
50. 4.0×10^{-6} m
52. 5.6%
54. 6.8, 11.9, 8.0, and 12.8×10^{10} N/m²
56. 8.1×10^5 N/m²
58. 4.3×10^3 m
60. 6.8×10^{-3}
62. 130° C
64. 3.8×10^5 N/m²
66. 2.8×10^{-6} m

Chapter 15

2. 1.7×10^4 m/s², 27 m/s; 2.0×10^4 N
4. (a) $A = 3.0$ m, $\nu = 0.32$ Hz, $\omega = 2.0$ radian/s, $T = 3.1$ s; (b) 0.26 s, 1.05 s; (c) at times larger and smaller by a multiple of one-half period
6. (a) $0.20 \cos 6\pi t + (4.0/6\pi) \sin 6\pi t$; (b) 0.043 s, -1.0×10^2 m/s²
8. 1.9×10^4 N
10. 1.13×10^{14} Hz
12. (a) 4.5 Hz; (b) 8.5 m/s, strong vibrations
14. (a) $x_1 = -x_2 = (-v_1/\omega) \cos \omega t$ with $\omega = \sqrt{3k/m}$; (b) $x_1 = x_2 = (-v_1/\omega) \cos \omega t$ with $\omega = \sqrt{k/m}$
16. (a) 6.6 J; (b) $t = 0.30$ s, $t = 0$; (c) $t = 0.15$ s
18. 18 eV; no
22. 0.19 Hz
24. (a) 0.73 min; (b) 1.0 mm
26. $4.189 \sqrt{l/g}$
28. (a) 1.3×10^{-3} J; (b) 0.14 m/s
30. (a) 9.4 J; (b) 1.6×10^{-5} W; (c) 6.0×10^{-3} N·m/radian
34. 1.988 s
36. 1.5 s
38. (b) $l = R/\sqrt{2}$
40. $\omega = \sqrt{(\kappa - mgR)/mR^2}$; $m = \kappa/gR$
42. 3.6 radians/s; 3.3 m/s
44. $2\pi\sqrt{4l/5g}$
46. (a) $T = mg[1 - \tfrac{1}{4}A^2 + \tfrac{3}{2}A^2 \sin^2(\sqrt{g/l}\, t)]$; (b) $t = \tfrac{1}{2}\pi/\sqrt{g/l}$ gives $T = mg(1 + A^2)$
48. $\gamma = 2.5 \times 10^{-3}$/s
50. 18

Chapter 16

2. 0.114 Hz, 0.716 radian/s, 0.0524/m, 13.7 m/s
4. (a) 6.0×10^{-3} m, 0.31 m, 20/m, 0.64 Hz, 4.0 radian/s, negative x, 0.20 m/s; (b) -0.26 s
6. 7.8 m/s
8. 5.5×10^{-7} m; 4.2×10^{-7} m
12. 0.69 s
14. 1.6 s
16. 0.026 s
18. μv^2
22. 1:2.4
24. 2.8 cm
26. $A = 5.0$; $\delta = -\tan^{-1} \tfrac{4}{3}$
28. 0.007 Hz
30. 1.00/day, 0.92/day; 1.00 day, 1.09 day; first is due to Sun, second due to Moon
32. 392 Hz, 588 Hz, 784 Hz, 980 Hz
34. 1.6 Hz, 3.2 Hz
36. 71 N
38. 1.25×10^{-7} m; 2.5×10^{-7} m
40. 3.7 m/s; 6.8×10^3 m/s²
42. (a) 9.6×10^3 N; (b) 12 Hz
44. 0.018 J/m $\times (\sin^2 \omega t \cos^2 kx + \cos^2 \omega t \sin^2 kx)$, where $\omega = 3.7 \times 10^3$ radian/s and $k = 28$/m; location of maxima depends on time: maxima are at 0.057 m, 0.17 m, 0.28 m when $\sin^2 \omega t < \cos^2 \omega t$, and maxima are at 0 m, 0.11 m, 0.23 m, 0.34 m when $\sin^2 \omega t > \cos^2 \omega t$; 0.018 J/m $\times \sin^2 \omega t$ or 0.018 J/m $\times \cos^2 \omega t$

Chapter 17

2. 55 Hz, 4187 Hz
4. 1.9, 1.8, 1.7, 1.6, 1.5, 1.5, 1.3, 1.3, 1.2, 1.1, 1.1, and 1.0 cm
6. 123 dB
8. 3.1 m
10. 130 times; 21 dB
12. 9.1 s; use visual signal
14. 150 m
16. (a) 1.2×10^{-4} s; (b) on the glass
18. 9 cm, 4000 Hz
20. 337 m/s
24. (a) 0.63 m; (b) 3.5, 3.3, 3.2, 3.0, 2.8, 2.7, 2.5, 2.4, 2.2, 2.1, 2.0, and 1.9 cm

28. 619 Hz
30. 22 m/s, 0.22 Hz
32. 594 Hz, 596 Hz
34. 476 Hz
36. 481 Hz
38. (a) 33.4°; (b) 30.4 s
40. 4.0 m/s, 0.40 Hz; 12 m/s, 0.12 Hz
42. 2.1 m/s^2; 4.9 m/s
44. 1.7×10^2 m
46. (a) 2.3/h; (b) 320 km, 740 km/h
48. (a) 31 m/s; (b) 0.11/h, 9.0 h; (c) probably
50. (b) 1.6 m, 3.2 m
52. 4.0×10^3 km
54. 7.2
56. (a) 6.3 m/s; (b) 1.0 m

Chapter 18

2. (a) 11 m; (b) 8.1 cm, 11.7 cm
4. 6.1×10^3 N; layer of air under the paper exerts an opposite pressure force
6. 130 cm^2
8. 2.34×10^5 N/m^2
10. 5.1×10^4 N
12. 52 atm
14. (a) 360 N; (b) 330 N
16. 0.86 m
18. 10.3 m
20. (b) 5.0×10^{11} N/m^2
22. (a) $\frac{1}{3}\rho g h \pi (2R_2^2 - R_1^2 - R_1R_2)$ up on sides, $\rho g h \pi R_2^2$ down on bottom, $\frac{1}{3}\rho g h \pi (R_2^2 + R_1^2 + R_1R_2)$ down; (b) $\frac{1}{3}\rho g h \pi (R_2^2 + R_1R_2 - 2R_1^2)$ down on sides, $\rho g h \pi R_1^2$ down on bottom, $\frac{1}{3}\rho g h \pi (R_2^2 + R_1^2 + R_1R_2)$ down
24. (a) 4.7×10^7 m^3; (b) 4.8×10^{10} kg
26. yes
28. (a) 9.5 m; (b) 28.4 m
30. 0.94 kg/m^3; 93 N/m^2
32. 2.2 m
34. 0.115%; 0.011%; 0.002%
36. $\rho = M/[\frac{4}{3}\pi R^3 + R'^2 (l - h)]$
38. 19 cm
40. 28 N/m^2
42. (a) 15 m/s; (b) 1.1×10^5 N/m^2 (overpressure)
44. (a) $v = \sqrt{2g(h_2 - h_1)}$; (b) $p_{atm} - \rho g h_2$; (c) $p_{atm}/\rho g$
46. 2.3 hp
50. $z = \frac{1}{2}h$; h
52. $(A/A')\sqrt{2l/g}$

Chapter 19

2. 58.0°C = 331.2 K; −88.3°C = 184.9 K
4. 2.69×10^{22} molecules
6. 3.4 atm
8. 0.060 kg/m^3
10. 1.4×10^{-9} N/m^2
12. 11.6 kg/m^3
14. 3.6×10^6 N/m^2
16. 1.3 cm
18. 4.5×10^7 N/m^2
20. 96.3 g
22. (a) 3.1×10^3 kg; (b) 2.8×10^3 m^3
24. (a) 1.2 m; (b) 2.5 kg at 2.5 atm
26. 29.0 g
30. 5.7×10^{-21} J
32. 428 m/s, 349 m/s; 4.9×10^{-21} J
34. 0.43%
36. 1.9×10^5 J; $\frac{3}{5}$; $\frac{2}{5}$
38. 284 K
40. 2.4×10^{-10} m
42. 4.4 cm
44. 17 m
46. 9.8 h

Chapter 20

2. 97 kW
4. 25°C
6. 97 W
8. 4.3 cal/s
10. 14.5 %
12. 3.5 h
14. 1.4 liters/min
16. 0.34 m
18. ±0.08°C
20. (a) 1.1 cm; (b) 47 cm
26. (a) 709 kg/m^3; (b) 41.1¢/kg, 42.3¢/kg
28. 9.4 cm
30. 100.32°C
32. 5.9×10^3 cal/s; 110 times as large
34. (a) 10 m^2 · °C · s/cal; (b) 13.6 ft^2 · °F · h/Btu
36. 0.085 kg/h
38. 1.8 cm
40. 0.45 cm/h
44. (a) 1.0×10^{11} kcal; (b) 5 A-bombs
46. 41 kcal/kg; the same
48. 22°C
50. 0.65 kg/h
52. 2.5×10^3 J; 50 m/s
54. 4.94 cal/K · mole, 6.93 cal/K · mole
56. 50 kcal; 119 kcal
58. 1.5×10^4 kcal
60. (a) 2.62 min; (b) 2.64 min, no
62. 30.8 atm; 507°C
64. −225°C
66. 1.0°C
68. 1.1×10^4 J

Chapter 21

2. $\Delta Q = 1.96$ J, $\Delta W = 1.96$ J, $\Delta E = 0$
4. 518 kcal/kg
6. 0.35, or 35%
8. 0.14, or 14%
10. 0.19 kg/s
14. (a) 2.0×10^4 J; (b) 3.9×10^4 J, 1.9×10^4 J
16. $1 - 3 \times 10^{-9}$, or 99.9999997%
18. 3.9×10^2 W
20. (a) 3.1×10^2 MW; (b) 1.0×10^3 MW
22. 7.1 kW
24. 1.7×10^2 J
26. 8.2×10^3 cal/K · s

28. 3.7×10^2 cal/K·day
30. -2.9×10^2 cal/K
32. 3.2 cal/K·s
34. (a) 6.9×10^3 cal/K; (b) 4.6×10^2 cal/K
36. (b) 0.19 kcal/K
38. 14.5 cal/K

Chapter 22

2. 2.89×10^{-9} N
4. 17 N; 2.6×10^{27} m/s^2
6. 6.3×10^{18} electrons
8. 0.045N; 4.9×10^{28} m/s^2
10. 9.4×10^{18} electrons/s
12. 8.4×10^{22}
14. 2.4×10^{28} of each
16. $(-6.9 \times 10^{-4}\,\hat{\mathbf{x}} + 1.7 \times 10^{-3}\,\hat{\mathbf{y}})$ N; $(6.9 \times 10^{-4}\,\hat{\mathbf{x}} - 1.7 \times 10^{-3}\,\hat{\mathbf{y}})$ N
18. 6.8×10^{32} electrons on Earth and 1.9×10^{32} electrons on Moon
20. 10^{-9}; 10^{-5}
22. 2.6×10^{-39} C; 1.3×10^{-6}:1; attractive
24. (c) 0.53×10^{-10} m
26. 4 electrons

Chapter 23

2. 6.2 m/s^2 upward
4. 1.3×10^{-13} N; 1.4×10^{17} m/s^2
6. 9.0×10^3 N/C
8. 6.2×10^{-7} m
10. 5.1×10^{11} N/C
12. 5.1×10^{12} N/C in positive x direction
14. (a) at $z = R/\sqrt{2}$
16. 2.9×10^4 N/C at 71° below horizontal
18. 2.3×10^4, 1.7×10^4, 0.88×10^4, 0.35×10^4, 0.11×10^4, and 0.017×10^4 N/C
20. $\sqrt{15}\lambda/(8\pi\varepsilon_0 d)$
22. $E_x = -\dfrac{1}{4\pi\varepsilon_0}\dfrac{\lambda}{\sqrt{x^2+y^2}}$, $E_y = \dfrac{1}{4\pi\varepsilon_0}\dfrac{\lambda}{y}\left(1+\dfrac{x}{\sqrt{x^2+y^2}}\right)$
24. 1.1×10^5 N/C up, 1.1×10^5 N/C down, 1.1×10^5 N/C down, 3.4×10^5 N/C down
26. 2.4×10^5 N/C
28. $E = \dfrac{\sigma}{2\varepsilon_0}\dfrac{z}{(z^2+R^2)^{1/2}}$
30. (a) $E = \dfrac{Q}{4\pi\varepsilon_0}\dfrac{1}{x(x+l)}$ in positive x direction
    (b) $E = \dfrac{Q}{2\pi\varepsilon_0}\dfrac{1}{y\sqrt{l^2+4y^2}}$ in positive y direction
32. $E = 2Q/(l^2\pi\varepsilon_0)$ toward lower right corner
34. $E = \dfrac{Q}{2\pi\varepsilon_0}\dfrac{1}{R^2}\left(1-\dfrac{z}{\sqrt{x^2+R^2}}\right)$ along axis
36. $E_x = \dfrac{\lambda}{4\pi\varepsilon_0}\left[\dfrac{1}{x}\left(1+\dfrac{y}{\sqrt{x^2+y^2}}\right)-\dfrac{1}{\sqrt{x^2+y^2}}\right]$
    $E_y = \dfrac{\lambda}{4\pi\varepsilon_0}\left[\dfrac{1}{y}\left(1+\dfrac{x}{\sqrt{x^2+y^2}}\right)-\dfrac{1}{\sqrt{x^2+y^2}}\right]$
38. $\dfrac{Q^2}{4\pi\varepsilon_0 L^2}\ln\left[\dfrac{(x+L)^2}{x(x+2L)}\right]$
40. (a) 2.0×10^{-17} C·m; (b) 1.2×10^{-11} N·m
42. (a) 8.5×10^{-30} C·m; (b) 0
44. 4.8×10^{-24} N·m

Chapter 24

2. 1.1×10^3
4. 1.1×10^5 N·m^2/C; 0 N·m^2/C; 1.1×10^5 N·m^2/C
6. 5.9×10^4 N/C
8. 4.4×10^{-11} C/m^3
10. 8.3×10^{-9} C
12. 4.8×10^{20} N/C; 3.5×10^3 N; 1.8×10^{28} m/s^2
14. 7×10^{-9} C/m
16. $\mathbf{E} = (\rho x/\varepsilon_0)\hat{\mathbf{x}}$ for $x < d/2$; $\mathbf{E} = \pm(\rho d/2\varepsilon_0)\hat{\mathbf{x}}$ for $\pm x > d/2$
18. $E = 0$ for $r < a$, $E = \dfrac{Q}{4\pi\varepsilon_0}\dfrac{r^3-a^3}{b^3-a^3}\dfrac{1}{r^2}$ for $a < r < b$, $E = \dfrac{Q}{4\pi\varepsilon_0}\dfrac{1}{r^2}$ for $r > b$
20. $E = \dfrac{Q}{4\pi\varepsilon_0}\left[\dfrac{1}{r^2}-\dfrac{1}{8(r-R/2)^2}\right]$ in radial direction
22. 0.5 Å
24. 1.2×10^7 m/s, 2.2×10^{-13} J, 1.1×10^{-34} J·s, 6.5×10^{20} Hz
26. 1.2×10^{21} N/C
30. (a) 1.9×10^{-25} kg·m/s; (b) 2.1×10^5 m/s; (c) 2.1×10^5 m/s
32. 10^{-6} C/$(4\pi\varepsilon_0 r^2)$; 4×10^{-6} C/$(4\pi\varepsilon_0 r^2)$
34. 8.8×10^{-10} C/m^2

Chapter 25

2. -1.5 eV
4. 4.5×10^3 V
6. 1.3×10^7 V/m
8. 1.9×10^3 V
10. 0.06 V; 0.03 V
12. 1.7×10^7 V; 2.5×10^7 V
14. 4.1×10^7 m/s; 5.0×10^7 m/s
16. 5.8×10^{-12} J
18. 7.5×10^7 V; 1.7×10^7 V
20. (c) -13.6 eV
22. $V = \dfrac{1}{4\pi\varepsilon_0}\dfrac{Q}{l}\ln\left(\dfrac{2+\sqrt{3}}{2-\sqrt{3}}\right)$
24. $V = \dfrac{1}{4\pi\varepsilon_0}\dfrac{Q}{l}\ln\left[\dfrac{x+l}{x}\dfrac{l+\sqrt{x^2+l^2}}{x}\dfrac{l+\sqrt{(x+l)^2+l^2}}{x+l}\right.$
    $\left.\dfrac{l+x+\sqrt{(x+l)^2+l^2}}{x+\sqrt{x^2+l^2}}\right]$
26. 3.1×10^7 m/s
28. $V = \dfrac{Q}{2\pi\varepsilon_0}\dfrac{1}{R^2}(\sqrt{R^2+z^2}-z)$
30. $V = 4k/(\varepsilon_0 r^{1/2})$
32. $\mathbf{E} = -8\hat{\mathbf{x}} - 4\hat{\mathbf{y}}$ in volt/meter
34. $V = \dfrac{1}{4\pi\varepsilon_0}\dfrac{Q}{l}\ln\left(\dfrac{x+l}{x}\right)$; $E = \dfrac{1}{4\pi\varepsilon_0}\dfrac{Q}{x(x+l)}$ in positive x direction
36. (a) $V = \dfrac{2Q}{3\pi\varepsilon_0 R^2}(\sqrt{z^2+R^2}-\sqrt{z^2+R^2/4})$
    (b) $E = \dfrac{2Qz}{3\pi\varepsilon_0 R^2}\left(-\dfrac{1}{\sqrt{z^2+R^2}}+\dfrac{1}{\sqrt{z^2+R^2/4}}\right)$
    along axis

38. (a) 6.4×10^7 V/m parallel to **p**;
    (b) 3.2×10^7 V/m antiparallel to **p**

40. (b) $E_x = \dfrac{Ql^2}{4\pi\varepsilon_0} \dfrac{x}{r^5} \left[2 + \dfrac{5(2z^2 - x^2 - y^2)}{r^2} \right]$

$E_y = \dfrac{Ql^2}{4\pi\varepsilon_0} \dfrac{y}{r^5} \left[2 + \dfrac{5(2z^2 - x^2 - y^2)}{r^2} \right]$

$E_z = \dfrac{Ql^2}{4\pi\varepsilon_0} \dfrac{z}{r^5} \left[-4 + \dfrac{5(2z^2 - x^2 - y^2)}{r^2} \right]$

where $r^2 = x^2 + y^2 + z^2$

Chapter 26

2. -8.1×10^{-18} J
4. $\dfrac{Q}{4\pi\varepsilon_0 d} (4 + \sqrt{2})$
6. 5.8×10^6 eV
8. -50 eV
10. $U = \dfrac{1}{4\pi\varepsilon_0} \dfrac{Q^2}{l^2} \left[x \ln \dfrac{(x+2l)x}{(x+l)^2} + 2l \ln \dfrac{x+2l}{x+l} \right]$
12. 1.1×10^{-4} J
14. 1.7 J
16. $e^2/8\pi\varepsilon_0 m_e c^2$ 1.41×10^{-15} m
18. 5.1×10^{31} J/m^3
20. 1.8×10^{31}, 4×10^{16}, 1.6×10^{12}, and 1.1×10^2 J/m^3; 2.0×10^{14}, 0.5, 1.8×10^{-5}, and 1.2×10^{-15} kg/m^3
22. 8.6×10^5 eV
24. 3.5×10^{-15} m
26. (b) 1.2×10^{29} J; $1 : 5.3 \times 10^{10}$
28. $0.980 \, e^2/8\pi\varepsilon_0 R = 7.1 \times 10^5$ eV

Chapter 27

2. (a) 2.0×10^{-11} F; (b) 4.0×10^{-6} C
4. 9.5 pF
6. (a) 2.1×10^{-8} C, 6.0×10^4 V/m;
   (b) 7.1×10^{-9} C, 4.0×10^4 V/m
8. 15.5 μF; 1.5 μF
10. $3\,\varepsilon_0 A/d$
12. 2.4×10^{-5} C, 7.2×10^{-5} C
14. 7.4 m^2
16. $C = \dfrac{\varepsilon_0 A}{d} \dfrac{2\kappa_1 \kappa_2}{\kappa_1 + \kappa_2}$
20. 0.11%
22. 0.20 C/m^2
24. (a) -2.6×10^{-6} C/m^2, 1.7×10^{-6} C/m^2;
    (b) 1.6×10^5 V/m, 1.1×10^5 V/m;
    (c) 3.0×10^5 V/m
26. 5.0×10^{-2} N
28. $C = \dfrac{2\pi\varepsilon_0 (\kappa_1 + \kappa_2)}{1/R_1 - 1/R_2}$
30. (a) 1.6×10^{-10} F; (b) 3.8×10^2 V;
    (c) 7.5×10^4 V/m; (d) 2.5×10^{-2} J/m^3;
    (e) 1.1×10^{-5} J
32. 0.2 C, 2×10^3 J
34. 2.0×10^3 V
36. 8.9×10^{-4} J

38. 1.1×10^8 m^3
40. 2400 V; 0.0012 C; 1.44 J

Chapter 28

2. 3.0 Ω; 2.5 A
4. 400 A
6. 5.7×10^{-14} J·s
8. (a) 3.6×10^{-2} V/m; (b) 1.7×10^{-4} m/s;
   (c) 46 years
10. 0.201, 0.256, 0.323, 0.407, and 0.514 Ω
12. 7.9 Ω
14. 2.2 cm
16. 9.9 V
18. 4.1×10^6 A/m^2; 6.9×10^{-2} V/m
20. 0.88 Ω; 14 A
22. 191 kg vs. 384 kg
24. 0.164 cm
26. 8×10^{-5} A
28. 33° C
30. 3.5 A, 2.5 A
32. (a) 2.1×10^2 A; (b) 2.7×10^{-19} A
34. 2.4%
36. 1.5 km from AB
38. 4.0 A, 2.4 A, 1.5 A; 7.9 A
40. 10.9 A, 7.1 A
42. 1.8 Ω
44. (a) 1.85 Ω; (b) 6.5 A;
    (c) $\Delta V_1 = \Delta V_2 = \Delta V_3 = 12$ V, $I_1 = 3.0$ A, $I_2 = 2.0$ A, $I_3 = 1.5$ A
46. (a) 2.7 Ω; (b) 0.56 A; (c) 1.5 V, 0.37 A; 1.12 V, 0.19 A; 0.38 V, 0.19 A
48. $(1 + \sqrt{3})\ \Omega$

Chapter 29

2. (a) 6.5×10^3 J, 2.4×10^6 J;
   (b) 1.4×10^2 J/cm^3, 2.0×10^2 J/cm^3;
   (c) 7.5×10^4 J/kg, 1.0×10^5 J/kg
4. (a) 2.9×10^3 J; (b) 48 s
6. $I_1 = 3.25$ A, $I_2 = 3.5$ A
8. 0.49 V, 0.22 Ω
10. 8.0 Ω
12. $R_i' \mathscr{E}/(RR_i + RR_i' + R_i R_i')$, $R_i \mathscr{E}/(RR_i + RR_i' + R_i R_i')$
14. 12.5 A, 25.5 A
16. (a) 2.6, 1.7, 1.3, 2.1, and 0.86 A; (b) 2.6 V
18. 5.1 A
20. 6.2×10^{12} protons per second; 7.0×10^2 W
22. 12.1 Ω, 48.4 Ω; two 24.2-Ω resistors connected in parallel or in series.
24. (a) 13.2 V; (b) 7.2 W
26. 6.7 h
28. 0.50 W
30. $\mathscr{E}^2 R/(R + R_i)^2$
32. (a) 1/30; (b) 1/3
34. (a) 1.8×10^2 V; (b) 1.8×10^6 W
36. 1.3×10^2 W; 0.022
38. 19 liters per minute
40. 433 Ω
42. 1.49936 V
44. 2.0 Ω

46. (a) 0, 4.7×10^{-5} C, 7.6×10^{-5} C; (b) 1.2×10^{-4} C; (c) 0.060 A
48. 0.011 s
50. 1.2×10^4 Ω
54. 5.3 s; 5.0×10^{-6} A

Chapter 30

2. 1.2×10^{-12} N, 0
4. $F = \dfrac{\mu_0}{4\pi} \dfrac{qq'}{2d^2} \dfrac{v^2}{\sqrt{2}}$
6. (a) $|\mathbf{F}| = 0.89 \times 10^{-13}$ N, $|\mathbf{F}'| = 1.78 \times 10^{-13}$ N; (b) 1.54×10^{-13} N in direction of $\mathbf{v}$
8. 8×10^{-16} N in direction of $-\mathbf{v} \times \mathbf{B}$
10. 8.2×10^{-8} N and 3.5×10^{-5} N; yes
12. 4.0×10^{-3} T
14. 34 V
16. 1.5×10^{13} m/s²
18. (a) 5.0×10^{-6} T; (b) 16°
20. $B = \dfrac{\mu_0}{2\pi} \dfrac{3\sqrt{2}}{2} \dfrac{I}{d}$
22. $B = \dfrac{\mu_0 I}{4\pi} \dfrac{\sqrt{x^2+y^2}+x+y}{xy}$ in positive z direction
24. $B = \dfrac{\sqrt{5}\,\mu_0 I}{2\pi L}$ into plane
26. $B = (\sqrt{5} + \sqrt{13}/3) \dfrac{\mu_0 I}{\pi h}$ out of plane
28. $B = \mu_0 I / 8R$ into plane
30. $B = \dfrac{\mu_0}{\pi} \dfrac{I}{b} \tan^{-1} \dfrac{b}{2z}$ perpendicular to z and to current
32. $B = \dfrac{\mu_0 I}{4\pi R}(3\pi/4 + 2)$ out of plane
34. $B = \dfrac{\mu_0 I}{2L}\left[\dfrac{z+L/2}{\sqrt{(z+L/2)^2+R^2}} - \dfrac{z-L/2}{\sqrt{(z-L/2)^2+R^2}}\right]$ in positive z direction
36. $B_x = 0$, $B_y = -\dfrac{\mu_0 I}{2\pi}\dfrac{Rz}{(z^2+R^2)^{3/2}}$
   $B_z = \dfrac{\mu_0 I}{2\pi}\left[\dfrac{\pi}{2}\dfrac{R^2}{(z^2+R^2)^{3/2}} + \dfrac{R}{R^2+z^2}\right]$
38. 1.9 T
40. 2.1×10^9 A; westward
42. $B = \dfrac{4\mu_0}{\pi}\dfrac{IL^2}{(4z^2+L^2)\sqrt{4z^2+2L^2}}$ in positive z direction
44. $Q\omega R^2/4$

Chapter 31

2. 6.4×10^{-16} N, 7.0×10^{14} m/s²; into wire
4. 4.0×10^{-8} T; 1.9×10^{-18} N
6. $\mathbf{B} = \dfrac{\mu_0 I}{2\pi}\left(\dfrac{1}{z+R} + \dfrac{1}{z-R}\right)\hat{\mathbf{y}}$ for $z > 2R$;
   $\mathbf{B} = \dfrac{\mu_0 I}{2\pi}\left(\dfrac{1}{z+R} + \dfrac{z-R}{R^2}\right)\hat{\mathbf{y}}$ for $2R \geq z \geq 0$;
   $\dfrac{2\mu_0 I}{3\pi R}$ at $z = 2R$
8. 0, $\mu_0 \sigma \hat{\mathbf{x}}$, 0, $-\mu_0 \sigma \hat{\mathbf{x}}$
10. 0.13 T
12. 6.9 T; 2.3 T

14. $B = \mu_0 \sqrt{n^2 I^2 + (I'/2\pi r)^2}$; helical
16. $B = \dfrac{\mu_0}{2\pi r}(NI_0 + I')$; $I' = NI_0$; $B = 0$
18. 1.1×10^{-17} kg·m/s
20. 3.4×10^{-17} kg·m/s
22. 3.3 T
24. 3.6×10^{-2} T
26. 0.39 A
28. 2×10^6 m/s eastward
30. 3.1×10^{-6} V
32. 0.12 N
34. 6.7×10^{-5} N
36. 7.2×10^{-4} N toward the straight wire
40. 2×10^5 A
42. 3.4×10^{-5} N·m
44. 2.0 N·m; 0.86 hp
46. (a) 4.0×10^{-3} N; (b) 3.2×10^{-4} N·m

Chapter 32

2. 0.75 V
4. 22 rev/s
6. 9×10^{17} V
8. 9.8×10^{-5} V, 7.5×10^{-5} V, 8.7×10^{-5} V
10. 40 rev/s
12. (a) 1.9×10^{-2} T/s; (b) 6.4×10^{-3} V
14. 3.0×10^2 A
16. 0.62 A
18. $\mathcal{E} = \dfrac{\mu_0 Iv}{2\pi}\dfrac{l}{(l+d)d}$
20. 1.8 Wb; 45 T
22. (a) 6.0 V/m; (b) 0 V/m; (c) 0.36 V
24. (a) 2.6×10^{-3} V/m; (b) 0 V/m; (c) 9.2×10^{-15} C/m²
26. (a) 0 Wb; (b) 90 V; (c) negative
28. 4.8×10^{-6} H; 9.5×10^{-4} V
30. (a) 6.3×10^{-3} H; (b) −1.9 V
32. $200\,\mu_0 n\pi R^2$; no
34. (a) $B = \dfrac{\mu_0}{2\pi}\dfrac{NI}{r}$; (b) $\Phi_B(1) = \dfrac{\mu_0 NI}{2\pi}(R_2 - R_1) \ln(R_2/R_1)$, $L = \dfrac{\mu_0 N^2}{2\pi}(R_2-R_1)\ln(R_2/R_1)$
38. 4×10^{11} J/m³
40. 7×10^{18} J
42. 3.6×10^7 J
44. 9.1×10^3 m³; about 40 m across
46. (a) $u = \dfrac{\mu_0}{8\pi^2}\left(\dfrac{NI}{r}\right)^2$; (b) $U = \dfrac{\mu_0}{4\pi}N^2 I^2 (R_2 - R_1)\ln(R_2/R_1)$; (c) $L = \dfrac{\mu_0}{2\pi}N^2(R_2 - R_1)\ln(R_2/R_1)$
48. (a) 0, 0.024 A, 0.038 A; (b) 0.060 A; (c) 3.0 A/s
52. 48 A; 14 s; 6.9 s
54. $\tfrac{1}{2}L\,(\mathcal{E}/R)^2$

Chapter 33

2. (a) 1.9 T; (b) -1.1×10^{-4} eV, 1.1×10^{-4} eV, parallel
4. (a) $\tan^{-1}[3zx/(2z^2 - x^2)]$; (b) 0°, 0°, 72°
6. 0.24 Hz
8. $(3/14)e\hbar/m_n = 2.2 \times 10^{-27}$ A·m²; opposite

10. (a) 13 T; (b) 3.3×10^9 radians/s
12. 3.2×10^{-6} T
14. 1.000912
18. 3.6×10^3 A
20. 2.0×10^{-6} N·m; no
22. 1.6×10^{28} electrons/m^3; 0.18 electron/atom
24. 4.0×10^4 A
26. 3.4×10^{-9} J·s
28. 2.1 T
30. (a) 4.4×10^{10} radians/s; (b) 2.3 m/s; (c) 4.7×10^{-24} J

Chapter 34

2. 400 W
4. (a) 3.1×10^3 A; (b) less loss in transmission line
6. 3.2×10^{-4} m/s; 0.12 m/s^2
8. (a) $Q = 6.5 \times 10^{-8} \sin(120\pi t)$; (b) 1/240 s, 0 s; (c) 5.8×10^{-8} J, 2.9×10^{-8} J
10. $I = \omega C \mathscr{E}_0 \cos \omega t$, $I = -\mathscr{E}_0/(\omega L) \cos \omega t$;
    $I = (\omega C - 1/\omega L) \mathscr{E}_0 \cos \omega t$,
    $P = (\omega C - 1/\omega L) \mathscr{E}_0^2 \cos \omega t \sin \omega t$
12. (a) 240 Ω; (b) 8.3×10^{-4} A; (c) -8.3×10^{-4} A, -5.9×10^{-4} A
14. 6.0×10^{-9} F to 6.6×10^{-10} F
16. 1.4×10^{-3} J; 7.9×10^{-4} s; 15.7×10^{-4} s
18. (a) circuit oscillates at decreased frequency with gradually diminishing amplitude; (b) circuit oscillates at original frequency, but with much diminished amplitude
20. (a) 5.8×10^{-3} J; (b) 40%; (c) 0.027s, 0.090s; (d) 0.035 s, 0.12 s
22. 1.0×10^{-2} H, 2.0×10^{-5} F
24. (a) $0.80 \sin(2\pi \times 2.2 \times 10^4 t)$; (b) $5.9 \times 10^{-2} \cos(2\pi \times 2.2 \times 10^4 t)$
26. 13.4×10^{-3} A, 6.6×10^{-3} A
28. (a) 0.20 A; (b) 6.3×10^{-4} s, 6.8×10^{-4} s
30. (a) 5.9×10^3 Hz; (b) 0.40 A; (c) 0.16 W
32. 0.37 A; $-43°$; 5.4×10^{-2} W
34. 0.15 W; 1.4×10^{-8} W
36. $\tfrac{1}{2} \mathscr{E}_{max}^2 (\omega C - 1/\omega L) \sin 2\omega t + (\mathscr{E}_{max}^2/R) \sin^2 \omega t$
38. 4550 turns
40. 9.1×10^4 A; 5.0×10^3 A
42. 1.96×10^3 A; 4.35×10^3 A

Chapter 35

2. (a) 1.0×10^6 V/m s; (b) 8.3×10^{-13} T, 1.7×10^{-12} T
6. (a) 2.0×10^{-3} A; (b) 1.0 s; (c) 0.89×10^{-9} T, 1.8×10^{-9} T, 2.7×10^{-9} T
8. (a) $(V_0/R) \sin \omega t$; (b) $(\varepsilon_0 A \omega V_0/d) \cos \omega t$; (c) $(V_0/R) \sin \omega t + (\varepsilon_0 A \omega V_0/d) \cos \omega t$; (d) $(\mu_0/2\pi) [(V_0/rR) \sin \omega t + (\varepsilon_0 \pi r \omega V_0/d) \cos \omega t]$
12. $\oint \mathbf{E} \cdot d\mathbf{S} = Q/\varepsilon_0$, $\oint \mathbf{B} \cdot d\mathbf{S} = q_m$,
    $\oint \mathbf{E} \cdot d\mathbf{l} = -I_m - d\Phi_B/dt$,
    $\oint \mathbf{B} \cdot d\mathbf{l} = \mu_0 I + \mu_0 \varepsilon_0 d\Phi/dt$
14. 5.7 cm
16. (a) $(2\pi/\mu_0 \omega^2) E_0 \sin \omega t$; (b) $(2\pi/\mu_0 \omega) E_0 \cos \omega t$
18. 5.6×10^{-8} V/m
20. 0.11 V/m; 6.7×10^{-10} s

22. (a) 7.4×10^{17} m/s^2; (b) 7.9×10^{-9} V/m
24. (a) 4.8×10^{-15} m, 6.0×10^{27} m/s^2; (b) 3.2×10^{11} V/m; (c) 0
26. 4.7×10^4 m/s^2
28. (a) 2.5×10^{-12} V/m, 8.3×10^{-21} T
30. (a) 1.1×10^{29} m/s^2; (b) 7.1×10^{13} V/m, 2.4×10^5 T
32. (a) 6.8×10^9 m/s^2; (b) 2.7×10^{-17} V/m, 9.1×10^{-26} T; (c) 6.1×10^5 V/m, 2.1×10^{-3} T; (d) 3.8×10^{-9} T
34. (a) $q^2 v a/(12\pi \varepsilon_0 c^3)$; (b) $q^2 v a/(12\pi \varepsilon_0 c^3)$

Chapter 36

2. 0.067 s
4. 5.3×10^{-6} A/m
6. 2.0×10^{-9} T, northward
8. (a) negative z; (b) 63°, 27°, 90°; (c) $2\hat{\mathbf{x}}(E_0/c) \sin(\omega t + \omega z/c) - \hat{\mathbf{y}}(E_0/c) \sin(\omega t + \omega z/c)$
10. 1.85 m
12. 5500 Å; 5000 Å and 6000 Å; 4800 Å and 6200 Å
14. 1.0×10^3 V/m, 3.4×10^{-6} T
16. 2.9×10^{-20} W; 3.6×10^{27} W
18. 6.9%
20. 1.9×10^9 W/m^2; 1.2×10^6 V/m, 4.0×10^{-3} T
22. 5.5×10^{-2} V/m, 1.8×10^{-10} T
24. 2.0×10^4 W
26. 51
30. (a) 50 W/m^2; (b) 1.9×10^2 V/m, 6.5×10^{-7} T; (c) 1.7×10^{-7} N/m^2
32. (a) 6.2×10^{-18} N; (b) 3.7×10^{-18} N, gravity is larger; (c) 7.7×10^{-19} N, 9.2×10^{-19} N, radiation pressure is larger
34. 5.8×10^{-4} kg·m/s
38. $0.81c$
40. 2203.13 MHz
42. (a) 1.8×10^{-7}; (b) 1.4×10^3 Hz
44. 2.2×10^3 m/s, 1.50×10^5 m/s
46. 2.75×10^8 m/s

Chapter 37

2. 0.50 m × 0.42 m
4. 29°
6. 50°
8. 39° 0′ 50″
10. l, $l + 2d/n$, $l + 4d/n$
12. (b) 0.67 mm; (c) 2.67 mm
14. 48°
16. 47 m
18. 5.7 m
20. 4.8°
22. 0.25
24. 53.1°; 36.9°
28. 0.38; 0.27; 10 or more
30. $I_0 \cos^2\phi/(\cos^2\phi + \sin^2\phi \cos^2\alpha)$

Chapter 38

2. R
4. 120 cm; concave

6. 60 cm behind mirror
8. (a) 60 cm in front of first mirror; (b) 30 cm in front of second mirror
10. 8.3 mm
12. 21 cm
14. 10 cm
18. (a) 36 cm to right of lens; (b) 130 cm to left of mirror
22. 475 cm to left of lens
24. 10.7 cm beyond third lens; -0.35
28. 21.4 cm from right surface; 22.9 cm from left surface
30. 1.3
32. 3.87 cm beyond curved surface of rod; 6.0 cm beyond curved surface of rod (that is, at its actual position)
34. 49 cm to infinity
36. 0.022 s
38. (a) -22 cm, divergent; (b) 0.90 cm
40. 10
42. 64
44. 1344

Chapter 39

2. 3×10^8 m/s
4. (b) 9.48×10^5, 9.48×10^5; (c) 1582 Å, 3164 Å
6. (a) 996 Å; (b) no
8. $2d = \frac{1}{2}\lambda, \frac{3}{2}\lambda, \frac{5}{2}\lambda, \ldots$
10. (c) 1.2 mm
12. 5.8 mm/s
14. 1.000277
16. 7.4 mm
18. $\lambda/d \cong 0.20$ implies nodal lines at 12°, 24°, 37°, and 53°
20. $\pm 1.3 \times 10^{-4}$ radian, $\pm 2.6 \times 10^{-4}$ radian, $\pm 3.9 \times 10^{-4}$ radian, etc.
24. 61°, 22°, $-7°$, $-39°$
26. 0.45
28. 5.7°
30. 1.0 mm; 2.6 mm
32. 8.7×10^4
34. 0.3 mm
36. 10 cm, $\frac{1}{2} \times 10$ cm, $\frac{1}{3} \times 10$ cm, ...; 3310 Hz, 2×3310 Hz, 3×3310 Hz, ...
38. $N - 2 = 8.7 \times 10^4$

Chapter 40

2. $\lambda/a \cong 0.44$ implies nodal lines at 26° and 63°
4. 47°
6. (a) no; (b) yes; (c) no
8. (a) 4.2 cm; (b) 2.1 cm
10. (b) $n - 1$
12. 0.040 mm; 5200 Å
14. (a) 1.9×10^{-4} radian; (b) 4.4×10^{-3} mm; (c) 2.3
16. 2.8°; 2.4×10^2 m
18. 3.4×10^{-3} radian, or 12'
20. (a) 3×10^{-7} radian; (b) 2 m
22. 1.3×10^{-5} radian; 2.4×10^{-4} cm
24. 1.1×10^{-5} radian; 6 cm
26. 13 km
28. (a) 1.3×10^{-5} radian; (b) 1.4×10^{-5} radian, eye
32. 6.4 cm

Chapter 41

2. (b) 2.4, $v'_x = 0.42$
4. (a) $x = 2.3 \times 10^8$ light-seconds, $y = 4.0 \times 10^8$ light-seconds, $z = 0$, $t = -1.7 \times 10^8$ s; (b) before
8. 15.5 min; 17.3 min; 34.4 min
10. $(1 + 3.4 \times 10^{-10})$; 3.0×10^{-5} s
12. 3.7×10^2 m/s; $(1 + 7.5 \times 10^{-13})$
14. $V = (1 - 1.0 \times 10^{-11})$; 2.2×10^6 years
16. (a) 8.2×10^8 s; (b) 7.6×10^8 s
18. 4×10^{-13}%
20. 0.87 m, 0.50 m; 0.62 m, 0.50 m; 39°
22. (a) 6.7×10^{-7} s; (b) 4.2×10^{-6} s
24. $0.999c$
26. $0.98c$
28. $0.61c$; 61°
30. $0.87c$
34. 1.6×10^{-22} kg·m/s; 1.3×10^{-14} J
36. 4.0×10^{-3}%
38. 130 m/s; 5.3×10^{-16} kg·m/s
40. 1.4×10^{-15} m/s
42. (a) 9.0×10^{13} m/s^2; (b) 1.8×10^{-24} kg; (c) 1.6×10^{-10} N
44. $0.83c$
48. (a) $0.99962c$; (b) $0.973c$; (c) 6.5×10^{10} J
50. (a) 1.0×10^{-3} m/s; (b) -4.3×10^4 C/m, $+4.3 \times 10^4$ C/m; (c) $(-4.3 \times 10^4 + 2.6 \times 10^{-19})$ C/m, $(4.3 \times 10^4 + 2.6 \times 10^{-19})$ C/m; 5.2×10^{-19} C/m

Chapter 42

2. (a) 1 mm; (b) 2×10^9 W
4. 1.3×10^7 m
6. 0.78; 0.87
10. 44 K
12. 7.5×10^{-20} J
14. 0.87 Hz; 5.8×10^{-34} J; 2.0×10^{-16} m
16. (a) 7.2×10^{15} Hz
18. 9×10^{24}
22. 1.2×10^6/s
24. 2.5×10^3/m^3
28. 1.8 eV
30. none; K; K, Cr, Zn
32. 17.5 eV
34. 2.8×10^{-15} J, 9.2×10^{-24} kg·m/s
36. 85°
38. 63 eV
40. 1.5×10^8 eV
42. 5.8×10^3 eV; 1.2×10^3 eV
44. (a) 2×10^{-39} kg·m/s; (b) 1×10^3 Hz

Chapter 43

2. first ($n = 5$); 34,076 Å
4. (a) 4102.9 Å and 4341.7 Å in Balmer series; (b) 1.1×10^6 m/s
6. (a) $p = -0.0407$, $d = -0.0015$; (b) Lyman and Balmer

8. 3.22×10^7 eV
12. $7.30 \times 10^{-3}\, c$
14. 1215.7 Å
16. $n = 3$ to $n = 1$
18. 2.12×10^{-10} m, 1.09×10^6 m/s, 2.10×10^{-34} J·s, 5.66×10^{21} m/s^2
20. 0.028 mm; -2.5×10^{-5} eV
22. 7.0×10^5 m/s
24. 0.176 Å; -122 eV
26. $r = \dfrac{n^2\hbar^2}{GMm^2}$, $E = -\dfrac{(GM)^2 m^3}{2\hbar^2}\dfrac{1}{n^2}$; $n = 330$, -5.6×10^{-17} eV
28. 0.69 Å, 0.58 Å
30. $r = (a_0 R^3 n^2)^{1/4}$, $E = \dfrac{n\hbar e}{\sqrt{4\pi\varepsilon_0 m_e R^3}} - \dfrac{e^2}{4\pi\varepsilon_0}\dfrac{3}{2R}$, $\nu = \dfrac{n_2 - n_1}{2\pi}\dfrac{e}{\sqrt{4\pi\varepsilon_0 m_e R^3}}$; $4a_0$
32. 1.06 Å, 2431 Å
36. 0.087 Å, 5.3 Å, 3.3 Å, 0.041 Å
38. 2.1 Å, 0.16 Å
40. 1.1×10^{-32} m
42. 0.39 Å
44. ± 7 cm
46. 9.2×10^{-20} kg·m/s, 5.5×10^7 m/s
48. (a) $\lambda = 2L/n$; (b) $E = h^2 n^2 / 8m_e L^2$; (c) 38 eV, 151 eV, 339 eV

Chapter 44

2. $\tfrac{5}{2}\hbar$; $\tfrac{3}{2}e\hbar/m_e$
4. 8 levels; $(1.90 \times 10^{-5}\text{ eV}) \times (\tfrac{7}{2}, \tfrac{5}{2}, \tfrac{3}{2}, \tfrac{1}{2}, -\tfrac{1}{2}, -\tfrac{3}{2}, -\tfrac{5}{2},$ and $-\tfrac{7}{2})$
6. 0.34 T
8. $n = 1, l = 0, m = 0, m_s = -\tfrac{1}{2}$; 1, 0, 0, $+\tfrac{1}{2}$; 2, 0, 0, $-\tfrac{1}{2}$; 2, 0, 0, $+\tfrac{1}{2}$; 2, 1, $-1, -\tfrac{1}{2}$
10. $n = 1, l = 0, m = 0, m_s = -\tfrac{1}{2}$; 1, 0, 0, $+\tfrac{1}{2}$; 2, 0, 0, $-\tfrac{1}{2}$; 2, 0, 0, $+\tfrac{1}{2}$; 2, 1, $-1, -\tfrac{1}{2}$; 2, 1, $-1, +\tfrac{1}{2}$; 2, 1, 0, $-\tfrac{1}{2}$; 2, 1, 0, $+\tfrac{1}{2}$; 2, 1, $+1, -\tfrac{1}{2}$; 2, 1, $+1, +\tfrac{1}{2}$; 3, 0, 0, $-\tfrac{1}{2}$; 3, 0, 0, $+\tfrac{1}{2}$
14. 0, $\tfrac{1}{2}\hbar$; $e\hbar/2m_e$
16. (a) completely filled shells; (b) completely filled shells plus one extra electron of zero orbital angular momentum outside of filled shell
18. (a) 1.33 eV; (b) -5.1 eV
20. (a) 4.9×10^2 N/m; (b) 6.1×10^{13} Hz, 5.0×10^{13} Hz
22. $3:5:7:(2n-1)$
24. 3.47×10^{-45} kg·m^2; 9.94×10^{-6} eV, 39.8×10^{-6} eV, 89.5×10^{-6} eV
26. $\lambda = (2.0\text{ Å})/n$, $K = 38n^2$ eV
28. 8.9 V
32. Use two circuit diagrams as in Fig. 44.9 and connect the output of the first circuit to the input of the second
34. 9%

Chapter 45

2. 7.7 fm
4. 4.3×10^{-15}
6. 6.8 MeV
8. 0.44×10^{-26} A·m^2
10. (a) antiparallel; (b) 5.9×10^{-6} eV
12. 490 MeV; 55.93 u
14. 0.03037 u; 28.29 MeV
16. 10.56 MeV
18. (a) 2.43 MeV per bond; (b) 122.8 MeV predicted vs. 127.6 MeV actual
20. 10.42 MeV
22. 5.4 MeV
24. 0.351; 0.492
26. 7.7×10^{15} Bq
28. 0.81×10^{-6} g
30. 130 days
34. 0.00083 u; atoms would disappear
36. (a) -0.353 MeV; (b) 0.471 MeV; (c) 1.41 MeV
38. 0.99703 u
40. (a) 17.59 MeV; (b) 11.7 MeV (with $v_1' = -4.72 \times 10^7$ m/s)

Chapter 46

2. -815 MeV; no
4. Y(11020); 11.83 u
6. no
8. yes; no; no
14. $\bar{u}d$
16. 8 more quarks created
18. 9 (however, the gluon combination red-antired + blue-antiblue + green-antigreen is equivalent to white, and does not count; this effectively leaves 8 gluons)

PHOTOGRAPH CREDITS

PRELUDE
- page 1 — James L. Mairs
- page 2 — (top) James L. Mairs; (bottom) James L. Mairs
- page 3 — (top) James L. Mairs; (center) Lockwood, Kessler, and Barlett, Inc.; (bottom) NASA
- page 4 — (top) U.S. Geological Survey; (center) General Electric Space Systems; (bottom) NASA
- page 5 — NASA
- page 9 — (top) Hale Observatories; (bottom) Hale Observatories
- page 10 — (top) Hale Observatories; (bottom) Hale Observatories
- page 11 — (top) J. R. Kuhn, Dept. of Physics, Princeton Univ.; (bottom) James L. Mairs
- page 12 — (top) Zeiss; (center) Dr. Ronald Radius, The Eye Institute, Milwaukee, Wis.; (bottom) From Tissue and Organs: A Test Atlas of Scanning Electron Microscopy by Richard G. Kessel and Randy H. Kardon, W. H. Freeman & Co., San Francisco, © 1979
- page 13 — (top) From Tissue and Organs: A Test Atlas of Scanning Electron Microscopy by Richard G. Kessel and Randy H. Kardon, W. H. Freeman & Co., San Francisco, © 1979. (bottom) Professor T. T. Tsong, Pennsylvania State Univ.
- page 14 — A. V. Crewe and M. Utlaut, Univ. of Chicago
- page 15 — L. S. Bartell, Univ. of Michigan

CHAPTER 1
- Fig. 1.7 — BIPM-Picture
- Fig. 1.9 — U.S. National Bureau of Standards
- Fig. 1.10 — BIPM-Picture
- Fig. 1.11 — U.S. National Bureau of Standards
- Fig. 1.12 — U.S. National Bureau of Standards
- Fig. 1.13 — Photo courtesy of Hewlett-Packard Company
- Fig. 1.14 — U.S. National Bureau of Standards
- Fig. 1.15 — U.S. National Bureau of Standards

CHAPTER 2
- Fig. 2.10 — PSSC Physics, 2nd ed., 1965, D. C. Heath & Co. with the Educational Development Center, Newton, Mass.
- Fig. 2.11 — BIPM-Picture
- Fig. 2.15 — NASA
- page 37 — (Galileo Galilei) AIP Niels Bohr Library

INTERLUDE I
- Fig. I.1 — Reproduced by gracious permission of Her Majesty Queen Elizabeth II; from the Royal Library, Windsor Castle, Berkshire
- Fig. I.2 — Duvall Corporation
- Fig. I.3 — R. Gronsky, National Center for Electron Microscopy, Lawrence Berkeley Laboratory
- Fig. I.4 — Martin J. Buerger, Institute Professor, MIT
- Fig. I.5 — Professor T. T. Tsong, Pennsylvania State Univ.
- Fig. I.6 — R. P. Goehner, General Electric
- Fig. I.7 — J. M. Karansinski, IBM Watson Research Center
- Fig. I.8 — David Scharf, 1977, through Stockton Books, Inc.
- Fig. I.9 — Earth Scenes, by Breck P. Kent
- Fig. I.10 — W. A. Bentley; reproduced from Snow Crystals by W. A. Bentley and W. J. Humphreys, Dover, New York, 1962
- Fig. I.11 — Hans C. Ohanian
- Fig. I.12 — Courtesy H. B. Huntington, RPI
- Fig. I.13 — © Beeldrecht, Amsterdam/VAGA, New York, Collection Haags Gemeentemuseum, The Hague
- Fig. I.14 — © Beeldrecht, Amsterdam/VAGA, New York, Collection Haags Gemeentemuseum, The Hague
- Fig. I.15 — © Beeldrecht, Amsterdam/VAGA, New York, Collection Haags Gemeentemuseum, The Hague
- Fig. I.16 — © Beeldrecht, Amsterdam/VAGA, New York, Collection Haags Gemeentemuseum, The Hague
- Fig. I.26 — C. S. Smith, MIT
- Fig. I.32 — Otsuka Kogeisha & Co., Ltd.
- Fig. I.33 — C. S. Smith, MIT
- Fig. I.34 — Metropolitan Museum of Art

CHAPTER 4
- Fig. 4.5 — PSSC Physics, 2nd ed., 1965, D. C. Heath & Co. with the Educational Development Center, Newton, Mass.
- Fig. 4.7 — © Smithsonian Institution
- Fig. 4.8 — Dr. Harold E. Edgerton, MIT
- Fig. 4.11 — (a) Photo courtesy of The Schenectady Gazette, Schenectady, N.Y.
- Fig. 4.16 — NASA
- Fig. 4.27 — Photo by Jeffrey Wood, PH1

CHAPTER 5
- Fig. 5.2 — The Smithsonian Institution
- Fig. 5.4 — NASA
- Fig. 5.5 — Road & Track
- Fig. 5.20 — Honeywell, Inc., Sperry Commercial Flight System Group
- Fig. 5.22 — The Science Museum, London

Photograph Credits **A-47**

page 99 (Sir Isaac Newton) Burndy Library, courtesy AIP Niels Bohr Library
page 102 (Ernst Mach) AIP Niels Bohr Library

CHAPTER 6
Fig. 6.1 (a) H. E. Edgerton; (b) H. E. Edgerton; (c) The Warder Collection; (d) Photo CERN
Fig. 6.3 NASA
Fig. 6.4 NASA
Fig. 6.11 S. J. Calabrese, RPI
Fig. 6.12 S. J. Calabrese, RPI
Fig. 6.31 Leo De Wys, Inc. © Larry Miller
Fig. 6.35 U.S. National Bureau of Standards
page 123 (Pierre Simon, Marquis de Place) AIP Niels Bohr Library
page 136 (Leonardo da Vinci) NMAH Archives Center, Smithsonian Institution

CHAPTER 7
Fig. 7.21 Neil Liefer, Sports Illustrated
Fig. 7.23 Six Flags Great Adventure Family Entertainment Center
page 161 (James Prescott Joule) © British Museum
page 174 (Christiaan Huygens) The Bettmann Archive

CHAPTER 8
Fig. 8.4 Courtesy Calspan Corp., Buffalo, N.Y.
Fig. 8.10 U.S. Air Force Photo
page 186 (Joseph Louis LaGrange) The Granger Collection
page 196 (Hermann von Helmholtz) AIP Niels Bohr Library
page 198 (James Watt) © British Museum

CHAPTER 9
Fig. 9.5 (a) From Cavendish, Philosophical Transactions of the Royal Society 18 (1798): 388
Fig. 9.14 Sovfoto Agency
Fig. 9.15 NASA
Fig. 9.34 NASA
Fig. 9.37 NFB Phototheque, photos by G. Blouin, 1949
Fig. 9.38 © 1987 Esselte Map Service AB Stockholm
Fig. 9.40 NASA
Fig. 9.45 National Optical Astronomy Observatory
page 215 (Henry Cavendish) © British Museum
page 216 (Nicolas Copernicus) National Portrait Gallery, Smithsonian Institution
page 219 (Johannes Kepler) AIP Niels Bohr Library

INTERLUDE II
Fig. II.1 (a) Palomar Observatory Photograph; (b) Palomar Observatory Photograph
Fig. II.2 Hale Observatories
Fig. II.3 Hale Observatories
Fig. II.4 Palomar Observatory Photograph
Fig. II.5 Palomar Observatory Photograph
Fig. II.6 Palomar Observatory Photograph
Fig. II.7 Palomar Observatory Photograph
Fig. II.8 Palomar Observatory Photograph
Fig. II.9 Palomar Observatory Photograph
Fig. II.11 Palomar Observatory Photograph
Fig. II.12 AT&T Bell Laboratories

CHAPTER 10
Fig. 10.6 PSSC Physics, 2nd ed., 1965, D. C. Heath & Co., with the Educational Development Center, Newton, Mass.
Fig. 10.12 The Bettmann Archive
Fig. 10.15 (a) NASA
Fig. 10.17 NASA
Fig. 10.18 Duomo Photography, Inc.

CHAPTER 11
Fig. 11.1 Mercedes-Benz of North America, Inc.
Fig. 11.9 Reprinted from Philosophical Magazine, R. H. Brown et al., vol. 40, 1949
Fig. 11.10 © Fundamental Photographs, 1972
Fig. 11.11 W. B. Hamilton, U.S. Geological Survey
Fig. 11.18 C. F. Powell and G. P. S. Occhialini, Nuclear Physics in Photographs, Oxford Univ. Press, New York

INTERLUDE III
Fig. III.4 Adapted from L. C. Lundstrom, SAE Transactions 75, 1967; reprinted with permission © 1967 Society of Automotive Engineers, Inc.
Fig. III.5 From Modern Automotive Structural Analysis by J. A. Wolf, Van Nostrand Reinhold, New York, 1982
Fig. III.6 Adapted from K. H. Lin, M. M. Kamal, J. W. Justusson, SAE paper 750117, 1975; reprinted with permission © 1975 Society of Automotive Engineers, Inc.
Fig. III.7 Volvo Cars of North America
Fig. III.12 Humanoid Systems, Humanetics, Inc.
Fig. III.13 Insurance Institute for Highway Safety
Fig. III.16 From The Prevention of Highway Injury, Marvin L. Selzer, Paul W. Gikas, and Donald F. Huelke, eds., Highway Safety Research Institute, 1967
Fig. III.17 Bob Costanzo, Daytona International Speedway
Fig. III.18 Mercedes-Benz of North America, Inc.
Fig. III.19 Courtesy Colonel John P. Stapp
Fig. III.22 Minicars, Inc.
Fig. III.24 Volvo Cars of North America

CHAPTER 12
Fig. 12.1 Dr. Harold E. Edgerton, MIT

CHAPTER 13
Fig. 13.14 Sperry Flight Systems
Fig. 13.21 PSSC Physics, 2nd ed., 1965, D. C. Heath & Co. with the Educational Development Center, Newton, Mass.
Fig. 13.27 Wide World Photos
Fig. 13.32 Chicago Historical Society
Fig. 13.41 NASA
Fig. 13.44 Lufthansa Photo

CHAPTER 14
Fig. 14.9 Golden Gate Bridge, Highway & Transportation District
Fig. 14.26 The Port Authority of NJ & NY
Fig. 14.37 Joseph Daniel, Wildpic

CHAPTER 15
Fig. 15.6 NASA
Fig. 15.8 Adapted from Energy, A Sequel to IPS, by Uri Haber-Schaim, Prentice-Hall, Englewood Cliffs, N.J., 1983
Fig. 15.14 © Lester Lefkowitz
Fig. 15.17 U.S. National Bureau of Standards
Fig. 15.25 Portescap, La Chaux-de-Fonds, Suisse; Service de Presse, Nicole Rouge

CHAPTER 16
- Fig. 16.22 — UPI © Bashford Thompson, Tacoma, Wash.

CHAPTER 17
- Fig. 17.1 — Fundamental Photographs
- Fig. 17.2 — Engineering Applications of Lasers and Holography by Winston E. Kock, Plenum Publishing Co., New York, 1975
- Fig. 17.5 — William B. Joyce
- Fig. 17.8 — S. Gilmore, General Electric Research and Development Center
- Fig. 17.9 — Courtesy of C. F. Quate and L. Lam, Hansen Laboratories, Stanford Univ.
- Fig. 17.18 — Dr. Harold E. Edgerton, MIT
- Fig. 17.23 — Mrs. Harry A. Simms, Sr.
- Fig. 17.25 — California Historical Society, San Francisco; Photographer: Arnold Genthe, New York, FN-27239
- Fig. 17.26 — R. Such, Lamont-Doherty Geological Observatory of Columbia University
- Fig. 17.28 — Courtesy of John S. Shelton
- Fig. 17.29 — PSSC Physics, 2nd ed., 1965, D. C. Heath & Co. with the Educational Development Center, Newton, Mass.
- Fig. 17.30 — PSSC Physics, 2nd ed., 1965, D. C. Heath & Co. with the Educational Development Center, Newton, Mass.
- Fig. 17.31 — Educational Development Center, Newton, Mass.
- Fig. 17.32 — Educational Development Center, Newton, Mass.
- page 446 — (Christian Doppler) AIP Niels Bohr Library

INTERLUDE IV
- Fig. IV.2 — Susan Leavines, Photo Researchers, Inc.
- Fig. IV.3 — Reprinted from the Proceedings of the Royal Society, P. M. S. Blackett and D. S. Lees, 1932
- Fig. IV.8 — Reprinted from the Proceedings of the Royal Society, P. M. S. Blackett and D. S. Lees, 1932
- Fig. IV.13 — Nuclear Medicine Section, Dept. of Radiology, Hospital of the Univ. of Pennsylvania
- Fig. IV.14 — Michael G. Velchik, M.D., Nuclear Medicine Section, Dept. of Radiology, Hospital of the Univ. of Pennsylvania
- Fig. IV.15 — Michael G. Velchik, M.D., Nuclear Medicine Section, Dept. of Radiology, Hospital of the Univ. of Pennsylvania
- Fig. IV.16 — Micheal G. Velchik, M.D., Nuclear Medicine Section, Dept. of Radiology, Hospital of the Univ. of Pennsylvania
- Fig. IV.17 — M. D. Anderson, Hospital and Tumor Institute, Univ. of Texas, Houston

CHAPTER 18
- Fig. 18.1 — Boston Museum of Science photo by Bradford Washburn
- Fig. 18.8 — A. D. Moore, Univ. of Michigan, from Introduction to Electric Fields by W. E. Rogers, McGraw-Hill Book Co., New York, 1954
- Fig. 18.9 — D. C. Hazen and R. F. Lehnert, Subsonic Aerodynamics Laboratory, Princeton
- Fig. 18.10 — D. C. Hazen and R. F. Lehnert, Subsonic Aerodynamics Laboratory, Princeton
- Fig. 18.11 — Photo courtesy of MCO Properties, Inc.
- Fig. 18.14 — (a) Naval Photographic Center
- Fig. 18.23 — The Balloon Works, Statesville, N.C.
- Fig. 18.31 — Hans C. Ohanian
- Fig. 18.33 — NASA
- Fig. 18.35 — U.S. Navy Photo
- Fig. 18.41 —
- page 473 — (Blaise Pascal) National Portrait Gallery, Smithsonian Institution
- page 480 — (Daniel Bernoulli) National Portrait Gallery, Smithsonian Institution

CHAPTER 19
- Fig. 19.12 — Physical Science Laboratory, National Scientific Balloon Facility
- page 495 — (Lord Kelvin) Wellcome Institute Library, London
- page 496 — (Robert Boyle) © British Museum
- page 506 — (Ludwig Boltzmann) AIP Niels Bohr Library

CHAPTER 20
- Fig. 20.5 — The Warder Collection
- Fig. 20.6 — AP, Wide World Photos
- Fig. 20.9 — U.S. National Bureau of Standards
- Fig. 20.10 — (b) U.S. National Bureau of Standards
- Fig. 20.12 — Photo CERN
- Fig. 20.17 — Robert Azzi, Woodfin Camp & Associates
- Fig. 20.18 — VANSCAN® thermogram by Daedalus Enterprises, Inc.
- page 515 — (Benjamin Rumford) © British Museum

CHAPTER 21
- Fig. 21.22 — Vorpal Galleries: New York, San Francisco, Laguna Beach, Calif.
- page 550 — (Sadi Carnot) AIP Niels Bohr Library
- page 555 — (Rudolph Clausius) National Portrait Gallery, Smithsonian Institution
- page 564 — (Walther Nernst) NMAH Archives Center, Smithsonian Institution

INTERLUDE V
- Fig. V.6 — UPI, Bettmann Archives
- Fig. V.7 — Los Alamos National Laboratory
- Fig. V.8 — Bureau of Reclamation
- Fig. V.9 — NASA
- Fig. V.10 — French Embassy Press & Information Division
- Fig. V.11 — Southern California Edison Co.
- Fig. V.12 — (b) Advanced Energy Technologies
- Fig. V.13 — NASA
- Fig. V.14 — GM Hughes Electronics
- Fig. V.15 — PG&E Photographs
- Fig. V.16 — PG&E Photographs
- Fig. V.17 — French Embassy Press & Information Division
- Fig. V.18 — New York Power Authority
- Fig. V.19 — PG&E Photographs
- Fig. V.20 — Novosti, Gamma
- Fig. V.22 — Kansas City Power & Light, Kansas Gas & Electric

CHAPTER 22
- Fig. 22.2 — L. S. Bartell, Univ. of Michigan
- Fig. 22.7 — National Center for Atmospheric Research, National Science Foundation
- Fig. 22.10 — Leybold-Heraeus

page 572	(Benjamin Franklin) National Portrait Gallery, Smithsonian Institution	
page 573	(Charles Augustin de Coulomb) The Bettmann Archive	

CHAPTER 23
| Fig. 23.22 | Robert Matthews, Princeton Univ. |
| Fig. 23.23 | Robert Matthews, Princeton Univ. |

CHAPTER 24
| Fig. 24.19 | Robert Matthews, Princeton Univ. |

CHAPTER 25
| page 628 | (Alessandro Volta) Lande Collection, AIP Niels Bohr Library |

CHAPTER 27
| Fig. 27.4 | (a) G. Holton, *The Project Physics Course.* Orlando, Fla.: Holt, Rinehart & Winston, 1970. |
| Fig. 27.15 | Earl Williams and Chatham Cook, High Voltage Laboratory, MIT |

CHAPTER 28
Fig. 28.1.	Jefimenko, *American Journal of Physics* 30, no. 19, 1962
Fig. 28.2.	Jefimenko, *American Journal of Physics* 30, no. 19, 1962
Fig. 28.6	(a) ©Richard Megna, Fundamental Photographs, 1980
page 687	(George Ohm) The Bettmann Archive

CHAPTER 29
Fig. 29.4	(b) NASA
Fig. 29.6	ARCO-Solar, Inc.
page 708	(Gustav Kirchhoff) The Bettmann Archive

INTERLUDE VI
Fig. VI.1	National Center for Atmospheric Research, National Science Foundation, Dave Baumhefner
Fig. VI.6	Hans C. Ohanian
Fig. VI.10	National Severe Storms Lab/NOAA, W. David Rust
Fig. VI.17	National Center for Atmospheric Research, Dave Baumhefner
Fig. VI.18	E. P. Krider, The Univ. of Arizona
Fig. VI.19	U.S. Department of Agriculture, Forest Division

CHAPTER 30
Fig. 30.12	Education Development Center, Newton, Mass.
Fig. 30.16	Education Development Center, Newton, Mass.
page 730	(Hans Christian Oersted) ©Smithsonian Institution
page 735	(Nikola Tesla) ©Smithsonian Institution
page 737	(Karl Friedrich Gauss) The Bettmann Archive
page 738	(Jean Baptiste Biot) The Bettmann Archive

CHAPTER 31
Fig. 31.6	Manfred Kage, Peter Arnold, Inc.
Fig. 31.10	(c)
Fig. 31.13	Leybold-Heraeus
Fig. 31.15	Lawrence Berkeley Laboratory
Fig. 31.16	Lawrence Berkeley Laboratory
Fig. 31.17	Brookhaven National Laboratory
Fig. 31.24	(a) "Lent to Science Museum, London," by the late J. J. Thompson, Trinity College, Cambridge, in the case of Cathode Ray Tube of J. J. Thomson
Fig. 31.41	Lawrence Berkeley Laboratory
page 753	(André Marie Ampère) AIP Niels Bohr Library
page 763	(Ernest Lawrence) Lawrence Berkeley Laboratory
page 766	(Sir Joseph John Thomson) The Bettmann Archive

CHAPTER 32
page 780	(Michael Faraday) AIP Niels Bohr Library
page 781	(Wilhelm Weber) NMAH Archives Center, Smithsonian Institution
page 793	(Joseph Henry) ©Smithsonian Institution

INTERLUDE VII
Fig. VII.1	NASA
Fig. VII.3	Hale Observatories
Fig. VII.22	Princeton Plasma Physics Laboratory
Fig. VII.23	Princeton Plasma Physics Laboratory
Fig. VII.25	Lawrence Livermore Laboratory
Fig. VII.26	Argonne National Laboratory

CHAPTER 33
| Fig. 33.6 | H. J. Williams, R. M. Bozorth, and W. Shockley, *Physical Review* 75 (1949): 155 |
| Fig. 33.9 | Education Development Center, Newton, Mass. |

CHAPTER 34
| Fig. 34.26 | Westinghouse Electric Corp. |

INTERLUDE VIII
Fig. VIII.2	A. Leitner, RPI
Fig. VIII.3	Cyrogenic Technology, Inc.
Fig. VIII.10	A. D. Little, Inc. Cambridge, Mass.
Fig. VIII.13	Intermagnetics General Corporation
Fig. VIII.14	General Electric
Fig. VIII.15	Argonne National Laboratory
Fig. VIII.16	Fermilab Photo
Fig. VIII.17	Intermagnetics General Corporation
Fig. VIII.18	General Electric
Fig. VIII.19	Brookhaven National Laboratories
Fig. VIII.20	Walter Sullivan, NYT Pictures

CHAPTER 35
Fig. 35.11	Stanford Linear Accelerator Center
Fig. 35.13	Hans C. Ohanian
Fig. 35.17	Education Development Center, Newton, Mass.
Fig. 35.18	Education Development Center, Newton, Mass.
page 849	(James Maxwell) AIP Niels Bohr Library

CHAPTER 36
Fig. 36.12	Hale Observatories
Fig. 36.15	Hale Observatories
Fig. 36.16	Polaroid Corporation
page 874	(Heinrich Hertz) Berlin Univ., WestphalCollection, Courtesy AIP Niels Bohr Library

CHAPTER 37
Fig. 37.10	Davis Instruments Corporation
Fig. 37.11	NASA
Fig. 37.18	AT&T Bell Laboratories
Fig. 37.31	Polaroid Corporation
Fig. 37.32	Polaroid Corporation
Fig. 37.33	John Shelton
page 899	(Christiaan Huygens) Museum Boerhaave
page 914	(Étienne Malus) Library of the Academy of Sciences, Paris, courtesy AIP Niels Bohr Library

CHAPTER 38

Fig. 38.24	C. C. Jones, Union College
Fig. 38.31	Hale Observatories
Fig. 38.32	Public Information Office, MMT Observatory
Fig. 38.33	Drawing by Chas. Addams ©1983, *The New Yorker Magazine*
Fig. 38.34	Hans C. Ohanian
Fig. 38.36	Point Reyes National Seashore, Point Reyes, Calif.

CHAPTER 39

Fig. 39.5	Bausch & Lomb
Fig. 39.11	Educational Development Center, Newton, Mass.
Fig. 39.12	(a) C. C. Jones, Union College
Fig. 39.20	C. C. Jones, Union College
Fig. 39.23	National Radio Astronomy Observatory, operated by Associated Universities, Inc., under contract with the National Science Foundation
Fig. 39.25	(b) Bausch & Lomb
page 947	(Thomas Young) NMAH Archives Center, Smithsonian Institution
page 953	(Albert Michelson) Univ. of Chicago

CHAPTER 40

Fig. 40.1	F. W. Sears, *Optics* © 1949, Addison-Wesley, Reading, Mass., Fig. 9.8. Reprinted with permission.
Fig. 40.5	C. C. Jones, Union College
Fig. 40.9	C. C. Jones, Union College
Fig. 40.11	C. C. Jones, Union College
Fig. 40.12	C. C. Jones, Union College
Fig. 40.13	Space Telescope Science Institute, Baltimore, Md.
Fig. 40.14	Cornell Univ.
Fig. 40.15	Robert L. Mutel, Univ. of Iowa
Fig. 40.19	C. C. Jones, Union College
Fig. 40.20	Sears, Zemansky and Young, *University Physics*, © 1982, Addison-Wesley, Reading, Mass. Reprinted with permission. Page 789, Fig. 41–18.
Fig. 40.21	R. Gronsky, Lawrence Berkeley Laboratory
Fig. 40.22	R. Olberg, Union College
Fig. 40.23	Education Development Center, Newton, Mass.

page 972	(Augustin Fresnel) AIP Niels Bohr Library
page 977	(John William Strutt Rayleigh, 3rd Baron) The Royal Society, London, courtesy AIP Niels Bohr Library

CHAPTER 41

page 989	(Albert Einstein) AIP Niels Bohr Library
page 1001	(Hendrik Lorentz) The Bettmann Archive

INTERLUDE IX

Fig. IX.3	NASA
Fig. IX.10	From Arthur Eddington, *Space, Time and Gravitation*, courtesy Cambridge Univ. Press
Fig. IX.23	Hewlett-Packard Company
Fig. IX.25	R. F. Vassot and M. L. Levine, Smithsonian Astrophysical Observatory
Fig. IX.30	Riccardo Giacconi, Harvard Smithsonian Center for Astrophysics
Fig. IX.31	Hale Observatories
Fig. IX.32	After Eardley and Press, *Annual Review of Astronomy and Astrophysics*, 1975

CHAPTER 42

Fig. 42.13	(a,b,c) E. R. Huggins, New York, 1968; (d) C. C. Jones, Union College
page 1031	(Max Planck) AIP Niels Bohr Library
page 1037	(Arthur Compton) AIP Niels Bohr Library
page 1041	(Max Born) AIP Niels Bohr Library
page 1042	(Werner Heisenberg) AIP Niels Bohr Library

CHAPTER 43

Fig. 43.5	Hale Observatories Photograph
Fig. 43.6	Hans C. Ohanian, *Modern Physics*, Prentice-Hall
page 1051	(Joseph von Fraunhofer) NMAH Archives, Smithsonian Institution
page 1055	(Sir Ernest Rutherford) The Bettman Archive
page 1058	(Niels Bohr) Princeton Univ., Courtesy of AIP Niels Bohr Library

page 1064	(Louis Victor, Prince de Broglie) AIP Meggers Gallery of Nobel Laureates
page 1066	(Erwin Schrödinger) Ullstein, courtesy AIP Niels Bohr Library

INTERLUDE X

Fig. X.4	Hughes Aircraft
Fig. X.7	The Warder Collection
Fig. X.8	NASA
Fig. X.9	Bell Laboratories
Fig. X.10	Arnold Zann ©1980 Black Star
Fig. X.11	United Technologies
Fig. X.13	U.S. National Bureau of Standards
Fig. X.14	Lawrence Livermore National Laboratory
Fig. X.15	Lawrence Livermore National Laboratory
Fig. X.16	Lawrence Livermore National Laboratory
Fig. X.17	Lawrence Livermore National Laboratory
Fig. X.19	From H. M. Smith, *Principles of Holography*, 2nd ed., ©John Wiley & Sons, Inc., 1975
Fig. X.24	Museum of Holography, New York City, ©Ronald R. Erickson, 1983
Fig. X.25	Karl Stetson
Fig. X.26	RW

CHAPTER 44

Fig. 44.9	Hans C. Ohanian, *Modern Physics*, Prentice-Hall
Fig. 44.22	Bell Laboratories
Fig. 44.26	The Warder Collection
page 1075	(Arnold Sommerfield) AIP Niels Bohr Library, Physics Today Collection
page 1078	(Wolfgang Pauli) The Bettmann Archive

CHAPTER 45

Fig. 45.7	(b) Amperex Electronic Corp.
Fig. 45.9	(left) The MIT Museum; (right) Cavendish Laboratory, Cambridge, England
Fig. 45.10	Lawrence Berkeley Laboratory
page 1095	(James Chadwick) AIP Meggers Gallery of Nobel Laureates
page 1098	(Enrico Fermi) The Bettmann Archive
page 1104	(Antoine Becquerel) AIP Niels Bohr Library

INTERLUDE XI

page 1106	(Marie and Pierre Curie) AIP Niels Bohr Library
Fig. XI.1	Los Alamos National Laboratory
Fig. XI.8	Los Alamos National Laboratory
Fig. XI.9	Los Alamos National Laboratory
Fig. XI.11	Los Alamos National Laboratory
Fig. XI.17	Dr. Harold E. Edgerton, MIT
Fig. XI.18	NASA
Fig. XI.19	From *The Effects of Nuclear War*, Office of Technology Assessment, U.S. Congress
Fig. XI.24	General Electric

CHAPTER 46

Fig. 46.1	Fermilab Photo
Fig. 46.2	Fermilab Photo
Fig. 46.4	Photo CERN
Fig. 46.5	Photo CERN
Fig. 46.6	Photo CERN
Fig. 46.7	(a) The MIT Bubble Chamber Group
Fig. 46.8	Photo CERN
Fig. 46.9	Photo CERN
Fig. 46.10	(left) Photo CERN
Fig. 46.11	Lawrence Berkeley Laboratory
Fig. 46.15	Photo CERN
Fig. 46.20	Photo CERN
page 1135	(Richard Feynman) Caltech Photo
page 1136	(Paul Dirac) AIP Meggers Gallery of Nobel Laureates

INDEX

aberration:
  chromatic, 932
  spherical, 923–24, 932
A-bomb, XI4–XI5
absolute acceleration, 101
absolute motion, 989
absolute temperature scale, 495–96, 523
absolute thermodynamic temperature scale, 499, 554
absolute time, 4–5, 87
absolute zero, entropy at, 564
absorbed dose, IV7
accelerated charge:
  electric field of, 859–63
  induced magnetic field of, 866
  magnetic field of, 863–66
  polarization of radiation of, 891
  radiation field of, 859, 861
acceleration, 32–42, 51
  absolute, 101
  angular, see angular acceleration
  average, 32
  average, in three dimensions, 76
  of center of mass, 257–58
  centripetal, 85–86, 101–2, 146–50
  conversion factors for, A24
  g as, 37–38, 40–42
  instantaneous, 33–34
  instantaneous, in three dimensions, 76
  motion with constant, 34–37, 77–78
  transformation equations for, 88–89
acceleration of gravity, 37–42, 214
  formula for, 51
  measurement of, 41–42, 395
  universality of, 37
  variation of, with altitude, 214
  variation of, with latitude, 102
accelerators:
  at Fermilab, 323, VIII8
  linear, 624, 857, 1123, 1139, 1143
  for particles, 528, 762–64, 1111–12, 1122–24, VIII9
accelerometers, 115
acceptor impurities, 693, 1086
accidents:
  automobile, 288, III1–III16
  nuclear, V12–V16
  reactor, V12–V16
AC circuits, 823–48
  simple, with an external electromagnetic force, 823–28
accretion, by black hole, IX12–IX13

AC current, 823
  hazards of, 721–23
acid rain, V11
acoustic micrograph, 443
action and reaction, 109–12
action-at-a-distance, 125, 213, 587, 590–91
action-by-contact, 587, 591
activity, radioactive, IV3–IV5
AC voltage, 788–89
Adams, J. C., 211
addition law for velocity, Galilean, 990
addition of vectors, 54–57
  associative law of, 56–57
  commutative law of, 56
  by components, 59–60, 61
  graphical method for, 61
  trigonometric method for, 61, 109
addition of velocities, relativistic, 1008–9
adiabatic equation, for gas, 533–34
adiabatic process, 531
age:
  of chemical elements, II5–II6
  of globular clusters, II5–II6
  of the universe, II5–II6
air bag restraint systems, III11, III12, III14
air burst, of bomb, XI6, XI7
air conditioner, 554
air cushion, 543
airfoil, flow around, 470–71, 480–81
air friction, 37, 38
air pollution, V10–V11
air resistance, in projectile motion, 83–84
Alamagordo, New Mexico, XI5
Alcator, VII9
Alfvén wave, VII6
Alpha Centauri, Prelude-8
alpha decay, 1104–5
alpha particles, scattering of, 1055–56
alpha rays, 1103–4, IV1–IV2
  radiation damage by, IV5
alternating current, 823
  hazards of, 721–23
alternating emf, 788–89
alternative energy sources, V5–V10
ammeter, 715, 723, 724, 773
ampere (A), 15, A20
Ampère, André Marie, 733, 753
Ampère's Law, VII4, 753–56
  displacement current and, 849–51

electric flux and, 850
in magnetic materials, 811
modified by Maxwell, 851, 853
right-hand rule for, 754
amplitude:
  of motion, 380
  of wave, 415
Angers, France, bridge collapse at, 402
angle:
  bearing, 2
  Brewster's, 912–13
  conversion factors for, A22
  definition of, A5
  elevation, 82–83
  of incidence and of reflection, 901–2
  latitude, 2
  longitude, 2
angular acceleration:
  average, 300
  constant, equations for, 301–2
  instantaneous, 300
  rotational motion with constant, 301–3
angular frequency, 380–81
  of physical pendulum, 395–96
  of simple harmonic oscillator, 385
  of torsional pendulum, 397
  of wave, 416
angular magnification:
  of magnifier, 935
  of microscope, 936
  of telescope, 937
angular momentum, 219
  central forces and, 312
  component of, along axis of rotation, 315–16
  cross product and, 312
  of Earth, 312, 323–24
  in elliptical orbit, 226
  of particle, 310–12, 327–31
  quantization of, 807–8
  rate of change of, 327–29
  rate of change of, relative to center of mass, 342
  of a rigid body, 312–17
  of a system of particles, 310–12, 327–31
  for uniform circular motion, 312
angular momentum, conservation of:
  of free particle, 311
  law of, 329
  for motion with central force, 312

In the two-volume edition, all pages after 570 are in Volume Two.
Interlude V is the last one in Volume One.

angular momentum (*continued*)
  in planetary motion, 219–20
  in rotational motion, 335–39
angular-momentum quantum number, 1058, 1058–59, 1076–77
angular resolution, of telescope, 977–80
angular velocity:
  average, 299
  instantaneous, 299
antibaryons, 1128
antiderivative, A14
antimatter-matter annihilation, 578
antimesons, 1128
antineutrino, 1105–7
antinodes and nodes, 429
antiparticles, 1128
antiquarks, 1139–40
anvil cloud, in thunderstorm, VI5
aphelion, 218
  of comets, 226
  of planets, 220
apogee, Prelude-5
  of artificial satellites, 221
apparent weight, 149–50
arch bridge, 370
Archimedes' Principle, 478–79
area, A4
  conversion factors for, A23
areas, law of, 219, 312
Arecibo radiotelescope, 894, 979–80
artificial satellites, 217–18, 220–22, 1090
  apogee of, 221
  perigee of, 221
  period of, 221
associative law of vector addition, 56–57
astigmatism, 934
astrology, 237–38
atmosphere, 473
  as electric machine, VII1
atmospheric electric circuit, VII1
atmospheric electric current, VII1–VII2
atmospheric electric field, 635–36, VII1–VII3
  time dependence of, VII3
atmospheric electricity, VII1–VII12
atmospheric pressure, 476–77
atom, 1121
  electron configuration of, 1079–81
  electron distribution in, 571
  nuclear model of, 1056
  nucleus of, *see* nucleus
  photograph of, Prelude-14, Prelude-15, X12
  quantum structure of, 1074–86
  stationary states of, 1060
  structure of, 571
  table of, Prelude-14
atomic clock, Cesium, 10
atomic clocks, portable, 11
atomic constants, A27–A28
atomic magnetic moments, 807–9
atomic mass unit, 13, 1097–98
atomic standard of mass, 12–13
atomic standard of time, 10
atomic states, quantum numbers of, 1080–81
atomic structure, 1049–57
atom smashers, 1121
Atwood's machine, 154, 331
Aurora Borealis, 765, VIII1

automobile accidents, 288, III1
automobile battery, 578–79, 707
automobile collision, 273, 288–89, III2–III14
  energy transfer in, III5
  force-deformation characteristics in, III3–III4, III7–III8
  injury to occupant in, III6–III11
  secondary, III6–III9
automobiles:
  automobiles vs., III4–III6
  barriers vs., III2–III4
  crash tests of, 190, 288, III2, III5, III8–III14
  electric fields and, 634
  energy conversion in, V1–V2
  federal safety standards for, III2
  solar, V9
average acceleration, 32
  in three dimensions, 76
average angular acceleration, 300
average angular velocity, 299
average power, 198
average speed, 26–27
average velocity, 27–29
  in three dimensions, 74–75
Avogadro's number, 12, 497
axis, instantaneously fixed, 339–40
axis of symmetry, I5

Babinet, Jacques, 981
Babinet's Principle, 981–82
back emf, 794
background radiation, cosmic, II6
balance, 102, 106–7, 127
  beam, 106, 127
  Cavendish, 215–16, 397–98, 576, IX1
  Coulomb's, 576
  spring, 106, 127
balance wheel, 397
ballistic curve, 83–84
ballistic missile, 83–84, 143
ballistic pendulum, 280–81
ball lightning, VII11
Balmer, Johann, 1052–53
Balmer series, 1052–53
band spectrum, 1083
bandwidth, 838
banked curve, 148
Bardeen, John, VIII6n
Bardeen-Cooper-Schrieffer (BCS) theory, VIII5–VIII7
barometer, mercury, 477
Barton, O., 473
baryon, 1128–30
baryon number, 1128, 1132
  conservation law for, 1112–13, 1132
base units, 14
  of SI system, A20
Basov, Nicolai G., X2n
bassoon, 445
bathyscaphe, 486
bathysphere, 473–74
battery:
  automobile, 578–79, 705
  dry cell, 706
  internal resistance of, 708–9
  lead-acid, 578–79, 705, 707–8
BCS (Bardeen-Cooper-Schrieffer) theory, VIII5–VIII7
beam balance, 106, 127

bearing angle, 2
beat frequency, 426–27
beats, of a wave, 426–27
becquerel (Bq), 1110, IV4
Becquerel, Antoine Henri, 1103–4, IV4n
Bednorz, J. Georg, VIII2n
Beebe, C. W., 473, 474
bending of light rays, 4, IX3–IX6
Bernoulli, Daniel, 480
Bernoulli's equation, 480–82
best values, of fundamental constants, A27–A29
beta decay, 1105–7
beta rays, 1103–4, IV1–IV2
  radiation damage by, IV5
betatrons, 786–87
Bethe, Hans A., XI6
Big Bang, 1137, II1, II5–II6
  element formation during, II5–II6
  radiant heat of, II6
bimetallic strip thermometers, 500, 522
binary star system, 223, 239
  resolution of telescope and, 978–79, 980
binding energy of nucleus, 1099–1103, XI1
  curve of, 1101, 1102, XI6
binoculars, 942
Biot, Jean Baptiste, 738
Biot-Savart Law, 738–42
birefringence, 913
blackbody, spectral emittance of, 1030
blackbody radiation, 1027–29, 1031–33
black hole, II8, IX10–IX13
  accretion by, IX12–IX13
  singularity in, IX12
blackout, 152
blast waves, nuclear, XI6
block and tackle, 364
blowhole, 445
blue, 1141–43
body-centered cube, I9
body-mass measurement device, 103–4, 383–84
Bohr, Niels, 323–24, 1058, 1075
Bohr magnetron, 808
Bohr radius, 1059
Bohr's Complementarity Principle, 1068
Bohr's Correspondence Principle, 1063
Bohr's postulates, 1058
boiling-water reactor, XI11
Boltzmann, Ludwig, 506
Boltzmann's constant, 498
bomb:
  A-, XI4–XI5
  air burst of, XI6
  fission, XI4–XI5
  H-, XI6
  nuclear, XI4–XI6
  plutonium, XI4–XI5
  uranium, XI4–XI5
bonds, interatomic, 1081
boom, sonic, 450
Born, Max, 1041, 1064
bottom quark, 1143
boundary conditions, 429
bound charges, 667
bound orbit, 193
Boyle, Robert, 496
Boyle's law, 496

In the two-volume edition, all pages after 570 are in Volume Two.
Interlude V is the last one in Volume One.

Brackett series, 1053
Braginsky, V. B., IX2
brake:
  hydraulic, 475
  power, 374–75
brake horsepower, 335
branch method, for circuits, 711
breeder reactors, V5, XI12
Breeding, E., III13
Bremsstrahlung, 863, 882
Brewster's angle, 912–13
Brewster's Law, 912–13
bridge collapse:
  at Angers, France, 402
  at Tacoma Narrows, 430–31
bridges:
  arch, 370
  suspension, 359–60
  thermal expansion and, 521
  truss, 358–59
British system of units, 8–9, 13, 26
British thermal unit (Btu), 516
Brown's Ferry, Alabama, reactor accident at, V13
bubble chamber, 528–29, 763–64, 1125–26
bulk modulus, 366–67
bullet, measuring speed of, 291
buoyant force, 478–79
Bureau International des Poids et Mesures (BIPM), 6–7, 12, 41
Bureau Internationale de l'Heure, 11

cable, superconducting, 701, VIII9–VIII10
calcite, birefringence displayed by, 913
calculus, 35, A10–A19
Callysto, 221
caloric, 515
calorie, 516, 517
camera:
  photographic, 932–33
  pinhole, IV10–IV11, 933
candela, 15, A20
capacitance, 662–67
  conversion factors for, A26
  of the earth, 663–64
  of a single conductor, 663
capacitive reactance, 826
capacitors, 662–67
  circuit with, 826
  cylindrical, 671
  energy in, 674–76
  guard rings for, 677
  multiplate, 678
  in parallel, 665–66
  parallel-plate, 664–65, 678
  in series, 665–67
  spherical, 671
Capella, 1034
Caph, spectrum of, 1051
capillary wave, 451
carbon:
  isotopes of, 1096–98
  mass of, 1098
carbon dioxide-nitrogen laser, X4
Carnot, Sadi, 550
Carnot cycle, 549–50, 552–55, 559–60
Carnot engine, 549–54
  efficiency of, 552–54
  Second Law of Thermodynamics and, 560
Carnot's theorem, 559–60
Cartesian diver, 486
cat, landing on feet by, 339
cathode rays, 766–67
cathode-ray tube, 766
Cavendish, Henry, 215
Cavendish balance, 215–16, 397–98, IX1
cavity oscillation, 854–58
  electromagnetic, 856–57
cavity radiation, 1029
celestial mechanics, 211
  accuracy of, 211
cell:
  thunderstorm, VI4
  triple-point, 499
Celsius temperature scale, 500
center of force, 143
center of mass, 251–59
  acceleration of, 257–58
  of continuous mass distribution, 254
  gravitational force acting on, 354–55
  motion of, 257–59
  rate of change of angular momentum relative to, 342
  rotation about, 342
  velocity of, 257, 279
central forces, angular momentum and, 312
centrifugal forces, 149
centrifuge, 86
centripetal acceleration, 85–86, 101–2, 146–50
  Newton's Second Law and, 147
centroid, 254
ceramic superconductors, VIII1–VIII2
Čerenkov counters, 1124
CERN (Organisation Européenne pour la Recherche Nucléaire), 528, 762, 1005, 1009, 1122–25, 1127, 1138, 1142
Cerro Prieto (Mexico), V9
Cesium atomic clock, 10
Cesium standard of time, 10
cgs system of units, 582
Chadwick, James, 1095
chain reactions, XI3–XI4
Chalk River, Canada, reactor accident in, V13
changes of state, 527–29
characteristic X rays, 1062
charge, electric, 573–79
  bound, 667–68
  bound, in dielectrics, 667–68
  conservation of, 578–79
  conversion factors for, A26
  of electron, 573
  of electron, measurement of, 603
  elementary, 574
  of elementary particles, 1128–29, 1140
  generation of, in thunderstorms, VI5–VI6
  image, VI7
  motion of, in magnetic field, 759–60
  of particles, 573–74, 577, 1140
  point, 575
  of proton, 573, 374
  quantization of, 577
  SI unit of, 769
  static equilibrium of, 617–18
  surface, on dielectric, 670–71
charged surface, flat, potential of, 628
Charles' Law, 496
charm, of quarks, 1142–43
chemical composition of universe, II5–II6
chemical elements, A30–A32
  age of, II6
  during Big Bang, II5–II6
  Periodic Table of, Prelude-14, 1079–81, A30
  transmutation of, IV4
chemical reactions, conservation of charge in, 579
Chernobyl, USSR, reactor accident in, V12–V14
China syndrome, V15
chromatic aberration, 932
chromatic musical scale, 441
circuit, electric:
  AC, 823–48
  atmospheric, VI1
  branch method for, 711
  with capacitor, 826
  with inductor, 827–28
  LC, 828–32
  LCR, *see* LCR circuit
  loop method for, 710
  multiloop, 710–12
  RC, 717–21
  with resistor, 824
  RL, 796–98
  single-loop, 707–9
circular aperture:
  diffraction by, 976–80
  minimum in diffraction pattern of, 977
circular orbits, 216–18, 1075
  energy for, 226
  in magnetic field, 761
clarinet, sound wave emitted by, 440–41
classical electron radius, 1068
classical mechanics, quantum mechanics vs., 1050
classical physics, 4
Clausius, Rudolph, 555
Clausius statement of Second Law of Thermodynamics, 560
Clausius' theorem, 555–56, 561
clipper ship, power of, V1
clock:
  Cesium atomic, 10
  pendulum, 392, 395
  portable atomic, 11
  synchronization of, 5, 11, 992–95
close packing, I7–I10
cloud chamber, 529
coal:
  production of, V4
  reserves of, V4
Cockcroft, J. D., 1111–12
coefficient of cubical thermal expansion, 519–20
coefficient of kinetic friction, 135–37
coefficient of linear thermal expansion, 518, 521
coefficient of restitution, 293
coefficient of static friction, 136, 138–39
coefficients of friction, 134–40
coherence of light, 954, X1–X2, X6
coherent radiation, X1

In the two-volume edition, all pages after 570 are in Volume Two.
Interlude V is the last one in Volume One.

colliding beams, 1124
collimator, IV10
collisions, 272–95
    automobile, 273, 288–89, III2–III14
    impulsive forces and, 272–76
    of nuclei, 285–88
    of particles, 285–88
collisions, elastic, 276–79
    conservation of energy in, 277, 282
    conservation of momentum in, 277, 282
    of deuteron and proton, 282–83
    in one dimension, 276–79
    one-dimensional, speeds after, 277–79
    in two dimensions, 281–83
collisions, inelastic, 279
    internal kinetic energy in, 287–88
    rest-mass energy in, 286–87
    totally, 279, 284
    in two dimensions, 283–85
color:
    of quarks, 1141–42
    of visible light, 883
combination of velocities, relativistic, 1008–9
combination principle, Rydberg-Ritz, 1054
comets, 227
    Halley's, 240
    orbits of, 226
    perihelion of, 227
    period of, 227
    tails of, 888
common logarithm, A3
communication satellites, 217–18
commutative law of vector addition, 56
compass needle, 730
Complementarity Principle, 1068
complementary plate, Babinet's Principle and, 981–82
component of angular momentum along axis of rotation, 315–16
components, 59–66
compression, 365, 367–68
Compton, Arthur Holly, 1037
Compton effect, 1037–39
Compton wavelength, 1068
concave spherical mirror, 923
Concorde SST, sonic boom of, 450
concrete, thermal expansion of, 521
conduction band, 1085
conduction of heat, 524–27
conductivity, 690
conductivity, thermal, 525–26
conductors, 579
    in electric field, 616–19
    energy of a system of, 652–54
    insulators vs., 692
conical pendulum, 159
conservation laws, 160–61
    for baryon number, 1112–13, 1132
    elementary particles and, 1132–33
    for mass number, 1112–13, 1132
    symmetry and, 161
conservation of angular momentum:
    of free particle, 311
    law of, 329
    for motion with central force, 312
    in planetary motion, 219–20
    in rotational motion, 335–39

conservation of electric charge, 578–79
    in chemical reactions, 579
conservation of energy, 161, 174–75, 184–210, 225
    general law of, 196
    in Lenz' Law, 788
    in nuclear reactions, 1113–14
    in one-dimensional elastic collision, 277
    in simple harmonic motion, 388, 392
    in two-dimensional elastic collision, 282
conservation of mass, 160
conservation of mechanical energy, 173–74, 187–88
    law of, 173–74, 187–88
    in rotational motion, 333
conservation of momentum, 113–14, 246–49, 277, 279
    in elastic collisions, 276–77, 282
    in fields, 591
    law of, 114, 246
conservative electric field, 633–34
conservative force, 184–90
    gravity as, 224
    path independence and, 185
    potential energy of, 186–90
conserved quantities, 113–14
constant:
    fine-structure, 1068
    fundamental, best values of, A27–A29
constant-volume gas thermometer, 499
constructive and destructive interference, 424–25, 947
    in Michelson interferometer, 953
    in standing electromagnetic wave, 949
    for wave reflected by thin film, 950–52
"contact" forces, 572–73
contact potential, 618–19
containment shell, V14
contamination, radioactive, V15–V16
continuity equation, 470
contraction of universe, II7
control rod, in nuclear reactor, V13, V14, XI10–XI11
convection, 526
conversion factors, 18–20, A22–A26
conversion of energy, see energy conversion
conversion of units, 17–18
converter reactor, XI11
convex spherical mirror, 924
cooling towers, V12
Cooper, Leon N., VIII6n
Cooper pairs, VIII6
Coordinated Universal Time (UTC), 10–11
coordinate grid, 2–3, 5–6, 86–88
coordinates:
    cylindrical, 2–3
    Galilean transformation of, 66n, 87, 88, 990, 998
    Lorenz transformation of, 66n, 1003
    origin of, 2, 5–6
    rotation of, 66–68
    transformation of, 66n, 67
Copernicus, Nicolas, 216
core:
    of Earth, 454
    of nuclear reactor, V13–V14

corner reflector, 902–3
    on Moon, 903, X5
corona discharge, VI8
Correspondence Principle, 1063–64
$\cos^{-1}$, A7
cosecant, A6
cosine, A6
    formula for derivatives of, 383
cosines, law of, A9
cosmic background radiation, II6
cosmic ray, 286, IV2–IV3
    primary, IV3
    secondary, IV2–IV3
    trajectory of, 760
Cosmological Principle, II4
cosmological red shift, 897
cosmology, II1
cotangent, A6
coulomb (C), 574
Coulomb, Charles Augustin de, 573
Coulomb's Law, 574–77, 599
    accuracy of, 576–77
    in vector notation, 575
critical angle, for total internal reflection, 908
critical magnetic field, VIII3
critical mass, XI4
critical radii, for nuclear explosion, XI8–XI9
crossed electric and magnetic fields, 765–67
cross product, 63–66, 600
    angular momentum and, 310–12, 328
    by components, 538
    in definition of torque, 325
    of unit vectors, 64–65
crush distance, III2, III3, III6, III13
crystal, I4–I14
    close packing and, I7–I10
    defects in, I10–I112
    of virus particles, I2
crystal lattice, I2
crystallites, I4
crystallography, I1
crystal structure, I2–I4
cube: I8, I9
curie (Ci), 1111, IV4–IV5
Curie, Marie Sklowdowska, 1106, IV4n
Curie, Pierre, 1106, IV4n
Curie temperature, 814
current, electric, 683–85
    AC, 823
    AC, hazards of, 721–23
    atmospheric, VI1–VI2
    in capacitor circuit, 826
    conversion factors for, A26
    DC, 703–28
    DC, hazards of, 721–23
    displacement, 849–52
    generating magnetic field, 738–39
    in inductor circuit, 827–28
    "let-go," 722
    magnetic force between, 733
    in parallel LCR circuit, 839–40
    persistent, VIII2
    in series LCR circuit, 835
    SI unit of, 769
    time-dependent, 717–21
current density, 685
current loop:

In the two-volume edition, all pages after 570 are in Volume Two.
Interlude V is the last one in Volume One.

potential energy of, 770–71
  torque on, 769–71
current resistor circuit, 824
curvature of space, IX3–IX6
  perihelion precession and, IX7
curve:
  banked, 148
  of binding energy, 1101, 1102, XI6
  of potential energy, 192–94
curved spacetime, IX7–IX8
cutoff frequency, in plasma, VII7
cutoff wavelength, 1047
cycloid, 340
cyclotron, 761–62, 1111–12
cyclotron emission, 882
cyclotron frequency, 760–61
Cygnus X-1, 239, IX12–IX13
cylindrical capacitor, 671
cylindrical coordinates, 2–3

damped harmonic motion, 399–400
damped oscillations, 399–402
damped oscillator:
  driving force on, 402
  energy loss of, 401
  Q of, 401–2
  resonance of sympathetic oscillation of, 402
  ringing time of, 401–2
dart leader, in lightning, VI9
daughter material, 1104–5, IV4
da Vinci, Leonardo, 135–36
Davisson, C. J., 1064
day:
  mean solar, 9
  sidereal, 101n
  solar, 9, 10n
DC current, 703–28
  hazards of, 721–23
  instruments used in measurement of, 715–17
DC-10 airliner crash near Orly, 1064, 1084
de Broglie, Louis Victor, Prince, 1064
de Broglie wavelength, 1064, 1084
Debye distance, VII2
decay of particles, 1104–8
decay rates of radioisotopes, 1109–11
deceleration, 34
decibel, 440
dees, 761
defects, in crystals, I10–I12
deflection of light, by Sun, 4, IX3–IX6
degree absolute, 495
degree-days, V17
degree of arc, definition of, A5
degree of freedom, 506n
degrees, A5
density, 14, 253–54
  conversion factors for, A24
  critical, for universe, II7–II8
  of field lines, 597–98
  of fluid, 467–68
  of nucleus, 1099
deoxyribonucleic acid, Prelude-13, Prelude-15
depth finder, 458
derivative:
  definition of, A10
  partial, 191–92

  of the potential, 634–35
  rules for, A12
derived unit, 14–15
  of SI system, A21
destructive and constructive interference, 424–25, 947–48
  in Michelson interferometer, 953
  in standing electromagnetic wave, 949
  for wave reflected by thin film, 950–52
determinant, 65
determinism, 123
deuterium:
  abundance of, in universe, II8
  as nuclear fuel, V6
deuterium-deuterium reaction, VII8–VII9
deuterium-tritium reaction, VII8–VII9
deuteron:
  constants for, A28
  elastic collision of proton and, 282–83
Dialogues (Galileo), 44
diamagnetic materials, 814–17
  permeabilities of, 817
diamagnetism, 814–17
diamond, I9
diatomic gas, energy of, 505–6
diatomic molecule, 192
  vibrations of, 304–5
Dicke, R. H., experiments on universality of free fall by, IX2
dielectric, 662, 667–76
  electric field in, 668–69
  energy density in, 675
  Gauss' Law in, 672–74
  linear, 668
  surface charge on, 671
dielectric constant, 669
differential equation, numerical solution of, 142–46
differentiation, A10
  rules for, A12
diffraction, 455–56, 971–88
  at a breakwater, 455
  by circular aperture, 976–80
  by a single slit, 971–76
  of sound waves, 456
  of water waves, 455
  X-ray, I3
diffraction pattern, 456
  for separation of isotopes, XI4–IX5
  of single slit, 973, 975
dimensions, 16–17
diodes, 1087–88, I12
Dione, 239
dip angle, 74
dip needle, 746
dipole, 599–601
  electric field of, 637–38
  magnetic, 743–45
  potential of, 638
  potential energy of, 600
  torque on, 600
dipole moment, 600–601
  magnetic, 743–44, 770
  permanent and induced, 703–28
Dirac, Paul Adrien Maurice, 1064, 1136
direct current, 703–28
  hazards of, 721–23
dislocation, I10–I12
dispersion, of light, 909

dispersive medium, 420
displacement, 8
displacement current, 849–52
displacement field, electric, 674
displacement vector, 52–54
dissipation of energy:
  in U.S., V3
  in U.S., per capita, V3
distributive law, 62n
DNA, Prelude-13, Prelude-15
domains, magnetic, 812
donor impurities, semiconductors with, 693, 1086
doping, I12
Doppler, Christian, 446
Doppler shift, 446–50, IX10
  of light, 888–90
doses, radiation, IV9
  effects of acute, IV9
dot product, 62
  components, 62–63
  in definition of work, 165–68
  of vectors, 62–64, 165–66
double-exposure hologram, X12
double refraction, 913
"doubling the angle on the bow," 70
down quark, 1139–40
dowsing, 818
drift velocity:
  of free-electron gas, 686, 688–89
  of free electrons, 689
driving force, on damped oscillator, 402
dry cell battery, 706
dynamic imbalance, 313–14
dynamics, 99–159
  fluid, 479–83
  of rigid body, 325–53
dynodes, 1036, 1037

Earnshaw's theorem, 640
Earth, 220
  angular momentum of, 312, 323–24
  capacitance of, 663–64
  core of, 454
  escape velocity from, 228
  magnetic moment of, 744–45
  mantle of, 454
  moment of inertia of, 315–17, 320, 322
  orbital speed of, 219–20
  perihelion of, 237
  reference frame of, 101–2
  rotation of, 9–10, 101
  spin of, 316–17
earthquake, 453–55
  in Alaska, 453
  in San Francisco, 454–55
Eddington, Arthur, 1039
eddy currents, 799
effects of acute radiation doses, IV9
efficiency:
  of Carnot engine, 552–54
  of energy conversion, V2–V3
  of engines, 548–49
eigenfrequencies, 430
Einstein, Albert, 4, 196, 212, 234, 989, 1034–35, 1137, IX1, IX3
  mass-energy equation of, 892
  photoelectric equation of, 1036
elastic body, 140, 365

In the two-volume edition, all pages after 570 are in Volume Two.
Interlude V is the last one in Volume One.

elastic body (*continued*)
  elongation of, 365–68
elastic collision, 275–79
  conservation of energy in, 277, 282
  conservation of momentum in, 277, 282
  of deuteron and proton, 282–83
  in one dimension, 276–79
  speeds after one-dimensional, 277–79
  in two dimensions, 281–83
elastic force, 140–42
elasticity of materials, 365–69
elastic limit, 141, 368–69
electrical measurements, 715–17
electric charge, 573–79
  bound, 667
  bound vs. free, in dielectrics, 667
  conservation of, 578–79
  conversion factors for, A26
  of electron, 573
  of electron, measurement of, 603
  elementary, 574
  of elementary particles, 1128, 1140
  generation of, in thunderstorms, V15–VI6
  image, VI7
  motion of, in magnetic field, 759–60
  of particles, 573–74, 577, 1140
  point, 575
  of proton, 573, 574
  quantization of, 577
  SI unit of, 769
  static equilibrium of, 617–18
  surface, on dielectric, 670–71
electric circuit, *see* circuit, electric
electric dipole, 599–601
electric displacement field, 674
electric energy, 650–61
  in capacitors, 674–76
  in nucleus, XI1
  of spherical charge distribution, 656–57
  in uranium nucleus, 656–57
electric energy density in dielectric, 675
electric field, 587, 590–608
  of accelerated charge, 859–63
  atmosphere, 635–36, VI1–VI3
  conductors in, 616–19
  conservative, 633–34
  conversion factors for, A26
  definition of, 591–92
  in dielectric, 668–69
  of dipole, 637–38
  electric dipole in, 599–601
  electric force and, 591–92
  energy density of, 654–57
  fair-weather, VI1–VI3
  as fifth state of matter, 591
  of flat sheet, 593–94
  induced, 779, 789–92, 855
  induced, in solenoid, 791–92
  of plane wave, 878–79
  of point charge, 592
  of spherical charge distribution, 616
  of thundercloud, VI6–VI8
  in uniform wire, 684
electric field lines, 595–99
  made visible, 601
  made visible, with grass seeds, 683

  of point charge, 595–97, 599, 736
  sources and sinks of, 597
electric flux, 609–11
  Ampère's Law and, 850
electric force, 571–74
  compared to gravitational force, 572, 575
  Coulomb's law for, 573–77
  electric field and, 591–92
  in nucleus, 1095–96, 1100, XI1
  qualitative summary of, 572–73
  superposition of, 588–90
electric fringing field, 654, 676–77
electric generators, 706
electric ground, 629
electricity:
  atmospheric, VI1–VI12
  frictional, 572, 580–81
  Gauss' Law for, 853
electric motor, 771
electric resistance thermometer, 500
electric shock, 721–23
electrolytes, 580
electromagnet, 758, 766
electromagnetic constants, A27
electromagnetic flowmeter, 800
electromagnetic force, 124–25, 572, 1130
electromagnetic generator, 788–89
electromagnetic induction, 779–805
electromagnetic interactions, 1130
electromagnetic launcher, 777–78
electromagnetic oscillation of cavity, 856–57
electromagnetic pulse (EMP), XI10
electromagnetic radiation:
  kinds of, 881–83
  wavelength and frequency bands of, 882–83
electromagnetic wave, 849, 859, 873–74
  energy flux in, 884–85
  generation of, 881–83
  momentum of, 885–88
  speed of, 865–66, 877–78
  standing, 948–50
electromagnetic wave pulse, 874–78
electromotive force, *see* emf (electromotive force)
electron:
  charge of, 573
  constants for, A27–A28
  discovery of, 766–67
  distribution of, in atoms, 571
  electric and magnetic fields of, 744
  elliptical orbit of, 1075–76
  free, 579–80, 687–89, 693
  free of gas, 580, 686, I9
  magnetic moment of, 744, 1077, 1096
  mass of, 107, 1096
  measurement of charge of, 604
  neutron and proton vs., 1096
  quantum behavior of, 1074–89
  spin of, 744, 1076–77, 1080, 1096
electron avalanche, VI8
electron configuration:
  of atoms, 1079–81
  of solids, 1084–86
electron field, 1135
electron-holography microscope,

  Prelude-15, 571, X12
electron microscope, Prelude-12, I2–I3
electron scanning microscope, I2
electron-volts (eV), 195, 630, 1064, 1085, 1099, 1121–22
electrostatic equilibrium, 617
  approach to, 618
electrostatic force, 572–73
electrostatic induction, 581–82
electrostatic potential, 627
  calculation of, 629–33
electroweak force, 1137
elementary charge, 574
elementary particles, 1121–48
  collisions between, 275–76, 282–83
  conservation laws and, 1132–33
  electric charges of, 1128, 1140
  masses of, 1128
  reaction rates among, 1131–32
  spins of, 1128
elements, chemical, A30–A32
  age of, II6
  during Big Bang, II5–II6
  Periodic Table of, Prelude-14, A30
  transmutation of, IV4
elevation angle, 82–84
elevator with counterweight, 133–34
ellipse, semimajor axis of, 218
elliptical galaxies, II2
elliptical orbits, 218–24
  angular momentum in, 226
  of electron, 1075–76
  energy in, 226–27
  vs. parabolic orbit for projectile, 223–24
elongation, 365–68
emergency cooling systems, V15
emf (electromotive force), 703–7
  alternating, 788–89
  induced, 779, 783–89, 792–94
  LCR circuit with, 832–40
  motional, 779–83
  power delivered by, 713
  in primary and secondary circuits of transformer, 840–41
  sources of, 705–7
emission, stimulated, 882, X2
empirical evidence, 3
emulsion, photographic, 286
endoergic reactions, 1113–14
energy, 160–210
  alternative sources of, V5–V10
  alternative units for, 195–96
  in capacitor, 674–76
  for circular orbit, 226
  conservation of mechanical, in rotational motion, 333
  conversion factors for, A24–A25
  of diatomic gas, 505–6
  dissipation of, in U.S., V3
  electric, *see* electric energy
  in elliptical orbit, 226
  entropy and, V2–V3
  equivalent to atomic mass unit, 1102
  geothermal, V9–V10
  gravitational potential, of a body, 256–57
  heat, 260
  of ideal gas, 504–6

In the two-volume edition, all pages after 570 are in Volume Two.
Interlude V is the last one in Volume One.

internal, 505
internal kinetic, 260–61, 279
internal kinetic, in inelastic
   collision, 287–88
kinetic, *see* kinetic energy
in LC circuits, 830–31
Lorentz transformations for, 1001–2
magnetic, 794–96
magnetic, in inductor, 795
mass and, 196–98
mechanical, *see* mechanical energy
momentum and, relativistic
   transformation for, 1013
nuclear, XI1
nuclear, from fission, V5–V6
nuclear, from fusion, V6
of photon, 1034–35
of point charge, 630–31
potential, *see* potential energy
rate of dissipation of, 206
released in fission, XI2–XI3
resources of, V4
rest-mass, 197, 1013
rest-mass, in inelastic collision,
   286–87
rotational, 1082–83
of rotational motion, 332–33
of rotational motion of gas, 505–6
in simple harmonic motion, 387–88
of simple pendulum, 391
solar, V6–V9
of stationary states of hydrogen, 1060
of system of conductors, 652–54
of system of particles, 259–61
of system of point charges, 650–52
thermal, 515
threshold, 288, 1114
tidal, V10
total relativistic, 1013
vibrational, 1082
in wave, 421–23, 883–85
energy, conservation of, 161, 174–75,
   184–210, 225
  general law of, 196
  in Lenz' Law, 788
  in nuclear reactions, 1113–14
  in one-dimensional elastic collision,
     277
  in simple harmonic motion, 388, 392
  in two-dimensional elastic collision,
     282
energy, conservation of mechanical,
   173–74, 187–88
  in rotational motion, 333
energy bands, 1084–85
energy banks, in solids, 1083–86
energy conversion, V1–V3
  in automobiles, V1–V2
  entropy and, V2–V3
  in thunderstorm, VI4
energy conversion efficiencies, V2–V3
energy density:
  in dielectric, 675
  in electric field, 654–57
  in magnetic field, 796
  of wave, 422–23
energy dissipation, in U.S., per capita, V3
energy distribution:
  in fission, XI2

  in nuclear explosion, X16–X17
energy-equivalent speed (EES), III2
energy flux, in electromagnetic wave,
   884–85
energy level, 193–94
  of hydrogen, 1060
  in molecules, 1081–83
energy-level diagram, 1060, 1061,
   1082–83, 1084
energy loss, of damped oscillator, 401
energy quantization, of oscillator, 1032
energy quantum, 1029–34
energy storage, VIII10–VIII11
energy use, in U.S., V3
energy-work theorem, 171
engine:
  efficiency of, 548–49
  heat, 548
enriched uranium, XI10
entropy, 554–59
  at absolute zero, 564
  change, in free expansion of gas, 558
  change, in isothermal expansion of
    gas, 557
  of different kinds of energy, V2–V3
  disorder and, 563–64
  energy conversion and, V2–V3
  of ideal gas, 556, 558
  irreversible process, 562–63
  negative, 567
Eötvös, Lorand von, 233, IX1
Eötvös experiments, 233–34, IX1–IX2
epicenter, 454
equation of motion, 123–24, 142–46
  for rotation, 329–30
  of simple harmonic oscillator, 384–85
  of simple pendulum, 390
  see also Newton's Second Law
equilibrium:
  electrostatic, 617, 618
  of fluid, 474
  stable and unstable, 394
  static, *see* static equilibrium
  thermal, 523–24
  thermodynamic, 547
equilibrium point, 194
equinox, precession of, 336*n*
equipartition theorem, 506
equipotential surface, 636–37
Equivalence, Principle of, 234, IX1–IX3
equivalence of inertial and
   gravitational mass, 233
equivalent absorbed dose, IV8
escape velocity, 227–28
  from Earth, 228
  from Sun, 228
  for universe, II7
Escher, M. C., I5–I7
etalon, 967
ether, 990–91
ether wind, 991
Euclidean geometry, IX5
  deviations from, IX6
Euclidean space, 3–5
Euler's theorem, 296*n*
Europa, 221
European Southern Observatory, 939
Evenson, K. M., 874
events, 1

excited states, 1060
Exclusion Principle, 1074, 1079–81,
   1084, 1106, 1140–41
exoergic reactions, 1113
expansion, free, of a gas, 546, 557
expansion, thermal:
  of concrete, 521
  linear, coefficient of, 519, 521
  of solids and liquids, 518–21
  of water, 520
expansion of universe, II1–II4
experimental safe vehicles (ESV),
   III13–III14
Explorer I, 221
Explorer III, 221, 240
Explorer X, 240
explosion, nuclear, *see* nuclear
   explosion
exponential function, A3
external emf, LCR circuit with, 832–40
external field, 599
external forces, 250
extremely low frequency (ELF) radio
   waves, 893
eye:
  astigmatic, 934
  compound, 984
  insect, 984
  myopic and hyperopic, 934
  nearsighted and farsighted, 934
  as optical instrument, 933–34

Fabry-Perot interferometer, 966
face-centered cube, I8
Fahrenheit temperature scale, 500
fair-weather electric field, VI1–VI3
fallout, from nuclear explosion, XI9
fallout pattern, for nuclear explosion,
   XI9
fallout shelters, XI9
farad (F), 663
Faraday, Michael, 779, 780
Faraday cage, 633–34
Faraday's constant, 584
Faraday's Law of induction, 783–84,
   791, 853, VII3
farsightedness, 934
fermi (fm), 1098
Fermi, Enrico, 1098, XI11
Fermi National Accelerator Laboratory
   (Fermilab), 762, 1122–23, 1143
  accelerator at, 323, VIII8
  Tevatron, 1122–23
ferromagnetic materials, 811–14
ferromagnetism, 811–14
Feynman diagram, 1135–36
Feynman, Richard P., 1135–36
fibers, optical, X5
fibrillation, 722
field, 126, 587
  electric, *see* electric field
  electron, 1135
  as fifth state of matter, 591
  magnetic, *see* magnetic field
  momentum in, 734
  quanta and, 1133–36
  radiation, of accelerated charge, 859,
    861

In the two-volume edition, all pages after 570 are in Volume Two.
Interlude V is the last one in Volume One.

8 Index

field (*continued*)
  radiation, of oscillating charge, 862
field lines, 595–99
  density of, 597–98
  made visible, 601
  of point charge, 595–97, 599
  sources and sinks of, 597
field meter, VI2
field mill, VI3
fifth state of matter, 591
filter, polarizing, 891, 911–14
fine-structure constant, 1068
fireball:
  of nuclear explosion, XI7–XI8
  primordial, II6
fireball radiation, cosmic, II6
firestorm, of nuclear explosion, XI9
First Law of Thermodynamics, 545–48
first overtone, 429–30
fission, 1096, XI1–XI3
  in alpha-decay reaction, 1104–5
  deformation of nucleus during, XI1–XI2
  distribution of energy in, XI2
  ejection of particles in, XI2
  energy released in, XI1, XI2–XI3
  multiplication factor for, XI4
  neutron-induced, XI3
  neutrons released in, XI2
  potential energy during, XI2–XI4
  reflector in, XI4
  as source of nuclear energy, V5–V6
  spontaneous, XI2
  symmetric, XI1–XI2
fission bomb, XI4–XI5
fission-fusion-fission device, XI6
flash burns, from nuclear explosion, XI8–XI9
flat sheet, electric field of, 593–94
flow:
  around an airfoil, 470–71, 480–81
  of free electrons, 687–89
  of heat, 524–25
  incompressible, 468–69
  methods for visualizing, 470–71
  in a nozzle, 482–83
  plastic, 466
  of rock, 466
  of solids, 466
  from source to sink, 470
  steady, 468–69
  velocity of, 467
  around a wing, 471, 480
flowmeter:
  electromagnetic, 800
  Venturi, 482, 492
fluid, 466
  density of, 467–68
  equilibrium of, 474
  incompressible, 469
  nuclear, 1100
  perfect, 467
  static, 474
fluid dynamics, 479–83
fluid mechanics, 466–93
fluorescent tube, flicker in, 843
flute, 445
flux:
  electric, 609–11
  magnetic, 780–81, 784

flux quantum, VIII5
f number, 933
focal length, of spherical mirror, 923
focal plane, 941
focal point:
  of lenses, 928
  of a mirror, 923
foot, 8–9
foot-pound-force (ftlbf), 162
force, 104–12
  buoyant, 478–79
  calculated from potential energy, 191–92
  center of, 143
  central, angular momentum and, 312
  centrifugal, 149
  color, 1141–42
  conservative, 184–90, 224
  contact, 572–73
  conversion factors for, A24
  as derivative of potential energy, 191–92
  elastic, 140–42
  electric, *see* electric force
  electromagnetic, 124–25, 572, 1130
  electromotive, *see* emf (electromotive force)
  electrostatic, 572–73
  electroweak, 1137
  equation of motion and, 123–24
  failure of Newton's Third Law for, 734
  fundamental, 124–25
  gravitational, *see* gravitational force
  impulsive, 272–76
  internal and external, 250
  inverse-square, 143
  Lorentz, 765, 853–54
  method of joints for calculation of, 359
  motion with constant, 129–34
  motion with variable, 142–46
  N as, 130–31
  net, 108
  nuclear, 1095
  power delivered by, 199
  between quarks, 1141–42
  range of, 126
  restoring, 140–42
  resultant, 108
  of seat belts, III9–III11
  of a spring, 140–42
  "strong," 125, 1095–96, 1099–1103, 1130, 1141, XI1
  tube of, 603
  units of, 105–6
  "weak," 125, 1107, 1130
force, magnetic, 572, 730–35
  between currents, 733
  failure of Newton's Third Law for, 734
  due to magnetic field, 735
  on moving point charge, 731, 733–34
  on relativistic correction, 732
  on wire, 768–69
forced oscillations, 399, 402
fossil fuel:
  pollutants from, V10–V11
  reserves of, V4
  supply of, V4
Foucault effect, 389*n*
Foucault pendulum, 101
Fourier series, 427

Fourier's theorem, 427, 440
frame of reference, 5–6, 86–89
  in calculation of work, 163
  of Earth, 101–2
  freely falling, 127–28
  inertial, 100–102, 991
  local inertial, IX2
  power and, 200
  for rotational motion, 297, 339
Franklin, Benjamin, 572, VII1
Fraunhofer, Joseph von, 1051
Fraunhofer approximation, 956, 973–74
Fraunhofer Lines, 1052
free body, 100
"free-body" diagram, 130, 134, 148, 153, 156
free charges, 675
freedom, degree of, 506*n*
free-electron gas, 580, I9
  drift velocity of, 696
  friction in, 686, 688–89
free electrons, 579–80, 693
  drift velocity of, 689
  flow of, 687–89
free expansion of a gas, 546
  entropy change in, 557
free fall, 233, 404
  impact speeds compared with, III7–III8
  universality of, 37, 234, 392, IX1–IX3
  weightlessness in, 127–28
free-fall motion, 37–39, 127–28
freon, 553–54
frequency, 299–300
  normal, 430
  proper, 430
  of simple harmonic motion, 380–81
  threshold, 1036
  of wave, 415–16
frequency bands, of electromagnetic radiation, 882–83
Fresnel, Augustin, 456, 972
Fresnel approximation, 973–74
Fresnel lens, 942, 944
friction, 134–40
  air, 37, 38–39
  coefficient of kinetic, 135–37
  coefficients of, 134–40
  in flow of free-electron gas, 686, 688–89
  on ice and snow, 136–37
  loss of mechanical energy by, 190
  microscopic and macroscopic area of contact and, 136
  sliding (kinetic), 134–38
  static, 138–40
  static, coefficient of, 136, 138–39
frictional electricity, 572, 580–81
fringes, 951
fringing field, 654, 676–77
fuel, fossil:
  pollutants from, V10–V11
  reserves of, V4
  supply of, V4
fuel cells, 706–7
  on Skylab, 707
fuel pellets, for laser fusion, X8, X9
fuel rods, of nuclear reactor, V13–V14, XI10

In the two-volume edition, all pages after 570 are in Volume Two.
Interlude V is the last one in Volume One.

Index

functions, inverse trigonometric, A7–A8
fundamental constants, best values for, A27–A29
fundamental forces, 124–26
fundamental mode, 429–30
Fundy, Bay of, 235
fusion, 528
   heat of, 527
   laser, X7–X9
   nuclear, V6, VII9–VII11, XI6
   as source of nuclear energy, V6
   thermonuclear, VII8–VII11, X7–X9, XI6
   triggered by laser, V6
fusion reactions, XI6
fusion reactor, V6, VII9–VIII11
future of universe, II6–II8

g, 37–38, 40–42
   measurement of, 41–42
   variation of, with latitude, 102
G (gravitational constant), 212
   measurement of, 214–16
G (unit of acceleration), 40, 41
Gabor, Dennis, X12
gain factor, 1088
galaxies, II1–II4
   elliptical, II3
   spiral, II3
Galaxy of the Milky Way, Prelude-9, II1
Galilean addition law for velocity, 990
Galilean coordinate transformations, 66n, 87, 88, 990, 998
Galilean Relativity, 114–15
Galilean telescope, 941
Galilean transformation, 88–89, 114
   for momentum, 1010
   for velocity, 88–89, 990
Galileo Galilei, 37, 44, 99, 115n
   experiments on universality of free fall by, 233, 392, 404, IX2
   isochronism of pendulum and, 408
   pendulum experiments by, 403
   tide theory of, 91
Galvani, Luigi, 628
gamma emission, 1107–8
gamma rays, 882, 1103–4, IV1–IV2
   Ganymede, 221
   radiation damage by, IV6
gas:
   adiabatic equation for, 533–34
   Boyle's law for, 496
   distribution of molecular speeds in, 503–4
   energy of ideal, 504–6
   entropy change in free expansion of, 558
   entropy change in isothermal expansion of, 557
   of free electrons, 580, 686, 688–89, I9
   free expansion of, 546
   Gay-Lussac's Law for, 496
   ideal, see ideal gas
   interstellar, Prelude-7
   natural, V4
   root-mean-square speed of, 502–3
   specific heat of, 529–31
gas constant, universal, 495
gas-cooled reactor, XI2

gas state, Prelude-2
gas thermometer, constant-volume, 499
gauge blocks, 7
gauge pressure, 477–78
gauss, 735
Gauss, Karl Friedrich, 737
Gaussian surface, 611, 612
Gauss' Law, 609, 611–16
   in dielectrics, 672–74
   for electricity, 853
   for magnetic field, 737
   for magnetism, 853
Gay-Lussac's Law, 496
Gedankenexperiment, 223–24
gee suit, 152
Geiger, H., 1055
Geiger counter, 1108
Gell-Mann, Murray, 1139
general law of conservation of energy, 196
General Relativity theory, 4, 212, 234, II1, IX1, IX6–IX8
general wavefunction, 413–14
generation of electric charge in thunderstorm, VI5–VI6
generator:
   electric, 706
   electromagnetic, 788–89
   homopolar, 782
   magnetohydrodynamic, 782–83
   superconducting, VIII9–VIII10
geometric optics, 922
geometrodynamics, IX6
geometry, deviations from Euclidean, IX6
geostationary orbit, 217–18
geostationary satellite, 218
geothermal energy, V9–V10
Germer, L., 1064
GeV, 195, 1122–24, 1138
Geysers, the (California), V9
G force, 40–41, 150
Giaever, Ivar, VIII7n
gimbals, 335
Glashow, Sheldon Lee, 1137
glide line, I7
globular clusters, age of, II5–II6
gluons, 1142
Golden Gate Bridge, 360, 370
Goudsmit, Samuel, 1076–77
gradient, of the potential, 634–37
graphite, I9–I10
grating, 960–62
   principal maximum of, 960–61
   reflection, 961
   resolving power of, 961
gravimeter, Jaeger, 41
gravitation, 211–44
   law of universal, 212–14
gravitational constant, 212
   measurement of, 214–16
gravitational force, 124–25, 212–14, 1130
   acting on center of mass, 354–55
   compared to electric force, 572, 575
   exerted by spherical mass, 229–32
   motion of recession of galaxies and, II7
gravitational interactions, 1130
gravitational mass, 232–34
gravitational potential energy, 172–76, 224–29
   of a body, 256–57
   of spherical mass, 231
gravitational red shift, IX8–IX10
gravitational singularity, IX12
gravitational time dilation, IX8–IX10
   experiments on, IX8–IX9
gravity:
   action and reaction and, 110
   as conservative force, 224
   of Earth, see weight
   formula for acceleration of, 51
   work done by, 168–69
gravity, acceleration of, 37–42, 214
   measurement of, 41–42, 395
   universality of, 37
   variation of, with altitude, 214
   variation of, with latitude, 102
gravity wave, 451
   speed of, 451–52
green, 1141–43
greenhouse effect, V11
   Earth's temperature increase and, V11
Greenwich Mean Time, 10
ground, electric, 629
ground state, 1060
group velocity, of a wave, 420
guard rings, for capacitors, 677
Guericke, Otto von, 117
guidance system, inertial, 115, 335
guitar, 445
Gulf Stream, 89
gun device, XI4–XI5
gyroscope, 335
   directional, 335
   precession of, 343–45

hadron, 1130
hairsprings, 397
half-life, 1109–10, IV4
Hall effect, 767–68
Halley's comet, 240
Hanford, Washington, XI11
harmonic function, 380–81
harmonic motion:
   damped, 399–400
   see also simple harmonic motion
harmonic wave, 414–17
harmonic wavefunction, 416
H-bomb, XI6
headlight method, II4
hearing, threshold of, 439
heat, 195, 515–43
   of evaporation, in water vapor, VI3–VI4
   of fusion, 527
   mechanical equivalent of, 517
   specific, see specific heat
   temperature changes and, 517
   transfer, by convection, 526
   transfer, by radiation, 526–27
   of transformation, 527
   of vaporization, 527–28
   waste, V11–V12
heat capacity, specific, 516
heat conduction, 524–27
   equation of, 525–26
heat current, 524
heat energy, 260

In the two-volume edition, all pages after 570 are in Volume Two.
Interlude V is the last one in Volume One.

heat engine, 548
heat flow, 524
heat pump, 553
heat reservoir, 548
heavy water, II8
heavy-water reactor, XI12
height, maximum, of projectiles, 81–82
Heisenberg, Werner, 1042, 1064
Heisenberg's uncertainty relation, 1042
helical motion, in magnetic field, 764
heliocentric system, 216
helium and hydrogen, abundance of, in universe, II6
helium-neon laser, X3–X4
helium nuclei, 1104–5, 1111
Helmholtz, Hermann von, 196
henry, 793
Henry, Joseph, 793
hertz (Hz), 381
Hertz, Heinrich, 874, 1035
High-Altitude Research Project (HARP) gun, 51
high-voltage transmission line, power dissipated in, 714
Hiroshima, Japan, XI5
"holes," in semiconductors, 692–93, 1085–86, 1087–88
hologram, X9–X12
  double-exposure, X12
holographic images, X10
holographic three-dimensional images, X10
holography, X9–X12
homopolar generator, 782
Hooke, Robert, 140
Hooke's law, 140–42, 365, 383–84
horizon, IX11
horsepower (hp), 198
hot rod, 152
Hubble, Edwin, II1
Hubble's Law, II3–II4
Hubble time, II5
human body:
  force-deformation characteristics of, III7–III8
  lever-like motion of, 363, 370, 375
  tolerance limits of, III11–III14
Huygens, Christiaan, 174, 403, 899
Huygens' Construction, 899–900
Huygens-Fresnel Principle, 971–72, 981
Huygens' tilted pendulum, 403
Huygens' wavelets, 900
Hyatt Regency hotel, collapse of "skywalks" at, 370
hydraulic brake, 475
hydraulic press, 474–75
hydroelectric power, V7
hydrogen:
  atom, stationary states of, 1060, 1061
  burning, in Sun, XI6
  energy levels of, 1061
  and helium, abundance of in universe, II6
  isotopes of, 1097, XI6
  molecule, oscillations of, 388–89
  quantum numbers for stationary states of, 1078, 1079
  spectral series of, 1052–54, 1060, 1076
hydrogen spectrum, 1052–53
  produced by grating, 960

hydrometer, 491
hydrostatic pressure, 476
hyperbola, 226–27
hyperbolic orbit, 226–27
hyperopic eye, 934
hysteresis, 369, 813
hysteresis loop, 818

Iapetus, 239
Idaho Falls, Idaho, reactor accident at, V13
ideal gas, 496
  energy of, 504–6
  entropy of, 556, 558
  kinetic theory and, 494–515
ideal-gas law, 495–98
ideal-gas temperature scale, 498–500, 524, 554
ideal particle, 1–2
ideal solenoid, 756–57
image:
  method of, VI7–VI8
  real, 926
  three-dimensional, from hologram, X10
  virtual, 902
image charges, VI7
image orthicon, 1037
imbalance, dynamic, 313–14
impact parameter, 1056
impact speeds, III3–III14
  heights of fall compared with, III7–III8
  minimum stopping distance vs., III13
impedance:
  for LCR circuit, 834
  for parallel LCR circuit, 840
implosion device, XI5
implosion of universe, II7
impulse, 273–74
impulsive force, 272–76
impurities, acceptor, 693, 1086
incidence, angle of, 901–2
incompressible flow, 468–69
incompressible fluid, 469
indeterminacy relation, 1042
index of refraction, 904, 908–9
induced dipole moments, 601
induced electric and magnetic fields between capacitor plates, 855
induced electric field, 779, 789–92
  in solenoid, 791–92
induced emf, 779, 783–89
  in adjacent coils, 792–94
  in solenoids, 785–86
induced magnetic field, 851–52
  of accelerated charge, 866
  in capacitor, 852
inductance, 792–94
  conversion factors for, A26
  mutual, 793
  self-, 794
induction:
  electromagnetic, 779–805
  electrostatic, 581–82
  Faraday's Law of, 783–84, 791
inductive reactance, 827–28
inductor:
  circuit with, 827–28
  current in, 827–28

inelastic collision:
  internal kinetic energy in, 287–88
  rest-mass energy in, 286–87
  totally, 279, 284
  in two dimensions, 283–85
inertia, law of, 100
inertia, moment of, see moment of inertia
inertial confinement, VII10–VII11
inertial guidance system, 115, 335
inertial mass, 233–34
inertial reference frames, 100–102
  local, IX2
infrared radation, 881–82
infrared slavery, 1142
inner product, of vectors, 62–64
instantaneous acceleration, 33–34
  in three dimensions, 76
instantaneous angular acceleration, 300
instantaneous angular velocity, 299
instantaneously fixed axis, 339–40
instantaneous power, 198
instantaneous velocity, 29–32
  analytic method for, 31–32
  as derivative, 32
  graphical method for, 31
  as slope, 30–31
  in three dimensions, 75–76
instantaneous velocity vector, 75–76
insulating material, see dielectric
insulators, 579
  conductors vs., 692
  electron configurations of, 1084–85
  resistivities of, 692
integral:
  definition of, A14
  for work, 164
integrated circuit, 1089
integration, 164, A12–A18
  rules for, A16–A18
intensity:
  of sound waves, 439
  for two-slit interference pattern, 958
intensity level of sound, 440
interaction:
  electromagnetic, 1130
  gravitational, 1130
  "strong," 1130
  "weak," 1107, 1130
interaction region, 282
interatomic bonds, 1081
intercontinental ballistic missile (ICBM), 84
interference, 947–70
  constructive and destructive, 424–25, 947, 949–53
  maxima and minima for, 955–56
  maxima and minima, for multiple slits, 960
  from multiple slits, 958–62
  in standing electromagnetic wave, 948–50
  in thin films, 950–52
  two-slit, pattern for, 955, 958
  from two slits, 954–58
interferometer, 7
  Fabry-Perot, 966
  Michelson, 952–54, X6
interferometry:
  laser, X6–X7
  very-long-baseline, 980, 985

intergalactic space, mass in, II8–II9
internal energy, 505
internal forces, 250
internal kinetic energy, 260–61, 279, III4
  in inelastic collision, 287–88
internal resistance, of batteries, 708–9
International Geodesy Association, 51
international standard meter bar, 6–7
International System of Units (SI), 6, 14–15, A20
  see also SI units
interstellar gas, Prelude-7
interstitial defects, I10
interstitial space, I10
invariance of speed of light, 992
inverse-square force, 143
inverse-square law, 576, 599
inverse trigonometric functions, A7–A8
inversion symmetry, I5
Io, 221
ionization, IV5
ionizing radiation, IV5
ion microscope, Prelude-13, I2–I3
ionosphere, VI1–VI2, VI7
ions, in electrolytes, 580
irreversible process, 561–62
  entropy change in, 562–63
isochronism, 385
  of simple pendulum, limitations of, 392
isospin, 1132
isothermal expansion of gas, entropy change in, 557
isotopes, Prelude-17, 1096–99, IV3
  of carbon, 1096–98
  chart of, Prelude-17, 1097–98
  in decay reactions, 1106–8
  of hydrogen, 1097, XI6
  radioactive, 1109–11, IV3–IV4
  of uranium, 1103, X14–X16, XI1

J/Ψ meson, 1143
Jaeger gravimeter, 41
Jodrell Bank radiotelescope, 987
Jordan, P., 1064
Josephson, Brian D., VIII7*n*
Josephson junction, VIII7
joule (J), 162, 198
Joule, James Prescott, 161
Joule heat, 713–14
Joule's experiment, 517
Jupiter, 220
  main moons of, 220–21

Keck Telescope, 939
"keeper," of magnet, 818
kelvin, 15, 495–96, A20
Kelvin, William Thomson, Lord, 374, 495
Kelvin-Planck statement of Second Law of Thermodynamics, 559–60
Kelvin temperature scale, 495–96, 499–500, 554
Kepler, Johannes, 222
Kepler's First Law, 218
Kepler's Laws, 218–24, 218–24
  of areas, 218–19
  limitations of, 224
  for motion of moons and satellites, 220–22

  for motion of stars, 223–24
Kepler's Second Law, 218–19, 312
Kepler's Third Law, 220
KeV, 195
killer smog, V10
kilocalorie, 196
kilogram, 6, 12, 14–15, A20
  multiples and submultiples of, 13
  standard, 102
kilowatt-hours, 196
kinematics, 25
  in one dimension, 25–51
  of rigid body, 296–324
  in three dimensions, 74–98
kinetic energy, 169–72
  in automobile collisions, III2, III4–III5
  as function of momentum, 172
  of ideal monatomic gas, 505–6
  internal, 260–61, 279, III4
  relativistic, 1011–12
  of rotation, 303–9
  in simple harmonic motion, 387–88
  of simple pendulum, 391–92
  of a system of particles, 259–61
kinetic friction, 134–38
  coefficient of, 135–37
kinetic pressure, 501–4
kinetic theory, ideal gas and, 494–515
Kirchhoff, Gustav Robert, 708
Kirchoff's first rule, 711
Kirchoff's second rule, 708–9, 717–18
klystron, 857–58
Krüger 60, 223
krypton wavelength standard, 7–8
Kugelblitz, VI11

Lagoona Beach, Michigan, reactor accident at, V13
Lagrange, Joseph Louis, Comte, 186
lap-and-diagonal-shoulder belt, III9–III10, III14
lap-and-double-shoulder harness, III10–III11
lap belt, III9–III10
Laplace, Pierre Simon, Marquis de, 123
  black holes and, IX11
  deflection of light and, IX4
La Rance river, V10
Larmor frequency, 816
laser, 882, X2–X4
  carbon dioxide-nitrogen, X4
  definition of speed of light and, X7
  fusion triggered by, V6
  helium-neon, X3–X4
  medical use of, X6
  neodymium glass, X3, X8
  resonant cavity in, X3–X4
  ruby, X2–X3
  solid-state, X5
  stabilized, 7–8, X7
  as torch, X6
  tunable, X7
laser fusion, X7–X9
laser-fusion reactors, X7–X8
laser interferometry, X6–X7
laser light, 954, X1–X14
laser spectroscopy, X7
latitude angle, 2
lattice, crystal, I2
lattice points, I6

Laue spots, I3
law of areas, 219, 312
law of conservation of angular momentum, 329
law of conservation of mechanical energy, 173–74, 187–88
law of conservation of momentum, 114, 246
law of cosines, A9
law of inertia, 100
law of Malus, 914
law of radioactive decay, 1109–11
law of reflection, 900–901
law of refraction, 906
law of sines, A9
Lawrence, Ernest Orlando, 761, 762, 763, 1111–12
Lawrence Berkeley Laboratory, 763
laws of planetary motion, 218–24
LC circuit, 828–31
  energy in, 830–31
  freely oscillating, 828–32
  natural frequency of, 829
LCR circuit:
  bandwidth of, 836
  current in, 835, 839–40
  with external emf, 832–40
  impedance for, 834, 840
  phase angle for, 834, 839
  power absorbed in, 837
  Q of, 832
  resonance in, 833
  steady state of, 833
lead-acid battery, 578–79, 705, 707–8
leader, in lightning, VI8–VI9
  dart, VI9
leap second, 10–11
length, 6–9
  conversion factors for, A22
  precision of measurement of, 8
  standard of, 6–8
length contraction, 995–96, 1007–8
  visual appearance and, 1007–8
lens:
  focal point of, 928
  magnification produced by, 930
  objective, 936–37
  ocular, 936–37
  sign conventions for, 932
  "Tessar," 932
  thin, 926–32
lens equation, 929
lens-maker's formula, 926–28
Lenz, Emil Khristianovich, 787
Lenz' Law, 787, 815, VII3
lepton, 1127–28, 1130
lepton number, 1132
"let-go" current, 722
Leverrier, U. J. J., 211
levers, 362–64
  human bones acting as, 363, 370, 375
  mechanical advantage and, 362, 364
Lichtenstein, R., 236
light:
  coherent, 954, X1–X2, X6
  deflection of, by Sun, 4, IX3–IX6
  dispersion of, 909
  Doppler shift of, 888–90
  from laser, 954, X1–X14
  polarized, 910–15
  quanta of, 1027–48

In the two-volume edition, all pages after 570 are in Volume Two.
Interlude V is the last one in Volume One.

12 Index

light (*continued*)
  reflection of, 900–903
  refraction of, 903–10, 912–13
  retardation of, by Sun, IX5–IX6
  spectral lines of, 910
  spectrum of, 910, 960, 1052–54
  as standard of speed, X7
  ultraviolet, 881–82
  unpolarized, 911
  visible, 881–83
light, speed of, 8, 860, 864–66, 990–91
  invariance of, 992
  laser standard for, X7
  in material medium, 904
  measurement of, 874, 892
  universality of, 992
light-emitting diode (LED), 1089
lightning, VI8–VI11
  ball, VI11
  dart leader in, VI9
  data on, VI10
  leader in, VI8–VI9
  radio waves from, VI10
  return stroke in, VI9
  step leader in, VI9
  thunder and, VI10
  in tornado, VI10
lightning bolt, VI10
light rays, bending of, by Sun, 4, IX3–IX4
light-second, 998–99
light-water reactor, XI12
light-wave communications, X5
light waves, 873–98
  incoherent, 882
linear accelerator, 1123
linear dielectric, 668
lines of electric field, 595–99
  made visible, 601, 683
liquid drop model of nucleus, 1100
liquids:
  bulk moduli for, 366–67
  thermal expansion of, 518–21
liquid state, Prelude-2
lithium, spectral series of, 1069
Lloyd's mirror, 968–69
Local Group, Prelude-10
local inertial reference frame, IX2
Local Supercluster, Prelude-10
logarithmic function, A3
longitude angle, 2
longitudinal wave, 413
long wave, 881
looping the loop, 149–50, 182
loop method, for circuits, 710
Lopes, Carlos, 15
Lorentz, Hendrik Antoon, 1001
Lorentz force, 765, 853–54
Lorentz transformations, 997–1004, 1137
  for coordinates, 66n, 1003
  for momentum and energy, 1001–2
  for velocity, 1008
Los Alamos National Laboratories, fusion experiments at, V6
lunar tides, 235
Lyman series, 1053, 1077–78

MacDonald Observatory, X5
Mach, Ernst, 102

Mach cone, 450
Mach front, of nuclear explosion, XI8
macroscopic and microscopic parameters, 494
Magdeburg hemispheres, 117
magic nuclei, 1117
magnet:
  "keeper" of, 818
  permanent, 811
  superconducting, VIII7–VIII8
magnetic bottles, VII4
magnetic dipole, 743–45
  field of, 743
magnetic dipole moment, 743–44, 770
  of earth, 744–45
  of electron, 744
magnetic domains, 812
magnetic energy, 794–96
  in inductor, 795
magnetic field, 735–45
  of accelerated charge, 863–66
  circular orbit in, 761
  conversion factors for, A26
  critical, VIII3
  derived from relativity, 1014–18
  of electron, 744
  energy density in, 796
  Gauss' Law for, 737
  generated by current, 738–39
  helical motion in, 764
  induced, in capacitor, 852
  induced, of accelerated charge, 866
  made visible with iron filings, 740
  of magnetic dipole, 743
  magnetic force and, 735
  motion of charge in, 759–60
  of plane wave, 879–80
  plasma and, VII3–VII6
  of point charge, 736
  of point charge, represented by field lines, 736–37
  principle of superposition for, 737–38
  relativity and, 1014–18
  of solenoid, 757–58
  of straight wire, 739–40
magnetic field lines:
  of circular loop, 742
  made visible with iron filings, 742, 756, 757
magnetic flux, 780–81, 784
magnetic flux quantum, VIII5
magnetic force, 572, 730–35
  between currents, 733
  failure of Newton's Third Law for, 734
  magnetic field and, 735
  on moving point charge, 731, 733–34
  on relativistic correction, 732
  on wire, 768–69
magnetic materials, 806–22
magnetic mirrors, VII5
magnetic moment, 1096
  of Earth, 744–45
  of electron, 744, 1077
  nuclear, 807–9
  orbital, 807
  spin, 808
magnetic monopole, 745
magnetic pressure, VII4
magnetic quantum number, 1076, 1078, 1079

magnetic suspension, VIII11
magnetism, Gauss' Law for, 853
magnetization, 809
magnetohydrodynamic (MHD) generator, 782–83
magnetosonic waves, VII6
magnetosphere, VII4
magnification, 930
magnifier, 934–35
Maiman, T. H., X2, X3
Malus, Étienne, 914
Malus, law of, 914
mandolin, 445
Manhattan Engineer District, XI5
Manned Spacecraft Center, 85–86
manometer, 477
mantle of Earth, 454
many-body problem, 224
Mariner 6, IX5
Mariner 7, IX5
Marly, France, waterworks at, V1
Mars, 220–21
Marsden, E., 1055
maser, X2
mass, 12–14
  atomic standard of, 12–13
  of carbon, 1098
  center of, *see* center of mass
  conservation of, 160
  conversion factors for, A23
  critical, XI4
  definition of, 102–4, 111
  of electron, 107
  of elementary particles, 1128
  energy and, 196–98
  equivalence of inertial and gravitational, 233
  gravitational, 232–34
  inertial, 233–34
  in intergalactic space, II8–II9
  moment of inertia of continuous distribution of, 305–6
  of neutron, 107, 1096
  of proton, 107, 1096
  rest, 197
  standard of, 12, 102
  supercritical, XI4
  weight vs., 127
mass defect, 1102, XI1–XI2
mass density, in universe, II7–II8
mass-energy relation, 892, 1013
mass number, 1097, IV3
  binding energy per nucleon vs., 1102
  conservation law for, 1112–13, 1132
mass spectrometer, 107, 767, 776
matter-antimatter annihilation, 578
maximum height, of projectiles, 81–82
Maxwell, James Clerk, 849, 874
  formula for speed of light by, 878
Maxwell-Ampère's Law, 851, 853
Maxwell distribution, 503–4
Maxwell's demon, 566
Maxwell's equations, 853–54
Mayer, Robert von, 537
mean free path, 507–9
mean lifetime, 1109
mean solar day, 9
mean-value theorem, 639–42
mechanical advantage, 362, 364–65
mechanical energy, 173–76, 187–88

In the two-volume edition, all pages after 570 are in Volume Two.
Interlude V is the last one in Volume One.

conservation of, in rotational motion, 333
law of conservation of, 173–74, 187–88
loss of, by friction, 190
mechanical equivalent of heat, 517
mechanics, 15
  celestial, 211
  classical vs. quantum, 1050
  fluid, 466–93
median speed, 510–11
medium, dispersive, 420
medium wave, 881
Meissner effect, VIII3–VIII5
Mercury, 220–21
  perihelion precession of, 211–12, IX1, IX7
mercury barometer, 477
mercury-bulb thermometers, 500, 522
meson, 1128–30, 1143
metal:
  resistivities, 691–92
  temperature dependence of resistivity of, 690–91
meteoroid incidents, 242
meter, 6–8, 14–16, A20
  cubic, 14
method of images, VI7–VI8
method of joints, 359
metric system, 6, 26, A20
MeV, 195, 762, 787, 1099, 1101–2, 1122, 1143, VII8, XI1–XI3, XI6
MHD (magnetohydrodynamic generator), 782–83
Michelson, Albert Abraham, 874, 953, 991
Michelson interferometer, 952–54, X6
Michelson-Morley experiment, 953–54, 991
microballoon, X8
microcrystal, I4
microfarad, 663
micrograph, acoustic, 443
microscope, 936
  angular magnification of, 936
  electron, Prelude-12, I2–I3
  electron-holographic, Prelude-15, 571, X12
  electron scanning, I2
  holographic, Prelude-15, X12
  ion, Prelude-13, I2–I3
  two-wavelength, I2
  ultrasonic, 443
microscopic and macroscopic parameters, 494
microwaves, 881, 911
middle C, 441
millibar, 473
Millikan, R. A., 1036
Millikan's experiment, 604
Minas Basin, Canada, V10
Minkowski diagram, 997–98
mirror:
  focal point of, 923
  image formed by, 902
  Lloyd's, 968–69
  magnetic, VII5
  magnification produced by, 930
  paraboloidal, 924
  plane, 902
  reflection by, 902–3
  sign conventions for, 932
  spherical, 922–26
  in telescope, 937–39
mirror equation, 925–26
mirror nuclei, 1115
missile, ballistic, 83–84, 143
mode, fundamental, 429–30
moderator, in nuclear reactor, XI10–XI11
modulation, of wave, 426
mole, 12–13, A20
molecular speeds, distribution of, in gas, 503–4
molecules:
  energy levels in, 1081–83
  quantum structure of, 1074, 1081–83
  rotational energy of, 1082–83
  vibrational energy of, 1082
  water, 320, 601
moment, magnetic dipole, 743–45, 770
moment arm, 326
moment of force, 325
moment of inertia:
  of continuous mass distribution, 305–6
  of Earth, 315–17, 320, 322
  of nitric acid molecule, 320
  of oxygen molecule, 320
  of system of particles, 303–9
  of water molecule, 320
momentum, 113–14
  angular, see angular momentum
  of an electromagnetic wave, 885–88
  energy and, relativistic transformation for, 1013–14
  in fields, 734
  Galilean transformations for, 1010
  kinetic energy as function of, 172
  Lorentz transformations for, 1001–2
  Newton's Laws and, 113–14
  of a photon, 1038
  rate of change of total, 250–51
  relativistic, 1010–11
  of a system of particles, 245–51, 257
momentum, conservation of, 113–14, 246–49, 277, 279
  in elastic collisions, 276–77, 282
  in fields, 591
  law of, 114, 246
monatomic gas, kinetic energy of ideal, 505–6
Moon-Earth distance measurement, X5
Moon simulator, 128–29
moons of Saturn, 239
Morley, E. W., 953–54, 991
most probable speed, 504
motion, 27–51
  absolute, 989
  amplitude of, 380
  of ballistic missile, 83–84
  of center of mass, 257–59
  of charges in magnetic field, 759–60
  with constant acceleration, 34–37
  with constant acceleration, in three dimensions, 77–78
  with constant angular acceleration, 401–4
  with constant force, 129–34
  equation of, see equation of motion
  free-fall, 37–39, 127–28
  harmonic, 399–400
  helical, in magnetic field, 764
  in one dimension, 27–28
  parabolic, 81
  periodic, 379
  planetary, 218–24
  of projectiles, 78–84
  as relative, 27, 86, 89, 989
  of rigid bodies, 296–98
  of rockets, 261–64
  rolling, 339–43
  rotational, see rotational motion
  simple harmonic, see simple harmonic motion
  three-dimensional, 74
  translational, 25, 296, 330–31, 339
  uniform circular, 84–86, 146–50, 312
  with variable force, 142–46
  wave, 412
motional emf, 779–83
motion of recession, II4
  of galaxies, gravitational force and, II7
Mt. Hopkins, Arizona, 938
Mt. Palomar telescope, 938
  resolution of, 979
Müller, K. Alex, VIII2n
multiloop circuits, 710–12
multiplate capacitor, 678
Multiple Mirror Telescope (MMT), 938–39
multiplication factor, for fission, XI4
multiwire chambers, 1126–27
muon, 1127–28, IV3
  constants for, A28
mushroom, of nuclear explosion, XI1, XI8
mutual inductance, 793
muzzle velocity, 82–83, 223–24, 266, 267
myopic eye, 934

N (normal force), 130
Nagasaki, Japan, XI5
National Bureau of Standards, 10–12
natural gas, V4
natural logarithmic function, A3
Naval Observatory, 11
near point, 934
nearsightedness, 934
negative velocity, 33
negentropy, 567
neodymium glass laser, X3, X8
Neptune, 220
  discovery of, 211
Nernst, Walther Hermann, 564
net force, 108
neutral-beam injectors, VII10
neutral equilibrium, 394
neutrino, 1106–7, 1128
neutron, Prelude-17, 1095, 1121
  beta-decay reaction and, 1105–6
  constants for, A28
  fission induced by, XI3
  isotopes and, 1096
  mass of, 107, 1096
  nuclear binding energy and, 1100–1103
  in nuclear reactor, XI10
  quark structure of, 1140

In the two-volume edition, all pages after 570 are in Volume Two.
Interlude V is the last one in Volume One.

14 Index

neutron (continued)
  radiation damage by, IV6–IV7
  released in fission, XI2
  thermal, XI10
neutron sources, IV3
newton (N), 105
Newton, Isaac, 4, 99, 223, 1098
  on action-at-a-distance, 587
  bucket experiment of, 484
  experiments on universality of free fall by, 234, 392, 404, IX1
  pendulum experiments by, 404
Newtonian physics, 4
Newtonian relativity, 114–15
newton-meter, 326–27
Newton's First Law, 99–102
  momentum and, 113
Newton's Law of Universal Gravitation, 212–16
Newton's Laws, 99–122
  angular momentum and, 328–29
  in rotational motion, 325
Newton's rings, 952
Newton's Second Law, 102–9, 134, 161, 328, 418
  centripetal acceleration and, 147
  empirical tests of, 105
  momentum and, 113
  see also equation of motion
Newton's theorem, 214, 229–32
Newton's Third Law, 109–12, 328
  failure of, for magnetic force, 734
  momentum and, 113–14
Nicol prism, 921
nitric acid molecule, moment of inertia of, 320
nodes and antinodes, 429
noise, white, 440
noise reduction, 457
normal force (N), 130
normal frequencies, 430
Northern Lights, 765, VIII1
Nova, X8
n-p-n junction transistor, 1088
n-type semiconductor, 692–93, 1085–88, 1089–90
nuclear accidents, V12–V16
nuclear bomb, 1096, XI4–XI10
  air burst of, XI6, XI7
nuclear density, 1099
nuclear energy, XI1
  from fission, V5–V6
  from fusion, V6
nuclear explosion, XI6–XI10
  blast wave of, XI6
  chronological development of, XI6–XI8
  critical radii of, XI8–XI9
  distribution of energy released in, XI6, XI7
  above Earth's atmosphere, XI9–XI10
  fallout from, XI9
  fallout pattern for, XI9
  fireball of, XI7–XI8
  firestorm of, XI9
  flash burns from, XI8–XI9
  Mach front of, XI8
  mushroom of, XI1, XI8
  underground, 455
nuclear fission, see fission

nuclear fluid, 1100
nuclear force, see "strong" force
nuclear fuel:
  deuterium as, V6
  reprocessing of, XI12
  supply of, V5
  uranium as, V5
nuclear fusion, V6, VII9–VII11, XI6
nuclear magnetic moments, 807–9
nuclear model of atom, 1056
nuclear power plant, XI11
  safety of, V12–V16
nuclear radius, 1098
nuclear reactions, 1111–14
  conservation of energy in, 1113–14
  endoergic, 1113–14
  exoergic, 1113
  Q value of, 1113–14
nuclear reactor, V5–V6, XI10–XI12
  accidents in, V12–V16
  accidents in, scenarios for, V15
  boiling-water, XI11
  breeder, V5, XI12
  control rods in, V13, V14, XI10–XI11
  converter, XI11
  core of, V13–V14
  fuel rods in, V13–V14, XI10
  fusion, V6, VII9–VIII11
  gas-cooled, XI12
  heavy-water, XI12
  laser-fusion, X7–X8
  light-water, XI12
  moderator in, XI10–XI11
  neutrons in, XI10
  plutonium (^{239}Pu) produced in, V5, XI11–XI12
  power, XI11
  pressurized-water, XI11
  rods in, XI10–XI11
  in submarines, XI12
nuclear reactor vessel, XI11–XI12
nuclear weapons, XI4–XI6
  effects of, XI6–XI10
  see also nuclear explosions
nuclear winter, XI12–XI13, XI14
nucleons, 1096, 1099
  see also neutron; proton
nucleus, Prelude-16, 1095–1120
  binding energy of, 1099–1103, XI2
  collisions of, 285–88
  deformation of, during fission, XI1–XI2
  density of, 1099
  electric energy in, XI1
  electric force and, 1095–96, 1100, XI2
  of helium, 1104–5, 1111
  liquid-drop model of, 1100
  magic, 1117
  mirror, 1115
  proton bombardment of, 1111–13
  stable and unstable, 1100
  table of, Prelude-17
numerical solution, of differential equation, 142–46

objective lens, 936–37
observable universe, II5
ocean waves:
  diffraction of, 455

  speed of, 451–52
octave, 441
ocular lens, 936–37
Odeillo, France, solar furnace at, V8
Oersted, Hans Christian, 730
ohm (Ω), 690
Ohm, Georg Simon, 687
ohmmeter, 723
Ohm's Law, 687
oil, production of, V4
oil reserves, V4
oil shales, V4
ommatidia, 984
one-way membrane, IX11
Onnes, Heike Kamerlingh, VIII1
optical fibers, X5
optical instruments, 932
optical pumping, X2
optics:
  geometric, 922
  geometric vs. wave, 947
  wave, 947
orbit:
  bound, 193
  circular, see circular orbits
  of comets, 226
  elliptical, 218–24, 226, 1075
  geostationary, 217–18
  hyperbolic, 226–27
  parabolic, 226–27
  period of, 216–17
  planetary, 216–17, IX7–IX8
  planetary, data on, 220–21
  synchronous, 217–18
  unbound, 194
orbital magnetic moment, 807
orbital quantum number, 1075, 1078, 1079
organ, 445
Organisation Européenne pour la Recherche Nucléaire (CERN), 528, 762, 1005, 1009, 1122–25, 1127, 1138, 1142
origin of coordinates, 2, 5–6
Orion Spiral Arm, Prelude-9
orthicon, 1037
oscillating charge, radiation field of, 862
oscillating systems, 393–99
oscillations, 379–411
  electromagnetic, of a cavity, 856–57
  forced, 399, 402
oscillator:
  damped, see damped oscillator
  energy quantization of, 1032
  simple harmonic, 383–87
overpressure, 477–78
overtones, 429–30
oxygen molecule, moment of inertia of, 320

parabola, 226–27
parabolic motion, 81
parabolic orbit, 226–27
parabolic vs. elliptical orbit for projectile, 223–24
paraboloidal mirror, 924
parallel-axis theorem, 307
parallel-plate capacitor, 664–65, 678
paramagnetic material, 809–11
  permeabilities of, 810–11

In the two-volume edition, all pages after 570 are in Volume Two.
Interlude V is the last one in Volume One.

paramagnetism, 809–11
parapsychology, 150–51
parent material, 1104, IV4
parity, 1133
partial derivative, 191–92
particle:
 accelerators for, 528, 762–64, 1111–12, 1122–24, VIII9
 alpha, 1055–56
 angular momentum of system of, 310–12, 327–31
 center of mass of system of, 251–59
 collisions of, 285–88
 decays of, 1104–8
 ejection of, in fission, XI2
 electric charges of, 573–74, 577, 1140
 elementary, *see* elementary particles
 free, 311
 ideal, 1–2
 kinetic energy of system of, 259–61
 moment of inertia of system of, 303–9
 momentum of system of, 245–51, 257
 in quantum mechanics, 1064–67
 scattering of, 1055–56
 stable, 1128–29
 subatomic, 316
 system of, *see* system of particles
 unstable, 1128–29
 W, 1135–38
 wave vs., 1039–42
 Z, 1135–38
pascal, 473
Pascal, Blaise, 473, 484
Pascal's Principle, 474–75
Paschen series, 1053
Passamaquody Bay, V10
Pauli, Wolfgang, 1076, 1078, 1079–80, 1107
pendulum, 175–76, 403, 408
 ballistic, 280–81
 conical, 159
 Foucault, 101
 Huygens' tilted, 403
 isochronism of, 392
 physical, 394–96
 reversible, 395
 simple, *see* simple pendulum
 torsional, 396–98
pendulum clocks, 392, 395
penetrating radiation, IV1–IV3
 sources of, IV2–IV3
Penzias, Arno A., II6
perfect fluid, 467
perigee, Prelude-5
 of artificial satellites, 221
perihelion, 218
 of comets, 227
 of Earth, 237
 of planets, 220
perihelion precession, 224, IX7–IX8
 curvature of space and, IX7–IX8
 of Mercury, 211–12, IX1, IX7
perimeters, A4
period, 299
 of artificial satellites, 221
 of comets, 227
 increase of, for simple pendulum, 392
 of orbit, 216–17
 of planets, 220
 of simple harmonic motion, 381

 of wave, 415
periodic motion, 379
Periodic Table of Chemical Elements, Prelude-14, 1079–81, A30
periodic waves, 414–18
permanent dipole moments, 601
permanent magnet, 811
permeability constant, 731
 relative, 810–11
permittivity constant, 574–75
perpendicular-axis theorem, 308
perpetual motion machine:
 of the first kind, 544–45
 of the second kind, 545
persistent current, VIII2
Pfund series, 1053
phase, 381
phase angle:
 for parallel LCR circuit, 839
 for series LCR circuit, 834
phase constant, 381–82
phase velocity, of a wave, 420–21
phasor, 835
phasor diagram, 835
photoelectric effect, 1035–37
photoelectric equation, 1036
photographic camera, 932–33
photographic emulsion, 286
photomultiplier tube, 1036–37, 1108
photons, 1027, 1034–37
 in Compton effect, 1038–39
 emitted in atomic transition, 1061–62
 energy of, 1034–35
 momentum of, 1038
 virtual, 1136
physical pendulum, 394–96
 angular frequency of, 395–96
physico-chemical constants, A29
picofarad, 663
pinch effect, VII5
pinhole camera, IV10–IV11, 933
pion, 1125–26
pion field, 1135
pitch, 297
Pitot tube, 492
Planck, Max, 559, 1031–33
Planck's constant, 317, 807n, 808, VIII8, 1031
Planck's Law, 1030–31
plane mirror, 902
planetary motion:
 conservation of angular momentum in, 219–20
 Kepler's laws of, 218–24
planetary orbit, 216–17
 data on, 220
 perihelion of, 220
 perihelion precession and, IX7–IX8
 period of, 220
plane wave, 439
 electric and magnetic fields of, 878–80
plane wave pulse, electromagnetic, 874–78
plasma, 580, VIII1–VIII12
 cutoff frequency in, VII7
 in fusion reactor, V6
 magnetic field and, VII3–VII6
 in universe, VII1
 waves in, VII6–VII8
plasma confinement, VII9–VIII11

plasma state, Prelude-6
plastic flow, 466
Pluto, 220
 mass of, 236
plutonium (239_{Pu}):
 in chain reaction, XI4
 produced in nuclear reactor, V5, XI11–XI12
plutonium bomb, XI4–XI5
p-n junctions, 1087–88, 1089–90
point charge, 575
 electric field of, 592
 electric field lines of, 595–97, 599, 736
 energy of, 630–31
 energy of a system of, 650–52
 magnetic field of, 736
 moving, magnetic force and, 731, 733–34
 potential of, 629–30
 self-energy of, 652
point discharge, VI8
point groups, I7n
Poisson spot, 984
polarization, 668–69, 880, 910–15
 of accelerated charge radiation, 891
 by reflection, 912–14, 915
polarized light, 910–15
polarized plane waves, electric and magnetic fields of, 879–80
polarizing filter, 891, 911–14
Polaroid, 911, 914, 915
pollution, V10–V12
 thermal, V11–V12
polycrystalline solids, I4
polymer, I2
population inversion, X2
portable atomic clocks, 11
position vector, 58
positive velocity, 33
potential:
 conversion factors for, A26
 derivatives of, 634–35
 of dipole, 638
 electrostatic, 627, 629–33
 of flat charged surface, 628
 gradient of, 634–37
 maxima and minima of, 640
 numerical method for calculation of, 640–42
 of point charge, 629–30
 of spherical charge distribution, 631–32
potential difference, 628
potential energy, 172
 of conservative force, 186–90
 of current loop, 770–71
 curve of, 192–94
 of dipole, 600
 during fission, XI2–XI4
 force calculated from, 191–92
 gravitational, 172–76, 224–29
 gravitational, of a body, 256–57
 gravitational, of spherical mass, 230–31
 mutual, 650–51
 in simple harmonic motion, 387–88
 of simple pendulum, 391
 of a spring, 188
 turning points and, 193–94
potentiometer, 716–17

In the two-volume edition, all pages after 570 are in Volume Two.
Interlude V is the last one in Volume One.

Potsdam, determination of acceleration of gravity at, 395
pound (lb), 13
pound-force (lbf), 105–6
pound-force-foot (lbf–ft), 327
power, 198–201
  absorbed by resistor, 825, 825
  absorbed in LCR circuit, 837
  average, 198
  of clipper ship, V1
  conversion factors for, A25
  delivered by force, 199
  delivered by source of emf, 713
  dissipated in high-voltage transmission line, 714
  dissipated in resistor, 71314
  dissipation of, in U.S., V3
  hydroelectric, V7
  instantaneous, V7
  from nuclear fission, V5–V6
  from nuclear fusion, V6
  reference frame and, 200
  in rotational motion, 333–34
  from solar cells, V8–V9
  transported by a wave, 423
  from windmills and waterwheels, V1, V7
power brake, 374–75
power factor, 837
power plant, nuclear, XI11
  safety of, V12–V16
power reactor, XI11
power transmission line, superconducting, VIII10
Poynting, John Henry, 884
Poynting vector, 884–85
precession, 344
  of the equinoxes, 336n
  of a gyroscope, 343–45
  perihelion, 224
  perihelion, of Mercury, 211–12
precession frequency, 344
prefixes for units, A21
pressure, 467, 472–74, 487
  atmospheric, 476–77
  conversion factors for, A25
  gauge, 477–78
  hydrostatic, 476
  kinetic, 501–4
  magnetic, VII4
  of radiation, 887
  standard temperature and, 498
  in a static fluid, 474–78
pressurized-water reactor, XI11
primary, of a transformer, 840–41
primary cosmic rays, IV3
primitive cell, I6
primitive vectors, I6
primordial fireball, II6
primordial universe, II6
Princeton Tokamak Fusion Test Reactor (TFTR), VII9–VII10
principal maximum, of grating, 960–61
principal quantum number, 1075, 1078, 1079
principal rays, 924
  of lens, 928–29
Principia Mathematica (Newton), 4
Principle of Equivalence, 234, IX1–IX3
principle of relativity, 992–97
principle of superposition, 107–8, 588

for magnetic field, 737–38
for waves, 423
prism, 908, 910, 921
probability interpolation of wave, 1041
Prochorov, Alexander M., X2n
production:
  of coal, V4
  of oil, V4
projectile:
  maximum height of, 81–82
  parabolic vs. elliptical orbit for, 223–24
  range of, 223–24
  time of flight of, 81
  trajectory of, 82–84
projectile motion, 78–84
  with air resistance, 83–84
prominence, on the Sun, VIII1
Prony dynamometer, 334
proper frequencies, 430
proton, Prelude-17, 269, 1095
  charge of, 573, 574
  charge distribution of, 624
  constants for, A28
  elastic collision of deuteron and, 282–83
  isotopes and, 1096
  mass of, 107, 1096
  nuclear binding energy and, 1100–1103
  nuclear bombardment by, 1111–13
  quark structure of, 1140
proton field, 1135
proton-proton chain, 1119–20
Proxima Centauri, Prelude-7
pseudo-gravity, 236
psychokinesis, 150–51
Ptolemaic system, 216
p-type semiconductor, 692–93, 1085–88, 1089–90
pulleys, 363–65
  mechanical advantage and, 364–65
p-V diagram, 550
P wave, 454–55

Q (change of rest-mass energy), 1113–14
Q (ringing time):
  of damped oscillator, 401–2
  of freely oscillating LCR circuit, 832
quadratic equation, A2
quantization:
  of angular momentum, 807–8, 1058–59
  of electric charge, 577
  of energy, 1032
  space, 1076
quantum:
  of energy, 1032
  fields and, 1133–36
  of flux, VIII5
  of light, 1027–48
quantum chromodynamics, 1142
quantum electrodynamics (QED), 1136
quantum jump, 1058
quantum mechanics, 1050. 1064–67
quantum numbers, 1032
  angular-momentum, 1058–59, 1076–77
  of atomic states, 1080–81
  electron configuration and, 1079–80
  magnetic, 1076, 1078, 1079

orbital, 1075, 1078, 1079
principal, 1075, 1078, 1079
spin, 1076, 1078, 1079
for stationary states of hydrogen, 1078, 1079
quantum structure, 1074–1120
quark, Prelude-18, 1121, 1138–43
  bottom, 1143
  charm of, 1142–43
  charmed, 1142–43
  color of, 1141–42
  down, 1139–40
  electric charges of, 577, 1140
  force between, 1141–42
  strange, 1139–40
  top, 1143
  up, 1139–40
quark fields, 1135
quark structure of proton and neutron, 1140
quasar, deflection of radio waves and, IX4
quasar Q1208+1011, 22, 898, II3
quasar 3C273, 894
quasar 3C279, IX4

rad, IV7
radar echo, from Venus, IX5
radian, A5, A20
  definition of, A5
radiant heat, of Big Bang, II6
radiation, 526–27, IV1–IV14
  blackbody, 1027–29, 1031–33
  cavity, 1029
  coherent, X1
  cosmic background, II6
  cosmic fireball, II6
  electromagnetic, 881–83
  electromagnetic, wavelength and frequency bands of, 882–83
  heat transfer by, 526–27
  infrared, 881–82
  ionizing, IV5
  penetrating, IV1–IV3
  physiological effects of, IV8–IV10
  pressure of, 887
  thermal, 1027, 1028–29
radiation damage:
  by alpha rays, IV5
  by beta rays, IV5
  by gamma rays, IV6
  in molecules and living cells, IV7–IV8
  by neutrons, IV6–IV7
radiation doses, IV9–IV10
  acute, affects of, IV9
radiation field:
  of accelerated charge, 859, 861
  of oscillating charge, 862
radiation pressure, 887
radioactive contamination, V15–V16
radioactive decay, IV3–IV5
  law of, 1109–11
radioactive isotopes, IV3–V4
radioactive materials, IV1
radioactive series, 1105
radioactivity, 1103–8, IV4–IV5
radioisotopes, IV10–IV12
  decay rates of, 1109–11
radio receiver, circuit diagram for, 665
radio station CHU, 10n
radio station WWV, 10

In the two-volume edition, all pages after 570 are in Volume Two.
Interlude V is the last one in Volume One.

radio telescope, 961–62, 980–81
　at Arecibo, 894, 979–80
　at Jodrell Bank, 987
　Very Large Array, 962
radio waves, 873–98
　extremely low frequency, 893
　from lightning, VI10
　quasars and, IX4
rail gun, 777–78
railroad tracks, thermal expansion and, 521
random walk, 508–9
range:
　of a force, 126
　of fundamental forces, 125
　of projectiles, 81
Rasmussen Report, V15
rate of change:
　of angular momentum, relative to center of mass, 342
　of momentum, 250–51
Rayleigh, John William Strutt, Lord, 977, 1029–30
Rayleigh's criterion, 978, 987
rays, 901
　alpha, 1103–4, IV1–IV2, IV5
　beta, 1103–4, IV1–IV2, IV5
　cathode, 766–67
　cosmic, 286, 760, IV2–IV3
　gamma, 882, 1103–4, IV1–IV2, IV6
　principal, 924, 928–29
　X, 882, 1062, I3, IV1
RBE (Relative Biological Effectiveness), IV8
RC circuit, 717–21
reactance:
　capacitive, 826
　inductive, 827–28
reaction and action, 109–12
reactor, nuclear, V5–V6, XI10–XI12
　boiling-water, XI11
　breeder, V5, XI12
　control rods in, V13, V14, XI10–XI11
　converter, XI11
　core of, V13–V14
　fuel rods in, V13–V14, XI10
　fusion, V6, VII9–VIII11
　gas-cooled, XI12
　heavy-water, XI12
　laser-fusion, X7–X8
　light-water, XI12
　moderator in, XI10–XI11
　neutrons in, XI10
　plutonium (239_{Pu}) produced in, V5, XI11-XI12
　power, XI11
　pressurized-water, XI11
　rods in, XI10–XI11
　in submarines, XI12
reactor accidents, V12–V16
　scenarios from, V15
real image, 926
recession, motion of, II4
recoil, 247–48, 266
rectangular coordinates, 2
rectifier, 1087–88, I12
red, 1141–43
redout, 152
red shift:
　cosmological, 897
　gravitational, IX8–IX10

red-shift experiments, IX9–IX10
red-shift method, II3
reference frame, 5–6, 86–89
　in calculation of work, 163
　of Earth, 101–2
　freely falling, 127–28
　inertial, 100–102
　local inertial, IX2
　power and, 200
　for rotational motion, 297, 339
reflection, 900–903
　angle of, 901–2
　critical angle for total internal, 908
　law of, 900–901
　by mirror, 902–3
　polarization by, 912–14, 915
　total internal, 908
reflection grating, 961
reflection symmetry, I5
reflector:
　corner, 902–3, X5
　in fission, XI4
　on Moon, 903, X5
refraction, 903–10
　double, 913
　index of, 904, 908–9
　law of, 906
　polarizers and, 912–13
refrigerators, use of freon and, 553–54
Relative Biological Effectiveness (RBE), IV8
relative coordinates, $5n$
relative measurement, 5
relative permeability constant, 810–11
relativistic kinetic energy, 1011–12
relativistic momentum, 1010–11
relativistic total energy, 1013
relativistic transformation:
　for momentum and energy, 1013–14
relativity:
　general theory of, 4, 212, 234, II1, IX1, IX6–IX8
　magnetic field and, 1014–18
　of motion, 27, 86, 89, 989
　Newtonian, 114–15
　principle of, 992–97
　of simultaneity, 992–94
　special theory of, 4, 989–1026
　of speed, 27
　of synchronization of clocks, 994–95
renormalization rules, 652
reserves:
　of coal, V4
　of fossil fuel, V4
　of oil, V4
resistance, 687
　in combination, 693–95
　conversion factors for, A26
　of human skin, 721–22
　internal, of batteries, 708–9
　of wires connecting resistors, 695
resistance thermometers, 522–23, 697
resistivity, 687
　conversion factors for, A26
　of insulators, 692
　of materials, 690–93
　of materials, temperature dependence of, 691
　of metals, 690–92
　of semiconductors, 692–93
　temperature coefficient of, 697

resistors, 693–95
　circuit with, 824
　in parallel, 694
　power absorbed by, 825
　power dissipated in, 713–14
　resistance of wires connecting, 695
　in series, 694
resolution, angular, of telescope, 977–80
resolving power, of grating, 961
resonance, 1129
　of damped oscillator, 402
　in LCR circuit, 833
　in musical instruments, 445–46
resonant cavity, in laser, X3–X4
resources:
　of energy, V4
　limits of, V4
　of nuclear fuels, V5–V6
rest frame, 5
restitution, coefficient of, 293
rest mass, 197
rest-mass energy, 197, 1013
　in inelastic collision, 286–87
restoring force, 140–42
resultant, of vectors, 55
resultant force, 108
retardation of light, by Sun, IX5–IX6
return stroke, in lightning, VI9
reversible pendulum, 395
reversible process, 549
　entropy for, 556
Rhea, 239
Richter, Burton, 1143
Richter magnitude scale, 454–55
right-hand rule, 64–65, 600
　for Ampère's Law, 754
　in angular momentum of particle, 310–11
　for induced emf, 784
rigid body, 296–324
　angular momentum of, 312–17
　component of angular momentum along axis of rotation of, 315–16
　dynamics of, 325–53
　kinematics of, 295–324
　kinetic energy of rotation of, 303–9
　moment of inertia of, 303–9
　motion of, 296–98
　parallel-axis theorem for, 307
　perpendicular-axis theorem for, 308
　rotational motion of, 296
　some moments of inertia for, 309
　spin of, 316, 316
　statics of, 354–55
　translational motion of, 296
ringing time (Q):
　of damped oscillator, 401–2
　of freely oscillating LCR circuit, 832
RL circuit, 796–98
rock, flow of, 466
rocket:
　motion of, 261–64
　thrust of, 263
rocket equation, 263
Roemer, Ole, measurement of speed of light and, 892
roll, 297
rolling motion, 339–43
Röntgen, Wilhelm Konrad, IV1n
root-mean-square speed, of gas, 502–3
root-mean-square voltage, 825

In the two-volume edition, all pages after 570 are in Volume Two.
Interlude V is the last one in Volume One.

rotation:
   about center of mass, 342
   of coordinates, 66–68
   of the Earth, 9–10, 101
   frequency of, 299–300
   kinetic energy of, 303–9
   noncommutative, 298
   period of, 299
rotational energy, of molecule, 1082–83
rotational motion, 29996–303
   conservation of angular momentum in, 335–39
   conservation of mechanical energy in, 333
   with constant angular acceleration, 301–3
   energy in, 332–33
   equation of, 329–30
   about a fixed axis, 298–301
   of gas, 505–6
   Newton's laws in, 325
   power in, 333–34
   reference frame for, 297, 339
   of rigid body, 296
   translational motion as analogous to, 330
   translation and, in rolling motion, 339
   work in, 332–33
rotation symmetry, I5–I7
Rowland, H., 969, 970
Rubbia, Carlo, 1138n
ruby laser, X2–X3
Rumford, Benjamin Thompson, Count, 515
Rutherford, Sir Ernest, 1049, 1055–57, 1095, 1104, 1111, XI13
Rutherford scattering, 1055–56
R-value, 539
Rydberg constant, 1053
   measurement of, X7
Rydberg-Ritz combination principle, 1054

St. Elmo's fire, VI8
Salam, Abdus, 1137
satellites:
   artificial, 217–18, 220–22, 1090
   communication, 217–18
   geostationary, 218
Saturn, 220
   moons of, 239
Saturn V rocket, 153, 270
scalar, 54
scalar product, 62
scale:
   chromatic musical, 441
   Richter magnitude, 454–55
scans:
   brain, IV11
   thyroid, IV11
scattering, 285
   of alpha particles, 1055–56
Schawlow, Arthur L., X2
Schick stabilizer, 346
Schrieffer, J. Robert, VIII6n
Schrödinger, Erwin, 1064, 1065
Schrödinger equation, 1065–66
Schwarzschild radius, IX11
Schwinger, Julian, 1136
scientific notation, 15–16

scintillation counters, IX11
screw axis, I7
Scylla, VII9
Scyllac, VII9
seat belts, III1–III2, III9–III11, III14
secant, A6
second, 6, 9–11, 14–15, A20
   multiples and submultiples of, 11
secondary, of a transformer, 840–41
secondary collisions, III6–III9
secondary cosmic rays, IV2–IV3
Second Law of Thermodynamics, 559–64
second overtone, 429–30
seiche, 458
seismic waves, 454–55
seismometer, 459
selection rule, 1082
self-energy, of point charge, 652
self-inductance, 794
semiconductors, 692–93, 1087–90, I12
   with donor and acceptor impurities, 693, 1086
   electron configurations of, 1084–86
   "holes" in, 692–93, 1085–86, 1087–88
   n-type, 692–93, 1085–88, 1089–90
   p-n junctions of, 1087–88, 1089–90
   p-type, 692–93, 1085–88, 1089–90
   resistivities of, 692–93
   temperature dependence of resistivity of, 1085
semimajor axis of ellipse, 218
   related to energy, 226
series limit, 1053
Sèvres, France, 6
shear, 365, 367–68
shear modulus, 366–68
shells, 1080–81
Shiva, X8
shock waves, X12
Shortt pendulum clock, 392, 395
short wave, 881
sidereal day, 101n
sign conventions, 932
significant figures, 15–16
silicon, p-type and n-type, 707
silicon solar cells, 707
simple harmonic motion, 379–83
   conservation of energy in, 388, 392
   frequency of, 380–81
   kinetic energy in, 387–88
   period of, 381
   phase of, 381
   potential energy in, 387–88
simple harmonic oscillator, 383–87
   angular frequency of, 385
   equation of motion of, 384–85
simple pendulum, 389–92
   energy of, 391
   equation of motion for, 390
   increase of period of, 392
   isochronism of, 392
   kinetic energy of, 391–92
   potential energy of, 391
simultaneity, relative, 992–94
$\sin^{-1}$, A7–A8
sine, A6
   formula for derivatives of, 383
   law of, A9

single-loop circuits, 707–9
single slit:
   diffraction by, 971–76
   diffraction pattern of, 973, 975
   minima in diffraction pattern of, 973–74
singularity, gravitational, IX12
sinks, of field lines, 597
siphon, 492
SI units, 6, 14–15, A20
   of current, 769
   derived, A21
   of electric charge, 769
   for radioactive decay rate, 1110–11
skin, human, resistance of, 721–22
sky diver, 42
Skylab mission, 102, 338–39, 234, 237, IX2–IX3
   body-mass measurement device on, 103–4, 383–84
   fuel cell on, 707
   solar panel on, V9
sliding friction, 134–38
slope, 20
   instantaneous velocity as, 30–31
   of worldline, 30
slowing of light, by Sun, IX5–IX6
slug, 106n
smog, V10–V11
Snell's Law, 906
Sobral, Brazil, eclipse at, IX4
solar cells, 707, 1089–90
   as power source, V8–V9
   silicon, 707
solar collectors, V8
solar day, 9, 101n
solar energy, V6–V9
   indirect, V6–V8
solar heating system, V8
Solar One power station, V8
solar panels, V9
solar power station, V8–V9
Solar System, 212, 218, 221
   data on, 220–21
solar tides, 235
solar wind, VII3–VII4
solenoid, 756–59
   induced electric field, 791–92
   induced emf in, 785–86
   magnetic field of, 757–58
solid:
   compression of, 365, 367–68
   elasticity of, 365–69
   elastic limit of, 368–69
   elastic moduli for, 366–68
   electron configuration of, 1084–86
   elongation of, 365–68
   energy banks in, 1083–86
   flow of, 466
   quantum structure of, 1074, 1083–90
   shear of, 365, 367–68
   thermal expansion of, 518–21
   ultimate tensile strength of, 366, 368–69
solid solutions, I10
solid state, Prelude-2
solid-state electronics, I12
solid-state laser, X5
Sommerfeld, Arnold, 1075–76
sonic boom, 450

In the two-volume edition, all pages after 570 are in Volume Two.
Interlude V is the last one in Volume One.

Index

sound:
  intensity level of, 440
  speed of, 443–46, 461
sound waves:
  in air, 439–43
  diffraction of, 456
  intensity of, 440
sources:
  alternative, for energy, V5–V10
  of electromotive force, 705–7
  of field lines, 597
  of penetrating radiation, IV2–IV3
space, curvature of, IX3–IX6
  perihelion precession and, IX7–IX8
space, interstitial, I10
space groups, I7n
space quantization, 1076
Space Shuttle, 128
space station, 235
Space Telescope, 979, 987
spacetime, curved, IX7–IX8
spacetime diagram, 997–98
spark chambers, 1126–27
Special Relativity theory, 4, 732, 989–1026
specific heat:
  at constant pressure, 529–30
  at constant volume, 529–30
  of a gas, 529–31
specific heat capacity, 516
speckle effect, IX4
spectra, color plate, 1051
spectral emittance, 1028
  of blackbody, 1030
spectral lines, 910, 1050–53
  splitting of, 1076–78
spectral series:
  of hydrogen, 1052–54, 1060, 1076
  of lithium, 1069
spectroscopy, laser, X7
spectrum, 1051
  band, 1083
  of Caph, 1051
  of hydrogen, 1052, 1053
  produced by grating, 960
  produced by prism, 910
speed:
  average, 26–27
  of bullet, 291
  conversion factors for, A24
  of Earth, 219–20
  of electromagnetic wave, 865–66, 877–78
  energy-equivalent, III2
  of impact, in automobile collisions, III7–III8, III13–III14
  median, 510–11
  molecular, in gas, 503–4
  most probable, 504
  after one-dimensional elastic collision, 277–79
  as relative, 27
  of sound, 443–46
  standard of, 8, IX7
  terminal, 42
  unit of, 26
  velocity vs., 29, 76
  of water waves, 451–52
  of waves, on a string, 418–21
speed of light, 8, 860, 864–66, 990–91

invariance of, 992
laser standard for, X7
in material medium, 904
measurement of, 874, 892
universality of, 992
Sperry C-12 directional gyroscope, 335
spherical aberration, 923–24, 932
spherical capacitor, 671
spherical charge distribution:
  electrical field of, 616
  electric energy of, 656–57
  potential of, 631–32
spherical coordinates, 2–3
spherical mass:
  gravitational force exerted by, 229–32
  gravitational potential energy of, 231
spherical mirrors, 922–26
  concave, 923
  convex, 924
  focal length of, 923
spin, 316
  of Earth, 316–17
  of electron, 744, 1076–77, 1080, 1096
  of elementary particles, 1128
  of rigid body, 316
  of subatomic particles, 316–17, 1078
spin magnetic moment, 808
spin quantum number, 1076, 1078, 1079
spin-spin force, 812
spiral galaxies, II3
spring balance, 106, 127
spring constant, 141–42
springs, 140–42, 383–84
  force of, 140–42
  potential energy of, 188
spring tides, 238
Sputnik I, 221–22, 242
Sputnik II, 221
Sputnik III, 221
SQUID (superconductive quantum interference device), VIII7
stabilized laser, 7–8
stable equilibrium, 394
stable particle, 1128–29
standard kilogram, 102
standard meter bar, international, 6–7
standard of length, 6–8
standard of mass, 12, 102
standard of speed, 8
  established with laser, X7
standard of time, 10
Standards, National Bureau of, 10–12
standard temperature and pressure (ST), 498
standing electromagnetic wave, 948–50
standing wave, 428–31, 948–49
  in a tube, 445
Stanford Linear Accelerator (SLAC), 624, 857, 1123, 1139, 1143
Stapp, J., III12–III13
states of atoms, stationary, 1060
states of matter, Prelude-2, Prelude-6
  fourth, VII1–VII3
  fifth, 591
static equilibrium, 354–62
  condition for, 354
  of electric charge, 617–18
  examples of, 355–62
static fluid, 474
static friction, 138–40

coefficient of, 136, 138–39
statics, of rigid body, 354–55
stationary state, 1058
  of hydrogen, 1060, 1061
steady flow, 468–69
steady state, 833
steel:
  crystal structure of, I11
  hard and soft, I12
Stefan-Boltzmann Law, 1033–34
Stellarator, VII9
step leader, in lightning, VI9
steradian, A20
stimulated emission, 882, X2
stopping potential, 1035
storage, of energy, VIII10–VIII11
strain, 367–69
strangeness, 1133
strange quark, 1139–40
streamlines, 468–71
  velocity along, 480–81
stream tube, 469–70
stress, 367–69
  thermal, 378
stringed instruments, 445–46
"strong" force, 125, 1130, 1141
  nuclear binding energy and, 1099–1103
  in nucleus, 1095–96, XI1
strong interactions, 1130
strontium, 1106
  radioactive decay of, 1109–11
subatomic particles, spin of, 316–17, 1078
sublimation, 527
substitutional impurities, I10
suction pump, 489
Sun:
  eclipse of, IX4
  escape velocity from, 228
  generation of energy in, XI6
  hydrogen burning in, XI6
  light rays and, 4, IX3–IX6
  prominence on, VIII1
sunglasses, Polaroid, 914, 915
Supercluster, Local, Prelude-10
superconducting cable, 701, VIII9–VIII10
superconducting generators, VIII9–VIII10
superconducting magnet, VIII7–VIII8
superconducting power transmission line, VIII10
Superconducting Super Collider (SSC), 1123–24
superconductivity, 691–92, VIII1–VIII2
superconductors, VIII1–VIII4
  of the second kind, VIII5
supercooling, 529
supercritical mass, XI4
superfluid, VIII1
superheating, 528–29
supernovas, II5
superposition:
  of electric forces, 588–90
  of waves, 423–27
superposition principle, 107–8, 588
  for magnetic field, 737–38
  for waves, 423

In the two-volume edition, all pages after 570 are in Volume Two.
Interlude V is the last one in Volume One.

supersonic aircraft, sonic boom and, 450
supply of fossil fuels, V4
surface charge, on dielectric, 671
surface wave, 454
suspension, magnetic, VIII11
suspension bridge, 359–60
S wave, 454–55
sword blades, Japanese, I12
symmetric fission, XI1–XI2
symmetry, in physics, I1, I5–I7
   conservation laws and, I61
   of a cube, I8
   inversion, I5
   reflection, I5
   rotation, I5–I7
   translation, I6–I7
symmetry axis, I5
sympathetic oscillation, of damped oscillator, 402
synchrocyclotron, 762
synchronization of clocks, 5, 11, 992–95
   relativity of, 994–95
synchronous (geostationary) orbit, 217–18
synchrotron, 762–63, 1122
system of particles, 245–71
   center of mass in, 251–59
   energy of, 259–61
   kinetic energy of, 259–61
   moment of inertia of, 303–9
   momentum of, 245–51, 257
system of units (SI), 6, 14–15, A20
   see also SI units

Tacoma Narrows, 430–31
tails of comets, 888
$\tan^{-1}$, A7
tangent, A6
tangent galvanometer, 773
Taylor series, A18–A19
telescope, 936–39
   angular magnification of, 937
   angular resolution of, 977–80
   Arecibo radio-, 979–80
   Galilean, 941
   Jodrell Bank radio-, 987
   Keck, 939
   mirror, 937–39
   Mt. Palomar, 938, 979
   Multiple Mirror, 938–39
   radio-, 894, 961–62, 979–81, 987
   Space, 979, 987
   Very Large, 939
   Very Large Array radio-, 962
teletherapy unit, V12
television cameras, 1037
temperature:
   coefficient of resistivity, 697
   lowest attained, 564
   standard pressure and, 498
temperature scales:
   absolute, 495–96, 523
   absolute thermodynamic, 499, 554
   Celsius, 500
   comparison of, 500
   Fahrenheit, 500
   ideal-gas, 498–500, 523, 554
   Kelvin, 495–96, 499–500, 554
tensile strength, of solids, 366, 368–69

tension, 111–12
   of seat belts, III9–III11
terminal speed, 42
tesla (T), 735
Tesla, Nikola, 735
"Tessar" lens, 932
Tethys, 239
TeV, 1122–23
Tevatron, Fermilab, 1122–23
thermal conductivity, 525–26
thermal energy, 515
thermal engine, efficiency of, 548–49
thermal equilibrium, 523–24
thermal expansion:
   of concrete, 521
   linear, coefficient of, 519, 521
   of solids and liquids, 518–21
   of water, 520
thermal neutrons, XI10
thermal pollution, V11–V12
thermal radiation, 1027, 1028–29
thermal stress, 378
thermal units, 196
thermocouples, 500, 523
thermodynamic equilibrium, 547
thermodynamics, 544–70
   First Law of, 545–48
   Second Law of, 559–64
   Third Law of, 564
   Zeroth Law of, 523n, 565
thermodynamic temperature scale, absolute, 499, 554
thermometer:
   bimetallic strip, 500, 522
   constant-volume gas, 499
   electric resistance, 500
   mercury-bulb, 500, 522
   resistance, 522–23, 697
   thermocouple, 500, 523
thermonuclear fusion, VII8–VII11, X7–X9, XI6
thermonuclear fusion reactor, VII9–VIII11
thermonuclear reactions, V6, VII8–VII9, X8, XI6
thermos bottle, 1029
thin films, interference in, 950–52
thin lenses, 926–32
Third Law of Thermodynamics, 564
Thomson, Sir Joseph John, 766–67, 1054–55
thorium, as nuclear fuel, V5
three-axis accelerometer, 110
three-dimensional space, 3
Three Mile Island, Pennsylvania, reactor accident at, V13
threshold energy, 288, 1114
threshold frequency, 1036
threshold of hearing, 439
thrust, 263
thunder, VII0
thundercloud, electric field of, VI6–VI8
thunderhead, VI5
thunderstorm, VI1, VI3–VI11
   anvil cloud in, VI5
   energy conversion in, VI4
   generation of electric charge in, VI5–VI6
   turrets in, VI5
thunderstorm cells, VI4

tidal energy, V10
tidal wave, 453
tides, 91, 235
   lunar, 235
   solar, 235
   spring, 238
time:
   absolute, 4–5, 87
   atomic standard of, 10
   Cesium standard of, 10
   Coordinated Universal, 10–11
   conversion factors for, A23
   standard of, 10
   unit of, 9–11
time constants:
   of RC circuit, 720
   of RL circuit, 797–98
time dilation, 995, 1004–7
   gravitational, IX8–IX10
   gravitational, experiments on, IX8–IX9
time of flight, of projectiles, 81
time signals, 10
Ting, Samuel Chao Chung, 1143
Titan, 239
Tokamak, VII9–VII10
Tomonaga, Sin-Itiro, 1136
top, 343
top quark, 1143
torch, laser as, X6
tornado:
   lightning in, VI10
   pressure in, 487
toroid, 759
torque, 325–27
   angular momentum and, 328–29
   cross product in definition of, 325
   on a current loop, 769–71
   on dipole, 600
   static equilibrium and, 354–56
torr, 473
Torricelli's equation, 481
torsional constant, 396–98
torsional pendulum, 396–98
torsional suspension fiber, 396–97
torsion balance, 215–16, 576
total angular momentum, rate of change of, 328
total internal reflection, 908
totally inelastic collision, 279, 284
Townes, Charles H., X2
trajectory of projectiles, 82–84
transfer of heat, 524–25, 526–27
transformation, heat of, 527
transformation equation, 67, 89
   Galilean, 66n, 88–89, 114, 990, 998, 1010
   Lorentz, 66n, 997–1004, 1008
transformation of coordinates, 66n, 67, 87, 88, 990, 998, 1003
transformer, 840–42
transistors, 1087, 1088–89, I12
translational motion, 25, 296, 331
   rotational motion as analogous to, 330
   rotation and, in rolling motion, 339
translation symmetry, I6–I7
transmission line:
   power dissipated in, 714
   superconducting, VIII10
transmutation of elements, IV4

In the two-volume edition, all pages after 570 are in Volume Two.
Interlude V is the last one in Volume One.

transverse radiation field of accelerated charge, 860
transverse wave, 413
très grande vitesse (TGV) train, 48
Trieste, 486
trigonometric functions, A5–A8
trigonometric identities, A8–A9
trigonometry, A5–A9
triode, 1087, 1088–89, I12
triple-point cell, 499
triple point of water, 499
tritium, as nuclear fuel, V6
trumpet, sound wave emitted by, 440–41
truss bridge, 358–59
  method of joints for forces in, 359
tsunamis, 453
tube of force, 603
tumors, IV11
tunable laser, X7
turbulence, 469n
turning points, 193
turrets, in thunderstorm, VI5
TV waves, 881
twin paradox, 1006
two-slit interference, 954–58
  pattern for, 955, 958
two-wavelength microscope, I2

UFOs, 458
Uhlenbeck, George, 1076–77
ultimate tensile strength, 366, 368–69
ultrasonic microscope, 443
ultrasound, 439
ultraviolet catastrophe, 1030
ultraviolet freedom, 1142
ultraviolet light, 881–82
unbound orbit, 194
uncertainty relation, 1042
underground nuclear explosions, 455
unified field theory, 1137
unified theory of weak and electromagnetic forces, 1136–38
uniform circular motion, 84–86, 146–50
  angular momentum for, 312
United States:
  energy dissipation in, V3
  power dissipation in, V3
unit of length, 6–9
units:
  consistency of, 16
  conversions of, 17–18
Units, International System of (SI), 6, 14–15
  base, A20
  of current, 769
  derived, A21
  of electric charge, 769
units of force, 105–6
unit vector, 58–59
  cross product of, 64–65
  dot product of, 62, 166
universal constants, A27
universal gas constant, 495
universal gravitation, law of, 212–14
universality of acceleration of gravity, 37
universality of free fall, 37, 234, 392, IX1–IX3
universality of speed of light, 992

universe:
  age of, II5–II6
  chemical composition of, II5–II6
  contraction of, II7
  critical density for, II7–II8
  deuterium abundance in, II8
  escape velocity for, II7
  expansion of, II1–II4
  future evolution of, II6–II8
  hydrogen and helium abundance of, II6
  implosion of, II7
  mass density in, II7–II8
  observable, II5
  plasma in, VII1
  primordial, II6
unpolarized light, 911
unstable equilibrium, 394
up quark, 1139–40
uranium, 1096
  alpha decay of, 1104–5
  in chain reaction, X13–X14
  enriched, XI10
  isotopes of, 1103, XI4–XI6
  natural, V5
  as nuclear fuel, V5
  nucleus of, electrical energy in, 656–57, XI1
  "weapons grade," XI6
uranium bomb, XI4–XI5
Uranus, 220
UX Arietis, 980

vacancies, I10
valence band, 1085
Van Allen belts, 764, 765
Van de Graaff, R. J., 1111–12
van der Meer, Simon, 1138n
Vanguard I, 221–22, 240
vaporization, 527–28
vector addition, 54–57
  associative law of, 56–57
  commutative law of, 56
  by components, 59–60, 61
  graphical method for, 61
  trigonometric method for, 61, 109
vector product, 63–66
vectors, 52–73
  angular-momentum, 310–12, 314
  components of, 59–66
  cross product of, 63–66
  definition of, 54
  displacement, 52–54
  dot (scalar, inner) product of, 62–64, 165–66
  instantaneous velocity, 75–76
  multiplication of, 58, 61–66
  position, 58
  Poynting, 884–85
  primitive, 6
  resultant of, 55
  rigorous definition of, 67
  subtraction of, 57
  unit, *see* unit vector
vector triangle, 55
velocity:
  angular, average and instantaneous, 299
  average, 27–29
  average, in three dimensions, 74–75

  of center of mass, 257, 279
  of flow, 467
  Galilean addition law for, 990
  instantaneous, *see* instantaneous velocity
  Lorentz transformations for, 1008
  muzzle, 82–83, 223–24, 266, 267
  negative, 33
  positive, 33
  relativistic combination of, 1008–9
  speed vs., 29, 76
  along streamlines, 480–81
  transformation equations for, 88–89, 990
velocity field, 468
velocity filters, 504
velocity selectors, 766
velocity transformation, Galilean, 990
Venturi flowmeter, 482, 492
Venus, 220–21
  radar echo from, IX5
Verne, Jules, 241
Very Large Array (VLA) radiotelescope, 962
Very Large Telescope (VLT), 939
very-long-baseline interferometry (VLBI), 980, 985
vibrational energy, of molecules, 1082
vibrations, of diatomic molecule, 304–5
violin, 446
  sound wave emitted by, 440–41
Virgo Cluster, Prelude-10
virtual image, 902
virtual photon, 1136
virus particles, arranged in crystal, I2
viscosity, 469n
visible light, 881–83
  colors and wavelengths of, 882–83
volt (V), 628
Volta, Alessandro, Conte, 628
voltage, 704–5
  AC, 788–89
voltmeter, 715–16, 723, 724
volume, 14, A4
  compression of, 365, 367–68
  conversion factors for, A23–A24
vortex, 445
vortex lines, VIII5

Wairakei (New Zealand), V9
Walton, E. T. S., 1111–12
WASH-740 Report, V15
waste heat, V11–V12
water:
  heavy, II8
  thermal expansion of, 520
  triple point of, 499
water molecule, 601
  moment of inertia for, 320
water waves, 451–53
  diffraction of, 455
  speed of, 451–52
waterwheel, 171, 337–38
  power from, V1, V7
watt (W), 198
Watt, James, 198
wave, 412–65
  Alfvén, VII6
  amplitude of, 415
  angular frequency of, 416

In the two-volume edition, all pages after 570 are in Volume Two.
Interlude V is the last one in Volume One.

Index

wave (continued)
  beats of, 426–27
  capillary, 451
  constructive and destructive interference for, 424–25
  on Earth's surface, 454–55
  electromagnetic, see electromagnetic wave
  energy in, 421–23, 883–85
  energy density of, 422–23
  frequency of, 415–16
  gravity, 451–52
  group velocity of, 420
  harmonic, 414–17
  light and radio, 873–98
  long, 881
  longitudinal, 413
  magnetosonic, VII6
  measurement of momentum and position of, 1041–42
  medium, 881
  modulation of, 426
  ocean, 451–52, 455
  P, 454–55
  particle vs., 1039–42
  period of, 415
  periodic, 414–18
  phase velocity of, 420–21
  plane, 439, 878–80
  in a plasma, VII6–VII8
  power transported by, 423
  probability interpolation and, 1041
  radio, 873–98
  radio, from lightning, VI10
  radio, quasars and, IX4
  S, 454–55
  seismic, 454–55
  short, 881
  sound, see sound waves
  speed of, on a string, 418–21
  standing, 428–31, 948–49
  standing, in a tube, 445
  superposition of, 423–27
  surface, 454
  tidal, 453
  transverse, 413
  TV, 881
  water, see water waves
wave crests, 414–15
wave equation, 418
wave fronts, 438–39, 441
  circular, 438–39
  plane, 439
  spherical, 438–39
wavefunction, 413–14
  harmonic, 416
waveguide, 858
wavelength, 7, 414–17
  cutoff, 1047
  of visible light, 882–83
wavelength bands, of electromagnetic radiation, 882–83
wavelength shift, of photon, in Compton effect, 1038–39
wavelets, 900, 1039
wave motion, 412
wave number, 415
wave optics, 947
wave pulse, 412–14
  electromagnetic, 874–78
wave trough, 414–15
wavicle, 1027, 1039–40
  measurement of, 1041–42
"weak" force, 125, 1107, 1130
"weak" interactions, 1107, 1130
"weapons grade" uranium, XI5
weber (Wb), 780
Weber, Wilhelm Eduard, 781
weight, 12, 126–29
  apparent, 149–50
  mass vs., 127
weightlessness, 127–28
  simulated, 485
Weinberg, Stephen, 1137
welding, by laser, X6
Wheatstone bridge, 717
Wheeler, John A., IX6
whistlers, VII7–VII8
white noise, 440
Wien's Law, 1033
Wilson, Robert W., II6
winches, 362
wind instruments, 445
windmills, power from, V1, V7
Windscale, England, reactor accident at, V13
wing, flow around, 471, 480
wire:
  magnetic force on, 768–69
  straight, magnetic field of, 739–40
  uniform, electric field in, 684
work, 160–245
  definition of, 161
  done by constant force, 161–62, 165
  done by gravity, 168–69
  done by variable force, 164, 168
  dot product in definition of, 165–69
  frame of reference in calculation of, 163
  as integral, 164
  internal, in muscles, 163
  in one dimension, 161–65
  in rotational motion, 332–33
  in three dimensions, 165–69
work energy theorem, 171
work function, 1036
worldline, 27–30
  in secondary collision, III6
  slope of, 30
W particles, 1135–38

X-ray diffraction, I3
X rays, 882, I3, IV1
  characteristic, 1062
  from Cygnus X-1, IX12, IX13

yaw, 297
ying-yang coil, VII10
Young, Thomas, 456, 947, 954
Young's modulus, 366–67

Zeeman effect, 1077–78
Zeiss "Tessar" lens, 932
zero G, 127–28
zero-point motion, $564n$
zero resistance, VIII1–VIII3
Zeroth Law of Thermodynamics, $523n$, 565
Z particles, 1135–38
Zweig, G., 1139

In the two-volume edition, all pages after 570 are in Volume Two.
Interlude V is the last one in Volume One.